Handbook of Matching and Weighting Adjustments for Causal Inference

An observational study infers the effects caused by a treatment, policy, program, intervention, or exposure in a context in which randomized experimentation is unethical or impractical. One task in an observational study is to adjust for visible pretreatment differences between the treated and control groups. Multivariate matching and weighting are two modern forms of adjustment. This handbook provides a comprehensive survey of the most recent methods of adjustment by matching, weighting, outcome modeling, and their combinations. Three additional chapters introduce the steps from association to causation that follow after adjustments are complete.

When used alone, matching and weighting do not use outcome information, so they are part of the design of an observational study. When used in conjunction with models for the outcome, matching and weighting may enhance the robustness of model-based adjustments. The book is for researchers in medicine, economics, public health, psychology, epidemiology, public program evaluation, and statistics, who examine evidence of the effects on human beings of treatments, policies, or exposures.

Chapman & Hall/CRC

Handbooks of Modern Statistical Methods

Series Editor
Garrett Fitzmaurice, *Department of Biostatistics, Harvard School of Public Health, Boston, MA, U.S.A.*

The objective of the series is to provide high-quality volumes covering the state-of-the-art in the theory and applications of statistical methodology. The books in the series are thoroughly edited and present comprehensive, coherent, and unified summaries of specific methodological topics from statistics. The chapters are written by the leading researchers in the field and present a good balance of theory and application through a synthesis of the key methodological developments and examples and case studies using real data.

Published Titles
Handbook of Quantile Regression
Roger Koenker, Victor Chernozhukov, Xuming He, and Limin Peng

Handbook of Statistical Methods for Case-Control Studies
Ørnulf Borgan, Norman Breslow, Nilanjan Chatterjee, Mitchell H. Gail, Alastair Scott, and Chris J. Wild

Handbook of Environmental and Ecological Statistics
Alan E. Gelfand, Montserrat Fuentes, Jennifer A. Hoeting, and Richard L. Smith

Handbook of Approximate Bayesian Computation
Scott A. Sisson, Yanan Fan, and Mark Beaumont

Handbook of Graphical Models
Marloes Maathuis, Mathias Drton, Steffen Lauritzen, and Martin Wainwright

Handbook of Mixture Analysis
Sylvia Frühwirth-Schnatter, Gilles Celeux, and Christian P. Robert

Handbook of Infectious Disease Data Analysis
Leonhard Held, Niel Hens, Philip O'Neill, and Jacco Walllinga

Handbook of Meta-Analysis
Christopher H. Schmid, Theo Stijnen, and Ian White

Handbook of Forensic Statistics
David L. Banks, Karen Kafadar, David H. Kaye, and Maria Tackett

Handbook of Statistical Methods for Randomized Controlled Trials
KyungMann Kim, Frank Bretz, Ying Kuen K. Cheung, and Lisa Hampson

Handbook of Measurement Error Models
Grace Yi, Aurore Delaigle, and Paul Gustafson

Handbook of Multiple Comparisons
Xinping Cui, Thorsten Dickhaus, Ying Ding, and Jason C. Hsu

Handbook of Bayesian Variable Selection
Mahlet Tadesse and Marina Vannucci

Handbook of Matching and Weighting Adjustments for Causal Inference
José Zubizarreta, Elizabeth A. Stuart, Dylan S. Small, Paul R. Rosenbaum

For more information about this series, please visit: https://www.crcpress.com/Chapman--HallCRC-Handbooks-of-Modern-Statistical-Methods/book-series/CHHANMODSTA

Handbook of Matching and Weighting Adjustments for Causal Inference

Edited by
José R. Zubizarreta
Elizabeth A. Stuart
Dylan S. Small
Paul R. Rosenbaum

CRC Press is an imprint of the
Taylor & Francis Group, an **informa** business
A CHAPMAN & HALL BOOK

First edition published 2023
by CRC Press
6000 Broken Sound Parkway NW, Suite 300, Boca Raton, FL 33487-2742

and by CRC Press
4 Park Square, Milton Park, Abingdon, Oxon, OX14 4RN

CRC Press is an imprint of Taylor & Francis Group, LLC

ISBN: 978-0-367-60952-8 (hbk)
ISBN: 978-0-367-60953-5 (pbk)
ISBN: 978-1-003-10267-0 (ebk)

DOI: 10.1201/9781003102670

Typeset in Nimbus Roman
by KnowledgeWorks Global Ltd.

Publisher's note: This book has been prepared from camera-ready copy provided by the authors.

To our families, students, and collaborators,
and the causal inference community.

Contents

Contributors

Joseph Antonelli
University of Florida
Gainsville, FL

Mike Baiocchi
Stanford University
Stanford, CA

Eli Ben-Michael
University of California
Berkeley, CA

Matteo Bonvini
Carnegie Mellon University
Pittsburgh, PA

Zach Branson
Carnegie Mellon University
Pittsburgh, PA

Matias Cattaneo
Princeton University
Princeton, NJ

Ting-Hsuan Chang
Bloomberg School of Public Health, Johns Hopkins University
Baltimore, MD

Donna Coffman
Temple University
Philadelphia, PA

Eric Cohn
Harvard T. H. Chan School of Public Health, Harvard University
Boston, MA

Vincent Dorie
Code for America

Avi Feller
University of California
Berkeley, CA

Colin B. Fogarty
University of Michigan
Ann Arbor, MI

Laura Forastiere
Yale University
New Haven, CT

Mark Fredrickson
University of Michigan
Ann Arbor, MI

Robert Greevy, Jr.
Vanderbilt University
Nashville, TN

Ben B. Hansen
University of Michigan
Ann Arbor, MI

Erin Hartman
University of California
Berkeley, CA

Jennifer Hill
Columbia University
New York, NY

Jesse Hsu
University of Pennsylvania
Philadelphia, PA

Hyunseung Kang
University of Wisconsin
Madison, WI

Bikram Karmakar
University of Florida
Gainsville, FL

Kosuke Imai
Harvard University
Cambridge, MA

Luke Keele
University of Pennsylvania
Philadelphia, PA

Edward Kennedy
Carnegie Mellon University
Pittsburgh, PA

Kwonsang Lee
Seoul National University
Seoul, Korea

Fan Li
Duke University
Durham, NC

Bo Lu
Ohio State University
Columbus, OH

Alessandra Mattei
University of Florence
Florence, Italy

Daniel F. McCaffrey
ETS
Princeton, NJ

Alec McClean
Carnegie Mellon University
Pittsburgh, PA

Fabrizia Mealli
University of Florence
Florence, Italy

George Perrett
New York University
New York

Samuel Pimentel
University of California
Berkeley, CA

Sherri Rose
Stanford University
Stanford, CA

Paul R. Rosenbaum
Wharton School, University of Pennsylvania
Philadelphia, PA

Megan S. Schuler
RAND Corporation
Pittsburg, PA

Dylan S. Small
Wharton School, University of Pennsylvania
Philadelphia, PA

Elizabeth Stuart
Johns Hopkins University
Baltimore, MD

Elizabeth Tipton
Northwestern University
Evanston, IL

Rocío Titiunik
Princeton University
Princeton, NJ

Trang Q. Nguyen
Johns Hopkins University
Baltimore, MD

Mark van der Laan
University of California
Berkeley, CA

Stefan Wager
Stanford University
Stanford, CA

Ruoqi Yu
University of Illinois
Urbana-Champaign, IL

José R. Zubizarreta
Harvard Medical School, Harvard University
Boston, MA

About the Editors

José Zubizarreta, PhD, is professor in the Department of Health Care Policy at Harvard Medical School and in the Department of Biostatistics at Harvard T.H. Chan School of Public Health. He is also a Faculty Affiliate in the Department of Statistics at Harvard University. He is a Fellow of the American Statistical Association and is a recipient of the Kenneth Rothman Award, the William Cochran Award, and the Tom Ten Have Memorial Award.

Elizabeth A. Stuart, PhD, is Bloomberg Professor of American Health in the Department of Mental Health, the Department of Biostatistics and the Department of Health Policy and Management at Johns Hopkins Bloomberg School of Public Health. She is a Fellow of the American Statistical Association and the American Association for the Advancement of Science, and is a recipient of the the Gertrude Cox Award for applied statistics, Harvard University's Myrto Lefkopoulou Award for excellence in Biostatistics, and the Society for Epidemiologic Research Marshall Joffe Epidemiologic Methods award.

Dylan S. Small, PhD, is the Universal Furniture Professor in the Department of Statistics and Data Science of the Wharton School of the University of Pennsylvania. He is a Fellow of the American Statistical Association and an Institute of Mathematical Statistics Medallion Lecturer.

Paul R. Rosenbaum is emeritus professor of Statistics and Data Science at the Wharton School of the University of Pennsylvania. From the Committee of Presidents of Statistical Societies, he received the R. A. Fisher Award and the George W. Snedecor Award. He is the author of several books, including Design of Observational Studies and Replication and Evidence Factors in Observational Studies.

Part I

Conceptual Issues

1

Overview of Methods for Adjustment and Applications in the Social and Behavioral Sciences: The Role of Study Design

Ting-Hsuan Chang and Elizabeth A. Stuart

CONTENTS

1.1 Introduction to Causal Effects and Non-experimental Studies

Determining the causal, rather than just associational, relationship between an exposure and outcome is the ultimate goal of many studies in the social, behavioral, and health sciences. The "exposures" of interest might correspond, for example, to policies (such as a state-wide opioid prescribing policy), programs (such as an early childhood program to increase literacy), treatments, and environmental factors. Outcomes of interest might include those measured in the short-term (e.g., reading levels at the end of a program) or longer-term outcomes (e.g., college graduation rates). In this chapter we provide an overview of methods to adjust for factors that may confound the relationship between the exposure and outcome(s) of interest, including discussion of some of the particular challenges that emerge in the social and behavioral sciences.

To formalize the goal of inference for the sorts of studies discussed, the causal effect of an exposure at the individual level is commonly defined based on Rubin's causal model as the difference between an individual's outcome under the exposure and their outcome without the exposure [1,2]. Formally, let Y_{1i} and Y_{0i} denote the "potential" outcome under exposure and no exposure for

DOI: 10.1201/9781003102670-1

individual i, respectively. The causal effect for individual i is defined as $Y_{1i} - Y_{0i}$. This definition of the individual causal effect entails the assumptions that an individual's outcome is unaffected by whether or not other individuals are exposed and that the exposure of interest is well-defined – these assumptions together form the stable unit treatment value assumption, or SUTVA for short [3]. In epidemiology the second piece of this is sometimes known as "consistency" and is encapsulated by the assumption that the outcome observed when someone receives the treatment is actually Y_{1i} and the outcome observed when someone receives the comparison condition is actually Y_{0i}: $Y_i = T_i \times Y_{1i} + (1 - T_i) \times Y_{0i}$, where Y denotes the observed outcome and T denotes the exposure status ($T = 1$ if exposed and $T = 0$ if not exposed).

The fundamental problem of causal inference refers to the fact that $Y_{1i} - Y_{0i}$ can never be directly observed because for each individual, only their potential outcome corresponding to their actual exposure status is observed [2]. That is, if individual i were exposed, we would only observe Y_{1i} and not Y_{0i} and vice versa for those who receive the comparison condition. This inability to directly observe the quantity of interest is what distinguishes causal inference from standard statistical inference. And because individual causal effects are inherently unobservable directly, the goal of estimation is instead often an average treatment effect, for example, the average across all individuals (average treatment effect; ATE), defined as $E[Y_{1i} - Y_{0i}]$. Other estimands are also sometimes of interest, such as the average treatment effect on the treated (ATT), defined as $E[Y_{1i} - Y_{0i} | T_i = 1]$.

Randomized controlled trials (RCT) are often seen as the gold standard for estimating causal effects because the ATE can be readily estimated from the study sample as the difference in the mean outcome between the exposed and unexposed. This is justified by the property that in expectation randomization produced two groups (exposed and unexposed, sometimes termed treated and control) comparable with respect to all pre-exposure characteristics. Thus, the outcome of the exposed group is expected to be representative of the outcome of the unexposed group (or all individuals in the study) had they all been exposed, and vice versa; this property is often referred to as "exchangeability" in the epidemiologic literature [4, 5]. $E[Y_{1i}]$ can thus be estimated by the mean outcome among the exposed, and similarly for $E[Y_{0i}]$, and the difference in outcomes observed in the two groups yields an unbiased estimate of the causal effect.

However, randomization is often hindered by ethical concerns and feasibility; for example, it is unethical to randomize individuals to potentially harmful exposures, such as childhood maltreatment. When randomization is not possible, non-experimental studies are often conducted to estimate causal effects, but require addressing a set of challenges first. The main challenge, and also the most common criticism of non-experimental studies, is that of confounding. In the presence of confounding, the outcome among the exposed is not representative of the outcome among the unexposed had they been exposed (for more formal definitions of confounding, see VanderWeele [5] and VanderWeele and Shpitser [6]). This is driven by differences between the exposed and unexposed groups in the distributions of covariates that predict both the exposure and outcome. For example, prognostic factors often predict whether a patient would receive treatment as well as their outcome; thus, treated individuals may be on average in worse health than untreated individuals and the contrast of outcomes between the two groups may not accurately reflect the effect of treatment.

Non-experimental studies thus must grapple with confounding and use strategies to adjust for potential confounders. Such approaches for dealing with confounding often rely on some degree of substantive knowledge of the causal structure, specifically covariates that associate with both the exposure and outcome. Causal graphs (or "causal directed acyclic graphs") can be a useful visual tool for identifying sources of confounding. A causal graph encodes the investigator's assumptions of the underlying relationships among the exposure, outcome, and other covariates (either measured or unmeasured) relevant to the research question. In a causal graph, covariates are represented as nodes and arrows represent causal relationships. Assuming the graph is correct, a set of rules have been developed and mathematically proven to identify confounders (i.e., covariates that need to be measured and adjusted for in order to obtain unconfounded effect estimates) from causal graphs [7, 8].

A common assumption, then, of non-experimental studies is that of unconfounded treatment assignment, also known as conditional exchangeability or "no unmeasured confounding" [1]. Formally, this is stated as $(Y_0, Y_1) \perp\!\!\!\perp T|\boldsymbol{X}$, i.e., that treatment assignment is independent of the potential outcomes given the observed covariates $\boldsymbol{X}$. This assumption essentially implies that the observed confounders are sufficient to address confounding – that once the observed confounders are "dealt with" (more on that below), there are no unobserved confounders that would bias the treatment effect estimate. Of course a key challenge is that it is impossible to test this assumption – due to the unobserved potential outcomes – and so researchers use design and analysis strategies to make the assumption more plausible, and sensitivity analyses to assess robustness to an unobserved confounder (see Section 1.5 for more details on those sensitivity analyses).

Approaches to confounding adjustment in non-experimental studies can be loosely categorized as either "analysis-based" or "design-based" [9]. Analysis-based approaches typically involve direct modeling of the outcome; for example, a regression model of the outcome with the exposure and potential confounders as predictors. Design-based approaches include methods that attempt to create or adjust a sample (a subset of the original sample obtained by matching/grouping individuals, or a weighted sample) with covariate balance between the exposed and unexposed, mimicking an RCT at least with respect to the observed covariates [9–11]. Importantly, the "design" of non-experimental studies involves "all contemplating, collecting, organizing, and analyzing of data that takes place prior to seeing any outcome data" [10] (p. 21), and design strategies might include some of the statistical methods discussed further in this chapter, as well as careful design elements such as careful subset selection by restricting analysis to individuals in the same geographic area as those in the exposed group (see [12–14] for more examples of clever design elements).

Design strategies are particularly crucial in non-experimental studies because reducing confounding is a first-order concern. In a randomized experiment, for example, researchers may be able to optimize design elements to reduce variance because unbiasedness is provided by the randomization, whereas in a non-experimental study, reducing bias due to confounders is of key importance. Because of this, some intuition about the design of non-experimental studies may differ from that for experiments – for example, subsetting the data to a smaller sample with better covariate balance (as discussed further below) may be a very appropriate choice in a non-experimental study.

A focus on non-experimental study design – as discussed in this chapter – has three key benefits. First, the design stage is done without use of the outcome data, setting up a clear separation between the design stage and the effect estimation, as happens naturally in an experiment. As stated by Rubin [15], "Arguably, the most important feature of experiments is that we must decide on the way data will be collected before observing the outcome data. If we could try hundreds of designs and for each see the resultant answer, we could capitalize on random variation in answers and choose the design that generated the answer we wanted! The lack of availability of outcome data when designing experiments is a tremendous stimulus for honesty in experiments and can be in well-designed observational studies as well" (p. 169). This attitude is echoed in the education literature by Murnane and Willett [16]: "The better your research design, the simpler your data analysis" (p. 48). Also, as Ho et al. [17] stated, the design stage can be viewed as a nonparametric preprocessing stage for effect estimation, which can often make the effect estimates less sensitive to the outcome analysis choice.

A second key benefit of a design approach is that there are clear diagnostics for its success. That is, after a matched or weighted sample (methods discussed further below) is created, covariate balance of this sample can be checked prior to using the outcome data to estimate the exposure effect [9].

Finally, a third key benefit is a reduction in potential extrapolation between the exposed and unexposed groups and a resulting reliance on a specific regression model form [17]. In particular, if the exposed and unexposed groups differ substantially on a covariate, the resulting effect estimate can be very sensitive to the underlying regression model, which essentially extrapolates from one group to the other. For example, if the exposed group is much younger than the unexposed group, then the implicit imputation of the missing potential outcomes under exposure for the unexposed group is going to be based on a model of that potential outcome fit among people much younger than the people for whom the prediction is being made (and vice versa for imputing the missing potential outcome for the exposed group). While researchers are generally cautioned against out of sample prediction and extrapolation, this can often happen in a causal inference context without researchers being fully aware (see Imai et al. [18], for more discussion of this point). Rubin [15] lays out the conditions under which regression adjustment for confounders is "trustworthy" (i.e., when a design approach is needed in addition to an analysis approach) and shows that analysis methods are sufficient only if the covariates differ across groups by less than 0.1 or 0.2 standard deviations.

This idea of a clear separation of design and outcome analysis traces back to the concepts discussed by William G. Cochran [19], who stressed that non-experimental studies should be planned carefully as if a RCT were conducted. These ideas have re-emerged in the epidemiology literature through emphasis on "trial emulation," which extends some of these basic study design ideas to longitudinal settings, such as when using electronic health record data to study the effects of healthcare interventions on outcomes [20].

It is also the case that combining the analysis- and design-based approaches – for example, regression adjustment on the resulting sample of a design-based method – has been shown to work better than either approach alone [21, 22]. This is similar to the concept of "doubly robust" methods [23–25], in that regression adjustment may help reduce any remaining imbalance in the sample after a design-based method is implemented [26]. Common analysis- and design-based methods are presented in 1.3.

We now turn to more specifics on the approaches that can be used to adjust for confounders in a design stage.

1.2 Metrics for Adjustment

The first step for adjustment is to define a distance used to determine the similarity between two individuals. A distance can be defined using all of the covariates individually or through a single scalar that summarizes the covariates. The simplest version of the former is exact matching, which defines "similar" as having the exact same values on all of the covariates. Exact matching ensures no remaining imbalance between the exposed and unexposed; however, it is often difficult to find exact matches, especially when there are many covariates or when the covariates are on a continuous scale. Coarsened exact matching (CEM) is a less restrictive alternative that defines "similar" as falling within certain ranges of the covariates (e.g., in the same age group rather than having the exact same age) [27]. Another approach, fine balance matching, aims to achieve exact balance on a set of covariates between the exposed and unexposed groups, but does not require individuals to be matched exactly [28].

Two common covariate summary measures are the Mahalanobis distance and the propensity score. The Mahalanobis distance is essentially the standardized difference between two covariate vectors. The propensity score is defined as the probability of being exposed given the measured covariates [29]. Propensity scores are often obtained by generating the fitted values from a logistic regression model of exposure on the observed covariates, but nonparametric methods such as generalized boosted models [30, 31], neural networks [32], and Super Learner [33] have been shown to be promising alternatives for propensity score estimation given their flexibility.

A key property of the propensity score that justifies it as a measure for similarity is that it is a "balancing score," meaning that conditional on the propensity score, the distribution of the covariates entered in the propensity score model is balanced between the exposed and unexposed [29] (see Chapter 2 for more details on propensity scores). When there are many covariates, matching individuals using propensity scores often leads to better matched samples than using the Mahalanobis distance [34]. However, because the propensity scores can only be estimated in non-experimental studies – they are not directly observed or known –, investigators should check the resulting balance from the proposed propensity score model and choose a model that achieves adequate balance. For recent reviews on the use of propensity scores for adjustment in non-experimental studies, see for example, Desai and Franklin [35] and Pan and Bai [36]. A more recent approach, entropy balancing, uses an algorithm to find individual weights such that the weighted exposed and unexposed groups satisfy a set of prespecified balance conditions, typically involving equalization of covariate moments [37]. Thus, unlike the propensity score approach, entropy balancing does not require iteratively checking covariate balance and refining models.

1.3 Design Strategies for Adjustment

In this section we briefly summarize the main ways of using design strategies to adjust for confounders: matching, weighting, and stratification, and contrast those design approaches with covariate adjustment in an outcome model.

1.3.1 Matching

Matching methods generally involve selecting a subset of the sample such that the observed covariates are balanced across exposure groups in this subset. The most common method is k:1 nearest neighbor matching, which is usually done by matching k controls to an exposed individual based on the chosen distance measure and discarding unmatched controls [26,38]. Thus, k:1 nearest neighbor matching is mostly used to estimate the ATT [26]. In addition to choosing a distance measure, there are several additional specifications to consider with k:1 nearest neighbor matching, including the number of matches (i.e., the value of k), the closeness of matches, the matching algorithm, and matching with or without replacement. Here we provide a brief discussion of these specifications; for more details we refer the reader to Stuart [26] and Greifer and Stuart [9]. We also refer the reader to Greifer [39] for a comprehensive and updated introduction to the R package MatchIt, which provides implementation of several matching methods as well as balance assessment [40].

1:1 nearest neighbor matching selects for each exposed individual a single control individual who is most similar to them (e.g., has the closest propensity score value), but this may result in many control individuals being discarded. Increasing k retains a larger sample size, which may improve precision of the treatment effect estimate, but may lead to larger bias since some of the matches will likely be of lower quality than the closest match [41,42]. One way to mitigate this bias-variance tradeoff is to perform variable ratio matching, which allows each exposed individual to have a different number of matches [43]. To ensure the quality of matches, one may also impose a prespecified caliper distance, usually on the difference in the propensity scores of matched individuals (e.g., 0.2 of the standard deviation of the logit of the propensity score [44]); exposed individuals with no controls falling within the caliper are discarded. This then will estimate a restricted ATT – the effect of the treatment among the treated group with matches – which will affect for whom inferences can be drawn.

One question is how exactly the matches are obtained. Two commonly used matching algorithms are greedy matching and optimal matching. Greedy matching forms matched pairs/sets one exposed individual at a time, and the order in which the exposed individuals are matched may affect the quality of matches [26,45]. Optimal matching is "optimal" in the sense that the matched sample is formed such that the average distance across all matched pairs/sets is minimized [46], but it is more computationally intensive and may be challenging to implement in large samples. Studies have shown that the resulting matched sample usually does not differ much between the two matching algorithms [34,45].

Matching without replacement requires control individuals to be matched with at most one exposed individual, whereas matching with replacement allows control individuals to be reused in matching. Once again investigators are faced with a bias-variance tradeoff as matching with replacement may decrease bias by selecting the closest match for each exposed individual, but the matched sample may only retain a small number of controls [26]. Matching with replacement may also complicate subsequent analyses since the repeated use of controls should be taken into account [26,47].

Additionally, optimization methods can be used to improve matching. A common optimization-based matching method is cardinality matching, which maximizes the size of the matched sample under covariate balance constraints specified by the investigator [48,49]. Some of the recently developed optimization-based matching methods can be found in Cho [50] and Sharma et al. [51].

1.3.2 Weighting

Weighting adjustments typically involve direct use of the estimated propensity scores to assign weights for each individual. For estimating the ATE, inverse probability of treatment weights (IPTW) are commonly used [52]; these are $\frac{1}{\hat{e}_i}$ for those exposed and $\frac{1}{1-\hat{e}_i}$ for those unexposed, where $\hat{e}_i$ represents the estimated propensity score for individual i. These weights are applied to weight both exposure groups to resemble the combined group with respect to the observed covariates. For estimating the ATT, exposed individuals receive a weight of 1, and unexposed individuals are assigned a weight equal to $\frac{\hat{e}_i}{1-\hat{e}_i}$, which weights the unexposed group to resemble the exposed group.

A potential drawback of the inverse probability weighting approach is that extreme weights (i.e., extremely small or extremely large propensity scores) may lead to bias and large variance in the effect estimate [15,26]. Extreme weights may signal that the propensity score model is severely misspecified [24,53], or that the exposed and exposed groups have a lack of overlap in the observed covariates (i.e., exposed and exposed groups are characteristically very different) [54].

To address the problems from extreme weights, a common solution is weight trimming. Weight trimming may refer to the exclusion of individuals with weights falling outside a pre-specified range [55,56], or the reduction of weights larger than some pre-specified maximum to that maximum [57]. Reducing large weights to some maximum value has been shown to help with reducing bias and variance when the weights are estimated using logistic regression [53]. However, there is currently little guidance on the optimal way to trim weights in the context of extreme propensity scores, and it is suggested that more attention should be paid on the specification of the propensity score model, with some evidence that some of the machine learning based estimation approaches are less likely to lead to extreme weights – and thus less need for weight trimming, as compared to logistic regression [53].

Overlap weighting is a more recently developed method to address some of the issues of inverse probability weighting [54,58]. The overlap weight is defined as the estimated probability of being in the exposed group ($\hat{e}_i$) for unexposed individuals, and being in the unexposed group ($1 - \hat{e}_i$) for exposed individuals. Details on overlap weighting are discussed in Chapter 14. In brief, the estimand of overlap weighting estimators is the average treatment effect in the overlap population (i.e., the population that has both exposed and unexposed individuals), and overlap weights have been shown to be advantageous in the presence of extreme propensity scores compared to other weighting methods [54,58]. If propensity score model misspecification (which may fail to achieve adequate balance and lead to biased results) is of great concern, an alternative approach is to directly solve for weights that target the balance of the sample at hand [59]. Details on such balancing approach to weighting are discussed in Chapter 16.

1.3.3 Stratification

Stratification refers to dividing the sample into mutually exclusive strata of individuals who are similar, usually based on quantiles of the propensity scores. The distribution of the observed covariates is thus expected to be roughly identical between the exposure groups within each strata. The number of strata should be chosen such that the strata are small enough to have adequate balance, but large enough to get stable within-stratum treatment effect estimates [26]. Five has been the most common number of strata in applications [60, 61], but more work on stratification approaches is needed; recent studies on the optimal number of strata can be found in, e.g., Neuhäuser et al. [62] and Orihara and Hamada [63].

A method that overcomes the difficulty in choosing the number of strata for stratification, and which generally provides better confounder adjustment, is full matching [64]. Unlike most matching methods, which may discard many unmatched individuals, full matching forms subsets that contain at least one treated and at least one control using all individuals in the data [65, 66]. Full matching also differs from stratification in that it forms subsets in an optimal way, such that the number of stratum and the stratification of individuals are optimized to minimize the average within-stratum distance between exposure groups. Interestingly, full matching can be thought of as an approach somewhat in between matching, weighting, and stratification, in that it is implemented by creating many (small) strata, it is called "matching," and effect estimation following full matching often involves a weighted analysis, using weights determined by the full matching strata. More discussions on full matching are presented in Chapter 5.

1.3.4 Covariate adjustment in a model

The most common analysis-based method involves constructing an outcome model with the exposure indicator and the observed covariates as predictors. Propensity scores can also be used in the outcome model as a replacement for the observed covariates. These approaches, however, may result in model extrapolation if there is insufficient overlap in the covariates between the exposed and unexposed individuals [26]. Moreover, it is often more difficult to correctly specify the outcome model than it is to correctly specify the treatment model. A "doubly robust" approach, where one would use the aforementioned adjustment methods to form a balanced sample and model the outcome using this sample, is recommended instead [67]. More recently, machine learning techniques, such as Bayesian additive regression trees (BART), have been utilized to model the outcome flexibly with the covariates and have been shown to be more effective than traditional regression adjustment [68]. See Chapter 20 for more details on such approaches.

1.3.5 The role of balance checking

A key aspect of all of the design approaches described above is that of balance checking – assessing whether the design strategy was effective at creating groups similar on the observed covariates. A variety of balance measures exist – basically the idea is to compare the distribution of the covariates between the adjusted exposed and unexposed groups, such as comparing covariate means in the matched/stratified samples, or weighted means if weighting is used. A common metric used to summarize covariate balance is the absolute standardized mean difference, defined as the absolute difference in means of a covariate divided by the standard deviation: $|\frac{\bar{X}_1 - \bar{X}_0}{s}|$, where $\bar{X}_1$ and $\bar{X}_0$ are the covariate means (after matching, stratification, or weighting) for the exposed and unexposed groups, respectively, and s is typically the pooled standard deviation from the exposed and unexposed groups combined. An absolute standardized mean difference below 0.1 is generally considered adequate balance [69–71].

1.4 Applications of Design Approaches for Covariate Adjustment

Our discussion on adjustment methods has focused on their use in non-experimental settings for comparing exposed and unexposed groups. It is important to note that adjustment methods can be used more broadly under different contexts, wherever there is interest in comparing two groups who have been equated on a set of characteristics. To illustrate, we present an example for each context below.

1.4.1 Non-experimental studies

In non-experimental studies, adjustment methods are used to induce comparability between the exposed and unexposed. In Amoah et al. [72], the investigators seeked to answer the question "Does transitioning patients with gram-negative bloodstream infections from intravenous to oral therapy impact 30-day mortality?" They used observational data from a cohort of patients with monomicrobial Enterobacteriales bloodstream infections who were hospitalized at one of three selected hospitals during a specified period of time. The "exposed" patients were those who switched from intravenous therapy to oral therapy and the "unexposed" patients were those who remained on intravenous therapy. Interest was in comparing the 30-day mortality between the two groups. The investigators collected data on potential confounders, i.e., variables that likely affected both the treatment switch decision and 30-day mortality, for each patient. The adjustment methods they used included propensity score matching, propensity score weighting (IPTW), propensity score stratification, and regression adjustment (a multivariable regression model of 30-day mortality, with the exposure status and measured confounders as predictors). For the design-based methods, the propensity scores were estimated using multivariable logistic regression. For matching, the investigators performed 1:1 nearest neighbor matching without replacement, based on the propensity score and with a caliper of 0.2 standard deviation of the logit of the propensity score. For weighting, in order to avoid issues with extreme weights, the investigators excluded patients whose propensity scores were higher than the 99th percentile of the unexposed group and smaller than the first percentile of the exposed group. After a matched or weighted sample was formed, the comparison in 30-day mortality between the two groups was done by modeling the outcome on the matched or weighted sample, with additional adjustments for covariates. For stratification, the investigators created five strata based on quintiles of the propensity score. Comparison of the two groups was done within each stratum using the same "doubly robust" analysis strategy as the matching and weighting methods, and the final effect estimate was obtained by pooling the stratum-specific comparisons.

1.4.2 Non-response adjustments

Non-response adjustments are commonly used in studies involving surveys to "equate" the responders and non-responders on baseline characteristics [73, 74]. The purpose is to improve the representativeness of the survey sample so that results derived from the survey responses are not biased by differences between the responders and non-responders ("nonresponse bias"). In Simonetti et al. [75], the goal was to assess whether nonresponse bias affected estimates of burnout prevalence from a nationwide survey of Veterans Health Administration (VHA) primary care employees. Administrative data on all employees was available, including demographics and clinic type. Responders and nonresponders were found to differ on several characteristics (e.g., responders were more likely to be female). In this context the propensity score is an employee's probability of responding to the survey, which was estimated using multivariable logistic regression with variables obtained from the administrative data included as predictors. The investigators then performed propensity score stratification with five strata and calculated the response rate in each stratum. To estimate the burnout prevalence among all employees (including both responders and non-responders), the survey responses were weighted by the inverse of this stratum-specific response rate.

1.4.3 Generalizability of randomized trial results to target populations

Randomized trials are often considered ideal for estimating exposure effects. However, similar to the context of survey analysis, trial participants may be characteristically different from non-participants, which could mean that the effect seen in the trial does not reflect the effect in some target population of interest – a problem of external validity. This is especially concerning if there is effect heterogeneity by subgroups that are disproportionately represented in the trials [76, 77]. In Susukida et al. [78], the investigators were interested in comparing the effect estimates of a substance use disorder (SUD) treatment obtained from randomized trials ("sample treatment effects") to the "population treatment effects" (i.e., treatment effects in the target populations). To make trial participants representative of patients in the target populations, they weighted each trial participant by the weight $\frac{1-\hat{p}}{\hat{p}}$, where $\hat{p}$ is their estimated probability of participating in the trial conditional on a set of variables measured comparably between the trials and target sample datasets. The weighted analyses aimed to estimate population treatment effects, whereas unweighted analyses could only estimate sample treatment effects.

1.4.4 Mediation analysis

Mediation analysis attempts to characterize how an exposure affects outcomes; specifically, how much of the exposure effect is explained by a mediating factor [79]. Jo et al. [80] proposed the use of propensity scores in mediation analyses, which we briefly describe here. The example in Jo et al. [80] investigated how training sessions for job search (exposure) among high-risk individuals reduced depression (outcome) through improving sense of mastery (mediator). For simplicity, the authors categorized the individuals into two types: "treatment improvers" were those who improved their sense of mastery when receiving the training sessions; and "treatment non-improvers" were those who did not improve their sense of mastery when receiving the training sessions. To estimate the effect of training sessions among treatment improvers (the same approach applies to treatment non-improvers), the investigators first estimated each individual's probability of being a treatment improver given covariates, which is the "propensity score" in this setting. Since the study was a randomized trial (participants were randomly assigned to receive training sessions), the propensity score model was fit using the treatment group members only. Next, individuals in the control group were weighted by their estimated propensity score to resemble the treatment improvers. Treatment improvers in the treatment group received a weight of 1, whereas treatment non-improvers in the treatment group received a weight of 0. The treatment effect among treatment improvers was then estimated using the weighted sample.

1.5 Sensitivity Analyses to an Unobserve Confounder

While the adjustment methods described above can address confounding by observed covariates, causal inferences from non-experimental studies are inevitably threatened by unmeasured confounding as the assumption of no unmeasured confounding can never be directly tested. Designing studies in certain ways can increase robustness to unobserved confounders, a concept termed "design sensitivity" [81, 82]. Rosenbaum [12] listed several design choices that would affect the quality of effect estimates – this includes choices of the research hypothesis, exposed and unexposed groups, and more. According to Rosenbaum [83], a well-designed study is one that "does more to address the inevitable critical discussion of possible bias from unmeasured covariates" (p. 145).

Sensitivity analysis assesses the robustness of effect estimates to unobserved confounders. Various techniques for sensitivity analysis have been developed over the years (e.g., [84–92]). Sensitivity analysis considers how strong the associations between a hypothetical unobserved confounder and the exposure and outcome would need to be to change the study conclusion. These techniques typically require a number of assumptions (e.g., that the unobserved confounder is binary [84]) or user-specified parameters such as sensitivity parameters (the associations between an unobserved confounder and the exposure and outcome) and the prevalence of the unobserved confounder in the exposed and unexposed groups (e.g., [85, 86, 90]). Newer approaches have been developed to accommodate more general scenarios and require fewer parameter specifications. For example, the E-value is a measure that represents the minimum strength of association between an unobserved confounder and the exposure and outcome for the unobserved confounder to explain away the observed effect [93]; it makes no assumptions regarding the type, distribution, or number of unobserved confounders and can be computed with a variety of outcome types [94]. Another recent approach is one that seeks to find the minimum proportion of confounded individuals such that the results indicate a null average treatment effect, which has the benefit of capturing unmeasured confounding heterogeneity [92]. See Chapter 25 for more discussion of these and related methods.

1.6 Conclusion

Although much focus has been on confounding in existing literature on non-experimental studies, the validity of results from non-experimental studies hinges on many factors including the specification of the causal question, design of the study, analysis and modeling choices, and data quality/usefulness [95].

There are also a number of complications that arise in practice, such as missing data, multilevel structures, and measurement challenges. Missing values can occur in the exposure, outcome, and covariates. Although missing data is common in randomized studies as well, non-experimental studies tend to be more vulnerable to missing data in that measurements of covariates are necessary for confounding adjustments [96]. Causal diagrams can be extended to encode assumptions about the missingness mechanisms (e.g., missing at random, missing not at random [97–99]) and studies have shown the importance of choosing the appropriate method for dealing with missing data (e.g., complete case analysis or multiple imputation) based on different missingness mechanisms [96, 100].

In non-experimental studies with a multilevel structure, individuals are nested within one or more clusters (e.g., patients within hospitals, students within classrooms) and covariates are often measured at both the individual and cluster levels (see Chapter 10). The exposure may be assigned at either the cluster or individual level, which would have different implications for adjustment approaches as well as the underlying assumptions. Chapter 10 provides an introduction to matching methods in multilevel settings, with a special focus on the effects of cluster-level exposures on individual-level outcomes. Some studies have extended adjustment methods using propensity scores, including matching [101, 102] and weighting [103], to multilevel settings with individual-level exposures.

Measurement can also be a particular challenge in the social and behavioral sciences, and propensity score methods need to be adjusted for use when some of the covariates are latent variables, measured through a set of observed items [104, 105].

The validity or plausibility of the key assumptions and the need to think about these assumptions may differ across fields. For example, in non-experimental studies in fields such as psychology, economics, and education, participants often self-select into the exposure condition (e.g., a job training program), whereas in medical contexts, the treatment decision process is often based on a set of factors (e.g., lab tests, patient age) in compliance with certain medical guidelines. In the latter case, investigators may have more information regarding the assignment mechanism and thus the "no unmeasured confounding" assumption may be more likely to hold compared to the former case. The availability of data may also differ across fields. For example, when using administrative data in educational settings to make causal inferences, covariate adjustment is often limited by the variables that are available in the data. Researchers should also use their substantive knowledge about the setting to assess the validity of the underlying assumptions. In short, different choices and assumptions may need to be made depending on the scientific context and the data that is typically available in that context, and it is crucial to assess the underlying assumptions within the context of each specific study.

References

[1] Donald B Rubin. Estimating causal effects of treatments in randomized and nonrandomized studies. *Journal of Educational Psychology*, 66(5):688–701, 1974.

[2] Paul W Holland. Statistics and causal inference. *Journal of the American Statistical Association*, 81(396):945–960, 1986.

[3] Donald B Rubin. Randomization analysis of experimental data: The fisher randomization test comment. *Journal of the American Statistical Association*, 75(371):591–593, 1980.

[4] Sander Greenland and James M Robins. Identifiability, exchangeability, and epidemiological confounding. *International Journal of Epidemiology*, 15(3):413–419, 1986.

[5] Tyler J VanderWeele. Confounding and effect modification: Distribution and measure. *Epidemiologic Methods*, 1(1):55–82, 2012.

[6] Tyler J VanderWeele and Ilya Shpitser. On the definition of a confounder. *Annals of Statistics*, 41(1):196–220, 2013.

[7] Judea Pearl. Causal diagrams for empirical research. *Biometrika*, 82(4):669–688, 1995.

[8] Sander Greenland, Judea Pearl, and James M Robins. Causal diagrams for epidemiologic research. *Epidemiology*, 10(1):37–48, 1999.

[9] Noah Greifer and Elizabeth A Stuart. Matching methods for confounder adjustment: An addition to the epidemiologist's toolbox. *Epidemiologic Reviews*, 43(1):118–129, 2021.

[10] Donald B Rubin. The design versus the analysis of observational studies for causal effects: Parallels with the design of randomized trials. *Statistics in Medicine*, 26(1):20–36, 2007.

[11] Donald B Rubin. For objective causal inference, design trumps analysis. *The Annals of Applied Statistics*, 2(3):808–840, 2008.

[12] Paul R Rosenbaum. Choice as an alternative to control in observational studies. *Statistical Science*, 14(3):259–304, 1999.

[13] Paul R Rosenbaum. *Observational Studies.* Springer, New York, NY, 2002.

[14] Paul R. Rosenbaum. *Design of Observational Studies.* Springer-Verlag, New York, 2010.

[15] Donald B Rubin. Using propensity scores to help design observational studies: Application to the tobacco litigation. *Health Services and Outcomes Research Methodology*, 2(3):169–188, 2001.

[16] Richard J Murnane and John B Willett. *Methods Matter: Improving Causal Inference in Educational and Social Science Research.* New York, NY, Oxford University Press, 2010.

[17] Daniel E Ho, Kosuke Imai, Gary King, and Elizabeth A Stuart. Matching as nonparametric preprocessing for reducing model dependence in parametric causal inference. *Political Analysis*, 15(3):199–236, 2007.

[18] Kosuke Imai, Gary King, and Elizabeth A Stuart. Misunderstandings between experimentalists and observationalists about causal inference. *Journal of the Royal Statistical Society: Series A (Statistics in Society)*, 171(2):481–502, 2008.

[19] William G Cochran and S Paul Chambers. The planning of observational studies of human populations. *Journal of the Royal Statistical Society. Series A (General)*, 128(2):234–266, 1965.

[20] Miguel A Hernán and James M Robins. Using big data to emulate a target trial when a randomized trial is not available. *American Journal of Epidemiology*, 183(8):758–764, 2016.

[21] Donald B Rubin. The use of matched sampling and regression adjustment to remove bias in observational studies. *Biometrics*, 29(1):185–203, 1973.

[22] Donald B Rubin and Neal Thomas. Combining propensity score matching with additional adjustments for prognostic covariates. *Journal of the American Statistical Association*, 95(450):573–585, 2000.

[23] James M Robins and Andrea Rotnitzky. Semiparametric efficiency in multivariate regression models with missing data. *Journal of the American Statistical Association*, 90(429):122–129, 1995.

[24] Joseph DY Kang and Joseph L Schafer. Demystifying double robustness: A comparison of alternative strategies for estimating a population mean from incomplete data. *Statistical Science*, 22(4):523–539, 2007.

[25] Michele Jonsson Funk, Daniel Westreich, Chris Wiesen, Til Stürmer, M Alan Brookhart, and Marie Davidian. Doubly robust estimation of causal effects. *American Journal of Epidemiology*, 173(7):761–767, 2011.

[26] Elizabeth A Stuart. Matching methods for causal inference: A review and a look forward. *Statistical Science: A Review Journal of the Institute of Mathematical Statistics*, 25(1):1–21, 2010.

[27] Stefano M Iacus, Gary King, and Giuseppe Porro. Causal inference without balance checking: Coarsened exact matching. *Political Analysis*, 20(1):1–24, 2012.

[28] Paul R Rosenbaum, Richard N Ross, and Jeffrey H Silber. Minimum distance matched sampling with fine balance in an observational study of treatment for ovarian cancer. *Journal of the American Statistical Association*, 102(477):75–83, 2007.

[29] Paul R Rosenbaum and Donald B Rubin. The central role of the propensity score in observational studies for causal effects. *Biometrika*, 70(1):41–55, 1983.

[30] Daniel F McCaffrey, Greg Ridgeway, and Andrew R Morral. Propensity score estimation with boosted regression for evaluating causal effects in observational studies. *Psychological Methods*, 9(4):403–425, 2004.

[31] Brian K Lee, Justin Lessler, and Elizabeth A Stuart. Improving propensity score weighting using machine learning. *Statistics in Medicine*, 29(3):337–346, 2010.

[32] Soko Setoguchi, Sebastian Schneeweiss, M Alan Brookhart, Robert J Glynn, and E Francis Cook. Evaluating uses of data mining techniques in propensity score estimation: A simulation study. *Pharmacoepidemiology and Drug Safety*, 17(6):546–555, 2008.

[33] Romain Pirracchio, Maya L Petersen, and Mark Van Der Laan. Improving propensity score estimators' robustness to model misspecification using super learner. *American Journal of Epidemiology*, 181(2):108–119, 2015.

[34] Xing Sam Gu and Paul R Rosenbaum. Comparison of multivariate matching methods: Structures, distances, and algorithms. *Journal of Computational and Graphical Statistics*, 2(4):405–420, 1993.

[35] Rishi J Desai and Jessica M Franklin. Alternative approaches for confounding adjustment in observational studies using weighting based on the propensity score: A primer for practitioners. *BMJ*, 367, 2019.

[36] Wei Pan and Haiyan Bai. Propensity score methods for causal inference: An overview. *Behaviormetrika*, 45(2):317–334, 2018.

[37] Jens Hainmueller. Entropy balancing for causal effects: A multivariate reweighting method to produce balanced samples in observational studies. *Political Analysis*, 20(1):25–46, 2012.

[38] Donald B Rubin. Matching to remove bias in observational studies. *Biometrics*, 29(1): 159–183, 1973.

[39] Noah Greifer. Matchit: Getting started. `http://cran.r-project.org/web/packages/MatchIt/vignettes/MatchIt.html`. Accessed: 2022-03-27.

[40] Daniel Ho, Kosuke Imai, Gary King, and Elizabeth A Stuart. Matchit: Nonparametric preprocessing for parametric causal inference. *Journal of Statistical Software*, 42(8):1-28, 2011.

[41] Peter C Austin. Statistical criteria for selecting the optimal number of untreated subjects matched to each treated subject when using many-to-one matching on the propensity score. *American Journal of Epidemiology*, 172(9):1092–1097, 2010.

[42] Jeremy A Rassen, Abhi A Shelat, Jessica Myers, Robert J Glynn, Kenneth J Rothman, and Sebastian Schneeweiss. One-to-many propensity score matching in cohort studies. *Pharmacoepidemiology and Drug Safety*, 21:69–80, 2012.

[43] Kewei Ming and Paul R Rosenbaum. Substantial gains in bias reduction from matching with a variable number of controls. *Biometrics*, 56(1):118–124, 2000.

[44] Peter C Austin. Optimal caliper widths for propensity-score matching when estimating differences in means and differences in proportions in observational studies. *Pharmaceutical Statistics*, 10(2):150–161, 2011.

[45] Peter C Austin. A comparison of 12 algorithms for matching on the propensity score. *Statistics in Medicine*, 33(6):1057–1069, 2014.

[46] Paul R Rosenbaum. Optimal matching for observational studies. *Journal of the American Statistical Association*, 84(408):1024–1032, 1989.

[47] Peter C Austin and Guy Cafri. Variance estimation when using propensity-score matching with replacement with survival or time-to-event outcomes. *Statistics in Medicine*, 39(11):1623–1640, 2020.

[48] José R Zubizarreta, Ricardo D Paredes, and Paul R Rosenbaum. Matching for balance, pairing for heterogeneity in an observational study of the effectiveness of for-profit and not-for-profit high schools in chile. *The Annals of Applied Statistics*, 8(1):204–231, 2014.

[49] Giancarlo Visconti and José R Zubizarreta. Handling limited overlap in observational studies with cardinality matching. *Observational Studies*, 4(1):217–249, 2018.

[50] Wendy K Tam Cho. An evolutionary algorithm for subset selection in causal inference models. *Journal of the Operational Research Society*, 69(4):630–644, 2018.

[51] Dhruv Sharma, Christopher Willy, and John Bischoff. Optimal subset selection for causal inference using machine learning ensembles and particle swarm optimization. *Complex & Intelligent Systems*, 7(1):41–59, 2021.

[52] Keisuke Hirano, Guido W Imbens, and Geert Ridder. Efficient estimation of average treatment effects using the estimated propensity score. *Econometrica*, 71(4):1161–1189, 2003.

[53] Brian K Lee, Justin Lessler, and Elizabeth A Stuart. Weight trimming and propensity score weighting. *PloS One*, 6(3):e18174, 2011.

[54] Fan Li, Laine E Thomas, and Fan Li. Addressing extreme propensity scores via the overlap weights. *American Journal of Epidemiology*, 188(1):250–257, 2019.

[55] Richard K Crump, V Joseph Hotz, Guido W Imbens, and Oscar A Mitnik. Dealing with limited overlap in estimation of average treatment effects. *Biometrika*, 96(1):187–199, 2009.

[56] Til Stürmer, Kenneth J Rothman, Jerry Avorn, and Robert J Glynn. Treatment effects in the presence of unmeasured confounding: Dealing with observations in the tails of the propensity score distribution—a simulation study. *American Journal of Epidemiology*, 172(7):843–854, 2010.

[57] Michael R Elliott. Model averaging methods for weight trimming in generalized linear regression models. *Journal of Official Statistics*, 25(1):1–20, 2009.

[58] Fan Li, Kari Lock Morgan, and Alan M Zaslavsky. Balancing covariates via propensity score weighting. *Journal of the American Statistical Association*, 113(521):390–400, 2018.

[59] Eli Ben-Michael, Avi Feller, David A. Hirshberg, and José R. Zubizarreta. The balancing act in causal inference. *arXiv preprint*, 2021.

[60] Paul R Rosenbaum and Donald B Rubin. Reducing bias in observational studies using subclassification on the propensity score. *Journal of the American Statistical Association*, 79(387):516–524, 1984.

[61] Jared K Lunceford and Marie Davidian. Stratification and weighting via the propensity score in estimation of causal treatment effects: A comparative study. *Statistics in Medicine*, 23(19):2937–2960, 2004.

[62] Markus Neuhäuser, Matthias Thielmann, and Graeme D Ruxton. The number of strata in propensity score stratification for a binary outcome. *Archives of Medical Science: AMS*, 14(3):695–700, 2018.

[63] Shunichiro Orihara and Etsuo Hamada. Determination of the optimal number of strata for propensity score subclassification. *Statistics & Probability Letters*, 168:108951, 2021.

[64] Elizabeth A Stuart and Kerry M Green. Using full matching to estimate causal effects in nonexperimental studies: Examining the relationship between adolescent marijuana use and adult outcomes. *Developmental Psychology*, 44(2):395–406, 2008.

[65] Paul R Rosenbaum. A characterization of optimal designs for observational studies. *Journal of the Royal Statistical Society: Series B (Methodological)*, 53(3):597–610, 1991.

[66] Ben B Hansen. Full matching in an observational study of coaching for the sat. *Journal of the American Statistical Association*, 99(467):609–618, 2004.

[67] Heejung Bang and James M Robins. Doubly robust estimation in missing data and causal inference models. *Biometrics*, 61(4):962–973, 2005.

[68] Jennifer L Hill. Bayesian nonparametric modeling for causal inference. *Journal of Computational and Graphical Statistics*, 20(1):217–240, 2011.

[69] Sharon-Lise T Normand, Mary Beth Landrum, Edward Guadagnoli, John Z Ayanian, Thomas J Ryan, Paul D Cleary, and Barbara J McNeil. Validating recommendations for coronary angiography following acute myocardial infarction in the elderly: A matched analysis using propensity scores. *Journal of Clinical Epidemiology*, 54(4):387–398, 2001.

[70] Muhammad Mamdani, Kathy Sykora, Ping Li, Sharon-Lise T Normand, David L Streiner, Peter C Austin, Paula A Rochon, and Geoffrey M Anderson. Reader's guide to critical appraisal of cohort studies: 2. assessing potential for confounding. *BMJ*, 330(7497):960–962, 2005.

[71] Peter C Austin. Balance diagnostics for comparing the distribution of baseline covariates between treatment groups in propensity-score matched samples. *Statistics in Medicine*, 28(25):3083–3107, 2009.

[72] Joe Amoah, Elizabeth A Stuart, Sara E Cosgrove, Anthony D Harris, Jennifer H Han, Ebbing Lautenbach, and Pranita D Tamma. Comparing propensity score methods versus traditional regression analysis for the evaluation of observational data: A case study evaluating the treatment of gram-negative bloodstream infections. *Clinical Infectious Diseases*, 71(9):e497–e505, 2020.

[73] Roderick JA Little. Survey nonresponse adjustments for estimates of means. *International Statistical Review/Revue Internationale de Statistique*, 54(2):139–157, 1986.

[74] Robert M. Groves, Don Dillman, Don A. Dillman, John L. Eltinge, and Roderick J. A. Little *Survey Nonresponse*. Hoboken, NJ, Wiley, 2002.

[75] Joseph A Simonetti, Walter L Clinton, Leslie Taylor, Alaina Mori, Stephan D Fihn, Christian D Helfrich, and Karin Nelson. The impact of survey nonresponse on estimates of healthcare employee burnout. *Healthcare*, 8(3):100451, 2020.

[76] Stephen R Cole and Elizabeth A Stuart. Generalizing evidence from randomized clinical trials to target populations: The actg 320 trial. *American Journal of Epidemiology*, 172(1):107–115, 2010.

[77] Elizabeth A Stuart, Stephen R Cole, Catherine P Bradshaw, and Philip J Leaf. The use of propensity scores to assess the generalizability of results from randomized trials. *Journal of the Royal Statistical Society: Series A (Statistics in Society)*, 174(2):369–386, 2011.

[78] Ryoko Susukida, Rosa M Crum, Cyrus Ebnesajjad, Elizabeth A Stuart, and Ramin Mojtabai. Generalizability of findings from randomized controlled trials: Application to the national institute of drug abuse clinical trials network. *Addiction*, 112(7):1210–1219, 2017.

[79] Tyler VanderWeele. *Explanation in Causal Inference: Methods for Mediation and Interaction*. New York, NY: Oxford University Press, 2015.

[80] Booil Jo, Elizabeth A Stuart, David P MacKinnon, and Amiram D Vinokur. The use of propensity scores in mediation analysis. *Multivariate Behavioral Research*, 46(3):425–452, 2011.

[81] Paul R Rosenbaum. Design sensitivity in observational studies. *Biometrika*, 91(1):153–164, 2004.

[82] José R Zubizarreta, Magdalena Cerda, and Paul R Rosenbaum. Effect of the 2010 chilean earthquake on posttraumatic stress reducing sensitivity to unmeasured bias through study design. *Epidemiology*, 24(1):79–87, 2013.

[83] Paul R Rosenbaum. Modern algorithms for matching in observational studies. *Annual Review of Statistics and Its Application*, 7:143–176, 2020.

[84] Paul R Rosenbaum and Donald B Rubin. Assessing sensitivity to an unobserved binary covariate in an observational study with binary outcome. *Journal of the Royal Statistical Society: Series B (Methodological)*, 45(2):212–218, 1983.

[85] Sander Greenland. Basic methods for sensitivity analysis of biases. *International Journal of Epidemiology*, 25(6):1107–1116, 1996.

[86] Danyu Y Lin, Bruce M Psaty, and Richard A Kronmal. Assessing the sensitivity of regression results to unmeasured confounders in observational studies. *Biometrics*, 54(3):948–963, 1998.

[87] Joseph L Gastwirth, Abba M Krieger, and Paul R Rosenbaum. Dual and simultaneous sensitivity analysis for matched pairs. *Biometrika*, 85(4):907–920, 1998.

[88] David J Harding. Counterfactual models of neighborhood effects: The effect of neighborhood poverty on dropping out and teenage pregnancy. *American Journal of Sociology*, 109(3):676–719, 2003.

[89] Onyebuchi A Arah, Yasutaka Chiba, and Sander Greenland. Bias formulas for external adjustment and sensitivity analysis of unmeasured confounders. *Annals of Epidemiology*, 18(8):637–646, 2008.

[90] Tyler J VanderWeele and Onyebuchi A Arah. Bias formulas for sensitivity analysis of unmeasured confounding for general outcomes, treatments, and confounders. *Epidemiology*, 22(1):42–52, 2011.

[91] Liangyuan Hu, Jungang Zou, Chenyang Gu, Jiayi Ji, Michael Lopez, and Minal Kale. A flexible sensitivity analysis approach for unmeasured confounding with multiple treatments and a binary outcome with application to seer-medicare lung cancer data. *arXiv preprint arXiv:2012.06093*, 2020.

[92] Matteo Bonvini and Edward H Kennedy. Sensitivity analysis via the proportion of unmeasured confounding. *Journal of the American Statistical Association*, 117(539):1540-1550, 2022.

[93] Tyler J VanderWeele and Peng Ding. Sensitivity analysis in observational research: Introducing the e-value. *Annals of Internal Medicine*, 167(4):268–274, 2017.

[94] Maya B Mathur, Peng Ding, Corinne A Riddell, and Tyler J VanderWeele. Website and r package for computing e-values. *Epidemiology*, 29(5):e45–e47, 2018.

[95] Steven N Goodman, Sebastian Schneeweiss, and Michael Baiocchi. Using design thinking to differentiate useful from misleading evidence in observational research. *JAMA*, 317(7):705–707, 2017.

[96] Neil J Perkins, Stephen R Cole, Ofer Harel, Eric J Tchetgen Tchetgen, BaoLuo Sun, Emily M Mitchell, and Enrique F Schisterman. Principled approaches to missing data in epidemiologic studies. *American Journal of Epidemiology*, 187(3):568–575, 2018.

[97] Rhian M Daniel, Michael G Kenward, Simon N Cousens, and Bianca L De Stavola. Using causal diagrams to guide analysis in missing data problems. *Statistical Methods in Medical Research*, 21(3):243–256, 2012.

[98] Karthika Mohan, Judea Pearl, and Jin Tian. Graphical models for inference with missing data. In *Proceedings of the 26th International Conference on Neural Information Processing Systems*, pages 1277–1285, 2013.

[99] Felix Thoemmes and Karthika Mohan. Graphical representation of missing data problems. *Structural Equation Modeling: A Multidisciplinary Journal*, 22(4):631–642, 2015.

[100] Rachael A Hughes, Jon Heron, Jonathan AC Sterne, and Kate Tilling. Accounting for missing data in statistical analyses: Multiple imputation is not always the answer. *International Journal of Epidemiology*, 48(4):1294–1304, 2019.

[101] Bruno Arpino and Fabrizia Mealli. The specification of the propensity score in multilevel observational studies. *Computational Statistics & Data Analysis*, 55(4):1770–1780, 2011.

[102] Bruno Arpino and Massimo Cannas. Propensity score matching with clustered data. an application to the estimation of the impact of caesarean section on the apgar score. *Statistics in Medicine*, 35(12):2074–2091, 2016.

[103] Fan Li, Alan M Zaslavsky, and Mary Beth Landrum. Propensity score weighting with multilevel data. *Statistics in Medicine*, 32(19):3373–3387, 2013.

[104] Hwanhee Hong, David A Aaby, Juned Siddique, and Elizabeth A Stuart. Propensity score–based estimators with multiple error-prone covariates. *American Journal of Epidemiology*, 188(1):222–230, 2019.

[105] Trang Quynh Nguyen and Elizabeth A Stuart. Propensity score analysis with latent covariates: Measurement error bias correction using the covariate's posterior mean, aka the inclusive factor score. *Journal of Educational and Behavioral Statistics*, 45(5):598–636, 2020.

2

Propensity Score

Paul R. Rosenbaum

CONTENTS

2.1 Goal of this Chapter

In several articles in the 1980s, Donald B. Rubin and I proposed the propensity score as an aid in adjusting for observed covariates in observational studies of causal effects [1–6]. The propensity score is often used for matching, stratification, and weighting both in causal inference and in adjustments for nonresponse. This chapter presents a concise review of some of this material from the 1980s; see also [7–10]. For a survey of recent developments in multivariate matching, not necessarily related to propensity scores, see [11–13].

DOI: 10.1201/9781003102670-2

2.2 Preliminaries

2.2.1 Example: nursing and surgical mortality

The following recent observational study will be mentioned at various points to make general concepts tangible. In this study a propensity score was used in conjunction with other matching techniques. For instance, the match also used a covariate distance, a network optimization algorithm, and an externally estimated risk-of-death or prognostic score; see [5, 14, 15].

Is mortality following general surgery lower at hospitals that have superior nursing environments? Silber et al. [16] compared mortality in 35 hospitals with superior nursing and 293 hospitals with inferior nursing. They took the patients undergoing general surgery at the 35 superior hospitals and matched each one with a different patient at one of the 293 inferior hospitals, forming 25,076 matched pairs. The matching paired exactly for 130 surgical procedures (i.e., 4-digit ICD-9 codes), and it balanced a total of 172 covariates, including the patient's other recorded health problems, such as congestive heart failure or diabetes, plus age, sex, emergency admission or not, and so on. The data were from the US Medicare program.

By definition, a hospital had superior nursing if it was so designated by the Magnet Nursing Services Recognition Program and also had a nurse-to-bed ratio of at least 1; see Aiken et al. [17]. A hospital had inferior nursing if it had neither of these attributes. For brevity, surgery at a hospital with superior nursing is called "treated," while surgery at a hospital with inferior nursing is called "control." Mortality refers to death within 30 days of surgery.

The question being asked is: Would your risk of death be lower if your operation were performed at one of the 35 superior hospitals, rather than one of the 293 inferior hospitals? It is the effect of going to one existing hospital in lieu of going to another existing hospital. The question is not about the effect of rebuilding hospitals to have attributes that they do not currently have. An important subquestion of the main question is whether patients are being channeled to the appropriate hospitals, in particular, whether high-risk patients are being channeled to the most capable hospitals, and whether that matters for surgical mortality.

Hospitals with superior nursing disproportionately have other desirable attributes as well, attributes that are not part of the explicit definition of a superior hospital. Nonetheless, these are genuine features of a superior hospital, and for the question under study these features should not be removed by adjustments. Superior hospitals tended to have a larger proportion of nurses with advanced degrees, and a larger proportion of registered nurses. Superior hospitals were more often major teaching hospitals (21.5% for the superior group versus 5.7% for the inferior group). Superior hospitals were somewhat larger on average (595 beds versus 430 beds).

2.2.2 Adjustments for covariates in observational studies

In his review of observational studies, Cochran [18, §2.2] wrote:

> Blocking (or matching) and adjustments in the analysis are frequently used. There is, however, an important difference between the demands made on blocking or adjustments in controlled experiments and in observational studies. In controlled experiments, the skillful use of randomization protects against most types of bias arising from disturbing variables. Consequently, the function of blocking or adjustment is to increase precision rather than to guard against bias. In observational studies, in which no random assignment of subjects to comparison groups is possible, blocking and adjustment take on the additional role of protecting against bias. Indeed, this is often their primary role.

A precursor of propensity scores was proposed by Cochran and Rubin [19, §6.3]. They observed that if covariates had a multivariate Normal distribution in treated and control groups, with different expectations but the same covariance matrix, then it was possible to linearly transform the covariates so that all of the bias is captured by the first coordinate – the linear discriminant – and the remaining coordinates have the same distribution in treated and control groups. In this multivariate Normal case, Cochran's "primary role" for blocking and matching, namely, bias reduction, involves a unidimensional random variable, while the secondary role of blocking and matching, namely, increasing precision, involves all of the coordinates. In a sense, the basic results about propensity scores say that these properties are true in general and require none of the following: Normal distributions, linear transformations, equal covariance matrices. In the Normal case with equal covariance matrices, the linear discriminant reflects the likelihood ratio – the ratio of the density of the covariate in treated and control groups – and in all cases, by Bayes theorem, this likelihood ratio is a strictly monotone function of the propensity score.

2.2.3 Dawid's notation for conditional independence

Dawid [20] proposed the notation $A \perp\!\!\!\perp B \mid C$ for $\Pr(A \mid B, C) = \Pr(A \mid C)$. If C is not present, $A \perp\!\!\!\perp B$ is written if $\Pr(A \mid B) = \Pr(A)$.

This notation is obviously meant to imply that A is a random event, or random variable, or random object of some sort. If B is also a random variable, then $A \perp\!\!\!\perp B \mid C$ says A is conditionally independent of B given C, and this implies $B \perp\!\!\!\perp A \mid C$. However, the notation is deliberately noncommittal about the status of B, which may be a fixed quantity, perhaps a fixed parameter, rather than a random variable, in which case $\Pr(A \mid B, C) = \Pr(A \mid C)$ says the fixed parameter drops out of the conditional distribution. If $A \perp\!\!\!\perp B \mid C$ when A is all the data, C is a function of A, and B is a fixed parameter, then C is a sufficient statistic for the parameter B; see Dawid [20, §3.2]. Indeed, B could have several components, some of which are fixed while others are random. A Bayesian and a frequentist might agree that a certain argument proves $A \perp\!\!\!\perp B \mid C$, while disagreeing about whether parameters are random variables. Dawid's notation has a diplomatic ambiguity that allows it to speak to people who speak in different ways.

Dawid's [20] Lemma 4 is a quick calculus that permits rapid recognition of simple arguments, particularly when A, B and C are complex expressions whose complexity is irrelevant. For instance, Dawid's [20] Lemma 4b says: if $A \perp\!\!\!\perp B \mid C$ and D is a function of A, then $D \perp\!\!\!\perp B \mid C$ and $A \perp\!\!\!\perp B \mid (C, D)$.

2.2.4 The effects caused by treatments

There is interest in the effect caused by a treatment, where individuals are assigned to either treatment with $Z = 1$ or control with $Z = 0$. Each individual has two potential outcomes, the response r_T seen under treatment if $Z = 1$ or the response r_C seen under control if $Z = 0$, so the response actually observed from an individual is $R = Z\, r_T + (1 - Z)\, r_C$, and the effect caused by the treatment, namely $r_T - r_C$ is not observed for any individual. This potential outcomes notation for treatment effects, (r_T, r_C), is due to Neyman [21] and Rubin [22]; see also [23, §5], [24] and [25, §6]. In §2.2.1, for one individual, r_T is the binary indicator of survival for 30 days after surgery if treated at a hospital with superior nursing, and r_C is the indictor of survival for this same person if treated at a hospital with inferior nursing. For any one patient, we see either r_T or r_C, never both.

The expected treatment effect is $\mathrm{E}\,(r_T - r_C)$. It seems natural to estimate $\mathrm{E}\,(r_T - r_C)$ by the average response among treated individuals minus the average response among controls, but, in general, that difference estimates something else, not the effect caused by the treatment. To see this, note that the expected response of a treated individual is $\mathrm{E}(R \mid Z=1) = \mathrm{E}(r_T \mid Z=1)$, because a treated individual has $Z=1$ and observed response $R=r_T$ if $Z=1$. In parallel, the expected response of a control is $\mathrm{E}(R \mid Z=0) = \mathrm{E}(r_C \mid Z=0)$. In general, the expected treatment effect is not the difference in expectations in the treated and control groups; that is, in general, $\mathrm{E}(r_T - r_C) \neq \mathrm{E}(r_T \mid Z=1) - \mathrm{E}(r_C \mid Z=0)$. The two quantities are equal if treatments are assigned by flipping a fair coin; then $Z \perp\!\!\!\perp (r_T, r_C)$, precisely because the coin is fair and the probability of a head does not change with (r_T, r_C). Because $Z \perp\!\!\!\perp (r_T, r_C)$ in a completely randomized experiment, $\mathrm{E}(r_T \mid Z=1) = \mathrm{E}(r_T)$ and $\mathrm{E}(r_C \mid Z=0) = \mathrm{E}(r_C)$, so we *can* estimate the expected treatment effect by comparing the mean outcomes in treated and control groups; that is, in this case, $\mathrm{E}(r_T \mid Z=1) - \mathrm{E}(r_C \mid Z=0) = \mathrm{E}(r_T) - \mathrm{E}(r_C) = \mathrm{E}(r_T - r_C)$. This is one of the simplest ways of saying that randomization plays a critical role in causal inference in randomized experiments. If people picked their own treatments, deciding Z for themselves, then there is no reason to expect $Z \perp\!\!\!\perp (r_T, r_C)$. In §2.2.1, if high-risk patients were channeled to more capable hospitals, then there is no reason to expect $Z \perp\!\!\!\perp (r_T, r_C)$.

2.2.5 Covariates

A covariate describes a person prior to treatment assignment, so that, unlike an outcome, there is only one version of a covariate. Changing the treatment to which a person is assigned does not change the value of a covariate.

Each individual is described by an observed covariate $\mathbf{x}$. Central to observational studies is the concern that people who appear comparable in terms of the observed $\mathbf{x}$ may not be comparable in terms of a covariate $\mathbf{u}$ that was not observed. In §2.2.1, matching balanced an $\mathbf{x}$ containing 172 covariates, but inevitably critical discussion focused on the 173rd covariate, the one that was not measured.

2.3 The Propensity Score

2.3.1 Definition of the propensity score

The propensity score is the conditional probability of treatment given the observed covariate $\mathbf{x}$, that is, $\lambda = \lambda(\mathbf{x}) = \Pr(Z=1 \mid \mathbf{x})$. The propensity score is defined in terms of the observed covariates, $\mathbf{x}$, whether or not important unmeasured covariates $\mathbf{u}$ have been omitted from $\mathbf{x}$. The propensity score is a function of $\mathbf{x}$, so if $\mathbf{x}$ is a random variable, then $\lambda = \lambda(\mathbf{x})$ is also a random variable. Some properties of the propensity score depend upon whether important unmeasured covariates are omitted from $\mathbf{x}$, while other properties do not. It is important to distinguish these two types of properties.

The propensity score describes the relationship between two quantities that were observed, the assigned treatment Z and the observed covariate $\mathbf{x}$. Although the function $\lambda(\mathbf{x})$ is typically unknown, no special issues arise in estimating $\lambda(\mathbf{x})$, because all of the relevant quantities are observed. For instance, it is common to model or fit the propensity score using a logit model, perhaps with interactions, transformations, splines, and so on, involving $\mathbf{x}$, perhaps with some form of regularization.

2.3.2 Balancing properties of the propensity score

Matching for the propensity score tends to balance the observed covariates, $\mathbf{x}$, whether or not there are also unobserved covariates $\mathbf{u}$ that also require attention. In §2.2.1, matching for a scalar,

$\lambda = \lambda(\mathbf{x}) = \Pr(Z = 1 \mid \mathbf{x})$, tends to balance the 172-dimensional covariate $\mathbf{x}$. Two individuals with the same λ, one treated with $Z = 1$, the other a control with $Z = 0$, may have different values of $\mathbf{x}$, say $\mathbf{x}^*$ and $\mathbf{x}^{**}$ with $\lambda(\mathbf{x}^*) = \lambda(\mathbf{x}^{**})$, but it is just as likely that the treated individual has $\mathbf{x}^*$ and the control has $\mathbf{x}^{**}$ as it is that the treated individual has $\mathbf{x}^{**}$ and the control has $\mathbf{x}^*$; so, over many pairs, $\mathbf{x}$ tends to balance. Stated informally, taking care of $\lambda(\mathbf{x})$ takes care of all of $\mathbf{x}$, but if there are important unmeasured covariates, then taking care of all of $\mathbf{x}$ may not be enough.

Proposition 2.1 is from Rosenbaum and Rubin [3], and it states the balancing property.

Proposition 2.1. *If* $\lambda = \lambda(\mathbf{x}) = Pr(Z = 1 \mid \mathbf{x})$, *then*

$$Z \perp\!\!\!\perp \mathbf{x} \,\Big|\, \lambda, \tag{2.1}$$

and if $f(\cdot)$ is a function then

$$Z \perp\!\!\!\perp \mathbf{x} \,\Big|\, \{\lambda, f(\mathbf{x})\}. \tag{2.2}$$

Remark 2.1. *Condition (2.1) can be read in two ways. As is, (2.1) says that the propensity score $\lambda(\mathbf{x})$ contains all of the information in the observed covariates $\mathbf{x}$ that is useful in predicting treatment assignment Z. Rewriting (2.1) as $\mathbf{x} \perp\!\!\!\perp Z \,\big|\, \lambda$, it says $Pr(\mathbf{x} \mid Z = 1, \lambda) = Pr(\mathbf{x} \mid Z = 0, \lambda)$, so treated and control individuals with the same λ have the same distribution of $\mathbf{x}$. As is, (2.1) is agnostic about whether $\mathbf{x}$ and λ are random variables; that is, perhaps only Z is random.*

Remark 2.2. *Condition (2.2) says that if you match on the propensity score plus other aspects of $\mathbf{x}$, then the balancing property is preserved. In the example in §2.2.1, in asking whether high-risk patients benefit more from more capable hospitals, it was important for pairs to be close not only on the propensity score, $\lambda(\mathbf{x})$, but also on the externally estimated risk-of-death score or prognostic score [14], which is another function $f(\mathbf{x})$ of $\mathbf{x}$; see [16, Table 5]. Condition (2.2) says $\mathbf{x}$ will balance if you match for both scores.*

Proof The proof is simple. It suffices to prove (2.2). Trivially, $\lambda = \Pr(Z = 1 \mid \mathbf{x}) = \Pr\{Z = 1 \mid \mathbf{x}, \lambda, f(\mathbf{x})\}$ because $\lambda = \lambda(\mathbf{x})$ and $f(\mathbf{x})$ are functions of $\mathbf{x}$. Also,

$$\begin{aligned} \Pr\{Z = 1 \mid \lambda, f(\mathbf{x})\} &= \mathrm{E}[\Pr\{Z = 1 \mid \mathbf{x}, \lambda, f(\mathbf{x})\} \mid \lambda, f(\mathbf{x})] \\ &= \mathrm{E}[\lambda \mid \lambda, f(\mathbf{x})] = \lambda = \Pr(Z = 1 \mid \mathbf{x}), \end{aligned}$$

proving (2.2). □

After matching for the propensity score, it is standard and recommended practice to check whether the covariates $\mathbf{x}$ are balanced, whether the treated and matched control groups look comparable in terms of $\mathbf{x}$, that is, to check condition (2.1) or (2.2). This check for balance of $\mathbf{x}$ can be viewed as a diagnostic check of the model for the propensity score $\lambda(\mathbf{x})$. For the example in §2.2.1, the check was done in Silber et al. [16, Table 2 and Appendix]. A general approach to checking covariate balance is discussed by Yu [26].

2.3.3 Estimating treatment effects using propensity scores

Proposition 2.1 says propensity scores balance observed covariates, $\mathbf{x}$. When does that suffice to estimate the effects caused by treatments?

Write $\zeta = \Pr(Z = 1 \mid \mathbf{x}, r_T, r_C)$ and call it the principal unobserved covariate [13, §6.3]. The principal unobserved covariate is so named because it conditions upon the potential outcomes, (r_T, r_C), viewing them as stratifying variables, consistent with Frangakis and Rubin [27]. Unlike the propensity score, $\lambda = \Pr(Z = 1 \mid \mathbf{x})$, we cannot estimate or calculate ζ because we never jointly observe (r_T, r_C). If we could calculate ζ, and if it satisfied $0 < \zeta = \Pr(Z = 1 \mid \mathbf{x}, r_T, r_C) < 1$, then causal inference would be straightforward. As will be seen, adjustments for the propensity score suffice if the propensity score and the principal unobserved covariate are equal.

Replacing $\mathbf{x}$ in the proof of Proposition 2.1 by $(\mathbf{x}, r_T, r_C)$, we have immediately:

Proposition 2.2. *If* $\zeta = Pr(Z = 1 \mid \mathbf{x}, r_T, r_C)$ *then:*

$$Z \perp\!\!\!\perp (\mathbf{x}, r_T, r_C) \Big| \zeta, \tag{2.3}$$

$$Z \perp\!\!\!\perp (\mathbf{x}, r_T, r_C) \Big| \zeta, f(\mathbf{x}) \tag{2.4}$$

for any $f(\cdot)$, *and*

$$Z \perp\!\!\!\perp (r_T, r_C) \Big| \zeta, \mathbf{x}. \tag{2.5}$$

Suppose that we sampled a value of ζ, then sampled a treated individual, $Z = 1$, and a control individual, $Z = 0$, with this value of ζ, and we took the treated-minus-control difference in their responses. That is, we sample ζ from its marginal distribution, then sample r_T from $\Pr(R \mid Z = 1, \zeta) = \Pr(r_T \mid Z = 1, \zeta)$ and r_C from $\Pr(R \mid Z = 0, \zeta) = \Pr(r_C \mid Z = 0, \zeta)$, taking the treated-minus-control difference of the two responses matched for ζ. Given ζ, the expected difference in responses is

$$\begin{aligned} \mathrm{E}(r_T \mid Z = 1, \zeta) - \mathrm{E}(r_C \mid Z = 0, \zeta) &= \mathrm{E}(r_T \mid \zeta) - \mathrm{E}(r_C \mid \zeta) \\ &= \mathrm{E}(r_T - r_C \mid \zeta), \end{aligned} \tag{2.6}$$

using $(r_T, r_C) \perp\!\!\!\perp Z \Big| \zeta$ from (2.3). However, ζ was picked at random from its marginal distribution, so the unconditional expectation of the treated-minus-control difference in the matched pair is the expectation of $\mathrm{E}(r_T - r_C \mid \zeta)$ with respect to the marginal distribution of ζ, so it is $\mathrm{E}(r_T - r_C)$, which is the expected treatment effect.

In brief, if we could match for ζ, instead of matching for the propensity score λ, then we would have an unbiased estimate of the expected effect caused by the treatment, $\mathrm{E}(r_T - r_C)$. If we could match for ζ, causal inference without randomization would be straightforward. There is the difficulty, alas not inconsequential, that we have no access to ζ, and so cannot match for it.

In [3], treatment assignment is said to be (strongly) ignorable given $\mathbf{x}$ if

$$0 < \zeta = \Pr(Z = 1 \mid \mathbf{x}, r_T, r_C) = \Pr(Z = 1 \mid \mathbf{x}) = \lambda < 1. \tag{2.7}$$

Condition (2.7) says two things. First, (2.7) says the principal unobserved covariate and the propensity score are equal, or equivalently that $(r_T, r_C) \perp\!\!\!\perp Z \Big| \mathbf{x}$. Second, (2.7) adds that treated and control individuals, $Z = 1$ and $Z = 0$, occur at every $\mathbf{x}$. That is, (2.7) says:

$$(r_T, r_C) \perp\!\!\!\perp Z \Big| \mathbf{x} \quad \text{and} \quad 0 < \Pr(Z = 1 \mid \mathbf{x}) < 1. \tag{2.8}$$

Again, form (2.8) takes (r_T, r_C) to be random variables, while form (2.7) is agnostic about whether (r_T, r_C) are random variables, or fixed quantities as in Fisher's [28] randomization inferences.

If treatment assignment is ignorable given $\mathbf{x}$ in the sense that (2.7) holds, then matching for the propensity score is matching for the principal unobserved covariate, and the above argument then shows that matching for the propensity score provides an unbiased estimate of the expected treatment effect, $\mathrm{E}(r_T - r_C)$. In brief, if treatment assignment is ignorable given $\mathbf{x}$, then it suffices to adjust for all of $\mathbf{x}$, but it also suffices to adjust for the scalar propensity score $\lambda(\mathbf{x})$ alone.

There are many minor variations on this theme. If (2.7) holds, matching for $\mathbf{x}$ instead of ζ yields an unbiased estimate of $\mathrm{E}(r_T - r_C)$. If (2.7) holds, matching for $\{\lambda,\ f(\mathbf{x})\}$ yields an unbiased estimate of $\mathrm{E}\{r_T - r_C \mid \lambda,\ f(\mathbf{x})\}$ and of $\mathrm{E}\{r_T - r_C \mid f(\mathbf{x})\}$. In §2.2.1, a treated individual was sampled, her value of λ was noted, and she was matched to a control with the same value of λ, leading to an estimate of the average effect of the treatment on the treated group, $\mathrm{E}(r_T - r_C \mid Z = 1)$, if (2.7) holds [6]. If (2.7) holds, the response surface $\mathrm{E}(r_T \mid Z = 1,\ \mathbf{x})$ of R on $\mathbf{x}$ in the treated group equals $\mathrm{E}(r_T \mid \mathbf{x})$, the response surface $\mathrm{E}(r_C \mid Z = 0,\ \mathbf{x})$ of R on $\mathbf{x}$ in the control group estimates $\mathrm{E}(r_C \mid \mathbf{x})$, and differencing yields the causal effect $\mathrm{E}(r_T - r_C \mid \mathbf{x})$; however, all of this is still true if $\mathbf{x}$ is replaced by $\{\lambda,\ f(\mathbf{x})\}$.

In brief, in estimating expected causal effects, say $\mathrm{E}(r_T - r_C)$ or $\mathrm{E}\{r_T - r_C \mid f(\mathbf{x})\}$, if it suffices to adjust for $\mathbf{x}$ then it suffices to adjust for the propensity score, $\lambda = \Pr(Z = 1 \mid \mathbf{x})$. In §2.2.1, it if suffices to adjust for 172 covariates in $\mathbf{x}$, then it suffices to adjust for one covariate, λ. Adjustments for all of $\mathbf{x}$, for λ alone, or for λ and $f(\mathbf{x})$, work if treatment assignment is ignorable give $\mathbf{x}$, that is, if (2.7) or (2.8) are true, and they work only by accident otherwise. The problem is that (2.7) or (2.8) may not be true.

2.3.4 Unmeasured covariates and the propensity score

One useful way to express the possibility that treatment assignment is not ignorable given $\mathbf{x}$ is to entertain the alternative possibility that it would have been ignorable given $(\mathbf{x}, \mathbf{u})$ but not given $\mathbf{x}$ alone, where $\mathbf{u}$ is some unobserved covariate. Saying this more precisely, we entertain the possibility that

$$0 < \Pr(Z = 1 \mid \mathbf{x},\ \mathbf{u},\ r_T,\ r_C) = \Pr(Z = 1 \mid \mathbf{x},\ \mathbf{u}) < 1 \tag{2.9}$$

is true, but (2.7) is false. Why is it useful to express failure of (2.7) in terms of (2.9)? First, it is common for scientists to say that adjustments for $\mathbf{x}$ are inadequate because $\mathbf{x}$ omits a specific $\mathbf{u}$, so (2.7) and (2.9) give precise form to a conversation that is going on anyway. Second, (2.9) is prescriptive: to fix the problem, measure $\mathbf{u}$. So, thinking of failures of (2.7) in terms of (2.9) challenges the critic of (2.7) to be specific about what $\mathbf{u}$ might be. The more fanciful the critic is in suggesting that a specific $\mathbf{u}$ should have been measured, the more unreasonable is his criticism of (2.7); see Bross [29]. Third, it may be possible to collect additional data that tells us something about $\mathbf{u}$; see [30–34]. Fourth, we are able to say what $\mathbf{u}$ would have to be like if failure of (2.7) is to alter the conclusions of the study; that is, we may conduct a sensitivity analysis of the type first done by Cornfield et al. [35].

Can we always get from $0 < \zeta = \Pr(Z = 1 \mid \mathbf{x},\ r_T,\ r_C) < 1$ to (2.9) for some unobserved $\mathbf{u}$? Indeed we can.

Proposition 2.3. *If* $0 < \zeta = Pr(Z = 1 \mid \mathbf{x},\ r_T,\ r_C) < 1$, *then there exists a scalar* u *with* $0 < u < 1$ *such that (2.9) is true, namely* $u = \zeta$.

Proof As $\zeta = \Pr(Z = 1 \mid \mathbf{x},\ r_T,\ r_C)$ is a function of $(\mathbf{x},\ r_T,\ r_C)$, conditioning on $(\mathbf{x},\ \zeta,\ r_T,\ r_C)$ is the same as conditioning on $(\mathbf{x},\ r_T,\ r_C)$, so

$$\Pr(Z = 1 \mid \mathbf{x},\ \zeta,\ r_T,\ r_C) = \Pr(Z = 1 \mid \mathbf{x},\ r_T,\ r_C) = \zeta. \tag{2.10}$$

Using (2.10) twice yields

$$\begin{aligned} \Pr(Z = 1 \mid \mathbf{x},\ \zeta) &= \mathrm{E}\{\Pr(Z = 1 \mid \mathbf{x},\ \zeta,\ r_T,\ r_C) \mid \mathbf{x},\ \zeta\} \qquad (2.11) \\ &= \mathrm{E}(\zeta \mid \mathbf{x},\ \zeta) = \zeta = \Pr(Z = 1 \mid \mathbf{x},\ \zeta,\ r_T,\ r_C), \end{aligned}$$

which is (2.9) when $\mathbf{u} = \zeta$. □

Frangakis and Rubin [27] refer to conditioning upon (r_T, r_C) as forming principal strata. In recognition of this, we might reasonably refer to the $u = \zeta$ in Proposition 2.3 as the principal unobserved covariate. Proposition 2.1 says that the aspects of an observed covariate $\mathbf{x}$ that bias treatment assignment may be summarized in a scalar covariate, namely the propensity score, $\lambda = \Pr(Z = 1 \mid \mathbf{x})$. Proposition 2.3 says that the aspects of unobserved covariates that bias treatment assignment may be summarized in a scalar, namely, the principal unobserved covariate.

The sensitivity analysis in Rosenbaum [7, 36] refers to the principal unobserved covariate. The principal unobserved covariate equals $u = \zeta = \Pr(Z = 1 \mid \mathbf{x}, u)$ using (2.11) and satisfies $0 \leq u \leq 1$ because u is a probability. The sensitivity analysis quantifies the influence of u on Z by a parameter $\Gamma \geq 1$ expressed in terms of the principal unobserved covariate u. Specifically, it says: any two individuals, i and j, with the same value of the observed covariate, $\mathbf{x}_i = \mathbf{x}_j$, may differ in their odds of treatment by at most a factor of Γ:

$$\frac{1}{\Gamma} \leq \frac{\Pr(Z = 1 \mid \mathbf{x}_i, u_i) \Pr(Z = 0 \mid \mathbf{x}_j, u_i)}{\Pr(Z = 1 \mid \mathbf{x}_j, u_i) \Pr(Z = 0 \mid \mathbf{x}_i, u_j)} = \frac{u_i(1 - u_j)}{u_j(1 - u_i)} \leq \Gamma \text{ if } \mathbf{x}_i = \mathbf{x}_j. \tag{2.12}$$

In [36], the parameter $\gamma = \log(\Gamma)$ is the coefficient of u in a logit model that predicts treatment Z from $\mathbf{x}$ and u. For any sensitivity analysis expressed in terms of the principal unobserved covariate u in (2.12), the amplification in Rosenbaum and Silber [37] and Rosenbaum [38, Table 9.1] reexpresses the sensitivity analysis in terms of other covariates with different properties that have exactly the same impact on the study's conclusions.

2.3.5 Kullback-Leibler information and propensity scores

The Kullback and Leibler [39] divergence distinguishing the distribution of observed covariates $\mathbf{x}$ in treated, $Z = 1$, and control, $Z = 0$, groups is:

$$J(1:0;\mathbf{x}) = \int \{\Pr(\mathbf{x} \mid Z = 1) - \Pr(\mathbf{x} \mid Z = 0)\} \log\left\{\frac{\Pr(\mathbf{x} \mid Z = 1)}{\Pr(\mathbf{x} \mid Z = 0)}\right\} d\mathbf{x}.$$

As noted by Kullback and Leibler, $J(1:0;\mathbf{x}) \geq 0$ with equality if and only if the distribution of $\mathbf{x}$ is the same in treated and control groups. Using Bayes theorem, $J(1:0;\mathbf{x})$ may be rewritten in terms of the propensity score:

$$J(1:0;\mathbf{x}) = \int \{\Pr(\mathbf{x} \mid Z = 1) - \Pr(\mathbf{x} \mid Z = 0)\} \log\left\{\frac{\lambda(\mathbf{x})}{1 - \lambda(\mathbf{x})}\right\} d\mathbf{x},$$

so $J(1:0;\mathbf{x})$ is the difference in the expectations of the log-odds of the propensity score in treated and control groups. Yu et al. [40] propose estimating $J(1:0;\mathbf{x})$ by the difference in sample means of the estimated log-odds of the propensity score, and they use $J(1:0;\mathbf{x})$ as an accounting tool that keeps track of magnitudes of bias before matching and in various matched samples.

As a measure of covariate imbalance, $J(1:0;\mathbf{x})$ has several attractions. It is applicable when some covariates are continuous, others are nominal, and others are ordinal. If the covariates are partitioned into two sets, then the total divergence may be written as the sum of a marginal divergence from the first set and a residual or conditional divergence from the second set. The divergence in all of $\mathbf{x}$ equals the divergence in the propensity score $\lambda(\mathbf{x})$ alone; that is, $J(1:0;\mathbf{x}) = J(1:0;\lambda)$. The divergence reduces to familiar measures when $\Pr(\mathbf{x} \mid Z = z)$ is multivariate Normal with common covariance matrix for $z = 1$ and $z = 0$. These properties are discussed by Yu et al. [40].

2.4 Aspects of Estimated Propensity Scores

2.4.1 Matching and stratification

Matching and stratification are done before outcomes are examined, so they are part of the design of an observational study; see Rubin [41]. That is, matching or stratification use Z and $\mathbf{x}$ without examining R. In that sense, matching or stratification resemble selecting one experimental design from a table of designs in a textbook, and they do not resemble fitting a covariance adjustment model that predicts R from Z and $\mathbf{x}$. In a traditional least squares regression of R on Z and $\mathbf{x}$, the design matrix is viewed as fixed. Viewed in this way, matching and stratification, in all of their various forms, are analogous to other transformations of the design matrix prior to examining outcomes.

Viewed in this way, Proposition 2.1 predicts that certain uses of the propensity score will produce a matched or stratified design that balances $\mathbf{x}$; however, we can and should check whether the desired balance is achieved. So far as matching and stratification are concerned, the propensity score and other methods are a means to an end, not an end in themselves. If matching for a misspecified and misestimated propensity score balances $\mathbf{x}$, then that is fine. If by bad luck, the true propensity score failed to balance $\mathbf{x}$, then the match is inadequate and should be improved. Various methods can improve covariate balance when a matched design is not adequately balanced [12, 13].

In a randomized trial in which treatments are assigned by flipping a fair coin, the true propensity score is $1/2$, but a propensity score estimated by a logit model will vary somewhat with $\mathbf{x}$ due to overfitting. For example, if by luck the coin flips happen to put relatively more women into 35(4): the treated group than men, then the estimated propensity scores for women will be a little higher than for men. Is this a problem? It might be a problem if the objective were to obtain the true propensity score; however, that is not the objective. Matching for the estimated propensity score will tend to correct for some of the imbalance in gender that occurred by chance. The logit model cannot distinguish an imbalance in covariates that is due to luck from an identical imbalance in covariates that is due to bias, and in either case, it will tend to remove imbalances in observed covariates. In §2.2.1, there we 172 covariates in $\mathbf{x}$ and more than 88,000 patients prior to matching, so some imbalances in $\mathbf{x}$ are expected by chance, and the estimated propensity works to remove these along with systematic imbalances.

In brief, the propensity score may be used in the design, without access to outcomes, to match or stratify to balance covariates. In this case, we are concerned with the degree of covariate balance that is achieved, with whether a good design has been built, not with whether the propensity score is well-estimated or poorly estimated.

The situation is different when the propensity score is used with outcome data. Some estimators or tests directly use estimated propensity scores and outcomes. Two ways to think about this are discussed in §2.4.2 and §2.4.3.

2.4.2 Conditional inference given a sufficient statistic

Consider I independent individuals, $i = 1, \ldots, I$, with treatment assignments $\mathbf{Z} = (Z_1, \ldots, Z_I)^T$, observed outcomes $\mathbf{R} = (R_1, \ldots, R_I)^T$, potential outcomes $\mathbf{r}_T = (r_{T1}, \ldots, r_{TI})^T$ and $\mathbf{r}_C = (r_{C1}, \ldots, r_{CI})^T$, and observed covariates $\mathbf{x}_i$, where $\mathbf{X}$ is the matrix with N rows whose ith row is $\mathbf{x}_i^T$. The vector of logits of the propensity scores is then $\boldsymbol{\kappa} = (\kappa_1, \ldots, \kappa_I)^T$ where $\kappa_i = \log\{\lambda_i/(1-\lambda_i)\}$ with $\lambda_i = \Pr(Z_i = 1 \mid \mathbf{x}_i)$, and treatment assignment is ignorable given $\mathbf{X}$ if $0 < \lambda_i = \Pr(Z_i = 1 \mid \mathbf{x}_i) = \Pr(Z_i = 1 \mid r_{Ti}, r_{Ci}, \mathbf{x}_i) < 1$ for each i. Fisher's [28] hypothesis of no treatment effect asserts $H_0 : r_{Ti} = r_{Ci}$ for each i, or equivalently $\mathbf{r}_T = \mathbf{r}_C$.

If the propensity score follows a logit model, $\boldsymbol{\kappa} = \mathbf{X}\boldsymbol{\beta}$ with unknown $\boldsymbol{\beta}$, then the propensity score is unknown. How can H_0 be tested when $\boldsymbol{\beta}$ is unknown? Two closely related approaches are discussed, an exact test that works when the coordinates of $\mathbf{x}_i$ are discrete and coarse, and a large-sample test that applies generally. It is assumed that $\mathbf{X}$ "contains" a constant term, meaning that the I-dimensional vector of 1s is in the space spanned by the columns of $\mathbf{X}$.

For the exact method, what are examples of a discrete, coarse $\mathbf{X}$? The simplest example is M mutually exclusive strata, so that $\mathbf{X}$ has M columns, and $x_{im} = 1$ if individual i is in stratum m, $x_{im} = 0$ otherwise, with $1 = \sum_{m=1}^{M} x_{im}$ for each i, say $m = 1$ for female, $m = 2$ for male. Instead, the first two columns of $\mathbf{X}$ could be a stratification of individuals by gender, followed by four columns that stratify by ethnicity, so $1 = \sum_{m=1}^{2} x_{im}$ and $1 = \sum_{m=3}^{6} x_{im}$ for each i. The seventh column of $\mathbf{X}$ might be age rounded to the nearest year. For the exact test, the columns of $\mathbf{X}$ need not be linearly independent. If $\mathcal{C}$ is a finite set, then write $|\mathcal{C}|$ for the number of elements of $\mathcal{C}$.

Under the logit model, $\mathbf{S} = \mathbf{X}^T\mathbf{Z}$ is a sufficient statistic for $\boldsymbol{\beta}$, so the conditional distribution of $\mathbf{Z}$ given $\mathbf{X}^T\mathbf{Z}$ does not depend upon the unknown $\boldsymbol{\beta}$. If treatment assignment is ignorable given $\mathbf{X}$, then using both ignorability and sufficiency,

$$\Pr\left(\mathbf{Z} = \mathbf{z} \mid \mathbf{X}, \mathbf{r}_T, \mathbf{r}_C, \boldsymbol{\beta}, \mathbf{X}^T\mathbf{Z} = \mathbf{s}\right) = \Pr\left(\mathbf{Z} = \mathbf{z} \mid \mathbf{X}, \mathbf{X}^T\mathbf{Z} = \mathbf{s}\right), \tag{2.13}$$

and $\Pr\left(\mathbf{Z} = \mathbf{z} \mid \mathbf{X}, \mathbf{X}^T\mathbf{Z} = \mathbf{s}\right)$ is constant on the set $\mathcal{C}_{\mathbf{s}} = \left\{\mathbf{z} : \mathbf{X}^T\mathbf{z} = \mathbf{s}\right\}$ of vectors of binary treatment assignments $\mathbf{z}$ that give rise to $\mathbf{s}$ as the value of the sufficient statistic, so

$$\Pr\left(\mathbf{Z} = \mathbf{z} \mid \mathbf{X}, \mathbf{r}_T, \mathbf{r}_C, \boldsymbol{\beta}, \mathbf{X}^T\mathbf{Z} = \mathbf{s}\right) = \frac{1}{|\mathcal{C}_{\mathbf{s}}|} \text{ for each } \mathbf{z} \in \mathcal{C}_{\mathbf{s}}. \tag{2.14}$$

Under Fisher's hypothesis H_0 of no effect, $\mathbf{R} = \mathbf{r}_T = \mathbf{r}_C$ is fixed by conditioning in (2.13) and (2.14), so that $t(\mathbf{Z}, \mathbf{R}) = t(\mathbf{Z}, \mathbf{r}_C)$ for any test statistic $t(\mathbf{Z}, \mathbf{R})$. The distribution of $t(\mathbf{Z}, \mathbf{R}) = t(\mathbf{Z}, \mathbf{r}_C)$ under H_0 is a known, exact permutation distribution,

$$\Pr\left\{t(\mathbf{Z}, \mathbf{r}_C) \geq k \mid \mathbf{X}, \mathbf{r}_T, \mathbf{r}_C, \boldsymbol{\beta}, \mathbf{X}^T\mathbf{Z} = \mathbf{s}\right\} = \frac{\left|\left\{\mathbf{z} \in \mathcal{C}_{\mathbf{s}} : t(\mathbf{z}, \mathbf{r}_C) \geq k\right\}\right|}{|\mathcal{C}_{\mathbf{s}}|}. \tag{2.15}$$

The hypothesis H_τ of an additive constant effect τ asserts $r_{Ti} = r_{Ci} + \tau$ for each i. If H_τ is true, then $\mathbf{R} - \tau\mathbf{Z} = \mathbf{r}_C$ satisfies the null hypothesis of no effect, so H_τ may be tested by applying (2.15) with $\mathbf{r}_C = \mathbf{R} - \tau\mathbf{Z}$. By testing every H_τ and retaining the τ's that are not rejected at level α, a $1 - \alpha$ confidence set for τ is obtained. In brief, there is an exact theory of inference about causal effects when the parameter $\boldsymbol{\beta}$ of the propensity score is unknown.

In the special case in which $\mathbf{X}$ represents S strata, this procedure reproduces the exact randomization distribution of familiar stratified randomization tests. For instance, it reproduces Birch [42]'s exact distribution for the Mantel-Haenszel test and the exact distribution of stratified rank tests, such as the Hodges-Lehmann [43] aligned rank test.

If the statistic testing H_τ is $t(\mathbf{Z}, \mathbf{r}_C) = \mathbf{Z}^T\mathbf{r}_C$ with $\mathbf{r}_C = \mathbf{R} - \tau\mathbf{Z}$, then the resulting test based on (2.15) is numerically the same as the uniformly most powerful unbiased test of $H_* : \omega = 0$ in the linear logit model $\kappa_i = \mathbf{x}_i^T\boldsymbol{\beta} + \omega\, r_{Ci}$; see Cox [44, §4.2]. Because $\mathbf{X}$ contains a constant term, this test is the same as the test based on the treated-minus-control difference in means,

$$t(\mathbf{Z}, \mathbf{r}_C) = \frac{\mathbf{Z}^T\mathbf{r}_C}{\mathbf{Z}^T\mathbf{Z}} - \frac{(\mathbf{1} - \mathbf{Z})^T\mathbf{r}_C}{(\mathbf{1} - \mathbf{Z})^T(\mathbf{1} - \mathbf{Z})},$$

in the sense that both tests either reject H_τ at level α or both tests accept H_τ.

In practice, when testing hypotheses about a coefficient in a linear logit model, it is customary to approximate the uniformly most powerful unbiased test of $H_* : \omega = 0$ by the likelihood ratio test. This leads to the large sample test of H_τ, namely, fit $\kappa_i = \mathbf{x}_i^T\boldsymbol{\beta} + \omega\, r_{Ci}$ by maximum likelihood and use the likelihood ratio test of $H_* : \omega = 0$ as the test of H_τ. The large sample test is computationally straightforward and it does not require that $\mathbf{X}$ be discrete and coarse. Indeed, the large sample test is sufficiently easy to compute that it may be inverted to obtain a confidence interval for τ. A rank test may be obtained by replacing r_{Ci} in $\kappa_i = \mathbf{x}_i^T \boldsymbol{\beta} + \omega \, r_{Ci}$ by the rank of r_{Ci}, ranking from 1 to I. Whether r_{Ci} or its rank is used, ω equals zero when treatment assignment is ignorable and H_τ is true. Instead of ranks, when testing H_τ, r_{Ci} may be replaced by residuals of $\mathbf{R} - \tau\mathbf{Z} = \mathbf{r}_C$ when robustly regressed on $\mathbf{x}_i$; see [45] and [46, §2.3].

The material in this subsection is from [1]. See also [47], [7, §3.4-§3.6] and [45].

2.4.3 Inverse probability weighting and post-stratification

The average treatment effect is $\delta = N^{-1} \sum_{i=1}^{N} (r_{Ti} - r_{Ci})$, but it cannot be calculated because r_{Ti} and r_{Ci} are not jointly observed. Suppose that treatment assignment is ignorable given $\mathbf{X}$, so that $0 < \lambda_i = \Pr(Z_i = 1 \mid \mathbf{x}_i) = \Pr(Z_i = 1 \mid r_{Ti}, r_{Ci}, \mathbf{x}_i) < 1$. Is there an unbiased estimate of δ? If the propensity scores λ_i were known, then

$$\delta^* = I^{-1} \sum_{i=1}^{I} \frac{Z_i \, r_{Ti}}{\lambda_i} - \frac{(1 - Z_i) \, r_{Ci}}{1 - \lambda_i}$$

would be unbiased for δ in the sense that $\mathrm{E}(\delta^* \mid \mathbf{X}, \mathbf{r}_T, \mathbf{r}_C) = \delta$, simply because $\mathrm{E}(Z_i \mid \mathbf{X}, \mathbf{r}_T, \mathbf{r}_C) = \lambda_i$. However, δ^* depends upon the unknown λ_i, so δ^* is not an estimator.

Is there an unbiased estimate of δ when, as in §2.4.2, the propensity score satisfies a linear logit model, $\boldsymbol{\kappa} = \mathbf{X}\boldsymbol{\beta}$, with unknown $\boldsymbol{\beta}$? Let $\overline{\lambda}_i = \mathrm{E}\left(Z_i \mid \mathbf{X}, \mathbf{r}_T, \mathbf{r}_C, \mathbf{X}^T\mathbf{Z} = \mathrm{s}\right)$, so that $\overline{\lambda}_i$ does not depend upon $\boldsymbol{\beta}$ because $\mathbf{X}^T\mathbf{Z}$ is a sufficient statistic; then,

$$\overline{\lambda}_i = \frac{|\{\mathbf{z} \in \mathcal{C}_\mathbf{s} : z_i = 1\}|}{|\mathcal{C}_\mathbf{s}|}.$$

That is, $\overline{\lambda}_i$ is known, even though $\boldsymbol{\beta}$ is not. Note also that $\overline{\lambda}_i \geq 1/|\mathcal{C}_\mathbf{s}| > 0$ whenever $Z_i = 1$, and similarly $\overline{\lambda}_i < 1$ whenever $Z_i = 0$, because $\mathbf{Z} \in \mathcal{C}_\mathbf{s}$ for $\mathrm{s} = \mathbf{X}^T\mathbf{Z}$. If $0 < \overline{\lambda}_i < 1$,

$$\mathrm{E}\left(\frac{Z_i \, r_{Ti}}{\overline{\lambda}_i} - \frac{(1 - Z_i) \, r_{Ci}}{1 - \overline{\lambda}_i} \,\middle|\, \mathbf{X}, \mathbf{r}_T, \mathbf{r}_C, \mathbf{X}^T\mathbf{Z} = \mathrm{s} \right) = r_{Ti} - r_{Ci}.$$

If $0 < \overline{\lambda}_i < 1$ for $i = 1, \ldots, I$, then

$$\widehat{\delta} = I^{-1} \sum_{i=1}^{I} \frac{Z_i \, r_{Ti}}{\overline{\lambda}_i} - \frac{(1 - Z_i) \, r_{Ci}}{1 - \overline{\lambda}_i}$$

is a function of the observed data with

$$\mathrm{E}\left(\widehat{\delta} \,\middle|\, \mathbf{X}, \mathbf{r}_T, \mathbf{r}_C, \mathbf{X}^T\mathbf{Z} = \mathrm{s} \right) = \delta;$$

so $\widehat{\delta}$ is unbiased for δ.

In practice, these calculations work for discrete, coarse covariates, and even then computation of $\overline{\lambda}_i$ can be difficult, so the maximum-likelihood estimate $\widehat{\lambda}_i$ of λ_i is commonly used instead. With this substitution, strict unbiasedness is lost. Here, as before, it is assumed that treatment assignment is ignorable and that λ_i follows a linear logit model, $\boldsymbol{\kappa} = \mathbf{X}\boldsymbol{\beta}$; so, both $\overline{\lambda}_i$ and $\widehat{\lambda}_i$ are estimators of λ_i that are functions of the sufficient statistic, $\mathbf{X}^T\mathbf{Z} = \mathrm{s}$ and $\mathbf{X}$. In this case, $\mathrm{E}(Z_i \mid \mathbf{x}_i) = \Pr(Z_i = 1 \mid \mathbf{x}_i) = \Pr(Z_i = 1 \mid r_{Ti}, r_{Ci}, \mathbf{x}_i) = \lambda_i$, so Z_i is a poor but unbiased estimate of λ_i, and $\overline{\lambda}_i = \mathrm{E}\left(Z_i \mid \mathbf{X}, \mathbf{r}_T, \mathbf{r}_C, \mathbf{X}^T\mathbf{Z} = \mathrm{s}\right)$ is the corresponding Rao-Blackwell estimator of λ_i.

In the simplest case, $\mathbf{X}$ encodes precisely M mutually exclusive strata, where M is small, as in the simplest case in §2.4.2, and the estimator $\widehat{\delta}$ is called poststratification; see [48]. In this simplest case, the logit model logit model, $\boldsymbol{\kappa} = \mathbf{X}\boldsymbol{\beta}$, says that the propensity score λ_i is constant for all individuals in the same stratum, and $\overline{\lambda}_i$ is the sample proportion of treated individuals in that stratum. In this simplest case, if each $\overline{\lambda}_i$ is neither 0 nor 1, then $\overline{\lambda}_i$ is also the maximum-likelihood estimate $\widehat{\lambda}_i$ of λ_i. When the true λ_i are constant, $\lambda_1 = \cdots = \lambda_I$, it is known that the poststratified estimate, $\widehat{\delta}$, which allows $\overline{\lambda}_i = \widehat{\lambda}_i$ to vary among strata, tends to be more efficient than δ^*, because $\widehat{\delta}$ corrects for chance imbalances across strata while δ^* does not; see [48]. In other words, in this simplest case, estimated propensity scores, $\overline{\lambda}_i = \widehat{\lambda}_i$, often outperform the true propensity score λ_i.

This subsection is based on [2]. The example in that paper uses maximum likelihood to estimate not just the expectations of $I^{-1}\sum_{i=1}^{I} r_{Ti}$ and $I^{-1}\sum_{i=1}^{I} r_{Ci}$, but also the cumulative distributions of r_{Ti} and r_{Ci} adjusting for biased selection. The adjustments are checked by applying them to the covariates in $\mathbf{X}$, where there is no effect, so the estimator is known to be trying to estimate an average treatment effect of zero.

2.5 Limitations of Propensity Scores

It is generally impossible to match closely for a high dimensional covariate $\mathbf{x}_i$, but it is often possible to balance $\mathbf{x}_i$. Propensity scores are helpful in building matched pairs or strata that balance many observed covariates. However, they do have several limitations.

Two individuals, i and i', with the same propensity score, $\lambda(\mathbf{x}_i) = \lambda(\mathbf{x}_{i'})$, may be very different in terms of the covariates themselves, $\mathbf{x}_i \neq \mathbf{x}_{i'}$. For this reason, matched pairs or strata are often constructed to control for both the propensity score $\lambda(\mathbf{x}_i)$ and a few of the most important features $f(\mathbf{x}_i)$ of $\mathbf{x}_i$; see [4, 5, 49].

Propensity scores use probability – the law of large numbers – to balance many low-dimensional summaries of high-dimensional covariates $\mathbf{x}_i$. In medium-size samples, the law of large numbers is of little help in balancing a coordinate of $\mathbf{x}_i$ that represents an exceeding rare binary attribute, or a nominal covariate with an enormous number of levels, such as Zip codes or ICD-10 Principal Procedure Codes. In this case, it is often helpful to match for $\lambda(\mathbf{x}_i)$ while also enforcing balance for rare categories, a technique known as fine balance; see [15, §3.2] and [12, 13, 50–54].

Because the propensity score is constructed to separate treated and control groups, it can be difficult to balance the propensity score. A directional caliper on the propensity score can help: it penalizes imbalances in the propensity score that reinforce the natural direction of the bias; see [55].

Propensity scores address biases from measured covariates, $\mathbf{x}_i$. Other methods are required to address biases from unmeasured covariates, that is, biases that arise when treatment assignment is not ignorable given observed covariates; see [13, 46, 56].

2.6 Nursing and Surgical Mortality, Concluded

Table 2.1 shows the covariate balance achieved in the example in §2.2.1 from [16] for 25,076 matched pairs. The match balanced 172 covariates, including 130 4-digit ICD-9 Principal Procedure codes that indicated the type of surgery. The 130 Principal Procedure codes were matched exactly, so matched treated and control patients always had the same type of surgery. The match controlled an estimated propensity score, and also a prognostic score, the probability of death within 30 days of surgery based on covariates from a logit model fitted to a different data set; see [14]. The match also minimized the total of the 25,076 within-pair Mahalanobis distances; see [15].

TABLE 2.1
Balance for selected covariates in 25076 matched pairs. The matching controlled for 172 covariates, including 130 4-digit ICD-9 Principal Procedure codes that were matched exactly. The standarized difference (Std. Dif.) is the difference in means or percentages after matching divided by a pooled estimate of the standard deviation within groups before matching.

Covariate	Treated	Control	Std. Dif.
N of Pairs	25076	25076	
Age	76.3	76.3	0.00
Male %	39.4	40.2	−0.02
Probability of death	0.04	0.04	−0.02
Propensity score	0.32	0.32	0.04
	Type of Admission		
Emergency admission %	38.1	40.2	−0.04
Transfer in %	3.0	2.3	0.05
	History of Health Problems		
Diabetes %	27.9	27.8	0.00
Arrhythmia %	25.7	25.4	0.01
COPD %	22.4	22.8	−0.01
Congestive heart failure %	21.7	22.6	−0.02
Myocardial infarction %	8.2	7.9	0.01
Dementia %	6.4	6.7	−0.01
Renal failure %	5.8	5.9	0.00
Angina %	3.3	3.6	−0.01
130 Principle Procedures	Every pair, matched exactly		

Table 2.1 shows covariate means or percentages in matched pairs. The standardized difference (Std. Dif.) is a traditional measure of covariate imbalance: it is the difference in means after matching divided by a pooled within-group standard deviation before matching [19]. The pooled standard deviation is the square root of the equally weighted average of the two within-group variances. Of course, the Std. Dif. is zero for the 130 Princpal Procedure codes, as the difference in percentages for each code is zero. Additional detail is found in the Appendix to [16].

This example is typical. The propensity score is one of several mutually supporting techniques used in matching. In addition to matching for the propensity score, other techniques used were exact matching for principal procedures, an externally estimated prognostic score, and optimization of a covariate distance.

Matching for the prognostic score played an important role in the analysis: pairs were separated into five groups of pairs based on the quintile of the estimated probability of death. The treatment effect was estimated for all pairs, and for pairs in each quintile of risk.

For all 25,076 pairs, the 30-day mortality rate was 4.8% in hospitals with superior nursing and 5.8% in hospitals with inferior nursing. As discussed in §2.2.1, this is an attempt to estimate the effect of having surgery in one group of existing hospitals rather than another group of existing hospitals, not the effect of changing the nursing environment in any hospital. The difference in mortality rates is not plausibly due to chance, with P-value less than 0.001 from McNemar's test. The within-pair odds ratio was 0.79 with 95% confidence interval 0.73 to 0.86. An unobserved covariate that doubled the odds of death and increased by 50% the odds of treatment is insufficient to produce this association in the absence of a treatment effect, but larger unmeasured biases could explain it; see [16, eAppendix 10].

Dividing the 25,076 pairs into five groups of roughly 5,015 pairs based on the quintile of the prognostic score yields five estimates of treatment effect for lower and higher risk patients;

see [16, Table 5]. For the lowest three quintiles, the difference in mortality rates was small and not significantly different from zero, although the point estimate was negative, with slightly lower mortality rates in hospital with superior nursing. The two highest risk quintiles had lower rates of death in hospitals with superior nursing. In the highest risk quintile, the mortality rates were 17.3% at hospitals with superior nursing and 19.9% at hospitals with inferior nursing, a difference of −2.6%, with P-value less than 0.001 from McNemar's test. This analysis raises the possibility that mortality rates in the population might be lower if high-risk patients were directed to have surgery at hospitals with superior nursing.

2.7 Summary

This chapter has reviewed, with an example, the basic properties of propensity scores, as developed in the 1980's in joint work with Donald B. Rubin [1–6]. The discussion has focused on role of propensity scores in matching, stratification, inverse probability weighting, and conditional permutation inference. Section 2.5 discussed limitations of propensity scores, and various methods for addressing them.

References

[1] Paul R Rosenbaum. Conditional permutation tests and the propensity score in observational studies. *Journal of the American Statistical Association*, 79(387):565–574, 1984.

[2] Paul R Rosenbaum. Model-based direct adjustment. *Journal of the American Statistical Association*, 82(398):387–394, 1987.

[3] Paul R Rosenbaum and Donald B Rubin. The central role of the propensity score in observational studies for causal effects. *Biometrika*, 70(1):41–55, 1983.

[4] Paul R Rosenbaum and Donald B Rubin. Reducing bias in observational studies using subclassification on the propensity score. *Journal of the American Statistical Association*, 79(387):516–524, 1984.

[5] Paul R Rosenbaum and Donald B Rubin. Constructing a control group using multivariate matched sampling methods that incorporate the propensity score. *The American Statistician*, 39(1):33–38, 1985.

[6] Paul R Rosenbaum and Donald B Rubin. The bias due to incomplete matching. *Biometrics*, 41:103–116, 1985.

[7] Paul R Rosenbaum. *Observational Studies*. Springer, New York, 2nd edition, 2002.

[8] Donald B Rubin. Estimating causal effects from large data sets using propensity scores. *Annals of Internal Medicine*, 127(8):757–763, 1997.

[9] Donald B Rubin. *Matched Sampling for Causal Effects*. Cambridge University Press, New York, 2006.

[10] Donald B Rubin and Richard P Waterman. Estimating the causal effects of marketing interventions using propensity score methodology. *Statistical Science*, 21:206–222, 2006.

[11] Noah Greifer and Elizabeth A Stuart. Matching methods for confounder adjustment: An addition to the epidemiologist's toolbox. *Epidemiologic Reviews*, 43(1):118-129, 2021.

[12] Paul R Rosenbaum. *Design of Observational Studies*. Springer, New York, 2nd edition, 2020.

[13] Paul R Rosenbaum. Modern algorithms for matching in observational studies. *Annual Review of Statistics and Its Application*, 7:143–176, 2020.

[14] Ben B Hansen. The prognostic analogue of the propensity score. *Biometrika*, 95(2):481–488, 2008.

[15] Paul R Rosenbaum. Optimal matching for observational studies. *Journal of the American Statistical Association*, 84(408):1024–1032, 1989.

[16] Jeffrey H Silber, Paul R Rosenbaum, Matthew D McHugh, Justin M Ludwig, Herbert L Smith, Bijan A Niknam, Orit Even-Shoshan, Lee A Fleisher, Rachel R Kelz, and Linda H Aiken. Comparison of the value of nursing work environments in hospitals across different levels of patient risk. *JAMA Surgery*, 151(6):527–536, 2016.

[17] Linda H Aiken, Donna S Havens, and Douglas M Sloane. The magnet nursing services recognition program: A comparison of two groups of magnet hospitals. *AJN The American Journal of Nursing*, 100(3):26–36, 2000.

[18] William G Cochran. The planning of observational studies of human populations. *Journal of the Royal Statistical Society, A*, 128(2):234–266, 1965.

[19] William G Cochran and Donald B Rubin. Controlling bias in observational studies: A review. *Sankhyā, Series A*, 35(4): 417–446, 1973.

[20] A Philip Dawid. Conditional independence in statistical theory. *Journal of the Royal Statistical Society B*, 41(1):1–15, 1979.

[21] Jerzy Neyman. On the application of probability theory to agricultural experiments. (English translation of Neyman (1923). *Statistical Science*, 5(4): 465–472, 1990.

[22] Donald B Rubin. Estimating causal effects of treatments in randomized and nonrandomized studies. *Journal of Educational Psychology*, 66(5):688, 1974.

[23] David R Cox. The interpretation of the effects of non-additivity in the latin square. *Biometrika*, 45:69–73, 1958.

[24] B L Welch. On the z-test in randomized blocks and latin squares. *Biometrika*, 29(1/2):21–52, 1937.

[25] Martin B Wilk and Oscar Kempthorne. Some aspects of the analysis of factorial experiments in a completely randomized design. *Annals of Mathematical Statistics*, 27(4):950–985, 1956.

[26] Ruoqi Yu. Evaluating and improving a matched comparison of antidepressants and bone density. *Biometrics*, 77(4):1276-1288, 2021.

[27] Constantine E Frangakis and Donald B Rubin. Principal stratification in causal inference. *Biometrics*, 58(1):21–29, 2002.

[28] R.A. Fisher. *Design of Experiments*. Oliver and Boyd, Edinburgh, 1935.

[29] Irwin DJ Bross. Statistical criticism. *Cancer*, 13(2):394–400, 1960.

[30] Paul R Rosenbaum. From association to causation in observational studies: The role of tests of strongly ignorable treatment assignment. *Journal of the American Statistical Association*, 79(385):41–48, 1984.

[31] Paul R Rosenbaum. The role of a second control group in an observational study. *Statistical Science*, 2(3):292–306, 1987.

[32] Paul R Rosenbaum. On permutation tests for hidden biases in observational studies. *Annals of Statistics*, 17(2): 643–653, 1989.

[33] Paul R Rosenbaum. The role of known effects in observational studies. *Biometrics*, 45:557–569, 1989.

[34] Eric Tchetgen Tchetgen. The control outcome calibration approach for causal inference with unobserved confounding. *American Journal of Epidemiology*, 179(5):633–640, 2014.

[35] Jerome Cornfield, William Haenszel, E Cuyler Hammond, Abraham M Lilienfeld, Michael B Shimkin, and Ernst L Wynder. Smoking and lung cancer: Recent evidence and a discussion of some questions. (Reprint of a paper from 1959 with four new comments). *International Journal of Epidemiology*, 38(5):1175–1201, 2009.

[36] Paul R Rosenbaum. Sensitivity analysis for certain permutation inferences in matched observational studies. *Biometrika*, 74(1):13–26, 1987.

[37] Paul R Rosenbaum and Jeffrey H Silber. Amplification of sensitivity analysis in matched observational studies. *Journal of the American Statistical Association*, 104(488):1398–1405, 2009.

[38] Paul R Rosenbaum. *Observation and Experiment: An Introduction to Causal Inference*. Harvard University Press, Cambridge, MA, 2017.

[39] Solomon Kullback and Richard A Leibler. On information and sufficiency. *Annals of Mathematical Statistics*, 22(1):79–86, 1951.

[40] Ruoqi Yu, Jeffrey H Silber, and Paul R Rosenbaum. The information in covariate imbalance in studies of hormone replacement therapy. *Annals of Applied Statistics*, 15, 2021.

[41] Donald B Rubin. The design versus the analysis of observational studies for causal effects: Parallels with the design of randomized trials. *Statistics in Medicine*, 26(1):20–36, 2007.

[42] M W Birch. The detection of partial association, i: The 2$\times$ 2 case. *Journal of the Royal Statistical Society B*, 26(2):313–324, 1964.

[43] J L Hodges and E L Lehmann. Rank methods for combination of independent experiments in analysis of variance. *The Annals of Mathematical Statistics*, 33:482–497, 1962.

[44] David R Cox. *Analysis of Binary Data*. Methuen, London, 1970.

[45] Paul R Rosenbaum. Covariance adjustment in randomized experiments and observational studies. *Statistical Science*, 17(3):286–327, 2002.

[46] Paul R Rosenbaum. *Replication and Evidence Factors in Observational Studies*. Chapman and Hall/CRC, New York, 2021.

[47] James M Robins, Steven D Mark, and Whitney K Newey. Estimating exposure effects by modelling the expectation of exposure conditional on confounders. *Biometrics*, 48:479–495, 1992.

[48] David Holt and T M Fred Smith. Post stratification. *Journal of the Royal Statistical Society A*, 142(1):33–46, 1979.

[49] Donald B Rubin and Neal Thomas. Combining propensity score matching with additional adjustments for prognostic covariates. *Journal of the American Statistical Association*, 95(450):573–585, 2000.

[50] Samuel D Pimentel, Rachel R Kelz, Jeffrey H Silber, and Paul R Rosenbaum. Large, sparse optimal matching with refined covariate balance in an observational study of the health outcomes produced by new surgeons. *Journal of the American Statistical Association*, 110(510):515–527, 2015.

[51] Paul R Rosenbaum, Richard N Ross, and Jeffrey H Silber. Minimum distance matched sampling with fine balance in an observational study of treatment for ovarian cancer. *Journal of the American Statistical Association*, 102(477):75–83, 2007.

[52] Ruoqi Yu, Jeffrey H Silber, and Paul R Rosenbaum. Matching methods for observational studies derived from large administrative databases. *Statistical Science*, 35(3):338–355, 2020.

[53] José R Zubizarreta. Using mixed integer programming for matching in an observational study of kidney failure after surgery. *Journal of the American Statistical Association*, 107(500):1360–1371, 2012.

[54] José R Zubizarreta, Ricardo D Paredes, and Paul R Rosenbaum. Matching for balance, pairing for heterogeneity in an observational study of the effectiveness of for-profit and not-for-profit high schools in chile. *The Annals of Applied Statistics*, 8(1):204–231, 2014.

[55] Ruoqi Yu and Paul R Rosenbaum. Directional penalties for optimal matching in observational studies. *Biometrics*, 75(4):1380–1390, 2019.

[56] Paul R Rosenbaum. How to see more in observational studies: Some new quasi-experimental devices. *Annual Review of Statistics and Its Application*, 2:21–48, 2015.

3

Generalizability and Transportability

Elizabeth Tipton and Erin Hartman

CONTENTS

In practical uses of causal inference methods, the goal is to not only estimate the causal effect of a treatment in the data in hand, but to use this estimate to infer the causal effect in a broader population. In medicine, for example, doctors or public health officials may want to know how to apply findings of a study to their population of patients or in their community. Or, in education or social welfare, school officials or policy-makers may want to know how to apply the findings to students in their schools, school district, or state. If unit specific treatment effects are all identical, these inferences

DOI: 10.1201/9781003102670-3

from the sample to the population are straightforward. However, when treatment impacts vary, the causal effect estimated in a sample of data may *not* directly generalize to the target population of interest.

In the causal inference literature, this is referred to as the *generalizability* or *transportability* of causal effects from samples to target populations. The distinction between these two concepts has to do with the relationship between the sample and target population. If the sample can be conceived of as a sample *from* the target population, this inference is referred to as generalization. If, on the other hand, the sample is *not* from the target population, this inference is referred to as one of transportation. Of course, we note that in practice this distinction can be difficult, since it is not always clear how a sample was generated, making questions of its target population of origin difficult to surmise.

The goal of this chapter is to provide an introduction to the problems of generalization and transportation and methods for addressing these concerns. We begin by providing an overview of validity concerns, then introduce a framework for addressing questions of generalization and transportation, methods for estimating population-level causal effects, extensions to other estimands, and approaches to planning studies that address these concerns as part of the study design. Note that this chapter is not comprehensive, and that there are several other field specific reviews that are also relevant [1–6].

3.1 The Generalization and Transportation Problem

We begin this chapter by situating the problems of generalization and transportation in the broader research *validity* literatures.

3.1.1 Validity concerns

The field of causal inference is one that, at its core, focuses on improving *internal validity*: the extent to which a study can establish a trustworthy cause-and-effect relationship between a treatment and outcome. Internal validity, however, is only one of four validity types that affect the claims that can be made from scientific studies. These other three validity types are *statistical conclusion* validity, *external* validity, and *construct* validity [7]. Overviews of each are provided in Table 3.1. While statistical conclusion validity – which has to do with estimation and hypothesis testing < is important, we put it aside here since it is not the focus of this chapter.

Both external and construct validities speak to concerns regarding how causal effects estimated in samples relate to the effects expected in target populations. External validity focuses, in particular, on the extent to which a causal effect varies across *units*, *treatments*, *outcomes*, and *settings* [8]. In this formulation, this is a question of the extent to which the *utos* of a study generalizes to its intended target population *UTOS* or to a new $UTOS^+$ beyond this target population (what we call "transportation"). Here *units* refer to characteristics of the unit of analysis – e.g., patients, students – while *settings* refer to the context in which these are situated – e.g., hospitals, schools, or time periods. Similarly, *treatments* refer to the fact that the version of a treatment found in a study may differ from that delivered in a target population, and similarly, the *outcomes* of concern may differ in the sample versus the target population. Importantly, following the broader literature, the focus of this chapter will be on methods for addressing concerns with generalizations across *units* and *settings*, more so than across treatments and outcomes.

Construct validity is a deeper issue, however, and deserves consideration as well, though it will not be the focus of this chapter. Construct validity has to do with the extent to which the variables measured in a study map onto higher order constructs that they represent. It is typical in causal studies, for example, to require the exact version of a *treatment* to be clearly defined, but these concerns are broader and apply to the other components of *UTOS* as well. For example, the *units* in a study may be described as coming from "low socioeconomic status households" ("low-SES"), but how this is operationalized matters. While various complex measures of SES exist – including family income, assets and parental education - in practice the operationalization that is far more common in U.S. education studies is an indicator that a student receives a Free or Reduced Priced Lunch (FRPL); unlike these more complex measures, this only takes into account income and household size and requires families to apply. Similarly, these same concerns hold with *outcomes* in a study. For example, the purpose of a treatment may be to reduce depression. However, the construct of "depression" also requires operationalization, often including answering 21 questions from Beck's Depression Inventory.

In the remainder of this chapter, we will focus on methods for making inferences from the variety of *units* and *settings* found in a study to those found in a target population. Throughout, we will assume that the *treatment* and *outcomes*, as defined in the study, are the focus in the target population. Similarly, we will put aside deeper questions of construct validity in this work. However, we do so only because these have not been the focus of this literature, not because these issues are not important.

3.1.2 Target validity

To understand potential external validity bias, it is essential that researchers clearly define their targets of inference. Assuming the target treatment and outcome of interest are defined as in the study, this means we must clearly define the target population and target setting of inference. We begin by assuming that each unit $i = 1, \dots, n$ in a sample S has a treatment effect,

$$\delta_i(c) = Y_i(1, c) - Y_i(0, c) \tag{3.1}$$

where we let c denote the experimental setting. As is common, we assume $c_i = c \; \forall \; i$, however multiple settings of interest may be included within the sample (e.g. both public and private schools). As a result of the Fundamental Problem of Causal Inference [9], δ_i is unobservable. For this reason, we typically focus instead on the Sample Average Treatment Effect (SATE),

$$\tau_S = SATE = \frac{1}{n}\sum_{i=1}^{n} \delta_i(c). \tag{3.2}$$

In a randomized experiment, for example, τ_S can be estimated unbiasedly using the simple difference in means estimator.

Suppose we are interested instead in the Population Average Treatment Effect (PATE),

$$\tau_P = PATE = \frac{1}{N}\sum_{i=1}^{N} \delta_i(c^*) \tag{3.3}$$

where this target population P includes N units all with context c^*. Note that if the sample $S \subset P$ and $c = c^*$ then this is a question of generalizability, while if $S \not\subset P$ or c^* is not included in the sample, this is a question of transportability.

Now, assume a general estimator $\hat{\tau}$ of this PATE. The total bias in this estimator (Δ) can be decomposed into four parts [10]:

$$\begin{aligned} \Delta &= \hat{\tau} - \tau_P \\ &= \Delta_S + \Delta_T \\ &= \Delta_{S_x} + \Delta_{S_u} + \Delta_{T_x} + \Delta_{T_u} \end{aligned}$$

where $\Delta_S = \tau_S - \tau_P$ is the difference between the SATE and the PATE and $\Delta_T = \hat{\tau} - \tau_S$ is the bias of the estimator for the SATE. In the third line, note that each of these biases are further divided into

TABLE 3.1
Overview of validity types

Validity Type	Inferences about:	Threats to Validity:
Internal	Whether observed covariation between A and B reflects a causal relationship from A to B in the form in which the variables were manipulated or measured.	(a) ambiguous temporal precedence, (b) selection, (c) history, (d) maturation, (e) regression artifacts, (f) attrition, (g) testing, and (h) instrumentation.
External	Whether the cause–effect relationship holds over variation in persons, settings, treatment variables, and measurement variables.	(a) interaction of the causal relationship with units, (b) interaction of the causal relationship over treatment variations, (c) interaction of the causal relationship with outcomes, and (d) interaction of the causal relationship with settings.
Statistical Conclusion	Whether the validity of conclusions that the dependent variable covaries with the independent variable, as well as that of any conclusions regarding the degree of their covariation.	(a) low statistical power, (b) violated assumptions of statistical tests, (c) "fishing" and the error rate problem, (d) unreliability of measures, (e) restriction of range, (f) unreliability of treatment implementation, (g) extraneous variance in the experimental setting, (h) heterogeneity of units, and (i) inaccurate effect-size estimation.
Construct	The higher order constructs that represent sampling particulars.	(a) inadequate explication of constructs, (b) construct confounding, (c) mono-operation bias, (d) mono-method bias, (e) confounding constructs with levels of constructs, (f) treatment-sensitive factorial structure, (g) reactive self-report changes, (h) reactivity to the experimental situation, (i) novelty and disruption effects, (j) experimenter expectancies, (k) compensatory equalization, (l) compensatory rivalry, (m) resentful demoralization, and (n) treatment diffusion.

two parts – those based on observed covariates X and those resulting from unobserved covariates U. Olsen and colleagues provide a similar decomposition, though from a design-based perspective [11].

By focusing on total bias – also called *target validity* [12] – it is clear that bias results both from how units selected into or were assigned to treatment (Δ_T) and from how these units selected or were sampled into the study (Δ_S). This also clarifies the benefits and costs of different study designs. For example, in a randomized experiment with no attrition, we can expect that $\Delta_T = 0$, but, depending upon study inclusion criteria and the degree of heterogeneity in treatment effects across units, Δ_S is likely non-zero and could be large. In converse, in an observational study using large administrative data on the entire population of students in a school system, it is likely that $\Delta_S = 0$, while, since treatment as not randomly assigned, depending upon the adjustments and covariates accounted for, Δ_T is likely non-zero. This target validity framework makes clear that there are strong trade-offs between internal and external validity and that focusing on one without the other can severely limit the validity of inferences from the study to its target population.

3.2 Data and Assumptions

In this section, we provide an overview to the data and assumptions required in order to estimate a PATE from a sample. These methods apply to both questions of generalization and transportation, as well as to both inferences made based upon randomized studies and observational data.

3.2.1 Data

To begin, we require data on both a sample S of units $i = 1, \ldots, n$ and on a target population P with units $i = 1, \ldots, N$. In the sample, let Z_i indicate if unit i receives the treatment (versus comparison), and let $Y_i = Y_i(1,c)Z_i + Y_i(0,c)(1 - Z_i)$ be the observed outcome. In the target population, we do not observe Z_i nor do we require that the same Y_i is observed (though when it is observed, this information is useful; see Section 3.3.5).

Let $P^* = P \cup S$ be the combined data across the sample and target population. When the sample is a subset of the target population ($S \subset P$), this combined data includes $N' = N$ rows and moving from the sample to the target population is referred to as *generalization*. When the sample is not a subset of the target population ($S \not\subset P$), the combined data has $N' = N + n$ rows and moving from the sample to the target population is referred to as *transportation* [13, 14] or *synthetic* generalization [15].

For all units in P^*, let W_i indicate if unit i is in the sample. Finally, for each unit in this combined P^*, a vector of covariates is also required $\mathbf{x}_i = (x_{1i}, x_{2i}, \ldots, x_{pi})$. Importantly, these covariates need to be measured the same way in the sample and target population, a criteria that can be difficult to meet in many studies (see [16] for a case study).

3.2.2 Assumptions

In order for an estimator $\hat{\tau}$ to be unbiased for the causal effect τ_P in the target population, the following assumptions are required.

Assumption 1 (Strongly ignorable treatment assignment [15, 17]). *For all units with $W = 1$, let $\mathbf{X} = (X_1, X_2, \ldots, X_p)$ be a vector of covariates that are observed for each. Let Z indicate if each of these units receives the treatment. Then the treatment assignment is strongly ignorable if,*

$$[Y(1,c), Y(1,c)] \perp\!\!\!\perp Z \mid W = 1, \mathbf{X} \textit{ and } 0 < Pr(Z = 1 \mid W = 1, \mathbf{X}) < 1.$$

In practice, this condition is met without covariates in a randomized experiment without attrition, or in an observational study in which $\mathbf{X}$ includes all covariates that are both related to the outcome and to treatment selection. Moving forward, we will assume that Assumption 1 is met, since the remainder of the chapter focuses on generalization and transportation.

Assumption 2 (Strongly ignorable sample selection [15, 17]). *In the combined population P^*, the sample selection process is strongly ignorable given these covariates $\mathbf{X}$ if*

$$\delta = [Y(1,c) - Y(0,c)] \perp\!\!\!\perp W \mid \mathbf{X} \textit{ and } 0 < Pr(W = 1 \mid \mathbf{X}) \leq 1.$$

Notice that this condition is of similar form to Assumption 1 but differs in two regards. First, it is the unit-specific treatment effects ($\delta = Y(1,c) - Y(0,c)$) that must be conditionally orthogonal to selection, not the vector of potential outcomes ($Y(1,c), Y(0,c)$). This means that we are worried here with identifying covariates that explain variation in treatment effects in the population, or put another way, the subset of those covariates $\mathbf{X}$ that are related to the outcome Y that *also* exhibit a different relationship with Y under treatment and comparison. The second difference here is with regard to the final inequality. Whereas in order to estimate the SATE, every unit needs to have some probability of being in either treatment or control, when estimating the PATE, it is perfectly fine for some population units to be deterministically included in the sample.

In practice, meeting Assumption 2 requires access to a rich set of covariates. Here the covariates that are essential are those that *moderate* or explain variation in treatment effects across units in the target population. The identification of potential moderators requires scientific knowledge and theory regarding the mechanism through which the intervention changes outcomes. This is a place in which substantive expertise plays an important role.

Researchers can often exert more control over what variables are collected in the sample than which variables are available in the target population. Egami and Hartman [18] provide a data-driven method for selecting covariates for generalization that relies only on sample data. Maintaining the common assumption that the researcher can specify the variables related to how the sample was selected, they estimate a Markov random field [19, 20] which is then used to estimate which pre-treatment covariates are sufficient for meeting Assumption 2. The algorithm can include constraints on what variables are measurable in the target population, which allows researchers to determine if generalization is feasible, and if so, what variables to use in the estimation techniques for the PATE we describe in Section 3.3.4. By focusing on estimating a set that is sufficient for meeting Assumption 2, the method allows researchers to estimate the PATE in scenarios where methods that require measurement of all variables related to sampling cannot.

Regardless of approach, researchers should keep in mind that in many cases – particularly in randomized experiments – the samples are just large enough to have adequate statistical power for testing hypotheses regarding the SATE and are quite under-powered for tests of treatment effect moderators. Thus, moderators that are detected should clearly be included, but those that are not precisely estimated should not be excluded based on this alone. Furthermore, in highly selected samples, it is possible that there are some moderators that simply do not vary within the sample at all and thus cannot be tested.

All of this means that it may be hard to empirically verify that this condition has been met, and instead the warrant for meeting Assumption 2 may fall on the logic of the intervention and theoretical considerations. In this vein, one approach to determining what variables to adjust for relies on a directed acyclic graphical (DAG) approach [14, 21]. Whereas the method in Egami and Hartman [18] empirically estimates the variables sufficient for meeting Assumption 2, Pearl and Bareinboim [14, 21] and related approaches first fully specify the DAG and then analytically select variables that address Assumption 2. The graphical approach also allows for alternative approaches to identification, under the assumption the DAG is fully specified.

Finally, regardless of the approach taken when identifying potential moderators, it is wise to conduct analyses regarding how sensitive the results are to an unobserved moderator. See Section 3.3.5 for more on this approach.

Assumption 3 (Contextual Exclusion Restriction [18]). *When generalizing or transporting to settings or contexts different from the experimental setting, we must assume*

$$Y(Z = z, \mathbf{M} = \mathbf{m}, c) = Y(Z = z, \mathbf{M} = \mathbf{m}, c = c^*)$$

where we expand the potential outcome with a vector of potential context-moderators $\mathbf{M}(c)$*, and fix* $\mathbf{M}(c) = m$.

When we change the setting or context, the concern is that the causal effect is different across settings even for the same units. For example, would the effect for student i be the same or different in a public vs. a private high school? At its core, this is a question about differences in mechanisms across settings. To address concerns about changes to settings, researchers must be able to adjust for context-moderators that capture the reasons why causal effects differ across settings. Assumption 3 states that, conditional on these (pre-treatment) context-moderators, the treatment effect for each unit is the same across settings. Much like the exclusion restriction in instrumental variables, to be plausible the researcher must assume that, conditional on the context-moderators $\mathbf{M} = \mathbf{m}$, there are no other pathways through which setting directly affects the treatment effect. As with instrumental variables, this assumption cannot be guaranteed by the design of the study, and must be justified with domain knowledge. This strong assumption emphasizes the role of strong theory when considering generalizability and transportability across settings [22, 23].

Assumption 4 (Stable Unit Treatment Value Assumptions (SUTVA) [15]). *There are two SUTVA conditions that must be met, SUTVA(S) and SUTVA(P). Again, let W indicate if a unit in P^* is in the sample and let Z indicate if a unit receives the treatment.*

1. *SUTVA (S): For all possible pairs of treatment assignment vectors $\mathbf{Z} = (Z_1, \ldots, Z_{N'})$ and $\mathbf{Z}' = (Z'_1, \ldots, Z'_{N'})$, SUTVA (S) holds if when $\mathbf{Z} = \mathbf{Z}'$ then $Y(\mathbf{Z}, W = 1, c) = Y(\mathbf{Z}', W = 1, c)$.*

2. *SUTVA (P): Let $\delta(1) = Y(Z = 1, W = 1, \mathbf{M} = \mathbf{m}, c) - Y(Z = 0, W = 1, \mathbf{M} = \mathbf{m}, c)$ and $\delta(0) = Y(Z = 1, W = 0, \mathbf{M} = \mathbf{m}, c^*) - Y(Z = 0, W = 0, \mathbf{M} = \mathbf{m}, c^*)$ be the unit specific treatment effects when included in the study sample versus not. For all possible pairs of sample selection vectors $\mathbf{W} = (W_1, \ldots, W_{N'})$ and $\mathbf{W}' = (W'_1, \ldots, W'_{N'})$, SUTVA (P) holds if when $\mathbf{W} = \mathbf{W}'$ then $\delta(\mathbf{W}) = \delta(\mathbf{W}')$ and when $\delta = \delta(1) = \delta(0)$.*

Here it is important that both of these stability conditions are met. SUTVA (S) is the "usual" SUTVA condition introduced by Rubin [24, 25], while SUTVA (P) is a condition specific to generalization and transportation introduced by Tipton [15]. SUTVA (P) requires two conditions to be met. First, it requires that there is no interference of units included (or not) in the sample and that there is only one version of the treatment. This condition may not be met if the treatment effects are a function of the proportion of the population receiving treatment, often referred to as the general equilibrium or scale-up problem. Second, there is the requirement that $\delta = \delta(1) = \delta(0)$, which may not be met when being in the sample itself affects the treatment effect. When the contextual exclusion restriction holds, this assumption allows for transportation across contexts where the context-moderators are sufficient. However, it rules out some specific patterns of transportation, for example, if simply being in a study may induce larger or smaller treatment effects.

3.3 Estimation

In order to estimate the PATE, there are five steps, which we describe here. First, sample selection probabilities must be estimated, and based upon the distributions of these probabilities, the ability to generalize needs to be assessed. In many cases problems of common support lead to a need to redefine the target population; this is a nuanced problem, since doing so involves changing the question of practical interest. Then the PATE can be estimated using one of several approaches. Finally, the PATE estimate needs to be evaluated based upon its ability to remove bias.

In this section we focus on estimation when generalizing to a broader, or different, population of *units*, and we assume that our target setting is the same as that in the experimental setting, i.e. $c = c^*$. In effect this means that Assumption 3 holds for all units i, which allows us to collapse our potential outcomes as $Y_i(w), w \in (0,1)$, which are no longer dependent on context. We return to estimation under changes to settings at the end of the section.

3.3.1 Estimate sample selection probabilities

For units $i = 1, \ldots, N'$ in the combined population $P^* = S \cup P$, let W_i indicate if unit i is in the sample and observe the covariate vector $\mathbf{x}_i = (x_{i1}, \ldots, x_{ip})$. Then define the sample selection probabilities as

$$s_i(\mathbf{x}) \propto \Pr(W_i = 1 \mid \mathbf{x}) \tag{3.4}$$

In this formulation we assume that the units are independent. We define this as "proportional to" since, in the case of transportability, the combined data is an artificial combination of the sample and target population (i.e., $N' > N$).

In practice, these sample selection probabilities must be estimated. Here, the same statistical methods available for estimating the treatment probabilities – propensity scores – can be applied as well. The simplest approach, of course, is to estimate these using a logistic regression, though other more modern methods can be used as well. Just as in the treatment selection case, the right-hand side of this equation may include not only main effects for these moderators, but also squared terms and interactions (if the moderation effect is nonlinear). Theoretically, this model could be extended as well in cases in which the sample selection process is multi-stage – e.g., selecting schools and then students – though to date there has been little work on this problem.

As later steps will illustrate, these selection probabilities are themselves not of direct interest. Indirectly, these selection probabilities are used to define weights or quantiles, depending upon the estimator of the PATE. Others have proposed, therefore, that there is no need to estimate the selection probabilities at all, and that instead weights should be developed directly. For example, one approach, balancing weights [26], are developed directly to minimize variance and balance moments using an optimization framework.

3.3.2 Assess similarity

Meeting the strong ignorability requirement (Assumption 2) involves determining if the positivity condition has been met. This can be re-written in terms of the sample selection probability as

$$0 < \Pr(W = 1 \mid s(\mathbf{x})) \leq 1 \tag{3.5}$$

Here the threat is that there are some units in the target population P that have zero probability of being in the sample S. Tipton [15] refers to this problem as a *coverage error*, or *under-coverage*. This can be diagnosed by comparing the distributions of the estimated sample selection probabilities (or their logits) in the sample S and the target population P. Here it can be helpful to focus on the 5th and 95th percentiles of these distributions, instead of on the full range, given sampling error.

Beyond determining if there is under-coverage, there are many situations in which it is helpful to summarize the similarity of these distributions. When these distributions are highly similar to one another, this suggests that the estimator of the PATE may not differ much from the SATE and may have similar precision. However, when these are highly different, considerable adjustments may be required, resulting in possible extrapolations and significantly less precision. This dissimilarity can arise either because the sample and population are truly very different or because the treatment effect heterogeneity is not strongly correlated with sample selection (see [27]). Thus, diagnostic measures are helpful when determining if it is reasonable to proceed to estimate the PATE or if, instead, a more credible sub-population (or different covariate set) needs to be defined (see Step 3).

One such metric for summarizing similarity is called the B-index or the *generalizability index* [28]. This index is defined as the geometric mean of the densities of the logits of the sample selection probabilities (i.e., $l(\mathbf{x}) = logit(s(\mathbf{x}))$) in the sample $f_s = f_s(l(\mathbf{x}))$ and target population $f_p = f_p(l(\mathbf{x}))$,

$$\beta = \int \sqrt{f_s f_p} df_p. \tag{3.6}$$

This index is bounded $0 \leq \beta \leq 1$, where larger values indicate greater similarity between the sample and target population. Tipton shows that the value of β is highly diagnostic of the degree of adjustment required by the PATE estimator and its effect on the standard error. Simulations indicate that when $\beta > 0.7$ or so, the sample is similar enough to the target population that adjustments have minimal effect on precision, and – in the case of cluster-randomized field trials – when $\beta > 0.9$ish the sample is as similar to the target population as a random sample of the same size [29]. Similarly, when $\beta \leq 0.5$, typically there is significant under-coverage and the sample and target population exhibit large differences, thus requiring extrapolations in order to estimate the PATE [28]. In cases in which the distributions of the logits are normally distributed, this reduces to a function of the standardized mean difference of the logits – another diagnostic – defined as

$$SMD = \frac{\overline{\hat{l}(\mathbf{x}|W=0)} - \overline{\hat{l}(\mathbf{x}|W=1)}}{sd}. \tag{3.7}$$

where here sd indicates the standard deviation of the logits distribution in the target population [17].

The approaches described so far for diagnosing similarity only require covariate information in both the sample and target population. When the same outcome measure, Y, is available in both the sample and target population; however, this information can also be used to assess similarity through the use of placebo tests. Stuart and colleagues [17] describe a test that relies only on the sample outcomes observed by those units in the control condition (i.e., $Y_i(0, c)$) and comparing these values with those observed for all units in the target population, assumed to be unexposed to treatment, under different estimation strategies. Hartman and colleagues [26] describe a placebo test for verifying the identifying assumption for the population average treatment effect on the treated (PATT) that compares the adjusted sample treated units to treated units in the target population. We will return to this option at the end of Step 4.

3.3.3 When positivity fails

When estimating the PATE, restricting inferences to the region of common support (so as to meet the positivity part of Assumption 2) comes at great cost, particularly when the study is a randomized experiment. The selection process into the study sample typically involves several steps and actors: First, the unit must be eligible to be in the study, then they must have been approached to be recruited, and then they must agree to be in the study. These initial selection steps have to do with the researchers conducting the study, and their goals for the study may not have been with the same target population in mind. Furthermore, this means that units with a zero probability of being in the study sample may be interested in and / or may have benefited greatly from the treatment yet were excluded from the study for practical reasons.

Put another way, restricting the target population to the region of common support means fundamentally changing the question and estimand of interest [6]. For example, Tipton [15] shows a case in which the initial goal is to estimate the PATE for the entire population of public middle schools in Texas, but that this is not possible for 8% of the schools (which are outside the region of common support). Concluding the study with an estimate of the PATE for the 92% of schools that have some probability of being in the experiment is not particularly informative in practice, since it is not at all clear which schools are in this target population and which are not. A simple approach that can be useful is to supplement these findings with maps or lists of units that are included in the newly defined target population [30].

Tipton and colleagues [31] provides a case study from a situation in which the resulting samples from two separate randomized trials were quite different than the target populations the studies initially intended to represent. In these studies, the eligibility criteria used by the recruitment team were not clear. They then explored possible eligibility criteria – based on one, two, or three covariates – to define a series of possible sub-populations. The sample was then compared to each of these sub-populations, and for each a generalizability index was calculated. The final target populations were then selected based on this index. This enabled the target population for the study to be clearly defined – using clear eligibility criteria that practitioners could apply – and for PATEs to be estimated.

Finally, an alternative to changing the estimand is to rely on stronger modeling assumptions. For example, balancing weights do not rely on a positivity assumption for identification of the PATE, rather the research must assume that either the treatment effect heterogeneity model is linear or the sample selection model is link-linear in the included $\mathbf{X}$ used to construct the weights [32, 33], allowing researchers to use weighting to interpolate or extrapolate beyond the area of common support. Cross-design synthesis, which combine data treatment and outcome data collected in both trial and observational target population data using model-based adjustments to project onto stratum with no observations, provides an alternative modeling approach to estimating generalized effects [34]. These approaches are particularly useful when considering issues of transportability, where the study might have had strict exclusion criterion for some units. Of course, the stronger the degree of extrapolation, such as in the studies described above, the more a researcher must rely on these modeling assumptions. Researchers should carefully consider the trade offs between restricting the target population and relying on modelling assumptions when considering failures of the positivity assumption. As discussed in Section 3.5, better study design, where feasible, can mitigate some of these issues.

3.3.4 Estimate the PATE

The PATE can be estimated using a wide variety of estimators, each of which mirrors its use in estimating the SATE in observational studies. Here we describe several estimators.

3.3.4.1 Subclassification

The first is the *subclassification* estimator [15, 30], defined as

$$\hat{\tau}_{ps} = \sum_{j=1}^{k} w_j \hat{\tau}_j \tag{3.8}$$

where w_j is the proportion of the population P in stratum j (so that $\Sigma w_j = 1$) and $\hat{\tau}_j$ is an estimate of the SATE in stratum j. In a simple randomized experiment, for example, $\hat{\tau}_j = \overline{Y}_{t_j} - \overline{Y}_{c_j}$. Here typically strata are defined in relation to the distribution of the estimated sampling probabilities in the target population P so that in each stratum $w_j = 1/k$; as few as $k = 5$ strata can successfully reduce most, but not all, of the bias (Cochran, 1968). The variance of this estimator can be estimated using,

$$\hat{V}(\hat{\tau}_{ps}) = \sum_{j=1}^{k} w_j^2 SE(\hat{\tau}_j)^2 \tag{3.9}$$

There are several extensions to this estimator that have been proposed, including the use of full matching instead [17] and estimators that combine subclassification with small area estimation [35].

3.3.4.2 Weighting

Another approach is *weighting-based*, in which weights are constructed that make the sample representative of the target population. When treatment assignment probabilities are equal across units, the weighting estimator for the PATE is defined as:

$$\hat{\tau}_w = \sum_{i=1}^{N} \frac{\hat{w}_i W_i Z_i Y_i}{\hat{w}_i W_i Z_i} - \sum_{i=1}^{N} \frac{\hat{w}_i W_i (1 - Z_i) Y_i}{\hat{w}_i W_i (1 - Z_i)} \tag{3.10}$$

where recall that W_i indicates inclusion in the sample, Z_i indicates a unit is in the treatment group, and $\hat{w}_i$ is the estimated probability that a unit is in the sample if generalizing, or odds of inclusion if transporting. To improve precision and reduce the impact of extreme weights, the denominator normalizes the weights to sum to 1 [27]. This also ensures that the outcome is bounded by the convex hull of the observed sample outcomes.

For generalizability, in which the sample is a subset of the target population, weights are estimated as *inverse probability weights* [12, 17, 36],

$$w_{i,ipw} = \frac{1}{\hat{s}_i} \tag{3.11}$$

When transporting to a target population disjoint from the experimental sample, i.e. with $Z_i = 0$, the weights are estimated as the *odds of inclusion*.

$$w_{i,odds} = \frac{1}{\hat{s}_i} \times \frac{1 - \hat{s}_i}{\Pr(W_i = 0)} \tag{3.12}$$

A benefit of this weighting-based estimator, $\hat{\tau}_w$ is that it weights the outcomes in the treatment and comparison groups separately; in comparison, the subclassification estimator cannot be calculated if one or more strata have either zero treatment or control units. Variance estimation is trickier with the IPW estimator, however. Buchanan and colleagues [36] provide a sandwich estimator for the variance of this IPW estimator, as well as code for implementing this in R. The bootstrap can also be used to construct variance estimators [37].

A related method for estimating the weights for generalizability and transportability is through calibration, or balancing weights [26, 38], which ensure the sample is representative of the target population on important descriptive moments, such as matching population means. This can be beneficial for precision, as well an alternative approach to estimation when positivity fails.

3.3.4.3 Matching

A third strategy is to *match* population and sample units. One approach would be to caliper match with replacement based on the estimated sampling. In this approach the goal would be to find the "nearest" match to each target population unit in the sample, keeping in mind that the same sample unit would likely be matched to multiple population units. Another approach, building off of optimization approaches for causal inference more generally uses mixed integer programming to directly match sample units to each target population unit. This approach does not require propensity scores, instead seeking balance (between the sample and target population) directly [39].

3.3.4.4 Modeling

In the previous approaches, the strata, weights, or matches are designed without any outcome data. Another approach is to model the outcome directly. The simplest version of this is linear regression, using a model like,

$$E\left[Y_i | Z_i, \mathbf{x}_i\right] = \beta_0 + \gamma_0 Z_i + \Sigma \beta_p x_{pi} + \Sigma \gamma_p Z_i X_{pi}, \tag{3.13}$$

which explicitly models the heterogeneity in treatment effects via interaction terms. This model could be fit using either only data from the sample, or, when data on the outcome Y is available in the target population, this can be used as well [40]. Based on this model, predicted outcomes are generated for each unit in the target population and these are then averaged and differenced to estimate the PATE.

In addition to this approach Kern and colleagues [41] investigated two additional approaches, with a focus on transporting a causal effect from a study to a new population. One of these approaches combines weighting (by the odds) with a specified outcome model, and they explore several different approaches to estimating the sampling probabilities (e.g., random forests). The other approach they investigate is the use of Bayesian Additive Regression Trees (BART), which does not require the assumptions of linear regression. Their simulations suggest that BART or IPW+RF work best, when sample ignorability holds, though linear regression performed fairly similarly. More recent work in this area further investigates the use of Bayesian methods using multilevel modeling and post-stratification in combination [42].

3.3.4.5 Doubly robust

A final approach, "doubly robust" estimators combine the approaches discussed above by modeling both the sampling weights and the treatment effects. These estimators are consistent if either the sampling weights or the outcome model are correctly specified, providing flexibility for the researcher in estimation. Kern and colleagues [41], Dahabreh and colleagues [43], Egami and Hartman [18] and Dong and colleagues [44] investigate the use of "doubly robust" estimators (that employ covariates to predict both the sample selection process and treatment effects), which may offer several benefits, particularly when samples are large. Egami and Hartman [45] discuss practical guidance for selecting between weighting-based, outcome-based, and doubly robust estimators depending on data availability and research goals. Estimation approaches that simultaneously incorporate information on the sampling propensity and the outcome include an augmented-IPW estimator [43], a targeted maximum likelihood estimator (TMLE) [46], and an augmented calibration estimator [44].

3.3.4.6 Changes to setting

So far, in this section, we have discussed how to estimate the PATE with changes to the *units*. The estimation approaches focus on adjusting for differences in the distribution of pre-treatment characteristics $\mathbf{X}$, either through weighting or modeling. When the target of inference also includes changes to setting or context, such that $c \neq c^*$, we must invoke Assumption 3, the contextual exclusion restriction. Recall that this requires that we adjust for variables that relate to the mechanisms affecting treatment effect heterogeneity differentially across settings. For this to be plausible, researchers must assume that, conditional on context-moderators $\mathbf{M}$, the treatment effects for units are the same across settings.

Estimation under the contextual exclusion restriction is very similar to the approaches we describe in *Step 3*, except we include both individual characteristics $\mathbf{X}$ and context-moderators $\mathbf{M}$ in the weights or outcome models. As discussed by Egami and Hartman [45], often times when we change settings we simultaneously change units, as well. The authors provide analogous estimators to those described here for simultaneous changes in units and setting, relying on both the sampling ignorability and the contextual exclusion restriction assumptions.

3.3.5 Evaluate PATE estimation

Given that there are several possible strategies for estimating the PATE, it is important to investigate how well the approach selected works in a given dataset. This is especially important since Tipton and colleagues [29] and Kern and colleagues [41] both find – via simulation studies – that no one approach dominates the others in all data analyses. At a minimum, when using subclassification, weighting, or matching, *balance* on the included moderators should be explored both before and after estimation. For example, this could include comparisons of standardized mean differences (standardized with respect to the target population) for the sample versus target population (before) and for the weighted sample versus target population (after). Ideally, the same assessment strategies used for the use of propensity scores in observational studies can be applied as well, e.g., that these values should be 0.25 or smaller.

Additionally, when the same outcome Y is available in the sample and target population, balance can also be assessed in the same way on with regard to the outcome, a form of placebo test. To do so, apply the estimator (e.g., weights) to units in the treatment group ($W = 1$) in the sample ($W = 1$) and then compare this estimate to that of the population average of the treated units [26]. A similar placebo test that relies only on control units is described by Stuart and colleagues [17].

Importantly, the approaches given focus only on establishing balance with regard to observed moderators. Even when balance has been achieved on these moderators, it is possible that there remains an unobserved moderator for which balance is not achieved. One way to approach this problem is by asking how *sensitive* the estimate is to an unobserved moderator. Nguyen and colleagues [47] discuss a sensitivity analysis for partially observed confounders, in which researchers have measured important treatment effect modifiers in the experimental sample but not in the target population. Researchers can assess sensitivity to exclusion of such a variable from either the weights or outcome model by specifying a sensitivity parameter, namely plausible values for the distribution of the partially observed modifier in the target population. Dahabreh and colleagues [48] provide a sensitivity analysis that bypasses the need for knowledge about partially or fully unmeasured confounders by directly specifying a bias function. Andrews and colleagues [49] provide an approximation to possible bias when researchers can assume that residual confounding from unobservables is small once adjustment for observables has occurred.

3.4 Extensions

The previous section focused on point estimation of the PATE in either the target population of origin (generalization) or in a new target population (transportation). But when treatment effects vary, other estimands may also be of interest.

3.4.1 Interval estimation

In order to identify and estimate the PATE, sampling ignorability and positivity assumptions must hold. These are strong assumptions which can be difficult to meet in practice. Another approach focuses instead on relaxing these assumptions by shifting focus from a point estimate to an interval estimate based; these are referred to as *bounds* [50]. While bounds can be developed with no assumptions, Chan shows that these bounds are never informative about the sign of the PATE. If it is reasonable to assume that treatment effects within the sample are *always larger* than those units not in the sample ('monotone sample selection') then, we can bound the PATE between (τ_p^L, τ_p^U), where

$$\tau_p^L = \tau_s \text{Pr}(W = 1) + (Y^L - Y^U)\text{Pr}(W = 0) \tag{3.14}$$

$$\tau_p^U = \tau_s. \tag{3.15}$$

where τ_s is the SATE. Thus, if the outcome is dichotomous, $Y \in (0, 1)$ and $\text{Pr}(W = 1) = n/N = 0.05$ then $\tau_p \in (0.05\tau_s - 0.95, \tau_s)$. However, these bounds are also not typically very tight and, for this reason, Chan suggests that they are perhaps most useful when implemented in combination with other approaches such as subclassification.

3.4.2 Sign-generalization

In between the point- and interval-estimation approaches is a focus on the *sign* of the PATE. In this approach, researchers hypothesize the direction, but not the magnitude, of the average treatment effect in a different population or setting. Sign-generalization is more limited in the strength of the possible claim about the PATE, but it answers an important aspect of external validity for many researchers. It may also be a practical compromise when the required assumptions for point-identification are implausible.

Egami and Hartman [45] provide a statistical test of sign-generalization. The authors describe the *design of purposive variations*, where researchers include variations of units, treatments, outcomes, or settings/contexts in their study sample. These variations are designed to meet an overlap assumption, which states that the PATE in the target population and setting lies within the convex hull of the average treatment effects observed in the study across purposive variations. These average treatment effect estimates are combined using a partial conjunction test [51,52] to provide statistical evidence for how many purposive variations support hypothesized sign of the PATE. An advantage of this method, in addition to weaker assumptions, is that it allows researchers to gain some leverage on answering questions regarding target validity with respect to changes in treatments and outcomes.

3.4.3 Heterogeneity and moderators

Methods for addressing generalization and transportation both require meeting a sampling ignorability assumption. This involves identifying the full set of covariates that moderate the treatment impact, which thus requires methods for identifying and estimating potential moderator effects in studies. Furthermore, if treatment impacts vary, in many instances the PATE may not be the parameter of focus since it could be averaging across a significant amount of heterogeneity. For example, the PATE could be close to zero – suggesting an ineffective intervention in this target population – and yet for some subset of the target population, the effect might be quite large and effective (for an example of this, see [53]).

Clearly, the simplest approach to exploring heterogeneity is to use linear regression including treatment by covariate interactions. A problem, then, is identifying which of these estimated effects is truly a moderator and which are noise, particularly when there are *many* possible moderators to be explored. This problem is similar to the estimation of the PATE (using model-based approaches) and to estimation of propensity scores (e.g., using machine learning). One strand of these methods use regression trees [54] or support vector machines [55], while several other approaches use versions of Bayesian Additive Regression Trees (BART) [56–59]. Others have focused on non-parametric tests [60] and randomization tests [61]. A head-to-head comparison of many of these procedures was conducted at the 2017 Atlantic Causal Inference Conference [62], with the BART-based approaches proving to be most effective.

Finally, analysts using these approaches should keep in mind that generalization and transportation problems apply just as much to estimating and testing treatment effect moderators as they do to the PATE. This is because tests of moderators are functions not only regarding the strength of the relationship, but also the degree of variation in the moderator covariate itself in the sample. This concern particularly applies to randomized experiments, which often include far more homogeneous samples and include both explicit and implicit exclusion criteria [63,64].

3.4.4 Meta-analysis

A final approach, often used for generalizing causal effects, is to do so by combining the results of multiple studies using meta-analysis [65]. In this approach, the goal is to estimate the ATE (and variation) in the population of *studies* not people or other units. To do so, outcomes are converted to *effect sizes* and weights for each are calculated (typically inverse-variance). By conceiving of the observed studies as a random sample of the super-population of possible studies, the *study average effect size* and the degree of variation in these effect sizes within- and across-studies can be estimated. When covariates are also encoded for each effect size (or study), *meta-regression* models can be used to estimate the degree to which these moderate the treatment effects.

When individual participant data is available for all units in each of these primary studies, [66] provides an approach that allows for the PATE to be estimated for the population of units, as in the remainder of this chapter. This requires each study to include the same set of moderator data, though it does not require the positivity assumption to hold in every study.

3.5 Planning Studies

As the previous sections have made clear, many of the problems of generalization and transportation result because the data needed is unavailable. For example, it may be that important moderator data is not collected in the sample or target population, or that certain subgroups of units or settings found in the target population are not represented in the sample. This suggests that the best methods for generalizing and transporting causal effects involve *planning* studies with this goal in mind from the beginning.

3.5.1 Planning a single study

In order to ensure that generalization is possible, the first step is to design a study to be focused on determining the causal effect in clearly defined target populations. This involves three parts [6, 31]. First, the target population of interest – including key eligibility criteria – should be defined with reference to the treatment under study. This determination should involve multiple parties – investigators, stakeholders, funders – and will likely be constrained by the resources available for the study. Second, a population frame is developed which enumerates the units in the population. Finally, set of potential moderators and variables related to sampling are pre-specified and measured in the target population. Certainly this list of moderators and sampling variables should include those identified in the population frame, but can also involve additional moderators or sampling variables only available in the sample. Ideally, at this stage, the target population definition and the set of selected moderators and sampling variables are preregistered.

At this point, study design – including sample selection – can be addressed using many of the methods found in the design of sample surveys. This includes a wide range of probability sampling methods [67], though it should be assumed that non-response will likely be high and that non-response adjustments will be required. Another approach is to use a model-based sampling framework and instead select the sample using purposive methods focused on achieving balance between the sample and target population. One approach Tipton for generalizing causal effects is to divide the population into strata of similar units (for a given set of moderators) using k-means [68,69]. Tipton and Olsen [6] provide an overview of each of these planning process steps, situated in the education research context.

When the goal from the outset is to transport the causal effect to a different target population, one approach is to divide the target population (for transportation) into strata and then to sample similar units (that are not in the target population) for inclusion in the study [31]. In many cases, however, determining the optimal study design for transporting causal effects is made more difficult by the fact that there are *many* possible target populations of interest. The best advice here is to try to anticipate, to the extent possible, *all* of the possible PATEs that may be of interest and then to narrow these to the most important. The best approach – in terms of the average precision of these PATE estimates – is to define the broadest possible target population [70]. While this broad target population may itself not be of direct interest, doing so may result in better estimates for each of the actual target populations of interest (i.e., subsets of the broad population).

If treatment effects vary, when possible it may be important to not only design the study to estimate a PATE, but also to predict unit level treatment effects or to test hypotheses about moderators. Like the considerations for transportability, this means that there are multiple estimands of interest – e.g., the PATE, moderators, and CATEs. Tipton and colleagues [71] provide an example in which the goal was to estimate both an overall PATE and several subgroup PATEs. The problem here is that the optimal sampling probabilities or design may be different for different estimands. Similarly, estimands other than averages may also be of interest, including tests of treatment effect heterogeneity and moderators. Tipton [64] provides an approach to sample selection in this case using a response surface model framework.

3.5.2 Prospective meta-analysis

The same general idea of planning a single study can also be applied to planning a suite of studies, with the goal of combining them meta-analytically. Here the biggest difference is with respect to the degree of coordination required to ensure that the data can be easily combined. For example, each of the elements of *utos*, as well as the moderators and the treatment, needs to be measured in the same way across studies. Tipton and colleagues [72] provide a general framework and considerations for this case.

3.6 Conclusion

In this chapter, we have provided an overview of methods for generalizing and transporting causal effects from a sample to one or more target populations. These involve approaching bias from the sample selection process in a similar vein to the bias from treatment selection. Like those for addressing treatment selection, these require careful consideration of assumptions, the right covariate (moderator) data to be available, and the need to clearly identify the estimand of interest.

We close by reminding readers that these approaches address but one part of the problems of external validity that apply to all research. Questions of generalization and transportation call to question the reasons that research is conducted, the questions asked, and the ways in which research will be used for decision-making for both individuals and policies. For this reason, perhaps the most important step researchers can take – whether using the methods described in this chapter or not – is to clearly define characteristics of the sample, the limits to their study, and where and under what conditions they expect results might generalize (or not), and to hypothesize the mechanism through which the treatment may work.

3.7 Glossary

Internal validity: describes the extent to which a cause-and-effect relationship established in a study cannot be explained by other factors.

External validity: is the extent to which you can generalize the findings of a study to other situations, people, settings and measures.

Construct validity: concerns the extent to which your test or measure accurately assesses what it's supposed to,

Statistical conclusion: validity concerns the degree to which the conclusions drawn from statistical analyses of data are accurate and appropriate.

Generalization: concerns inferences about the value of a parameter estimated in a sample to a target population that the sample is a part of.

Transportation: concerns inferences about the value of a parameter estimated in a sample to a target population that the sample is not a part of.

References

[1] Benjamin Ackerman, Ian Schmid, Kara E Rudolph, Marissa J Seamans, Ryoko Susukida, Ramin Mojtabai, and Elizabeth A Stuart. Implementing statistical methods for generalizing randomized trial findings to a target population. *Addictive Behaviors*, 94:124–132, 2019.

[2] Irina Degtiar and Sherri Rose. A review of generalizability and transportability, 2021.

[3] Michael G. Findley, Kyosuke Kikuta, and Michael Denly. External validity. *Annual Review of Political Science*, 24(1):365–393, 2021.

[4] Hartman, Erin. "Generalizing Experimental Results." *Advances in Experimental Political Science*, edited by James N. Druckman and Donald P. Green, Cambridge University Press, Cambridge, 2021, pp. 385–410.

[5] Catherine R Lesko, Ashley L Buchanan, Daniel Westreich, Jessie K Edwards, Michael G Hudgens, and Stephen R Cole. Generalizing study results: A potential outcomes perspective. *Epidemiology (Cambridge, Mass.)*, 28(4):553, 2017.

[6] Elizabeth Tipton and Robert B Olsen. A review of statistical methods for generalizing from evaluations of educational interventions. *Educational Researcher*, 47(8):516–524, 2018.

[7] Thomas D Cook, Donald Thomas Campbell, and William Shadish. *Experimental and quasi-experimental designs for generalized causal inference*. Houghton Mifflin, Boston, MA, 2002.

[8] Lee J Cronbach and Karen Shapiro. *Designing evaluations of educational and social programs*. Jossey-Bass, San Francisco, CA, 1982.

[9] Paul W Holland. Statistics and causal inference. *Journal of the American statistical Association*, 81(396):945–960, 1986.

[10] Kosuke Imai, Gary King, and Elizabeth A Stuart. Misunderstandings between experimentalists and observationalists about causal inference. *Journal of the Royal Statistical Society: Series A (Statistics in Society)*, 171(2):481–502, 2008.

[11] Robert B Olsen, Larry L Orr, Stephen H Bell, and Elizabeth A Stuart. External validity in policy evaluations that choose sites purposively. *Journal of Policy Analysis and Management*, 32(1):107–121, 2013.

[12] Daniel Westreich, Jessie K Edwards, Catherine R Lesko, Stephen R Cole, and Elizabeth A Stuart. Target validity and the hierarchy of study designs. *American Journal of Epidemiology*, 188(2):438–443, 2019.

[13] Elias Bareinboim and Judea Pearl. A general algorithm for deciding transportability of experimental results. *Journal of Causal Inference*, 1(1):107–134, 2013.

[14] Judea Pearl and Elias Bareinboim. External validity: From do-calculus to transportability across populations. *Statistical Science*, 29(4):579–595, 2014.

[15] Elizabeth Tipton. Improving generalizations from experiments using propensity score subclassification: Assumptions, properties, and contexts. *Journal of Educational and Behavioral Statistics*, 38(3):239–266, 2013.

[16] Elizabeth A Stuart and Anna Rhodes. Generalizing treatment effect estimates from sample to population: A case study in the difficulties of finding sufficient data. *Evaluation Review*, 41(4):357–388, 2017.

[17] Elizabeth A Stuart, Stephen R Cole, Catherine P Bradshaw, and Philip J Leaf. The use of propensity scores to assess the generalizability of results from randomized trials. *Journal of the Royal Statistical Society: Series A (Statistics in Society)*, 174(2):369–386, 2011.

[18] Naoki Egami and Erin Hartman. Covariate selection for generalizing experimental results: Application to a large-scale development program in uganda*. *Journal of the Royal Statistical Society: Series A (Statistics in Society)*, n/a(n/a), n/a.

[19] Steffen L Lauritzen. *Graphical models*. Clarendon Press, Oxford, 1996.

[20] Jonas Haslbeck and Lourens J Waldorp. mgm: Estimating time-varying mixed graphical models in high-dimensional data. *Journal of Statistical Software*, 93(8):1–46, 2020.

[21] Elias Bareinboim and Judea Pearl. Causal inference and the data-fusion problem. *Proceedings of the National Academy of Sciences*, 113(27):7345–7352, 2016.

[22] Wilke, Anna M. and Macartan Humphreys. Field experiments, theory, and external valid- ity. In: *The SAGE Handbook of Research Methods in Political Science and International Relations. 2:* 55 London: SAGE Publications Ltd, 2020.

[23] Carlos Cinelli and Judea Pearl. Generalizing experimental results by leveraging knowledge of mechanisms. *European Journal of Epidemiology*, 36(2):149–164, 2021.

[24] Donald B Rubin. Bayesian inference for causal effects: The role of randomization. *The Annals of Statistics*, 6(1): 34–58, 1978.

[25] Donald B Rubin. Formal mode of statistical inference for causal effects. *Journal of Statistical Planning and Inference*, 25(3):279–292, 1990.

[26] Erin Hartman, Richard Grieve, Roland Ramsahai, and Jasjeet S Sekhon. From sample average treatment effect to population average treatment effect on the treated: Combining experimental with observational studies to estimate population treatment effects. *Journal of the Royal Statistical Society: Series A (Statistics in Society)*, 178(3):757–778, 2015.

[27] Luke W. Miratrix, Jasjeet S. Sekhon, Alexander G. Theodoridis, and Luis F. Campos. Worth weighting? how to think about and use weights in survey experiments. *Political Analysis*, 26(3):275–291, 2018.

[28] Elizabeth Tipton. How generalizable is your experiment? an index for comparing experimental samples and populations. *Journal of Educational and Behavioral Statistics*, 39(6):478–501, 2014.

[29] Elizabeth Tipton, Kelly Hallberg, Larry V Hedges, and Wendy Chan. Implications of small samples for generalization: Adjustments and rules of thumb. *Evaluation Review*, 41(5):472–505, 2017.

[30] Colm O'Muircheartaigh and Larry V Hedges. Generalizing from unrepresentative experiments: A stratified propensity score approach. *Journal of the Royal Statistical Society: Series C: Applied Statistics*, 63(2), 195–210, 2014.

[31] Elizabeth Tipton, Lauren Fellers, Sarah Caverly, Michael Vaden-Kiernan, Geoffrey Borman, Kate Sullivan, and Veronica Ruiz de Castilla. Site selection in experiments: An assessment of site recruitment and generalizability in two scale-up studies. *Journal of Research on Educational Effectiveness*, 9(sup1):209–228, 2016.

[32] Qingyuan Zhao and Daniel Percival. Entropy balancing is doubly robust. *Journal of Causal Inference*, 5(1): 20160010, 2016.

[33] Erin Hartman, Chad Hazlett, and Ciara Sterbenz. Kpop: A kernel balancing approach for reducing specification assumptions in survey weighting. *arXiv preprint arXiv:2107.08075*, 2021.

[34] Eloise E Kaizar. Estimating treatment effect via simple cross design synthesis. *Statistics in Medicine*, 30(25):2986–3009, 2011.

[35] Wendy Chan. The sensitivity of small area estimates under propensity score subclassification for generalization. *Journal of Research on Educational Effectiveness*, 15(1), 178-215, 2021.

[36] Ashley L Buchanan, Michael G Hudgens, Stephen R Cole, Katie R Mollan, Paul E Sax, Eric S Daar, Adaora A Adimora, Joseph J Eron, and Michael J Mugavero. Generalizing evidence from randomized trials using inverse probability of sampling weights. *Journal of the Royal Statistical Society. Series A,(Statistics in Society)*, 181(4):1193, 2018.

[37] Ziyue Chen and Eloise Kaizar. On variance estimation for generalizing from a trial to a target population, 2017.

[38] Kevin P Josey, Fan Yang, Debashis Ghosh, and Sridharan Raghavan. A calibration approach to transportability with observational data. *arXiv preprint arXiv:2008.06615*, 2020.

[39] Magdalena Bennett, Juan Pablo Vielma, and José R Zubizarreta. Building representative matched samples with multi-valued treatments in large observational studies. *Journal of Computational and Graphical Statistics*, 29(4):744–757, 2020.

[40] Melody Huang, Naoki Egami, Erin Hartman, and Luke Miratrix. Leveraging population outcomes to improve the generalization of experimental results, 2021.

[41] Holger L Kern, Elizabeth A Stuart, Jennifer Hill, and Donald P Green. Assessing methods for generalizing experimental impact estimates to target populations. *Journal of Research on Educational Effectiveness*, 9(1):103–127, 2016.

[42] Lauren Kennedy and Andrew Gelman. Know your population and know your model: Using model-based regression and poststratification to generalize findings beyond the observed sample. *Psychological Methods*, 48(6), 3283-3311, 2021.

[43] Issa J Dahabreh, Sarah E Robertson, Jon A Steingrimsson, Elizabeth A Stuart, and Miguel A Hernan. Extending inferences from a randomized trial to a new target population. *Statistics in Medicine*, 39(14):1999–2014, 2020.

[44] Nianbo Dong, Elizabeth A Stuart, David Lenis, and Trang Quynh Nguyen. Using propensity score analysis of survey data to estimate population average treatment effects: A case study comparing different methods. *Evaluation Review*, 44(1):84–108, 2020.

[45] Naoki Egami and Erin Hartman. Elements of external validity: Framework, design, and analysis. *American Political Science Review* (2022) 1–19.

[46] Kara E Rudolph and Mark J van der Laan. Robust estimation of encouragement-design intervention effects transported across sites. *Journal of the Royal Statistical Society. Series B, Statistical Methodology*, 79(5):1509, 2017.

[47] Trang Quynh Nguyen, Benjamin Ackerman, Ian Schmid, Stephen R Cole, and Elizabeth A Stuart. Sensitivity analyses for effect modifiers not observed in the target population when generalizing treatment effects from a randomized controlled trial: Assumptions, models, effect scales, data scenarios, and implementation details. *PloS One*, 13(12):e0208795, 2018.

[48] Issa J Dahabreh, James M Robins, Sebastien JP Haneuse, Iman Saeed, Sarah E Robertson, Elisabeth A Stuart, and Miguel A Hernán. Sensitivity analysis using bias functions for studies extending inferences from a randomized trial to a target population. *arXiv preprint arXiv:1905.10684*, 2019.

[49] Isaiah Andrews and Emily Oster. Weighting for external validity. *NBER working paper*, (w23826), 2017.

[50] Wendy Chan. Partially identified treatment effects for generalizability. *Journal of Research on Educational Effectiveness*, 10(3):646–669, 2017.

[51] Yoav Benjamini and Ruth Heller. Screening for Partial Conjunction Hypotheses. *Biometrics*, 64(4):1215–1222, 2008.

[52] Bikram Karmakar and Dylan S. Small. Assessement of the extent of corroboration of an elaborate theory of a causal hypothesis using partial conjunctions of evidence factors. *Annals of Statistics*, 2020.

[53] David S Yeager, Paul Hanselman, Gregory M Walton, Jared S Murray, Robert Crosnoe, Chandra Muller, Elizabeth Tipton, Barbara Schneider, Chris S Hulleman, Cintia P Hinojosa, et al. A national experiment reveals where a growth mindset improves achievement. *Nature*, 573(7774):364–369, 2019.

[54] Susan Athey and Guido Imbens. Recursive partitioning for heterogeneous causal effects. *Proceedings of the National Academy of Sciences*, 113(27):7353–7360, 2016.

[55] Kosuke Imai and Marc Ratkovic. Estimating treatment effect heterogeneity in randomized program evaluation. *The Annals of Applied Statistics*, 7(1):443–470, 2013.

[56] Hugh A Chipman, Edward I George, and Robert E McCulloch. Bart: Bayesian additive regression trees. *The Annals of Applied Statistics*, 4(1):266–298, 2010.

[57] Donald P Green and Holger L Kern. Modeling heterogeneous treatment effects in survey experiments with bayesian additive regression trees. *Public Opinion Quarterly*, 76(3):491–511, 2012.

[58] P Richard Hahn, Jared S Murray, and Carlos M Carvalho. Bayesian regression tree models for causal inference: Regularization, confounding, and heterogeneous effects (with discussion). *Bayesian Analysis*, 15(3):965–1056, 2020.

[59] Jennifer L Hill. Bayesian nonparametric modeling for causal inference. *Journal of Computational and Graphical Statistics*, 20(1):217–240, 2011.

[60] Richard K Crump, V Joseph Hotz, Guido W Imbens, and Oscar A Mitnik. Nonparametric tests for treatment effect heterogeneity. *The Review of Economics and Statistics*, 90(3):389–405, 2008.

[61] Peng Ding, Avi Feller, and Luke Miratrix. Randomization inference for treatment effect variation. *Journal of the Royal Statistical Society: Series B: Statistical Methodology*, 78(3), 655-671, 2016.

[62] P Richard Hahn, Vincent Dorie, and Jared S Murray. Atlantic causal inference conference (acic) data analysis challenge 2017. *arXiv preprint arXiv:1905.09515*, 2019.

[63] Elizabeth Tipton and Larry V Hedges. The role of the sample in estimating and explaining treatment effect heterogeneity. *Journal of Research on Educational Effectiveness*, 10(4):903–906, 2017.

[64] Elizabeth Tipton. Beyond generalization of the ate: Designing randomized trials to understand treatment effect heterogeneity. *Journal of the Royal Statistical Society: Series A (Statistics in Society)*, 184(2):504–521, 2021.

[65] Larry V Hedges and Ingram Olkin. *Statistical methods for meta-analysis*. Academic Press, Orlando, FL, 1985.

[66] Issa J Dahabreh, Lucia C Petito, Sarah E Robertson, Miguel A Hernán, and Jon A Steingrimsson. Towards causally interpretable meta-analysis: Transporting inferences from multiple studies to a target population. *arXiv preprint arXiv:1903.11455*, 2019.

[67] Robert B Olsen and Larry L Orr. On the "where" of social experiments: Selecting more representative samples to inform policy. *New Directions for Evaluation*, 2016(152):61–71, 2016.

[68] Elizabeth Tipton. Stratified sampling using cluster analysis: A sample selection strategy for improved generalizations from experiments. *Evaluation Review*, 37(2):109–139, 2014.

[69] Elizabeth Tipton and Laura R Peck. A design-based approach to improve external validity in welfare policy evaluations. *Evaluation Review*, 41(4):326–356, 2017.

[70] Elizabeth Tipton. Sample selection in randomized trials with multiple target populations. *American Journal of Evaluation*, 43(1), 70-89, 2022. Forthcoming.

[71] Elizabeth Tipton, David S Yeager, Ronaldo Iachan, and Barbara Schneider. Designing probability samples to study treatment effect heterogeneity. *Experimental methods in survey research: Techniques that combine random sampling with random assignment*, pages, edited by Lavrakas, Paul J., Michael W. Traugott, Courtney Kennedy, Allyson L. Holbrook, Edith D. de Leeuw, and Brady T. West, John Wiley & Sons, 435–456, 2019.

[72] Elizabeth Tipton, Larry Hedges, David Yeager, Jared Murray, and Maithreyi Gopalan. Global mindset initiative paper 4: Research infrastructure and study design. *SSRN*, 2021. https://ssrn.com/abstract=3911643.

Part II

Matching

4

Optimization Techniques in Multivariate Matching

Paul R. Rosenbaum and José R. Zubizarreta

CONTENTS

DOI: 10.1201/9781003102670-4

4.1 Introduction

4.1.1 Goal of this chapter

Optimization techniques for multivariate matching in observational studies are widely used and extensively developed, and the subject continues to grow at a quick pace. The typical research article published today presumes acquaintance with some technical background that is well-known in computer science and operations research, but is not standardly included in doctoral education in statistics. The relevant technical material is interesting and not especially difficult, but it is not curated in a form that permits relatively quick study by someone who wishes to develop new techniques for matching in statistics. The goal of this chapter is to review some of this technical background at a level appropriate for a PhD student in statistics. This chapter assumes no background in combinatorial optimization, and the chapter may serve as an entry point to the literature on combinatorial optimization for a student of statistics. In contrast, a scientist seeking a review of methods for constructing a matched sample in an empirical article might turn to references [1,2].

Matching is used to achieve a variety of goals. Mostly commonly, it is used to reduce confounding from measured covariates in observational studies [3]. Matching may be used to strengthen an instrument or instrumental variable [4–7]. In addition, it may be used to depict the processes that produce racial or gender disparities in health or economic outcomes [8,9]. Close pairing for covariates highly predictive of the outcome can also reduce sensitivity of causal inferences to biases from unmeasured covariates [10,11].

For instance, Neuman and colleagues [6] used matching to reduce confounding from measured covariates, while also strengthening an instrument intended to address certain unmeasured covariates, in a study of the effects of regional versus general anesthesia in hip fracture surgery. Some hospitals commonly use regional/local anesthesia for hip fracture surgery, while other hospitals commonly use general anesthesia. Someone who breaks their hip is likely to taken to an emergency room, and to receive during hip fracture surgery the type of anesthesia commonly used in that hospital. A possible instrument for regional-versus-general anesthesia is the differential distance to the nearest hospital that commonly uses each type of anesthesia. In New York State as a whole, this instrument is weak: in the densely populated parts of the State, such as New York City with its many hospitals in a small space, geography exerts only a slight nudge, pushing a patient towards one type of anesthesia or the other. In contrast, in the sparsely populated parts of upstate New York, there may be only one hospital in a mid-sized town, and geography strongly influences the type of anesthesia that a patient receives. This distinction is vividly displayed on a map of New York State in Figure 2 of Neuman et al. [6]. Matching sought to form pairs close in term of measured covarates in which one person was much closer to a hospital using regional anesthesia, and the other person was much closer to a hospital using general anesthesia. In this way, a large study with a weak instrument is replaced by a smaller study with a strong instrument [4,5]. Using data from New York's Statewide Planning and Research Cooperative System, Neuman and colleges [6] built a closely matched study with a strong instrument consisting of 10,757 matched pairs of two patients undergoing hip fracture surgery, one close only to a hospital using regional anesthesia, the other close only to a hospital using general anesthesia. Numerous covariates were controlled by matching.

4.1.2 The role of optimization in matching

Matching is part of the design, not the analysis, of an observational study, and it is completed before outcomes are examined for the first time [12]. Just as one might consider and compare several experimental designs before selecting one for a particular experiment, so too one typically compares several matched samples before selecting one for a particular observational study. In both experiments and observational studies, the design is completed before a single, preplanned primary analysis is conducted when outcomes are examined for the first time. A single primary analysis does not preclude secondary analyses or exploratory analyses; rather, it distinguishes among such analyses.

Optimization is a framework within which matched samples may be constructed and compared. An optimal match is best in terms of a certain criterion subject to certain requirements or constraints. However, optimization of one criterion is not recommendation; rather, human judgement is commonly required to compare multiple criteria. An optimal match is one study design that may be compared with other study designs optimal in terms of altered criteria or constraints, as a recommended design is chosen. It is difficult to compare study designs that lack definite properties, and optimization gives each study design certain definite properties.

In truth, we rarely need exact optimality except as a foundation for comparison. Wald [13] proved the consistency of maximum likelihood estimators under extremely weak conditions by showing that all estimators with sufficiently high likelihoods are consistent. Huber [14] showed that there are estimators that are very nearly the equal of maximum likelihood estimators when the underlying model is true, and that are much better when the model is false in particular ways. Both Wald and Huber asserted that a suboptimal estimator was adequate or in some way superior, but they could only say this because they could compare its performance with an optimal procedure. Optimality is a benchmark, not a recommendation.

Typically, an investigator has several objectives for a matched sample, some of which cannot be measured by quantitative criteria. For instance, the investigator wants the match to be correctly and appropriately persuasive to its target audience – perhaps scientists or medical researchers in a particular field, or a regulatory agency, or a judge and jury – but it is difficult give a mathematical definition of persuasive. A persuasive study design is public, open to view, open to assessment, and open to criticism by its target audience. The design has to be intelligible to its target audience in a way that permits criticism by experts in that audience. It can be a powerful endorsement of a study and its findings if the design is intelligible to experts and the only criticisms that arise seem inconsequential. The audience must understand the design if it is to judge certain criticisms as inconsequential. Because a study design is proposed before outcomes are examined, the design may be shared with its target audience before it is used, so that reasonable criticisms may be addressed before any analysis begins.

In brief, optimality is a framework within which competing study designs may be compared. Within such a framework, one quickly discovers new ways to improve study designs, replacing a design optimal in an impoverished sense by a new design optimal in an enriched sense.

4.1.3 Outline

Section 4.2 begins with an old problem that had been well-solved in 1955. This optimal assignment problem forms matched treated-control pairs to minimize the total covariate distance within the pairs. This problem is not yet a practical problem from the perspective of statistics, but it is a good place to clarify certain issues. Section 4.3 discusses the way computer scientists traditionally measure the speed of an algorithm, including the limitations in statistics of this traditional measure. The assignment algorithm generalizes to the problem of minimum cost flow in a network, as discussed in §4.4, leading to many practical methods for statistical problems. Optimal assignment and minimum cost flow problems of the sizes commonly encountered can be solved quite quickly by the traditional standard of computational speed in §4.3. Network techniques are easily enhanced by various tactics, as discussed in §4.5. Section 4.6 greatly expands the set of possible matched designs using mixed integer programming. Unlike minimum cost flow problems, integer programs can be difficult to solve by the traditional standard of computational speed in §4.3; however, this may be a limitation of the traditional standard when used in this context, not a limitation of matching using integer programming; see §4.6.

4.2 Optimal Assignment: Pair Matching to Minimize a Distance

4.2.1 Notation for the assignment problem

There are T treated individuals, $\mathcal{T} = \{\tau_1, \ldots, \tau_T\}$, and $C \geq T$ potential controls, $\mathcal{C} = \{\gamma_1, \ldots, \gamma_C\}$. There is a distance, $\delta_{tc} \geq 0$, between τ_t and γ_c based on observed covariates for individuals τ_t and γ_c. For instance, δ_{tc} might be some variant of the Mahalanobis distance. In practice, a better distance sets $\delta_{tc} = \infty$ if τ_t and γ_c differ substantially on the propensity score (described in Chapter 2) – that is, if a caliper on the propensity score is violated – and otherwise δ_{tc} is a variant of the Mahalanobis distance. The exact form of the distance is not a concern at the moment.

The assignment problem is to pair each τ_t with a different γ_c to minimize the sum of the T within pair distances. Write $a_{tc} = 1$ if τ_t is paired with γ_c, and $a_{tc} = 0$ otherwise. The assignment problem is to make an optimal choice of the a_{tc}. The assignment problem is

$$\min \sum_{t=1}^{T} \sum_{c=1}^{C} \delta_{tc}\, a_{tc} \tag{4.1}$$

subject to constraints

$$\sum_{c=1}^{C} a_{tc} = 1 \text{ for each } t, \text{ and } \sum_{t=1}^{T} a_{tc} \leq 1 \text{ for each } c, \text{ with each } a_{tc} \in \{0, 1\}. \tag{4.2}$$

In (4.2), $\sum_{c=1}^{C} a_{tc} = 1$ insists that τ_t is paired with exactly one control γ_c, and $\sum_{t=1}^{T} a_{tc} \leq 1$ insists that each control is used at most once.

Two misconceptions are common. The assignment problem can appear to be either trivial or impossible, but it is neither. It can seem trivial if you assign to each τ_t to the control γ_i who is closest, the one with $\delta_{ti} = \min_{c \in \mathcal{C}} \delta_{tc}$; however, that fails because it may use the same control many times, and might even produce a control group consisting of a single control rather than T distinct controls. Upon realizing that the assignment problem is not trivial, it can seem impossible: there are $C!/(C-T)!$ possible solutions to the constraints (4.2), or 3.1×10^{93} solutions for a tiny matching problem with $T = 50$ and $C = 100$. It seems impossible to evaluate (4.1) for $C!/(C-T)!$ solutions to find the best solution.

4.2.2 Greedy algorithms can produce poor solutions to the assignment problem

A greedy algorithm pairs the two closest individuals, i and j, with $\delta_{ij} = \min_{t \in \mathcal{T}} \min_{c \in \mathcal{C}} \delta_{tc}$, and removes i and j from further consideration. From the individuals who remain, the greedy algorithm pairs the two closest individuals, and so on. There is a small but interesting class of problems for which greedy algorithms provide optimal solutions [15, Chapter 7]; however, the assignment problem is not one of these problems.

Greedy and optimal matching are compared in Table 4.1, where $T = C = 3$, and $\beta \gg 5\varepsilon > \varepsilon > 0$. Here, β is for a "big" distance, much larger than 5ε. Greedy would form pairs (τ_1, γ_2) and (τ_2, γ_3) with a total within-pair distance of $\delta_{12} + \delta_{23} = 0 + 0 = 0$, but then would be forced to add the pair (τ_3, γ_1), making the total distance $\delta_{12} + \delta_{23} + \delta_{31} = 0 + 0 + \beta = \beta$. In Table 4.1, greedy matching starts strong, with very small distances, but paints itself into a corner, and its final match is poor. The optimal matching is (τ_1, γ_1), (τ_2, γ_2), (τ_3, γ_3), with a total within pair distance of $\varepsilon + \varepsilon + \varepsilon = 3\varepsilon \ll \beta$. As β can be any large number or even ∞, it follows that $\beta - 3\varepsilon$ can be any large number or ∞, so the total distance produced by greedy matching can be vastly worse than the total distance produced by optimal matching. The optimization in optimal matching is needed to recognize that τ_1 had to be paired with γ_1 to avoid pairing τ_2 or τ_3 with γ_1.

TABLE 4.1
A $T \times C = 3 \times 3$ distance matrix, δ_{tc}.

	γ_1	γ_2	γ_3
τ_1	ε	0	$5 \times \varepsilon$
τ_2	β	ε	0
τ_3	β	$5 \times \varepsilon$	ε

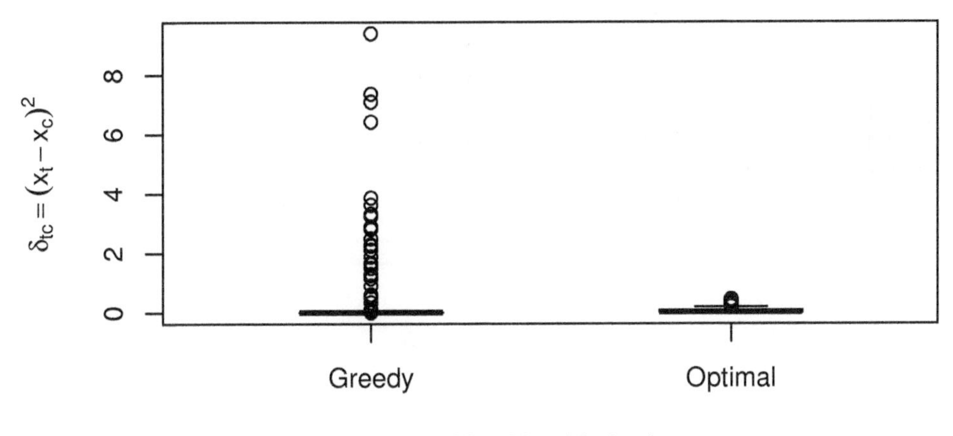

FIGURE 4.1
Boxplots of 300 within pair distances, $\delta_{tc} = (x_t - x_c)^2$, for greedy and optimal matching. Thirty greedy distances, 30/300 = 10%, exceed the maximum distance, 0.4373, for optimal matching, and in all 30 pairs, the covariate difference is positive, $(x_t - x_c) > 0$.

The typical distance matrix in current use is broadly similar in structure to Table 4.1. There are some big distances, like β, and some small distances, like ε or 5ε. The big distances may reflect penalties imposed to try to enforce some constraint, while the small distances reflect close pairs that satisfy the constraint. For instance, [3] recommended imposing a caliper on the propensity score and, within that caliper, matching using the Mahalanobis distance. Large distances like β would then refer to individuals whose propensities were far apart, violating the caliper, while small distances like ε or 5ε would distinguish people with similar propensity scores but differing patterns of covariates. In this context, in Table 4.1, greedy matching would produce one pair of people with very different propensity scores, while optimal matching would form no pairs with very different propensity scores.

Figure 4.1 is a small simulated example, in which $T = 300$ treated individuals are matched with $C = 600$ potential controls to form 300 matched pairs. There is a single covariate x drawn from $N(0.75, 1)$ for treated individuals and from $N(0, 1)$ for controls. The distance is $\delta_{tc} = (x_t - x_c)^2$, and 300 within-pair distances are shown in Figure 4.1 for greedy and for optimal matching. Prior to matching, for a randomly chosen treated individual and a control, $x_t - x_c$ is $N(0.75, 2)$ so $(x_t - x_c)^2$ has expectation $0.75^2 + 2 = 2.56$, and has sample mean 2.59 is the simulation. As in Table 4.1, in Figure 4.2 greedy matching starts strong, finding many close pairs, but paints itself into a corner and finally accepts some very poor pairs. These poorly matched pairs could have been avoided and were avoided by optimal matching. The largest distance δ_{tc} for a pair matched by the greedy method is 9.39, while the largest distance by the optimal method is 0.44. In total, $30/300 = 10\%$ of the greedy pairs have larger distances than the largest distance, .44, produced by optimal matching. This is shown in greater detail in Table 4.2 which lists the total of 300 within-pair distances, their mean, median, and upper percentiles.

TABLE 4.2
Summary of 300 within-pair distances, $\delta_{tc} = (x_t - x_c)^2$, for greedy and optimal matching. Also included is the summary of the 300 × 600 distances before matching.

	Total	Mean	Median	75%.	90%	95%	99%	max
Before Matching		2.59	1.18	3.42	7.03	10.03	17.24	47.60
Greedy	85.62	0.29	0.00	0.00	0.35	2.12	7.09	9.39
Optimal	18.87	0.06	0.01	0.08	0.23	0.28	0.36	0.44

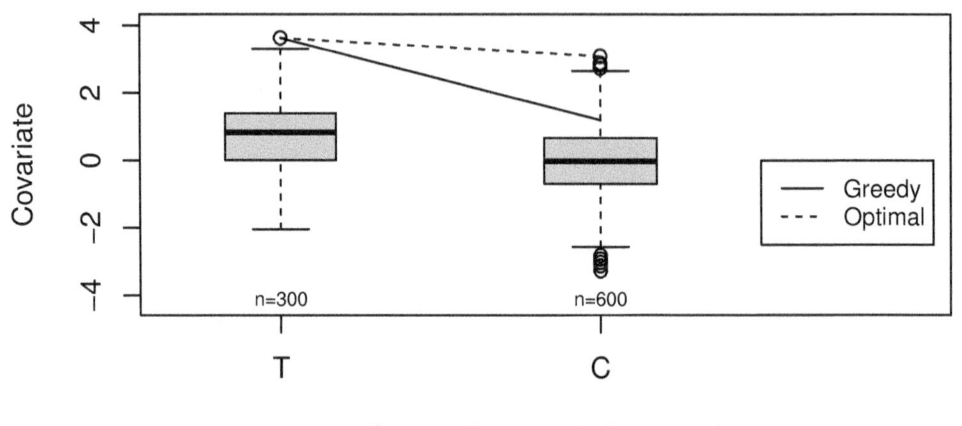

FIGURE 4.2
The 300th of 300 pairs formed by greedy matching (solid line) compared to the pair for the same treated individual formed by optimal matching (dashed line).

Also important in Figure 4.1 are the signs of the 300 within-pair differences, $x_t - x_c$. Before matching, the bias was positive, $\mathrm{E}(x_t - x_c) = 0.75$. In greedy matching, when the distance, $\delta_{tc} = (x_t - x_c)^2$, is large, the difference $x_t - x_c$ tends to be a large positive number, not a large negative number, so the original bias is intact. All of the $30/300 = 10\%$ largest $\delta_{tc} = (x_t - x_c)^2$ from greedy matching are positive differences, $x_t - x_c > 0$, often with differences larger than 0.75.

Figure 4.2 shows where greedy matching went wrong. It shows the 300th pair formed by greedy matching, and the corresponding match for optimal matching. The treated individual with the highest x_t was matched last by greedy matching, when its good controls were gone, whereas optimal matching paired the highest x_t to the highest x_c.

Are Figures 4.1 and 4.2 and Table 4.2 typical? Yes and no. When there are many covariates, x is replaced by the propensity score, so Figures 4.1 and 4.2 and Table 4.2 remain relevant. However, some matching problems are impossible, say $x_t \sim N(5, 1)$ and $x_c \sim N(0, 1)$, and in impossible problems neither greedy nor optimal matching is useful. Similarly, some matching problems are easy, say pair matching with $x_t \sim N(0, 1)$ and $x_c \sim N(0, 1)$ and $C/T = 5$; then, greedy and optimal matching both work well. Optimal matching outperforms when matching is neither impossible nor easy, when there is serious competition among treated individuals for the same control, as in Figure 4.2 and Table 4.2. Even in the easy case just described, with $C/T = 5$, we might prefer to match two or three controls to each treated individual, so an easy pair matching problem becomes a more challenging 3-to-1 matching problem, and optimal matching would outperform greedy matching, perhaps substantially. For instance, a poor 3-to-1 greedy match may lead an investigator to settle for a 2-to-1 match, when a good 3-to-1 optimal match is possible. Perhaps equally important, an investigator using greedy with a good 2-to-1 greedy match and a poor 3-to-1 greedy match does not know whether a good 3-to-1 match exists, precisely because no criterion of success has been optimized.

4.2.3 Nearest neighbor matching with replacement can waste controls

Although greedy matching made some poor decisions in Figures 4.1 and 4.2, it did form 300 pairs with 300 distinct controls. In contrast, so-called nearest neighbor matching *with replacement* uses one or more nearby controls, even if the control has been used before. If implemented in an unwise way, matching with replacement can be inefficient, using fewer controls when it could use more controls. The issue is that one person does not become two people by virtue of being used twice. In the same simulated example, nearest neighbor pair matching reuses controls, so it formed 300 pairs containing only 184 distinct controls. As discussed at the end of this section, most of what is imagined to be attractive about matching *with replacement* can be had using full matching *without replacement* [16–21].

Imagine outcomes for controls are independent, perhaps with different expectations, but with constant variance σ^2. In matching with replacement, using a control twice increases the variance of the estimator in two different ways. First, there are 184 controls, not 300 controls, so even the unweighted mean of 184 control observations has variance $\sigma^2/184$ rather than $\sigma^2/300$; therefore, even the unweighted mean has variance that is $\left(\sigma^2/184\right) / \left(\sigma^2/300\right) = 1.63$ times larger, or 63% larger.

Second, matching is intended to remove bias, and to remove bias, the 184 controls need to be weighted to reflect that one control may stand for several controls. The use of unequal weights further increases the variance. In the simulated example from Figure 4.1, Table 4.3 shows the duplication of controls in nearest neighbor pair-matching with replacement. One control was paired to seven treated individuals, and so receives weight 7. Five controls were each used 5 times, and so each receive weight 5. The weighted mean of the controls has variance

$$\sigma^2 \frac{\left(7^2 \times 1 + 5^2 \times 5 + \cdots + 2^2 \times 44 + 1^2 \times 114\right)}{300^2} = \frac{686\sigma^2}{300^2}$$

as opposed to

$$\sigma^2 \frac{\left(1^2 \times 300\right)}{300^2} = \frac{300\sigma^2}{300^2} = \frac{\sigma^2}{300}$$

for pair matching without replacement, so matching with replacement has a variance that is $686/300 = 2.29$ times larger than matching without replacement. The 2.29 reflects the use of 184 controls rather than 300 controls plus the need for unequal weights.

TABLE 4.3
Reuse of the same control in 300 matched pairs by nearest neighbor matching with replacement. The 300 controls are represented by only 184 distinct controls.

Times Used	1	2	3	4	5	6	7	Total
Number of Controls	114	44	14	6	5	0	1	184

Unlike randomized treatment assignment, matching with replacement focuses on the closeness of matched individuals, ignoring the comparability of treated and matched control groups as a whole. Like randomized experiments, modern matching methods are more concerned with the comparability of treated and matched control groups as whole groups, not with close pairing for many covariates; see §4.5.2, §4.6.2 and §4.6.3. Close pairing for a few important predictors of the outcome does affect design sensitivity; however, close pairing for all covariates is not required to pair closely for a few important predictors of the outcome while also balancing many more covariates [10, 11].

A practical alternative to using fewer than 300 controls in matching with replacement is to use more than 300 controls in full matching. Full matching without replacement can match to use every available control, and even when it is allowed to discard some distant controls, it can often use many more controls than pair matching. In full matching, each matched set contains either one treated individual and one or more controls, or one control and one or more treated individuals. The sets do not overlap, unlike matching with replacement. Full matching can produce smaller distances and use more controls than pair matching, but weights are needed. The increased sample size may partially offset the need for weights. For discussion of full matching, see [16–25].

4.2.4 Finding an optimal assignment by the auction algorithm

There are several algorithms for finding an optimal assignment. Kuhn's Hungarian algorithm dates from 1955. Here, we briefly sketch one good method, the auction algorithm, due to Bertsekas [26]. For statisticians, the auction algorithm has two advantages. First, it can be described in familiar economic language as holding an auction in which controls are sold to treated individuals in an auction with variable prices. As the auction proceeds, the prices rise for each control wanted by several treated individuals, so eventually a treated individual with several good controls available is unwilling to pay the high price for that individual's closest control. Second, as discussed in §4.4, a version of the auction algorithm called RELAX IV due to Bertsekas and Tseng [27] is available inside R. The brief description that follows is based on Bertsekas' texbook [28].

The auction algorithm is described first for $T = C$, with other cases reduced to this case in §4.2.5. The auction maximizes a nonnegative benefit, rather than minimizing a nonnegative distance, but they are equivalent. Let $\delta_{\max} = \max_{t \in \mathcal{T},\, c \in \mathcal{C}} \delta_{tc}$. Define the benefit δ'_{tc} from matching t to c to be $\delta'_{tc} = \delta_{\max} - \delta_{tc} \geq 0$. The assignment that maximizes the benefit also minimizes the distance, and the maximum benefit is $T\delta_{\max}$ minus the minimum distance.

All $T = C$ controls, γ_c, start with price $p_c = 0$ for $c = 1, \ldots, C$. As the auction proceeds, the prices p_c increase. The rules of the auction require that a new bid increases a price by at least $\epsilon > 0$. As in a real auction, a minimum bid is needed to ensure that prices actually increase every time someone makes a bid.

At every stage of the auction, the net benefit to treated individual τ_t of being matched to γ_c is the benefit, δ'_{tc}, minus the price, p_c, so the net benefit is $\delta'_{tc} - p_c$. At current prices, τ_t would like to be matched to the control γ_j with maximum net benefit, $\delta'_{tj} - p_j = \max_{c \in \mathcal{C}} \delta'_{tc} - p_c = v_t$, say, but it is possible that γ_j is also the preferred control for several treated individuals, so bids will increase prices to settle the question of who finally gets γ_j. Let γ_k be second best after γ_j for τ_t, so $\delta'_{tk} - p_k = \max_{c \neq j, c \in \mathcal{C}} \delta'_{tc} - p_c = w_t$, say. If the price of γ_j increased from p_j to $p_j^* = p_j + v_t - w_t$, then τ_t would be indifferent between γ_j and γ_k, because the net benefits would be the same, $\delta'_{tj} - p_j^* = \delta'_{tj} - (p_j + v_t - w_t) = w_t = \delta'_{tk} - p_k$. The idea is that to alter τ_t's preference for γ_j, someone else would have to offer to pay more than p_j^* for γ_j. So, the price increases not to p_j^* but to $p_j^* + \epsilon$: if anyone else offered to pay $p_j^* + \epsilon$ for γ_j, then τ_t would no longer prefer γ_j and would prefer γ_k instead.

At each stage of the auction, one of the unmatched treated individuals τ_t is selected to make a bid. That individual is tentatively paired with his or her preferred control, namely γ_j in the previous paragraph. If someone else, say τ_ℓ, had previously been paired with γ_j, then τ_ℓ becomes unpaired. The price of γ_j increases from p_j to $p_j + v_t - w_t + \epsilon = p_j^* + \epsilon$. This process continuous until everyone is paired.

As in a real auction, this minimum bid ϵ means that the highest prices may not be exactly attained. Perhaps I am willing to pay \$50 more than the current price, but the minimum bid increase is \$100 more than the current price, so the sale takes place to someone else at the current price, not at the current price plus \$50. That is, τ_ℓ might have been willing to pay $p_j + \epsilon/2$ for γ_j, but does not ultimately take γ_j from τ_t, because τ_ℓ is not willing to pay $p_j + \epsilon$.

Bertsekas [28, Proposition 2.6, p. 47] shows that the auction terminates in an assignment that minimizes the total within pair distance with an error of at most $T\epsilon$. If the δ_{tc} are nonnegative integers and $\epsilon < 1/T$, then the auction finds an optimal assignment.

The description of the auction algorithm given here is a conceptual sketch. There are many technical details omitted from this sketch that are important to the performance of the auction algorithm. As discussed in §4.4, a good implementation is available inside R.

4.2.5 Simple variations on optimal assignment

A typical strategy for solving a new combinatorial optimization problem is to reduce it to a previously solved problem. Suppose that $T = 300$, $C = 1000$, and we wish to match each treated individual to 2 controls to minimize the total of the 600 treated-control distances within the 300 matched sets. Start with the 300×1000 distance matrix $\boldsymbol{\delta}$ with entries δ_{tc}, let $\mathbf{0}$ be a 400×1000 matrix of zeros, and define

$$\boldsymbol{\Delta} = \begin{bmatrix} \boldsymbol{\delta} \\ \boldsymbol{\delta} \\ \mathbf{0} \end{bmatrix}$$

to be the 1000×1000 distance matrix. Solve the minimum distance assignment problem for $\boldsymbol{\Delta}$, regarding the first matched set as composed of τ_1 and the two controls paired with rows 1 and 301, and so on for the remaining matched sets. Many small changes in the optimal assignment problem solve useful statistical problems; e.g, [29].

The matrix $\boldsymbol{\Delta}$ solves the problem of matching 2-to-1 by reducing that problem to matching 1-to-1. Reductions of this form have advantages and disadvantages. The advantages tend to be theoretical, and the disadvantages tend to be computational. It takes only a second to see that the reduction yields an algorithm for optimal 2-to-1 matching that has the same computational complexity as matching 1-to-1. The disadvantage is that the matrix $\boldsymbol{\Delta}$ is an inefficient way to represent the information it contains. A better approach to representing this information is to use a network, as described in §4.4.

4.3 The Speed of Algorithms

It is possible to sort C numbers into nondecreasing order in at most $\kappa C \log(C)$ steps for some positive number κ, that is, in $O\{C \log(C)\}$ steps as $C \to \infty$; see [30, Chapter 1]. The constant κ tends to depend on details of the computer code and on what one calls "one step," but the rate of growth, $O\{C \log(C)\}$, often does not. Consequently, the speed of an algorithm is characterized by the rate of growth.

The bound, $\kappa C \log(C)$ or $O\{C \log(C)\}$, is a worst-case bound. That means: no matter what order the C numbers are in, they can be sorted into nondecreasing order in at most $O\{C \log(C)\}$ steps. In the theory of algorithms, the typical measure of the performance of an algorithm is a rate of growth, say $O\{C \log(C)\}$, for the worst possible input of size C.

In §4.2, we had $C \geq T$. There is an algorithm for the optimal assignment problem that finds T pairs to minimize the total within pair distance $\boldsymbol{\delta}$ in a $T \times C$ distance matrix in $O(C^3)$ steps. If sufficiently many pairings are forbidden — say, if $\delta_{tc} = \infty$ for sufficiently many entries in $\boldsymbol{\delta}$ — then the time bound can improve to $O(C^2 \log C)$. For both bounds, see [31, Theorem 11.2]. In practical terms, quite large optimal assignment problems can be solved. The 2-to-1 matching problem in §4.2.5 can also be solved in at most $O(C^3)$ steps, simply because it has been reduced to a square optimal assignment problem of size C.

Some combinatorial optimization problems are exceedingly difficult, in the sense that they are widely believed to have no worst case bound of the form $O\left(C^{\lambda}\right)$ for any finite λ, where C is the size of the problem. This collection of combinatorial optimization problems is called NP-hard. Integer programming is a combinatorial optimization problem in which a linear program is restricted to solutions that have integer coordinates. Integer programming arises in various contexts: it is, for example, unhelpful to be told that the best navy that can be built from limited resources is composed of 12.5 submarines and 1.5 battleships. In general, integer programming is NP-hard [31, Theorem 15.42, p. 405]. Again, this is not a statement about your integer program, but rather about the worst cases in a sequence of integer programs of growing size.

A worst case bound on computation time can be very conservative for statistical problems. Often, the worst possible input to an algorithm is bizarre, and unlikely to arise in a data set that arises naturally. As we discuss in §4.6, many algorithms that have terrible worst case bounds perform competitively without them.

4.4 Optimal Matching by Minimum Cost Flow in a Network

4.4.1 Minimum cost flow and optimal assignment

The minimum cost flow problem is one of the basic combinatorial optimization problems [28, 31]. Because there are several good algorithms for finding a minimum cost flow in a network, it would be common to regard some new combinatorial optimization problem as "solved" once it has been expressed as a minimum cost flow problem. Moreover, some of these good solutions have computational time bounds, as in §4.3, that are low order polynomials in the size of the network.

In one sense, the assignment problem in §4.2 is a simple special case of the minimum cost flow problem: every assignment problem can be written as, and solved as, a minimum cost flow problem. In another sense, the minimum cost flow problem can be transformed into an assignment problem and solved as one; see [28, p. 15]. So, viewed from a certain theoretical angle, the two problems are equivalent.

The representation of a statistical matching problem as a minimum cost flow problem has several advantages over its representation as an assignment problem. Although the minimum cost flow representation involves elementary notions from graph theory, once those elementary notions are mastered, the resulting diagrammatic representation of the problem aids intuition and description. For several reasons, the minimum cost flow representation is more concise: it tends to make more efficient use of computer storage, and encourages consideration of sparse networks that can be optimized more quickly. Finally, in `R`, by loading the `optmatch` and `rcbalance` packages, one has access to [27]'s RELAX-IV code for the minimum cost flow problem using the function `callrelax`; so, once a problem is expressed as a minimum cost flow problem, it is in a form that can be solved in `R`.

4.4.2 The minimum cost flow problem

There is a finite, nonempty set of nodes, $\mathcal{N}$. In the typical statistical matching problem, each person – treated or control – is a node, but $\mathcal{N}$ includes some additional nodes that achieve specific objectives; see §4.5. For any finite set $\mathcal{S}$, write $|\mathcal{S}|$ for the number of elements of $\mathcal{S}$. When a network is represented as a diagram on paper, the nodes are drawn as points.

Some pairs of nodes are connected by a directed edge, drawn on paper as an arrow from the first node to the second. An edge e is an ordered pair of distinct nodes, $e = (n_1, n_2)$ with $n_1 \in \mathcal{N}$, $n_2 \in \mathcal{N}$ and $n_1 \neq n_2$, and is described as pointing from n_1 to n_2. The set of edges $\mathcal{E}$ in a network is, therefore, a subset of the ordered pairs of nodes, $\mathcal{E} \subset \mathcal{N} \times \mathcal{N}$. The containment $\mathcal{E} \subset \mathcal{N} \times \mathcal{N}$ is strict because nodes never point to themselves; e.g., $(n_1, n_1) \in \mathcal{N} \times \mathcal{N}$, but $(n_1, n_1) \notin \mathcal{E}$.

In the first applications of minimum cost flow methods in operations research, the nodes might be factories and warehouses, and an edge $e = (n_1, n_2)$ indicated the possibility of moving units of something – say televisions – from location n_1 to location n_2. A flow indicates how many units were moved from one node n_1 to another n_2 along each edge $e = (n_1, n_2)$; that is, for each $e \in \mathcal{E}$, the flow f_e is a nonnegative integer. An integer is required because televisions cannot be cut in half for transport. The flow $\mathbf{f}$ in a network is a vector, $(f_e,\ e \in \mathcal{E})$ of dimension equal to the number $|\mathcal{E}|$ of edges, and we will find a flow $\mathbf{f}$ that minimizes cost subject to various constraints. In statistical matching problems, the requirement that f_e be an integer will say that intact people, not fractional people, must be matched.

Each edge has a nonnegative integer capacity to carry flow, $\text{cap}\,(e) \geq 0$, and a nonnegative cost per unit flow, $\text{cost}\,(e) \geq 0$, so the total cost of a flow $\mathbf{f}$ is $\sum_{e \in \mathcal{E}} f_e \,\text{cost}\,(e)$, and the flow must satisfy the constraint $0 \leq f_e \leq \text{cap}\,(e)$ for all $e \in \mathcal{E}$.

For a flow $\mathbf{f}$, the total flow into node n is a sum of f_e over all edges $e = (m, n) \in \mathcal{E}$ that point to n, or $\sum_{m:e=(m,n)\in\mathcal{E}} f_e$. The total flow out from node n is a sum of f_e over all edges $e = (n, m) \in \mathcal{E}$ that point from n to some other node m, or $\sum_{m:e=(n,m)\in\mathcal{E}} f_e$. For each node n, the flow out minus the flow in, $s_n = \sum_{m:e=(n,m)\in\mathcal{E}} f_e - \sum_{m:e=(m,n)\in\mathcal{E}} f_e$, is called the supply. For instance, a factory that manufactures televisions has a positive supply, $s_n > 0$, a retail outlet that sells televisions has a negative supply, $s_n < 0$, an a train depot that immediately passes along whatever televisions arrive has supply $s_n = 0$. In the minimum cost flow problem, the supplies are fixed, and the only question is how much should be transported to which nodes to meet the supply while minimizing transport costs.

In brief, the minimum cost flow problem is to find a flow $\mathbf{f} = (f_e,\ e \in \mathcal{E})$ of minimum cost

$$\min_{\mathbf{f}} \sum_{e \in \mathcal{E}} f_e \,\text{cost}\,(e) \tag{4.3}$$

subject to constraints

$$f_e \in \{0,\ 1,\ 2, \ldots\},\quad 0 \leq f_e \leq \text{cap}\,(e) \quad \text{for all} \quad e \in \mathcal{E}, \tag{4.4}$$

$$s_n = \sum_{e=(n,m)\in\mathcal{E}} f_e - \sum_{e=(m,n)\in\mathcal{E}} f_e \quad \text{for each} \quad n \in \mathcal{N}. \tag{4.5}$$

A flow $\mathbf{f} = (f_e,\ e \in \mathcal{E})$ is feasible if it satisfies the constraints (4.4) and (4.5). A feasible flow may or may not exist. A feasible flow is optimal if it minimizes the cost (4.3) among all feasible flows. There may be more than one optimal feasible flow.

4.4.3 Simple examples of matching by minimum cost flow

The simplest example of the minimum cost flow problem is the assignment problem considered in §4.2. The nodes are the people, treated and control together, plus one more ω called a sink, $\mathcal{N} = \mathcal{T} \cup \mathcal{C} \cup \{\omega\} = \{\tau_1, \ldots, \tau_T, \gamma_1, \ldots, \gamma_C, \omega\}$; so, $|\mathcal{N}| = T + C + 1$. The edges $e \in \mathcal{E}$ are all of the possible pairs, $e = (\tau_t, \gamma_c)$ for $t = 1, \ldots, T$ and $c = 1, \ldots, C$, plus an edge connecting each control to the sink, $e = (\gamma_c, \omega)$ for $c = 1, \ldots, C$; so, $|\mathcal{E}| = TC + C$. See Figure 4.3.

Every edge has capacity $\text{cap}\,(e) = 1$, so (4.4) implies $f_e = 0$ or $f_e = 1$ for every edge $e \in \mathcal{E}$. Each treated node τ_t has supply 1, each control node γ_c has supply 0, and the sink ω has supply $-T$; that is, $s_{\tau_t} = 1$, $s_{\gamma_c} = 0$, $s_\omega = -T$. Each treated node τ_t supplies one unit of flow, each control γ_c passes along to the sink whatever flow it receives from treated units, and the sink absorbs all T units of flow collected from T controls. Because $\text{cap}\,(e) = 1$ for each $e = (\gamma_c, \omega)$, each control can accept at most one unit of flow from at most one treated unit τ_t.

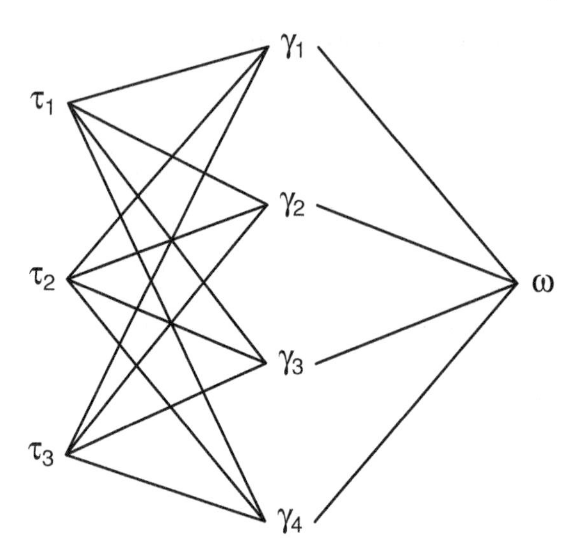

FIGURE 4.3
Network representation of the pair-match or assignment problem. Every edge has capacity 1. The three τ_t each supply one unit of flow, the sink ω absorbs three units of flow, and the γ_c pass along all flow they receive. The node supplies and edge capacities mean that a feasible flow pairs each treated τ_t to a different control γ_c. A minimum cost flow discards one control and pairs the rest to minimize the total of the three within-pair distances.

A moment's thought shows that every feasible flow has precisely T edges $e = (\tau_t, \gamma_c)$ with $f_e = 1$, and these comprise the T matched pairs, and it has $TC - T = T(C-1)$ edges $e = (\tau_t, \gamma_c)$ with $f_e = 0$. Moreover, every feasible flow has matched C distinct controls, specifically the controls γ_c with $f_e = 1$ for $e = (\gamma_c, \omega)$, and it has discarded $C - T$ controls with $f_e = 0$ for $e = (\gamma_c, \omega)$. So each feasible flow corresponds with one of the $C!/(C-T)!$ possible pair matchings in §4.2.

If we set $\text{cost}(e) = \delta_{tc}$ for $e = (\tau_t, \gamma_c)$ and $\text{cost}(e) = 0$ for $e = (\gamma_c, \omega)$, then the cost of a feasible flow is the total within pair distance, as in §4.2. A minimum cost feasible flow is the solution to the assignment problem, that is, a minimum distance pair matching.

If $C \geq 2T$, to match two controls to each treated individual, as in §4.2.5, then retain the same network but change the supplies to $s_{\tau_t} = 2$ for $t = 1, \ldots, T$, $s_{\gamma_c} = 0$, for $c = 1, \ldots, C$, $s_\omega = -2T$. Unlike the assignment representation of the 2-to-1 matching problem in §4.2.5, the network representation of 2-to-1 matching does not involve duplication of distance matrices or treated individuals.

4.4.4 Solving the minimum cost flow problem in R

A variant of the Bertsekas' auction algorithm in §4.2.4 solves the minimum cost flow problem (4.3)-(4.5); see [28]. The RELAX-IV code of [27] is accessible in `R` using `callrelax` in Samuel Pimentel's `rcbalance` package, which requires Ben Hansen's `optmatch` package. The input to `callrelax` is simple: (i) the edges $\mathcal{E}$ are represented by two vectors of the same length, whose jth coordinates give start node n_i and the end node n_i' of the jth edge, $e_j = \left(n_i, n_i'\right)$; (ii) a vector of costs, $\text{cost}(e_i)$, (iii) a vector of capacities, $\text{cap}(e_i)$, and (iv) a vector of supplies for nodes, b_n, $n \in \mathcal{N}$. Then `callrelax` reports whether the problem is feasible, and if it is, then it returns a minimum cost flow. Treated individual τ_t is matched to control γ_c if edge $e = (\tau_t, \gamma_c)$ has flow $f_e = 1$.

4.4.5 Size and speed of minimum cost flow problems

This subsection continues the discussion of computational speed begun in §4.3. Recall that $T \leq C$ so that $T = O(C)$ and $T + C = O(C)$. In typical statistical matching problems, the number of nodes, $|\mathcal{N}|$, is slightly larger than the number of people, $T + C$; so, as more people are added to the match, as $T + C \to \infty$, in typical cases the number of nodes grows as $|\mathcal{N}| = O(C)$. Because $\mathcal{E} \subset \mathcal{N} \times \mathcal{N}$, it follows that $|\mathcal{E}| < |\mathcal{N} \times \mathcal{N}| = |\mathcal{N}|^2$, so $|\mathcal{E}| = O\left(|\mathcal{N}|^2\right)$. Because typical statistical problems have $|\mathcal{N}| = O(C)$, they also have $|\mathcal{E}| = O(C^2)$. Finally, in typical statistical matching problems, $\sum_{n \in \mathcal{N}} |s_n| = O(C)$.

There is an algorithm that solves the minimum cost flow problem in

$$O\left\{ |\mathcal{N}| \cdot |\mathcal{E}| + (|\mathcal{E}| + |\mathcal{N}| \log |\mathcal{N}|) \sum_{n \in \mathcal{N}} |s_n| \right\} \tag{4.6}$$

steps; see [31, Theorem 9.13]. Simplifying for statistical matching problems, this is at most $O(C^3)$, but it can be smaller if $|\mathcal{E}|$ is much smaller than C^2. For example, if we include in $\mathcal{E}$ only treatment-control edges $e = (\tau_t, \gamma_c)$ for which τ_t and γ_c have very similar propensity scores in such a way that $|\mathcal{E}| = O(C \log C)$, then the time bound improves to $O(C^2 \log C)$.

4.5 Tactics for Matching by Minimum Cost Flow

4.5.1 What are tactics?

Within the framework of minimum cost flow, a tactic adjusts the network to produce some desired effect. Some tactics adjust the costs, cost (e_i), others add or delete nodes and edges, and still others adjust the supplies s_n. Once it has been shown that this adjustment produces the desired effect, the problem is again solved using the minimum cost flow algorithm. The current section briefly mentions a few tactics.

4.5.2 Fine, near-fine balance and refined balance

Fine balance is a constraint on the marginal distributions of a covariate in treated and control groups. In the simplest form, fine balance constrains the marginal distributions of a nominal covariate to be equal in treated and control groups, without constraining who is matched to whom. If the treated and control groups are both 50% female, then gender is finely balanced, even if women are often not matched to women, and men are often not matched to men. Fine balance means that an imbalance in one pair is counterbalanced by an opposing imbalance in some other pair. This simplest form of fine balance was proposed in [32, §3.2].

Fine balance can be implemented by splitting the one sink in Figure 4.3 into two sinks in Figure 4.4, a sink for women and a sink for men. Female controls are connected to the sink for women, and male controls are connected to the sink for men. The sink for women has a supply that is the negative of the number of women in the treated group. The sink for men has a supply that is the negative of the number of men in the treated group. A flow is feasible in this network if and only if the number of female controls equals the number of treated females and the number of male controls equals the number of treated males. A minimum cost feasible flow is a minimum distance match subject to the constraint that gender is finely balanced.

Fine balance is not always feasible. Although $C \geq T$, it may happen that the number of treated women exceeds the number of women in the pool of potential controls. Near-fine balance means

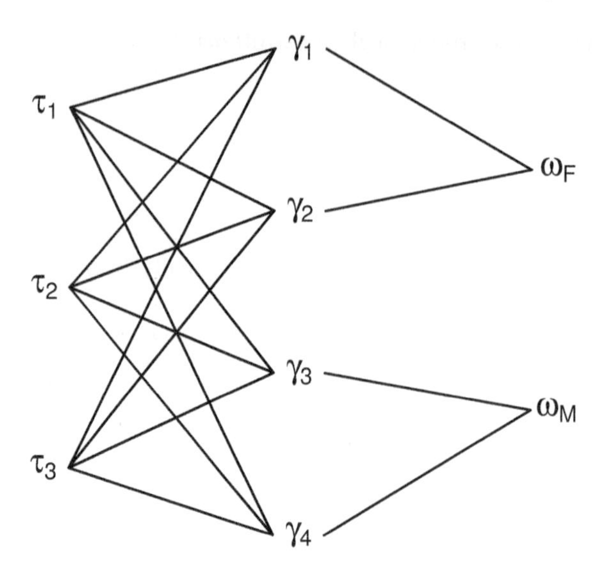

FIGURE 4.4
Fine balance for gender is achieved by splitting the sink into a female sink, ω_F and a male sink ω_M, and controls γ_c are connected to the sink that represents their own gender. If two treated individuals were female and one was male, the supply of ω_F would be -2 and the supply of ω_M would be -1, forcing the selection of two female controls and one male control, without forcing women to be paired with women, or men to be paired with men.

that the deviation from fine balance, though not zero, is minimized; see [33]. Near-fine balance is implemented by adding one more sink to Figure 4.4. Because there is a deficit of women, the female sink has supply equal to the negative of the number of women in the control group, while the male sink has supply equal to the number of men in the treated group. An additional sink absorbs precisely the flow that cannot pass through the category sinks: it has supply equal to the negative of the total deficit in the category sinks.

In forming 38,841 matched pairs of two children, one receiving surgery at a children's hospital, the other at a conventional hospital, Ruoqi Yu and colleagues [34] constrained the match to achieve near-fine balance for 973 Principal Diagnoses, while matching exactly for 463 Principal Surgical Procedures. That is, children were paired for surgical procedure, but diagnoses were balanced without necessarily being paired. Subject to those constraints, a covariate distance was minimized.

When fine balance is infeasible, it may not be enough to minimize the extent of the imbalance. Often, nominal categories have a structure, so that some categories are similar to one another. In this case, if fine balance is infeasible, we prefer to counterbalance with a control in a near-by category. Refined balance replaces the two sinks in Figure 4.4 by a tree-structure of sinks, whose categories are the leaves of the tree, preferring to counterbalance with categories that are close together on the same branch; see Pimentel and colleagues [35] who balance 2.8 million categories.

Fine balance and its variants apply in a similar way when matching in a fixed ratio, say 1-to-2. Pimentel, Yoon and Keele [36] extend the concept of fine balance to matching with a variable number of controls.

Bo Zhang and colleagues [37] extend the concept of fine balance. In fine balance, a match must meet two criteria, one defined in terms of balancing a nominal covariate, the other in terms of close pairing of treated and control individuals. The balance criterion refers to the control group as a whole, not to who is paired with whom. In parallel, Zhang and colleagues [37] have two criteria, but the balance criteria need not be confined to a nominal covariate. In their approach, the balance criteria do not affect who is paired with whom, but they do affect who is selected for the control group as a whole. For instance, the balance criteria might include a narrow caliper on the propensity score, ensuring that the entire distribution of propensity scores is similar in treated and matched control groups, but individual pairs might have only a loose caliper on the propensity score, so that pairs are also close in terms of a few covariates highly predictive of the outcome.

4.5.3 Adjusting edge costs to favor some types of pairs

Changing the costs, cost (e), of edges $e = (\tau_t, \gamma_c)$ can produce a wide variety of desired effects. If a sufficiently large constant – a large penalty – is added to cost (e) of $e = (\tau_t, \gamma_c)$ whenever τ_t and γ_c differ in gender, then a minimum cost flow will maximize the number of pairs matched exactly for gender, and among matches that do that, it will minimize the total of the unpenalized costs. Various simple tactics of this kind are discussed in [1, Chapters 9 and 10].

Small directional penalties adjust the cost (e) of $e = (\tau_t, \gamma_c)$ in a way that views τ_t and γ_c asymmetrically [38]. For instance, because treated individuals tend to have higher propensity scores than controls, this tendency might be opposed by slightly increasing the cost (e) of $e = (\tau_t, \gamma_c)$ whenever τ_t has a higher propensity score than γ_c. Directional penalties come in many forms, and some of these forms constitute a Lagrangian for linear constraints on the optimal flow; see §4.6.2.

4.5.4 Using more controls or fewer treated individuals

In Figure 4.3, each treated unit supplies the same number of units of flow, 1 for pair matching, 2 for matching 2-to-1, and so on. With 4 controls in Figure 4.3, it is not possible to match 2-to-1. Matching in a variable ratio can use more controls. This is implemented by adding a new node, ς, called a source to Figure 4.3. The source has an edge pointing to each treated individual, τ_t, where that edge has zero cost and specified capacity; see [32]. For instance, in Figure 4.3, each τ_t might supply one unit of flow, and the source might supply one additional unit of flow, making four units of flow in total. A feasible flow would form two matched pairs and one set matched 1-to-2, thereby using all four controls. A minimum cost flow would minimize the total of the four within-set covariate distances.

In some contexts, there are some treated individuals who are unlike all controls in terms of observed covariates. For example, we may plot the propensity score and see that the highest propensity scores in the treated group far exceed all of the propensity scores in the control group. In this case, we may wish to match in the region of the covariate space – equivalently, the region of propensity scores – that exhibits reasonable overlap. Of course, this changes the treated population. The treated population now consists of those treated individuals who had some realistic prospect of receiving the control instead. This can be implemented in Figure 4.3 by adding an additional sink, ω_U. Each treated unit τ_t has an edge pointing to ω_U with capacity one and cost $\lambda > 0$. For large enough λ, the cost of sending flow to ω_U is prohibitive, and all treated T treated individuals are matched to controls in pairs. As λ decreases, some pairs that do the most to increase the average within-pair cost are shunted to ω_U. For properties of this subset matching method, see [39]. To emphasize limited overlap on the propensity score when deciding to omit a treated individual, the cost or distance should emphasize the propensity score, say using a caliper. Samuel Pimentel's `R` package `rcsubset` combines subset matching with fine or refined covariate balance.

4.5.5 Threshold algorithms that remove edges

A network with fewer edges takes less time to construct, less time to optimize; see §4.4.5. Often, a glance at τ_t and γ_c – say a glance at their propensity scores – indicates that they should never be matched, so $e = (\tau_t, \gamma_c)$ might reasonably be removed from the network as a candidate match. Removing an edge changes the optimization problem, but it may do so in a reasonable way.

A problem with removing edges – even very bad edges – from a network is that, if too many edges are removed, or the wrong edges are removed, then there may be no feasible flow, hence no possible match. Let $\nu : \mathcal{T} \times \mathcal{C} \rightarrow \mathbb{R}$ assign a real number, $\nu(e)$, to each $e = (\tau_t, \gamma_c)$. A threshold algorithm starts with a candidate threshold, say $\varkappa$, and removes all edges with $\nu(e) > \varkappa$, and finds a minimum cost flow in the resulting network. If the problem is infeasible, it increases $\varkappa$ and tries again. If the problem is feasible, the algorithm reduces $\varkappa$ and tries again. A binary search for the smallest feasible $\varkappa$ quickly produces a short interval of values of $\varkappa$ such that feasibility is assured for the upper endpoint of the interval and is lacking at the lower endpoint. If the number of edges is $O(C \log C)$ for each $\varkappa$, then a dozen iterations of the threshold algorithm may take much less time and space than a single optimization of a dense network with $O(C^2)$ edges; see §4.4.5. Garfinkel [40] used a threshold algorithm for the so called bottleneck assignment problem.

A simple version adds a constraint to the match in Figure 4.3. The constraint finds the minimum total cost pair-match among all pair-matches that minimize the maximum value $\nu(e)$ for all $e = (\tau_t, \gamma_c)$ who are matched; see [41].

The threshold algorithm is particularly fast when the requirement that $\nu(e) \leq \varkappa$ creates a doubly convex bipartite graph from $\mathcal{T} \times \mathcal{C}$. For example, this occurs when $\nu(e)$ is the absolute difference in propensity scores. It also occurs if $\nu(e) = \infty$ for $e = (\tau_t, \gamma_c)$ when τ_t and γ_c belong to different categories of a nominal covariate, but if they are in the same category then $\nu(e)$ is the absolute difference in propensity scores. In these and similar cases, Glover's [42] algorithm may be applied in each iteration of the binary search for the smallest feasible $\varkappa$; see also [43]. Ruoqi Yu and colleagues [34] use this method to determine the smallest feasible caliper for the propensity score in large matching problems that also employ exact match requirements, a covariate distance and fine balance. An extension of her method can discard edges using complex distances [44].

4.6 Optimal Matching using Mixed Integer Programming

4.6.1 Working without a worst-case time bound

Worst-case time bounds were discussed in §4.3. A polynomial time-bound says: the most difficult computational problem of size C can be solved in $O(C^\lambda)$ steps, for some fixed number λ as $C \rightarrow \infty$. As noted in §4.3, there are algorithms for minimum distance matching and related problems that possess a worst-case time bound that is a polynomial, $O(C^3)$, in the number, C, of controls available, where there are $T \leq C$ treated individuals. Specifically, this is true when the problem is expressed as an appropriate minimum cost flow problem; see §4.4.2. Even for minimum cost flow problems, some of the algorithms that are highly competitive – i.e., very fast in practice – lack a time bound of the form $O(C^\lambda)$; for instance, the time-bound for Bertsekas' [26] auction algorithm is not of this form, as it involves more information than just the size, C, of the problem. Other algorithms often found to work well in practice lack any kind of known worst-case time bound.

There are useful matching problems that cannot be expressed as minimum cost flow problems of the type in §4.4.2. Many of these problems may be expressed as mixed-integer programs; that is, as linear programs in which some of the variables are required to take integer values. It is believed that there can be no algorithm that solves the general mixed-integer program in time $O(C^\lambda)$ no matter what value is given to λ; that is, these problems are NP-hard, as discussed by Schrijver [45]. Again, this is not a statement about your matching problem, but rather a statement about the most difficult problem of size C as $C \rightarrow \infty$. People solve large mixed integer programs all the time, but they do so without a guarantee that a solution can be produced in reasonable time, and in some instances no solution can be found in reasonable time. The existence of a worst-case bound $O(C^\lambda)$ is often a useful guide to the difficulty of a general class of problems, but sometimes the worst-case situation is quite peculiar and unlikely to correspond with the situation that you happen to have.

For a statistician with a matching problem, ask: How important is it to have a worst-case bound on computation time of the form $O(C^{\lambda})$ as $C \to \infty$? Arguably, the statistician is concerned with a particular matching problem with a particular size C, and if the computations are feasible for that problem, then that is all that the statistician requires. If the computations are not feasible for that problem, then perhaps the original matching problem can be replaced by a different matching problem that is computationally feasible. The worst that can happen if you attempt to build a match without a worst-case time bound is that the algorithm will run for longer than you can tolerate, you will cancel the algorithm's execution and reformulate the problem as one that is computationally easier. Indeed, the same thing can happen when there is a worst-case time bound like $O(C^{\lambda})$ if C is large and λ is not small. In practice, worst-case time bounds tend to label a computation problem as tractable or not; however, they do not identify the best algorithm for solution of that problem.

In practice, in designing an observational study, it is common to build several matched samples – always without access to outcomes – and then select the one matched sample that best balances various objectives. Because the investigator does not have access to outcomes – only to treatments and observed covariates – this process of considering alternative designs is analogous to considering several designs for an experiment before selecting one for actual use: it cannot bias the study, because outcomes are considered in only one study design. In this context it is not a major inconvenience if there are one or two failed attempts to build certain matched samples, because the matching algorithm did not complete its task in a tolerable amount of time. Commonly, with $T + C$ equal to a few thousand, an optimal match is completed in the time it takes to leave your computer to refill your coffee mug. With $T + C$ equal to tens or hundreds of thousands, greater attention to computational efficiency is required [34, 37, 44, 46].

There are several methods in common use for matching problems that are not minimum cost flow problems. First, certain matching problems are very similar to minimum cost flow problems, but they have some extra linear constraints. This class of problems has been extensively studied; see Ahuja, Magnanti and Orlan [47, Chapter 16] and Bertsekas [48, Chapter 10]. Second, there are some theorems for very specific situations that show that solving a linear program, rather than an integer program, and randomly rounding the solution to obtains integers, can be only slightly worse than finding the optimal integer solution [49, 50]. Finally, there are several commercial solvers, including CPLEX and Gurobi, that use a variety of techniques to solve mixed integer programs. These commercial solvers are often free to academics and highly effective, the principle downside being that they are black boxes so far as the user is concerned – you cannot look at the code or see how the solution was produced. The `R` packages `designmatch` and `mipmatch` use these commercial solvers to produce matched samples, and they have been widely used in practice [6, 9, 51–58].

4.6.2 Linear side constraints for minimum distance matching

Suppose that there are $T = 100$ treated individuals, where 50 treated individuals are older than 42.5 years and 50 are younger than 42.5 years, so 42.5 years is a median age in the treated group. In a pair match, perhaps we want 50 of the matched controls to be older 42.5 years and 50 of the matched controls to be younger than 42.5 years, so 42.5 years is also a median age in the matched control group. How can we achieve this?

The condition just described – the condition on the median age – is an additional linear constraint on the minimum cost flow problem in (4.3)-(4.5). The constraint says that 50 equals the sum of the flows f_e over all edges $e = (\tau_t, \gamma_c)$ such that control γ_c is over 42.5 years old.

There are many analogous linear constraints on flows f_e. One can constrain the upper and lower quartiles of age in the same way the median of age was constrained, simply by adding two more linear constraints on the flows f_e. One can do the same for income, using three constraints for age and three constraints for income. One can cut age and income at their quartiles, making a 4×4

table of frequencies for the joint distribution of cut-age and cut-income in the treated group; then, linear constraints on the flows f_e can force the same counts in the 4×4 table for the matched control group. Indeed, any multiway contingency table describing the treated group can be balanced in the matched control group. Obviously, if too many side constraints are imposed upon the flows, then there may be no flow f_e that satisfies them all; that is, (4.3)–(4.5) plus the side constraints may define an infeasible optimization problem. If you request the impossible, you will have to settle for less than you requested.

Notice that the constraints in the previous paragraph are constraints on the marginal distributions of the covariates in treated and control groups as two whole groups, not constraints on who is paired with whom. In this sense these linear side constraints resemble fine balance constraints in §4.5.2. The important difference is that a fine balance constraint in §4.5.2 can be implemented as part of a minimum cost flow problem (4.3)-(4.5) for a network without side constraints – see Figure 4.4 – so the $O(C^3)$ time bound applies to matching with fine balance, but that time bound does not apply to minimum cost flow problems with general linear side constraints.

Suppose that there are K binary covariates. Within the framework of a minimum cost flow optimization (4.3)-(4.5), one could finely balance the 2^K categories formed from these K covariates. If $K = 30$, then 2^K is more than a billion, so exact fine balance with a billion categories may be infeasible even with millions of people. Some of the variants of fine balance in §4.5.2, such as refined balance, might be practical in this case. However, an attractive alternative is to avoid balancing the enormous 2^K table for $K = 30$, and instead to balance many of its marginal tables. For example, linear side constraints on the flows f_e can balance each of the K covariates one at a time. Or, one could balance the $K(K-1)/2$ tables of size 2×2 recording the joint distribution of pairs of covariates [59].

In a pair match, the mean age in the matched control group is T^{-1} times the sum of f_e times the age of control γ_c over edges $e = (\tau_t, \gamma_c)$. So, the mean age in the control group is another linear function of the flows f_e. It is usually unwise to constrain the equality of the means of a continuous covariate in treated and control groups – in all likelihood that is infeasible. However, an inequality constraint is possible. For instance, one can insist that the mean age in the control group be at most 1 year older than the treated group and at most one year younger than the treated group – that becomes two linear inequality constraints. By constraining the means of squares of centered ages, one can constrain the variances of age to be close in treated and matched control groups. In parallel, one can constrain correlation or skewness to be similar in treated and matched control groups.

A direct and effective way to implement all of these ideas uses a commericial solver such as CPLEX or Gurobi, as proposed by Zubizarreta [60] and as implemented in the `R` package `designmatch`. Although no worst-case time bound is available, experience suggests that `designmatch` using Gurobi solves certain constrained matching problems about as fast as minimum cost flow algorithms solve unconstrained problems.

A common tactic for solving minimum cost flow problems with linear side constraints uses Lagrangian relaxation; see Fisher [61], Ahuja, Magnanti and Orlan [47, Chapter 16] and Bertsekas [48, Chapter 10]. In this approach $\text{cost}(e)$ in (4.3) is altered to penalize violations of a linear constraint, and the altered minimum cost flow problem (4.3)-(4.5) is solved. At successive iterations, the penalties are adjusted to gradually enforce the constraint. For instance, if the treated group has a higher median age than the control group, then a penalty is added to $\text{cost}(e)$ for $e = (\tau_t, \gamma_c)$ whenever control γ_c has an age below the treated group median age. By solving (4.3)-(4.5) while adjusting the size of the penalty, one can often force the medians to be the same. This tactic has long been used informally in matching under the name "directional penalty" [38]. A Lagrangian penalty is directional in that it pushes in a direction; here, it pushes to select older controls. If the directional penalty is too large, the control group may become too old, but if the directional penalty is not large enough, it may not equate the median ages. The penalty must be adjusted to produce the desired effect. It is quite possible that, inside their black boxes, commercial solvers, such as CPLEX or Gurobi, have automated the use of Lagrangians and other standard techniques of integer programming.

4.6.3 Cardinality matching: Matching without pairing

Fine balance in §4.5.2 and the method in §4.6.2 both constrain a minimum distance match to balance certain covariates. Cardinality matching carries this idea to its logical extreme: it does all of the work in selecting a control group using balance constraints, and makes no effort to form matched pairs [23,46,62–65]. Having selected the control group to balance covariates, these selected controls may subsequently be paired to treated individuals for a few covariates highly predictive of outcomes [11]. Close pairing for covariates highly predictive of outcomes increases insensitivity to unmeasured biases by increasing design sensitivity [10, 11]. The `cardmatch` function of the `designmatch` package in `R` implements cardinality matching. When `designmatch` uses Gurobi as a solver, the method works quickly in large problems [46, 64].

More precisely, cardinality matching creates a list of balance constraints describing the comparability of treated and matched control groups, such as a constraint that both groups have median age 42.5, as in §4.6.2. In the usual formulation, cardinality matching finds the largest control group that satisfies those constraints; that is, it maximizes the size or cardinality of the control group subject to balance constraints.

Cardinality matching has also been used to make a personalized control group for a single patient [62]. It is difficult to match one person to several controls who are each close on many covariates. Cardinality matching can instead create a group of controls that balance or average to resemble a single person: one control who is a little too old is balanced by another who is a little too young, and so on.

A related method is due to Katherine Brumberg and colleagues [66]. Various strata are defined by their importance or interest. In each stratum, a control group of fixed size is picked from controls in the same stratum to minimize the total of K absolute standardized imbalances in K covariates. For example, one might pick three controls for each treated individual in each stratum, minimizing the absolute treated-minus-control difference in mean ages in units of the standard deviation of age. Typically, K is moderately large – there are quite a few covariates – some of which describe individual strata, while others describe the treated group as a whole. As in cardinality matching, there is no pairing of individuals, and the size of the control group need not be an integer multiple of the size of the treated group. Unlike cardinality matching, which maximizes the size of a balanced control group, Brumberg's method minimizes the total covariate imbalance over K covariates for a control group of fixed size. The method solves a linear program whose solution is guaranteed to have very few non-integer coordinates, and these few coordinates are subjected to randomized rounding [49, 50]. In effect, the linear programming solution wants to cut a few people into pieces, using parts of several people rather than intact people; however, it only wants to do this to very few people. For these few people, randomized rounding flips biased coins to select intact people, but the result is barely different from the linear programming solution because there are provably few fractional coordinates in the linear programming solution. The linear programming solution produces a smaller total absolute imbalance in covariates than any integer solution; so, if the randomly rounded solution is only a bit worse – because few people are rounded – then the algorithm has demonstrated that there is no point in burning the computer searching for the optimal integer solution. After all, the total absolute covariate imbalance for the optimal integer solution must fall between its value for the fractional linear programming solution and the value for randomly rounded integer solution. As illustrated in Brumberg's example, the method is practical in very large problems.

4.7 Discussion

As mentioned in §4.1.2, optimization is a framework within which statistical matching problems may be framed, solved and the solutions compared. In practice, a matched study design has several or many objectives, so we do not automatically pick the best design for a single objective. Nonetheless, possessing the optimal designs formulated in terms of several different objectives is almost always helpful in selecting one design for implementation. For example, the mean within-pair covariate distance will be smaller with 1-to-1 matching than with 1-to-2 matching, but under simple models the standard error of the mean difference in outcomes will be larger [24], and the design sensitivity will be worse [67]. Having both the best 1-to-1 match, the best 1-to-2 match, and the best match with variable numbers of controls in hand is helpful in making an informed choice between them.

References

[1] Paul R Rosenbaum. *Design of Observational Studies*. Springer, New York, 2 edition, 2020.

[2] Paul R Rosenbaum. Modern algorithms for matching in observational studies. *Annual Review of Statistics and Its Application*, 7:143–176, 2020.

[3] Paul R Rosenbaum and Donald B Rubin. Constructing a control group using multivariate matched sampling methods that incorporate the propensity score. *The American Statistician*, 39(1):33–38, 1985.

[4] Mike Baiocchi, Dylan S Small, Scott Lorch, and Paul R Rosenbaum. Building a stronger instrument in an observational study of perinatal care for premature infants. *Journal of the American Statistical Association*, 105(492):1285–1296, 2010.

[5] Ashkan Ertefaie, Dylan S Small, and Paul R Rosenbaum. Quantitative evaluation of the trade-off of strengthened instruments and sample size in observational studies. *Journal of the American Statistical Association*, 113(523):1122–1134, 2018.

[6] Mark D Neuman, Paul R Rosenbaum, Justin M Ludwig, Jose R Zubizarreta, and Jeffrey H Silber. Anesthesia technique, mortality, and length of stay after hip fracture surgery. *Journal of the American Medical Association*, 311(24):2508–2517, 2014.

[7] José R Zubizarreta, Dylan S Small, Neera K Goyal, Scott Lorch, and Paul R Rosenbaum. Stronger instruments via integer programming in an observational study of late preterm birth outcomes. *The Annals of Applied Statistics*, 7: 25–50, 2013.

[8] Paul R Rosenbaum and Jeffrey H Silber. Using the exterior match to compare two entwined matched control groups. *American Statistician*, 67(2):67–75, 2013.

[9] Jeffrey H Silber, Paul R Rosenbaum, Amy S Clark, Bruce J Giantonio, Richard N Ross, Yun Teng, Min Wang, Bijan A Niknam, Justin M Ludwig, Wei Wang, and Keven R Fox. Characteristics associated with differences in survival among black and white women with breast cancer. *Journal of the American Medical Association*, 310(4):389–397, 2013.

[10] Paul R Rosenbaum. Heterogeneity and causality: Unit heterogeneity and design sensitivity in observational studies. *The American Statistician*, 59(2):147–152, 2005.

[11] José R Zubizarreta, Ricardo D Paredes, Paul R Rosenbaum, et al. Matching for balance, pairing for heterogeneity in an observational study of the effectiveness of for-profit and not-for-profit high schools in chile. *Annals of Applied Statistics*, 8(1):204–231, 2014.

[12] Donald B Rubin. The design versus the analysis of observational studies for causal effects: Parallels with the design of randomized trials. *Statistics in Medicine*, 26(1):20–36, 2007.

[13] Abraham Wald. Note on the consistency of the maximum likelihood estimate. *Annals of Mathematical Statistics*, 20(4):595–601, 1949.

[14] Peter J Huber. Robust estimation of a location parameter. *The Annals of Mathematical Statistics*, 35(1):73–101, 1964.

[15] Eugene L Lawler. *Combinatorial Optimization: Networks and Matroids*. Dover, Mineola, NY, 2001.

[16] Peter C Austin and Elizabeth A Stuart. Optimal full matching for survival outcomes: A method that merits more widespread use. *Statistics in Medicine*, 34(30):3949–3967, 2015.

[17] Peter C Austin and Elizabeth A Stuart. The performance of inverse probability of treatment weighting and full matching on the propensity score in the presence of model misspecification when estimating the effect of treatment on survival outcomes. *Statistical Methods in Medical Research*, 26(4):1654–1670, 2017.

[18] Ben B Hansen. Full matching in an observational study of coaching for the sat. *Journal of the American Statistical Association*, 99(467):609–618, 2004.

[19] Ben B Hansen and Stephanie Olsen Klopfer. Optimal full matching and related designs via network flows. *Journal of Computational and Graphical Statistics*, 15(3):609–627, 2006.

[20] Paul R Rosenbaum. A characterization of optimal designs for observational studies. *Journal of the Royal Statistical Society: Series B (Methodological)*, 53(3):597–610, 1991.

[21] Elizabeth A Stuart and Kerry M Green. Using full matching to estimate causal effects in nonexperimental studies: Examining the relationship between adolescent marijuana use and adult outcomes. *Developmental Psychology*, 44(2):395, 2008.

[22] Ben B Hansen. Optmatch: Flexible, optimal matching for observational studies. *New Functions for Multivariate Analysis*, 7(2):18–24, 2007.

[23] Cinar Kilcioglu, José R Zubizarreta, et al. Maximizing the information content of a balanced matched sample in a study of the economic performance of green buildings. *The Annals of Applied Statistics*, 10(4):1997–2020, 2016.

[24] Kewei Ming and Paul R Rosenbaum. Substantial gains in bias reduction from matching with a variable number of controls. *Biometrics*, 56(1):118–124, 2000.

[25] Fredrik Sävje, Michael J Higgins, and Jasjeet S Sekhon. Generalized full matching. *Political Analysis*, 29(4):423–447, 2021.

[26] Dimitri P Bertsekas. A new algorithm for the assignment problem. *Mathematical Programming*, 21(1):152–171, 1981.

[27] Dimitri P Bertsekas and Paul Tseng. The relax codes for linear minimum cost network flow problems. *Annals of Operations Research*, 13(1):125–190, 1988.

[28] Dimitri P Bertsekas. *Linear Network Optimization*. MIT Press, Cambridge, MA, 1991.

[29] Kewei Ming and Paul R Rosenbaum. A note on optimal matching with variable controls using the assignment algorithm. *Journal of Computational and Graphical Statistics*, 10(3):455–463, 2001.

[30] Robert Sedgewick and Philippe Flajolet. *An Introduction to the Analysis of Algorithms*. Addison-Wesley, New York, 1996.

[31] Bernhard H Korte and Jens Vygen. *Combinatorial Optimization*. New York: Springer, 5 edition, 2012.

[32] Paul R Rosenbaum. Optimal matching for observational studies. *Journal of the American Statistical Association*, 84(408):1024–1032, 1989.

[33] Dan Yang, Dylan S Small, Jeffrey H Silber, and Paul R Rosenbaum. Optimal matching with minimal deviation from fine balance in a study of obesity and surgical outcomes. *Biometrics*, 68(2):628–636, 2012.

[34] Ruoqi Yu, Jeffrey H Silber, and Paul R Rosenbaum. Matching methods for observational studies derived from large administrative databases (with Discussion). *Statistical Science*, 35(3):338–355, 2020.

[35] Samuel D Pimentel, Rachel R Kelz, Jeffrey H Silber, and Paul R Rosenbaum. Large, sparse optimal matching with refined covariate balance in an observational study of the health outcomes produced by new surgeons. *Journal of the American Statistical Association*, 110(510):515–527, 2015.

[36] Samuel D Pimentel, Frank Yoon, and Luke Keele. Variable-ratio matching with fine balance in a study of the peer health exchange. *Statistics in Medicine*, 34(30):4070–4082, 2015.

[37] Bo Zhang, Dylan S Small, Karen B Lasater, Matt McHugh, Jeffrey H Silber, and P R Rosenbaum. Matching one sample according to two criteria in observational studies. *Journal of the American Statistical Association*, DOI:10.1080/01621459.2021.1981337, 2022.

[38] Ruoqi Yu and Paul R Rosenbaum. Directional penalties for optimal matching in observational studies. *Biometrics*, 75(4):1380–1390, 2019.

[39] Paul R Rosenbaum. Optimal matching of an optimally chosen subset in observational studies. *Journal of Computational and Graphical Statistics*, 21(1):57–71, 2012.

[40] Robert S Garfinkel. An improved algorithm for the bottleneck assignment problem. *Operations Research*, 19(7):1747–1751, 1971.

[41] Paul R Rosenbaum. Imposing minimax and quantile constraints on optimal matching in observational studies. *Journal of Computational and Graphical Statistics*, 26(1):66–78, 2017.

[42] Fred Glover. Maximum matching in a convex bipartite graph. *Naval Research Logistics Quarterly*, 14(3):313–316, 1967.

[43] George Steiner and Julian Scott Yeomans. A linear time algorithm for maximum matchings in convex, bipartite graphs. *Computers & Mathematics with Applications*, 31(12):91–96, 1996.

[44] Ruoqi Yu and Paul R Rosenbaum. Graded matching for large observational studies. *Journal of Computational and Graphical Statistics*, 2022.

[45] Alexander Schrijver. *Theory of Linear and Integer Programming*. John Wiley & Sons, New York, 1998.

[46] Magdalena Bennett, Juan Pablo Vielma, and José R Zubizarreta. Building representative matched samples with multi-valued treatments in large observational studies. *Journal of Computational and Graphical Statistics*, 29(4):744–757, 2020.

[47] Ravindra K Ahuja, Thomas L Magnanti, and James B Orlin. *Network Flows*. Prentice Hall, Upper Saddle River, New Jersey, 1993.

[48] Dimitri Bertsekas. *Network Optimization: Continuous and Discrete Models*, volume 8. Athena Scientific, Nashua, NH, 1998.

[49] Vijay V Vazirani. *Approximation Algorithms*. Springer, New York, 2001.

[50] David P Williamson and David B Shmoys. *The Design of Approximation Algorithms*. Cambridge University Press, New York, 2011.

[51] Wenqi Hu, Carri W Chan, José R Zubizarreta, and Gabriel J Escobar. Incorporating longitudinal comorbidity and acute physiology data in template matching for assessing hospital quality. *Medical Care*, 56(5):448–454, 2018.

[52] Rachel R Kelz, Caroline E Reinke, José R Zubizarreta, Min Wang, Philip Saynisch, Orit Even-Shoshan, Peter P Reese, Lee A Fleisher, and Jeffrey H Silber. Acute kidney injury, renal function, and the elderly obese surgical patient: A matched case-control study. *Annals of Surgery*, 258(2):359, 2013.

[53] Neel Koyawala, Jeffrey H Silber, Paul R Rosenbaum, Wei Wang, Alexander S Hill, Joseph G Reiter, Bijan A Niknam, Orit Even-Shoshan, Roy D Bloom, Deirdre Sawinski, and Peter P. Reese. Comparing outcomes between antibody induction therapies in kidney transplantation. *Journal of the American Society of Nephrology*, 28(7):2188–2200, 2017.

[54] Caroline E Reinke, Rachel R Kelz, Jose R Zubizarreta, Lanyu Mi, Philip Saynisch, Fabienne A Kyle, Orit Even-Shoshan, Lee A Fleisher, and Jeffrey H Silber. Obesity and readmission in elderly surgical patients. *Surgery*, 152(3):355–362, 2012.

[55] Jeffrey H Silber, Paul R Rosenbaum, Wei Wang, Justin M Ludwig, Shawna Calhoun, James P Guevara, Joseph J Zorc, Ashley Zeigler, and Orit Even-Shoshan. Auditing practice style variation in pediatric inpatient asthma care. *JAMA Pediatrics*, 170(9):878–886, 2016.

[56] Jeffrey H Silber, Paul R Rosenbaum, Richard N Ross, Joseph G Reiter, Bijan A Niknam, Alexander S Hill, Diana M Bongiorno, Shivani A Shah, Lauren L Hochman, ORIT Even-Shoshan, and Keven R Fox. Disparities in breast cancer survival by socioeconomic status despite medicare and medicaid insurance. *The Milbank Quarterly*, 96(4):706–754, 2018.

[57] Giancarlo Visconti. Economic perceptions and electoral choices: A design-based approach. *Political Science Research and Methods*, 7(4):795–813, 2019.

[58] José R Zubizarreta, Magdalena Cerdá, and Paul R Rosenbaum. Effect of the 2010 chilean earthquake on posttraumatic stress reducing sensitivity to unmeasured bias through study design. *Epidemiology (Cambridge, Mass.)*, 24(1):79, 2013.

[59] Jesse Y Hsu, José R Zubizarreta, Dylan S Small, and Paul R Rosenbaum. Strong control of the familywise error rate in observational studies that discover effect modification by exploratory methods. *Biometrika*, 102(4):767–782, 2015.

[60] José R Zubizarreta. Using mixed integer programming for matching in an observational study of kidney failure after surgery. *Journal of the American Statistical Association*, 107(500):1360–1371, 2012.

[61] Marshall L Fisher. The Lagrangian relaxation method for solving integer programming problems. *Management Science*, 27(1):1–18, 1981.

[62] Eric R Cohn and Jose R Zubizarreta. Profile matching for the generalization and personalization of causal inferences. *Epidemiology*, 33(5):678–688, 2022.

[63] María de los Angeles Resa and José R Zubizarreta. Evaluation of subset matching methods and forms of covariate balance. *Statistics in Medicine*, 35(27):4961–4979, 2016.

[64] Bijan A Niknam and Jose R Zubizarreta. Using cardinality matching to design balanced and representative samples for observational studies. *Journal of the American Medical Association*, 327(2):173–174, 2022.

[65] José R Zubizarreta and Luke Keele. Optimal multilevel matching in clustered observational studies: A case study of the effectiveness of private schools under a large-scale voucher system. *Journal of the American Statistical Association*, 112(518):547–560, 2017.

[66] Katherine Brumberg, Dylan S Small, and Paul R Rosenbaum. Using randomized rounding of linear programs to obtain unweighted natural strata that balance many covariates. *Journal of the Royal Statistical Society, Series A*, 2022. DOI:10.1111/rssa.12848

[67] Paul R Rosenbaum. Impact of multiple matched controls on design sensitivity in observational studies. *Biometrics*, 69(1):118–127, 2013.

5

Optimal Full Matching

Mark M. Fredrickson and Ben Hansen

CONTENTS

In an observational study the distribution of outcomes and covariates for units receiving the treatment condition may differ greatly from the distribution of outcomes and covariates for units receiving the control condition. Simple estimates of treatment effects or tests of hypotheses for causal effects may be confounded if appropriate adjustment is not applied. In this chapter, we discuss *optimal full matching*, a particular form of statistical matching to control for background imbalances in covariates or treatment probabilities. Ideally, the matched sets would be homogeneous on background variables and the probability of receiving the treatment condition. In practice, such precise matching is frequently not possible, which suggests matches that minimize observed discrepancies. Optimal full matches can be found efficiently in moderate sized problems on modern hardware, and the algorithm permits researcher control of matched sets that can be used to control which units are matched and also reduce the complexity of large problems. The algorithms for optimal full matches are readily available in existing software packages and are easily extended to include additional constraints on acceptable matches. We conclude with a brief discussion of the use of matches in outcome analysis, a simulation study, and an overview of recent developments in the field.

Additional material on matching is available in the reviews by Sekhon [1], Stuart [2], Gangl [3], and Rosenbaum [4]. Chapters on optimal matching can be found in the book by Imbens and Rubin [5] and on optimal full matching particularly in the books by Rosenbaum [6, 7].

5.1 Adjusting Outcomes for Confounding with Matching

Consider the setting of n total units exposed to a binary treatment W, writing $N_1 = \sum_{i=1}^{n} W_i$ and $N_0 = n - N_1$ for the number of treated and control units, respectively. We posit the existence

DOI: 10.1201/9781003102670-5

of potential outcomes Y_1 and Y_0 representing the responses to each of the treatment levels. The observed outcome is one of the two potential outcomes, depending only on the treatment assignment: $Y = WY_1 + (1 - W)Y_0$. Inherent in this notation is the assumption that for each unit in the study, the potential outcomes depend only on the treatment level received by the unit itself and not any spillover effect from other units, the Stable Unit Treatment Value Assumption (SUTVA) [8]. In the following discussion this assumption is not required for the key results, but it does greatly simplify the exposition.[1] One particular implication of SUTVA is limiting *treatment effects* to comparisons between unit level potential outcomes under different treatment levels, such as $\tau = Y_1 - Y_0$.

In the observational context the potential outcomes and treatment assignment may not be independent, and the conditional distributions of $Y_1 \mid W = 1$ and $Y_0 \mid W = 0$ may be quite different from the marginal distributions of Y_1 and Y_0, imparting bias to impact estimates and hypothesis tests that do not employ corrective steps. In addition to the observed outcomes $\{Y_{1i} \mid i \leq n, W_i = 1\}$ and $\{Y_{0i} \mid i \leq n, W_i = 0\}$, we also observe *covariates* $\boldsymbol{X}$, attributes fixed at the time that units are assigned to treatment conditions. Classical approaches to causal inference employ a parametric model of connecting the covariates to the potential outcomes $(Y_1, Y_0) \mid \boldsymbol{X}$, but the validity of this approach depends on the correctness of the parametric model. More generally, if treatment assignment can be assumed *strongly ignorable given covariates* $\boldsymbol{X} = \boldsymbol{x}$ for all $\boldsymbol{x}$, $W \perp\!\!\!\perp (Y_0, Y_1) \mid \boldsymbol{X} = \boldsymbol{x}$ and $\Pr W = 1 \mid \boldsymbol{X} = \boldsymbol{x} \in (0, 1)$, then conditioning on $\boldsymbol{X} = \boldsymbol{x}$ is sufficient to establish unbiased estimates of treatment effects and valid hypothesis tests – with or without a model for $(Y_1, Y_0) \mid \boldsymbol{X}$ [14]. This insight suggests that researchers could *stratify* units by $\boldsymbol{X}$ to make inferences about treatment effects for observational data. The practical difficulty is that unless $\boldsymbol{X}$ is rather coarse, there may be no treated or control units for certain levels of $\boldsymbol{X}$, making such an endeavor infeasible [15].

One solution to this problem is to forgo perfect stratification for an approximate stratification in which all units within each stratum do not differ very much on the observed characteristics $\boldsymbol{X}$ (or some suitable transformation of $\boldsymbol{X}$) and each stratum includes at least one treated and one control unit. Define a *match* $\mathcal{M} = \{\mathcal{S}_1, \mathcal{S}_2, \ldots, \mathcal{S}_s\}$ as a partition of $n_{\mathcal{M}} \leq n$ units in the sample into s disjoint strata that include at least one treated unit ($W = 1$) and at least one control unit ($W = 0$), writing $i \in \mathcal{S}_b$ if unit i is in stratum b. In some situations, in may be necessary to discard units in the original sample so that $n_{\mathcal{M}} < n$. Define a *distance function* $d(\boldsymbol{X}_i, \boldsymbol{X}_j)$ such that $d(\boldsymbol{X}_i, \boldsymbol{X}_j)$ is small for units with similar covariates. Overall, the total distance between treated and control units is given in Equation (5.1):

$$d(\mathcal{M}) = \sum_{b=1}^{s} \sum_{i,j \in \mathcal{S}_b} W_i(1 - W_j) d(\boldsymbol{X}_i, \boldsymbol{X}_j). \tag{5.1}$$

One commonly used distance function is the *Mahalanobis distance* between units i and j:

$$d(\boldsymbol{X}_i, \boldsymbol{X}_j) = (\boldsymbol{X}_i - \boldsymbol{X}_j)' S(\boldsymbol{X})^{-1} (\boldsymbol{X}_i - \boldsymbol{X}_j), \tag{5.2}$$

where $S(\boldsymbol{X})$ is the sample variance-covariance matrix [16]. Mahalanobis distance rescales the observed data such that all variables are uncorrelated and have unit variance and can be viewed as finding Euclidean distance on the principle component scores. Mahalanobis distance matching provides for *equal percent bias reduction*, meaning matched samples have comparable reductions in the difference of means between treated and control groups across all covariates [17].

As an alternative to matching directly on covariate values, Rosenbaum and Rubin [14] show conditioning on the one-dimensional *propensity score*, the conditional probability of treatment assignment $e(\boldsymbol{x}) = P(W = 1 \mid \boldsymbol{X} = \boldsymbol{x})$, is equivalent to conditioning on $\boldsymbol{X} = \boldsymbol{x}$ directly. Therefore stratifying units such that $e(\boldsymbol{X}_i) = e(\boldsymbol{X}_j)$ for all i and j in the same stratum achieves the

[1]For discussions of relaxing this assumption see, Rosenbaum [9], Hudgens and Halloran [10], Bowers et al. [11], Aronow and Samii [12], and Athey et al. [13].

same purpose as stratifying on $\boldsymbol{X}$, suggesting the use of a *propensity score distance*,

$$d(\boldsymbol{X}_i, \boldsymbol{X}_j) = |\log(e(\boldsymbol{X}_i)/(1 - e(\boldsymbol{X}_i))) - \log(e(\boldsymbol{X}_j)/(1 - e(\boldsymbol{X}_j)))|. \quad (5.3)$$

With propensity score matching, units sharing a stratum may have quite different observed characteristics, even if the probability of treatment is comparable. In practice, the true propensity score will not be known and must be estimated, typically using logistic regression, but errors induced by this estimation have not been found to be problematic [18, 19]. Of course, it may still be difficult to find units with exactly the same (estimated) propensity scores. Practical propensity score matching generally allows pairs (i, j) with $e(\boldsymbol{X}_i) \neq e(\boldsymbol{X}_j)$ with the intention that $e(\boldsymbol{X})$-variation will be far greater between than within strata.

If matched sets are homogeneous in $\boldsymbol{X}$ or $e(\boldsymbol{X})$ and strong ignorability holds, treatment effect estimates will be unconfounded. It may, however, be the case that one or both of these conditions are not met: sets may have variation in $\boldsymbol{X}$ and $e(\boldsymbol{X})$ in a way that correlates with treatment status, and conditioning on $\boldsymbol{X}$ may not be sufficient to make the potential outcomes independent of the treatment assignment. The degree to which the study design is robust to violations of these assumptions is termed the *design sensitivity* [20]. Observational studies supporting inferences that are stable under modest deviations from homogeneity in treatment probabilities are said to be less sensitive. An important factor of design sensitivity rests on minimizing variation in the units under study [21], so even in the presence of unmeasured confounders and difficult to compare subjects, finding ways to arrange subjects such that Equation 5.1 is minimized remains an important goal.

5.2 Optimal Full Matching

In minimizing Equation (5.1) nothing precludes sets that include multiple treated units and multiple control units. In fact, this might seem most natural from the perspective of viewing the matched strata as blocks in a randomized trial in which there may be multiple treated and multiple control units. Interestingly, however, for any optimal match with a stratum that contains at least two treated and at least two control units, one could always further partition that stratum using the one treated and control pair with the minimum distance. For example, consider that a stratum contains treated units A and B and control units X, Y, and Z. Placing A and X into one stratum and B, Y, and Z into a second stratum would have total distance no larger than the combined stratum, and in the cases where the distances from A to Y and Z and B to X are non-zero, the two strata would have strictly smaller total distance. Applying this logic to the entire match, the result remains a valid match with distance no greater than the original match and with sets composed entirely of pairs, one treated unit and several controls, or one control unit and several treatment. Matches of this form are known as *full matches*. If $\mathcal{M}'$ is an *optimal match*, in the sense of minimizing Equation (5.1), there must be a full match $\mathcal{M}$ with distance no greater than $\mathcal{M}'$ [22]. There is, therefore, no loss to considering exclusively full matches when seeking to minimize Equation (5.1) as the *optimal full match* would have distance no greater than the optimal stratification allowing multiple treated and control units per stratum.

While any pair match is a full match, a minimum distance pair match can easily be sub-optimal among full matches, as a simple toy example demonstrates. For modest sized problems, it can be useful to enumerate all possible treatment-control distances in the form of a n_1 by n_0 matrix.[2] Table 5.1 shows a toy matching problem with three treated units (A, B, C) and three control units (X, Y, Z). For this matching problem, a valid full match could create either three pairs or two triples. In creating pairs it may be tempting to match B to Y as the distance between these units is zero.

[2]For large problems, this matrix can represent a substantial computation and memory requirement. Yu et al. [23] discuss techniques for matching without explicitly computing all distances when $d(\boldsymbol{X}_i, \boldsymbol{X}_j)$ is suitably well behaved.

TABLE 5.1
A distance matrix for three treated units (A, B, C) and three control units (X, Y, Z).

	X	Y	Z
A	8	1	3
B	5	0	1
C	9	2	9

Doing so, however, forces A and C to be matched either to X and Z or to Z and X, respectively, with corresponding distances 8 and 9 or 3 and 9, respectively. The resulting total distance (5.1) is no less than 12. However, foregoing the choice to begin "greedily" – joining the most favorable pair without regard to its effect on later pairings – allows for matching C to Y, A to Z and B to X, for a total distance of 10. While an improvement over the greedy match, this optimal pair match is less desirable than the match that places A, C, and Y in one group and B, X, and Z in another. The total distance of this match is only 9. With two treated matched to one control in one set and two controls matched to one treated unit in the second set, the best solution in Table 5.1 represents an optimal full match.

The earliest matching literature employed greedy or nearest neighbor matching as a necessity [16, e.g.,]. Rosenbaum [24] introduced a tractable algorithm for finding guaranteed optimal solutions based on a *minimum cost network flow* (MNCF) that provided either a pair match, necessarily discarding some units, or a match with one treated per set and multiple controls. Hansen and Klopfer [25] updated the algorithm to allow for both variable matching ratios for both treated and control units, that is strata composed of one treated unit matched to several controls units or one control unit matched to several treated units, and restrictions on the magnitudes of these ratios. This has been found to decrease the bias of the match [26]. Kallus [27] considers matching as a method of creating linear estimators for treatment effects and finds that optimal matching, as compared to greedy approaches, is optimal in the sense of minimizing mean squared error under a particular model.

Optimal full matching has been shown to work quite well in practice. Simulation studies on survival, binary, and continuous outcomes find that optimal full matching performs better than competing techniques with respect to mean squared error compared to other methods [28–30]. Chen and Kaplan [31] report similar findings in the context of a Bayesian framework. Augurzky and Kluve [32] compares optimal full matching to greedy full matching and finds that optimal matching is generally preferred. Hill et al. [33] investigate several matching techniques and found that full matching performed particularly well in achieving balance in observed covariates.

5.3 Algorithms for Optimal Full Matching

While an optimal match does not need to be a full match, an important attraction of full matching lies in the fast algorithms available to find optimal full matches by constructing a minimum cost network flow, a well-studied problem in the fields of computer science and operations research. In these problems, a directed network is composed of a vertex set V and edge set E. Each vertex in the network may contribute to the total supply of flow in the network (a source), may subtract from the total supply (a sink), or neither (a balanced vertex). Each edge has a minimum and maximum amount of flow it is allowed carry, with an associated cost per unit of flow carried. A *feasible flow* is a subset of the edges such that all supply and demand is balanced and no edge carries more than its capacity.

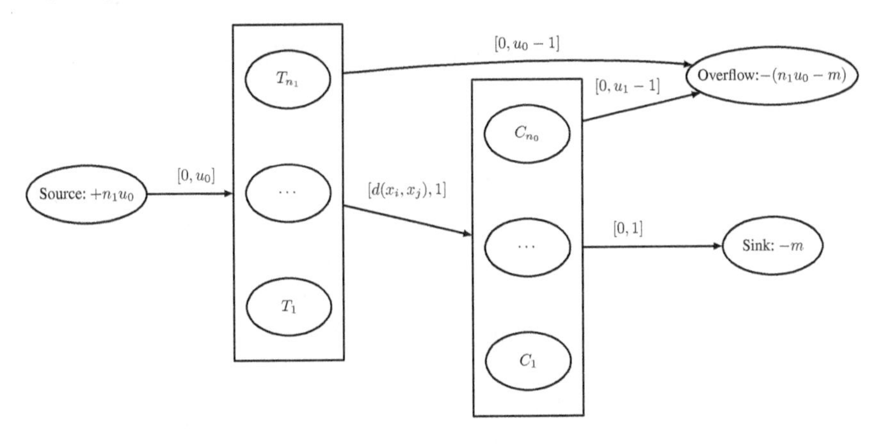

FIGURE 5.1
Edges labeled by [cost, capacity]. Edges to/from boxes represent one edge from each node in the box and from each node in the box.

A *minimum cost flow* is a feasible flow where the total cost to move all units of supply is minimized, though this solution need not be unique. Throughout, we only consider integer capacity and costs.

The problem of finding a minimum distance full match can be shown to be equivalent to the problem of finding a minimal cost flow in a network [24, 25]. Nodes in the network can be either classified as representing the units in the study (treated and control nodes) or nodes required to make the solution a valid matching problem (bookkeeping nodes). The edges can be similarly classified as having direct reference to the distance matrix of the matching problem and edges required to achieve a feasible flow. Each edge has a capacity, the maximum amount of flow allowed to pass over the edge, and a cost per unit of flow carried.

Figure 5.1 provides an overview of the network. Broadly, we see treatment nodes, control nodes, and three bookkeeping nodes. The method of Rosenbaum [24] only included the source and sink nodes and was restricted to solutions of pair matches or single treated units matched to fixed numbers of control units. Hansen and Klopfer [25] introduced the overflow node, which permits varying numbers of treated and control units, subject to full match constraints, and discarding a specified number of treated or control units. In this version of the algorithm, in addition to the distances $d(\boldsymbol{X}_i, \boldsymbol{X}_j)$, the researcher must supply the number of control units to match m, the maximum number of control nodes that can share a single treated unit u_0, and the maximum number of treated units that can share a control unit u_1. There are some natural constraints on these values to maintain feasibility of the solution, though we omit a detailed discussion.

Flow starts at the source node. As each treated unit can match up to u_0 control units, the flow contains n_1u_0 total flow, which gets divided across the n_1 treated units equally. All edges connecting treated units to control units have capacity one and cost equal to the distance $d(\boldsymbol{X}_i, \boldsymbol{X}_j)$, assumed positive for each i and j. (This can be ensured by adding a small increment to each distance.) Edges that move flow from treated units to control units indicate units matched in the optimal solution. Any flow not used to match treated units to control units goes to the overflow node. By allowing no more than $u_0 - 1$ flow over the edges to the overflow, all treated units must be matched to at least one control unit, but no more than u_0 control units.

To ensure that m control units are matched, each control unit is connected by a capacity-1 edge to the sink node which absorbs m total flow. For control units matched to more than one treated unit, there will be additional flow that cannot be sent to sink, so it must go to the overflow node. By limiting the amount of such over flow to $u_1 - 1$, control nodes can never be matched to more than u_1 treated units.

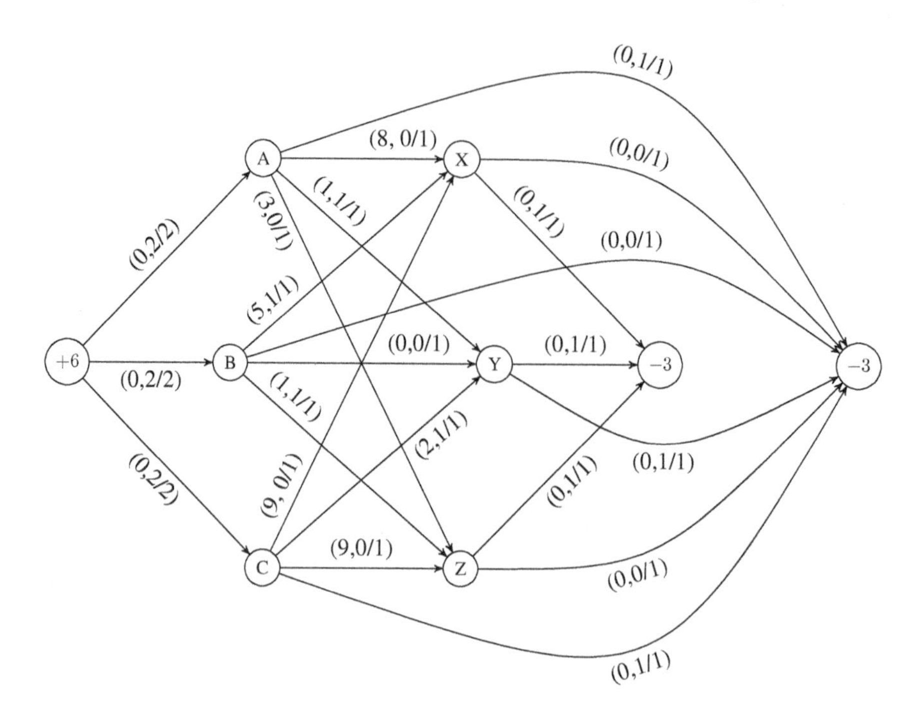

FIGURE 5.2
Network flow diagram for a toy matching problem described in Table 5.1. Edges are labeled with (cost, flow/capacity) for the optimal full match $\{\{A, C, Y\}, \{B, X, Z\}\}$.

All of the edges connecting the bookkeeping nodes to the treated and control nodes having a cost of zero, the only edges with a positive cost are the edges between treated and control nodes. Therefore, the total cost of the flow is equal to the total cost of pairing each treated and control unit joined by an edge carrying positive flow. It can be shown that for any minimum cost flow, the arrangment of pairs that is induced in this way must be a full matching; Hansen and Klopfer [25] provide details. Pair matches and matches that restrict sets to have no more than a given number of treated or control units can be enforced using careful selection of m, u_1, and u_0, and solutions will be optimal within these restrictions.

To be more concrete, Figure 5.2 shows the network corresponding to the matching problem of Table 5.1. Each edge shows its cost and the amount of flow, out of the maximum, that flows along the edge. Overall, six units of flow enter the network and must pass through the nodes for the treated units A, B, and C, though to ensure all treated units are matched, each treated node receives two units of flow. The between the treated and control nodes have cost equal to the distance between the units and a maximum capacity of 1. With a maximum capacity of one, precisely three of these edges will carry one unit of flow each, which then be sent to the sink. Since A and C share control node Y, both A and C send one unit of flow to the overflow node. Likewise, Y must also send one unit of flow to the overflow.

The network based algorithms have several practical implications that can be useful for practitioners to understand. First, as flow can only travel in one direction along an edge, there is a subtle asymmetry between units designated as treatment and units designated as control. Software interfaces may require stating restrictions with respect to one group, typically the "treatment" group, even if they apply to both. For methods that allow variable matching ratios of treated to control units [26], the maximum allowed number of shared units for one group is the minimal number of allowed controls for the other group. In some settings it may be useful to swap the treated and control labels to make these calculations easier to parse.

Second, many of the algorithms used to find minimum cost network flows require *integer costs* [34]. If distances between nodes are provided in floating point values, they must be discretized, inducing some error. Specified tolerance levels can be used control how close the integer approximation is to the floating point distance, but smaller tolerances will negatively impact the run time of the algorithm.

Third, it is also valuable to understand how the *algorithmic complexity* of these algorithms scales with larger input sizes. For each new treated unit (control unit) added to a study, there are n_0 (n_1) new edges to connect the new node to the opposite assignment nodes, along with some small number of additional edges to properly connect to sinks and sources. As the algorithm must consider these edges, it should be intuitive that number of steps to achieve a MCNF grows much faster than linearly with the number of units in the study. In fact, the earliest solutions for MCNFs have an algorithm complexity of $O(|V|^3)$, and newer approaches frequently do no better than $O(|V||E|)$. In the discussions so far, $|E| \approx |V|^2$, so again the cubic rate of growth is typical. Researchers are cautioned that the difference between a modest sized problem, for which an optimal match can be found in a reasonable amount of time, and a large problem, which is quite difficult to solve, can be smaller than one might initially anticipate. We discuss some techniques for reducing the size of the matching problem in the next section. Additional discussion of the particular algorithms for solving MCNFs and the associated algorithmic complexity can be found in the books by Ahuja et al. [35], Bertsekas [34], Korte and Vygen [36], and Williamson [37], as well as the recent survey paper by Sifaleras [38]. For empirical comparisons on a variety of networks, see Kovács [39].

5.4 Controlling Local and Global Features of Matches

One reason for the success of network flow-based solutions is that they permit researchers to blend both local information about particular treated-control pairs, but can also be easily extended to include additional global constraints on the overall matched structure.

With respect to local criteria, one advantage of the network approach is that "distances" between treated and control observations need not follow any particular geometry. Two treated units may be "close" to a particular control unit, but arbitrarily far from others. The suitability of matching units i and j only depends on local information contained in the (i, j) entry of the matrix. As no constraints are placed on any of the distances, the researcher is free in include additional criteria in judging the suitability of matching any particular units. Penalty terms can be included to discourage the matching algorithm from selecting units with particular qualities. Matches can be expressly forbidden by making the distance between the units infinite. As a side benefit, eliminating possible matches improves computation time by decreasing the number of edges in the network graph. Two techniques used to forbid particular matches are *exact matching* and *calipers*.

Exact matching limits matches to units that have precisely the same value on a categorical variable, for example allowing matches of clinics in the same state, but forbidding matched sets including clinics from two or more states. Exact matching effectively breaks the overall problem into a series of independent matching problems, which can be solved in parallel. Exact matching can be particularly useful when researchers are interested in *conditional treatment effects* that restrict inference within a subset of the population. In unrestricted matching, matched sets may involve treated units with $\boldsymbol{X} = \boldsymbol{x}$, but no control units with $\boldsymbol{X} = \boldsymbol{x}$ (or vice versa). By ensuring that all units within each set have the same value of $\boldsymbol{X}$, inference for treatment effects can proceed using only the strata with $\boldsymbol{X} = \boldsymbol{x}$.

Calipers loosen the requirement that units match exactly on given covariates by permitting matches for which some comparison of the units does not exceed a given threshold. This may be implemented using the distance function and a fixed threshold for all units, $d(\boldsymbol{X}_i, \boldsymbol{X}_j) < c$, or using other measures and unit specific thresholds. Calipers can often improve the quality of the match while simultaneously improving computational efficiency. In the example of Table 5.1, neither the greedy nor the optimal solution considered matching A and C. Little would be lost if distances as large as 8 were eliminated from consideration. Some attention must be given to ensuring that selected caliper values do not eliminate some units from consideration entirely (or that doing so would be desirable to the researcher). For some well behaved distance functions, finding the minimal feasible caliper can be quite efficient [23, 40–42]. Calipers need not restrict matches based only on $d(\boldsymbol{X}_i, \boldsymbol{X}_j)$. A common technique blends Mahalanobis distance and propensity score matching by matching units primarily on the propensity score, but forbidding matches for which Equation (5.2) exceeds a given value [43]. For selecting caliper width, for propensity scores a width of 1/5 of a standard deviation on the scale of the linear predictors has been found to work well in simulation studies [44].

The network approach also makes additional global constraints relatively straight forward. Examples of extending the networks structure include constraints on the order statistics of the within matched set distances [45] and requiring certain covariates to have precisely the same marginal distributions in the treated and control groups [46–48]. For clustered studies, the multilevel structure can be accommodated by matching both clusters and units within clusters [49].

To illustrate the use of both local and global restrictions, we consider an example originally presented in Fredrickson et al. [42]. This analysis recreated a propensity score analysis of vascular closure devices (VCD) on patients undergoing percutaneous coronary intervention compared with patients with whom VCD was not used [50]. The data set included approximately 31,000 subjects in the VCD (treatment) group and 54,000 potential controls. Even on modern computing hardware, this problem was sufficiently large as to be difficult to solve in a reasonable amount of time.

The original analysis of Gurm et al. [50] included three propensity score models and 192 exact matching categories. Even with this structure to remove potential matches across exact match categories, the problem remained difficult due to its size. Additionally, treated and control units were quite different on estimated propensity scores, with a particular outlying treated observation having no control units within 17 standard deviations on the logit scale. Any caliper that applied to all treated units would either eliminate the outlying treatment member or not reduce the size of the matching problem. Using the flexibility of the full matching approach, we did not place a caliper on the 52 (out of 31,000) most extreme treated units. Using an algorithm to select the largest feasible caliper for the remaining units, details given in Fredrickson et al. [42], for the full matching problem resulted in a decrease of 99% in the number of arcs in the underlying network flow problem, making the solution relatively quick to compute.

Additionally, our solution enforced global constraints on the maximum size sets. Initial matches produced a significant number of sets with more than five members of a treatment level. Restricting every match to be a pair match would be infeasible with the selected caliper constraints. We iteratively increased the maximum number of units per set until a feasible full match could be constructed, resulting in at most one treated to three controls and vice-versa. Compared to the best possible pair matching, the full matched sets had an average within set distance that was smaller by a factor of ten, illustrating the gap in optimality between pair matching and full matching.

5.5 Inference for Treatment Effects After Matching

By creating discrete strata, optimal full matching seeks to follow the advice of Cochran [15] to design an observational study in the mold of a randomized controlled trial. After selecting a match $\mathcal{M}$, index the units by matched set $b = 1, \ldots, s$ and write $n_b = |\mathcal{S}_b|$. The event $\mathcal{F}_{\mathcal{M}}$ describes an appropriate conditioning scheme to facilitate inference:

$$\mathcal{F}_{\mathcal{M}} = \{\boldsymbol{X}_{bi} = \boldsymbol{x}_{bi}; \mathcal{M}; \sum_{i=1}^{n_b} W_{bi} = n_{b1}, b = 1, \ldots, s; n_{b0} = n_b - n_{b1}, b = 1, \ldots, s\}.$$

Conditioning on $\mathcal{F}_{\mathcal{M}}$ leads to precisely the same structure as a blocked RCT, opening the possibility of drawing on the strength of randomization based analysis, either estimating average treatment effects [5, 51, 52] or testing sharp null hypothesis using permutation techniques [53–56], permuting treatment assignments uniformly within strata.

For estimating the average treatment effect $\Delta = n^{-1} \sum_{i=1}^{n} Y_1 - Y_0$, if all strata contain units with either identical covariates or identical propensity scores, a stratified estimator is unbiased for the average treatment effect:

$$\hat{\Delta}_{\mathcal{M}} \mid \mathcal{F}_{\mathcal{M}} = \sum_{b=1}^{s} \frac{n_b}{n} \sum_{i=1}^{n_b} \frac{W_{bi} Y_i}{n_{b1}} - \frac{(1 - W_{bi}) Y_i)}{n_{b0}}. \tag{5.4}$$

Alternatively, researchers can use a linear regression of the outcome on the treatment assignment and matched set fixed effects. While the coefficient for the treatment assignment could be understood as an estimated treatment effect, in general it does not precisely coincide with $\hat{\Delta}_{\mathcal{M}}$; it combines treatment effect estimates for strata b using weights proportional to $n_{b1} n_{b0} / n_b$, not n_b [57]. Some authors suggest including interaction terms between the treatment indicator and the stratum indicators [58,59]. Covariate adjustment can also be implemented by interacting covariates with the treatment assignment. As within set units typically are more similar on covariates, matching plays a role in increasing the precision of estimates of Δ, though simulation studies find that additional covariance adjustment after matching can be quite beneficial [60]. Inference can proceed using sandwich estimators for the variance, clustered at the level of strata [61].

In matching scenarios that discard units, the definition of the set of units for which the average treatment effect is defined changes from all units to only those units retained in the match. In many cases the researcher will only be discarding units from one of the two treatment levels, as when a very large pool of controls exists to compare to a smaller treatment group. Recognizing that the entire treatment group will be retained, it may be sensible to select the conditional ATE for the treatment group only, the *average treatment effect on the treated* (ATT),

$$\Delta_{W=1} = \mathrm{E}\,(Y_1 - Y_0 \mid W = 1) \tag{5.5}$$

In this situation, weights $W_{bi} = n_{b1} / (n_{b1} W_{bi} + n_{b0}(1 - W_{bi}))$ estimate $\Delta_{W=1}$ in a regression of the outcome on the treatment assignment vector. (This is "weighting by the odds" [62], because control group members receive case weights of n_{b1}/n_{b0}, the odds of a random selection from the matched set's belonging to the treatment group.)

While the estimator of Equation (5.4) and related variations for the ATT are unbiased when $(Y_1, Y_0) \perp\!\!\!\perp W \mid \mathcal{F}_{\mathcal{M}}$, the variance of the estimator can be quite poor, particularly when the matched set sizes vary greatly [63]. Returning to the inherent weighting of the linear regression estimator, while not necessarily unbiased for average treatment effects, greatly improves the variance of the estimator and frequently contributes minimal bias, as demonstrated in the simulation study below.

Both estimators' variance is improved by full matching with structural restrictions, that is upper and lower limits on matched sets' ratios of treatment group members to controls. Using the linear regression estimator after full matching on the propensity score, Hansen [64] explored various restrictions on this ratio, finding that allowing ratios between half and twice the ratio that obtained prior to matching preserved essential benefits achieved by unrestricted full matching, while furnishing standard errors some 17% smaller. Analyzing a distinct example, Stuart and Green [65] also report good performance from full matching with treatment to control ratios restricted to fall between half and twice the ratios that obtained prior to matching (In either instance, full matching was combined with exact matching, with ratios of treatment to control units varying across levels of exact matching variables; by "ratios that obtained prior to exact matching" we mean the ratios of treatment to control units by exact matching category.). Yoon [66] explored setting upper and lower limits on the number of matched controls per treatment in terms of the "entire number," a transformation of the estimated propensity score; see also Pimentel et al. [67]. Fredrickson et al. [42] suggested structural restrictions allowing a sufficiently narrow range of matching ratios that it becomes feasible for the statistician to present a compact table giving the numbers of each matched configuration $\ell : 1, \ldots, 1 : u$ that were ultimately created, along with the corresponding weights assigned to strata of each of those types by the selected estimator (For matching configurations $(m_1, m_0) = \ell : 1, \ldots, 1 : u$, the average treatment effect estimator (5.4) has stratum weights proportional to $m_1 + m_0$, whereas the ATT estimator's are proportional to m_1 and the linear regression estimator's are proportional to $m_1 m_0/(m_1 + m_0)$.). Limiting the number of entries to such a table can reduce estimates' standard errors even as it enhances their interpretability.

5.6 Simulation Study

In this section we illustrate the benefits of full matching as compared with one-to-many and pair matching. In these simulations we consider a simple data generation process. Let X be a categorical covariate with three levels, a, b, and c. Conditional X, we draw the potential outcome to control from one of three Normal distributions, all with unit variance but with different means:

$$Y_0 \mid X = a \sim N(0,1), Y_0 \mid X = b \sim N(2,1), Y_0 \mid X = c \sim N(4,1).$$

In our simulations we fix $n = 100$, with 50 units for whom $X = a$, and 25 units for $X = b$ and $X = c$ each.

After drawing Y_0, conditional on X, we draw a noise variable $\epsilon = N(0,1)$ marginally, but constrain the overall collection of ϵ to have precisely a mean of one and a standard deviation of 1. We then construct $Y_1 = Y_0 + \epsilon + \tau$, where τ is the true treatment effect, held at 7 in these simulations.

To confound the treatment assignment W with the covariate X, we vary the conditional probability of treatment across the levels of X:

$$P(W = 1 \mid X = a) = 0.2, P(W = 1 \mid X = b) = 0.5, P(W = 1 \mid X = c) = 0.8.$$

We restrict treatment assignments such that each level always has at least one treated and at least one control unit in each simulation.

For 10,000 replications, we generate an observed $Y = WY_1 + (1 - W)Y_0$ for each unit. For each sample, we consider six different matching strategies.

1. No matching, effectively treating the sample as a single stratum.
2. Full matching, with exact matching constraints that forbid matches across the levels of X. All units will be matched, though set sizes may involve many control units for a treatment unit or vice versa.
3. Pair matching, with exact matching constraints. By necessity, within levels ℓ with unequal numbers of treatment and control units, some units are discarded. (After matching, such levels will have will have $\min(n_{\ell 1}, n_{\ell 0}) > 0$ pairs and $\max(n_{\ell 1}, n_{\ell 0}) - \min(n_{\ell 1}, n_{\ell 0}) > 0$ unmatched individuals.)
4. One treated to multiple control units, with exact matching constraints. All units will be matched when control units outnumber treated units, but for levels of X with more treated than control units, some portion of the treated units will be discarded.

TABLE 5.2
Results from a simulation of a single categorical covariate confounded with treatment and the potential outcomes. Estimates for bias and mean squared error (MSE) are based on 10,000 replications. 95% confidence intervals are included.

Match Type	Cross-Category Matching	Bias	MSE
None	No restriction	1.61 (1.60, 1.62)	2.71 (2.67, 2.75)
Pair	Forbidden	0.00 (−0.01, 0.01)	0.11 (0.10, 0.11)
One-Many	Forbidden	0.00 (−0.01, 0.01)	0.10 (0.10, 0.11)
Full	Forbidden	0.00 (−0.01, 0.01)	0.09 (0.09, 0.10)
Pair	Penalized	1.36 (1.34, 1.37)	2.01 (1.97, 2.04)
One-Many	Penalized	1.29 (1.27, 1.30)	1.83 (1.80, 1.87)
Full	Penalized	0.00 (−0.01, 0.01)	0.09 (0.09, 0.09)

5. Full matching, without the exact matching constraint but with a penalty for matching across the levels of X. As before, all units will be matched.
6. Pair matching, with a penalty for cross category matching. In this pair match, typically fewer units will be discarded as number of treated and control units will be close in the overall sample.
7. One to many matching, with the penalty. As with pair matching, more units will be matched, though some treated units may be discarded if treated units exceed control units in the entire sample.

After matching, we estimate treatment effects using a linear regression of the outcome on the treatment assignment, with fixed effects for matched sets. We compare the estimated value to the true average treatment effect. Table 5.2 provides the key results of the simulations, the bias and mean squared error (MSE) of each of the methods. Unsurprisingly, without addressing the confounding, the treatment effect estimates are significantly biased. Additionally, as the variance of potential outcomes within groups defined by X is smaller than the pooled variance, the difference of means estimate is both biased and suffers from a higher variance. With exact matching restrictions units with different values are never matched, so all three matches are effectively unbiased when estimating treatment effects and have much lower variance.

When the exact matching restriction is removed and replaced with a penalty for matching across levels of X, it is no longer the case that matches must be made between units with identical values of X. For full matching, however, since set sizes can be either many treated to one control or many controls to one treated, the minimum distance match continues to be one in which all sets have the same value of X. For pair matching and one to many matching, however, the requiremejts on the set sizes require that some matches are made between unit with different values of X. Consequently, treatment effect estimates based on pair matching and one-to-many matching become biased as demonstrated by Table 5.2.

5.7 Software

To implement the solution to the MNCF, a variety of packages exist. See Kovács [39] for a comparison of several open source and commercial implementations. The `optmatch` package [25] in `R` combines the RELAX-IV implementation from Bertsekas and Tseng [68] with a set of routines for creating distance matrices, including methods for Mahalanobis matching, propensity score matching based on logistic regression, exact matching, calipers, and combining distance matrices from different distance functions. Several of methods extension methods mentioned in this paper have implementations in R that rely in part on `optmatch`. the The `MatchIt` package [69] for `R` provides a simplified interface to `optmatch` with several useful features, including calculating appropriate weights for estimating the ATE and ATT.

5.8 Related Recent Developments

Optimal full matching is not the only technique available to minimize distances between treated and control units. In some instances faster algorithms can take advantage of particular structure of the distance measure to find the minimum distance match as defined by Equation 5.1. In the situation in which absolute distance is computed on a single scalar index such as the propensity score, finding the optimal full match simplifies to sorting the observations (within treatment condition) on x [70,71]. As sorting can be achieved in time approximately proportional to $n \log(n)$, this represents a substantial improvement over the MCNF solvers for the general case. Restrictions on the maximum set sizes induces a slight penalty, $O(n^2)$, but is still quite competitive [72]. Ruzankin [40] considers distances that are Lipschitz of a certain order for pair matching and one-to-many matching.

Other approaches add additional restrictions to allowed matches or other optimization goals. Zubizarreta [73] extends the objective function of Equation (5.1) to include an additional penalty term for other balancing statistics and provides solutions using a mixed integer programming approach. Zubizarreta and Keele [74] additionally extends the integer programming approach to match within and across clusters simultaneously, and Bennett et al. [75] demonstrates the feasibility of scaling the approach to large databases. Iacus et al. [76], Hainmueller [77], Tam Cho et al. [78], and Imai and Ratkovic [79] focus on minimizing observed imbalance between treatment and control groups directly as a global criterion rather than within set distances.

Approximate solutions often achieve results close to those of optimal full matching. Diamond and Sekhon [80] suggests using a genetic algorithm to find an approximate minimizer of $d(\mathcal{M})$. Sävje et al. [81] proposes a very fast approach by solving a non-optimal problem, though with bounded worst case performance compared to the optimal solution. Dieng et al. [82] uses with replacement matching to find control units with the nearly the same values of discrete covariates for each individual treated unit.

Optimal matching has been extended to cases with more than two treatment conditions. Lu et al. [83] discusses the case of non-bipartite matching, in which one treatment group is matched to more than one control group, and recent developments in this area can be found in Bennett et al. [75] and Nattino et al. [84].

Recent years have seen increasing development of in the theoretical behavior of matching-based estimators of treatment effects such as asymptotic behavior and optimality with respect to mean squared error [27,85–87].

5.9 Summary

For observational studies of treatment effects, the distributions of the potential outcomes for the observed treatment and control group, $Y_1 \mid W = 1$ and $Y_0 \mid W = 0$, may differ greatly from the marginal distributions, Y_1 and Y_0. Provided observable covariates $\boldsymbol{X}$ are sufficient to make $W \perp\!\!\!\perp (Y_1, Y_0) \mid \boldsymbol{X} = \boldsymbol{x}$ for all $\boldsymbol{x}$, units could be stratified to achieve valid hypothesis tests for and unbiased biased estimates of treatment effects. In practice, stratification can be difficult to implement, so close approximations are necessary in the form of matches that ensure at least one treated and control unit per stratum and strive to minimize differences in either observed characteristics or the probability of treatment assignment.

Optimal full matching provides solutions that have guaranteed minimum distance strata. For any minimum distance match, a full match exists with the same total distance, so nothing is lost considering only matches that have pairs, one treated to many control units, or one control and many treated units. Optimal full matches can be found using algorithms for minimum cost network flows. The flexibility of these approaches allows for both local control of strata through the use of distance matrices, but also global control of several properties of the match overall, such as restricting to pair matches or one-to-many matches. Software packages make these algorithms widely available and straightforward to use. After matching, inference for treatment effects can follow techniques for analyzing randomized controlled trials, either estimating treatment effects or testing hypothesis using permutation tests.

References

[1] Jasjeet S. Sekhon. Opiates for the matches: Matching methods for causal inference. *Annual Review of Political Science*, 12(1):487–508, 2009.

[2] Elizabeth A. Stuart. Matching Methods for Causal Inference: A Review and a Look Forward. *Statistical Science*, 25(1):1 – 21, 2010.

[3] Markus Gangl. Matching estimators for treatment effects. In Henning Best and Christof Wolf, editors, *The SAGE Handbook of Regression Analysis and Causal Inference*. Sage Publications, London, 2013.

[4] Paul R. Rosenbaum. Modern algorithms for matching in observational studies. *Annual Review of Statistics and Its Application*, 7(1):143–176, 2020.

[5] Guido W. Imbens and Donald B. Rubin. *Causal Inference for Statistics, Social, and Biomedical Sciences: An Introduction*. Cambridge University Press, New York, NY, 2015.

[6] Paul R. Rosenbaum. *Observational Studies*. Springer, 2nd edition, 2002.

[7] Paul R. Rosenbaum. *Design of Observational Studies*. Springer, New York, second edition, 2020.

[8] Donald B. Rubin. Randomization analysis of experimental data: The Fisher randomization test comment. *Journal of the American Statistical Association*, 75(371):591–593, 1980.

[9] Paul R. Rosenbaum. Interference between units in randomized experiments. *Journal of the American Statistical Association*, 102(477):191–200, 2007.

[10] M. G. Hudgens and M. E. Halloran. Toward causal inference with interference. *Journal of the American Statistical Association*, 103(482):832–842, 2008.

[11] Jake Bowers, Mark M. Fredrickson, and Costas Panagopoulos. Reasoning about interference between units: A general framework. *Political Analysis*, 21(1):97–124, 2013.

[12] Peter M. Aronow and Cyrus Samii. Estimating average causal effects under general interference, with application to a social network experiment. *Annals of Applied Statistics*, 11(4):1912–1947, 12 2017.

[13] Susan Athey, Dean Eckles, and Guido W. Imbens. Exact p-values for network interference. *Journal of the American Statistical Association*, 113(521):230–240, 2018.

[14] Paul R. Rosenbaum and Donald B. Rubin. The central role of the propensity score in observational studies for causal effects. *Biometrika*, 70(1):41–55, 1983.

[15] W. G. Cochran. The planning of observational studies of human populations. *Journal of the Royal Statistical Society. Series A (General)*, 128(2):234–266, 1965.

[16] Donald B. Rubin. Bias reduction using mahalanobis-metric matching. *Biometrics*, 36(2):293–298, 1980.

[17] Donald B. Rubin. Multivariate matching methods that are equal percent bias reducing, ii: Maximums on bias reduction for fixed sample sizes. *Biometrics*, 32(1):121–132, 1976.

[18] Donald B. Rubin and Neal Thomas. Characterizing the effect of matching using linear propensity score methods with normal distributions. *Biometrika*, 79(4):797–809, 12 1992.

[19] Donald B. Rubin and Neal Thomas. Matching using estimated propensity scores: Relating theory to practice. *Biometrics*, 52(1):249–264, 1996.

[20] Paul R. Rosenbaum. Design sensitivity in observational studies. *Biometrika*, 91(1):153–164, 2004.

[21] Paul R. Rosenbaum. Heterogeneity and causality. *The American Statistician*, 59(2):147–152, 2005.

[22] Paul R. Rosenbaum. A characterization of optimal designs for observational studies. *Journal of the Royal Statistical Society. Series B (Methodological)*, 53(3): 597–610, 1991.

[23] Ruoqi Yu, Jeffrey H. Silber, and Paul R. Rosenbaum. Matching Methods for Observational Studies Derived from Large Administrative Databases. *Statistical Science*, 35(3):338 – 355, 2020.

[24] Paul R. Rosenbaum. Optimal matching for observational studies. *Journal of the American Statistical Association*, 84(408): 1024–1032, 1989.

[25] Ben B. Hansen and Stephanie Olsen Klopfer. Optimal full matching and related designs via network flows. *Journal of Computational and Graphical Statistics*, 15(3), 2006.

[26] Kewei Ming and Paul R. Rosenbaum. Substantial gains in bias reduction from matching with a variable number of controls. *Biometrics*, 56(1):118–124, 2000.

[27] Nathan Kallus. Generalized optimal matching methods for causal inference. *Journal of Machine Learning Research*, 21(62):1–54, 2020.

[28] Peter C. Austin and Elizabeth A. Stuart. Optimal full matching for survival outcomes: A method that merits more widespread use. *Statistics in Medicine*, 34(30):3949–3967, 2015.

[29] Peter C. Austin and Elizabeth A. Stuart. Estimating the effect of treatment on binary outcomes using full matching on the propensity score. *Statistical Methods in Medical Research*, 26(6):2505–2525, 2017.

[30] Peter C. Austin and Elizabeth A. Stuart. The effect of a constraint on the maximum number of controls matched to each treated subject on the performance of full matching on the propensity score when estimating risk differences. *Statistics in Medicine*, 40(1):101–118, 2021.

[31] Jianshen Chen and David Kaplan. Covariate balance in Bayesian propensity score approaches for observational studies. *Journal of Research on Educational Effectiveness*, 8(2):280–302, 2015.

[32] Boris Augurzky and Jochen Kluve. Assessing the performance of matching algorithms when selection into treatment is strong. *Journal of Applied Econometrics*, 22(3):533–557, 2007.

[33] Jennifer L. Hill, Jane Waldfogel, Jeanne Brooks-Gunn, and Wen-Jui Han. Maternal employment and child development: A fresh look using newer methods. *Developmental Psychology*, 41(6):833, 2005.

[34] Dimitri Bertsekas. *Network Optimization: Continuous and Discrete Models*. Athena Scientific, Belmont, MA, 1998.

[35] Ravindra K. Ahuja, Thomas L. Magnanti, and James B. Orlin. *Network Flows: Theory, Algorithms, and Applications*. Prentice Hall, Englewood Cliffs, New Jersey, 1993.

[36] Bernhard Korte and Jens Vygen. *Combinatorial Optimization: Theory and Algorithms*. Springer, Berlin, 2012.

[37] David P Williamson. *Network Flow Algorithms*. Cambridge University Press, New York, NY, 2019.

[38] Angelo Sifaleras. Minimum cost network flows: Problems, algorithms, and software. *Yugoslav Journal of Operations Research* 23:1, 2016.

[39] Péter Kovács. Minimum-cost flow algorithms: An experimental evaluation. *Optimization Methods and Software*, 30(1):94–127, 2015.

[40] Pavel S. Ruzankin. A fast algorithm for maximal propensity score matching. *Methodology and Computing in Applied Probability*, May 2019.

[41] Fredrik Sävje. Comment: Matching Methods for Observational Studies Derived from Large Administrative Databases. *Statistical Science*, 35(3):356 – 360, 2020.

[42] Mark M. Fredrickson, Josh Errickson, and Ben B. Hansen. Comment: Matching Methods for Observational Studies Derived from Large Administrative Databases. *Statistical Science*, 35(3):361 – 366, 2020.

[43] Paul R. Rosenbaum and Donald B. Rubin. Constructing a control group using multivariate matched sampling methods that incorporate the propensity score. *The American Statistician*, 39(1):33–38, 1985.

[44] Peter C. Austin. Optimal caliper widths for propensity-score matching when estimating differences in means and differences in proportions in observational studies. *Pharmaceutical Statistics*, 10(2):150–161, 2011.

[45] Paul R. Rosenbaum. Imposing minimax and quantile constraints on optimal matching in observational studies. *Journal of Computational and Graphical Statistics*, 26(1):66–78, 2017.

[46] Paul R Rosenbaum, Richard N Ross, and Jeffrey H Silber. Minimum distance matched sampling with fine balance in an observational study of treatment for ovarian cancer. *Journal of the American Statistical Association*, 102(477):75–83, 2007.

[47] Dan Yang, Dylan S. Small, Jeffrey H. Silber, and Paul R. Rosenbaum. Optimal matching with minimal deviation from fine balance in a study of obesity and surgical outcomes. *Biometrics*, 68(2):628–636, 2012.

[48] Samuel D. Pimentel, Rachel R. Kelz, Jeffrey H. Silber, and Paul R. Rosenbaum. Large, sparse optimal matching with refined covariate balance in an observational study of the health outcomes produced by new surgeons. *Journal of the American Statistical Association*, 110(510):515–527, 2015. PMID: 26273117.

[49] Samuel D. Pimentel, Lindsay C. Page, Matthew Lenard, and Luke Keele. Optimal multilevel matching using network flows: An application to a summer reading intervention. *Annals of Applied Statistics*, 12(3):1479–1505, 09 2018. doi: 10.1214/17-AOAS1118. URL `https://doi.org/10.1214/17-AOAS1118`.

[50] Hitinder S. Gurm, Carrie Hosman, David Share, Mauro Moscucci, and Ben B. Hansen. Comparative safety of vascular closure devices and manual closure among patients having percutaneous coronary intervention. *Annals of Internal Medicine*, 159(10):660–666, 2013.

[51] Jerzy Splawa Neyman. On the application of probability theory to agricultural experiments. Essay on principles. Section 9. *Statistical Science*, 5(4):465–480, 1923. (Originally in Roczniki Nauk Tom X (1923) 1–51 (Annals of Agricultural Sciences). Translated from original Polish by Dambrowska and Speed.).

[52] Guido W. Imbens. Nonparametric estimation of average treatment effects under exogeneity: A review. *Review of Economics and Statistics*, 86(1):4–29, Feb 2004.

[53] Ronald A. Fisher. *The Design of Experiments*. Oliver and Boyd, Edinburgh, 1935.

[54] J. S. Maritz. *Distribution-Free Statistical Methods*. Chapman and Hall, London, 1981.

[55] Paul R. Rosenbaum. Conditional permutation tests and the propensity score in observational studies. *Journal of the American Statistical Association*, 79(387): 565–574, 1984.

[56] Phillip I. Good. *Permutation, Parametric and Bootstrap Tests of Hypotheses*. Springer, New York, third edition, 2005.

[57] G. Kalton. Standardization: A technique to control for extraneous variables. *Journal of the Royal Statistical Society. Series C (Applied Statistics)*, 17(2):118–136, 1968.

[58] Winston Lin. Agnostic notes on regression adjustments to experimental data: Reexamining freedman's critique. *The Annals of Applied Statistics*, 7(1):295–318, 2013.

[59] Peter Z. Schochet. Is regression adjustment supported by the neyman model for causal inference? *Journal of Statistical Planning and Inference*, 140(1):246–259, 2010.

[60] K. Ellicott Colson, Kara E. Rudolph, Scott C. Zimmerman, Dana E. Goin, Elizabeth A. Stuart, Mark van der Laan, and Jennifer Ahern. Optimizing matching and analysis combinations for estimating causal effects. *Scientific Reports*, 6(1):23222, 2016.

[61] Cyrus Samii and Peter M. Aronow. On equivalencies between design-based and regression-based variance estimators for randomized experiments. *Statistics and Probability Letters*, 82(2):365 – 370, 2012.

[62] Valerie S. Harder, Elizabeth A. Stuart, and James C. Anthony. Propensity score techniques and the assessment of measured covariate balance to test causal associations in psychological research. *Psychological Methods*, 15(3):234, 2010.

[63] Ben B. Hansen. Propensity score matching to extract latent experiments from nonexperimental data: A case study. In Neil Dorans and Sandip Sinharay, editors, *Looking Back: Proceedings of a Conference in Honor of Paul W. Holland*, chapter 9, pages 149–181. Springer, 2011.

[64] Ben B Hansen. Full matching in an observational study of coaching for the sat. *Journal of the American Statistical Association*, 99(467):609–618, 2004.

[65] Elizabeth A Stuart and Kerry M Green. Using full matching to estimate causal effects in nonexperimental studies: Examining the relationship between adolescent marijuana use and adult outcomes. *Developmental Psychology*, 44(2):395, 2008.

[66] Frank B. Yoon. *New methods for the design and analysis of observational studies*. PhD thesis, University of Pennsylvania, 2008.

[67] Samuel D. Pimentel, Frank Yoon, and Luke Keele. Variable-ratio matching with fine balance in a study of the peer health exchange. *Statistics in Medicine*, 34(30):4070–4082, 2015.

[68] Dimitri P. Bertsekas and Paul Tseng. Relax-iv: A faster version of the relax code for solving minimum cost flow problems. Technical Report LIDS-P-2276, Massachusetts Institute of Technology, November 1994.

[69] Daniel E. Ho, Kosuke Imai, Gary King, and Elizabeth A. Stuart. MatchIt: Nonparametric preprocessing for parametric causal inference. *Journal of Statistical Software*, 42(8):1–28, 2011.

[70] Justin Colannino, Mirela Damian, Ferran Hurtado, John Iacono, Henk Meijer, Suneeta Ramaswami, and Godfried Toussaint. An $O(n \log n)$-time algorithm for the restriction scaffold assignment problem. *Journal of Computational Biology*, 13(4): 979-989, 2006.

[71] Justin Colannino, Mirela Damian, Ferran Hurtado, Stefan Langerman, Henk Meijer, Suneeta Ramaswami, Diane Souvaine, and Godfried Toussaint. Efficient many-to-many point matching in one dimension. *Graphs and Combinatorics*, 23(1):169–178, 2007.

[72] Fatemeh Rajabi-Alni and Alireza Bagheri. An $O(n^2)$ algorithm for the limited-capacity many-to-many point matching in one dimension. *Algorithmica*, 76(2):381–400, Oct 2016.

[73] José R. Zubizarreta. Using mixed integer programming for matching in an observational study of kidney failure after surgery. *Journal of the American Statistical Association*, 107(500):1360–1371, 2012.

[74] José R. Zubizarreta and Luke Keele. Optimal multilevel matching in clustered observational studies: A case study of the effectiveness of private schools under a large-scale voucher system. *Journal of the American Statistical Association*, 112(518):547–560, 2017.

[75] Magdalena Bennett, Juan Pablo Vielma, and José R. Zubizarreta. Building representative matched samples with multi-valued treatments in large observational studies. *Journal of Computational and Graphical Statistics*, 0(0):1–29, 2020.

[76] Stefano M. Iacus, Gary King, and Giuseppe Porro. Causal inference without balance checking: Coarsened exact matching. *Political Analysis*, 1:1–24, 2012. Working paper.

[77] Jens Hainmueller. Entropy balancing for causal effects: A multivariate reweighting method to produce balanced samples in observational studies. *Political Analysis*, 20(1):25–46, 2012.

[78] Wendy K. Tam Cho, Jason J. Sauppe, Alexander G. Nikolaev, Sheldon H. Jacobson, and Edward C. Sewell. An optimization approach for making causal inferences. *Statistica Neerlandica*, 67(2):211–226, 2013.

[79] Kosuke Imai and Marc Ratkovic. Covariate balancing propensity score. *Journal of the Royal Statistical Society: Series B (Statistical Methodology)*, 76(1):243–263, 2014.

[80] Alexis Diamond and Jasjeet S. Sekhon. Genetic matching for estimating causal effects: A general multivariate matching method for achieving balance in observational studies. *The Review of Economics and Statistics*, 95(3):932–945, 2013.

[81] Fredrik Sävje, Michael J. Higgins, and Jasjeet S. Sekhon. Generalized full matching. *Political Analysis*, 29(4), 423-447, 2020.

[82] Awa Dieng, Yameng Liu, Sudeepa Roy, Cynthia Rudin, and Alexander Volfovsky. Interpretable almost-exact matching for causal inference. In Kamalika Chaudhuri and Masashi Sugiyama, editors, *Proceedings of the Twenty-Second International Conference on Artificial Intelligence and Statistics*, volume 89 of *Proceedings of Machine Learning Research*, pages 2445–2453. PMLR, 16–18 Apr 2019.

[83] Bo Lu, Robert Greevy, Xinyi Xu, and Cole Beck. Optimal nonbipartite matching and its statistical applications. *The American Statistician*, 65(1):21–30, 2011. PMID: 23175567.

[84] Giovanni Nattino, Chi Song, and Bo Lu. Polymatching algorithm in observational studies with multiple treatment groups. *Computational Statistics & Data Analysis*, 167:107364, 2022.

[85] Alberto Abadie and Guido W. Imbens. Matching on the estimated propensity score. *Econometrica*, 84(2):781–807, 2016.

[86] Alberto Abadie and Jann Spiess. Robust post-matching inference. *Journal of the American Statistical Association*, 0(0):1–13, 2021.

[87] Stefano M. Iacus, Gary King, and Giuseppe Porro. A theory of statistical inference for matching methods in causal research. *Political Analysis*, 27(1):46–68, 2019.

6

Fine Balance and Its Variations in Modern Optimal Matching

Samuel D. Pimentel

CONTENTS

6.1 The Role of Balance in Observational Studies and Randomized Experiments

In studies of causal effects convincing results require some demonstration that subjects receiving the condition or treatment of interest are comparable in other respects to subjects receiving the opposite control condition. In randomized studies this state of affairs is guaranteed with high probability in large samples because of the study designer's ability to allocate the treatment in a random way, blind to all attributes of the subjects receiving them. In particular, in randomized experiments, the empirical

DOI: 10.1201/9781003102670-6

distributions of covariates in the treated group and in the control group will be very similar. This similarity in distribution is known as covariate balance and is frequently demonstrated via a table of summary statistics for each covariate in the two groups, along with standardized differences in means between the groups on individual variables and the results of two-sample hypothesis tests. Here balance is a confirmatory signal that randomization has been conducted properly and that the particular allocation of treatments has avoided low-probability large discrepancies between groups

In observational studies the study designer has no control over who in the original sample receives treatment and who receives control. As a result, observed groups may differ substantially on variables besides treatment. In this setting the key problem is to design, by appropriately reweighting or taking subsets from the raw data, a new set of comparison groups that are comparable on observed variables, mimicking the type of data that might have arisen had a randomized experiment been performed [1]. In this case covariate balance takes on a much more central role: it is the primary goal of the matching or weighting procedure used to transform the raw data. Of course, covariate balance on observed variables in an observational study is not informative about the similarity of groups on unobserved covariates, and substantial bias may arise from assuming a matched or weighted design is identical to a randomized study when unobserved covariate differences are present.

The most common form of matching, propensity score matching, targets overall covariate balance only indirectly. When many pairs of subjects, each with identical propensity scores but opposite treatment status, are matched to one another, and when no unobserved covariates are present, the resulting study is equivalent to a paired randomized trial in which treatment is assigned uniformly at random within pairs [2, §3]. However, paired individuals need not have identical covariates, and balance is only guaranteed approximately and in large samples. Indeed, completely randomized trials themselves do not guarantee balance in any particular finite sample, since an unlucky random draw of treatment assignment can induce large imbalances.

A balance constraint is a requirement imposed on a matching problem, specifying that the match produced must guarantee some level of balance between the selected treated and control samples. In particular, a fine balance constraint requires that the empirical distributions of a nominal covariate be identical in the matched treated sample and the matched control sample. In contrast to approaches such as exact matching and propensity score calipers, which place constraints on the similarity of subjects who are paired to one another, fine balance and other balance constraints describe similarity between marginal distributions of covariates. This means that fine balance may be achieved on a variable even when the values of this variable differ between matched individuals in many cases.

6.1.1 Example: comparing patients of internationally trained and U.S.-trained surgeons

About 15% of surgeons practicing in the United States received their primary medical education in other countries before coming to the United for post-graduate residency training. These international medical graduates (IMGs) tend to have a different experience from U.S. medical graduates (USMGs) even after entering residency, since they are less likely to be placed in their top choice of postgraduate program and are more likely to transfer from one residency to another. Do the differences in training undergone by these two surgeon groups influence the health outcomes of their patients once they enter practice? Zaheer et al. [3] and Pimentel & Kelz [4] study this question using state health department records from New York and Florida, linked to surgeon records maintained by the American Medical Association. While the original data is bound by strict data use agreements, Pimentel & Kelz [4] used simulations based on summary statistics to generate a synthetic version of the original dataset that is very similar in structure and can be used to replicate their analyses. We consider a subset of this synthetic data, focusing on five hospitals containing a total of 1,315 surgical patients. 345 of these patients were treated by IMGs, and the remainder by USMGs; we aim to match each IMG patient to a similar USMG patient to permit comparisons of performance without confounding due

TABLE 6.1
Marginal balance on ER admission for two matches in the IMG-USMG data. Both matches retain all IMG patients; Match 1 matches only on the robust Mahalanobis distance, while Match 2 also imposes fine balance on ER admission.

Admitted	IMG	USMG Patients		
through ER	Patients	Unmatched	Match 1	Match 2
No	95	351	101	95
Yes	146	356	140	146

to systematic differences in the type of patients that IMGs and USMGs tend to treat. The outcome of interest is death within 30 days of surgery.

Table 6.1 describes two matches conducted in this data, and their relative level of success in balancing the rates of emergency room (ER) admission between IMG and USMG patients. ER admission is an important predictor of 30-day mortality for general surgery patients, and it is initially substantially imbalanced between the groups, with 61% of IMG patients admitted through the ER and only 50% of USMG patients so admitted. In both matches we retain all IMG patients and select a subset of USMG patients, so the ER admission rate of IMG patients remains the same across comparisons. In addition, following Pimentel & Kelz [4] we match patients only within hospitals.

Match 1 in Table 6.1 focuses purely on minimizing the robust Mahalanobis distance between paired units; for more discussion of the use of Mahalanobis distance in matching see Rubin [5] and Rubin & Thomas [6]. ER admission is one of eight variables included in the Mahalanobis distance and the resulting match improves balance on ER admission substantially, but the new distribution of ER admission still differs between groups, with 58% of matched USMG patients admitted through the ER compared to 61% of IMG patients. In contrast, Match 2 in Table 6.1 minimizes the robust Mahalanobis distance among all matches achieving fine balance on ER admission. The result is a match with identical empirical cumulative distribution functions of ER admission in the two groups: 146/241 patients admitted through the ER, 95/241 patients not admitted through the ER.

We note also that the match with fine balance does not match exactly on ER admissions. In 17 cases, an IMG patient not admitted through the ER is matched to a USMG patient who was; however, in exactly 17 other cases, an IMG patient admitted through the ER is matched to a USMG patient who was, ensuring that the marginal distributions balance. Matching exactly on ER admission is sufficient for fine balance, but not necessary. This is important because of situations where exact matching is not feasible but fine balance is. For instance, if one of the hospitals in our data contained more IMG ER admissions than USMG ER admissions, it would not be possible to match each IMG patient to a within-hospital USMG counterpart with the same ER admission status. However, if total USMG ER admissions in other hospitals still exceeded IMG ER admissions, it might still be possible to achieve fine balance on ER admissions over the match as a whole. Figure 6.1 provides a toy illustration of such a setting.

6.1.2 Outline

The IMG-USMG matches in Table 6.1 illustrate the basic definition of fine balance and hint at its value in practice. In what follows we describe fine balance and closely related balance and their implementation and application in greater detail. First, in Section 6.2 we provide more precise motivation for covariate balance as a method for bias reduction under a statistical model. In Section 6.3 we present a formulation of matching as a discrete optimization problem, and give formal definitions of several different balance constraints closely related to fine balance; these address many common situations in real observational studies, including cases in which variables cannot be balanced exactly, in which many variables must be balanced in a prioritized order, in which

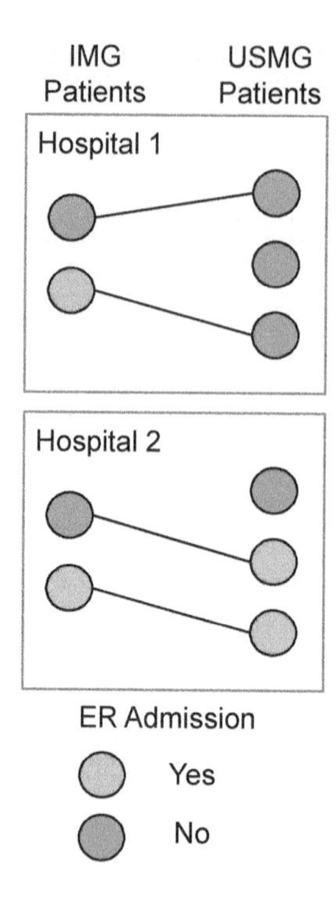

FIGURE 6.1
A hypothetical setting in which exact matching on ER admission is not possible within hospitals, but fine balance is still achieved. Lines indicate which IMG patients are matched to which USMG patients. In Hospital 1 exact matching is not possible since there are no USMG patients admitted through the ER, and the IMG ER patient must be matched to a non-ER USMG patient. However, in Hospital 2 this imbalance can be offset by matching an IMG non-ER patient to a USMG ER patient so that the overall counts of ER and non-ER patients in both matched samples is identical.

interactions between variables of a certain order must be balanced, and in which continuous variables must be balanced. In Section 6.4 we discuss algorithms for solving the matching optimization problem under the constraints of Section 6.3 and related tradeoffs between flexible constraints and efficient computation; we also review the compatibility of balance constraints with other alterations to the original matching problems such as increasing the number of control units matched to treated units, calipers and exact matching constraints, trimming of treated units, and changing the objective function.In Section 6.5 we consider how balance constraints can be useful in settings where the goal is not to achieve balance between the two samples at hand but to bring two or more samples into balance with an external population or sample. In Section 6.6 we discuss practical aspects of observational study design with balance constraints, including selection of tuning parameters. Finally, in Section 6.7 we conclude with discussion of connections between matching under balanced constraints and closely related weighting methods, as well as a survey of outstanding open problems related to matching under balance constraints.

6.2 Covariate Balance and Bias

As argued in Section 6.1, greater covariate balance makes designed observational studies appear more similar to designed randomized trials, enhancing the weight of the evidence they provide.

This heuristic explanation also has a more precise formulation in terms of a statistical model. The explicit link between balance after matching and covariate bias in a statistical model goes back at least to Rubin [5, 7], who discussed balance as a path to reduced bias in the context of a simple "mean-matching" algorithm that uses marginal covariate balance as an objective function rather than in constraints and that searches for solutions in a greedy manner rather than finding a global optimum. Here we consider the specific case of balance constraints in greater detail.

Suppose there are I matched pairs, with individual j in matched pair i characterized by treatment indicator W_{ij}, potential outcomes Y_{1ij} and Y_{0ij}, and a p-dimensional vector of observed binary or continuous covariates $\mathbf{X}_{ij} = (X_{1,ij}, \dots, X_{p,ij})$. In matched studies it is common to estimate the average effect of the treatment on the treated units using the average matched pair difference. We denote this estimand by SATT (where the S stands for "sample" and indicates that the average is taken over the units in the observed sample rather than over an infinite population) and denote the estimator by $\widehat{\text{SATT}}$. Formally:

$$\begin{aligned}
\text{SATT} &= \frac{1}{I}\sum_{i=1}^{I}\sum_{j=1}^{2} Z_{ij}(Y_{1ij} - Y_{0ij}) \\
\widehat{\text{SATT}} &= \frac{1}{I}\sum_{i=1}^{I}(Z_{i1} - Z_{i2})(Y_{i1} - Y_{i2}) \\
&= \text{SATT} + \frac{1}{I}\sum_{i=1}^{I}(Z_{i1} - Z_{i2})(Y_{0i1} - Y_{0i2}).
\end{aligned}$$

The bias of $\widehat{\text{SATT}}$ for SATT is characterized by the average matched pair difference in potential outcomes under control. Suppose Y_{0ij} values are sampled from a generalized additive model in the observed covariates, i.e. (writing expectation with respect to the sampling distribution):

$$E[Y_{0ij}] = \sum_{k=1}^{p} f_k(X_{k,ij}).$$

In this setting we can write the expected bias of $\widehat{\text{SATT}}$ for SATT (with expectation taking over the sampling distribution of the potential outcomes) as

$$\sum_{k=1}^{p}\left\{\frac{1}{I}\sum_{i=1}^{I}\sum_{j:Z_{ij}=1} f_k(X_{k,ij}) - \frac{1}{I}\sum_{i=1}^{I}\sum_{j:Z_{ij}=0} f_k(X_{k,ij})\right\}.$$

The quantity in the braces is a measure of treatment-control covariate balance on some transformation $f_k(\cdot)$ of the covariate $X_{k,ij}$. When $X_{k,ij}$ is discrete and fine balance has been achieved, this difference will be zero for any choice f_k, meaning that these observed variables contribute no bias on average over possible sampled realizations of the potential outcomes. When interactions of discrete variables are balanced, this same argument applies to a more flexible additive model that allows arbitrary nonlinear interactions between the discrete variables which are jointly balanced.

As such we can understand fine balance constraints and their variants as a method of bias reduction (at least for a particular estimator of a particular causal effect). Imposing these constraints in a problem such as cardinality matching, where more stringent balance constraints lead to smaller matched designs, can result in a form of bias-variance tradeoff, since smaller matched designs tend to produce test statistics with higher sampling variability (although for discussion of other factors influencing bias and variance of matched estimators see Aikens et al. [8]). Wang & Zubizarreta [9] studied the asymptotic impact of balance jointly on the bias and variance of the paired difference in means estimator under nearest-neighbor matching with replacement and found that imposing balance constraints removes sufficient bias to render the estimator $\sqrt{n}$-consistent for a population average causal effect, in contrast to previous results showing that the estimator fails to achieve $\sqrt{n}$-consistency in the absence of balancing constraints. Sauppe et al. [10] and Kallus [11] provide alternative motivations for balance-constrained matching and closely related methods as minimax solutions to a bias-reduction problem in which the goal is to control the worst possible bias obtained under a given family of models for potential outcomes under control.

Resa & Zubizarreta [12] also explored the bias-reduction potential of balancing constraints in finite samples through an extensive simulation study comparing several of the balance constraints described in Section 6.3. Their findings show that methods involving fine balance performed especially well relative to competitor matching methods without balance constraints in cases where the outcome was a nonlinear function of covariates, achieving root mean squared estimation error 5–10 times smaller.

6.3 Fine Balance and Its Variants Defined

At a high level, the findings reviewed in Section 6.2 suggest that scientific intuition or historical data about the data-generating process for the potential outcomes should be a primary consideration when deciding on which variables, or functions of variables, to impose balance. Assuming that one or more covariates of interest have been identified in advance, we now consider options for demanding balance on this chosen variable set. We begin by reviewing the broader framework of optimal matching in which balance constraints are imposed, and then discuss several specific constraints in detail.

6.3.1 Matching as an optimization problem

Matching in observational studies refers to constructing a design composed of many sets or pairs, each typically containing exactly one treated unit and one or more matched controls, although under some designs sets with exactly one control unit and one or more matched treated units may also be included. This is in contrast to stratified designs, which also group data points into homogeneous bins but which allow multiple individuals from each treatment arm within the same group. In what follows we focus specifically on matched designs with a single treated unit; in addition, we focus on optimal matching without replacement. *Optimal* matching signifies that the configuration of pairs is chosen by minimizing some objective function, rather than by a sequential greedy procedure. Matching without replacement stipulates that matched sets be disjoint, in contrast to matching with replacement, which allows the same control unit to appear in multiple matched sets.

We begin in a simple setting with a treated sample $\mathcal{T} = \{\tau_1, \ldots, \tau_T\}$ of size $|\mathcal{T}| = T$ and a control sample $\mathcal{C} = \{\gamma_1, \ldots, \gamma_C\}$ of size $|\mathcal{C}| = C$. The problem of selecting optimal non-overlapping matched pairs, each containing one treated unit and one control, can be formalized as follows:

$$\min \sum_{(\tau_i, \gamma_j) \in \mathcal{A}} D(\tau_i, \gamma_j) x_{ij} \tag{6.1}$$

s.t.

$$\sum_{j=1}^{C} x_{ij} = 1 \quad \forall i$$

$$\sum_{i=1}^{T} x_{ij} \leq 1 \quad \forall j$$

$$x_{ij} \in \{0, 1\} \quad \forall (i, j)$$

The first two constraints specify that each treated unit must be matched to exactly one control, and that each control may be matched to at most one treated unit. The $D(\tau_i, \gamma_j)$ terms in the objective function refer to some distance or dissimilarity between treated unit τ_i and control unit γ_j, typically calculated based on their respective covariate values and considered fixed; the decision variables x_{ij} are indicators for whether treated unit i is matched to control unit j. In addition, the set $\mathcal{A} \subseteq \mathcal{T} \times \mathcal{C}$ describes which treatment-control pairings are permitted; while sometimes $\mathcal{A} = \mathcal{T} \times \mathcal{C}$, it may instead be chosen as a strict subset, ruling out some possible pairings, in order to forbid matches differing by more than a fixed caliper on the propensity score or forbid matches that are not exactly identical on some important variable (such as hospital in the IMG-USMG match). In what follows we will denote the case $\mathcal{A} = \mathcal{T} \times \mathcal{C}$ as a fully dense matching, referencing the general classes of dense and sparse matches described in greater detail in Section 6.4.2. In Sections 6.4.1-6.4.3 we will discuss algorithms for solving this problem and in Section 6.4.4 we will discuss the impact of various changes to the constraints and the objective function.

Although we are primarily interested in the optimal solution to problem (6.1), any value of the decision variables $\mathbf{x} = (x_{11}, x_{12}, \ldots, x_{1C}, x_{21}, \ldots x_{TC})$ in the feasible set represents a possible match, and can equivalently be represented as a subset of treated-control ordered pairs $\mathcal{M} = \{(\tau_i, \gamma_j) \in \mathcal{T} \times \mathcal{C} : x_{ij} = 1\}$. For convenience, we also define $\mathcal{C}_{\mathcal{M}} = \{\gamma_j \in \mathcal{C} : \sum_{i=1}^{T} x_{ij} = 1\}$ as the set of controls selected to be in the match, and we define $\mathcal{F}$ as the set of all matches $\mathcal{M}$ feasible for problem (6.1).

6.3.2 Fine balance and near-fine balance

Suppose a nominal covariate with K categories is measured for all treated and control units in the sample. Formally, we can represent such a covariate as a mapping $\nu : \{\mathcal{T}, \mathcal{C}\} \longrightarrow \{\lambda_1, \ldots, \lambda_K\}$. We say that a match $\mathcal{M}$ exhibits fine balance on ν if and only if the treated sample and the matched control sample $\mathcal{C}_{\mathcal{M}}$ have identical counts of units in each category of ν. Formally:

$$|\{\tau_i \in \mathcal{T} : \nu(\tau_i) = \lambda_k\}| = |\{\gamma_j \in \mathcal{C}_{\mathcal{M}} : \nu(\gamma_j) = \lambda_k\}| \qquad \text{for all } k \in \{1, \ldots, K\}.$$

The term fine balance was first introduced in Rosenbaum et al. [13] in the context of a patient outcomes study involving patients from 61 different sites; although patients were not always matched to counterparts from the same site, each site contributed identical numbers of patients to each matched sample.

Fine balance on ν is not always feasible even when a feasible match $\mathcal{M}$ exists for problem (6.1). For example, suppose that the treated sample contains more subjects in some category λ_k of ν than are present in the entire control sample:

$$|\{\tau_i \in \mathcal{T} : \nu(\tau_i) = \lambda_k\}| > |\{\gamma_j \in \mathcal{C} : \nu(\gamma_j) = \lambda_k\}| \qquad \text{for all } k \in \{1, \ldots, K\}.$$

Since $\mathcal{C}_{\mathcal{M}} \subseteq \mathcal{C}$, there is no choice of $\mathcal{M}$ for which fine balance holds. This phenomenon occurs in the synthetic patient outcomes data with the procedure type variable. IMG surgeons performed 8 surgeries of procedure type 10 while USMG surgeons performed only 5. No matter how we select USMG surgeries in our match, we can obtain at most 5 in procedure type 10 which will be less than the total of 8 for that category in the IMG surgery group.

Motivated by a similar patient outcomes study, Yang et al. [14] defined a more general type of balance constraint, *near-fine balance*, that remains feasible even when treated and control counts in each category cannot be made identical. For any $\mathcal{M} \in \mathcal{F}$ and any $k \in \{1, \ldots, K\}$, let

$$\beta_k = \big|\, |\{\tau_i \in \mathcal{T} : \nu(\tau_i) = \lambda_k\}| - |\{\gamma_j \in \mathcal{C}_{\mathcal{M}} : \nu(\gamma_j) = \lambda_k\}| \,\big|. \tag{6.2}$$

Essentially β_k measures the absolute deviation from fine balance in category λ_k, or more intuitively the amount of "overflow" by which the count in category λ_k for one of the groups exceeds that of the other. In fact the quantity $\sum_{k=1}^{K} \beta_k/(2T)$ measures the total variation distance between empirical cumulative distribution functions of ν in the treated and the control group. Near-fine balance specifies that this quantity must be minimized. In formal terms (writing β_k as a function of $\mathcal{M}$ to emphasize its dependence on the match selected):

$$\sum_{k=1}^{K} \beta_k(\mathcal{M}) = \min_{\mathcal{M}' \in \mathcal{F}} \sum_{k=1}^{K} \beta_k(\mathcal{M}'). \tag{6.3}$$

When problem (6.1) is fully dense, so that all treatment-control pairings are allowed, the right-hand side of this constraint is equal to twice the sum of the "overflow" by which treated counts exceed control counts in each category in the raw data (zero when the control count is larger):

$$2\sum_{k=1}^{K} \max\left\{0, \{\tau_i \in \mathcal{T} : \nu(\tau_i) = \lambda_k\}| - |\{\gamma_j \in \mathcal{C} : \nu(\gamma_j) = \lambda_k\}\right\} \tag{6.4}$$

In cases where the match is not fully dense, this value provides a lower bound on the sum $\sum \beta_k(\mathcal{M}')$. It can be easily calculated by inspecting tables of the original samples prior to matching and identifying categories of the balance variable in which the number of treated units exceeds the number of controls. For the procedure type variable in the surgical example, there are three categories with more treated units than controls: category 4 (4 IMG patients, 2 USMG patients), category 10 (8 IMG, 5 USMG), and category 12 (7 IMG, 5 USMG). The total overflow is therefore $(4-2)+(8-5)+(7-5)=7$. Thus, solving problem (6.1) in the context of the hospital data under a near-fine balance constraint on procedure type ensures firstly that if possible all the USMG surgeries in each of procedure types 4, 10, 12 will be selected, minimizing the treated overflow to exactly 7 in these categories, and ensures secondly that the matched control overflow in all other procedure types is also limited to 7 if possible, and to its minimum feasible value otherwise. The "if possible" qualifier is necessary because the hospital match is not fully dense, forbidding matches across hospitals. In this particular case it is possible to limit the total overflow in both treated and control categories to 7, as shown in Table 6.2; however, if instead some hospital contained 4 controls with procedure type 10 and only 3 treated units in total, it would not be possible to include all the controls of procedure type 10 in the match and the best achievable overflow would be at least 8.

Near-fine balance as described so far minimizes the total variation distance between the empirical covariate distributions in the matched samples, essentially maximizing the overlap of the probability mass between the two distributions. However, it does not specify how the control overflows are to be allocated. In the context of the surgical example, a total of 7 extra controls must be distributed over the otherwise-balanced procedure types excluding categories 4, 10, and 12, but the basic near-fine balance constraint makes no distinction between distributing these controls relatively evenly across many categories or choosing to select them all from the same category. As the former is likely more desirable in practice than the latter, Yang et al. [14] define two stronger versions of near-fine balance that encourage more even distribution of control overflow across categories. One version minimizes the maximum overflow in any one control category; in the surgical example this corresponds to allowing exactly one unit of overflow in 7 of the 29 categories initially containing more controls than treated units. A second version minimizes the chi-squared distance between category counts in the two groups, essentially allocating overflow across control categories proportionally according to their relative prevalences. In what follows we use the term "near-fine balance" to refer specifically to constraint (6.3) as is typical in the literature.

6.3.3 Refined balance

While it is convenient to impose fine or near-fine balance on a single important variable, matching problems typically involve many variables on which we hope to achieve balance. An obvious

TABLE 6.2
Contingency table for procedure type in raw data and in data matched with a near-fine balance constraint. The first row gives counts for the IMG (treated) patients, who are fully retained in the matched dataset; the second row gives the counts in the USMG (control) group overall, and the third gives the counts in the matched USMG group. The final row is the difference between rows 1 and 3, which add to zero since the row totals in rows 1 and 3 are identical. The absolute values of the discrepancies are exactly the values β_k defined in (6.2), and dividing their sum by the sum of all numbers in rows 1 and 3 gives the total variation distance between marginal distributions of procedure type in the matched data.

	Procedure Type																
	1	2	3	4	5	6	7	8	9	10	11	12	13	14	15	16	17
IMG	1	2	0	4	1	1	0	4	4	8	4	7	5	12	4	7	9
USMG	1	2	3	2	2	4	13	13	15	5	13	5	14	35	13	20	27
Matched USMG	1	2	0	2	1	1	0	4	4	5	4	5	5	12	4	7	9
Discrepancy	0	0	0	2	0	0	0	0	0	3	0	2	0	0	0	0	0

	Procedure Type														
	18	19	20	21	22	23	24	25	26	27	28	29	30	31	32
IMG	12	11	4	12	6	8	14	11	20	21	19	8	6	8	8
USMG	34	21	31	28	53	17	39	29	39	35	34	58	37	40	25
Matched USMG	12	11	4	12	7	8	14	11	20	21	19	12	8	8	8
Discrepancy	0	0	0	0	−1	0	0	0	0	0	0	−4	−2	0	0

solution is to add multiple fine balance constraints to the matching problem. However, feasibility issues arise, since even if fine balance is possible for each of two variables individually it may not be simultaneously possible for both in the same match. For similar reasons, satisfying multiple near-fine balance constraints, at least in the form defined in (6.3), may not be simultaneously possible. Furthermore, even in settings where satisfying multiple constraints is possible, it may be computationally difficult for reasons discussed in Section 6.4.5.

Another option is to impose a single fine balance constraint on an interaction of several nominal covariates, i.e., a new nominal variable in which each unique combination of categories from the component variables becomes a new category. When fine balance on the interaction is feasible, it guarantees fine balance on all the individual component variables and the interactions involving any subset of variables. However, interactions tend to involve very large numbers of categories, many with small counts, and fine balance is unlikely to be possible in these cases. While near-fine balance is well-defined for the interaction, near-fine balance on an interaction does not guarantee near-fine balance on component variables. For example, matching in the IMG-USMG data with near-fine balance on the interaction of ER admission and procedure type limits the total variation distance between the two groups to 24/482, i.e., treated counts overflow control counts by a total of 12 units in some subset of categories while control counts overflow treated units by a total of 12 units in the complementary subset. However, in the match produced, there remains an overflow of 12 units on each side for procedure type alone, when by our previous match with near-fine balance we know that an overflow of at most 7 on each side is achievable for this variable, and there remains an overflow of 6 units on each side for emergency room, even though exact fine balance is possible in this dataset.

A second drawback to all the methods described above is that they necessarily treat all nominal variables involved as equally important. In practice some variables are often known a priori be more important to balance than others, either because they initially exhibit a very high degree of imbalance or because they are known to be prognostically important. In settings where not all variables can be balanced exactly, it is desirable to prioritize balance on the important variables over balance on others.

Refined balance is a modification of fine balance that offers a way to address balance on multiple covariates with heterogeneous levels of importance in a prioritized way. Assume we have L nominal covariates (possibly created by interacting other nominal covariates), $\nu_1, \ldots, \nu_L$ listed in decreasing order of priority for balance. In addition, suppose that for any $\ell \in \{1, \ldots, L-1\}$, covariate $\nu_{\ell+1}$ is nested inside covariate ν_ℓ in the sense that any two subjects with the same level for covariate $\nu_{\ell+1}$ also share the same level for covariate ν_ℓ. We say that a match $\mathcal{M}$ satisfies refined balance on the ordered list $\nu_1, \ldots, \nu_L$ if and only if:

1. $\sum_{k=1}^{K} \beta_k(\mathcal{M}) = \min_{\mathcal{M}' \in \mathcal{F}} \sum_{k=1}^{K} \beta_k(\mathcal{M}')$, i.e. $\mathcal{M}$ satisfies near-fine balance on ν_1.

2. $\sum_{k=1}^{K} \beta_k(\mathcal{M}) = \min_{\mathcal{M}' \in \mathcal{F}_1} \sum_{k=1}^{K} \beta_k(\mathcal{M}')$ where $\mathcal{F}_1$ is the subset of matches in $\mathcal{F}$ also satisfying constraint 1.

3. $\sum_{k=1}^{K} \beta_k(\mathcal{M}) = \min_{\mathcal{M}' \in \mathcal{F}_2} \sum_{k=1}^{K} \beta_k(\mathcal{M}')$ where $\mathcal{F}_2$ is the subset of matches in $\mathcal{F}$ also satisfying constraints 1 and 2.

   $\vdots$

L. $\sum_{k=1}^{K} \beta_k(\mathcal{M}) = \min_{\mathcal{M}' \in \mathcal{F}_{L-1}} \sum_{k=1}^{K} \beta_k(\mathcal{M}')$ where $\mathcal{F}_{L-1}$ is the subset of matches in $\mathcal{F}$ also satisfying constraints 1 through $L-1$.

Essentially, refined balance enforces the closest remaining balance (in the sense of the total variation distance) on each variable subject to the closest possible balance on the previous variables in the priority list. For matches that are fully dense, this is equivalent to achieving near fine balance on each level. Pimentel et al. [15] introduced refined balance and applied it in a study where the finest interaction had over 2.9 million different categories.

Typically, the nested structure of the list $\nu_1, \nu_2, \ldots, \nu_L$ is created by choosing either a single variable or an interaction of a small number variables as ν_1, then choosing ν_2 as an interaction of ν_1 with one or more additional variables, and so on, letting each new covariate be an interaction of the previous variable with some additional variables. For example, we may run a new match in the IMG-USMG data with refined balance on the following list of nested covariates in decreasing order of priority: ER admission, ER admission $\times$ procedure, ER admission $\times$ procedure $\times$ patient sex. In this match we achieve exact fine balance on ER admission so that all category counts agree exactly in the two groups, an overflow of 12 units on each side for the second interaction (for an overall total variation distance of 24/482), and an overflow of 19 units on each side in the third interaction (with total variation distance 38/482). In this case, the balance on the second two levels can be verified to be equivalent to near-fine balance by equation (6.4), although this is not guaranteed a priori in our case because the match is not fully dense.

6.3.4 Strength-k matching

While refined covariate balance is a natural constraint in cases where there is a clear way to divide covariates into groups of decreasing priority, it is less attractive in cases where balance on each individual variable is more important than balance on higher-order interactions. One example, described in detail in Hsu et al. [16], is problems in which substantial treatment effect variation may be present and in which investigators hope to search for partitions of the matched sample associated with differential effects in the data after matching. In this problem it is particularly important to have exact matches on a given nominal variable in order to partition matched pairs by its categories to examine its potential role in effect modification. While achieving exact matches on many variables at once simultaneously is not usually possible, if fine balance can be achieved simultaneously on many variables at once, then each variable may be separately tested as an effect modifier by rematching the treated units to the matched controls with an exact matching constraint (so that the individual pairs used vary across tests of individual variables, but not the overall samples compared). Since researchers may not want to rule out potential effect modifiers a priori and since interest in explaining effect heterogeneity is usually limited to fairly simple models without very high-order interactions, it is more important here to achieve marginal balance on each individual variable and possibly on low-order interactions than to give attention to high-order interactions as refined balance tends to do.

Strength-k matching, introduced in Hsu et al. [16], is a variant of fine balance designed for this setting. Assuming L nominal covariates are initially present in the data for balancing, fine balance constraints are enforced on all interactions of exactly $k < L$ variables. Strength-1 matching corresponds to requiring fine balance only on each variable individually; strength-2 matching corresponds to requiring fine balance on any interaction between two variables; and so on. While strength-k matching's relatively inflexible demands make it infeasible in many settings and appropriate only for certain kinds of optimization algorithms (see Section 6.4.3), it is very useful for guaranteeing perfect balance on many variables and interactions at once. In their simulation study Resa and Zubizarreta [12] found that strength-2 matching performed better in terms of mean-squared estimation error for causal effects than competitor methods such as distance-based matching and fine balance on individual variables only in cases where strong interactions were present in the outcome model. It is notable that the estimation error was smaller despite the fact that strength-2 matching also tended to result in much smaller sample sizes, and hence higher sampling variability under the outcome model, due to the stringency of the constraint.

Consider an example from the IMG-USMG data, in which we might wish to balance all of the following four variables: ER admission, patient sex, patient race, and comorbidity count. All of these variables are prognostically important and some could explain effect modification (if, for instance, IMGs and USMGs perform similarly on patients with few comorbidities but differ in their performance for patients with many). Matching with strength-k balance on these four variables ensures that all four of them are balanced exactly. In contrast, if we match with near-fine balance on the four-way interaction of these variables, we do not obtain perfect balance on any of them individually; refined balance could be used to ensure at least one of the individual variables is balanced perfectly, but not for all four variables at once since they are not nested.

6.3.5 Controlled deviations from balance

A slightly different framework for constraining balance on variables is described by Zubizarreta [17], in which a collection of variables $X_1, \ldots, X_p$ is assumed to be present, each mapping values from $\{\mathcal{T}, \mathcal{C}\}$ into some covariate space, possibly continuous or possibly nominal. Balance constraints of the following form are added to the optimization problem, where $f_1, \ldots, f_{J_p}$ are functions of the covariates to be balanced:

$$\left| \sum_{\tau \in \mathcal{T}} f_j\left(X_i(\tau)\right) - \sum_{\gamma \in \mathcal{C}} f_j\left(X_i(\gamma)\right) \right| < \delta_{ij} \qquad i = 1, \ldots, p, \quad j = 1, \ldots, J_p.$$

We refer to this constraint as controlled deviation from balance. For continuous variables X_i, the function f_j can be chosen as the identity to balance sample means; f_j can raise X_i to higher powers to induce balance on higher moments, or can indicate whether X_i lies above a threshold to balance distributions in the tails. For nominal covariates with K categories, one f_j function can be added for each category indicating whether the level for a given unit is in this category, so that the deviation in treated control counts in each category k is controlled at a corresponding level δ_{ik}.

In certain cases controlled deviation from balance, as described above for nominal covariates, is equivalent to fine or near-fine balance. When all δ_{ik} values are set to zero for a given nominal covariate, the K deviation constraints are equivalent to a fine-balance constraint. Furthermore, controlled deviation from balance is sufficient for near-fine balance (at least in a fully dense match) when

$$\sum_{k=1}^{K} \delta_{ik} = \min_{\mathcal{M}' \in \mathcal{F}} \sum_{k=1}^{K} \beta_k(\mathcal{M}').$$

where the $\beta_k(\mathcal{M}')$ terms are defined on variable i. Note that this constraint is not quite equivalent to near-fine balance however, since while near-fine balance places no constraint on the allocation off the overflow, controlled deviation dictates exactly where overflows are allowed. Beyond these special cases controlled deviation from balance need not guarantee fine or near-fine balance.

Controlled deviation from balance offers several advantages relative to fine and near-fine balance constraints. First and most importantly, it provides natural handling of continuous variables. Second, in large studies where tiny deviations in balance are unlikely to introduce bias of substantive importance (e.g. a deviation of 1 subject in category counts in a study with 100,000 matched pairs), controlled deviation can free the optimization algorithm from overly stringent restrictions, allowing users to specify a tolerance of similarity. On the other hand, controlled deviation requires users to specify tolerances δ_{ij} for each constraint, a practically tedious process, while near-fine balance and refined balance are designed to adaptively identify the best achievable balance subject to other constraints on the match. For discussion of the choice of δ_{ij} parameters in a closely related weighting problem, see Wang & Zubizarreta [18].

6.4 Solving Matching Problems under Balance Constraints

In Section 6.3 we defined an optimization problem in equation (6.1) and described a variety of constraints under which one might seek to solve it. We now turn to the problem of producing optimal solutions under the various balancing constraints. In particular, we describe three algorithmic approaches – the assignment method, the network flow method, and the integer programming method – in order of increasing generality. The first two approaches do not support all the balance constraints described above, but may offer computational advantages relative to the third approach. After describing the approaches we also consider how other modifications to problem (6.1) interact with the compatibility and performance of the algorithms.

6.4.1 Assignment method

The assignment method leverages algorithms designed to solve the classical assignment or marriage problem in operations research. In this problem there are two groups of units, call them $\mathcal{A}$ and $\mathcal{B}$, both of size N as well as an $N \times N$ matrix D whose ijth entry D_{ij} describes a distance or dissimilarity between unit i in $\mathcal{A}$ and unit j in $\mathcal{B}$. The problem is to pair each unit from A to a distinct unit in B such that the total within-pair distance is as small as possible. Algorithms to solve this problem have been a subject of interest for decades and are reviewed in detail by Burkard et al. [19].

The assignment problem coincides with problem (6.1) when the treated and control groups are equal in size. However, matching designs based on problem (6.1) are primarily useful only when the control group is substantially larger than the treated group, since improvement over a raw comparison between groups comes from excluding control units. Therefore, to use the assignment problem to solve problem (6.1) it is necessary to create additional false treated units so that the groups are equal in size. In particular, let $N = C$, let $\mathcal{B} = \mathcal{C}$, and let $\mathcal{A} = \mathcal{T} \cup \{\rho_1, \ldots, \rho_{N-T}\}$. We define D_{ij} as $D(\tau_i, \gamma_j)$ where $i \leq T$ and define $D_{ij} = 0$ for all $i > T$. Solving the assignment problem with these parameters and selecting the subset of pairs that include a member of $\mathcal{T}$ solves problem (6.1) exactly. Essentially the false treated units ρ_i can remove the most difficult-to-match controls from the problem and ensuring they contribute nothing to the objective function, leaving only the optimal set of T controls to be matched to the treated units so as to minimize the objective function.

The assignment method can be easily adapted to enforce a fine balance or a near-fine balance constraint. This is done by placing restrictions on which controls can be matched to the false treated units $\rho_1, \ldots, \rho_{C-T}$. In particular, for each level λ_k of fine balance covariate ν, define treated and control counts as follows:

$$M_k = |\{\gamma \in \mathcal{C} : \nu(\gamma) = \lambda_k\}|, \qquad \text{and} \qquad n_k = |\{\tau \in \mathcal{T} : \nu(\tau) = \lambda_k\}|.$$

Fine balance is possible exactly when $n_k \leq M_k$ for all $k = 1, \ldots, K$; in this setting is also the case that

$$\sum_{k=1}^{K} (M_k - n_k) = C - T.$$

To modify the assignment algorithm to enforce the fine balance constraint, instead of adding $C - T$ interchangeable false treated units ρ_i with zero distances to all controls, we add $M_k - n_k$ false treated units ρ_{ik} for each category $k = 1, \ldots, K$, each of which have zero distance to controls γ satisfying $\nu(\gamma) = \lambda_k$ and infinite distance to all other controls (in practice infinite distances can be replaced by any large distance exceeding the maximum of the other distances in the matrix). This modification forces exactly $M_k - n_k$ controls to be removed from consideration in each category, guaranteeing fine balance on the match produced at the end.

The strategy for enforcing near-fine balance is similar. As before, in all categories where $M_k \geq n_k$, we add $M_k - n_k$ false treated units ρ_{ik} to each category $k = 1, \ldots, K$, each with zero distance only to control units in the same category. However, if fine balance is not possible then we have $n_k > M_k$ in at least one category, and this also means that the total number of false treated units added satisfies:

$$\sum_{k:M_k \geq n_k} (M_k - n_k) = C - \left[T + \sum_{k:n_k > M_k} (n_k - M_k) \right] > C - T.$$

We now have more rows than columns in our matrix D and so we must add some additional columns to ensure that it is square. In particular, for each category k in which $n_k > M_k$ we add $n_k - M_k$ false control units κ_j, each of which has zero distance to false treated units and infinite distance to real treated units. Essentially, these false controls absorb excess false treated units to ensure that all treated units are matched.

The method just described enforces near-fine balance with respect to the total variation distance alone, but Yang et al. [14] describe further tweaks that can achieve near-fine balance in the sense of minimizing both total variation distance and maximum overflow in any category or of minimizing chi-squared distance in category counts. Essentially these tweaks involve changing a certain number of the zero distances for false controls back to infinities to ensure that the false treated units excluded by these false controls do not come disproportionately from certain groups rather than others.

Notice that the modifications described to enforce near-fine balance assume implicitly that the right-hand side of constraint (6.3) is known a priori. As such the assignment algorithm is best used for near-fine balance in cases where the match is fully dense (as is indeed assumed throughout [14]). For cases where the match is not fully dense, the network flow algorithm of the following section is often a more convenient approach.

6.4.2 Network flow method

The network flow method for solving optimal matching problems, like the assignment problem, transforms the original problem into a canonical problem in operations research, the minimum-cost network flow problem [20]. In minimum-cost network flow, a graph or network is assumed to be present, with nodes $\mathcal{N}$ and directed edges $\mathcal{E}$. Certain "source" nodes supply a commodity in positive integral quantities, other "sink" nodes exhibit positive integral demand for this commodity; the problem is to transport the commodity from sources to sinks over edges in the graph while paying the minimum transportation cost. In particular each edge has a positive integral upper bound governing how much commodity can be sent across it, as well as a nonnegative-real-valued cost per unit flow across the edge. Formally, we write the problem as follows where integer b_n represents the supply or demand (supplies being positive values and demands being negative values) at node n, u_e and c_e represent upper capacity and cost respectively for edge e, $\mathcal{E}_{\text{in}}(n)$ and $\mathcal{E}_{\text{out}}(n)$ represent the subsets of $\mathcal{E}$ directed into node n and out of node n, respectively, and x_e are the flow or decision variables.

$$\begin{aligned} &\min \sum_{e \in \mathcal{E}} c_e x_e \\ &\text{s.t.} \\ &\sum_{e \in \mathcal{E}_{\text{in}}(n)} x_e - \sum_{e \in \mathcal{E}_{\text{out}}(n)} x_e = b_n \quad \text{for all } n \in \mathcal{N} \\ &0 \leq x_e \leq c_e. \end{aligned} \tag{6.5}$$

The second constraint, which ensures that the amount of flow coming into a node differs from the amount coming out only by the demand or supply, is known as the "preservation of flow" constraint. One feature of problem 6.5 that is important to matching is that although the x_e decision variables are real-valued, some optimal solution with integer-valued decision must exist, and can be produced via polynomial time algorithms. Formally, this derives from the fact that network incidence matrices are totally unimodular; for more discussion, see Papadimitriou & Stieglitz [21, §13.2]. Algorithms for solving the minimum-cost network flow problem are reviewed by Bertsekas [22], and the RELAX-IV network flow solver [23] is accessible in R via the package `optmatch` [24]. Although `optmatch` itself does not implement any balance constraints, related R packages do so within the network flow framework. Package `bigmatch` [25] implements near-fine balance on a single nominal variable, and packages `rcbalance` and `rcbsubset` implement refined balance (as well as near-fine and fine balance on a single nominal variable as special cases). Pimentel [26] provides a user-friendly introduction to the `rcbalance` package.

To represent matching as a network flow problem (Figure 6.2(a)), first take a directed bipartite graph with nodes for each treated unit and each control unit, in which each treated unit τ_i connected to all control units to which it can be matched, or all γ_j such that $(\tau_i, \gamma_j) \in \mathcal{A}$ in the language of problem (6.1). Each edge is given a cost associated with the covariate distance between the units it connects and a capacity of one. Next, add an edge with zero cost and capacity one from each control node to an additional sink node. Finally, assign a supply of one to each treated node and a demand equal to the number of treated units T to the sink node. Intuitively, asking for the best network flow solution now asks for the best way to route the commodity produced at each treated node through a distinct control node to its destination at the sink; the choice of flow across the treated-control edges is bijective with the choice of an optimal match in problem (6.1).

How can balance constraints be represented in a network flow problem? The method described in the previous paragraph can be adapted to solve any assignment problem, including the methods for fine and near-fine balance described in Section 6.3. Here we describe another approach, more elegant in that it uses a graph with fewer nodes and edges. To implement a fine balance constraint, take the graph in the previous paragraph remove the control-sink edges, and add a node for each category of the fine balance variable. Finally, add edges of cost zero and upper capacity one from each control node to the fine balance node for the category to which it belongs, and add edges of cost zero and capacity n_k from each category node k to the sink node, where the n_k are the counts for each category k in the treated group (Figure 6.2(b)). These last edges enforce fine balance by ensuring that exactly the same number of controls as treated or chosen in each category. To impose

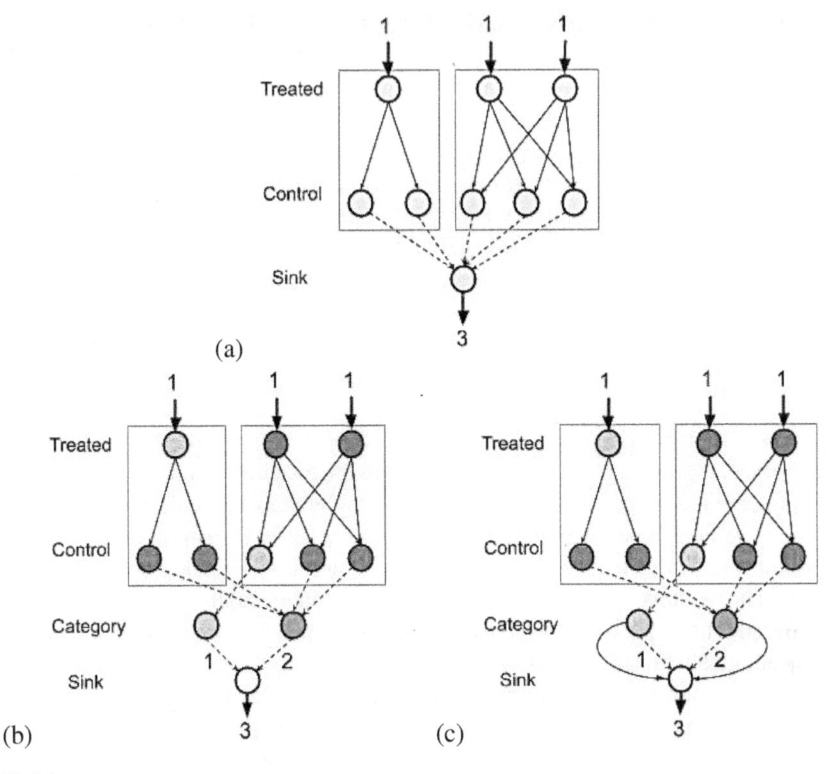

FIGURE 6.2
Network flow algorithms for optimal pair matching within blocks: without balance constraints (a), with a fine balance constraint (b), and with a near-fine balance constraint (c). Each algorithm is based on a directed bipartite graph containing nodes for each treated unit and each control unit, with edges connected treated units only to control units in the same block; additional sink and category nodes are added for housekeeping and do not refer to any unit in the original problem. Short boldface arrows represent a supply when directed into nodes and a demand when directed out; solid edges connecting nodes have some nonzero cost while dashed edges have zero cost.

near-fine balance instead, the capacity constraints on the category-sink edges can be changed from n_k to $\max(n_k, M_k)$ if the match is fully dense. More generally, if the match is not necessarily fully dense, an additional edge may be added between each category node and the sink node with a very large capacity but a high cost (Figure 6.2(c)). In cases where restrictions on who can be matched to whom make the minimal achievable imbalance larger than the n_k and M_k values suggest, a solution is still feasible in this network, since additional controls supplied in some category can always be routed through the new high-cost edge. However, for sufficiently high cost the number of units sent across these edges will be minimized. A generalization of this latter strategy is also useful for implementing refined balance. Here the network described above for near-fine balance is generalized by adding multiple layers of fine balance category nodes, each associated with a different variable in the refined balance hierarchy. The most granular layer with the most categories receives flow directly from the control nodes, then channels into category nodes of the next layer up in the hierarchy and so on. The costs associated with not meeting the balance constraints exactly increase dramatically with each additional level, enforcing the relative priority among the balance levels. For more discussion of this method and of the choice of costs to enforce balance, see Pimentel et al. [15].

In what settings will a match not be fully dense? In the IMG-USMG example, we have already seen one case, where matches are restricted to appear within certain blocks in the data; for additional discussion of matching with exact agreement on discrete variables and its statistical benefits see Iacus et al. [27], Hsu et al. [28], and Pimentel et al. [15]. Another common constraint on pairings is a propensity score caliper, which requires paired individuals' estimated propensity scores differ by no more than a fixed amount; such calipers are important for ensuring randomization inference after matching has a clear basis [29]. Calipers are an especially important case for network flow methods, since rather than partitioning the overall match into fully dense bipartite subgraphs they induce a complex structure of presence and absence among the edges that can make it very difficult to calculate the right-hand sides of constraints like (6.3) a priori. One workaround to the problems of non-fully dense matches is to assign very large edge costs $D(\tau, \gamma)$ to edges associated with pairings the researcher wishes to forbid; under this strategy the match essentially remains fully dense and the assignment algorithm can be used in place of network flow to solve for near-fine balance. However, there are two important drawbacks to this strategy. Firstly, the near-fine balance constraint, if implemented via the assignment algorithm described above or by setting exact edge capacities, will take precedence over the constraint forbidding certain edges to be used, so that propensity calipers or requirements to match exactly will be violated if necessary to achieve balance.

Secondly, the number of edges in a network flow problem has computational implications. The number of edges is the primary driver of worst-case complexity for state-of-the-art network flow solution algorithms and large edge counts tend also to be associated with long computation times in practice even when the worst case is not realized. As discussed in detail by Pimentel et al. [15], there is a major difference in worst-case complexity between dense problems, in which the number of edges scales quadratically in the number of nodes, and sparse problems, in which the scaling is linear. The latter case can be achieved by matching within natural blocks of a bounded size, among other ways. Sparse matches are a setting where balance constraints add an especially large amount of value; while the absence of most treatment-control edges greatly limits the options for achieving small within-pair covariate distances (and, in the case of blocks of bounded size, eliminates the possibility of within-pair covariate distances shrinking to zero asymptotically as is assumed in studies such as Abadie & Imbens [30]), the impact of edge deletions on balance tends to be much milder, and high degrees of covariate balance can often be achieved even in highly sparse matches [15]. Balance constraints in turn motivate the development of methods that can simultaneously match very large treatment and control pools; while in the absence of balance constraints exact matching within small blocks could be implemented by matching within each block separately in parallel, assessing balance across the many blocks requires doing all the within-block matches together as part of a coordinated algorithm.

Just as any assignment problem can be represented as a network flow problem via the transformation described several paragraphs ago, any network problem can also be transformed into an assignment problem [31, §4.1.4], so from a certain abstract perspective these two solution approaches are equivalent. In practice, however, it is meaningful to distinguish them for computational reasons, since dedicated network solvers produce solutions faster and in more convenient formats than could be achieved by first transforming network problems into assignment problems.

6.4.3 Integer programming method

While the network flow approach offers substantially flexibility in balance constraints via refined balance under a guarantee of low computational cost, not all interesting balance constraints can easily be incorporated into this framework. As a simple example, consider the problem of matching with fine balance on three or more different nominal covariates, none of them nested in any other, when fine balance on their interaction is not possible. A network flow problem with fine balance can be used to balance one of the several variables, or a refined balance constraint can balance several increasingly complex interactions between the variables as closely as possible subject to exact balance on one but no network flow method described so far, or known to exist, can guarantee fine balance on all the variables. In particular, this means that network flow methods cannot implement strength-k matching except in very special cases (such as when only k variables are to be balanced and fine balance is possibly on their full interaction).

To solve matching problems under such constraints, integer programming and mixed integer programming methods may instead be adopted [17]. For example, letting $\beta_1, \ldots, \beta_{K_1}$ and $\beta'_1, \ldots, \beta'_{K_2}$ represent the absolute discrepancies in counts after matching for the categories in the two variables to be balanced, we may write the problem as:

$$\min \sum_{(\tau_i,\gamma_j)\in\mathcal{A}} D(\tau_i,\gamma_j)x_{ij} \tag{6.6}$$

$$\begin{aligned}
&\text{s.t.}\\
&\sum_{j=1}^{C} x_{ij} = 1 \quad \forall i\\
&\sum_{i=1}^{T} x_{ij} \leq 1 \quad \forall j\\
&\beta_k = 0 \quad \forall k \in \{1,\ldots,K_1\}\\
&\beta'_k = 0 \quad \forall k \in \{1,\ldots,K_2\}\\
&x_{ij} \in \{0,1\} \quad \forall (i,j)
\end{aligned}$$

Alternatively, for sufficiently large costs ω_1 and ω_2 and additional decision variables z_1, z_2,we can represent the problem as follows [17] :

$$\min \sum_{(\tau_i,\gamma_j)\in\mathcal{A}} D(\tau_i,\gamma_j)x_{ij} + \omega_1 z_1 + \omega_2 z_2 \tag{6.7}$$

$$\begin{aligned}
&\text{s.t.}\\
&\sum_{j=1}^{C} x_{ij} = 1 \quad \forall i\\
&\sum_{i=1}^{T} x_{ij} \leq 1 \quad \forall j\\
&\sum_{k=1}^{K_1} \beta_k \leq z_1 \quad \forall k \in \{1,\ldots,K_1\}\\
&\sum_{k=1}^{K_2} \beta'_k \leq z_2 \quad \forall k \in \{1,\ldots,K_2\}\\
&z_1, z_2 \geq 0\\
&x_{ij} \in \{0,1\} \quad \forall (i,j)
\end{aligned}$$

Problem (6.6) is an integer program (all decision variables are integers) while problem (6.7) is a mixed integer program (having both real and integer-valued decision variables). The former problem explicitly represents the balance requirements as constraints, while the latter enforces them indirectly via a penalty term in the objective function. The differences between these two formulations recall the differences between the two networks for near-fine balance described earlier. The integer program has the advantage of simplicity, not requiring the penalties to be chosen, but the mixed integer program has the advantage of feasibility even in settings where fine balance is not achievable. Either of these problems may be solved by the application of a standard integer-program solver such as Gurobi, which implements integer programming techniques such as cutting planes and branch-and-bound algorithms. Unlike linear programming and network flow methods, these approaches lack strong guarantees of limited computation time in the worst case; more precisely, they do not generally have polynomial-time guarantees (see Section 6.4.5 for more discussion). However, in many settings in practice highly optimized solvers can be competitive with network flow methods, especially if one is willing to accept an approximation to the correct solution in place of an exactly optimal solution [32]. The R package `designmatch` [33] implements matching based on integer programming methods and supports all the varieties of balance constraint described in Section 6.3.

6.4.4 Balance and other aspects of the matching problem

So far we have focused on solving problem (6.1) under various balance constraints. However, balance constraints are not the only possible modification to the main optimization problem. In this section we discuss other changes to the constraints or objective functions in problem (6.1) and how they affect both the formulation of balance constraints and the viability of the solution algorithms just discussed.

Control:treated ratio. Solutions to problem (6.1) consist of matched pairs, each treated unit being matched to exactly one control unit. However, when the control pool is much larger than the treated pool, it may be desirable to match each treated unit to κ controls where $\kappa > 1$. Formally, this corresponds to changing the right-hand side of the first constraint in problem (6.1) to κ. Fine balance and all the variants discussed above extend easily to the $\kappa > 1$ case; total variation distances between treated and matched control empirical distributions are now rescalings of summed absolute differences between category counts in the treated group, multiplied by κ, and category counts in the control group. Assignment algorithms proceed by creating κ copies of each treated unit as a preprocessing step, and network flow methods proceed by changing the supply at each treated node from 1 to κ and multiplying the capacities on edges enforcing balance by κ.

Variable-ratio matching. Even when large numbers of controls are present, not all treated units may be similarly easy to match, and choosing the same control:treated ratio κ for all matched pairs may force some treated units to select dissimilar controls, or force other controls to use only a small number of available good matches, or both. A solution to this problem is variable-ratio matching, in which the optimization problem allows for different treated units to select different numbers of controls. In settings where there is heterogeneity across the treated units in their matchability, these methods may drive substantial improvements in bias and variance [34]. Unfortunately, while fully flexible variable-ratio matching in which control-treated ratios are chosen for each treated unit as part of the optimization can be implemented with the assignment problem [35] or with network flow methods [36], these solution methods are not compatible with existing fine or near-fine balance constraints. Intuitively, this problem arises because network flow formulations of balance rely on predefined network structures (such as the category nodes described in Section 6.4.2) that sum balance discrepancies across controls; in a variable-ratio problem, calculation of balance requires reweighting these quantities in a way that depends on the particular ratios selected in the given solution. A compromise solution, described by Pimentel et al. [37], suggests stratifying the data prior to matching on the entire number, a quantity that describes the approximate matching ratio achievable for each treated unit in large samples , and using fixed-ratio matching and fine balance constraints within each stratum, choosing a different matching ratio for each sample. Integer programming methods can also be used to enforce fine balance while allowing for variable treated-control ratios [38].

Subset and cardinality matching. The first constraint in problem (6.1) requires every treated unit to be included in the match. While this may be desirable in cases where the treated group is known to be a random sample from a larger population of interest, it is frequently the case that certain treated units cannot be matched to similar controls. For example, in the larger dataset from which the IMG-USMG sample used here is drawn, some hospitals have more IMG patients than USMG patients; under the constraint of matching only within hospitals, not all IMG patients can be paired to distinct USMG patients. Similar problems may arise with calipers. As described by Pimentel & Kelz [4], network flow methods may be modified to exclude the minimal number of treated individuals necessary to render the problem feasible by adding "bypass" edges to the network that connect treated units directly to the sink node; when these edges have sufficiently high cost (much higher than the highest individual treated-control dissimilarity in the matching problem) the number of treated units retained will be maximized. This innovation is also compatible with fine, near-fine, and refined balance; in this case the bypass edges go not directly to the sink but to the most granular fine balance category node. The penalties for the bypass edges must not only be much larger than the treated-control dissimilarities but also be much larger than the near-fine balance penalties, in order to ensure that the maximal number of controls is retained.

The problem as just described prioritizes keeping the largest sample possible and achieves optimal balance subject to this constraint. A different idea is to require exact balance, and maximize the sample size subject to this constraint. This method is called cardinality matching. Formally, it involves adding fine balance constraints to problem (6.1, relaxing the first constraint from an equality to an inequality, and replacing the cost-minimization objective with the following:

$$\max \sum_{i=1}^{T} \sum_{j=1}^{C} x_{ij}.$$

Cardinality matching, which must generally be solved by integer or mixed integer programming, always guarantees a specified degree of balance, at the potential cost of sacrificing sample size. In addition it is always feasible, although constraints that are too difficult to satisfy may result in unusably tiny matches. It works especially well with hard balance constraints such as strength-k balance and controlled deviations from balance which can frequently lead to infeasibility when enforced while also requiring all treated units to be retained. As described cardinality matching does not make any direct use of pair costs, but these can be optimized subject to balance and maximal sample size, sometimes as a second step after an initial balanced set of samples is obtained. In this latter case when the match is also fully dense, the initial optimization problem can use a much smaller set of decision variables, indicating simply whether each treated or control unit is in the matched sample and not which other unit is matched to it. For discussion of these ideas and some associated computational advantages see Zubizarreta et al. [39] and Bennett et al. [32].

Both cardinality matching and the network-flow-based method for excluding treated units just described work by imposing some strict order of priority among the competing priorities of minimizing marginal balance, maximizing sample size, and minimizing pair distances. The network flow method places top priority on sample size (retaining as many treated units as possible), followed by balance, followed by pair costs; cardinality matching prioritizes balance first, then sample size, then pair costs. In practice some happy medium among these different goals is desired, and strictly prioritizing them in some order may not manage the tradeoffs in an ideal manner. In these settings one may instead seek to construct matches that are Pareto optimal for multiple objective functions, not achieving optimality on any one of the objectives but neither being dominated by any other solution on all objective functions. For example, consider the procedure variable in the IMG-USMG study, which cannot be balanced exactly when all treated units are retained. Matching with near-fine balance produces a solution in which all 241 treated units are retained and in which treated categories overflow control categories by only 7 units (for a total variation distance of $14/482 \approx 0.029$); cardinality matching under a fine balance constraint would select a match with only 234 treated units but in which procedure is exactly balanced. There also exists a match with 236 treated units and a total variation distance of $4/482 \approx 0.008$. This match is neither quite as large as the near-fine balanced match nor quite as well-balanced as the cardinality match but is not dominated on both objectives by either option, so it is Pareto optimal with respect to sample size and total variation imbalance. In particular, if we are comfortable with any total variation distance below 0.01 this match may be preferable to either.

Rosenbaum [40] introduced Pareto optimality in the context of optimal matched designs with a specific formulation exploring tradeoffs between achieving minimal average pair distances and maximal sample size, and King et al. [41] explored the set of matches Pareto optimal for sample size and covariate balance, although only in the special case of matching with replacement. Pimentel & Kelz [4] generalized the notion to include any objective function that can be represented as a linear function in a network flow problem and provided network flow algorithm that finds Pareto optimal solutions in settings where total variation imbalance on some variable is one of the objective functions; this method could in principle be used to recover the 236-pair match mentioned above for the IMG-USMG data More generally, matches with controlled deviations from balance may be understood as Pareto optimal solutions for the balance measure controlled. Using this approach, the 236-pair match could be recovered via a cardinality match with a constraint specifying total variation distance must not exceed 0.01.

6.4.5 Computational complexity theory

Optimal matching under balance constraints can be computationally demanding, especially for large datasets, and an important question is to understand how computation time scales as problems grow in size. A standard tool for addressing this question is worst-case complexity bounds. These bounds count the maximal number of mathematical operations required for a given algorithm to produce a solution and describe this number as a function of the size of the problem. A key distinction is between problems for which the worst-case computation time or complexity is known to be bounded by a polynomial in the size of the problem, and between another class of problems known as NP-hard, for which no polynomial bounds are known to exist and for which any polynomial bound for one problem in the class must also yield a polynomial bound for all other problems in the class. The network flow and assignment problems mentioned above as solution algorithms for matching have polynomial time bounds in the number of units to be matched; integer programming methods, in contrast, do not have worst-case polynomial bounds, and certain problems that can be formulated as integer programs are NP-hard. Bertsekas commented, "for our purposes, ... given a problem that ... has nonpolynomial complexity ... we should give up hope of formulating it as a minimum cost flow problem" [31, §8]. However, just because a matching problem does not admit an obvious network flow solution does not immediately mean it is NP-hard.

In recent work, Hochbaum et al. [42] explore these issues for fully dense matches and provide definitive statements about the worst-case complexity of many matching problems with balance. In particular, they focus on minimum cost pair matching with fine balance (i.e. problem (6.1) with one or more fine balance constraints) and cardinality matching with fine balance, both in the case where no pair costs are considered and in the case where pair costs are minimized subject to maximum cardinality under the balance constraints. They require that all treated-control edges are present (i.e. no exact matching or caliper constraints) and they consider a regime in which the number of categories of the fine balance variable(s) grows with the number of units. Under these conditions, two key quantities play an important role in determining complexity: the number of covariates on which fine balance is required, and the control:treated matching ratio κ. When only one fine balance covariate is present and $\kappa = 1$, both minimum-cost matching and both forms of cardinality matching have polynomial time guarantees. When $\kappa > 1$ and one fine balance covariate is present, both minimum-cost pair matching and cardinality matching ignoring pair costs have polynomial time guarantees; however, cardinality matching with optimal pair costs is NP-hard for $\kappa \geq 3$ and the case with $\kappa = 2$ remains open. For two fine balance covariates, both forms of cardinality matching are NP-hard regardless of κ, while minimum-cost pair matching admits a polynomial-time solution if $\kappa = 1$ and is NP-hard for $\kappa \geq 3$ with the $\kappa = 2$ case open. Finally, if three fine balance covariates are present all the methods are NP-hard.

If the number of levels for each fine balance covariate is capped at some small value rather than being allowed to grow so that the parts of the computation that scale with the number of categories in play may be treated as a constant, Hochbaum et al. [42] argue that polynomial time bounds are available over a much broader array of settings. In particular, minimum-cost pair matching with balance constraints and cardinality matching without pair costs both generally have polynomial time algorithms for any fixed number of covariates. However, cardinality matching with minimum-cost pairing is still NP-hard with $\kappa \geq 3$ and remains open at $\kappa = 2$.

Of course, all of these results refer only to asymptotic worst-case performance, and may not provide perfect guidance about computational performance in practice. On the one hand, pathological worst-case settings are not always particularly plausible or common in problems arising from real-world data, which may be quicker to solve on average. On the other hand, polynomial time bounds may involve large constants which are ignored in the complexity analysis but may lead to long computation in practice. Furthermore, the worst-case results given here apply only to exact solutions, and near-optimal solutions may require very different levels of computational effort. We note in particular that the R package `designmatch` has been shown to solve certain integer programming formulations of matching problems with high worst-case complexity efficiently in practice, and can also produce approximate solutions for reduced effort [32].

6.5 Balancing to an External Population

When experimenters wish to prioritize internal validity above all else or when the existing treated sample is known to be representative of some larger population of interest, optimizing similarity between the matched controls and the treated sample in hand is reasonable. However, in some cases there are reasons to optimize similarity between the matched samples and some other distribution or set of distributions. In general, balance constraints can be adapted to do this.

One example is a setting where causal inference may benefit from dissimilarities between groups on a given variable. Pimentel et al. [43] discuss the role of "innocuous covariates" that are conditionally independent of outcomes given other variables in the study but have strong associations with treatment. In settings where unobserved confounding is also present, matched samples that differ substantially in the innocuous variable have reduced bias compared to those that match them exactly. In these settings it is desirable to select controls from an empirical distribution that resembles the treated group for most covariates but differs on covariates known or strongly suspected to be innocuous.

Another common setting where the "target" distribution for balance differs from the observed treated sample is problem where it is important to establish external validity with respect to some larger population of interest for which summary statistics are available. In Bennett et al. [32], effects of a natural disaster on school attendance in Chile was assessed by matching students in regions experiencing a massive earthquake to students outside that region on demographic and past educational performance variables and comparing their rates of attendance. However, to the degree effects of natural disasters on school attendance vary across regions of Chile, the researchers in this study are interested in learning not about the average effect in the regions that happened to be affected by this earthquake but in the average across all high school students in the entire country. As such, rather than taking the treated group as a given and choosing similar controls, it is desirable to select both the treated group and the control group to be similar in their empirical distributions to the overall distribution in the entire database of Chilean high school students.

The case of balancing treated and control groups to a third more general population is similar mechanically to a method called template matching, which was introduced not for purposes of measuring causal effects but of evaluating quality or performance among a large number of institutions such as hospitals which may differ systematically in the population of subjects or units they serve [44]. In template matching these differences are accounted for by taking a single template, usually drawn as a random sample from a larger population of interest (such as all the subjects across all the institutions) and matching subjects in each institution to those in the template, thereby comparing performance in each institution on a common set of subjects with a representative profile.

How can standard fine, near-fine, and related balance constraints be adapted to this setting? In the case of template matching or balancing both groups to a general population, this is simply done by repeated use of the standard bipartite matching approach, treating the template or a sample from the external population as the treated group and treating each of the other comparison groups in turn as the control pool. Bennett et al. [32] give a helpful mathematical formulation of this repeated optimization process. The situation is slightly different in the case where a difference on some variable is desired, since the treated group is held fixed, and since the desired target distribution to which to make controls similar is not necessarily observed directly. In this case the researcher first defines a perturbed version of the treated sample in which most covariates are held fixed but values of the innocuous variable are changed in the direction of anticipated bias reduction. Then standard fine or near-fine balance constraints are defined with respect to the new perturbed target population, by choosing appropriate constraints in a network flow problem; see online appendix to Pimentel et al. [43] for full details.

6.6 Practical Recommendations for Matching with Balance Constraints

The wealth of possible formulations of balance and the many possibilities offered by balancing interactions of discrete variables or transformations of continuous variables make matching with balance constraints a potentially challenging tuning problem. In general different matches are obtained under different sets of balance constraints. Furthermore, choosing one set of balance constraints may limit performance on another set of balance variables, so that the choice of which set of constraints to apply is consequential.

A useful heuristic is the notion of some fixed "budget" of flexibility in the matched sets chosen. Every time a balance constraint is imposed, some degree of flexibility is spent to enforce it, leaving less flexibility to enforce other constraints. This suggests the importance of establishing priority among which covariates to balance initially, as the refined balance framework explicitly requires. In practice, establishing priority among variables involves a mix of scientific insight into the mechanisms of possible confounding and of data-driven selection of covariates with strong potential for confounding.

We suggest two particularly important tools to aid in the data-driven part of selecting balance constraints. Firstly, where possible it is highly desirable to fit a prognostic model for the outcome variable among control units. Where historical controls not being used in the study are available these data can be used to fit the model. Alternatively, a random sample of the control data can be removed from the matching pool in order to fit the prognostic model, or a group of controls not selected for the match can be used [8, 45]. Variable importance can be assessed for the variables used in fitting the score, for example by comparing the size of standardized coefficients in a linear model, and those with the highest importance in the model can be prioritized; alternatively, the risk score itself, possibly discretized into quintiles, can be used as a top-priority balance variable.

Secondly, it is recommended to examine multiple specifications of the balance constraints to understand in-sample tradeoffs in balance on different variables, as well as on other important match quality metrics such as sample size. The tradeoffs algorithm of Pimentel & Kelz [4] provides a structured approach to exploring tradeoffs between two objective functions. The analysis given focuses on exploring tradeoffs by repeatedly solving network flow problems with different penalty parameters, and can provide insight into the impact of a balance constraint, under varying degrees of priority, on other aspects off the match; it could also be used to examine the impact of increasing priority on one discrete variable in the refined balance framework relative to balance on another nested variable. Similar principles may be applied in an integer programming context, for example by gradually increasing the tolerance on one set of controlled-deviation-from-balance constraints while holding other aspects of the matching problem fixed.

6.7 Discussion

Balance constraints of some kind are virtually always a good idea in matching. In the author's experience, a single balance constraint or two can almost always be added to a problem with only minimal impact on the average similarity of pairs. This practice both guarantees improvements in observed balance (as shown in a balance table) and tends to reduce bias under a wide variety of statistical models.

While this review has focused on the role of balance constraints in observational studies, covariate balance is also a topic of central interest in modern experimental design. In particular, rerandomized experimental designs offer balance guarantees for randomized trial similar to those achieved by balance constraints [46]. Algorithms for constructing rerandomized designs differ substantially from those for creating optimal matches because treatment assignments are not fixed in value, but results about inference under the two settings may be mutually relevant. Balance is also important in the gray area between classical observational studies and experiments occupied by natural experiments and instrumental variable designs; for an application of refined balance as a way to strengthen the credibility of an instrumental variable, see Keele et al. [47].

Matching methods that focus almost exclusively, or with top priority, on multiple balance constraints, such as cardinality matching as described in [32], have many similarities with modern optimal weighting methods such as entropy balancing [48], covariate balancing propensity scores [49], stable balancing weights [50], and overlap weights [51]. All of these methods find a continuous re-weighting of one group to match the distribution of the other under constraints that covariates or their transformations exhibit exact balance or controlled or minimized deviation from balance. Loosely speaking, matching can be understood as a restricted form of weighting in which weights are restricted to take on values of 0 and 1 only (for more in-depth discussion of this connection, see Kallus [11]). Indeed, theoretical analyses such as those in Wang & Zubizarreta [9] exploit this connection to prove asymptotic results about matching. As a general rule, we expect many of the documented benefits of matching under balance constraints to extend to optimal balancing weights approaches and vice versa.

Many important methodological problems involving matching with balance constraints remain to be solved. One direction is in providing solution algorithms and complexity analyses for various balance constraints in more complicated matching settings, such as nonbipartite matching [52], multilevel matching [53, 54], and matching in longitudinal settings including risk-set matching [55] and designs with rolling enrollment [56]. While individual studies have used balance constraints of some kind in these settings in an incidental manner, comprehensive studies of balance constraints in these settings are not extant.

In the area of statistical practice, there is also a need for improved software and guidance for practitioners in the use of balance constraints. For instance, most of the work reviewed above focuses on matching without replacement. While balance constraints are known to have theoretical benefits for matching with replacement as well, applied examples and software adding balance constraints to methods for matching with replacement are currently lacking.

At a more abstract level, another important question is understanding the limits and value of balance constraints in settings where the number of covariates is very large. One major selling point of balance constraints relative to near-exact matching in any given finite sample is that they are easier to satisfy; even in settings where it is impossible to pair individuals exactly on all measured covariates, it may still be possible to balance the marginal distributions of all or most measured covariates exactly by aligning inexact matches in a way so that they cancel. In this way balance constraints offer a partial solution, or at least a mitigation, for the curse of dimensionality. However, satisfying marginal balance on multiple covariates also becomes increasingly difficult as the number of covariates to be balanced grows, and it is not clear how much the balancing approach buys us in asymptotic regimes where p grows quickly with n.

Finally, matches with balance constraints pose interesting questions for matched randomization inference. While simulation studies and theoretical analyses such as those of Resa & Zubizarreta [12] and Wang & Zubizarreta [9] have focused on statistical guarantees over samples from a population model on the potential outcomes, a more common way to analyze matched designs with balance constraints in practice is to condition on the potential outcomes and the pairs selected and compare observed test statistics to those obtained by permuting treatment labels within pairs. This method, proposed originally by Fisher [57] for randomized experiments and adapted by Rosenbaum [2] for matched studies, frees researchers from the need to make strong assumptions about sampling or outcome models, and admits attractive methods of sensitivity analysis to assess the role of unobserved confounding. The key requirement needed to develop these methods in the absence of actual randomization is sufficiently similar propensity scores within matched pairs [29]. Balance constraints alone provide no such guarantee, so that the validity of randomization inference depends on other features of the match such as the presence of a propensity score caliper. Furthermore, since imbalance is a function of both treatment status and covariates, it is not preserved under permutations of treatment status within pairs; this means that when randomization inference is performed in a matched design with a balance constraint, the constraint need not hold in the null distribution draws obtained by permuting treatment labels. Li et al. [58] consider an analogous discrepancy between data configurations permitted by the design and those considered by the inferential approach in the randomized case and finds that analyzing a rerandomized design while ignoring the rerandomization preserves Type I error control and suffers only in terms of precision. However, it is not yet clear how the story plays out in the observational case.

6.8 R Code Appendix

```
###### PREPROCESS DATA #####
load('simulated_data.Rdata')
library(rcbalance)

#take a random sample of 5 hospitals in this dataset
set.seed(2021-4-29)
my.hosps <- sample(unique(my.slice$hospid), size = 5)
mini.dat <- my.slice[my.slice$hospid %in% my.hosps,]
mini.dat$hospid <- droplevels(mini.dat$hospid)

#create robust Mahalanobis distance for match within hospitals
match.covs <- c('proc_type', 'comorb', 'emergency', 'mf', 'age', 'exp',
                'med_income', 'race_cat')
my.dist <- build.dist.struct(mini.dat$img, mini.dat[match.covs], calip.option =
                             'none', exact =  mini.dat$hospid)

#### Section 1.1: fine balance example #####

#overall distribution of ER admission across group
table(mini.dat$emergency, mini.dat$img) #raw counts
table(mini.dat$img, mini.dat$emergency)/matrix(rep(table(mini.dat$img),2),
                                               ncol = 2) #percentages
```

```
#match with fine balance on emergency room status
match.fb.er <- rcbalance(my.dist, fb.list = c('emergency'), treated.info =
                         mini.dat[mini.dat$img == 1,], control.info =
                         mini.dat[mini.dat$img == 0,])
match.fb.er$fb.tables[[1]] #raw counts
match.fb.er$fb.tables[[1]]/nrow(match.fb.er$matches) #percentages

#count how many matches are exact on emergency room
match.fb.er$fb.tables[[1]]
emergency.comp <-paste(mini.dat$emergency[mini.dat$img == 1][
  as.numeric(rownames(match.fb.er$matches))],
  mini.dat$emergency[mini.dat$img == 0][match.fb.er$matches], sep = '.')
table(emergency.comp)

#repeat for match without fine balance
match.naive <- rcbalance(my.dist)
naive.dat <- rbind(mini.dat[which(mini.dat$img == 1)[
  as.numeric(rownames(match.naive$matches))],],
  mini.dat[which(mini.dat$img == 0)[match.naive$matches],])
table(naive.dat$emergency, naive.dat$img)
emergency.comp <-paste(mini.dat$emergency[mini.dat$img == 1][
  as.numeric(rownames(match.naive$matches))], mini.dat$emergency[mini.dat$img ==
          0][match.naive$matches], sep = '.')
table(emergency.comp)

###### Section 1.3.2: near-fine balance #####


match.proc <- rcbalance(my.dist, fb.list = c('proc_type'), treated.info =
                          mini.dat[mini.dat$img == 1,], control.info =
                          mini.dat[mini.dat$img == 0,])

#replicate Table 1.2
matched.ind <- rep(FALSE, nrow(mini.dat))
matched.ind[mini.dat$img == 1] <- TRUE
matched.ind[mini.dat$img==0][match.proc$matches] <- TRUE
proc_type_factor <- as.factor(mini.dat$proc_type)
print.tab <- table(mini.dat$img, proc_type_factor)
print.tab <- rbind(print.tab[2:1,], table(proc_type_factor[mini.dat$img == 0 &
                                                           matched.ind]))
print.tab <- rbind(print.tab, abs(print.tab[1,] - print.tab[3,]))
rownames(print.tab) <- c('IMG', 'USMG','Matched USMG',
                         'Discrepancy')
library(xtable)
xtable(print.tab[,1:16])
xtable(print.tab[,17:32])

##### Section 1.3.3: refined balance #####

#Example: near-fine balance on interactiond doesn't guarantee it on components
#create interaction
proc.er <- paste(mini.dat$proc_type, mini.dat$emergency, sep = '.')
mini.dat$proc_er <- proc.er
mytab <- table(proc.er, mini.dat$img)
sum(pmax(mytab[,2] - mytab[,1],0)) # 12 units of overflow

match.proc.er <- rcbalance(my.dist, fb.list = c('proc_er'), treated.info =
                           mini.dat[mini.dat$img == 1,], control.info =
                           mini.dat[mini.dat$img == 0,])

#check near-fine balance is achieved on interaction
sum(abs(match.proc.er$fb.tables[[1]][,1]-match.proc.er$fb.tables[[1]][,2]))/2

#near-fine balance not achieved on individual component variables
#create dataframe with only matched data
```

```
matched.data.nfb <- rbind(mini.dat[mini.dat$img == 1,][
  as.numeric(rownames(match.proc.er$matches)),],
  mini.dat[mini.dat$img == 0,][match.proc.er$matches,])
mytab <- table(matched.data.nfb$proc_type, matched.data.nfb$img)
sum(pmax(mytab[,2] - mytab[,1],0)) # 12 overflow whe we achieved 7 above
#near-fine balance not achieved on emergency either
table(matched.data.nfb$emergency, matched.data.nfb$img)
#6 overflow, when we achieved zero above.

#Now repeat with refined balance on:
# emergency
# emergency x procedure
# emergency x procedure x sex
match.rfb <- rcbalance(my.dist, fb.list = list(c('emergency'),
                                              c('emergency', 'proc_type'),
                                              c('emergency', 'proc_type','mf')),
                       treated.info = mini.dat[mini.dat$img == 1,],
                       control.info = mini.dat[mini.dat$img == 0,])
#check overflows at each level
laply(match.rfb$fb.tables, function(x) sum(pmax(x[,2]-x[,1],0)))

#check balance overall on gender
match.data.rfb <- rbind(mini.dat[mini.dat$img == 1,][
  as.numeric(rownames(match.rfb$matches)),],
  mini.dat[mini.dat$img == 0,][match.rfb$matches,])
table(match.data.rfb$mf, match.data.rfb$img)
#compare to balance in match with only near-fine balance on the interaction
table(matched.data.nfb$mf, matched.data.nfb$img)
#gender balance has improved from a discrepancy of 11 to only 7


##### Section 1.3.4: strength-k matching ######

library(designmatch)
#sort data in order of treatment variable as designmatch requires
mini.sort <- mini.dat[order(mini.dat$img, decreasing = TRUE),]
#convert my.dist to matrix
dist.mat <- matrix(1000, nrow = sum(mini.sort$img), ncol = sum(1 - mini.sort$img))
for(i in 1:length(my.dist)){
  dist.mat[i,as.numeric(names(my.dist[[i]]))] <- my.dist[[i]]
}
#strength-1 matching on sex, ER, race, and comorbidity count
my.match <- bmatch(t_ind = mini.sort$img, dist_mat = dist.mat,
                   total_groups = sum(mini.sort$img),
                  exact = mini.sort[c('hospid')],
                  fine = list(covs = mini.sort[c('mf','emergency','race_cat',
                     'comorb')]))

matched.data <- rbind(mini.sort[my.match$t_id,], mini.sort[my.match$c_id,])
#confirm we have fine balance on each variable
table(matched.data$mf, matched.data$img)
table(matched.data$emergency, matched.data$img)
table(matched.data$race_cat, matched.data$img)
table(matched.data$comorb, matched.data$img)
mytab <- table(paste(matched.data$mf, matched.data$comorb, matched.data$race_cat,
            matched.data$emergency, sep = '.'), matched.data$img)
sum(pmax(mytab[,2]-mytab[,1],0))


#check that you don't get this by near-fine balance on the interaction
my.match.alt <- rcbalance(my.dist, fb.list =
                            list(c('emergency','mf','race_cat','comorb')),
                          treated.info = mini.dat[mini.dat$img == 1,],
                          control.info = mini.dat[mini.dat$img == 0,])
matched.data <- rbind(mini.dat[mini.dat$img == 1,][
  as.numeric(rownames(my.match.alt$matches)),],
  mini.dat[mini.dat$img == 0,][my.match.alt$matches,])
```

```
table(matched.data$mf, matched.data$img)
table(matched.data$emergency, matched.data$img)
table(matched.data$race_cat, matched.data$img)
table(matched.data$comorb, matched.data$img)
```

References

[1] Paul R Rosenbaum. *Design of Observational Studies*. Springer, New York, NY, 2010.

[2] Paul R Rosenbaum. *Observational Studies*. Springer, New York, NY, 2002.

[3] Salman Zaheer, Samuel D Pimentel, Kristina D Simmons, Lindsay E Kuo, Jashodeep Datta, Noel Williams, Douglas L Fraker, and Rachel R Kelz. Comparing international and united states undergraduate medical education and surgical outcomes using a refined balance matching methodology. *Annals of Surgery*, 265(5):916–922, 2017.

[4] Samuel D Pimentel and Rachel R Kelz. Optimal tradeoffs in matched designs comparing us-trained and internationally trained surgeons. *Journal of the American Statistical Association*, 115(532):1675–1688, 2020.

[5] Donald B Rubin. Bias reduction using mahalanobis-metric matching. *Biometrics*, 36(2):293–298, 1980.

[6] Donald B Rubin and Neal Thomas. Affinely invariant matching methods with ellipsoidal distributions. *The Annals of Statistics*, 20(2):1079–1093, 1992.

[7] Donald B Rubin. Matching to remove bias in observational studies. *Biometrics*, 29:159–183, 1973.

[8] Rachael C Aikens, Dylan Greaves, and Michael Baiocchi. A pilot design for observational studies: Using abundant data thoughtfully. *Statistics in Medicine*, 39(30):4821–4840, 2020.

[9] Yixin Wang and José R Zubizarreta. Large sample properties of matching for balance. *arXiv preprint arXiv:1905.11386*, 2019.

[10] Jason J Sauppe, Sheldon H Jacobson, and Edward C Sewell. Complexity and approximation results for the balance optimization subset selection model for causal inference in observational studies. *INFORMS Journal on Computing*, 26(3):547–566, 2014.

[11] Nathan Kallus. Generalized optimal matching methods for causal inference. *Journal of Machine Learning Research*, 21(62):1–54, 2020.

[12] Marıa de los Angeles Resa and José R Zubizarreta. Evaluation of subset matching methods and forms of covariate balance. *Statistics in Medicine*, 35(27):4961–4979, 2016.

[13] Paul R Rosenbaum, Richard N Ross, and Jeffrey H Silber. Minimum distance matched sampling with fine balance in an observational study of treatment for ovarian cancer. *Journal of the American Statistical Association*, 102(477):75–83, 2007.

[14] Dan Yang, Dylan S Small, Jeffrey H Silber, and Paul R Rosenbaum. Optimal matching with minimal deviation from fine balance in a study of obesity and surgical outcomes. *Biometrics*, 68(2):628–636, 2012.

[15] Samuel D Pimentel, Rachel R Kelz, Jeffrey H Silber, and Paul R Rosenbaum. Large, sparse optimal matching with refined covariate balance in an observational study of the health outcomes produced by new surgeons. *Journal of the American Statistical Association*, 110(510):515–527, 2015.

[16] Jesse Y Hsu, José R Zubizarreta, Dylan S Small, and Paul R Rosenbaum. Strong control of the familywise error rate in observational studies that discover effect modification by exploratory methods. *Biometrika*, 102(4):767–782, 2015.

[17] José R Zubizarreta. Using mixed integer programming for matching in an observational study of kidney failure after surgery. *Journal of the American Statistical Association*, 107(500):1360–1371, 2012.

[18] Yixin Wang and Jose R Zubizarreta. Minimal dispersion approximately balancing weights: Asymptotic properties and practical considerations. *Biometrika*, 107(1):93–105, 2020.

[19] Rainer Burkard, Mauro Dell'Amico, and Silvano Martello. *Assignment problems: Revised reprint*. SIAM, 2012.

[20] Lester Randolph Ford, Jr and Delbert Ray Fulkerson. *Flows in Networks*. Princeton University Press, Princeton, NJ, 1962.

[21] Christos H Papadimitriou and Kenneth Steiglitz. *Combinatorial Optimization: Algorithms and Complexity*. Courier Corporation, North Chelmsford, MA, 1982.

[22] Dimitri P Bertsekas. *Linear Network Optimization: Algorithms and Codes*. MIT Press, Cambridge, MA, 1991.

[23] Dimitri P Bertsekas and Paul Tseng. Relax-iv: A faster version of the relax code for solving minimum cost flow problems. Technical report, Massachusetts Institute of Technology, Laboratory for Information and Decision Systems, Cambridge, MA, 1994.

[24] Ben B. Hansen and Stephanie Olsen Klopfer. Optimal full matching and related designs via network flows. *Journal of Computational and Graphical Statistics*, 15(3):609–627, 2006.

[25] Ruoqi Yu. *bigmatch: Making Optimal Matching Size-Scalable Using Optimal Calipers*, 2020. R package version 0.6.2.

[26] Samuel D Pimentel. Large, sparse optimal matching with r package rcbalance. *Observational Studies*, 2:4–23, 2016.

[27] Stefano M Iacus, Gary King, and Giuseppe Porro. Multivariate matching methods that are monotonic imbalance bounding. *Journal of the American Statistical Association*, 106(493): 345–361, 2011.

[28] Jesse Y Hsu, Dylan S Small, and Paul R Rosenbaum. Effect modification and design sensitivity in observational studies. *Journal of the American Statistical Association*, 108(501):135–148, 2013.

[29] Ben Hansen. Propensity score matching to recover latent experiments: Diagnostics and asymptotics. Technical Report 486, University of Michigan, 2009.

[30] Alberto Abadie and Guido W Imbens. Large sample properties of matching estimators for average treatment effects. *Econometrica*, 74(1):235–267, 2006.

[31] Dimitri P Bertsekas. *Network Optimization: Continuous and Discrete Models*. Athena Scientific, Belmont, MA, 1998.

[32] Magdalena Bennett, Juan Pablo Vielma, and José R Zubizarreta. Building representative matched samples with multi-valued treatments in large observational studies. *Journal of Computational and Graphical Statistics*, 29(4):744–757, 2020.

[33] Jose R. Zubizarreta, Cinar Kilcioglu, and Juan P. Vielma. *designmatch: Matched Samples that are Balanced and Representative by Design*, 2018. R package version 0.3.1.

[34] Kewei Ming and Paul R Rosenbaum. Substantial gains in bias reduction from matching with a variable number of controls. *Biometrics*, 56(1):118–124, 2000.

[35] Kewei Ming and Paul R Rosenbaum. A note on optimal matching with variable controls using the assignment algorithm. *Journal of Computational and Graphical Statistics*, 10(3):455–463, 2001.

[36] Paul R Rosenbaum. Optimal matching for observational studies. *Journal of the American Statistical Association*, 84(408):1024–1032, 1989.

[37] Samuel D Pimentel, Frank Yoon, and Luke Keele. Variable-ratio matching with fine balance in a study of the peer health exchange. *Statistics in Medicine*, 34(30):4070–4082, 2015.

[38] Cinar Kilcioglu and José R Zubizarreta. Maximizing the information content of a balanced matched sample in a study of the economic performance of green buildings. *The Annals of Applied Statistics*, 10(4):1997–2020, 2016.

[39] José R Zubizarreta, Ricardo D Paredes, and Paul R Rosenbaum. Matching for balance, pairing for heterogeneity in an observational study of the effectiveness of for-profit and not-for-profit high schools in chile. *The Annals of Applied Statistics*, 8(1):204–231, 2014.

[40] Paul R. Rosenbaum. Optimal Matching of an Optimally Chosen Subset in Observational Studies. *Journal of Computational and Graphical Statistics*, 21(1):57–71, 2012.

[41] Gary King, Christopher Lucas, and Richard A Nielsen. The balance-sample size frontier in matching methods for causal inference. *American Journal of Political Science*, 61(2):473–489, 2017.

[42] Dorit S Hochbaum, Asaf Levin, and Xu Rao. Algorithms and complexity for variants of covariates fine balance. *arXiv preprint arXiv:2009.08172*, 2020.

[43] Samuel D Pimentel, Dylan S Small, and Paul R Rosenbaum. Constructed second control groups and attenuation of unmeasured biases. *Journal of the American Statistical Association*, 111(515):1157–1167, 2016.

[44] Jeffrey H Silber, Paul R Rosenbaum, Richard N Ross, Justin M Ludwig, Wei Wang, Bijan A Niknam, Nabanita Mukherjee, Philip A Saynisch, Orit Even-Shoshan, Rachel R Kelz, et al. Template matching for auditing hospital cost and quality. *Health Services Research*, 49(5):1446–1474, 2014.

[45] Adam C Sales, Ben B Hansen, and Brian Rowan. Rebar: Reinforcing a matching estimator with predictions from high-dimensional covariates. *Journal of Educational and Behavioral Statistics*, 43(1):3–31, 2018.

[46] Kari Lock Morgan and Donald B Rubin. Rerandomization to improve covariate balance in experiments. *The Annals of Statistics*, 40(2):1263–1282, 2012.

[47] Luke Keele, Steve Harris, Samuel D Pimentel, and Richard Grieve. Stronger instruments and refined covariate balance in an observational study of the effectiveness of prompt admission to intensive care units. *Journal of the Royal Statistical Society: Series A (Statistics in Society)*, 183(4):1501–1521, 2020.

[48] Jens Hainmueller. Entropy balancing for causal effects: A multivariate reweighting method to produce balanced samples in observational studies. *Political Analysis*, 20:1, 25–46, 2012.

[49] Kosuke Imai and Marc Ratkovic. Covariate balancing propensity score. *Journal of the Royal Statistical Society: Series B: Statistical Methodology*, 76(1):243–263, 2014.

[50] José R Zubizarreta. Stable weights that balance covariates for estimation with incomplete outcome data. *Journal of the American Statistical Association*, 110(511):910–922, 2015.

[51] Fan Li, Kari Lock Morgan, and Alan M Zaslavsky. Balancing covariates via propensity score weighting. *Journal of the American Statistical Association*, 113(521):390–400, 2018.

[52] Bo Lu, Robert Greevy, Xinyi Xu, and Cole Beck. Optimal nonbipartite matching and its statistical applications. *The American Statistician*, 65(1):21–30, 2011.

[53] José R Zubizarreta and Luke Keele. Optimal multilevel matching in clustered observational studies: A case study of the effectiveness of private schools under a large-scale voucher system. *Journal of the American Statistical Association*, 112(518):547–560, 2017.

[54] Samuel D Pimentel, Lindsay C Page, Matthew Lenard, Luke Keele, et al. Optimal multilevel matching using network flows: An application to a summer reading intervention. *The Annals of Applied Statistics*, 12(3):1479–1505, 2018.

[55] Yunfei Paul Li, Kathleen J Propert, and Paul R Rosenbaum. Balanced risk set matching. *Journal of the American Statistical Association*, 96(455):870–882, 2001.

[56] Samuel D Pimentel, Lauren V Forrow, Jonathan Gellar, and Jiaqi Li. Optimal matching approaches in health policy evaluations under rolling enrollment. *Journal of the Royal Statistical Society: Series A (Statistics in Society)*, 2020. in press.

[57] Ronald A Fisher. *The Design of Experiments*. Oliver and Boyd, Edinburgh, 1935.

[58] Xinran Li, Peng Ding, and Donald B Rubin. Asymptotic theory of rerandomization in treatment–control experiments. *Proceedings of the National Academy of Sciences*, 115(37):9157–9162, 2018.

7

Matching with Instrumental Variables

Mike Baiocchi and Hyunseung Kang

CONTENTS

In this chapter we first introduce the assumptions and frequently used terminology in instrumental variables (IV). The introduction is meant to illustrate the basics of how an instrumental variables analysis works in the context of optimal matching. The literature on instrumental variables is large, spanning nearly 100 years and intersecting multiple disciplines, and we recommend the following literature for a more thorough perspective: [1–5], and [6].

After the introduction to instrumental variables, we explore how to design matched sets that incorporate IVs. At a high level, a matching-based instrumental variables analysis replaces the treatment variable discussed in previous chapters with an instrumental variable. The goal in IV matching is to group units with different values of the instrument but similar values of the observed covariates. This design produces sets wherein the salient difference between the units is their values of the instrument [7, 8]. We then compare the relative differences in both the treatment and the outcome variables within each matched set to assess the causal effect of the treatment on the outcome.

To make the discussions more concrete, throughout the chapter, we use a real example concerning the effects of malaria among children; the dataset was generously provided by Dr. Benno Kreuels and Professor Jürgen May. The example studies the effect of malaria on stunted growth among children in sub-Saharan Africa using a binary instrumental variable based on a sickle cell genotype. In addition to highlighting how to use matching-based instrumental variables, the real example points out some common, but important discussions about using an instrumental variables analysis, notably how to justify an instrumental variable's plausibility.

DOI: 10.1201/9781003102670-7

7.1 A Brief Background on Instrumental Variables

7.1.1 Motivation: The problem of unmeasured confounding in observational studies

As discussed in previous chapters, in observational studies, a principle challenge to estimating the casual effect of treatment on the outcome is the non-random sorting of study units into different levels of the treatment, which introduces both overt (i.e., measurable) and hidden (i.e., unmeasurable) biases; see Chapter 3.1.1 of [9] for more discussion in the context of matching. The propensity score [10] provides a scalar summary of how measured confounders are correlated with the sorting into treatment levels. In the special case where there is only overt bias in an observational study, matching study units with identical propensity scores can remove the overt bias; this special case is often referred to as strongly ignorable treatment assignment (SITA).

However, SITA will often be suspect in observational studies. Specifically, in the case of a binary treatment variable, a study unit's propensity to take the treatment (or control) may depend on other characteristics beyond measured confounders, say unmeasured confounders.

As a concrete example, consider the study by [11] and [12] on the effect of malaria on stunted growth among children in sub-Saharan Africa. Briefly, in 2013 alone, there were 128 million estimated cases of malaria in sub-Saharan Africa, with most cases occurring in children under the age of 5 [13], and stunting, defined as a child's height being two standard deviations below the mean for his/her age, is a key indicator of child development [14]. If malaria does cause stunted growth, several intervention strategies can be implemented to mitigate stunted growth, such as distribution of mosquito nets, control of the mosquito population during seasons of high malarial incidence, and surveillance of mosquito populations.

Analyzing the causal effect of malaria on stunted growth must address confounders, especially unmeasured confounders. For example, diet, specifically a child's intake of micronutrients such as vitamins, zinc, or iron has a critical impact a child's growth as well as his immune system's ability to fight off a malaria episode [15, 16]. Unfortunately, precisely measuring a child's nutrition is generally difficult and unmeasured confounders are likely present in any observational study concerning the effect of malaria on stunting.

7.1.2 Instrumental variables: Assumptions, definitions, and examples

Instrumental variables (IVs) are commonly used to estimate the causal effect of an exposure on the outcome when there is unmeasured confounding, provided that a valid instrument is available. The core assumptions for a variable to be a valid instrumental variable are that the variable (A1) is associated with the exposure, (A2) has no direct pathways to the outcome, and (A3) is not associated with any unmeasured confounders after controlling for the measured confounders; see Figure 7.1 and Section 7.3.3 for more concrete discussions.

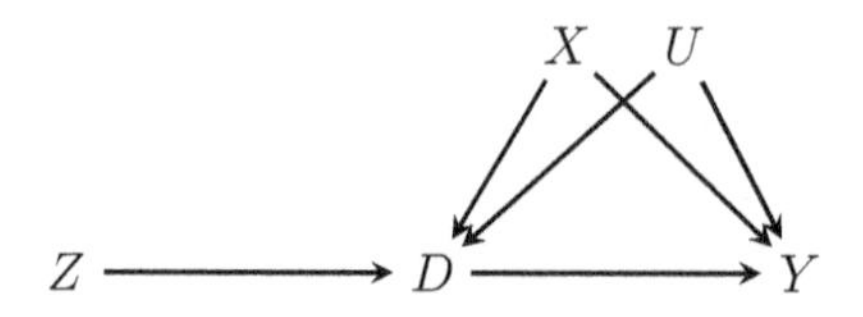

FIGURE 7.1
Causal directed acyclic graph with instrument Z, treatment D, outcome Y, measured confounder X, and unmeasured confounder U. Solid lines are non-zero causal paths. The graph represents the "prototypical' instrumental variable setting where there are no confounders associated with the instrument Z, say if Z is based on a randomized encouragement design.

TABLE 7.1
Compliance classes in a randomized encouragement design trial.

Compliance Class	Encourage to Take Treatment	Encouraged to Take Control
Complier	Take treatment	Take control
Always-taker	Take treatment	Take treatment
Never-taker	Take control	Take control
Defier	Take control	Take treatment

A randomized encouragement design [17] is often considered to be a prototypical example of an instrumental variable. Briefly, in a randomized encouragement design, the researcher encourages or nudges a participant to select a particular treatment, but the participant ultimately decides which level of the treatment is taken; in short, the nudge or encouragement to take a particular treatment remains unconfounded through randomization, but the actual treatment taken remains confounded. For example, consider a study of the causal effect of a weightlifting program on blood pressure. Under a randomized encouragement design, researchers may randomly encourage participants to one of two arms of the trial (i.e., either perform the weightlifting program or continue with their usual level of activity), say by randomly distributing discount coupons to the weightlifting program to some participants. If the researcher can precisely measure and quantify every aspect of the participant's decision to participate in the weightlifting program, then a standard SITA-based analysis where the propensity score includes all of these factors that contribute to the participant's propensity to participate in the weightlifting program can be used to study the casual effect of the weightlifting program. However, in many cases, researchers cannot measure all the factors contributing to participant's decision-making process. Instead, using the randomized encouragement as an instrument may be more appropriate to learn more about the causal effect of the treatment on the outcome.

To study the causal effect of the treatment on the outcome in a randomized encouragement design (or more generally, in an IV design), it's helpful to divide up the study participants into four, mutually exclusive types described in Table 7.1. Following the terminology from [1], compliers are group of individuals who take the treatment when encouraged to do so and who don't take the treatment when not encouraged to do so. Never-takers are participants who would never take the treatment regardless of the encouragement status. Always-takers are participants who would take the treatment regardless of the encouragement status. Defiers are participants that would take the opposite of their encouragement status. [1] showed that under assumptions (A1)–(A3) and if there are no defiers, an assumption known as monotonicity (or assumption (A4) for short), an IV analysis can identify the average causal effect among the compliers (CACE); see Section 7.2.3 for details.

The fact that we can only identify a subgroup's treatment effect presents an important caveat in an IV analysis. Usually, other subgroups will have meaningfully different causal responses to the treatment. Yet, the usual IV analysis produces an unbiased causal estimate only for the compliers subgroup. The causal effect estimated by the IV is not guaranteed to be useful in understanding the causal effect for other participants in the study or other members of the target population unless additional assumptions are met about treatment effect homoegeneity between the subgroups. More generally, this lack of identification of causal effects across different subgroups can be seen as a cost of allowing for unmeasured confounding in the treatment variable.

Finally, we remark that a randomized encouragement design is not the only study design in the IV literature. Far more frequent in practice is an instrumental variable analysis based on an observational study. In an observational study the instrument is not controlled by the researcher and therefore its exact characteristics tend to be less well understood than a researcher-directed randomization-based instrument from an encouragement design. Considerable care must be exercised when choosing an instrument in observational studies and our empirical example about malaria illustrates one example of justifying an instrument derived from an observational study. For example, in the malaria example above, genotypic variations are used as instruments, specifically the presence of a sickle cell genotype (HbAS) versus carrying the normal hemoglobin type (HbAA) as an instrument. Briefly, the sickle cell genotype (HbAS) is a condition where a person inherits from one parent a mutated copy of the hemoglobin beta (HBB) gene called the sickle cell gene mutation, but inherits a normal copy of the HBB gene from the other parent. We discuss in detail the validity of the sickle cell trait as an instrument in Section 7.3.3.

7.2 Instruments and Optimal Matching

7.2.1 Notation

To formalize the idea of using matching with an instrument, we introduce the following notation. Let $i = 1, \ldots, I$ index the I total matched sets that individuals are matched into. Each matched set i contains $n_i \geq 2$ subjects who are indexed by $j = 1, \ldots, n_i$ and there are a total of $N = \sum_{i=1}^{I} n_i$ individuals in the data. Let Z_{ij} denote a binary instrument for subject j in matched set i. In each matched set i, there are m_i subjects with $Z_{ij} = 1$ and $n_i - m_i$ subjects with $Z_{ij} = 0$. Let $\mathbf{Z}$ be a random variable that consists of the collection of Z_{ij}'s, $\mathbf{Z} = (Z_{11}, Z_{12}, ..., Z_{I,n_I})$. Note that implicit in our notation, we allow our matched sets to accommodate pair matching (i.e. $n_i = 2$, $m_i = 1$), matching with fixed controls (i.e. $n_i = k$, $m_i = 1$ for some fixed $k \geq 2$), matching with variable controls (i.e. n_i varies, but $m_i = 1$), or full matching (i.e. n_i varies, but $m_i = 1$ or $m_i = n_i - 1$).

Let $\mathbf{z}$ be one possible realization of $\mathbf{Z}$. Define Ω as the set that contains all possible values of $\mathbf{Z}$, so $\mathbf{z} \in \Omega$ if z_{ij} is binary and $\sum_{j=1}^{n_i} z_{ij} = m_i$ for all I matched sets. The cardinality of Ω denoted as $|\Omega|$ is $|\Omega| = \prod_{i=1}^{I} n_i m_i$. Denote $\mathcal{Z}$ to be the event that $\mathbf{Z} \in \Omega$. We remark that implicit in our definition of Ω is that the instrument is assigned individually to each study unit; for instruments that are assigned as a group, say in cluster randomized trials, see [18] and [19].

For individual j in matched set i, define $d_{ij}^{(1)}$ and $d_{ij}^{(0)}$ to be the potential exposure values under $Z_{ij} = 1$ or $Z_{ij} = 0$, respectively. With the malaria example, $d_{ij}^{(0)}$ and $d_{ij}^{(0)}$ represent the number of malaria episodes the child would have if the child carried the sickle cell trait, $Z_{ij} = 1$, and no sickle cell trait, $Z_{ij} = 0$, respectively. Also, define $r_{ij}^{(1,k)}$ to be the outcome individual i would have if she were assigned instrument value 1 and level $k \in \mathcal{K}$ of the exposure, and $r_{ij}^{(0,k)}$ to be the outcome individual i would have if she were assigned instrumental value 0 and level k of the exposure. Also, let $r_{ij}^{(1,d_{ij}^{(1)})}$ and $r_{ij}^{(0,d_{ij}^{(0)})}$ be the potential outcomes if the individual were assigned levels $Z_{ij} = 1$ and $Z_{ij} = 0$ of the instrument respectively and the exposure took its natural level given the instrument, resulting in exposures, $d_{ij}^{(1)}$ and $d_{ij}^{(0)}$, respectively. With the malaria example, $r_{ij}^{(1,d_{ij}^{(1)})}$ is a binary variable that represents whether the jth child in the ith matched set would be stunted (i.e. 1) or not (i.e. 0) if the child carried the sickle cell trait (i.e. if $Z_{ij} = 1$) and $r_{ij}^{(0,d_{ij}^{(0)})}$ is a binary variable that represents whether the child would be stunted or not if the child carried no sickle cell trait (i.e. if $Z_{ij} = 0$). The potential outcome notations assume the Stable Unit Treatment Value Assumption (SUTVA) where, among other things, (i) an individual's outcome and exposure depend only on her own value of the instrument and not on other people's instrument values and (ii) an individual's outcome only depends on her own value of the exposure and not on other people's exposure [20].

For individual j in matched set i, let R_{ij} be the binary observed outcome and D_{ij} be the observed exposure. Under SUTVA, the potential outcomes $r_{ij}^{(1,d_{ij}^{(1)})}, r_{ij}^{(0,d_{ij}^{(0)})}, d_{ij}^{(1)}$, and $d_{ij}^{(0)}$ and the observed values R_{ij}, D_{ij}, and Z_{ij} are related by the following equation:

$$R_{ij} = r_{ij}^{(1,d_{ij}^{(1)})} Z_{ij} + r_{ij}^{(0,d_{ij}^{(0)})}(1 - Z_{ij}) \qquad D_{ij} = d_{ij}^{(1)} Z_{ij} + d_{ij}^{(0)}(1 - Z_{ij}) \tag{7.1}$$

For individual j in matched set i, let $\mathbf{X}_{ij}$ be a vector of observed covariates and u_{ij} be the unobserved covariates. For example, in the malaria data, $\mathbf{X}_{ij}$ represents each child's covariates listed in Table 7.2 while u_{ij} is an unmeasured confounder, like diet. We define the set $\mathcal{F} = \{(r_{ij}^{(1,k)}, r_{ij}^{(0,k)}, d_{ij}^{(1)}, d_{ij}^{(0)}, \mathbf{X}_{ij}, u_{ij}), i = 1, ..., I, j = 1, ..., n_i, k \in \mathcal{K}\}$ to be the collection of potential outcomes and all confounders, observed and unobserved.

7.2.2 Matching algorithm and covariate balance

A matching algorithm controls the bias resulting from baseline covariates by creating I matched sets indexed by $i, i = 1, \ldots, I$ such that individuals within each matched set have similar covariate values $\mathbf{x}_{ij}$ and the only difference between individuals in each matched set is their instrument values, Z_{ij}. In a pair matching algorithm, each matched set i contains one individual with $Z_{ij} = 1$ and one individual with $Z_{ij} = 0$. In a full matching algorithm, each matched set i either contains $m_i = 1$ individual with $Z_{ij} = 1$ and $n_i - 1$ individuals with $Z_{ij} = 0$ or $m_i = n_i - 1$ individuals with $Z_{ij} = 1$ and 1 individual with $Z_{ij} = 0$.

[8, 9], [21], and [22] provide an overview of matching and a discussion on various tools to measure similarity of observed and missing covariates. In general, much of the same advice used in non-IV matching designs from previous chapters apply to create matched sets in IV matching designs. More concretely, for the malaria data which we analyze below, we used propensity score caliper matching with rank-based Mahalanobis distance to measure covariate similarity where the propensity score is an instrument propensity score, specifically the probability of $Z_{ij} = 1$ given a set of pre-instrument covariates $\mathbf{X}_{ij}$. Once we have obtained the distance matrix, we can use the same optimal matching software used for non-IV matching designs, say the R package *optmatch* developed by [23], to find the optimal match.

Once matched sets are found, covariate balance would be assessed between the different values of the instruments. The actual numerical exercise to assess covariate balance in IV settings is nearly identical to that in non-IV settings where we replace the treatment variable with the instrumental variable. Specifically, we would compare the standardized differences in covariate means between $Z_{ij} = 1$ and $Z_{ij} = 0$ before and after matching and assess where measured covariates are similar between individuals with $Z_{ij} = 1$ and $Z_{ij} = 0$; see Section 7.3.2 for a worked-out example. Also, similar to non-IV matched designs, depending on the accepted level of covariate balance, investigators can further tweak the matching algorithm to achieve better covariate balance. Notably, [21] discusses how the size of matched sets can be restricted to improve efficiency and/or covariate balance, and [24] analyzes different restrictions on matched sets in terms of balance and efficiency.

7.2.3 Parameter of interest: Effect ratio

A popular parameter to estimate in matched IV designs is called the *effect ratio*.

$$\lambda = \frac{\sum_{i=1}^{I} \sum_{j=1}^{n_i} r_{1ij}^{(1,d_{ij}^{(1)})} - r_{ij}^{(0,d_{ij}^{(0)})}}{\sum_{i=1}^{I} \sum_{j=1}^{n_i} d_{ij}^{(1)} - d_{ij}^{(0)}} \tag{7.2}$$

Roughly speaking, the effect ratio is the change in the outcome caused by the instrument divided by the change in the exposure caused by the instrument. Note that this is a parameter of the finite population of $N = \sum_{i=1}^{I} n_i$ individuals.

The effect ratio admits a well-known interpretation in the IV literature if the core IV assumptions (A1)-(A4) are satisfied. Specifically, in the context of matching, assumptions (A1)-(A4) formally translate to

$$\text{(A1)} \quad \sum_{i=1}^{I}\sum_{j=1}^{n_i} d_{ij}^{(1)} - d_{ij}^{(0)} \neq 0,$$

$$\text{(A2)} \quad \forall k \in \mathcal{K},\ r_{ij}^{(1,k)} = r_{ij}^{(0,k)} = r_{ij}^{(k)},$$

$$\text{(A3)} \quad \forall \mathbf{z} \in \mathcal{Z},\ P(\mathbf{Z} = \mathbf{z} \mid \mathcal{F}) = \prod_{i=1}^{I} P(Z_{i1} = z_{i1}, \ldots, Z_{in_i} = z_{in_i} \mid \mathcal{F}) = \prod_{i=1}^{I} \frac{1}{n_i m_i},$$

$$\text{(A4)} \quad d_{ij}^{(0)} \leq d_{ij}^{(1)}.$$

Additionally, suppose d_{1ij} and d_{0ij} are discrete values from 0 to M (i.e. $\mathcal{K} = \{0, 1, \ldots, M\}$), which is the case with the malaria example where d_{1ij} and d_{0ij} are the number of malaria episodes. Then, [24] showed that

$$\lambda = \sum_{i=1}^{I}\sum_{j=1}^{n_i}\sum_{k=1}^{M} (r_{ij}^{(k)} - r_{ij}^{(k-1)}) w_{ijk} \tag{7.3}$$

where

$$w_{ijk} = \frac{\chi(d_{1ij} \geq k > d_{0ij})}{\sum_{i=1}^{I}\sum_{j=1}^{n_i}\sum_{l=1}^{M} \chi(d_{1ij} \geq l > d_{0ij})}$$

and $\chi(\cdot)$ is an indicator function. In words, with the core IV assumptions (A1)–(A3) and the monotonicity assumption (A4), the effect ratio is interpreted as the weighted average of the causal effect of a one unit change in the exposure among individuals in the study population whose exposure would be affected by a change in the instrument. Each weight w_{ijk} represents whether an individual j in stratum i's exposure would be moved from below k to at or above k by the instrument, relative to the number of people in the study population whose exposure would be changed by the instrument. For example, if $\lambda = 0.1$ in the malaria example and we assume the said assumptions, 0.1 is the weighted average reduction in stunting from a one-unit reduction in malaria episodes among individuals who were protected from malaria by the sickle cell trait. Similarly, each weight w_{ijk} represents the jth individual in ith stratum's protection from at least k malaria episodes by carrying the sickle cell trait compared to the overall number of individuals who are protected from varying degrees of malaria episodes by carrying the sickle cell trait. We remark that the interpretation of λ is akin to Theorem 1 in [25], except that our result is for the finite-sample case and is specific to matching.

We conclude with some remarks about the subtle nuances of interpreting the effect ratio. First, only assumptions (A1) and (A3) are necessary to identify the "bare-bone" interpretation of λ in (7.2), the ratio of causal effects of the instrument on the outcome (numerator) and on the exposure (denominator). This interpretation can be useful, especially in the setting where the exposure is continuous. However, this interpretation based on the ratio of differences cannot identify the weighted average of effects of the treatment on the outcome as described in (7.3). Second, irrespective of whether the investigator prefers the "bare-bone" interpretation of λ or the usual interpretation of λ based on complier average causal effect, λ can be identified from the observed data by the following formula, i.e.

$$\lambda = \frac{\sum_{i=1}^{I}\sum_{j=1}^{n_i} E(R_{ij}|Z_{ij} = 1, \mathcal{F}, \mathcal{Z}) - E(R_{ij}|Z_{ij} = 0, \mathcal{F}, \mathcal{Z})}{\sum_{i=1}^{I}\sum_{j=1}^{n_i} E(D_{ij}|Z_{ij} = 1, \mathcal{F}, \mathcal{Z}) - E(D_{ij}|Z_{ij} = 0, \mathcal{F}, \mathcal{Z})} \tag{7.4}$$

In other words, assuming assumption (A2) or assumption (A4) does not change the identification strategy of λ. Third, recent works by [26] and [27] discuss heterogeneous effect ratios, i.e. the effect ratio among a subgroup of individuals defined by pre-instrument covariates. Heterogeneous effect ratios are useful to discover heterogeneity of the effect ratio within the study population.

7.2.4 Inference for effect ratio

Suppose an investigator wants to test the following hypothesis test about the effect ratio λ.

$$H_0 : \lambda = \lambda_0, \quad H_a : \lambda \neq \lambda_0 \tag{7.5}$$

To test the hypothesis in (7.5), consider the following test statistic

$$T(\lambda_0) = \frac{\frac{1}{I}\sum_{i=1}^{I} V_i(\lambda_0)}{S(\lambda_0)} \tag{7.6}$$

where

$$V_i(\lambda_0) = \frac{n_i}{m_i}\sum_{j=1}^{n_i} Z_{ij}(R_{ij} - \lambda_0 D_{ij}) - \frac{n_i}{n_i - m_i}\sum_{j=1}^{n_i}(1 - Z_{ij})(R_{ij} - \lambda_0 D_{ij})$$

$$S^2(\lambda_0) = \frac{1}{I(I-1)}\sum_{i=1}^{I}\{V_i(\lambda_0) - T(\lambda_0)\}^2$$

Each variable $V_i(\lambda_0)$ is the difference in adjusted responses, $R_{ij} - \lambda_0 D_{ij}$, of those individuals with $Z_{ij} = 1$ and those with $Z_{ij} = 0$. Under the null hypothesis in (7.5), these adjusted responses have the same expected value for $Z_{ij} = 1$ and $Z_{ij} = 0$ and thus, deviation of $T(\lambda_0)$ from zero suggests H_0 is not true.

[24] states that under regularity conditions, the asymptotic null distribution of $T(\lambda_0)$ is standard Normal. This result can be used to derive a point estimate as well as a confidence interval for the effect ratio. For the point estimate, in the spirit of [28], we find the value of λ that maximizes the p-value, Specifically, setting $T(\lambda) = 0$ and solving for λ gives an estimate for the effect ratio, $\hat{\lambda}$

$$\hat{\lambda} = \frac{\sum_{i=1}^{I}\frac{n_i^2}{m_i(n_i - m_i)}\sum_{j=1}^{n_i}(Z_{ij} - \bar{Z}_{i.})(R_{ij} - \bar{R}_{i.})}{\sum_{i=1}^{I}\frac{n_i^2}{m_i(n_i - m_i)}\sum_{j=1}^{n_i}(Z_{ij} - \bar{Z}_{i.})(D_{ij} - \bar{D}_{i.})}$$

where $\bar{Z}_{i.}$, $\bar{R}_{i.}$, and $\bar{D}_{i.}$ are averages of the instrument, response, and exposure, respectively, within each matched set i. For confidence interval estimation, say a two-sided, 95% confidence interval for the effect ratio, we can solve the equation $T(\lambda) = \pm 1.96$ for λ to get the two endpoints of the 95% confidence interval. In general, for any two-sided, $1 - \alpha$ level confidence interval, we can solve the equation $T(\lambda) = \pm z_{1-\alpha/2}$ where $z_{1-\alpha/2}$ is the $1 - \alpha/2$ quantile of the standard Normal distribution. A closed form solution for the confidence interval is provided in [24] and [18]. The software to implement the inferential procedure is described in [29].

Finally, we remark that the inference procedure we develop for the effect ratio allows for binary, discrete, or continuous outcomes and exposures, even though our malaria data have binary outcomes and whole-number exposures. However, as remarked earlier, if the exposure is continuous, interpreting λ based on the complier average causal effect can be challenging; see [30] for additional discussions.

7.2.5 Sensitivity analysis

Similar to sensitivity analysis done for the treatment variable, we can conduct an instrumental variables sensitivity analysis which quantifies how a violation of assumption (A3) in Section 7.3.3 would impact the inference on λ [9]. This section provides a high-level illustration of this type of sensitivity analysis; see [31] for another type of sensitivity analysis in instrumental variables.

Formally, under assumption (A3), the instrument is assumed to be free from unmeasured confounders or free after conditioning on observed confounders via matching. However, even after matching for observed confounders, unmeasured confounders may influence the viability of assumption

(A3). For example, with the malaria study, within a matched set i , two children, j and j', may have the same birth weights, be from the same village, and have the same covariate values ($\mathbf{x}_{ij} = \mathbf{x}_{ij'}$), but have different probabilities of carrying the HbAS genotype, $P(Z_{ij} = 1|\mathcal{F}) \neq P(Z_{ij'} = 1|\mathcal{F})$ due to unmeasured confounders, denoted as u_{ij} and $u_{ij'}$ for the jth and j'th unit, respectively. Despite our best efforts to minimize the observed differences in covariates and to adhere to assumption (A3) after conditioning on the matched sets, unmeasured confounders could still be different between the jth and j'th child, and this difference could make the inheritance of the sickle cell trait depart from randomized assignment, violating assumption (A3).

To model this deviation from randomized assignment due to unmeasured confounders, let $\pi_{ij} = P(Z_{ij} = 1|\mathcal{F})$ and $\pi_{ij'} = P(Z_{ij'} = 1|\mathcal{F})$ for each unit j and j' in the ith matched set. The odds that unit j will receive $Z_{ij} = 1$ instead of $Z_{ij} = 0$ is $\pi_{ij}/(1 - \pi_{ij})$. Similarly, the odds for unit j' is $\pi_{ij'}/(1 - \pi_{ij'})$. Suppose the ratio of these odds is bounded by $\Gamma \geq 1$

$$\frac{1}{\Gamma} \leq \frac{\pi_{ij}(1 - \pi_{ij'})}{\pi_{ij'}(1 - \pi_{ij})} \leq \Gamma \tag{7.7}$$

If unmeasured confounders play no role in the assignment of Z_{ij}, then $\pi_{ij} = \pi_{ij'}$ and $\Gamma = 1$. That is, child j and j' have the same probability of receiving $Z_{ij} = 1$ in matched set i. If there are unmeasured confounders that affect the distribution of Z_{ij}, then $\pi_{ij} \neq \pi_{ij'}$ and $\Gamma > 1$. For a fixed $\Gamma > 1$, we can obtain lower and upper bounds on π_{ij}, which can be used to derive the null distribution of $T(0)$ under $H_0 : \lambda = 0$ in the presence of unmeasured confounding and be used to compute a range of possible p-values for the hypothesis $H_0 : \lambda = 0$ [9]. The range of p-values indicates the effect of unmeasured confounders on the conclusions reached by the inference on λ. If the range contains α, the significance level, then we cannot reject the null hypothesis at the α level when there is an unmeasured confounder with an effect quantified by Γ. In addition, we can amplify the interpretation of Γ using [32] to get a better understanding of the impact of the unmeasured confounding on the outcome and the instrument; see [24] for the derivation of the sensitivity analysis and the amplification of Γ.

7.2.6 A continuous instrument

We briefly discuss using a continuous instrument in optimal matching. Throughout the chapter, we only considered a binary instrument, allowing us to formulate the problem as a matching problem between two disjoint sets. However, in practice, there may be an instrumental variable that is naturally continuous. For example, in healthcare policy, differential travel time of patients to different healthcare facilities is often used as an instrument; see [33] and [34]. Briefly, differential travel time can be calculated as the time to the closest "high-intensity" facility (measured in minutes) subtracting the time to the closest "low-intensity" facility. One patient might need to travel an additional 60 minutes to get care a high-intensity facility compared to a low-intensity facility, and therefore that additional travel time may encourage the patient to seek care at the closer low-intensity facility. Whereas another patient may only need to travel an additional 5 minutes to get to a nearby high-intensity facility. Assuming that differential travel time satisfies assumptions (A1)–(A4), we might say that low values of the continuous IV mean that the patients are encouraged to receive care at the high-intensity facilities and high values of the IV encourage treatment at the low-intensity facility. When the continuous IV is used in matching, we would like to create matched sets wherein the baseline covariates look as similar as possible, but the values of the continuous IV are quite dissimilar. That is, the more dissimilar the IV values the "stronger" the contribution of the IV to determining the treatment actually received is.

Roughly speaking, there are two ways to incorporate a continuous IV in matching. One naive, but popular way is to take the continuous IV and rework it into a binary IV. For example, an investigator might look at the histogram of the IV and then decide that all observations at the qth percentile of the

IV are now "high-IV" and all observations below the pth percentile are "low-IV." The advantage of doing this is that investigators can use existing software as described earlier to conduct an IV analysis. Also, in non-matching contexts, [35] provides some justification behind the dichotomiziation of a continuous instrument.

However, in some situations, it may be unwise to dichotomize the continuous instrument into a binary variable, notably when the values of the instrument are correlated with observed covariates, but are still uncorrelated with unobserved covariates. In that case, it might be useful to simultaneously consider the IV values and covariate balance within each matched set. That is, in some situations, we may accept slightly less than ideal matched sets in terms of covariate balance if we can achieve excellent differences in the IV values because the instrument's independence to unmeasured confounders is useful for achieving unbiased estimation. This kind of consideration arose in the analysis of neonatal intensive care units in [34] and technical aspects of choosing the strength of the instrument are well-described in [36]. Computationally, using a continuous IV in matching may require software that is, at the time of writing this, not standard. But, case-specific implementations are discussed in [34], [37], and [36].

7.3 Application: The Causal Effect of Malaria on Stunting

7.3.1 Data background

We analyze the study on the causal effect of malaria on stunting introduced in previous sections using an instrumental variables approach. Following [38], we only consider children with the heterozygous strand HbAS, the sickle cell trait, or wildtype HbAA and exclude children with the homozygous strand (HbSS), or a different mutation on the same gene leading to hemoglobin C (HbAC, HbCC, HbSC); this reduced the sample size from 1070 to 884. Among 884 children, 110 children carried HbAS and 774 children carried HbAA.

The instrument in this study was a binary variable indicating either the HbAS or HbAA genotype. The treatment of interest was the malarial history, which was defined as the total number of malarial episodes during the study. A malaria episode was defined as having a parasite density of more than 500 parasites/μl and a body temperature greater than 38°C or the mother reported a fever within the last 48 hours. The outcome was whether the child was stunted at the last recorded visit, which occurred when the child was approximately two years old. The difference in episodes of malaria between children with HbAS and HbAA is significant (Risk ratio: 0.82, p-value: 0.02, 95% CI: (0.70, 0.97)), indicating that the sickle cell trait satisfies Assumption (A1) of being associated with the exposure.

Table 7.2 summarizes the measured covariates in the study. All the covariates were collected at the beginning of the study, which is three months after the child's birth. We will match on all these covariates as each covariate may be associated with the outcome and the instrument. Also, matching a covariate that is associated with the outcome increases efficiency and reduces sensitivity to unobserved biases [39, 40]. In general, [41] argues for erring on the side of being inclusive when deciding which variables to match on (i.e. control for) in an observational study. Failure to match for a covariate that has an important effect on outcome and is slightly out of balance can cause substantial bias.

In terms of covariate balance, before matching, we see that there are a few significant differences between the HbAS and HbAA groups, most notably in birth weight, village of birth, and mosquito protection status. Children with the sickle cell trait (HbAS) tend to have high birth weights and lack any protection against mosquitos compared to HbAA children. Also, children living in the village of Tano-Odumasi tend to inherit HbAA more frequently than HbAS. Any one of these differences

TABLE 7.2
Characteristics of study participants at recruitment. P-values were obtained by doing a Pearson's chi-squared test for categorical covariates and two-sample t tests for numerical variables.

	HbAS ($n = 110$)	HbAA ($n = 774$)
Birth weight (Mean,(SD))	3112.44 (381.9) (32 missing)	2978.7 (467.9) (239 missing) ***
Sex (Male/Female)	46.4% Male	51.0% Male
Birth season (Dry/Rainy)	56.4% Dry	55.3% Dry
Ethnic group (Akan/Northerner)	86.4% Akan	88.8 % Akan (4 missing)
α-globin genotype (Norm/Hetero/Homo)	75.7% / 21.5% / 2.8% (3 missing)	74.4% / 23.1% / 2.6% (29 missing)
Village of residence:		
Afamanso	4.6 %	4.8%
Agona	10.0%	13.6%
Asamang	13.6%	11.1%
Bedomase	5.5%	4.5%
Bipoa	14.5 %	10.7%
Jamasi	15.5 %	13.8%
Kona	16.4 %	12.8%
Tano-Odumasi	4.5 %	12.3%**
Wiamoase	15.5 %	16.4%
Mother's occupation (Nonfarmer/Farmer)	79.0% Nonfarmer	78.0% Nonfarmer (11 missing)
Mother's education (Literate/Illiterate)	91.7% Literate (2 missing)	90.5% Literate (8 missing)
Family's financial status (Good/Poor)	69.1% Good (13 missing)	70.1% Good (84 missing)
Mosquito protection (None/Net/Screen)	55.7% / 32.0% / 12.4% (13 missing)	45.4%* / 35.1% / 19.5% (76 missing)
Sulphadoxine pyrimethamine (Placebo/SP)	49.1% Placebo	50.1% Placebo

*** corresponds to a p-value of less than 0.01, ** corresponds to a p-value between 0.01 and 0.05, and * corresponds to a p-value between 0.05 and 0.1.

can contribute to the violation of IV assumption (A3) if we do not control for these differences. For instance, it is possible that children with low birth weights were malnourished at birth, making them more prone to malarial episodes and stunted growth compared to children with high birth weights. We must control for these differences to eliminate this possibility, which we do through matching

7.3.2 Implementation details

We conduct full matching on all observed covariates. In particular, we group children with HbAS and without HbAS based on all the observed characteristics in Table 7.2 as well as match for patterns of missingness. To measure similarity of the observed and missing covariates, we use the rank-based Mahalanobis distance as the distance metric for covariate similarity [8]. Children with missing values in their covariates were matched to other children with similar patterns of missing data [8]. Once covariate similarity was calculated, an optimal matching algorithm implemented in the software package optmatch in R [23] matched children carrying HbAS with children carrying HbAA.

Figure 7.2 shows the covariate balance before and after full matching using absolute standardized differences. Absolute standardized differences before matching are computed by taking the difference of the means between children with HbAS and HbAA for each covariate, taking the absolute value of it, and normalizing it by the within group standard deviation before matching (the square root of the average of the variances within the groups). Absolute standardized differences after matching are computed by taking the differences of the means between children with HbAS and HbAA within each strata, averaging this difference across strata, taking the absolute value of it, and normalizing it by the same within group standard deviation before matching as before. As mentioned before, before matching, there are differences in birth weight, mosquito protection, and village of residence between children with HbAS and HbAA. After matching, these covariates are balanced. Specifically, the standardized differences for birth weight, village of residence, and mosquito protection, are under 0.1 indicating balance [42]. In fact, all the covariates are balanced after matching and the p-values used to test the differences between HbAS and HbAA in Table 7.2 are no longer significant after matching.

7.3.3 Conditions for sickle cell trait as a valid instrument

We now assess the validity of (A1)–(A4) for the sickle cell trait, the instrument for our analysis on the effect of malaria on stunting. For assumption (A1), there is substantial evidence that the sickle cell trait does provide protection against malaria as compared to people with two normal copies of the HBB gene (HbAA) [43–47]. For assumption (A2), this could be violated if the sickle cell trait had effects on stunting other than through causing malaria, for instance, if the sickle cell trait was pleiotropic [48]. We can partially test this assumption by examining individuals who carry the sickle cell trait, but who grew up in a region where malaria is not present. That is, if assumption (A2) were violated, heights between individuals with HbAS and HbAA in such a region would be different since there would be a direct arrow between the sickle cell trait and height. We examined studies among African American children and children from the Dominican Republic and Jamaica for whom the sickle cell trait is common, but there is no malaria in the area. These two regions also match nutritional and socioeconomic conditions that are closer to our study population in Ghana so that the populations (and subsequent subpopulations among them) are comparable. From these studies from the regions, we found no evidence that the sickle cell trait affected a child's physical development [49–52]. This supports the validity of assumption (A2).

For assumption (A3), this assumption would be questionable in our data if we did not control for any population stratification covariates. Population stratification is a condition where there are subpopulations, some of which are more likely to have the sickle cell trait, and some of which are more likely to be stunted through mechanisms other than malaria [48]. For example, in Table 7.2, we observed that the village Tano-Odumasi had more children with HbAA than HbAS. It is possible

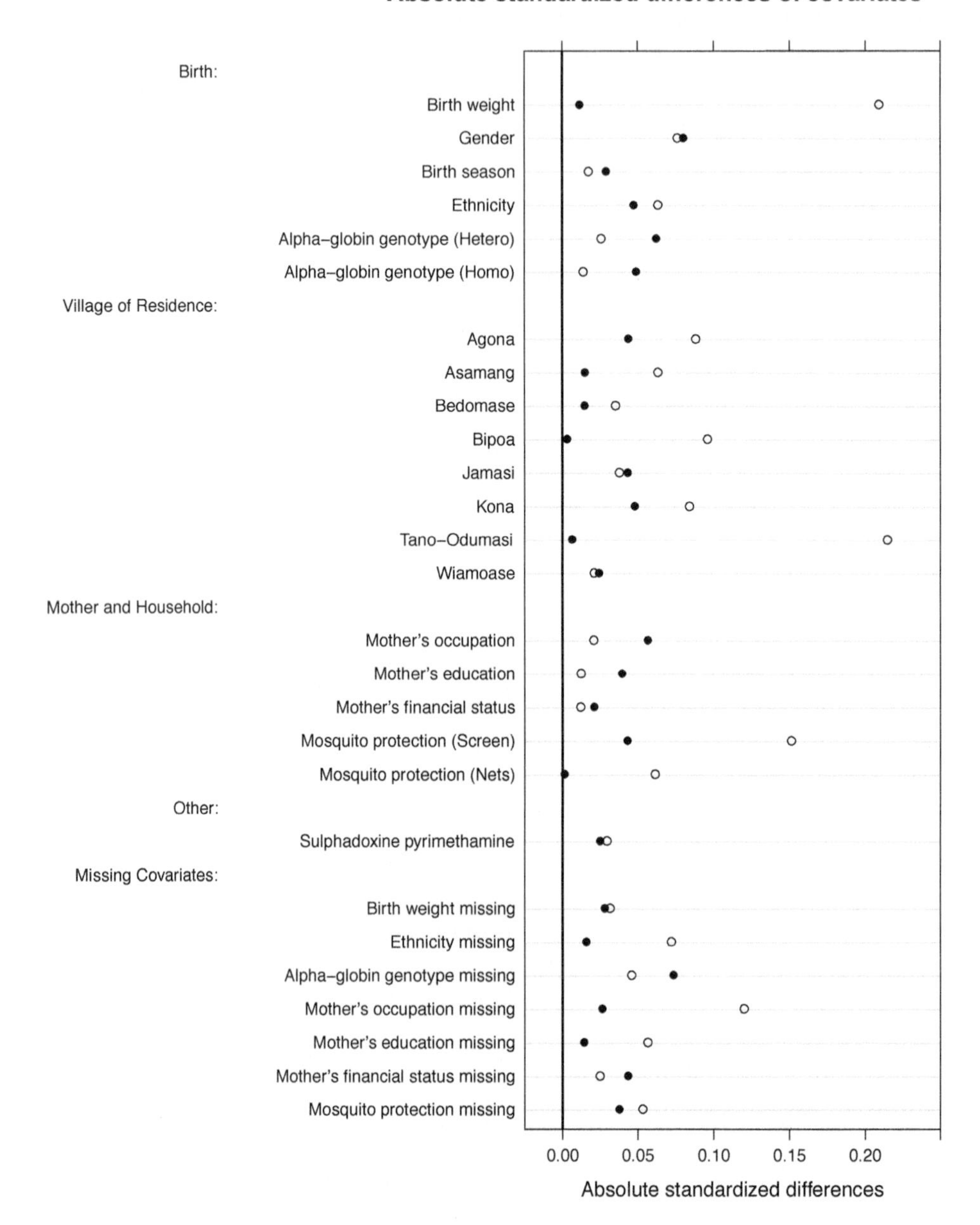

FIGURE 7.2
Absolute standardized differences before and after full matching. Unfilled circles indicate differences before matching and filled circles indicate differences after matching.

that there are other variables besides HbAA that differ between the village Tano-Odumasi and other villages and affect stunting. Hence, assumption (A3) is more plausible if we control for observed variables like village of birth. Specifically, within the framework of full matching, for each matched set i, if the observed variables $\mathbf{x}_{ij}$ are similar among all n_i individuals, it may be more plausible that the unobserved variable u_{ij} plays no role in the distribution of Z_{ij} among the n_i children. If (A3) exactly holds and subjects are exactly matched for X_{ij}, then within each matched set i, Z_{ij}

TABLE 7.3
Estimates of the causal effect using full matching compared to two-stage least squares and multiple regression.

Methods	Estimate	P-value	95% Confidence Interval
Full matching	0.22	0.011	(0.044, 1)
Two stage least squares	0.21	0.14	(−0.065, 0.47)
Multiple regression	0.018	0.016	(0.0034, 0.033)

is simply a result of random assignment where $Z_{ij} = 1$ with probability m_i/n_i and $Z_{ij} = 0$ with probability $(n_i - m_i)/n_i$ when we condition on the number of units in the matched set with $Z_{ij} = 1$ being m_i. In Section 7.2.5, we discuss a sensitivity analysis that allows for the possibility that even after matching for observed variables, the unobserved variable u_{ij} may still influence the assignment of Z_{ij} in each matched set i, meaning that assumption (A3) is violated.

For assumption (A4), there are known biological mechanisms by which the sickle cell genotype protects against malaria [44, 46, 53, 54] and no known mechanisms by which the sickle cell genotype increase the risk of malaria. Hence, there is no biological or epidemiological evidence to suggest that not carrying the sickle cell genotype would decrease the risk of getting malaria episodes compared to carrying it.

7.3.4 Estimate and inference of causal effect of malaria on stunting

Table 7.3 shows the estimates of the causal effect of malaria on stunting from different methods, specifically our matching-based method in Section 7.2.4, a popular method in instrumental variables known as two stage least squares (2SLS), and multiple regression. Briefly, 2SLS computed the estimate by regressing all the measured covariates and the instrument on the exposure and using the prediction from that regression and the measured covariates to obtain the estimated effect. Inference for 2SLS was derived using standard asymptotic Normality arguments [55]. Finally, the multiple regression estimate was derived by regressing the outcome on the exposure and the covariates and the inference on the estimate was based on a standard t test.

We see that the full matching method estimates λ to be 0.22. That is, the risk of stunting among children with the sickle cell trait is estimated to decrease by 0.22 times the average malaria episodes prevented by the sickle cell trait. Furthermore, we reject the hypothesis $H_0 : \lambda = 0$, that malaria does not cause stunting, at the 0.05 significance level. The confidence interval λ is $(0.044, 1.0)$. Even the lower limit of this confidence interval of 0.044 means that malaria has a substantial effect on stunting; it would mean that the risk of stunting among children with the sickle cell trait is decreased by 0.044 times the average malaria episodes prevented by the sickle cell trait.

The estimate based on 2SLS is 0.21, similar to our method. However, our method achieves statistical significance but 2SLS does not. Also, multiple regression, which does not control for unmeasured confounders, estimates a much smaller effect of 0.018.

Table 7.4 shows the sensitivity analysis in Section 7.2.5 due to unmeasured confounders. Specifically, we measure how sensitive our estimate and the p-value in Table 7.3 is to violation of assumption (A3) in Section 7.3.3, even after matching. We see that our results are somewhat sensitive to unmeasured confounders at the 0.05 significance level. If there is an unmeasured confounder that increases the odds of inheriting HbAS over HbAA by 10%, i.e. $\Gamma = 1.1$, then we would still have strong evidence that malaria causes stunting. But, if an unmeasured confounder increases the odds of inheriting HbAS over HbAA in a child by 20% (i.e. $\Gamma = 1.2$), the range of possible p-values includes 0.05, the significance level, meaning that we would not reject the null hypothesis of $H_0 : \lambda = 0$, that malaria does not cause stunting.

TABLE 7.4
Sensitivity analysis for instrumental variables with full matching. The range of significance is the range of p-values over the different possible distributions of the unmeasured confounder given a particular value of Γ, which represents the effect of unobserved confounders on the inference of λ.

Gamma	Range of Significance
1.1	(0.0082, 0.041)
1.2	(0.0034, 0.074)
1.3	(0.0015, 0.12)

7.4 Summary

Incorporating an IV into a matched provides an opportunity to quantify the effect of the treatment in the presence of unmeasured confounders. In particular, by using matching to adjust for measured confounders, a matching-based IV analysis provides a matching-based approach to study the causal effect of the treatment on an outcome. Similar to other matching-based designs discussed in the book, this IV design makes it explicitly clear how these covariates were adjusted by stratifying individuals based on similar covariate values. Importantly, as with other design-based studies, this analysis only looked at the outcome data once the balance was acceptable, i.e., once the differences in birth weight, village of residence, and mosquito protection between children with HbAS and HbAA were controlled for. If the balance was unacceptable, then comparing the outcomes between the two groups would not provide reliable causal inference since any differences in the outcome can be attributed to the differences in the covariates.

Of course, similar to a SITA-based analysis, an IV analysis carries untestable assumptions, which must be carefully assessed on a case-by-case basis; Section 6 of [6] discusses additional approaches to assessing IV assumptions in application. Also, in the binary setup discussed above, we are only able to identify the treatment effect among a subgroup of individuals known as the compliers. Other formulations of instrumental variables discussed in the overview allow identification of different treatment effects, including the treatment effect for the entire population or among those who are treated. However, in general, all formulations of instrumental variables must contend with the core IV assumptions (A1)-(A3). All investigators using an instrumental variables analysis must be attentive to these assumptions, providing readers with both empirical and application-specific arguments for their plausibility. If done well then an IV analysis can yield useful evidence about the causal effect of the treatment in the presence of unmeasured confounding.

References

[1] Joshua D. Angrist, Guido W. Imbens, and Donald B. Rubin. Identification of causal effects using instrumental variables. *Journal of the American Statistical Association*, 91(434):444–455, 1996.

[2] Miguel A. Hernán and James M. Robins. Instruments for causal inference: An epidemiologist's dream? *Epidemiology*, 17(4):360–372, 2006.

[3] M. Alan Brookhart and Sebastian Schneeweiss. Preference-based instrumental variable methods for the estimation of treatment effects: assessing validity and interpreting results. *The International Journal of Biostatistics*, 3(1):14, 2007.

[4] Jing Cheng, Jing Qin, and Biao Zhang. Semiparametric estimation and inference for distributional and general treatment effects. *Journal of the Royal Statistical Society: Series B (Statistical Methodology)*, 71(4):881–904, 2009.

[5] Sonja A. Swanson and Miguel A. Hernán. Commentary: How to report instrumental variable analyses (suggestions welcome). *Epidemiology*, 24(3):370–374, 2013.

[6] Michael Baiocchi, Jing Cheng, and Dylan S. Small. Instrumental variable methods for causal inference. *Statistics in Medicine*, 33(13):2297–2340, 2014.

[7] Amelia Haviland, Daniel S. Nagin, and Paul R. Rosenbaum. Combining propensity score matching and group-based trajectory analysis in an observational study. *Psychological Methods*, 12(3):247, 2007.

[8] Paul R. Rosenbaum. *Design of Observational Studies*. Springer Series in Statistics. Springer, New York, 2010.

[9] Paul R. Rosenbaum. *Observational Studies*. Springer Series in Statistics. Springer-Verlag, New York, second edition, 2002.

[10] Paul R Rosenbaum and Donald B Rubin. The central role of the propensity score in observational studies for causal effects. *Biometrika*, 70(1):41–55, 1983.

[11] Hyunseung Kang, Benno Kreuels, Ohene Adjei, Ralf Krumkamp, Jürgen May, and Dylan S. Small. The causal effect of malaria on stunting: A mendelian randomization and matching approach. *International Journal of Epidemiology*, 42(5):1390–1398, 2013.

[12] Hyunseung Kang, Benno Kreuels, Jürgen May, and Dylan S Small. Full matching approach to instrumental variables estimation with application to the effect of malaria on stunting. *The Annals of Applied Statistics*, 10(1):335–364, 2016.

[13] World Health Organization. World malaria report 2014. *World Health Organization*, 2014.

[14] WHO Multicentre Growth Reference Study Group. WHO child growth standards based on length/height, weight and age. *Acta Paediatrica. Supplement*, 450:76–85, 2006.

[15] Florie Fillol, Jean B. Sarr, Denis Boulanger, Badara Cisse, Cheikh Sokhna, Gilles Riveau, Kirsten B. Simondon, and Franck Remoué. Impact of child malnutrition on the specific anti-plasmodium falciparum antibody response. *Malaria Journal*, 8(1):116, 2009.

[16] Amare Deribew, Fessehaye Alemseged, Fasil Tessema, Lelisa Sena, Zewdie Birhanu, Ahmed Zeynudin, Morankar Sudhakar, Nasir Abdo, Kebede Deribe, and Sibhatu Biadgilign. Malaria and under-nutrition: A community based study among under-five children at risk of malaria, south-west ethiopia. *PLoS One*, 5(5):e10775, 2010.

[17] Paul W. Holland. Causal inference, path analysis, and recursive structural equations models. *Sociological Methodology*, 18(1):449–484, 1988.

[18] Hyunseung Kang and Luke Keele. Estimation methods for cluster randomized trials with noncompliance: A study of a biometric smartcard payment system in india. *arXiv preprint arXiv:1805.03744*, 2018.

[19] Bo Zhang, Siyu Heng, Emily J MacKay, and Ting Ye. Bridging preference-based instrumental variable studies and cluster-randomized encouragement experiments: Study design, noncompliance, and average cluster effect ratio. *Biometrics*, 2021. https://doi.org/10.1111/biom.13500

[20] Donald B. Rubin. Comment on "randomized analysis of experimental data: The fisher randomization test". *Journal of the American Statistical Association*, 75(371):591–593, 1980.

[21] Ben B. Hansen. Full matching in an observational study of coaching for the sat. *Journal of the American Statistical Association*, 99(467):609–618, 2004.

[22] Elizabeth A. Stuart. Matching methods for causal inference: A review and a look forward. *Statistical Science*, 25(1):1, 2010.

[23] Ben B. Hansen and Stephanie Olsen Klopfer. Optimal full matching and related designs via network flows. *Journal of Computational and Graphical Statistics*, 15(3):609–627, 2006.

[24] Hyunseung Kang, Laura Peck, and Luke Keele. Inference for Instrumental Variables. *Journal of the Royal Statistical Society: Series A (Statistics in Society)*, 181(4):1231–1254. https://doi.org/10.1111/rssa.12353

[25] Joshua D. Angrist and Guido W. Imbens. Two-stage least squares estimation of average causal effects in models with variable treatment intensity. *Journal of the American Statistical Association*, 90(430):431–442, 1995.

[26] Michael Johnson, Jiongyi Cao, and Hyunseung Kang. Detecting heterogeneous treatment effects with instrumental variables and application to the oregon health insurance experiment. *Annals of Applied Statistics*, 16 (2):1111–1129, 2022.

[27] Colin B. Fogarty, Kwonsang Lee, Rachel R. Kelz, and Luke J. Keele. Biased encouragements and heterogeneous effects in an instrumental variable study of emergency general surgical outcomes. *Journal of the American Statistical Association*, 116(536):1625–1636, 2021.

[28] J. L. Hodges and E. L. Lehmann. Estimation of location based on ranks. *Annals of Mathematical Statistics*, 34(2):598–611, 1963.

[29] Hyunseung Kang, Yang Jiang, Qingyuan Zhao, and Dylan S Small. Ivmodel: an r package for inference and sensitivity analysis of instrumental variables models with one endogenous variable. *Observational Studies*, 7(2):1–24, 2021.

[30] Joshua D. Angrist, Kathryn Graddy, and Guido W. Imbens. The interpretation of instrumental variables estimators in simultaneous equations models with an application to the demand for fish. *The Review of Economic Studies*, 67(3):499–527, 2000.

[31] Dylan S. Small. Sensitivity analysis for instrumental variables regression with overidentifying restrictions. *Journal of the American Statistical Association*, 102(479):1049–1058, 2007.

[32] Paul R. Rosenbaum and Jeffrey H. Silber. Amplification of sensitivity analysis in matched observational studies. *Journal of the American Statistical Association*, 104(488):1398–1405, 2009.

[33] Mark McClellan, Barbara J McNeil, and Joseph P Newhouse. Does more intensive treatment of acute myocardial infarction in the elderly reduce mortality?: Analysis using instrumental variables. *JAMA*, 272(11):859–866, 1994.

[34] Mike Baiocchi, Dylan S. Small, Scott Lorch, and Paul R. Rosenbaum. Building a stronger instrument in an observational study of perinatal care for premature infants. *Journal of the American Statistical Association*, 105(492):1285–1296, 2010.

[35] Joy Shi, Sonja A Swanson, Peter Kraft, Bernard Rosner, Immaculata De Vivo, and Miguel A Hernán. Instrumental variable estimation for a time-varying treatment and a time-to-event outcome via structural nested cumulative failure time models. *BMC Medical Research Methodology*, 21(1):1–12, 2021.

[36] Luke Keele and Jason W Morgan. How strong is strong enough? strengthening instruments through matching and weak instrument tests. *The Annals of Applied Statistics*, 10(2):1086–1106, 2016.

[37] José R. Zubizarreta, Dylan S. Small, Neera K. Goyal, Scott Lorch, and Paul R. Rosenbaum. Stronger instruments via integer programming in an observational study of late preterm birth outcomes. *The Annals of Applied Statistics*, 7(1):25–50, 2013.

[38] Benno Kreuels, Stephan Ehrhardt, Christina Kreuzberg, Samuel Adjei, Robin Kobbe, Gerd Burchard, Christa Ehmen, Matilda Ayim, Ohene Adjei, and Jürgen May. Sickle cell trait (hbas) and stunting in children below two years of age in an area of high malaria transmission. *Malaria Journal*, 8(1):16, 2009.

[39] Paul R. Rosenbaum. Heterogeneity and causality: Unit heterogeneity and design sensitivity in observational studies. *American Statistician*, 59(2):147–152, 2005.

[40] José R. Zubizarreta, Ricardo D. Paredes, and Paul R. Rosenbaum. Matching for balance, pairing for heterogeneity in an observational study of the effectiveness of for-profit and not-for-profit high schools in chile. *Annals of Applied Statistics*, (1):204–231, 2014.

[41] Donald B. Rubin. Should observational studies be designed to allow lack of balance in covariate distributions across treatment groups? *Statistics in Medicine*, 28(9):1420–1423, 2009.

[42] Sharon-Lise T. Normand, Mary Beth Landrum, Edward Guadagnoli, John Z. Ayanian, Thomas J. Ryan, Paul D. Cleary, and Barbara J. McNeil. Validating recommendations for coronary angiography following acute myocardial infarction in the elderly: A matched analysis using propensity scores. *Journal of Clinical Epidemiology*, 54(4):387–398, 2001.

[43] Michael Aidoo, Dianne J. Terlouw, Margarette S. Kolczak, Peter D. McElroy, Feiko O. ter Kuile, Simon Kariuki, Bernard L. Nahlen, Altaf A. Lal, and Venkatachalam Udhayakumar. Protective effects of the sickle cell gene against malaria morbidity and mortality. *The Lancet*, 359(9314):1311–1312, 2002.

[44] Thomas N. Williams, Tabitha W. Mwangi, David J. Roberts, Neal D. Alexander, David J. Weatherall, Sammy Wambua, Moses Kortok, Robert W. Snow, and Kevin Marsh. An immune basis for malaria protection by the sickle cell trait. *PLoS Medicine*, 2(5):e128, 2005.

[45] Jürgen May, Jennifer A. Evans, Christian Timmann, Christa Ehmen, Wibke Busch, Thorsten Thye, Tsiri Agbenyega, and Rolf D. Horstmann. Hemoglobin variants and disease manifestations in severe falciparum malaria. *Journal of the American Medical Association*, 297(20):2220–2226, 2007.

[46] Rushina Cholera, Nathaniel J. Brittain, Mark R. Gillrie, Tatiana M. Lopera-Mesa, Séidina A. S. Diakité, Takayuki Arie, Michael A. Krause, Aldiouma Guindo, Abby Tubman, Hisashi Fujioka, Dapa A. Diallo, Ogobara K. Doumbo, May Ho, Thomas E. Wellems, and Rick M. Fairhurst. Impaired cytoadherence of plasmodium falciparum-infected erythrocytes containing sickle hemoglobin. *Proceedings of the National Academy of Sciences*, 105(3):991–996, 2008.

[47] Benno Kreuels, Christina Kreuzberg, Robin Kobbe, Matilda Ayim-Akonor, Peter Apiah-Thompson, Benedicta Thompson, Christa Ehmen, Samuel Adjei, Iris Langefeld, Ohene Adjei, and Jürgen May. Differing effects of hbs and hbc traits on uncomplicated falciparum malaria, anemia, and child growth. *Blood*, 115(22):4551–4558, 2010.

[48] George Davey Smith and Shah Ebrahim. 'mendelian randomization': can genetic epidemiology contribute to understanding environmental determinants of disease? *International Journal of Epidemiology*, 32(1):1–22, 2003.

[49] Michael T. Ashcroft, Patricia Desai, and Stephen A. Richardson. Growth, behaviour, and educational achievement of jamaican children with sickle-cell trait. *British Medical Journal*, 1(6022):1371–1373, 1976.

[50] Michael S. Kramer, Yolanda Rooks, and Howard A. Pearson. Growth and development in children with sickle-cell trait: a prospective study of matched pairs. *New England Journal of Medicine*, 299(13):686–689, 1978.

[51] Michael T. Ashcroft, Patricia Desai, G. A. Grell, Beryl E. Serjeant, and Graham R. Serjeant. Heights and weights of west indian children with the sickle cell trait. *Archives of Disease in Childhood*, 53(7):596–598, 1978.

[52] N. Rehan. Growth status of children with and without sickle cell trait. *Clinical Pediatrics*, 20(11):705–709, 1981.

[53] Milton J. Friedman. Erythrocytic mechanism of sickle cell resistance to malaria. *Proceedings of the National Academy of Sciences*, 75(4):1994–1997, 1978.

[54] Milton J. Friedman and William Trager. The biochemistry of resistance to malaria. *Scientific American*, 244:154–155, 1981.

[55] Jeffrey M. Wooldridge. *Econometric Analysis of Cross Section and Panel Data*. MIT Press, Cambridge, Massachusetts, 2nd edition, 2010.

8

Covariate Adjustment in Regression Discontinuity Designs

Matias D. Cattaneo, Luke Keele, Rocío Titiunik

CONTENTS

8.1 Introduction

In causal inference and program evaluation, a central goal is to learn about the causal effect of a policy or treatment on an outcome of interest. There is a variety of methodological approaches for the identification, estimation, and inference for causal treatment effects, depending on the specific application. In experimental settings, where the treatment assignment mechanism is known, studying treatment effects is relatively straightforward because the methods are based on assumptions that are known to be true by virtue of the treatment assignment rule. In contrast, in observational studies, the assignment mechanism is unknown and researchers are forced to invoke identifying assumptions whose validity is not guaranteed by design. Well-known observational study methods include selection on observables, instrumental variables, difference-in-differences, synthetic controls, and regression discontinuity designs.[1]

In experimental and observational studies, baseline covariates generally serve different purposes. In experimental settings covariate adjustment based on pre-intervention measures is often used either for efficiency gains or for the evaluation of treatment effect heterogeneity. On the other hand, in many observational studies, the primary purpose of covariate adjustment is for the identification of causal effects. For example, under selection on observables, the researcher posits that, after conditioning

[1] See, for example, [1], [2] and [3] for reviews on causal inference and program evaluation, and [4] for further discussion on the role of natural experiments in observational research.

DOI: 10.1201/9781003102670-8

on a set of pre-intervention covariates, the assignment mechanism is not a function of potential outcomes, thereby mimicking an experimental setting after conditioning on those covariates. Under this assumption, researchers can obtain valid estimates of treatment effects by comparing groups of units whose covariate values are similar. Implementation of covariate adjustment methods in experimental and observational studies can be done in different ways, including via regression methods, inverse probability weighting, and matching. However, the validity of covariate adjustment methods is not automatically guaranteed because it requires additional (usually stronger) assumptions, if valid at all. This is particularly true in the case of the regression discontinuity (RD) design.

Covariate adjustment in RD designs has been employed in a variety of ways and for different purposes in recent years. Because the RD design has a unique feature relative to other non-experimental research strategies, covariate adjustment has distinctive implications for the analysis and interpretation of RD treatment effects. This has led to confusions and misconceptions about the role of auxiliary covariates in RD designs. The main purpose of this chapter is to review the different roles of covariate adjustment in the RD literature, and to offer methodological guidance for its correct use in applications.

To streamline the presentation, we do not provide a comprehensive review of RD designs and methods.[2] We focus our discussion on the canonical (sharp) RD design, where all units receive an observable score and a treatment is assigned to all units whose value of the score exceeds a known and fixed cutoff, and withheld from units whose score is below it. Under appropriate conditions, the abrupt change in the probability of receiving treatment that occurs at the cutoff justifies comparing units just above the cutoff to units just below it to learn about the causal effect of the treatment. For this canonical RD setup, and employing the taxonomy introduced by [10], we discuss the different roles of covariate adjustment within the two leading conceptual frameworks for the interpretation and analysis of RD designs: the continuity framework and the local randomization framework.

One of the most important roles of baseline covariates in the canonical RD design is for falsification or validation purposes. In the continuity framework for RD analysis, covariates are used to verify that the effect of the treatment on the covariates at the cutoff is zero, offering evidence in support of the assumption that only the probability of treatment status given the score changes discontinuously at the cutoff. In the local randomization approach, covariates are often used to select the window around the cutoff where treated and control units are similar to each other, and where the assumption of as-if random treatment assignment is assumed to hold.

The canonical RD setup is sometimes augmented with additional covariate information for other purposes beside falsification testing, and the role of these additional covariates varies depending on the setting and specific goals. For example, additional covariates may be used for efficiency gains or for the analysis of treatment effect heterogeneity. This use of pre-intervention covariates is motivated by the analogy between randomized experiments and local randomization in RD designs. Alternatively, covariates may be used to construct other treatment effects of interest or for extrapolation purposes (i.e., external validity). This occurs, for instance, when the additional covariates capture other features of the design such as geographic location, evolution over time, or group identity. Finally, sometimes covariates are also used in conjunction with statistical adjustment with the hope of "fixing" an otherwise invalid RD design. See [9] for a comprehensive review and references.

In this chapter we discuss the most common uses that have been proposed for covariate adjustment in the RD literature, including efficiency, heterogeneity, and extrapolation. In each case we discuss whether these methods are best justified from a continuity or a local randomization perspective, and how they methodologically relate to each other. Our main focus is conceptual, and mostly from an identification perspective, although we do discuss implementation issues when appropriate. Covariate adjustment in RD designs can be implemented by either local linear regression, inverse probability weighting, or "matching" methods more generally.

[2]For recent introductions and practical guides see, for example, [5–9], and references therein.

The rest of the chapter is organized as follows. The next section reviews the different roles of covariate adjustment in RD designs. Then, in Section 8.3, we discuss the important question of whether the use of covariate adjustment such as adding fixed effects for group of units can "restore" the validity of an RD design where the key identifying assumptions do not hold (e.g., settings where units are suspected to have sorted around the cutoff). Section 8.4 offers generic recommendations for practice, and Section 8.5 concludes.

8.2 Covariate Adjustment in RD Designs

For each type of covariate adjustment in RD designs, we discuss specific implications of using covariates. In particular we outline how covariate adjustment can lead to the unintended redefinition of the estimand to a quantity that does not coincide with a parameter of interest. We clarify whether the covariate adjustment applies to the continuity framework, the local randomization framework, or both. While most of the available proposals were introduced within one of these two conceptual frameworks, we also discuss the potential implications that each of the covariate adjustment approaches may have for the other. In this discussion it is important to remember that the core identifying assumptions differ between the two frameworks, and consequently so does the RD parameter of interest. For simplicity, we focus only on sharp RD designs (i.e., settings with perfect treatment compliance or intention-to-treat analysis), but the references given often include discussions of fuzzy, kink, and other related RD setups.

8.2.1 Overview of the canonical RD design

We assume that the observed data is (Y_i, D_i, X_i, Z_i), $i = 1, 2, \ldots, n$, where $Y_i = Y_i(0)(1 - D_i) + Y_i(1)D_i$, $Y_i(0)$ is the potential outcome under control, $Y_i(1)$ is the potential outcome under treatment, $D_i = \mathbb{1}(X_i \geq c)$ denotes the treatment assignment rule, X_i denotes the score variable, c denotes the cutoff, and Z_i denotes other pre-determined covariates. Depending on the framework considered, the potential outcomes are taken to be random or non-random, and the data can be seen either as a (random) sample from some large population or as a finite population. The core assumption in the RD design is that the probability of treatment assignment changes abruptly at the cutoff from zero to one, that is, $\mathbb{P}[D_i = 1|X_i = x] = 0$ if $x < c$ and $\mathbb{P}[D_i = 1|X_i = x] = 1$ if $x \geq c$.

In the continuity framework, the data is assumed to be a random sample and the potential outcomes are assumed to be random variables. In this context, the most common parameter of interest is the average treatment effect at the cutoff,

$$\tau_{\mathsf{SRD}} = \mathbb{E}[Y_i(1) - Y_i(0)|X_i = c],$$

and the key identifying assumption is the continuity of the conditional expectations of the potential outcomes given the score: $\mathbb{E}[Y_i(1) \mid X_i = x]$ and $\mathbb{E}[Y_i(0) \mid X_i = x]$ are continuous in x at c, which leads to the well-known continuity-based identification result [11]:

$$\tau_{\mathsf{SRD}} = \lim_{x \downarrow c} \mathbb{E}[Y_i|X_i = x] - \lim_{x \uparrow c} \mathbb{E}[Y_i|X_i = x].$$

Many methods are available for estimation, inference, and validation of RD designs within the continuity framework. The most common approach is to use (non-parametric) local polynomial methods to approximate the two regression functions near the cutoff. The implementation requires choosing a bandwidth around the cutoff, and then fitting two polynomial regressions of the outcome on the score – one above and the other below the cutoff – using only observations with scores within the chosen region around the cutoff as determined by the bandwidth. The bandwidth is typically chosen in an optimal and data-driven fashion to ensure transparency and to avoid specification searching. The most common strategy is to choose a bandwidth that minimizes the asymptotic mean squared error (MSE) of the RD point estimator (see [12], and references therein).

Although the implementation of local polynomial methods reduces to fitting two local, lower-order polynomials using least-squares methods, the interpretation and properties are different from the typical parametric least squares framework. Local polynomials are meant to provide a non-parametric approximation to the regression functions of the potential outcomes given the score. Because these functions are fundamentally unknown, the local polynomial approximation has an unavoidable misspecification error. When the bandwidth is chosen to be MSE-optimal, this approximation error appears in the standard distributional approximation and renders the conventional least squares confidence intervals invalid because conventional inference assumes that the approximation error is zero [13] developed valid confidence intervals based on a novel distributional approximation for a test statistic in which the RD point estimator is adjusted using bias correction and also the standard error is modified to account for the additional variability introduced by the bias estimation step. These robust bias-corrected confidence intervals lead to valid inferences when using the MSE-optimal bandwidth, and remain valid for other bandwidth choices, in addition to having several other optimal properties [12].

The canonical implementation of local polynomial estimation and inference in RD designs uses only the outcome and the score, without additional covariates. We discuss below how the local polynomial approach can be augmented to include pre-determined covariates in the two polynomial fits. In addition we discuss how the local polynomial estimation and inference methods are adapted to incorporate covariates for other purposes such as heterogeneity analysis or extrapolation.

In the local randomization framework for RD analysis, the parameter of interest is defined differently because the core identifying assumptions are different from those in the continuity framework [10, 14]. Because in the local randomization approach the central assumption is that the RD design creates conditions that resemble a randomized experiment in a neighborhood near the cutoff, the parameter is defined within this neighborhood—as opposed to at the cutoff as in the continuity approach. The local randomization parameter can be generically written as

$$\tau_{\text{SLR}} = \frac{1}{N_{\mathcal{W}}} \sum_{i:X_i \in \mathcal{W}} \mathbb{E}_{\mathcal{W}}[Y_i(1) - Y_i(0)]$$

where $\mathcal{W} = [c - w, c + w]$ is a symmetric window around the cutoff, $N_{\mathcal{W}}$ is the number of units with $X_i \in \mathcal{W}$, $\mathbb{E}_{\mathcal{W}}$ denotes an expectation computed conditionally for all units with $X_i \in \mathcal{W}$, and two conditions are assumed to hold within $\mathcal{W}$: (i) the joint distribution of scores in $\mathcal{W}$ is known; and (ii) the potential outcomes are unaffected by the score X_i in $\mathcal{W}$. Under these local randomization assumptions, estimation and inference can proceed using standard methods for the analysis of experiments, including Fisherian methods where the potential outcomes are assumed non-random and inferences are finite-sample exact, and Neyman or super-population methods where inferences are based on large-sample approximations with either non-random or random potential outcomes. In general the parameter of interest can be written as

$$\tau_{\text{SLR}} = \frac{1}{N_{\mathcal{W}}} \sum_{i:X_i \in \mathcal{W}} \mathbb{E}_{\mathcal{W}}\left[\frac{T_i Y_i}{\mathbb{P}_{\mathcal{W}}[T_i = 1]}\right] - \frac{1}{N_{\mathcal{W}}} \sum_{i:X_i \in \mathcal{W}} \mathbb{E}_{\mathcal{W}}\left[\frac{(1 - T_i) Y_i}{1 - \mathbb{P}_{\mathcal{W}}[T_i = 1]}\right],$$

where $\mathbb{P}_{\mathcal{W}}$ denotes a probability computed conditionally for all units with $X_i \in \mathcal{W}$, which indicates a natural estimator by virtue of the known assignment mechanism within $\mathcal{W}$ and the other assumptions imposed.

The window $\mathcal{W}$ where the local randomization assumptions are assumed to hold is unknown in most applications. The recommended approach to choose this window in practice is based on balance tests performed on one or more pre-determined covariates [14]. The main idea behind this method is that the treatment effect on any pre-determined covariate is known to be zero, and thus a non-zero effect is evidence that the local randomization assumptions do not hold. Assuming that the covariate is correlated with the score outside of $\mathcal{W}$ but not inside, a test of the hypothesis that the mean (or other feature of the distribution) of the pre-determined covariate is the same in the treated and control groups should be rejected in all windows larger than $\mathcal{W}$, and not rejected in $\mathcal{W}$ or any window contained in it. The method thus consists of performing balance tests in nested (often symmetric) windows, starting with the largest possible window and continuing until the hypothesis of covariate balance fails to be rejected in one window and in all windows contained in it.

The window $\mathcal{W}$ in the local randomization approach is analogous to the bandwidth in the continuity-based local polynomial analysis. An important distinction, however, is that the bandwidth can be chosen in a data-driven and optimal way using only data on the outcome and the score, while choosing the local randomization window $\mathcal{W}$ requires data on auxiliary pre-determined covariates. In this sense, the local polynomial analysis can be fully implemented without additional covariates, while the local randomization analysis typically cannot—unless $\mathcal{W}$ is known or chosen in an arbitrary manner. For further details on RD estimation and inference under the continuity-based and local randomization approaches, we refer the reader to [5–9], and references therein.

8.2.2 Efficiency and power improvements

In the analysis of randomized experiments, it is well known that including pre-intervention covariates in the estimation of treatment effects can increase precision and statistical power, as discussed in standard textbooks on the analysis of experiments [1, 2, 15]. Covariate adjustment in experimental analysis can be implemented both before and after randomization has occurred. Adjustment before randomization is typically done via blocking or stratification, where units are first separated into groups according to their values of one or more pre-treatment covariates, and the treatment is then assigned randomly within groups. These methods are effective at increasing precision, but unfortunately they are not available for RD designs because the assignment of treatment is not under the control of the researcher, who typically receives the data after the treatment has already been assigned.[3]

One popular ex-post method of covariate adjustment for experiments transforms the outcome to subtract from it a prior measure (also known as "pre-test outcome" or simply a "pre-test") that may come from a baseline survey or pilot study collected before the treatment was assigned. For example, instead of estimating the average treatment effect on the outcome, the researcher estimates the average treatment effect on the transformed outcome. The adjusted estimator is consistent for the average treatment effect, and leads to efficiency gains when the pre-test outcome is highly predictive of the post-treatment outcome. This idea can be easily implemented using modern non-parametric or machine learning methods.

Another popular method for ex-post covariate adjustment in experiments, and one of the most widely used, is to add pre-treatment covariates to a linear regression fit between the outcome and the treatment indicator. This regression-based covariate adjustment typically leads to efficiency gains when the added covariates are correlated with the outcome, and has been studied in detail in various contexts (see, for example, [2], and references therein). An interesting finding in this literature is that a regression model that incorporates the pre-intervention covariates and their interaction with the treatment variable is more efficient than a simple covariate adjusted estimator excluding the interaction term, unless the treatment and control groups have equal number of observations or the covariates are uncorrelated with the individual treatment effect.

[3]In recent years there has been interest in designing ex-ante surveys for participants in RD designs. At the moment, the RD literature only has methodological developments for simple power calculations and survey sample design ([16]), but no work is available taking into account covariates for how to use ex-ante survey information to increase efficiency in the ex-post analysis.

These ideas can be applied to the analysis of RD designs with suitable modifications. In the local randomization RD approach, the particular properties of covariate adjustment follow directly from applying the results from the literature on experiments to the window around the cutoff where the local randomization assumptions are assumed to hold. Depending on the assumptions imposed (Fisherian, Neymann, or super-population), and the specific estimation and inference methods considered, the efficiency gains may be more or less important.

Furthermore, regression-based covariate adjustment can also be used in the local randomization approach to relax some of the underlying assumptions. For example, [10] model the potential outcomes as polynomials of the running variable, to allow for a more flexible functional form of the potential outcomes inside the window $\mathcal{W}$ where the local randomization assumptions are assumed to hold. This model can be extended directly to include pre-determined covariates in the polynomial model, in addition to the running variable.

In the continuity-based framework, applying the same arguments is not immediate because, although the local polynomial RD estimator is based on linear regressions above and below the cutoff, these regressions are non-parametric and the average treatment effect is estimated by extrapolating to the cutoff point. The properties of regression-based covariate adjustment in this context were first studied by [17], who show that the inclusion of covariates other than the score in a local polynomial analysis can lead to asymptotic efficiency gains, if carefully implemented. The authors also show that covariate adjustment can result in unintended changes to the parameter that is being estimated depending on how the covariates are introduced in the estimation, a topic we revisit in Section 8.3.

The standard local linear estimator of the RD treatment is implemented by running the weighted least squares regression of Y_i on a constant, T_i, X_i, and X_iT_i using only units with scores inside the chosen bandwidth, $X_i \in [c-h, c+h]$, and applying weights based on some kernel function. This leads to a point estimator, which can be interpreted as MSE-optimal if the bandwidth employed is chosen to minimize the mean squared error of the RD point estimator. This estimator includes only the score X_i, and thus it is said to be "unadjusted" because it does not incorporate any additional pre-treatment characteristics of the units. As discussed above, under standard continuity assumptions, the unadjusted local linear estimator $\hat{\tau}_{\mathtt{SRD}}$ is consistent for the continuity-based RD treatment effect $\tau_{\mathtt{SRD}}$, and robust bias-corrected inference can be employed as is now standard in the literature.

In practice, it is common for researchers to adjust the RD estimator by augmenting the local polynomial specification with the additional pre-determined covariates Z_i. [17] show that augmenting the local polynomial specification by adding covariates in an additive-separable, linear-in-parameters way that imposes the same common coefficient for treated and control groups leads to a covariated-adjusted RD estimator that remains consistent for the canonical sharp RD treatment effect $\tau_{\mathtt{SRD}}$, while offering a reduction in its variance in large samples (see also [18]). The key required restriction, of course, is that the covariates are pre-intervention, and that they have non-zero correlation with the outcome. More specifically, whenever the effect of the additional pre-intervention covariates on the potential outcomes near the cutoff is (approximately) the same in the control and treatment groups, augmenting the specification with these covariates can lead to efficiency gains. Furthermore, the authors also show that other approaches commonly used in the applied RD literature for regression-based covariate adjustment can lead to inconsistent estimators of the parameter of interest $\tau_{\mathtt{SRD}}$.

More recently, [19] proposed a high-dimensional implementation of the covariate-adjusted local polynomial regression approach of [17], which allows for selecting a subset of pre-intervention covariates out of a potentially large covariate pool. Their method is based on penalization techniques (Lasso) and can lead to further efficiency improvements in practice. More generally, combining modern high-dimensional and machine learning methods can be a fruitful avenue for further efficiency improvements via covariate adjustment in RD settings. For example, non-parametric or machine learning adjustments can be applied directly to the outcome variable to then employ the residualized outcomes in the subsequent local polynomial analysis, a procedure that can lead to further efficiency and power improvements.

In sum, our discussion highlights several important issues. First, standard methods for covariate adjustment in the analysis of experiments can in principle be applied to RD designs with suitable modifications depending on the specific conceptual framework used. These methods and their properties apply immediately to RD estimation and inference in a local randomization framework, within the window $\mathcal{W}$ where the local randomization assumption is assumed to hold. The extension to continuity-based RD analysis is less immediate, as it requires considering the properties of the specific techniques at boundary points in a non-parametric local polynomial regression setup. However, there is a critical conceptual point: when using covariates for efficiency gains in RD designs, adjusting for those covariates should not change the RD point estimate. Regardless of the methods considered, covariate adjustment should not change the parameter of interest, and therefore unadjusted and adjusted point estimators should be similar in applications.

Second, pre-intervention covariates can also affect tuning parameter selection and implementation of the RD design more broadly. For example, in the local randomization framework, covariates can be used to select the window where the key assumptions hold and to relax those assumptions, while in the continuity-based framework they can be used to further refine bandwidth selection for local polynomial estimation and robust bias-corrected inference.

Finally, while augmenting RD analysis with covariates to increase efficiency is a principled goal, researchers should be careful to avoid using covariate adjustment for specification searches. In empirical studies, researchers should always report unadjusted RD results first, and covariate-adjusted results second. Covariate adjustment can be implemented in many different ways, and different adjustments might lead to different results and conclusions. Instead of trying multiple ways of covariate adjustment at the time of analysis, researchers should pre-register their preferred adjustment method, and perform only that specification as opposed to trying multiple regression models. We return to this important point when discussing recommendations for practice in Section 8.4.

8.2.3 Auxiliary information

In the context of RD designs, covariate adjustment has also been proposed to address (i) missing data and measurement error, and (ii) to incorporate prior information via Bayesian methods.

A common concern in RD designs is that the RD running variable may be measured with error [20], and several approaches employing covariates have been recently proposed to address this issue. Generally speaking, key ideas from the imputation literature can be applied with appropriate modifications to RD designs. In the continuity-based framework, [21] study the case where the RD score is measured with error and this error changes across different groups of observations, and the researcher has access to auxiliary data that can be used to learn about the group-specific measurement error. The general procedure is to replace the mismeasured running variable by a "corrected" running variable of the same dimension that is estimated using the auxiliary covariates. Similarly, [22] study a fuzzy RD design where the running variable is measured with error, but the researcher has access to an additional covariate – the true running variable – that is observed for a sample of treated individuals. With this auxiliary data, the RD treatment effect at the cutoff can be identified. The procedure thus leads to heterogeneity tests that can be directly employed in empirical applications to assess whether the treatment affects different populations differently. Finally, in the local randomization framework, conventional methods can be used based on the standard missing-at-random assumption, where the missingness indicator and the potential outcomes are assumed to be independent conditional on pre-determined covariates. This assumption and associated methods can be applied directly within the window $\mathcal{W}$ to justify the use of imputation approaches.

The use of covariates has also been proposed to incorporate information via Bayesian approaches to RD analysis – which fall in a separate category, different from both local randomization and continuity-based RD approaches. [23] propose an infinite-mixture Bayesian model that allows the density of the outcome variable to depend flexibly on covariates. At a minimum, these covariates include the running variable and the treatment indicator, but other covariates can be added to the specification. A somewhat more general use of pre-determined covariates in Bayesian approaches to RD estimation is to use them to inform priors. Here, researchers sometimes use pre-treatment information rather than specific covariates. For example, in their study of the effect of prescribing statins to patients with high cardiovascular disease risk scores on their future LDL cholesterol levels, [24] use data on previous experimental studies to formulate informative priors in a Bayesian approach to estimate RD effects. Finally, in a related approach, [25] use principal stratification in a Bayesian framework to estimate complier average treatment effects in fuzzy RD designs. They model the imperfect compliance near the cutoff using a discrete confounder variable that captures different types of subjects: compliers, never-takers, and always-takers. Their setup leads to a mixture model averaged over the unknown subject types. The compliers' potential outcomes are modeled as a function of observed pre-determined covariates and the running variable, plus an idisyoncractic shock that follows a t-distribution. Given a prior distribution on the type probabilities, the posterior distribution of the complier average treatment effect is obtained by Markov chain Monte Carlo (MCMC) methods.

These types of methods have not been widely adopted in practice, but they provide useful examples of the different roles that covariate adjustment methods can have in RD applications. Importantly, in all these methods, the goal is ultimately to identify, estimate and conduct inference on the canonical (sharp and other) RD treatment effects, that is, without changing the parameter of interest. In the upcoming sections, we discuss covariate adjustment methods that do change the estimand and therefore require careful interpretation.

8.2.4 Treatment effect heterogeneity

Another important use of covariates for RD analysis is to assess whether the treatment has different effects for different subgroups of units, where the subgroups are defined in terms of observed values of pre-determined covariates. The use of covariates to explore heterogeneity has a long tradition in the analysis of both experimental and non-experimental data, as researchers are frequently interested in assessing the effects of the treatment for different subpopulations. From a broader perspective, this analysis is equivalent to estimating treatment effects conditional on pre-treatment characteristics, which can typically be implemented using standard non-parametric or machine learning methods.

In RD designs, and regardless of the conceptual framework employed, the analysis of treatment effect heterogeneity necessarily changes the parameter of interest. For example, in the continuity-based framework, the parameter of interest is no longer $\tau_{\mathtt{SRD}}$, but rather

$$\tau_{\mathtt{CSRD}}(z) = \mathbb{E}[Y_i(1) - Y_i(0)|X_i = c, Z_i = z],$$

where Z_i denotes the additional pre-intervention covariate. The parameter $\tau_{\mathtt{CSRD}}(z)$ corresponds to the (conditional, sharp) RD treatment effect at the cutoff for the subpopulation with covariates $Z_i = z$. Naturally, under appropriate assumptions, the canonical sharp RD treatment effect $\tau_{\mathtt{SRD}}$ becomes a weighted average of the conditional RD treatment effects, with weights related to the conditional distribution of $Z_i|X_i$ near the cutoff. Similar ideas apply to the local randomization framework, depending on the specific assumptions invoked.

When the covariates Z_i are few and discrete, the simplest strategy to explore heterogeneity is to conduct the RD analysis for each subgroup defined by $Z_i = z$, for each value z, separately. For example, in a medical experiment, researchers may be interested in separately estimating treatment effects for patients in different age groups, e.g., patients between 45 and 64 versus patients 65 and above. In such cases separate analyses by subgroups have the advantage of being entirely non-parametric and involving no additional assumptions. One disadvantage is that the number of observations in each subgroup is lower than in the overall analysis, which reduces statistical power. Whether this is a limitation will depend on the number of observations and other features of the data generating process in each particular application. Practically, this type of analysis can also be implemented by generating indicator variables for each category of the pre-intervention covariates, and then conducting estimation and inference using a fully saturated, interacted local polynomial regression model; if the indicators cover all possible subgroups, this strategy is equivalent to estimating effects separately by subgroup. This approach is also valid with respect to estimation and inference, and can be deployed in both the continuity-based and local randomization frameworks. However, if the covariates are multi-valued or continuous, using an interactive model imposes additional parametric assumptions, local to the cutoff, and the validity of inferences will depend on the validity of these assumptions.

If the number of covariates is large and researchers wish to explore heterogeneity in multiple dimensions, estimating separate average treatment effects by subgroups may be infeasible or impractical. In this case, researchers are usually interested in exploring heterogeneity in a large number of covariate partitions without specifying subgroups a priori. Standard non-parametric or machine learning methods can be useful in this situation, as they allow researchers to learn the relevant subgroups from the data. Machine learning methods in the continuity-based RD framework require modifications and have begun to be explored only recently. In particular, [26] explores heterogeneity of the RD effect in subpopulations defined by levels of pre-determined covariates, creating a tree where each leaf contains an RD effect estimated on an independent sample. The approach assumes a parametric q-th order polynomial in each leaf of the tree; under this parametric assumption and additional regularity assumptions, the method leads to subgroup point estimates and standard errors that can be used for inference.

A different continuity-based approach to examine RD treatment effect heterogeneity on different subpopulations is proposed by [27]. The authors impose stronger continuity conditions than in the standard RD design, requiring (in the sharp RD case) continuity of the expectation of the potential outcomes conditional on both the running variable and the additional covariates, and continuity of the conditional distribution of the additional covariates given the running variable. Under these conditions, they derive conditional average treatment effects given the covariates, and propose tests for three hypothesis: (i) that the treatment is beneficial for at least some subpopulations defined by values of the pre-determined covariates, (ii) that the treatment has any impact on at least some subpopulations, and (iii) that the treatment effect is heterogeneous across all subpopulations.

[27] define the null and alternative hypotheses by conditional moment inequalities given both the running variable and the additional covariates, and convert these conditional moment inequalities into an infinite number of unconditional-on-covariates moment inequalities (that is, inequalities that are conditional only on the running variable and not on the additional covariates). Once the null and the alternative hypotheses are re-defined as instrumented moment conditions, these are estimated with local linear polynomials. Under regularity conditions, the asymptotic distribution of the local polynomial estimators of the moment conditions can then be used to derive the distribution of test statistics under the null hypothesis. In a follow-up paper, [28] employ similar methods to develop a monotonicity test to assess whether a conditional local average treatment effect in a sharp RD design or a conditional local average treatment effect for compliers in a fuzzy RD design has a monotonic relationship with an observed pre-determined covariate – that is, the null hypothesis is that the conditional local average treatment effect, seen as a function of a covariate z, is non-decreasing in z.

Covariate adjustments for heterogeneity analysis within the local randomization framework may be more challenging due to limited sample sizes. Since the local randomization framework is generically more appropriate for very small windows $\mathcal{W}$ around the cutoff, and these windows typically contain only a few observations, conducting estimation and inference for subsets of observations according to $Z_i = z$ can be difficult in practice, leading to limited statistical power.

In sum, there are several important issues when using covariate adjustment for the analysis of heterogenous RD treatment effects. First, focusing on treatment effect heterogeneity requires that researchers redefine the parameter of interest. Second, estimating different RD treatment effects for different subgroups defined by covariate values can be done in a non-parametric way by estimating fully saturated models – or equivalently, by analyzing each sub-group separately. When the covariate dimensionality is too high and sub-group analysis is not possible, machine learning methods may offer an attractive strategy to discover relevant subgroups. Finally, augmented models that include interactions between the RD treatment and non-binary covariates require additional parametric assumptions for identification and inference.

8.2.5 Other parameters of interest and extrapolation

Covariate adjustment has also been proposed in RD designs to identify additional parameters of interest. This includes cases where covariates are used to define new parameters for subpopulations, similar to the case of heterogeneity analysis discussed above, and cases where covariates are used to extrapolate treatment effects for score values far from the cutoff.

In some RD designs, researchers have access to special covariates, and in exploring heterogeneity along those covariates, they redefine the parameter of interest. For example, it is common in practice to see treatments that follow the RD assignment rule but use different cutoffs for different subpopulations of units, that is, $T_i = \mathbb{1}(X_i \geq c_j)$ for $j = 1, 2, \ldots, J$. In this case, the cutoffs $c_1, c_2, \ldots, c_J$ naturally define subpopulations of units, where each subpopulation is defined by the cutoff that those units are exposed to [29] introduced the Multi-cutoff RD design, and investigated how multiple cutoffs can be used to provide more information about RD effects. In particular the heterogeneity of the average treatment effect across cutoffs can be explored directly by performing RD analysis separately for each subgroup of units exposed to each cutoff value, as discussed in the prior section. The authors also discuss the role of the commonly used normalizing-and-pooling strategy, which leads to a causal interpretation of the canonical RD treatment effect as a weighted average of the cutoff-specific RD treatment effects under additional assumptions. See [30] for an empirical application.

In some cases researchers may prefer to use the information provided by the multiple cutoffs in combination with additional assumptions to define new parameters of interest. For example, [31] studies a Multi-Cutoff RD setup where each individual is exposed to a particular cutoff and a particular treatment dosage that changes as a function of that cutoff. Under additional assumptions, the author uses the changes in treatment dosage applied to individuals at various levels of the running variable to explore average treatment effects for counterfactual policies that set the treatment dosage to levels not observed in the data. [32] combine a Multi-Cutoff RD design with additional assumptions to define the average treatment effect at a particular cutoff for a subpopulation of units exposed to a different cutoff value. Both approaches lead to extrapolation of RD treatment effects – that is, to parameters that capture the average effect of the treatment at values of the score different from the original cutoff to which units were exposed.

Another common setup involves an RD design observed in two periods of time, where the treatment of interest is confounded by an additional treatment in one period but not the other [33] call this setup a differences in discontinuities design. In the first period, there are two treatments that are assigned with the same RD rule, that is, the probability of receiving each treatment goes from zero to one at the same cutoff. In the second period, only one treatment changes at the cutoff, but the other treatment – which is the treatment of interest—is not active. The authors propose an estimator that takes the difference between the first-period RD effect and the second-period RD effect. Whether the limit of this estimator is the standard RD parameter will depend on the particular assumptions imposed. Under the usual parallel trends assumption of difference-in-differences designs, the estimand will remain unchanged. However, researchers could also make different assumptions about the change of the regression functions before and after, and redefine the parameters accordingly.

When multiple cutoffs or time periods are not available, extrapolation of treatment effects can be done by using auxiliary pre-determined covariates. In this setting, with additional assumptions, researchers can learn about the treatment effect for units whose scores are not in the immediate neighborhood of the cutoff. For example, [34] propose a conditional independence assumption to study the effect of the treatment for observations far from the cutoff. This is formalized by assuming that, conditional on the covariates, the potential outcomes are mean-independent of the running variable. This "selection on observables" assumption (together with the standard common support assumption) immediately allows for extrapolation of the RD treatment effect, since it is assumed to hold for the entire support of the running variable. In this case covariates allow for the identification of new parameters that capture the effect of the treatment at different values or regions of the running variable.

A similar conditional independence assumption has been used by in the context of a geographic or multi-score RD designs. In a geographic RD design, a treatment is assigned to units in a geographic area and withheld from units in an adjacent geographic area. In this setup, units can be thought of as having a score (such as a latitude-longitude pair) that uniquely defines their geographic location and allows the researcher to calculate their distance to any point on the boundary between the treated and control areas. The assignment of the treatment can then be viewed as a deterministic function of this score, and the probability of receiving treatment jumps discontinuously at the border that separates the treated and control areas. As discussed by [35], seen in this way, the geographic RD design can be analyzed with the same continuity-based and local randomization tools of standard, one-dimensional RD designs [36] consider a geographic RD application where units appear to choose their location on either side of the boundary in a strategic way. They propose a conditional independence assumption according to which, for each point on the boundary, the potential outcomes are independent of the treatment assignment conditional on a set of covariates for units located in a neighborhood of that point (see also [37] for an extension to multiple dimensions). An important difference between their assumption and the conditional independence assumption proposed by [34] is that the former imposes conditional independence in a neighborhood of the cutoff and thus allows for extrapolation only within a neighborhood, while the latter imposes (mean) conditional independence along the entire support of the running variable and thus allows for extrapolation in the entire support of the running variable. Despite the differences, in both cases covariates are used for identification purposes, to define new parameters of interest that capture the average treatment effect at values of the running variable that are different from the cutoff.

Another use of auxiliary covariates for extrapolation purposes is proposed by [38], who augment the usual RD design with an exogenous measure of the outcome variable. Under the assumption that the exogenous outcome data can be used to consistently estimate the regression function of the actual outcome on the score in the absence of the treatment, treatment effects can be extrapolated to values of the score other than the actual cutoff used for assignment.

In sum, additional covariates can be used in the RD design to define new parameters of interest, including parameters capturing treatment effects for units with score values different from the original cutoff. These covariates are sometimes part of the original RD design, as in the case of multiple cutoffs or multiple time periods, while in other settings they are external to the design and must be collected separately, as in the case of auxiliary unit characteristics or exogenous outcomes.

8.3 Can Covariates Fix a Broken RD Design?

In a randomized experiment the fact that the treatment is randomly assigned implies that the treatment variable D_i is independent of both potential outcomes $(Y_i(1), Y_i(0))$ and predetermined covariates Z_i. This means that, in experimental settings, we have independence between potential outcomes and treatment assignment, $(Y_i(1), Y_i(0)) \perp\!\!\!\perp D_i$, and also conditional independence between potential outcomes and treatment assignment given the covariates, $(Y_i(1), Y_i(0)) \perp\!\!\!\perp D_i|Z_i$. At the same time, it is well known that the assumption of conditional independence given covariates alone, $(Y_i(1), Y_i(0)) \perp\!\!\!\perp D_i|Z_i$, does not imply $(Y_i(1), Y_i(0)) \perp\!\!\!\perp D_i$. In observational studies researchers often assume that $(Y_i(1), Y_i(0)) \perp\!\!\!\perp D_i|Z_i$ in order to identify treatment effects. This assumption directly justifies covariate adjustment via regression models, inverse probability weighting, and other matching methods.

This observation has motivated some researchers to employ covariate adjustment to "fix" invalid RD designs where the identifying assumptions are suspected not to hold. In these cases, it is common to encounter RD analyses that include fixed effects or other covariates via additive linear adjustments in (local) polynomial regression estimation. In the methodological RD literature, adjustment for covariate imbalances has been proposed using multi-step non-parametric regression [39] and inverse probability weighting [40].

In the continuity-based framework, covariate adjustment for imbalances in RD designs is done within a framework where $\mathbb{E}[Y_i(t)|X_i = x]$ is assumed to be discontinuous at the cutoff but $\mathbb{E}[Y_i(t)|X_i = x, Z_i = z]$, seen as a function of x, is assumed continuous in x, for $t = 0, 1$. These assumptions combined imply that the conditional distribution of $Z_i|X_i$ must be discontinuous at the cutoff, which in turn implies that canonical average RD treatment effects are necessarily not identifiable in general. Rather, in the absence of additional strong assumptions, the estimand that can be identified in this case is a weighted average of treatment effects where the weights depend on the conditional distributions of $Z_i|X_i$ for control and treatment units, which are different from each other [17, 39]. As a consequence, in imbalanced RD designs, canonical RD treatment effects are not identifiable in general, and by implication covariate adjustment cannot fix a broken RD design. In the local randomization framework, the situation is analogous.

In conclusion, for invalid RD designs where important covariates are imbalanced at the cutoff, covariate adjustment cannot restore identification of the canonical RD treatment effect. At best, covariate adjustment can identify some weighted average of the conditional treatment effects, $\mathbb{E}[Y_i(1)|X_i = x, Z_i = z] - \mathbb{E}[Y_i(0)|X_i = x, Z_i = z]$, with weights determined by the properties of the two distinct conditional distributions of $Z_i|X_i$ below and above the cutoff. Unfortunately, in most applications, this covariate-adjusted estimand is not a parameter of interest and requires strong, additional assumptions on the underlying data generating process to deliver a useful parameter of interest. In other words, the main advantages of canonical RD designs in terms of identification are lost once the key identifying assumptions fail, and cannot be restored by simply using covariate adjustment methods without additional untestable assumptions.

8.4 Recommendations for Practice

When employed correctly, baseline covariates can be useful for the analysis and interpretation of RD designs. When pre-intervention covariates are available, they can be used for two main purposes without affecting the main RD identification strategy: (i) to improve efficiency and power, and (ii) to define new parameters of interest. However, analysts should keep in mind that canonical RD designs do not necessitate covariate adjustments, and therefore researchers should always report unadjusted RD treatment effect estimates and associated unadjusted robust bias-corrected inference methods.

Covariate adjustment for efficiency and power improvement is a natural way of incorporating baseline covariates in the RD analysis. This can be done by using regression adjustments, or via other well-known approaches from the experimental literature such as outcome adjustments, inverse probability weighting, or matching methods. Importantly, because the treatment effect of interest is the same with and without covariate adjustment, researchers should always report both unadjusted and covariate-adjusted RD results. The main and only goal of this practice is to improve efficiency and power (e.g., shorten confidence intervals), but the RD point estimate should remain unaffected, which should be borne out by the empirical analysis. Furthermore, it is important to guard against p-hacking in this context. Researchers should be cognizant of the dangers of multiple hypothesis testing when incorporating covariate adjustment in the RD analysis. One possible solution to address this concern is pre-registration of the analysis, which requires practitioners to declare ex-ante the covariate adjustment approach to be used in the subsequent RD analysis in other to avoid ex-post specification searching.

Pre-intervention covariates can also be employed for identification of other useful RD treatment effects. In Section 8.2, we discussed different examples such as treatment effect heterogeneity, multi-cutoff, geographic and multi-score designs, and extrapolation. These approaches offer relatively principled ways of incorporating covariates into an RD analysis. In the case of treatment effect heterogeneity, no substantial additional assumptions are needed, particularly in cases where the covariates are discrete and low-dimensional, in which case subset analysis is a natural way to proceed empirically. In the other cases, additional assumptions are needed in order to exploit covariate adjustment for identification of other meaningful RD treatment effects. These covariate adjustment practices are reasonable, and can be used in applications whenever the necessary additional assumptions are clearly stated and judged to be plausible. Again, it is important to guard against model specification searching, just like when employing pre-interventation covariates for efficiency and power improvements.

Finally, an important misconception among some practitioners and methodologists is that covariates can be used to restore identification or "fix" an RD design where observations just above the cutoff are very different from observations just below it – in other words, a design where there is covariate imbalance at or near the cutoff and the RD assumptions are not supported by the empirical evidence. However, as we discussed in Section 8.3, covariate adjustment in such cases requires strong additional assumptions, and cannot in general recover canonical RD treatment effects. In the absence of additional assumptions, adjusting for imbalanced covariates will at best recover other estimands at the cutoff that are unlikely to be of substantive interest in applications.

8.5 Conclusion

We provided a conceptual overview of the different approaches for covariate adjustment in RD designs. We discussed benign approaches based on pre-intervention covariates, including methods for efficiency and power improvements, heterogeneity analysis, and extrapolation. Those methods are principled and generally valid under additional reasonable assumptions on the data generating process. However, we also highlighted a natural tension between incorporating covariates in an RD analysis and issues related to p-hacking and specification searching. As a consequence, researchers employing covariate adjustment in RD designs should always be cognizant of issues related to model/covariate selection when implementing those methods in practice. Last but not least, we addressed the common misconception that covariate adjustment can fix an otherwise invalid RD design, and discussed how those methods can at best recover other estimands that may not be of interest.

In conclusion, covariate adjustment can be a useful additional tool for the analysis and interpretation of RD designs when implemented carefully and in a principled way. In canonical RD settings it should never replace, only complement, the basic RD analysis based on the score and outcome variables alone. In other settings, covariate adjustment can offer additional insights in terms of heterogeneous or other treatment effects of interest, under additional design-specific assumptions.

8.6 Acknowledgments

We thank our collaborators Sebastian Calonico, Max Farrell, Michael Jansson, Xinwei Ma, Gonzalo Vazquez-Bare, and Jose Zubizarreta for stimulating discussions on the topic of this chapter. Cattaneo and Titiunik gratefully acknowledge financial support from the National Science Foundation (SES-2019432), and Cattaneo gratefully acknowledge financial support from the National Institutes of Health (R01 GM072611-16).

References

[1] Paul R. Rosenbaum. *Design of Observational Studies*. Springer-Verlag, 2010.

[2] Guido W Imbens and Donald B Rubin. *Causal Inference in Statistics, Social, and Biomedical Sciences*. Cambridge University Press, 2015.

[3] Alberto Abadie and Matias D. Cattaneo. Econometric methods for program evaluation. *Annual Review of Economics*, 10:465–503, 2018.

[4] Rocío Titiunik. Natural experiments. In J. N. Druckman and D. P. Gree, editors, *Advances in Experimental Political Science*, chapter 6, pages 103–129. Cambridge University Press, 2021.

[5] Matias D. Cattaneo, Rocío Titiunik, and Gonzalo Vazquez-Bare. The regression discontinuity design. In L. Curini and R. J. Franzese, editors, *Handbook of Research Methods in Political Science and International Relations*, chapter 44, pages 835–857. Sage Publications, 2020.

[6] Matias D. Cattaneo, Nicolás Idrobo, and Rocío Titiunik. *A Practical Introduction to Regression Discontinuity Designs: Foundations*. Cambridge University Press, 2020.

[7] Matias D. Cattaneo, Nicolás Idrobo, and Rocío Titiunik. *A Practical Introduction to Regression Discontinuity Designs: Extensions*. Cambridge University Press (to appear), 2023.

[8] Matias D. Cattaneo, Luke Keele, and Rocío Titiunik. A guide to regression discontinuity designs in medical applications. Working Paper, 2022.

[9] Matias D. Cattaneo and Rocío Titiunik. Regression discontinuity designs. *Annual Review of Economics*, 14:821–851, 2022.

[10] Matias D. Cattaneo, Rocío Titiunik, and Gonzalo Vazquez-Bare. Comparing inference approaches for rd designs: A reexamination of the effect of head start on child mortality. *Journal of Policy Analysis and Management*, 36(3):643–681, 2017.

[11] Jinyong Hahn, Petra Todd, and Wilbert van der Klaauw. Identification and estimation of treatment effects with a regression-discontinuity design. *Econometrica*, 69(1):201–209, 2001.

[12] Sebastian Calonico, Matias D. Cattaneo, and Max H. Farrell. Optimal bandwidth choice for robust bias corrected inference in regression discontinuity designs. *Econometrics Journal*, 23(2):192–210, 2020.

[13] Sebastian Calonico, Matias D. Cattaneo, and Rocío Titiunik. Robust nonparametric confidence intervals for regression-discontinuity designs. *Econometrica*, 82(6):2295–2326, 2014.

[14] Matias D. Cattaneo, Brigham Frandsen, and Rocío Titiunik. Randomization inference in the regression discontinuity design: An application to party advantages in the u.s. senate. *Journal of Causal Inference*, 3(1):1–24, 2015.

[15] Donald P. Green Green and Alan Gerber. *Field Experiments: Design, Analysis, and Interpretation*. W. W. Norton & Company, 2012.

[16] Matias D. Cattaneo, Rocío Titiunik, and Gonzalo Vazquez-Bare. Power calculations for regression discontinuity designs. *Stata Journal*, 19(1):210–245, 2019.

[17] Sebastian Calonico, Matias D. Cattaneo, Max H. Farrell, and Rocío Titiunik. Regression discontinuity designs using covariates. *Review of Economics and Statistics*, 101(3):442–451, 2019.

[18] Jun Ma and Zhengfei Yu. Empirical likelihood covariate adjustment for regression discontinuity designs. arXiv:2008.09263, 2022.

[19] Yoichi Arai, Taisuke Otsu, and Myung Hwan Seo. Regression discontinuity design with potentially many covariates. arXiv:2109.08351, 2021.

[20] Zhuan Pei and Yi Shen. The devil is in the tails: Regression discontinuity design with measurement error in the assignment variable. In Matias D. Cattaneo and Juan Carlos Escanciano, editors, *Regression Discontinuity Designs: Theory and Applications (Advances in Econometrics, volume 38)*, pages 455–502. Emerald Group Publishing, 2017.

[21] Otavio Bartalotti, Quentin Brummet, and Steven Dieterle. A correction for regression discontinuity designs with group-specific mismeasurement of the running variable. *Journal of Business & Economic Statistics*, 39(3):833–848, 2021.

[22] Laurent Davezies and Thomas Le Barbanchon. Regression discontinuity design with continuous measurement error in the running variable. *Journal of Econometrics*, 200(2):260–281, 2017.

[23] George Karabatsos and Stephen G Walker. A bayesian nonparametric causal model for regression discontinuity designs. In R. Mitra and P. Müller, editors, *Nonparametric Bayesian Inference in Biostatistics*, pages 403–421, 2015.

[24] Sara Geneletti, Aidan G O'Keeffe, Linda D Sharples, Sylvia Richardson, and Gianluca Baio. Bayesian regression discontinuity designs: Incorporating clinical knowledge in the causal analysis of primary care data. *Statistics in Medicine*, 34(15):2334–2352, 2015.

[25] Siddhartha Chib and Liana Jacobi. Bayesian fuzzy regression discontinuity analysis and returns to compulsory schooling. *Journal of Applied Econometrics*, 31(6):1026–1047, 2016.

[26] Agoston Reguly. Heterogeneous treatment effects in regression discontinuity designs. *arXiv preprint arXiv:2106.11640*, 2021.

[27] Yu-Chin Hsu and Shu Shen. Testing treatment effect heterogeneity in regression discontinuity designs. *Journal of Econometrics*, 208(2):468–486, 2019.

[28] Yu-Chin Hsu and Shu Shen. Testing monotonicity of conditional treatment effects under regression discontinuity designs. *Journal of Applied Econometrics*, 36(3):346–366, 2021.

[29] Matias D. Cattaneo, Luke Keele, Rocío Titiunik, and Gonzalo Vazquez-Bare. Interpreting regression discontinuity designs with multiple cutoffs. *Journal of Politics*, 78(4):1229–1248, 2016.

[30] Yasin Kursat Onder and Mrittika Shamsuddin. Heterogeneous treatment under regression discontinuity design: Application to female high school enrolment. *Oxford Bulletin of Economics and Statistics*, 81(4):744–767, 2019.

[31] Marinho Bertanha. Regression discontinuity design with many thresholds. *Journal of Econometrics*, 218(1):216–241, 2020.

[32] Matias D Cattaneo, Luke Keele, Rocío Titiunik, and Gonzalo Vazquez-Bare. Extrapolating treatment effects in multi-cutoff regression discontinuity designs. *Journal of the American Statistical Association*, 116(536):1941–1952, 2021.

[33] Veronica Grembi, Tommaso Nannicini, and Ugo Troiano. Do fiscal rules matter? *American Economic Journal: Applied Economics*, 8(3):1–30, 2016.

[34] Joshua D Angrist and Miikka Rokkanen. Wanna get away? regression discontinuity estimation of exam school effects away from the cutoff. *Journal of the American Statistical Association*, 110(512):1331–1344, 2015.

[35] Luke J Keele and Rocío Titiunik. Geographic boundaries as regression discontinuities. *Political Analysis*, 23(1):127–155, 2015.

[36] Luke Keele, Rocío Titiunik, and Jose R Zubizarreta. Enhancing a geographic regression discontinuity design through matching to estimate the effect of ballot initiatives on voter turnout. *Journal of the Royal Statistical Society. Series A (Statistics in Society)*, 178(1):223–239, 2015.

[37] Juan D Diaz and Jose R Zubizarreta. Complex discontinuity designs using covariates for policy impact evaluation. *Annals of Applied Statistics,* forthcoming, 2022.

[38] Coady Wing and Thomas D Cook. Strengthening the regression discontinuity design using additional design elements: A within-study comparison. *Journal of Policy Analysis and Management*, 32(4):853–877, 2013.

[39] Markus Frolich and Martin Huber. Including covariates in the regression discontinuity design. *Journal of Business & Economic Statistics*, 37(4):736–748, 2019.

[40] Sida Peng and Yang Ning. Regression discontinuity design under self-selection. In *International Conference on Artificial Intelligence and Statistics*, pages 118–126. PMLR, 2021.

9

Risk Set Matching

Bo Lu and Robert A. Greevy, Jr.

CONTENTS

When a treatment may be given at different time points, it is natural to examine the treatment delay effect, i.e., the impact of giving the treatment at an earlier time point in lieu of at some later time point. In such situation, it is crucial to form matched pairs or sets in which patients have similar covariate history up to the time point of treatment, but not matching on any post-treatment event. Risk set matching is a powerful design to match a cohort of subjects, in which the treatment was given at various times. The inference following such design may reveal the causal effect of a delayed treatment.

9.1 Treatment at Different Time Points

In a typical randomized experiment, every subject is assigned to treatment or control upon entry into the study. In some observational studies, however, the timing of treatment may vary substantially. Under such setting, everyone enters the study as untreated, and over the time, some of them may be given the treatment at different time points, and others never receive the treatment. To gauge the treatment effect, one might be tempted to make comparisons of outcomes between the groups – ever treated versus never treated. But this way of forming groups is quite problematic as it does not take into account the fact that the never-treated subjects might be too ill to survive long enough to get

DOI: 10.1201/9781003102670-9

the treatment. In some other situations, most or all of subjects may get the treatment eventually if the study followup is long enough. A naive way is to split the subjects into early- and late-treatment groups by an intermediate time point, then contrast their outcomes. But again this is a mistake as the timing of treatment could be substantially affected by underlying health status. For example, those who receive treatment earlier are much sicker and the rapid disease progression demands early treatment, while those who receive treatment later are healthier and they can afford longer waiting time.

A good illustration of this problem comes from early studies of the effects of transplantation [1]. A patient who needed an organ transplant was immediately enrolled into the study but had to wait until a suitable organ become available, depending on many factors including blood types and severity of illness. The early studies usually compared the duration of survival from entry into the study for patients who were transplanted to those who were not transplanted. People who died before a matching organ was available, were considered "control" by definition. In a critique of the early studies evaluating cardiac transplantation [2], Gail pointed out that "this assignment method biases the results in favor to the transplanted group." Patients with more severe conditions were likely to die before a heart became available, thus classified as controls. In contrast, transplanted patients survived at least long enough to receive a heart. This way of assigning transplant and control groups would introduce bias in comparison as the control group was expected to have a shorter overall survival, due to the fact that many of them failed to reach the time point when a transplant could be conducted.

To address the difficulty introduced by the potential treatment delay, waiting for a matching heart, Gail suggested a paired randomized experiment design that removes such bias. When a heart becomes available, two patients in waiting who are most compatible with that heart are identified, and one is chosen at random to receive it, while the other becomes a control. To assess the treatment effect of survival, the survival time is measured from the time of randomization. This design effectively remove the biases associated with the fact that hearts are not immediately available to patients upon entry into the study. Li and colleagues [3] suggested a slightly modified version of the design by allowing the control subject, who did not receive the available heart, to remain eligible to receive a suitable heart at a later time.

This modified design addresses a different problem of estimating the effect a treatment delay, that is, the effect of treating now as contrasted with waiting and possibly treating at a later time. In fact, the conventional treated-vs-control comparison can be regarded as a special case of the treatment delay problem, in the sense that the waiting time is forever. The effect of a treatment delay has important practical implication in terms of treatment management. Does treating immediately improve patient outcomes? Or shall we wait to see if there are better options later? If immediate treatment is deemed unnecessary, how long can we delay the treatment? Conceivably, longer delays are likely to have different effects than shorter delays. A randomized experiment with pre-specified delay times could address those questions. Risk set matching is a design tool for observational studies analogous to such randomized experiments. It takes advantage of the varying treatment times in an observational dataset to created matched pairs of similar subjects with one being treated immediately and the other being delayed. At the moment when someone receives treatment, that person is paired with another who has not yet treated and whose observed covariates are similar up to the treatment time point. It is crucial to emphasize that their similarity is solely determined by observed covariates history before treatment. The pairing should not match on any post-treatment variable, which may potentially remove part of the treatment effect. As seen in the old way of analyzing transplant patients, controls were defined as those who never got transplanted. Knowing who was control is nearly the same as knowing the future outcome, namely, that they died before a heart became available. This would likely bias the treatment effect estimation, while risk set matching matches on the past, not the future. Another important feature of risk set matching is that it can balance the time-varying covariates, as the conventional two-group matching usually does not address the temporal change of covariates.

The chapter is organized as following: section 2 introduces some theoretical background about risk set matching; section 3 describes the implementation of risk set matching via both bipartite and non-bipartite algorithms; section 4 presents three illustrative examples from the literature, including evaluation of surgery for interstitial cystitis, impact of premature infants staying longer on subsequent healthcare costs and outcomes, and drug effect for pregnant women with recurrent pre-term birth, followed by a summary of the chapter.

9.2 Methodology Overview

Suppose the study population contains I patients, $i = 1, \cdots, I$. For each patient, he/she may be treated at some time point W_i (using any time unit appropriate for the specific study) or not at all, where is signified as censored by denoting $W_i = c$. Assume patient i could receive the treatment once at any time between entry into the study to a terminal time point, s_i, then $W_i \in [0, s_i] \cup c$. Extending Rubin's notation on potential outcomes, every patient has a series of potential outcomes during the study period. Specifically, $y_{t,W_i,i}$ denotes the potential outcome for patient i at time t, given that patient received the treatment at W_i. Therefore, the treatment delay effect for patient i at month $t = 6$ of postponing the treatment from entry into study to three months later is well defined as $y_{6,3,i} - y_{6,0,i}$. The strong null hypothesis of no treatment effect for every patient asserts that the response patient i exhibits at time t is the same as the response without treatment, no matter when the patient receives the treatment, i.e.,

$$H_0 : y_{t,w,i} = y_{t,c,i} \text{ for all } t \in [0, s_i], i = 1, \cdots, I.$$

where w is an arbitrary time point when the treatment occurs with $w \in [0, s_i] \cup c$.

In risk set matching, a treated patient is matched to a not-yet-treated control who appears similar in terms of observed covariates up to the time point when the treated patient gets the treatment. Unlike the conventional studies with dichotomous treatment arms, the treatment decision becomes a time-varying variable here. Moreover, there could be time-varying confounders that impact the treatment decision. Assuming that treatment decisions for distinct patients are mutually independent, Li et al. [3] proposed to use the hazard of receiving treatment to model the treatment trajectory. The following model is assumed for the hazard of treatment reception at time t for patient i:

$$h_{ti} = exp\{\xi_t(\mathbf{a}_{ti}) + \gamma u_{ti}\} \tag{9.1}$$

where $\mathbf{a}_{ti}$ contains all the accumulated observed information about patient i until the instant before time t, which can include outcome information prior to time t, and $\xi_t(\cdot)$ is an unknown function for each t. u_{ti} denotes the potential unobserved confounder, and γ is an unknown scalar (sensitivity parameter). When $\gamma = 0$, the hazard of treatment at time t only depends on observed data up to time t, which is regarded as a time-dependent version of strong ignorability as discussed by Rosenbaum and Rubin [4].

Under the strong ignorability assumption, a successful risk set match recreates a randomization like scenario, in the sense, with each matched pair, one patient is "randomly" chosen to receive the treatment and the other patient is "randomly" chosen to wait. Therefore, the differences in their outcomes down the road can be interpreted as the causal effect of treatment delay. Given the time-dependent nature, a successful match means matching on the observed histories up to the moment of treatment. At first, all patients are unmatched. At time t, let $R_t(\mathbf{a})$ be the set of unmatched patients with a history of observed information of $\mathbf{a}$ up to time t. This is a subset of patients who have not received the treatment right before t, known as risk set for treatment in survival analysis terminology. Under model (1.1) for the hazard, the chance of any patient in risk set $R_t(\mathbf{a})$ receives the treatment at time t becomes $1/|R_t(\mathbf{a})|$, where $|\cdot|$ denotes the size of the risk set. Therefore, the conditional distribution of treatment assignments within matched sets is a randomization distribution, and conventional methods of inference, such as the signed-rank test or McNemar's test may be used.

To match on the observed histories, Li et al. [3] first picked important variables at different time points (prior to treatment) and stratified them by quantiles, then conducted optimal matching within each stratum using Mahalanobis distance computed based on relevant covariates. They also pointed out that exact matching on the entire covariate history is not needed and some dimension reduction strategy can be applied to obtain the same result. Lu [5] extended the idea by conducting risk set matching on time-dependent propensity scores. The propensity score is estimated with a Cox proportional hazards model with time-varying covariates. It is shown that matching on the covariates-related hazard component is sufficient to balance the covariates distribution, hence to justify the randomization based inference.

9.3 Implementation of Risk Set Matching

To accommodate the time-varying nature of treatment initiation, risk set matching implementation needs to create matched sets at different time points. There are two ways to accomplish this goal – sequential matching and simultaneous matching. For illustrative purpose, we only consider pair matching in this chapter. We follow notations in Hade et al. [6] to describe both matching procedures.

9.3.1 Sequential matching

Sequential matching creates matched pairs of treated and not-yet-treated subjects for each risk set chronologically. Suppose there are N subjects and they receive the treatment at s distinct time points, $t_0 = 0 < t_1 < t_2 < \cdots < t_s$, where t_0 represents the baseline. Without loss of generality, for those who never receive the treatment, we may denote their treatment time as $t_s = t_{end}^+$, where t_{end} is the end of the study period. Since t_{end}^+ is the last time point, those subjects will always be used as controls where possible. We also assume the number of subjects being treated at each time point are $n_1, n_2, \cdots, n_s$, where $n_1 + n_2 + \cdots + n_s = N$. To construct risk set matching, we may pair treated subjects with not-yet-treated subjects at each time point, sequentially in the time order. With the understanding that the risk here refers to the probability of receiving treatment, the risk set at time point t includes all subjects who have not yet received the treatment at $t - \epsilon$ for a tiny $\epsilon > 0$. Starting with t_1, the risk set at t_1 includes everyone and the size of the risk set is N. Then, we may create matches between the n_1 subjects who receive treatment at t_1 and the $N - n_1$ subjects who are not-yet-treated, based on a pre-specified distance metric (i.e., propensity score distance). With paired design, we get n_1^m matched pairs, where $n_1^m \leq n_1$ in case there are unmatchable treated subjects. Moving to the next treatment time point t_2, we exclude all subjects treated at t_1 and the corresponding matched controls (i.e., n_1^m not-yet-treated subjects matched at t_1). The risk set at t_2 includes $N - n_1 - n_1^m$ subjects, with n_2^* treated and $N - n_1 - n_1^m - n_2^*$ not-yet-treated subjects. Applying the same matching procedure between the two groups, we get n_2^m matched pairs at time t_2, where $n_2^m \leq n_2^*$. We continue the process for each subsequent time point whenever there are treated subjects available. Generally, for time point t_k, the risk set includes $N - (n_1 + n_2 + \cdots + n_{k-1} + n_1^m + n_2^m + \cdots + n_{k-1}^m)$ subjects, with n_k^* treated and $N - (n_1 + n_2 + \cdots + n_{k-1} + n_1^m + n_2^m + \cdots + n_{k-1}^m) - n_k^*$ not-yet-treated. We stop when there are no more treated subjects available or no more matches can be made based on the pre-specified matching rule. For each risk set, it is essentially a two-group matching problem, so bipartite matching algorithms can be used to implement it, such as *optmatch()* package in R [7]. Sequential risk set matching is easy to implement, however, it could be cumbersome when there are many risk set time points. For example, with 20 distinct time points, one needs to conduct the bipartite matching procedure 20 times.

9.3.2 Simultaneous matching

Unlike sequential matching, simultaneous matching only performs the matching procedure once. To do this, nonbipartite matching algorithms need to be used (see chapter 12 for details). Nonbipartite matching takes full advantage of the time-varying nature of the treatment initiation. For a patient treated at time point 2, he/she can be used either as a control for another patient treated at time point 1, or as a treated case for someone treated at time point 3. The algorithm compares all possible combinations of matched pairs at once and find the best way to use each subject, in the sense of minimizing a pre-specified distance measure. The distance matrix needs to be calculated before matching. For subject i, distance is computed individually with any other subject j. If both are treated at the same time, the distance is set to infinity to prohibit a match. If i is treated earlier than j, by default, i is regarded as the treated and j is regarded as the control in the (i, j) pair. Then the distance is computed based on covariate values at the time when i receives the treatment. If i is treated later than j, by default, j is regarded as the treated and i is regarded as the control in the (i, j) pair. Then the distance is computed based on covariate values at the time when j receives the treatment. Additional constraints can be built in the calculation to discourage certain matching structure, for example, a penalty can be added to discourage matching two subjects with adjacent treatment times as the treatment delay might not be meaningful for two treatment initiations very close in time. With such distance matrix incorporating all covariate information across different time points, non-bipartite matching only needs to match once to create matched pairs for all risk sets. Optimal nonbiparite matching can be implemented with R package *nbpmatch()* [8]. In the next subsection, we present a small example to illustrate how to accomplish both sequential and simultaneous matching in details.

9.3.3 A toy example

This example illustrates how the risk set matching is implemented with both sequential and simultaneous matching. For simplicity, there is only one time-varying variable, indicating the disease severity score. Before the treatment is applied, it serves as a time-varying covariate, which we try to balance. After treatment, it serves as an outcome. Without loss of generality, a higher value indicates more severe disease or symptom. The treatment is supposed to be effective and having the treatment will reduce the severity score. To demonstrate a harmful treatment delay effect, the data are generated to reflect a scenario that the treatment is not as effective when applied to a patient with extremely high severity score (8 or 9) in contrast to being applied to a patient with moderately high score (6 or 7). Suppose there are 12 patients in this study and they are evaluated every week during the study period of 5 weeks (Table 9.1). Week 0 is the baseline and treatment can only be applied starting week 1. Week 5 is the end of the study and no new treatment will be provided (though all patients will still be evaluated at week 5). Overall, each patient has a sequence of six weekly disease severity scores. All patients receive the treatment at some point during the study, as shown in Table 9.1. For example, patient A is treated at week 1 with all six scores listed in the parenthesis. His/her baseline score is 6 and week 1 score is 7. These are pre-treatment scores, which will be used in matching. His/her week 2-5's scores are 4, 4, 4, and 5, respectively. These are post-treatment outcomes, underlined to indicate that they will not be used in matching. The treatment is effective as the week 2 score (right after treatment) is 3 units lower than week 1 score (right before treatment). As in many chronic diseases, the treatment may not cure it completely. So the score may relapse after three weeks (week 5 score is one unit higher). In summary, two patients receive treatment at week 1, four at week 2, another four at week 3 and the last two at week 4.

In a naive early-vs-late comparison, it is tempting to create treatment groups by splitting treatment time right after week 2. Therefore, patients treated in weeks 1 and 2 are "Early" treated and those treated in weeks 3 and 4 are "Late" treated. The mean week 5 severity score difference between early

TABLE 9.1
Treatment time and time-varying covariate measurement for all twelve patients.

Trt Time	Patients Being Treated
0	Baseline
1	A (6, 7, 4, 4, 4, 5), B (6, 6, 3, 3, 3, 4)
2	C (6, 7, 8, 6, 6, 6), D (6, 6, 7, 4, 4, 4), E(5, 6, 7, 4, 4, 4), F(5, 5, 6, 3, 3, 3)
3	G (5, 6, 7, 8, 6, 6), H (5, 5, 6, 7, 4, 4), I (4, 5, 5, 6, 3, 3), J (3, 4, 5, 6, 3, 3)
4	K (4, 5, 5, 6, 7, 4), L (3, 4, 5, 6, 7, 4)
5	End of Study

and late groups is 0.33, indicating a slightly beneficial effect (probably not significant) of delaying the treatment.

It does not, however, take into account the fact that patients receiving treatment earlier are often those who are sicker and need the treatment immediately. As apparent in Table 9.1, patients A and B, who receive treatment at week 1, have severity scores of 6 or higher in the first two weeks, while patients K and L, who receive treatment much later, have scores 5 or lower. Therefore, a more valid comparison is to match patient scores up to the moment of treatment, then contrast the outcomes between the one immediately treated and the other with delayed treatment.

We first use the sequential matching design to construct matched pairs at each week. Starting from week 1, the risk set consists of two treated and ten not-yet-treated patients. For simplicity, the distance metric between any two patients is the average absolute difference of severity scores prior to treatment. For example, since the treatment is after week 1, the first two scores from each patient will be used to calculate the distance for week 1 risk set. The distance between A and C is zero since they have exactly the same scores at baseline and week 1. The distance between A and D is $\frac{|6-6|+|7-6|}{2} = 0.5$. The distances are set to ∞ between treated (or not-yet-treated) patients to prohibit matching. The full distance matrix is shown below and a conventional optimal bipartite matching algorithm produces pairs of (A, C) and (B, D).

$$\begin{bmatrix}
\infty & \infty & 0 & 0.5 & 1 & 2.5 & 1 & 2.5 & 4 & 9 & 4 & 9 \\
\infty & \infty & 0.5 & 0 & 0.5 & 1 & 0.5 & 1 & 2.5 & 6.5 & 2.5 & 6.5 \\
0 & 0.5 & \infty & & & & \cdots & \cdots & & & & \infty \\
0.5 & 0 & \infty & & & & \cdots & \cdots & & & & \infty \\
1 & 0.5 & \infty & & & & \cdots & \cdots & & & & \infty \\
2.5 & 1 & \infty & & & & \cdots & \cdots & & & & \infty \\
1 & 0.5 & \infty & & & & \cdots & \cdots & & & & \infty \\
2.5 & 1 & \infty & & & & \cdots & \cdots & & & & \infty \\
4 & 2.5 & \infty & & & & \cdots & \cdots & & & & \infty \\
9 & 6.5 & \infty & & & & \cdots & \cdots & & & & \infty \\
4 & 2.5 & \infty & & & & \cdots & \cdots & & & & \infty \\
9 & 6.5 & \infty & & & & \cdots & \cdots & & & & \infty
\end{bmatrix}$$

Prior to conducting match for week 2 risk set, matched patients (A, B, C, D) are removed to prevent matching the same patient more than once. Therefore, week 2 risk set consists of eight patients, two treated and six not-yet-treated. Now, three pre-treatment severity scores (baseline, week 1 and week 2) can be used in distance calculation. For example, the distance between patients E and H is $\frac{|5-5|+|6-5|+|7-6|}{3} = 0.67$. The full distance matrix for week 2 risk set is shown below and the

TABLE 9.2
Risk set matching process: Sequential vs. simultaneous.

Risk Set Week	Patients in the Risk Set	Matched Pairs
Sequential matching		
1	Treated: A, B	(A, C)
	Not-yet-treated: C, D, E, F, G, H, I, J, K, L	(B, D)
2	Treated: E, F	(E, G)
	Not-yet-treated: G, H, I, J, K, L	(F, H)
3	Treated: I, J	(I, K)
	Not-yet-treated: K, L	(J, L)
Simultaneous matching		
1-3	A, B, C, D, E, F, G, H, I, J, K, L	(A, C), (B, D) (E, G), (F, H) (I, K), (J, L)

optimal bipartite matching algorithm produces pairs of (E, G) and (F, H).

$$\begin{bmatrix} \infty & \infty & 0 & 0.67 & 2 & 4 & 2 & 4 \\ \infty & \infty & 0.67 & 0 & 0.67 & 2 & 0.67 & 2 \\ 0 & 0.67 & \infty & & \cdots & \cdots & & \infty \\ 0.67 & 0 & \infty & & \cdots & \cdots & & \infty \\ 2 & 0.67 & \infty & & \cdots & \cdots & & \infty \\ 4 & 2 & \infty & & \cdots & \cdots & & \infty \\ 2 & 0.67 & \infty & & \cdots & \cdots & & \infty \\ 4 & 2 & \infty & & \cdots & \cdots & & \infty \end{bmatrix}$$

After removing matched patients from week 2 risk set, week 3 risk set consists of only four patients, two treated and two not-yet-treated. Four pre-treatment severity scores can be used in distance calculation at week 3. The full distance matrix for week 2 risk set is shown below and the optimal bipartite matching algorithm produces pairs of (I, K) and (J, L).

$$\begin{bmatrix} \infty & \infty & 0 & 0.5 \\ \infty & \infty & 0.5 & 0 \\ 0 & 0.5 & \infty & \infty \\ 0.5 & 0 & \infty & \infty \end{bmatrix}$$

The whole process of sequential matching is summarized in the upper panel of Table 9.2. The bottom panel shows the process of simultaneous matching, which can be done in one step. This is possible because simultaneous matching uses nonbipartite algorithm, which regards patients treated at different time points as different groups, not confined to only two treatment groups. The key to implement nonbipartite matching is to calculate distance correctly at each time point. For example, patient C may be used as a control if he/she is matched to A or B, or C may be used as a treated if he/she is matched to G, H, I, J, K or L. When matching with A or B, the distance is calculated based on the first two time points since the treatment occurs at week 1. When matching with G to L, the distance is calculated based on the first three time points since the treatment occurs at week 2. The distance between C and D, E, F is set to ∞, because they are treated at the same time.

Distance for simultaneous matching:

$$\begin{bmatrix}
\infty & \infty & 0 & 0.5 & 1 & 2.5 & 1 & 2.5 & 4 & 9 & 4 & 9 \\
\infty & \infty & 0.5 & 0 & 0.5 & 1 & 0.5 & 1 & 2.5 & 6.5 & 2.5 & 6.5 \\
0 & 0.5 & \infty & \cdots & \cdots & \infty & 1 & 3 & 5.67 & 9 & 5.67 & 9 \\
0.5 & 0 & \infty & \cdots & \cdots & \infty & 0.33 & 1 & 3 & 4.67 & 3 & 4.67 \\
1 & 0.5 & \infty & \cdots & \cdots & \infty & 0 & 0.67 & 2 & 4 & 2 & 4 \\
2.5 & 1 & \infty & \cdots & \cdots & \infty & 0.67 & 0 & 0.67 & 2 & 0.67 & 2 \\
1 & 0.5 & 1 & 0.33 & 0 & 0.67 & \infty & \cdots & \cdots & \infty & 2.5 & 4 \\
2.5 & 1 & 3 & 1 & 0.67 & 0 & \infty & \cdots & \cdots & \infty & 0.75 & 1.75 \\
4 & 2.5 & 5.67 & 3 & 2 & 0.67 & \infty & \cdots & \cdots & \infty & 0 & 0.5 \\
9 & 6.5 & 9 & 4.67 & 4 & 2 & \infty & \cdots & \cdots & \infty & 0.5 & 0 \\
4 & 2.5 & 5.67 & 3 & 2 & 0.67 & 2.5 & 0.75 & 0 & 0.5 & \infty & \infty \\
9 & 6.5 & 9 & 4.67 & 4 & 2 & 4 & 1.75 & 0.5 & 0 & \infty & \infty
\end{bmatrix}$$

9.4 Illustrative Real-World Studies

In this section, we review three real-world data examples utilizing risk set matching to infer causal effects when treatments occur at different points over time. In the example of surgery for interstitial cystitis, a time-dependent propensity score based on the Cox proportional hazards model was used in risk set matching to ensure a valid causal evaluation of the surgery effect. In the example of premature infant care, multivariate matching created pairs of developmentally similar babies that were sent home at different days after born and subsequent health care costs and clinical outcomes were assessed. In the example of drug treatment for pregnant women with recurrent pre-term birth, time-dependent propensity score matching was applied to evaluate survival outcomes.

9.4.1 Evaluation of surgery for interstitial cystitis

Interstitial cystitis (IC) is a chronic bladder disorder, which resembles the symptoms of a urinary tract infection, but with no evidence of infection. A multi-center cohort study was carried out to evaluate the effects of a surgical intervention, cystoscopy and hydrodistention (C/H), on IC [9]. Patients, who had symptoms of IC for at least the previous six months, were evaluated at entry into the study and at intervals of approximately every 3 months thereafter for up to 4 years. Three measurements were taken at each evaluation: nocturnal frequency of voiding, pain, and urgency. The latter two were recorded on scale of 0 to 9, with higher numbers indicating greater intensity. This subsection summarizes the implementation of risk set matching and the results as described in Li et al. [3] and Lu [5].

In C/H study, some patients were treated with C/H after the enrollment, and the selection for surgery was by no means at random. Conceivably, a patient who can not tolerate the current symptoms is more likely to demand a surgery. The naive comparison that contrasts all patients who received surgery with those who did not receive surgery is problematic, because those never treated might be relatively healthier and their symptoms never became severe. Such comparison may introduce some kind of survival bias as many never treated may have survived their symptoms without needing a surgery. Ideally, we would like to see pairs of patients who were similar up to the time point that one of them received surgery. The untreated patient at the moment would serve as a control and what happened afterwards were outcomes. Therefore, a newly treated patient is paired with an untreated patient with similar characteristics, including symptoms, up to the moment of surgery. If everything is well matched, the decision of surgery is viewed as independent of any other factors, more or less like flipping a coin. It is important to point out that the treatment assignment looks random only at the moment when the first patient receives surgery, as we do not know what would happen afterwards. The control patient may receive surgery at the next follow-up time point or not at all until the end of the study. This paired design estimates the effect of having surgery now versus delaying surgery into the future. This is a typical question that physicians and patients face in practice for chronic diseases.

Unlike many drug studies or behavioral therapies [10], as an invasive procedure, C/H was given at the most once during the entire study period. This sets up the stage to answer the causal question about treatment delay given no multiple treatment assignments for the same patient. In this C/H study on IC patients, the time intervals between visits were approximately three months. By design, the control patient did not receive surgery in the three months interval immediately following surgery for the matched treated patient. So naturally, we can estimate the treatment delay effect of three months using all matched pairs.

There are many patient level characteristics to be balanced with matching, both fixed and time-varying. At baseline, there are five binary demographic variables: Race, Education Level, Full-Time Employment Status, Part-Time Employment Status, and Income. The symptom measures are arguably more important due to their potential strong association with surgery decision. Three measurements taken repeatedly every three months are Nocturnal Frequency of Voiding, Pain and Urgency. Nocturnal Frequency of Voiding is a daily count of nocturnal voids. Pain and Urgency are scores scaled on 0 to 9 with 9 indicating the most severe rating. Those three measurements will be considered as time-varying covariates. After removing cases with missing covariate information, the final analytical dataset consists of 424 patients with 273 controls and 151 patients who received surgery at some point during the course of the study.

A Cox proportional hazards model with time-varying covariates is used to estimate the hazard of being treated at a certain time point for each patient. Nocturnal Frequency of Voiding, Pain, and Urgency measured at the entry of the study are treated as baseline characteristics, along with five demographic variables. Symptom measures post baseline are treated as time-varying covariates. Since the goal is to balance all covariates up to the moment of surgery, all predictors are kept in the model regardless of their statistical significance. For the distance metric in matching, the Euclidean distance between any two patients is calculated based on the linear hazard component $\beta^T X_m(t)$, which serves as a linear propensity score [11]. The distance is computed at each follow-up time point whenever a surgery occurred. If two patients were treated at the same time, their distance would be set to ∞.

The goal of risk set matching is to balance the distribution of both fixed and time-varying covariates, hence to remove observed confounding bias. For time-varying symptoms, it is desirable to track their measurements at different time points. Two time points are selected in balance checking – baseline and the month of surgery based on the treated patient in each pair (denoted as "At Trt" in Table 1 from the original paper [5]; refer to the original paper for table content). For instance, in one pair, if the treated patient received surgery six months after entry into the study, then for both patients in this pair, the "At Trt" month meant six months post baseline. In another pair, if the treated patient received surgery 12 months after entry into the study, then for both patients, the "At Trt" month meant 12 months post baseline. Because all covariates are categorical, the Mantel-Haenszel test or Mante's extension test was applied to the stratified 2×2 or $2 \times k$ tables to check the independence between each covariate and treatment group. The strata were the risk sets at different time points. To measure the balance of covariates without matching, the Mantel–Haenszel test and Mantel's extension test with only one stratum were applied to the data before matching, as shown in the first column in Table 1 from the original paper [5]. Without matching, there is no "At Trt" time point, so the last three rows are marked as not available. P-values from corresponding tests are reported for each scenario. If a p-value of less than 0.1 indicates some imbalance in the covariates, Education, Part-Time Employment, and Frequency seem to be unbalanced, before matching. After either sequential matching or simultaneous matching, all covariates look balanced. It implies, barring any unmeasured confounder, within each matched pair, the decision to have surgery is more or less at random, and the outcome difference three months after the at-treatment time should reflect the effect of surgery.

Li et al. [3] reported effect estimates based on a paired risk set matching design, where only 100 pairs of patient data were used. Lu [5] included more patients with a 1–3 design. Ninety-one matched sets were formed, including 364 patients which is 86% of the full data. Matching with multiple controls is more efficient than pair matching, in terms of increasing the precision of treatment effect estimates. Specifically, 1–3 design would be 50% more efficient than the pair design. Following [5], the patients' outcomes for each symptom are compared at three different time points: at baseline, at the time of treatment, and 3 months after the treatment. Each matched set yields one value of the contrast for each measurement, which is defined using all three time points: $(T_3 - \frac{T_b+T_0}{2}) - (C_3 - \frac{C_b+C_0}{2})$, where T_b is the measurement at baseline for the treated patient in the matched set, T_0 is at the moment before the treatment was given, and T_3 is at 3 months after the treatment occurred. The same notations apply for the control patients. To accommodate multiple controls, the average of the measurements of the three controls is used. The Wilcoxon signed rank test is applied to the contrasts to test the hypothesis of no treatment effect. In Table 3 from the original paper [5], the measures of Frequency, Pain, and Urgency are compared at different time points. As a group, the treated patients and the control patients are quite comparable both at baseline and at the time of treatment. A significant drop is observed for Frequency, and the magnitude is about 0.75, which equals one quarter of the Nocturnal Frequency of Voiding before the treatment. No significant differences are observed for Pain or Urgency scores. These findings are consistent with the results in Li et al. [3].

9.4.2 Impact of premature infants staying longer on subsequent health care costs and outcomes

More complications may arise for premature babies, such as weakened immune system, bleeding in brain, etc. as babies mature best within the nurturing environment of mother's womb. An infant born 2 or 3 months premature may need to spend a substantial amount of time in the neonatal intensive care unit (NICU) to mature until additional physiology functioning makes discharge appropriate. It is also associated with a hefty price tag, as the estimated annual medical cost approaches $33,200 per premature infant. So it is important to know, when a premature baby stays a few days longer in the hospital, does the accompanying increased physiologic maturity reduce expenditures after discharge? Silber et al. [12] provided an answer to his question using the risk set matching design and their study is summarized as below. This study serves as a perfect example of one type of longitudinal observational studies, where everyone will receive the treatment eventually, as all infants will be discharged sooner or later.

The Infant Functional Status (IFS) Study dataset was used in this evaluation, which include eligible infants born at one of five Kaiser Permanente Medical Care Program (KPMCP) hospitals between 1998 and 2002. All infants surviving to discharge who were born at a gestational age (GA) of 32 weeks or less were included in the study cohort, plus a random sample of infants with a GA of 33 or 34 weeks. The final IFS cohort included 1402 infants with 246 having a GA of 28 weeks or below, after removing infants meeting exclusion criteria or with missing data. Information on physiologic maturity, including respirator and incubator settings, body temperature, notations of apnea and bradycardia, use of caffeine or methylxanthines, weight, feeding method, and requirements for intravenous fluids, was obtain through chart abstraction. KPMCP resource estimates were used to estimate costs related to hospital stays.

To answer the question of longer stay on subsequent health care costs, a risk set matching design was utilized by forming 701 pairs of "Early" and "Late" babies. Early babies were those who was discharged first within the pair, and Late babies were those who looked very similar on the day each Early baby went home, but who actually were discharged between 2 and 7 days later (use postmenstrual age as the time scale, PMA). The choice of 2–7 days represents a period of discretion on the part of neonatologists that has economic significance. Because the two babies were very alike at the time of the early discharge, in terms of multiple maturity and risk factors, the decision of letting one go home but not the other could be viewed as random. This served as the basis for a causal comparison to address whether the extra hospital stay would benefit babies who received them. Specifically, Silber and colleagues tried to answer two questions: (1) Are 6-month total costs comparable between the Early and Late babies? and (2) Are 6-month clinical outcomes similar or not?

Five types of costs were compared in their paper and only one of them, "Total Cost" (TC), is introduced here for illustrative purpose. TC is 180 days worth of resource consumption starting from the Early baby discharge. TC is an adequate measure for cost comparison between early and late babies in terms of PMA, as both babies were of the same age when the Early baby went home. Therefore, both babies had the same 180 days, in terms of PMA. There were five deaths among the 1402 babies, and deaths were counted as infinite costs. Silber and colleagues converted a variety of clinical outcomes after discharge into coherence rank scores, with higher scores to babies with worse outcomes. Due to the dimensionality of clinical outcomes, when two babies were compared, one might have a uniformly worse outcome than the other, or each of them had some worse outcomes. To break the tie, the score viewed death as the worst outcome, days in the ICU and total hospitalized days as the second and third most serious, then number of visits to the emergency department, and lastly, sick visits to a physician.

A large number of covariates, as listed in Table 1 from the original paper [12] (refer to the original paper for table content), were considered to ensure matched babies are comparable in every conceivable way. The distance metric was defined as a Mahalanobis distance on key covariates plus a time-dependent propensity score caliper. The propensity score was fitted with a Cox proportional hazards model with two time-varying covariates (daily maturity score and current weight) and other fixed covariates. The optimal nonbipartite matching algorithm was implemented to minimize the total covariate distance between babies discharged at different times. Table 1 from the original paper presents covariate summaries for early babies and late babies at the time early babies were discharged, and their standardized differences (DIFFAVE, mean differences in units of standard deviations). The standardized difference is a commonly used measure for assessing balance between two groups, with values smaller than 0.1 indicating good balance. All covariates are well balanced as shown, implying that the matched babies were indeed comparable on the day one baby went home and the other stayed in the hospital.

As the matching algorithm achieved its goal, recreating a randomization-like scenario within each pair, outcomes would be brought in for analysis. Median and its 95% nonparametric confidence interval were reported for each outcome by groups in Table 2 from the original paper [12]. Wilcoxon's signed rank test was applied to compare outcomes in matched pairs, with associated Hodges-Lehmann point estimate and confidence interval [13]. It turns out that TCs were higher in those who stayed longer in the hospital, with a typical Late-Early difference of $5016, which was highly significant. On the other hand, no significant differences were found for clinical outcomes. Within Late-Early matched pairs, a typical Late baby had outcomes that ranked, slightly but not significantly, worse than the early ones (p=0.21). Therefore, there is no evidence that early discharge could be harmful. Overall, using risk set matching design, the findings suggest that delaying discharge would increase the hospital costs significantly, and such increase cannot be counterbalanced by subsequent savings derived from babies being more mature at discharge.

9.4.3 Drug effect for pregnant women with recurrent pre-term birth

Survival time is an important type of outcomes in clinical studies, which measures the time until a specific event occurs, e.g., time until death after certain treatment or time until tumor relapse. With an observational longitudinal cohort of patients, people may receive treatments at different time points and the timing of treatment may have some effect on the survival outcomes. Patients with more rapid disease progression, hence shorter survival time without treatment, may choose to receive the treatment at an early time point. On the other hand, patients who are healthier may have longer survival time regardless of treatment. Hade et al. [6] examined the effect of a preterm-birth risk reducing drug on pregnant women with a history of preterm birth (birth before 37 weeks' gestation), in which the risk set matching design was applied to balance covariates at different treatment weeks during pregnancy.

For several decades, 17-alpha hydroxyprogesterone caproate (17P), a synthetic pro-gestin, was used to treat female hormone disturbances, to prevent recurrent abortion, and to treat uterine cancer, but the drug as withdrawn in 2000 because of better treatments for these conditions became available [14]. A randomized trial in 2003 found that weekly injection of 17P at 250 mg between 16 and 20 weeks' gestation have been found to reduce the risk of preterm birth and birth complications in a select group women with a history of preterm birth [15]. Since then, 17P has been available through local and national compounding specialty pharmacies, although it was never approved by the FDA to prevent preterm labor. 17P has been the standard of care for more than ten years, but the use of the drug is not without controversies. This is because the mechanisms by which it works have not been fully determined and it remains unclear why prophylactic treatment is not more widely effective in high risk women. It has been suggested that inconsistent findings of 17P in trials and observational studies could be due to the variability of initiation of 17P during pregnancy [16, 17].

Inconsistent effectiveness of 17P by timing of initiation has been reported previously; however, no studies have fully examined treatment delay effect beyond a naive "early" versus "late" two group comparison. Recent work by Ning et al. [17] reported improved outcomes with earlier initiation, before 17 weeks' gestation. However, other published results suggested no detrimental effect of late 17P initiation [16]. Erinn and colleagues [6] implemented risk set matching design to carefully examine the delayed timing of 17P treatment on the time to delivery, in a cohort of women who received prenatal care at a single academic medical center between 2011 and 2016.

Four hundred and twenty-one women with singleton pregnancies and a history of spontaneous preterm birth who initiated 17P therapy were included in the cohort. All women had cervical length (CL) measurements prior to initiation of 17P. All patient characteristics were measured at the time of initial visit to the prematurity clinic (baseline) which occurred one or more weeks prior to 17P initiation, with the exception of CL. The time-fixed covariates include demographics, insurance, and some clinical measures (the full list shown in Table 5 from the original paper [6]; refer to the original paper for table content). The time-varying CL measurements were obtained approximately every two weeks and were most often initiated at week 16. Initiation of 17P is recommended between 16 and 20 weeks' gestation and generally will not be initiated prior to 14 weeks' gestation or after 24 weeks.

To answer the causal question whether delayed 17P initiation had a detrimental effect (shorter) on time to delivery, risk set matching design was implemented to create pairs of treated and not-yet-treated patients with similar covariate histories up to the moment that one woman received 17P first. Patients in this cohort initiated treatment with 17P between 14 and 22 weeks' gestation, creating nine possible risk sets (one per week). The time-dependent propensity score was estimated via Cox PH model with a time-varying CL measurement and other time-fixed covariates, to estimate the hazard of receiving 17P at each gestational age week. Pairs were required to have an estimated propensity score within a caliper of 20% of the propensity score standard deviation. Moreover, to mimic a clinically meaningful delay pattern, treated and not-yet-treated patients were only allowed to be paired if they were at least two weeks apart. At 14 weeks' gestation, the first possible week of receiving treatment, 30 patients were treated with 17P (and 391 were not-yet-treated) and all 30 patients were paired with another not-yet-treated patient. Matching continued for each subsequent week among unused patients until no more pairs could be formed. Overall, a total of 126 matched pairs, 256 total patients, were created. As an alternative measure of assessing covariate balance (in addition to standardized difference of the mean), the effect of each covariate on the hazard of treatment and the associated Wald test p-value, were calculated through a Cox PH model for the time to treatment. Such measure was reported for both before and after matching in Table 5 from the original paper. Ideally, if covariates are well balanced, significant impacts of covariates on hazard ratios of treatment (i.e., a HR different form 1 and a p-value< 0.10) are not expected. As was shown before matching, insurance status and earliest gestational age of prior preterm birth were strongly associated with time to treatment. After matching, all covariates are sufficiently balanced.

The outcome of interest was time to delivery in weeks from the moment that the first treatment of 17P occurred in each pair. For survival outcomes under a matched design, stratified logrank (SLR) test or paired Prentice-Wilcoxon (PPW) test may be used [18, 19]. For estimating hazard ratios, a stratified Cox PH model may be used, as long as the PH assumption holds [18]. All three tests failed to reject the null hypothesis of no treatment delay effect in time to delivery, as shown in Table 6 from the original paper. The stratified Cox PH model also reported a non-significant hazard ratio of 0.95. Therefore, delaying 17P initiation in women with preterm birth history did not seem to cause any change to the time to delivery later.

9.5 Summary

In longitudinal studies with time varying covariates, conventional fixed two group matching designs may not produce matched sets that adequately incorporate time-related information. Because time-varying covariates may have substantial impact on the treatment decision over time, it is important to balance them at the time of treatment reception. Risk set matching provides a unique design to form matched pairs in which patients have similar covariate history up to the moment of treatment, mimicking a randomization mechanism of delaying treatment at certain time points. It is particularly useful for situations where the majority of the cohort would receive the treatment eventually, but the treatment might be delayed due to individual characteristics. The conventional pair design may compare a large group of ever-treated patients to a small group of never-treated patients, leading to biased treatment effect estimation.

There are several practical issues regarding applying risk set matching design to real data. First, the matching design can incorporate time-varying covariates, but it cannot handle repeated treatments. An individual may change her status from not-treated to treated at most once in entire study period. This scenario is reasonable for many surgical interventions, but may not be a reasonable framework for some drug studies when individuals' drug taking behavior may change often over time. Second, to obtain reasonable results, the number of treated subjects in each risk set should not vary dramatically over time. If the majority of treatment occurs at a particular time point, the matching will result in one huge risk set with many small ones. Then the effect estimate is probably a more accurate reflection for that particular time point, rather than the entire study period. Third, like any study with time-varying exposure, it requires some stability assumption about the treatment effect. For example, to pool matched risk sets together for estimating an overall causal effect, it may need an underlying assumption about homogeneous treatment delay effect, which assumes the same treatment effect regardless the delay occurs early or late in time.

It is well known that matching only balances observed confounders. Unmeasured confounding is a major concern in observational studies due to the lack of randomization. Rosenbaum [20] proposed a comprehensive framework to assess the impact of potential hidden bias on the observed significant association, based on matching design. Li et al. [3] specifically discussed the sensitivity analysis for potentially unmeasured time-varying confounding on continuous outcomes. Lu [19] implemented a sensitivity analysis for survival outcomes based on the PPW test in matched data.

More applications of risk set matching can be found in different subject areas. Haviland and colleagues [21] used a simpler form of risk set matching to balance the covariate trajectory prior to age 14 to examine if joining a gang at 14 initiate a violent career for boys. Nieuwbeerta et al. [22] combined group-based trajectory modeling with risk set matching to balance a variety of measurable indicator of criminal propensity, in order to assess the impact of first-time imprisonment on offenders' subsequent criminal career development. Apel et al. [23] adopted the risk set matching design to study the impact of imprisonment on marriage and divorce, as a means of learning whether imprisonment may impede people's ability to reintegrate into society. Zubizarreta et al. [24] linked risk set matching to a general class of devices that can extract a natural experiment from observational studies. Yoo and colleagues [25] created risk set matched pairs between kidney transplant patients and not-yet-treated patients to compare their survival outcomes. More recent use of risk set matching can be found in [26, 27].

9.6 Acknowledgement

This work was partially supported by grant DMS-2015552 from National Science Foundation. This work was also supported, in part, by the National Center for Advancing Translational Sciences of the National Institutes of Health under Grant Number UL1TR002733. The content is solely the responsibility of the authors and does not necessarily represent the official views of the National Institutes of Health.

9.7 Glossary

Confounding: A causal inference concept. Generally speaking, when a common variable impacts both the treatment assignment and the outcome, causing a spurious association, confounding is present.

Longitudinal Study: In a longitudinal study, subjects are measured for an extended period of time. Subjects usually have more than one observation; hence, the data are correlated over time.

Mahalanobis Distance: It is a multivariate measure of distance between two subjects, in units of standard deviation. It takes account of the correlations among variables.

Nonbipartite Matching: An unconventional matching algorithm that pairs subjects that do not come from a fixed two group setup. It is often used for matching with multiple groups or no clearly defined groups.

Potential Outcomes: An outcome is a variable measured after treatment. In causal inference with dichotomous treatment options, an outcome exists in two versions, one seen under treatment, and the other seen under control. Both of them cannot be observed simultaneously; hence, they are known as potential outcomes. Sometimes, also referred to as counterfactual outcomes.

Propensity Score: The conditional probability of treatment reception given observed covariates. Proper adjustment of propensity score leads to unbiased estimate of the treatment effect.

Risk Set Matching: When people receive treatments at different times, risk set matching pairs people who were similar up to the moment that one of them received treatment. Risk set matching controls for the past, not for the future.

Time-varying Variable: A variable whose values may change over time. It could be a covariate or an outcome.

References

[1] P.R. Rosenbaum. *Observation and Experiment: An Introduction to Causal Infernece*. Cambridge, Massachusetts, Havard University Press, 2017.

[2] M.H. Gail. Does cardiac transplantation prolong life? A reassessment. *Annals of Internal Medicine*, 76:815–817, 1972.

[3] Y. Li, K.J. Propert, and P.R. Rosenbaum. Balanced risk set matching. *Journal of the American Statistical Association*, 96(455):870–882, 2001.

[4] P.R. Rosenbaum and D.B. Rubin. The central role of propensity score in observational studies for causal effects. *Biometrika*, 70:41–55, 1983.

[5] B. Lu. Propensity score matching with time-dependent covariates. *Biometrics*, 61:721–728, 2005.

[6] E.M. Hade, G. Nattino, H.A. Frey, and B. Lu. Propensity score matching for treatment delay effects with observational survival data. *Statistical Methods in Medical Research*, 29(3):695–708, 2020.

[7] B.B. Hansen and J. Bowers. Covariate balance in simple, stratified and clustered comparitive studies. *Statistical Science*, 23(2):219–236, 2008.

[8] B. Lu, R. Greevy, X. Xu, and C. Beck. Optimal nonbipartite matching and its statistical applications. *The American Statistician*, 65:21–30, 2011.

[9] K.J. Propert, A. Schaeffer, C. Brensinger, J.W. Kusek, L.M. Nyberg, J.R. Landis, and ICDB Study Group. A prospective study of interstitial cystitis: Results of longitudinal follow-up of the interstitial cystitis diabetes cohort. *Journal of Urology*, 163:1434–1439, 2000.

[10] M. Joffe, D. Hoover, L. Jacobson, L. Kingsley, J. Chmiel, B. Visscher, and J. Robins. Estimating the effect of zidovudine on kaposi's sarcoma from observational data using a rank preserving structural failure time model. *Statistics in Medicine*, 17:1073–1102, 1998.

[11] G.W. Imbens and D.B. Rubin. *Causal Inference: For Statistics, Social, and Biomedical Sciences, and introduction*. Cambridge University Press, 2015.

[12] J.H. Silber, S.A. Lorch, P.R. Rosenbaum, B. Medoff-Cooper, S. Bakewell-Sachs, A. Millman, L. Mi, O. Even-Shoshan, and G.J. Escobar. Time to send the preemie home? additional maturity at discharge and subsequent health care costs and outcomes. *Hospital Services and Outcomes*, 44:444–463, 2009.

[13] M. Hollander and D.A. Wolfe. *Nonparametric Statistical Methods*. John Wiley & Sons, 1999.

[14] Y. Patel and M.M. Runmore. Hydroxyprogesteroe caproate injection (makena) one year later. *Pharmacology and Therapeutics*, 37:405–411, 2012.

[15] P.J. Meis, M. Klebanoff, and E. Thom. Prevention of recurrent preterm delivery by 17 alpha-hydroxyprogesterone caproate. *New England Journal of Medicine*, 348:2379–2385, 2003.

[16] H.Y. How, J.R. Barton, N.B. Istwan, D.J. Rhea, and G.J. Stanziano. Prophylaxis with 17 alpha-hydroxyprogesterone caproate for prevention of recurrent preterm delivery: does gestational age at initiation of treatment matter? *American Journal of Obstetrics and Gynecology*, 197(260):e1–e4, 2007.

[17] A. Ning, C.J. Vladutiu, S.K. Dotters-Katz, and W.H. Goodnight. Gestational age at initiation of 17-alpha hydroxyprogesterone caproate and recurrent preterm birth. *American Journal of Obstetrics and Gynecology*, 217(371):e1–e7, 2017.

[18] P.C. Austin. The use of propensity score methods with survival or time-to-event outcomes: reporting measures of effect similar to those used in randomized experiments. *Statistics in Medicine*, 33(7):1242–1258, 2014.

[19] B. Lu, D. Cai, and X. Tong. Testing causal effects in observational survival data using propensity score matching design. *Statistics in Medicine*, 37(11):1846–1858, 2018.

[20] P.R. Rosenbaum. *Observational Studies*. New York: Springer, 2002.

[21] A.M. Haviland, D.S. Nagin, P.R. Rosenbaum, and R.E. Tremblay. Combining group-based trajectory modeling and propensity score matching for causal inferences in nonexperimental longitudinal data. *Developmental Psychology*, 44:422–436, 2008.

[22] P. Nieuwbeerta, D.S. Nagin, and A.A.J. Blokland. Assessing the impact of first-time imprisonment on offenders' subsequent criminal career development: A matched samples comparison. *Journal of Quantitative Criminology*, 25:227–257, 2009.

[23] R. Apel, A.A.J. Blokland, P. Nieuwbeerta, and M. van Schellen. The impact of imprisonment on marriage and divorce:a risk set matching approach. *Journal of Quantitative Criminology*, 26:269–300, 2010.

[24] J.R. Zubizaretta, D.S. Small, and P.R. Rosenbaum. Isolation in the construction of natural experiments. *The Annals of Applied Statistics*, 8(4):2096–2121, 2014.

[25] K.D. Yoo, C.T. Kim, and J.P. Lee. Superior outcomes of kidney transplantation compared with dialysis. *Medicine*, 95:33, 2016.

[26] D. Watson, A.B. Spaulding, and J. Dreyfus. Risk-set matching to assess the impact of hospital-acquired bloodstream infections. *American Journal of Epidemiology*, 188(2):461–466, 2019.

[27] J.H. Silber, P.R. Rosenbaum, J.G. Reiter, A.S. Hill, S. Jain, D.A. Wolk, D.S. Small, S. Hashemi, B.A. Niknam, M.D. Neuman, L.A. Fleisher, and R. Echenhoff. Alzheimer's dementia after exposure to anesthesia and surgery in the elderly: a matched natural experiment using appendicitis. *Annals of Surgery*, 2020.

10

Matching with Multilevel Data

Luke Keele and Samuel D. Pimentel

CONTENTS

10.1 Multilevel Data Structures

A common feature of many data sets in the biomedical and social sciences is what is known as multilevel data structure. In a multilevel data structure, units are nested in one or more clusters and measurements are taken at both the unit and cluster level. Educational settings are perhaps the most well-known multilevel structure, in which we observe student level measures such as gender, but also school level covariates such as the proportion of female students and the size of the school [1]. Other multilevel data structures include patients within hospitals, households within villages, and firms within industries. Often, we are interested in the effects of a treatment administered at the group level on individual level outcomes. For example, a study may focus on the effect of a school level treatment on student level outcomes. A conventional approach to estimating treatment effects with multilevel data relies on hierarchical regression models. In this chapter, we review how matching methods can be applied to data with a multilevel structure. As we outline, the application of matching to multilevel

DOI: 10.1201/9781003102670-10

data depends critically on the treatment assignment mechanism. Different treatment assignment mechanisms lead to very different applications of matching with multilevel data. In what follows we describe two primary classes of multilevel observational studies, those with treatment assignment at the individual level, and those with treatment at the cluster level, the latter class designated as clustered observational studies (COSs). We briefly review matching in settings with individualistic treatment assignment, which is closely related to existing matching methods for single-level data. The balance of the chapter is devoted to a specific form of matching tailored to clustered treatment assignment, where two matches are typically performed within the same dataset, one at the cluster level and one at the individual. We also discuss implications for inference, sensitivity analysis, and open problems in this general area.

10.2 Models for treatment assignment in multilevel studies

We begin with some basic notation that we use to precisely define a multilevel data structure. We observe a population of units that are nested within clusters. There are J clusters and each cluster contains n_j units. We label the units by (i, j), $i = 1, \ldots, n_j$ and $j = 1, \ldots, J$. The total number of units is $N = \sum_{j=1}^{J} n_j$. For these units, we observe two types of covariates. We let $\mathbf{x}_{i,j}$ represent a matrix of covariates that vary across units. We let $\mathbf{w}_j$ represent a matrix of covariates that vary across clusters, but are fixed for units within the same cluster. In an education application, $\mathbf{x}_{i,j}$ would contain measures of student test scores and race; while $\mathbf{w}_j$ would contain measures of school size or type, but also aggregate measures of $\mathbf{x}_{i,j}$, such as average test scores. In a medical application, $\mathbf{x}_{i,j}$ would contain measures of such as patient comorbidities and income; while $\mathbf{w}_j$ would contain measures such as teaching status. Finally, we denote outcomes as $Y_{i,j}$.

In addition to outcomes and covariates, we observe treatment assignment Z_{ij} for each unit in the study, with $Z_{ij} = 1$ if the unit is assigned to treatment and $Z_{ij} = 0$ if the unit is assigned to control; we also define potential outcomes y_{zij} for $z \in \{0, 1\}$ each representing the realization of the outcome variable under a particular value of Z_{ij}. The mechanism by which treatment is assigned is perhaps the most critical aspect in designing the study and selecting an appropriate method of inference. In particular, there are substantial differences between studies in which treatment decisions are made separately for each individual unit (*individualistic* treatment assignment) and studies in which treatment decisions are made for entire clusters (*clustered* treatment assignment). Under individualistic treatment assignment, different units within the same cluster can be either treated or control. In contrast, when treatment assignment is clustered, all units in the same cluster necessarily share the same treatment assignment so that we may write $Z_{ij} = Z_{i'j} = Z_j$ for all units i, i' in cluster j. Moreover, we assume that units stay within the same cluster, which implies for each unit that cluster membership is fixed. The differences between the two treatment assignment mechanism can be easily understood by comparing them to the imagined randomized trial that would occur in each case. For example, imagine a new reading program for first grade students. Treatment assignment would be individualistic if we randomized the program as the student level. As result, we would observe that treatment assignment will vary within schools such that there are treated and control students in the same school. Treatment assignment would be clustered if we assigned the reading program at the school level. Now, we would observe that each student in a treated school is treated, and each students in a control school is untreated.

While many levels of hierarchy may be present in a study, it is generally possible to classify it as either a study with individualistic assignment (if treatment is assigned at the same level of granularity at which individual outcomes are measured) or clustered assignment (if treatment is assigned only to groupings of measurement units). For example, a study of educational outcomes may contain data on performance of students nested within classrooms which are in turn nested within schools.

If outcomes are reported at the student level and treatment is assigned at the classroom level, then this is a clustered study at the classroom level. The school-level groupings may be important to consider but are not fundamental to the design in the same way as the classroom-level groupings. The school labels play a role similar to other covariate explaining variation in student outcomes such as socioeconomic status or past performance. If instead we observe similar data in which treatments are assigned to individual students, then treatment assignment is individualistic, although classroom and school may both play important roles as covariates.

One reasonable question to ask is why analysts should apply multilevel matching rather than other forms of statistical adjustment, such as regression. One answer is that matching tends to be more robust—especially when treated and control covariate distributions do not have good overlap [2]. COSs in education often have relatively small numbers of treated and control schools which may exacerbate this problem. Recent research found that multilevel matching generally outperformed regression models [3]. Another potential solution might be to use inverse probability weighting. However, propensity score estimation in multilevel data faces special challenges in modeling the hierarchical relationships, and often suffers from convergence problems when the number of clusters is large. Multilevel matching provides an alternative that avoids the need to explicitly estimate and invert propensity scores in these settings.

10.3 Matching with Individualistic Treatment Assignment

10.3.1 Key assumptions

First we discuss methods for matching in multilevel data with individualistic treatment assignment. This setting is essentially a special case of the classical single-level causal inference setting in which we draw special distinctions among the measured covariates. Here each individual has two potential outcomes, $\mathbf{y}_{Tij}$ giving the outcome that would be observed under treatment and $\mathbf{y}_{Cij}$ giving the potential outcome under control. To develop tests or estimates for causal effects the stable unit treatment variation assumption (SUTVA) is generally assumed [4], which says that "the observation on one unit should be unaffected by the particular assignment of treatments to the other unit" [5][§2.4]. In formal notation, this requires $Y_{ij} = Z_{ij}y_{Tij} + (1 - Z_{ij}y_{Cij})$. In addition, unmeasured confounding is typically assumed absent at the individual level:

$$Pr(Z_{ij} = 1|\mathbf{y}_{Tij}, \mathbf{y}_{Cij}, \mathbf{x}_{i,j}, \mathbf{w}_j) = Pr(Z_{ij} = 1|\mathbf{x}_{i,j}, \mathbf{w}_j). \tag{10.1}$$

We also assume that each individual's propensity score $Pr(Z_{ij} = 1|\mathbf{x}_{i,j}, \mathbf{w}_j)$ lies on the interval $(0, 1)$. These latter assumptions are similar to other ignorability or no-unmeasured confounding assumptions; in particular, they differ from assumptions stated by Rosenbaum [6][§3.2] only in explicitly representing both individual-level covariates $\mathbf{x}_{i,j}$ and cluster-level covariates $\mathbf{w}_j$ separately. The primary methodological questions are how this special covariate structure shapes the causal estimands and the matched estimation strategy. We discuss implications of the presence of both $\mathbf{x}_{i,j}$ and $\mathbf{w}_j$ for estimation in the following section.

10.3.2 Propensity score formulation and matching approaches

When treatment assignment is individualistic, matching can be applied to multilevel data in two different ways. First, the analyst can simply apply matching to both $\mathbf{x}_{i,j}$ and $\mathbf{w}_j$. This could be done by simply including both sets of covariates in a propensity score model and matching on the propensity score. Alternatively, the investigator could apply optimal matching methods to both sets of covariates directly. As assumption 10.1 indicates, individualistic treatment assignment may depend on cluster-level covariates. As such, it is natural to include such cluster-level covariates among the variables for matching. Applying matching to the cluster level covariates will seek to ensure that cluster-level characteristics are similar across the treated and control groups. A match of this type will allow treated and control matches to occur across clusters.

With data of this type, however, we can implement an alternative type of match. If investigators have collected $\mathbf{w}_j$, then there are group indicators that indicate which units are nested within each cluster. We denote these group indicators as G_{ij}. If $G_{ij} = 1$ then unit i is a member of cluster j, and if $G_{ij} = 0$ unit i is not a member of cluster j. In this alternative match, the investigator would match on $\mathbf{x}_{i,j}$ and match exactly on G_{ij}. In this match, we would ensure that we only match treated and control units within the same cluster. For example, let's say we are interested in the effect of minimally invasive surgery compared to standard surgical methods, and that our sample of patients is grouped within hospitals. Here, we would match on patient characteristics like age and diagnosis, but we would also match exactly on hospital. That is, we would ensure that we only compare treated and control patients within the same hospital. Intuitively, such a match is appealing, since it ensures that we keep the cluster level factors fixed for all the treated and control units. Here, we ensure that all the matched pairs of patients are within the same hospital. In addition, if the confounders are suspected to be correlated within clusters, this type of cluster exact matching may remove the effects of unobserved cluster level confounders. How can this be the case? Specifically, controlling for G_{ij} via an exact match will remove the effect of unobserved cluster level confounders when the true model for the potential outcomes under the control condition is additive and linear in G_{ij}.

The relative desirability of across-cluster matching and within-cluster matching in any specific setting depends on several factors, including both the form of the true model for treatment and the number and size of the clusters. In general when clusters are large the consensus is in favor of within-cluster matching, since this method controls so strongly for the presence of cluster indicators or cluster-level variables in the true treatment model. Kim and Seltzer [7] emphasize that within-cluster matching is especially important when cluster-covariate interactions are present so that the form of the true propensity score differs across clusters in a manner more complex than a fixed shift in intercept, or in settings where there is interest in measuring treatment effect heterogeneity across clusters. When many small clusters are present, however, within-cluster matching may struggle to make full use of the data effectively. That is, when matching within clusters, it may be difficult to balance individual level characteristics. When conducting across-cluster matching on a propensity score in these settings, Arpino and Mealli [8] and Thoemmes and West [9] both emphasize the importance of accounting for clusters in the propensity score model by including either random or fixed cluster effects. In cases where fixed effects models cannot be estimated well due to the large number of parameters involved or where standard random effects assumptions are deemed implausible, Kim and Steiner [10] and Lee et al. [11] suggest a middle-ground approach in which groups of small clusters with similar treatment prevalences are formed and using the resulting groups to structure propensity score estimation, either by estimating separate propensity score models for each group or by using group-level random effects in an overarching model.

Finally, Zubizarreta et al. [12] present an example of conducting both within-cluster and across-cluster matching in the same dataset, yielding two independent tests of the null hypothesis. A major advantage of this approach is its implications for robustness to unmeasured bias. If across-cluster comparisons and within-cluster comparisons are subject to different types of unmeasured bias, then agreement between the answers produced by the different methods builds confidence that the observed effects are due to a common underlying true effect of treatment; in contrast, if the estimated effects disagree sharply there is clear evidence that one of the comparisons is biased.

10.3.3 Inference and sensitivity analysis

Hypothesis testing for multilevel designs with individualistic treatment assignment can either be conducted by comparing observed test statistics to null distributions constructed by viewing

$(\mathbf{y}_{Tij}, \mathbf{y}_{Cij}, \mathbf{x}_{i,j}, \mathbf{w}_j)$ as a random vector sampled from some population, or by considering only the conditional distribution $P(Z_{ij} \mid \mathbf{y}_{Tij}, \mathbf{y}_{Cij}, \mathbf{x}_{i,j}, \mathbf{w}_j)$. The latter approach is often described as randomization inference because of its close connections to Fisher randomization tests for designed experiments. Sampling inference is the more prevalent approach, but in the multilevel case it is complicated by the need to account for intracluster correlation of outcomes; typically some kind of hierarchical outcome model must be assumed; see for example the outcome models discussed by Kim and Seltzer [7] and Arpino and Mealli [8]. Furthermore, although sampling inference for various matching designs has been characterized for various versions of the classical setting without multilevel structure, sampling distributions have not been worked out for the hierarchical case.

In contrast, since the randomization inference framework conditions on the match constructed and the potential outcomes, methods of inference are essentially identical in the multilevel and the classical settings. Randomization inference can be implemented by repeatedly permuting treatment and control labels within each matched pair independently of other pairs, holding outcomes fixed. As described in detail by Rosenbaum [6][§2], the values of the test statistic under all such permutations constitute a null distribution under the sharp null hypothesis of no effect of treatment for any individual, and comparing the observed test statistic to this null distribution produces exact finite-sample p-values (under the model off equal probability of treatment within pairs). In practice randomization inference can be conducted either by Monte Carlo sampling of within-pair permutations or by using an appropriate large-sample approximation to the randomization distribution of the test statistic.

Randomization inference for matched designs also offers the benefit of a natural associated method of sensitivity analysis. While the initial model assumes equal probabilities of treatment are equal within matched pairs, this assumption is plausible only if pairs are matched closely both on observed covariates (or at least on a propensity score estimated with observed covariates) and also share similar values of unobserved covariates. The latter condition cannot be checked empirically; if it fails, probabilities of treatment may differ by some amount within pairs. A sensitivity analysis asks whether unobserved confounding of a certain degree would be sufficient to reverse the results observed under the equal-probability assumption. Specifically, for unit j in cluster i define:

$$\pi_{ij} = Pr(Z_{ij} = 1 \mid \mathbf{y}_{Tij}, \mathbf{y}_{Cij}, \mathbf{x}_{i,j}, \mathbf{w}_j)$$

The sensitivity analysis of Rosenbaum [6][§4] posits that for paired units (i, j) and (i', j') with and some fixed parameter $\Gamma > 1$,

$$\frac{1}{\Gamma} \leq \frac{\pi_{ij}/(1-\pi_{ij})}{\pi_{i'j'}/(1-\pi_{i'j'})} \leq \Gamma. \tag{10.2}$$

While this model admits a range of possible distributions with which to permute the treatment labels, the worst-case distribution (i.e. the one producing the largest p-value) can be determined and used for inference; if the test still rejects then unobserved confounding of strength Γ is not sufficient to explain the observed effect. This procedure may be repeated for larger and larger values of Γ until a value is found at which the test ceases to reject, which can be reported as a summary of a test's robustness to unmeasured bias. Again, for individualistic treatment assignment these methods apply directly to matched datasets in multilevel regimes with no special adaptations from the classical setting.

10.4 Matching with Clustered Treatment Assignment

10.4.1 Causal assumptions

As we outlined above, with multilevel data, the other possible treatment assignment mechanism is one that is clustered. Under this paradigm, treatment is applied or withheld to entire clusters. That is, treatment is allocated to either schools or hospitals such that all the units within those clusters are treated or not. We refer to this study design as a clustered observational study (COS). One can view the COS as the observational study counterpart to the clustered randomized trial. In a COS, the difference in the treatment assignment mechanism is reflected in the notation in that assignment no longer depends on the units. In general, in a COS assignment will occur such that there are N_1 treated clusters and N_2 control clusters. In an education application, the clusters would be entire schools. The key assumptions needed for the identification of causal effects in a COS mirror those when treatment assignment is individualistic. In parallel to assumption (10.1) for the individualistic treatment model, we require that there is some set of covariates such that treatment assignment is random conditional on these covariates. We write that assumption in the following way:

$$\pi_j = Pr(Z_j = 1 | \mathbf{y}_{Tji}, \mathbf{y}_{Cji}, \mathbf{x}_{i,j}, \mathbf{w}_j, \mathbf{u}_{i,j}) = Pr(Z_j = 1 | \mathbf{x}_{i,j}, \mathbf{w}_j).$$

In words, this says that after conditioning on observed characteristics, $\mathbf{x}_{i,j}$ and $\mathbf{w}_j$, a given cluster's probability of assignment to treatment is related neither to the potential outcomes of its units $(\mathbf{y}_{Tji}, \mathbf{y}_{Cji})$ nor to unobservables $(\mathbf{u}_{i,j})$. That is, we assume there are no unobservable differences between the treated and control groups. Next, we assume that all clusters have some probability of being treated or untreated such that $0 < \pi_j < 1$.

Finally, we assume a cluster-level version of SUTVA, in which an individual's outcome is affected only by treatments given to individuals in the same cluster and cannot be influenced by treatments assigned to individuals in other clusters. In contrast to the assumptions of Section 10.3.1, these assumptions do not require an absence of interference between units in the same cluster, merely across clusters. In principle this assumption allows for many more than two potential outcomes for each individual, one for each of the different possible vectors of treatment across individuals in a cluster. However, in a COS only two possible vectors of individual treatments are ever given within a cluster (all zeroes for a cluster assigned to control, and all ones for a cluster assigned to treatment), so it is still sufficient to express each observed outcome as a treatment-weighted sum of two potential outcomes for the individual in question: $Y_{ij} = Z_j y_{Tij} + (1 - Z_j) y_{Cij}$. In summary, arbitrary within-cluster interference is not problematic in a COS because the restricted nature of treatment assignment makes limits the number of potential outcomes that can be observed. For a thorough discussion of the implications of relaxing SUTVA in more complicated versions of the COS where not all units in a cluster receive the same treatment, see Hong and Raudenbush [13].

10.4.2 An aggregated design

One crude but simple method for the analysis of a COS is to aggregate all within-cluster outcomes to the cluster level, erasing all hierarchy below the level at which treatment is assigned. This produces a data structure that effectively resembles individualistic treatment assignment for clusters. Such an approach is generally undesirable because it ignores information about individual heterogeneity within clusters. One case in which accounting for individual-level information is particularly important is when the units observed within a cluster are not the census of units in that cluster but some subset, selected in a way that may vary from cluster to cluster. For example, suppose an intervention that is assigned to schools is then only administered to students who are struggling academically. As such, the treated units are a subset of the entire cluster population. Following Page et al. [1] we refer to the setting in which the observed within-cluster samples include all cluster members as Design 1, and the setting with within-cluster selection into the study as Design 2. When Design 2 is in operation, it opens a number of complexities that are not present in Design 1. Under Design 2, it may be the case that the within-cluster treatment assignment introduces a secondary treatment assignment mechanism that investigators would need to understand. That is, if schools operate a secondary treatment assignment mechanism, it opens up the possibility that treatment effects vary from treated cluster to treated cluster due to an interaction between school and student covariates. As such, under

Design 2 it may be critical to adjust for individual-level covariates to ensure that differences in selection within clusters do not introduce bias into causal comparisons. Moreover, under Design 2, it is possible that the intervention spillovers from the within-school treated subset of students to the untreated within school subgroup. The investigator may also be interested in understanding whether such spillover effects occur. See Hong and Raudenbush [13] for a discussion of spillovers of this type. In general, Design 2 may require critical refinements or tailoring of the research design depending on specifics of the application. However, even under Design 1, one may opt to match on individual level characteristics, since those covariates may be strongly correlated with cluster level covariates, which will tend to induce a pattern where the distribution of the individuals also differ across treated and control clusters.

10.4.3 Multilevel matching

Assuming the investigator decides not to aggregate the data to the cluster level, what approaches to matching are available? Since treatment is determined at the cluster level, a natural approach is to fit a propensity score for each cluster, or conduct matching between clusters, with reference only to cluster-level covariates $\mathbf{w}_j$. If the cluster-level propensity score is correctly specified in $\mathbf{w}_j$, this method will be sufficient to estimate effects. However, under Design 2 in particular, individual-level variables may remain poorly balanced in designs even where such a propensity score is correctly estimated, and some degree of matching at the individual level is desirable.

The primary alternative is multilevel matching. Multilevel matching is a method of matching specifically designed for the multilevel considerations that arise in a COS. Multilevel matching is designed to implement a process of matching at both the cluster and individual levels. The multilevel matching methods we describe do not require any explicit model fitting and tend to be robust across unobserved data-generating processes [3]. In addition, multilevel matching is compatible both with covariate balance constraints that enforce priority among variables and sample trimming, which provide the analyst with control and flexibility to increase treated-control comparability on covariates deemed to be of critical importance. For example, an investigator may choose to more closely balance past test scores relative to other covariates such as school size. Moreover, trimming allows the investigator to find the set of clusters and units with the highest levels of comparability. We provide more details on these aspects of the matching process in Sections 10.4.6.1 and 10.4.6.2 below.

How should one organize the process of matching at both the cluster and individual levels? First, we note that the "within-cluster" and "between-cluster" matching approaches recommended above for individualistic treatment assignment are both impossible and unattractive in the COS setting. Within-cluster matching does not work because in a COS all units within the same cluster share the same treated status. Between-cluster matching is inappropriate as it ignores matching at the cluster-level entirely and matches treated students in the same cluster to control students dispersed across multiple clusters. The pairs created in a match of this type do not reflect the structure of a CRT and complicate rather than streamline the process of accounting for cluster-level treatment. Instead, it is vital to match treated clusters to control clusters, and in addition it is desirable, particularly for studies following Design 2, to match individuals in each treated cluster to individuals in the corresponding paired control cluster. Even in Design 1 settings when pairing is not conducted within clusters, the overall framework described here is useful to guide design [14].

A natural approach for creating a match of this type is what Kim and Seltzer [7] refer to as "sequential matching": first conducting a cluster-level match, and then conducting the individual-level match within the resulting cluster pairs. The drawback to this approach is that the two-step sequential match involves an element of greedy search and thus is not guaranteed to find the optimal match among all designs that contain individual subject pairs nested within school pairs [20].

To fix ideas we now provide a brief conceptual overview of matching in a COS as an optimization problem. Let $\mathcal{T} = \{j : j \in \{1, \ldots, J\}, Z_j = 1\}$ and $\mathcal{C} = \{j : j \in \{1, \ldots, J\}, Z_j = 0\}$ be the index sets for the treated and control clusters respectively, and let $\mathcal{S}_j = \{(i, j) : i \in \{1, \ldots, n_j\}\}$

be the set of ij index tuples for individuals in cluster j. We will represent a cluster-level match as a set $\mathcal{M} \subset \mathcal{T} \times \mathcal{C}$ of paired indices, each giving one treated cluster and its paired control counterpart. Similarly, we can represent an individual-level match compatible with this cluster-level match as $\bigcup_{(j,j')\in\mathcal{M}} \mathcal{M}_{jj'}$ where $\mathcal{M}_{jj'} \subset \mathcal{S}_j \times \mathcal{S}_{j'}$. Let $g : \{\mathcal{W} : \mathcal{W} \subset \mathcal{T} \times \mathcal{C}\} \longrightarrow \mathbb{R}$ and $f_{jj'} : \{\mathcal{W} : \mathcal{W} \subset \mathcal{S}_j \times \mathcal{S}_{j'}\} \longrightarrow \mathbb{R}$ for any $j, j' \in \{1, \ldots, J\}, j \neq j'$ be objective functions measuring the desirability of the cluster-level and individual-level matches respectively and let the overall objective value for match $\mathcal{M}$ be given by:

$$g(\mathcal{M}) + \sum_{(j,j')\in\mathcal{M}} f_{jj'}(\mathcal{M}_{jj'}). \tag{10.3}$$

In words, this says that the overall quality of the match can be represented as a sum of some measure of quality in the cluster-level pairs plus some measure of quality in the individual-level pairs. For other more detailed examples of formulating matching as an optimization problem see Bennett et al. [15], Hansen and Klopfer [16], Pimentel et al. [17], Rosenbaum [18], and Zubizarreta [19].

Under this setup, the sequential matching procedure first searches over all cluster-level matches and chooses the configuration of pairs $\mathcal{M}$ minimizing $g(\mathcal{M})$. Then, subject to each $(j, j') \in \mathcal{M}$, individual-level pairs $\mathcal{M}_{jj'}$ are chosen to minimize $f_{jj'}$. This procedure may not find the optimum, since the choice of cluster-level match $\mathcal{M}^*$ that minimizes $g(\cdot)$ alone may not contain the cluster pairs (j, j') needed to minimize the overall objective function. In brief, because the cluster-level match rules out many possible individual-level matches without paying attention to their relative quality as characterized by the $f_{jj'}$ functions, it may miss the optimum. Keele and Zubizarreta [20] offer a formal proof of this sub-optimality, and Pimentel et al. [14] provide additional simulation evidence.

In contrast, consider the following procedure. For each unique pair of clusters (j, j'), compute the optimal match $\mathcal{M}_{jj'}$ with respect to objective function $f_{jj'}$, storing the match and the objective value obtained. Then select the optimal match $\mathcal{M}$ with respect to objective function (10.3), which is known since all of the individual-level matches were pre-computed. Formally, this procedure guarantees optimality because of the identity:

$$\min_{\mathcal{M},\{\mathcal{M}_{jj'}:(j,j')\in\mathcal{M}\}} \left\{ g(\mathcal{M}) + \sum_{(j,j')\in\mathcal{M}} f_{jj'}(\mathcal{M}_{jj'}) \right\}$$

$$= \min_{\mathcal{M}} \left\{ g(\mathcal{M}) + \sum_{(j,j')\in\mathcal{M}} \min_{\mathcal{M}_{jj'}} f_{jj'}(\mathcal{M}_{jj'}) \right\}.$$

Each of the terms in the sum is computed in advance, which works because the problem is separable: decision variables for individual-level matching within each cluster pair are disjoint from the decision variables for individual-level matching within any other cluster-pair in the same match.

For example, suppose in an educational COS, there are 3 treated schools and 5 control schools indexed by $k_t \in \mathcal{K}_t = \{1, ..., 3\}$ and $k_c \in \mathcal{K}_c = \{1, ..., 5\}$, respectively. We first evaluate the student-pair matches across all the possible pairs of treated and control schools. Since there are 3 treated schools and 5 control schools, we conduct $3 \times 5 = 15$ student-level matches. In general, where there are N_1 treated schools and N_2 control schools, the number of such possible pairings will be $N_1 \times N_2$. Although this involves assessing a large number of possible matches, the form of these matches is straightforward, since we are simply conducting standard student-to-student matches within each potential school pair. Next, we score the quality of each of these student matches by recording the value of function $f_{jj'}$ achieved in each, and the cluster-level match can now be conducted, searching over all possible pairings of the three treated schools to three of the five control schools, optimizing the linear combination of cluster-level quality $g(\mathcal{M})$ and the individual-level similarity of the associated student pairings $f_{jj'}(\mathcal{M}_{jj'})$.

One critical point to understand about multilevel matching is that student-level covariates are incorporated into the match in an unconventional fashion. Given the multilevel structure of the data, student-level covariates can be included as either student-level measures or as aggregate measures. For example, a covariate such as student sex can be used as a binary, student-level measure or as the proportion of female students in the school. The first stage of a multilevel match avoids the need to create aggregate measures from student-level covariates. All student-level covariates will inform the score, and this score is used directly in the school-level match in the next step. However, one can also include aggregate covariates during the school matching process. To make these concepts more concrete and address practical aspects of carrying out this procedure, we now consider two specific versions of this problem, each with a different choice of objective function and solution algorithm.

10.4.4 Multilevel cardinality matching and implementation via integer programming

Keele and Zubizarreta [20] first formalized multilevel matching as a generalization of cardinality matching. Zubizarreta [21] developed cardinality matching as a method for dealing with a lack of overlap for applications with individualistic treatment assignment. Under standard cardinality matching, the researcher specifies desired levels of post-matching balance on covariates, and then identifies the largest set of treated matched pairs that achieves those balance constraints. That is, it optimally trims the number of treated units to meet a specified degree of overlap [21]. For example, the analyst might specify that all treated control standardized differences to be at least 0.15. Cardinality matching will find the largest number of matched pairs which satisfies this balance constraint. Moreover, the smallest possible number of treated units are discarded from the match. That is, under cardinality matching, the objective function to be maximized is the post-match sample size, whic

To generalize this framework to multilevel matching, Keele and Zubizarreta [20] specified two distinct sets of balance constraints, one on the cluster-level covariates (to be achieved in the cluster-level match) and one for the individual-level matches (to be achieved in each within-cluster-pair individual match). In the specific case study they considered, they set the objective function $g(\cdot)$ for the cluster-level match to zero, and chose the individual-level objective functions $f_{jj'}(\cdot)$ to be the total number of samples included from cluster-pair (j, j'). Essentially, this objective function maximizes the number of individuals in the study, paying attention to the cluster level only in order to ensure that cluster-level covariates are properly balanced. In other cases one might want to pay attention to the number of clusters selected as well; for instance, in cases where cluster-to-cluster variability is much higher than within-cluster variability it might be preferable to have a large number of clusters of moderate sample size rather than a few large clusters. In this case $g(\cdot)$ could be chosen as the number of clusters selected and the quantities $f_{jj'}(\cdot)$ could all be scaled by a common parameter λ controlling the relative importance of cluster count and total number of individuals.

To implement multilevel cardinality matching, the recipe described in the previous section is followed: an individual-level match is performed for units across each unique combination of a treated cluster j and a control cluster j'; in each case sample size is maximized under the balance constraints by solving an integer program. Then, treating the minimal $f_{jj'}(\cdot)$ values produced as fixed numbers, the overall objective is maximized under cluster-level balance constraints. Since each individual match requires an integer program to be solved, the computation time associated with this procedure can be very high [20], but it has the advantage of guaranteeing specific levels of balance.

10.4.5 Multilevel minimum-distance matching and implementation via network flow algorithm

Now we consider an alternative approach, generalizing minimum-distance matching methods such as Mahalanobis distance matching and propensity-score matching to the multilevel setting. In minimum-distance matching, covariate distances are defined between each of the treated units and each of the control units being matched, based on some pre-specified multivariate distance between their respective covariate vectors. Matched sets or pairs are selected in such a way that the sum of covariate distances across the pairs is made as small as possible. In formal terms, the objective function for a match is simply the sum of the distances for paired units.

For multilevel matching, we require one covariate distance between treated clusters and control clusters, and another covariate distance, or possibly a set of covariate distances, between individuals in different clusters. The cluster-level distance can be chosen as the cluster-level propensity score or as some other multivariate distance; however, since no individual-level propensity score exists (since treatment is at the cluster level in a COS) some other distance must be used at the individual level, such as a Mahalanobis distance. The decision of whether to use a single global covariate distance for all individual-level comparisons, or instead to define a separate distance within each individual potential cluster pair, depends on the anticipated degree of heterogeneity in the covariate distributions and particularly in the outcome models from cluster to cluster. In general, we find overall Mahalanobis distance to be a good default [14]. The cluster-level objective function $g(\cdot)$ can be given as a measure of cluster match quality given by within-pair cluster distances. In particular, since this form of $g(\cdot)$ can be decomposed into a sum over individual cluster pairings, it is possible to implement the final cluster-level match whose distance matrix is a sum between the N_1 by N_2 cluster-level distance matrix and an N_1 by N_2 matrix storing the optimal objective values $f_{jj'}$ from the individual-level matches.

In principle one might wish to define objectives $f_{jj'}(\cdot)$ as sums of individual-level covariate distances, but some complexities arise in combining these quantities into one overall objective. Most problematic is the issue of sample size. In a single-level minimum-distance match all treated units are usually retained, which holds the overall sample size of the resulting match fixed. In a multilevel match, while all treated clusters may be retained at the cluster level, not all treated individuals can be retained unless control clusters are uniformly larger in size than treated clusters (or unless small control clusters are deleted a priori). The simplest solution is to retain either all control units or all treated units in an individual-level match for a given cluster pair, depending on which of the two clusters is smaller in size.

Returning to the problem of specifying the objective function with this aspect of the individual-level matches in mind, it is problematic to use sums of within-pair covariate distances for the $f_{jj'}$ terms because smaller matches will lead to smaller sums, even when the matches are not any better on average. Using average distances helps only a little, since small average distances can often be achieved when there is a gross mismatch in size between matched clusters even though this frequently results in throwing away larger portions of the large clusters than is warranted. One solution is to specify $f_{jj'}$ as a linear combination of a penalty inversely related to sample size and the average covariate distance achieved. Tuning the penalty allows more careful management of the bias-variance tradeoff.

In contrast to multilevel cardinality matching, each of the individual-level matches and the overall cluster match may be implemented in this case by solving a minimum-cost network flow problem or an assignment problem. In contrast to integer programs, both of these classes of optimization problems have strong computational guarantees ensuring that the worst-case computation time is polynomial in the size of the problem. More specifically, due to their special structure they can produce optimal integer solutions using solution methods usually used for linear programming problems. This is attractive in terms of scalability, especially since so many problems must be solved in the first step when calculating all possible individual matches within cluster pairs. See Hansen &

Klopfer [15], Pimentel & Kelz [16], Pimentel et al. [14], and Rosenbaum [22] for a more in-depth discussion of network flow methods and their value in matching.

Many other choices are possible. In particular, Pimentel et al. [14] define $f_{jj'}$ as a linear combination of a measure of matched sample size and a fixed balance penalty multiplied by the number of pre-specified covariates that do not meet predefined balanced benchmarks in the individual-level match. This mimics the cardinality matching method of Keele and Zubizarreta [20] by emphasizing sample size and balance rather than matched pair quality; following this same idea, they implement the cluster-level match without any cluster distance, instead using the $f_{jj'}$s alone as the distances but imposing refined balance constraints [18] on the cluster-level covariates. As discussed by Pimentel et al. [14], this choice of $f_{jj'}$s offers the benefit of being applicable in Design 1 settings where individual-level matches are not conducted but individual-level comparisons may still inform overall match quality; if you simply replace the matched sample size with the harmonic mean of the raw treated and control sample sizes, $f_{jj'}$ is calculable even without a pairing.

In fact, Pimentel et al. [14] conduct the individual-level matches as minimum-cost Mahalanobis distance matches, which means they may not succeed in fully optimizing the distinct objective function $f_{jj'}$ actually used for the overall objective criterion; to some extent. However, the minimum-cost Mahalanobis distance offers major computational advantages in exchange for this deviation from perfect optimality, since directly optimizing $f_{jj'}$ as defined would likely require an integer programming method.

Notably, multilevel matching based on network flows is available in the R package multiMatch. Keele et al [3] evaluate this form of multilevel matching in several different scenarios and present an applied example.

10.4.6 Additional features and practical advice

Next, we describe some additional features and variations on the process of multilevel matching as described above.

10.4.6.1 Balance prioritization

More standard approaches to matching tend to treat all covariates equally. That is, the goal is balance all covariates. However, in many applications, investigators may prefer to implement covariate balance prioritization. In matching, an investigator may prefer to balance one or a set of covariates more closely than others. This is referred to as covariate prioritization. Covariate prioritization is one key method for using subject matter expertise in matching. More specifically, subject matter expertise can guide the identification of which covariates should be prioritized for higher levels of balance. The simplest form of covariate prioritization is exact matching. Here, treated and control units are matched exactly on a nominal covariate. Exact matching is a form of covariate balance prioritization, since all imbalance is removed for the covariate that is exactly matched. Recent work has developed more generalized forms of covariate prioritization, such that analysts can more precisely select covariates for balance prioritization, sometimes by specifying multiple tiers of variables in descending order of importance, each of which is balanced subject to controlled balance on previous tiers [18, 21].

Both forms of multilevel matching describe above include covariate prioritization. Cardinality matching relies on balance constraints for individual variables, selected by the user to receive priority. In addition, the multilevel matching functions implemented in multiMatch allows investigators to prioritize balance on multiple tiers of school-level covariates. Covariate prioritization can be applied to individual covariates or to covariate sets. While prioritization of covariates does not guarantee individually optimal balance on all individual covariates, it does ensure that the covariates that are deemed to be of overriding importance are better balanced. This implies that selecting covariates for prioritization is a process of managing potential tradeoffs between improving balance for important covariates that may cause increased imbalance in other, less important covariates. In the case study presented later, we demonstrate key aspects of balance prioritization in the context of COS.

10.4.6.2 Trimming treated clusters and individuals

Often investigators find that, despite matching, treated and control units are not sufficiently comparable. This manifests itself as post-match balance statistics that show treated and control units as too dissimilar. In our experience, this is common in education settings—especially when the pool of control schools is fairly small. What can be done in this situation? Cardinality matching as described above is one elegant solution, maximizing sample size without ever sacrificing balance in order to include more samples. However, trimming options are also available for network-flow based matching methods. Two specific strategies are caliper matching and optimal subset matching. A caliper forbids matches between units that are not sufficiently close in terms of some continuous variable, often an estimated propensity score [23]; the size of the caliper controls the degree of trimming. Optimal subset matching [24] is similar, specifying a threshold parameter on the covariate distance scale; treated units are excluded until the average improvement in covariate distance per pair for the next treatment unit to be excluded is lower than the threshold.

In the matchMulti package, optimal subset matching is implemented in a user-friendly way via the $\underline{n}$ parameter, which specifies a minimum number of treated units that must be included in the match. For example, if there are 20 treated clusters and $\underline{n}$ is set to 19, the algorithm will discard the one treated cluster that contributes the most to the imbalance. Thus, it seeks to discard the treated cluster that is most problematic with respect to our balance criteria. Using this parameter, analysts can drop treated clusters until balance improves. In general, we recommend dropping clusters one-by-one until balance is deemed acceptable.

If the investigator decides to trim one or more clusters to improve balance, they need to be aware of the fact that this changes the interpretation of the estimated treatment effect. The estimated treatment effect is no longer the effect among the treated, but the treatment effect for a subset of the treated. Thus, some amount of generalizability is lost. That is, we can no longer describe the estimated effect as the treatment effect for the entire treated population. Instead, it is the treatment effect for the subsample of the treated clusters included in the match. When this occurs, it is critical that the analyst provide descriptive statistics for both the treated units retained in the match and those discarded by the matching. This provides important information on the population for which the treatment effect is relevant.

In addition, as mentioned above, trimming of individual treated units within treated clusters is often unavoidable due to disparities in cluster sizes. Here the trimming is not done to enforce balance or common support but is simply a byproduct of the structure of the pair match. Under a Design 2 match, the causal estimand is a cluster-level contrast for a set of individuals within the cluster who are at risk for the treatment. After matching under Design 2, analysts should again present descriptive statistics for both the individuals included in the match and those discarded from the match. In general, the causal effect in a COS is a cluster-level causal effect that may be altered by the matching process in various ways.

10.4.6.3 Control:treatment ratios and additional constraints

There is an extensive literature on modifications to the basic pair matching problem to improve the quality of the matches created or their suitability for various statistical tasks. Exact or near-exact matching may be required on certain variables [25]. In addition, in cases where controls are plentiful multiple controls may be matched to each treated unit, either the same number for each treated unit or differing numbers depending on the prevalence of comparable controls for each treated unit [26]. In full matching, multiple treated units may also be matched to a single control [27].

Conceptually, there is little difficulty in incorporating such modifications into the framework of multilevel matching. Exact or near-exact matching may be done for individual-level matches based on individual-level covariates, or for cluster-level matches based on cluster-level covariates, with no repercussions except for some potential trimming of treated units if they cannot be matched exactly. At the individual level, full matches and matches involving multiple controls may be conducted; the only change necessary here is to assess balance and sample size with the correct weights when constructing the objective functions $f_{jj'}$; choices of weights for test statistics and balance measurements are discussed in [27], and apply similarly to the question of what measure of sample size to penalize in minimum-cost matching. Of course, procedures such as variable-ratio matching and matching with near-fine balance are not always mutually compatible [28], nor are all combinations of these options supported by standard software such as matchMulti.

Full matching and matching with multiple controls at the cluster level presents complications in principle as well as practice. The notion of an individual-level match across a group of more than two clusters requires further refinement — for instance, should each individual in the treated cluster be matched to an individual in each of the control clusters, or should each treated individual match only to a single control individual in one off the clusters? — and depending on the specific refinement chosen the first step of the multilevel matching algorithm might need to be modified to consider all possible matched sets of treated-control clusters rather than merely all pairs, likely with substantial cost to the matching process in computational complexity. Such innovations have not been explored in the literature to date.

10.4.7 Clustered randomization inference and sensitivity analysis

As in the case of individualistic treatment assignment, methods of sampling-based inference for matching designs clustered observational studies remain underdeveloped and generally involve complications due to potential within-cluster dependence in the outcome model. In contrast, methods of randomization inference both exist specifically for such designs and avoid many complications associated with outcome modeling by focusing solely on random variation due to treatment, as suggested by Fisher in the case of randomized studies [29]. They also have associated sensitivity analyses based on model (10.2).

Hansen et al. [30] give an in-depth exploration of randomization inference and sensitivity analysis for clustered observational studies; we briefly review their approach in the context of multilevel matching. We focus on testing the sharp null hypothesis $y_{Cij} = y_{Tij}$ for all clusters i and individuals j; in other words, cluster-level treatment or control status has no impact on any individual's outcome, which would have been the same observed value regardless of which cluster-level treatment had been assigned. To assess evidence against this hypothesis, we will use a test statistic of the following form:

$$T = \sum_{(j,j')\in\mathcal{M}} q_{jj'}(2Z_j - 1) \cdot \frac{1}{|\mathcal{M}_{jj'}|} \sum_{(k,k')\in\mathcal{M}_{jj'}} (Y_{jk} - Y_{j'k'}) \tag{10.4}$$

Essentially, this test statistic takes the matched treated-control difference within each cluster pair and computes a weighted average of these differences across cluster pairs, with cluster-pair weights $q_{jj'}$. Common choices of weights are equal weights or weights proportional to the total number of individuals in the cluster pair; for more discussion of the best choice of weights under different settings see Hansen et al. [30][§3.5].

To conduct randomization inference, we consider the distribution of this test statistic under the null hypothesis, conditional on the configuration of matched cluster pairs and individual pairs, the values of the cluster-level and individual-level covariates, and the individual-level potential outcomes. Since the only random quantity remaining is treatment, and since under the sharp null hypothesis observed outcomes Y_{jk} are invariant to treatment status, it suffices to consider permuting $Z_j, Z_{j'}$ values independently within each cluster pair, with equal probabilities, and recomputing test statistic (10.4). This repeated process generates a null distribution to which the observed test statistic can be compared.

While randomization inference methods can be applied after matching, using regression models with the matched data is also useful strategy for estimating treatment effects. Under this approach, the outcome is regressed on the treatment indicator using the matched data set. While standard regression models can be used at this stage, one might opt to use a random intercept model, since it will account for within-school correlations in the standard error estimates [31,32]. Treatment effect estimation via regression models after matching is also useful, since it allows for additional bias correction. That is, any covariates that are not fully balanced by the match can also be included in the regression model to reduce bias while estimating treatment effects [33]. Using regression models to further reduce imbalance after matching is a well-known idea [34,35]. Moreover, when regression is applied after matching, it is applied to a subsample of the data that is well-balanced and in which covariate distributions overlap. Since regression is used locally in the covariate space, the corresponding results should be less sensitive to minor changes in the specification of the regression function [36].

For sensitivity analysis, one simply applies model (10.2) to entire cluster-level treatment indicators and replaces the uniform randomization inference just described with a worst-case randomization inference under which probabilities of treatment differ by up to a factor of Γ within matched pairs. The details of the worst-case analysis for test statistic (10.4) are slightly different from typical worst-case calculations in cases with individualistic treatment assignment, in ways that tend to make the inference less sensitive to unmeasured bias than would have been the case if treatments had been assigned individually. For more details, see Hansen et al. [30].

10.4.8 Case Study

Next, we present an empirical application to demonstrate some of the practical aspects of multilevel matching for a COS. Here, we review details from an analysis first reported in Keele et al. [3]. In the original analysis, the investigators sought to evaluate multilevel matching using a design that focused on recovering experimental benchmarks using within-study comparison (WSC) design [37]. In a WSC design, the investigator uses data from a randomized trial, sets aside the experimental control group, and replaces it with a new control group from the overall population that was not included in the original randomized trial. In their analysis, they applied multilevel matching to make the new control group comparable to the treated group on observed covariates. The original study focused exclusively on outcome estimates, we focus on aspects of the matching process. More specifically, we demonstrate how the match can be customized using substantive information from the application.

This WSC design used data from a CRT that evaluated Achieve3000, an adaptive literacy software program intended to increase student reading proficiency. In the CRT, 32 elementary schools participated in the study. These schools were sorted pairwise on 2012-13 composite reading scores and randomly assigned to the intervention within pairs. The schools that were selected for the treatment condition implemented Achieve3000 for all students in grades 2-5 and outcomes were measured in school years 2013-14, 2014-15, and 2015-16. In this WSC design, the control schools from the CRT were discarded and replaced with the 60 elementary schools that chose not to participate in the randomized trial. The data include both student and school level covariates. Student-level covariates included sex, race/ethnicity, English language learner status, gifted status, and prior achievement. School-level covariates included the percentage of Hispanic or African American students, students receiving free/reduced price lunch, students classified as English language learners, students proficient in reading and mathematics, novice teachers, (e.g., first-year teachers), white teachers, and an indicator for Title I school classification.

As the first step in the analysis, we review balance statistics. That is, we calculate treated and control differences for the baseline covariates. Specifically, we report treated-control means and the standardized differences in means, which is the difference in means divided by the standard deviation before matching, for the unmatched sample. Table 10.1 includes the balance statistics from the RCT. Given the small number of clusters in the CRT, we observe that there are notable imbalances. Specifically, treated schools had a lower percentage of African American students. A general rule of

TABLE 10.1
Balance from Achieve 3000 RCT. p-values from either Fisher exact test or Wilcoxon rank test.

	Treated Mean	Control Mean	Std. Diff.	p-value
% Proficient Reading 2011	76.35	77.02	−0.09	0.08
% Proficient Reading 2012	74.55	74.13	0.06	0.00
% Free Lunch	0.38	0.38	−0.00	0.91
% Limited English	0.13	0.10	0.41	0.00
Enrollment	770.70	700.38	0.50	0.00
% Teachers Beginners	0.15	0.14	0.16	0.00
% Teachers Whites	0.88	0.87	0.19	0.00
Title 1 School	0.78	0.60	0.40	0.00
Title 1 Focus School	0.25	0.19	0.14	0.00
% Students African-American	0.21	0.30	−0.68	0.00
% Students Hispanic	0.21	0.18	0.32	0.00

TABLE 10.2
Balance in the Constructed Observational Data. p-values from either Fisher exact test or Wilcoxon rank test.

	Treated Mean	Control Mean	Std. Diff.	p-value
% Proficient Reading 2011	76.35	81.48	−0.65	0.00
% Proficient Reading 2012	74.55	79.86	−0.72	0.00
% Free Lunch	0.38	0.29	0.58	0.00
% Limited English	0.13	0.10	0.50	0.00
Enrollment	770.70	772.52	−0.01	0.00
% Teachers Beginners	0.15	0.15	−0.06	0.00
% Teachers Whites	0.88	0.90	−0.14	0.00
Title 1 School	0.78	0.40	0.84	0.00
Title 1 Focus School	0.25	0.08	0.46	0.00
% Students African-American	0.21	0.19	0.17	0.00
% Students Hispanic	0.21	0.15	0.66	0.00

thumb is that matched standardized differences should be less than 0.20 and preferably 0.10 [38]. Table 10.2 contains the balance from the constructed observational data set. The differences are generally larger, with six standardized differences exceeding 0.50. As such, the control schools differ from treated schools in a number of observable ways.

The next step is to use multilevel matching to find the set of control schools that are most similar to the treated schools in terms of the observed covariates. However, as we outlined above, analysts have a number of different options in terms of how they implement multilevel matching. First, the investigator should decide whether students should be matched. In a WSC, we would argue, it makes little sense to match students, since we have no reason to suspect there was variation in terms of further student selection. That is, the treatment was applied to all students in each treated school. The next key question is whether or not to include any covariate prioritization. That is, should we seek to prioritize balance on some school level covariates? In the CRT, Achieve3000 was a literacy intervention, and the primary outcome measure is student level reading scores. As such, one might seek to prioritize balance for the school level reading scores. Alternatively, one might simply prioritize the school level covariates with the largest imbalances prior to matching. To that end, we implemented three different matches. The first match used the defaults in the matchMulti matching package. This match doesn't include any balance prioritization. In the second match, we

TABLE 10.3
Balance in RCT, Unmatched, and 3 Matched Groups. Cell entries are standardized differences in means. Match 1 includes no balance refinements. Match 2 prioritizes balance on prior reading test scores. Match 3 prioritizes overall balance

	RCT	Unmatched	Match 1	Match 2	Match 3
% Proficient Reading 2011	−0.09	−0.65	−0.30	−0.08	−0.04
% Proficient Reading 2012	0.06	−0.72	−0.42	−0.09	−0.30
% Free Lunch	−0.00	0.58	0.24	−0.04	−0.04
% Limited English	0.41	0.50	0.29	0.04	−0.06
Enrollment	0.50	−0.01	−0.09	0.25	0.36
% Teachers Beginners	0.16	−0.06	0.07	0.01	−0.09
% Teachers Whites	0.19	−0.14	−0.31	−0.12	−0.10
Title 1 School	0.40	0.84	0.41	0.14	0.00
Title 1 Focus School	0.14	0.46	0.33	0.16	0.00
% Students African-American	−0.68	0.17	−0.09	−0.38	−0.43
% Students Hispanic	0.32	0.66	0.40	0.06	0.11

prioritized school-level reading scores. That is, we sought to balance reading scores more than any other covariate. The final match prioritized balance on the covariates with the worst imbalances to produce the best overall balance. In the two final matches, we gave the lowest level of priority to the school enrollment covariate. For this covariate, the standardized difference is large, but the difference in school sizes across treated and control is not expected to matter as much for differences in outcomes.

Table 10.3 contains balance statistics for the RCT, the unmatched sample, and the three matched samples. First, matching at the defaults, improves balance substantially. In the second match, where we prioritized balance on reading scores, we find that the standardized differences for both reading score covariates are now both less than the 0.10 threshold. In the final match, one of the reading scores now displays higher imbalance, but we reduce imbalances on other covariates. Figure 10.1 includes a graphical comparison of the distribution of standardized differences. Here, it is clear that the match based on defaults does not improve on the balance from the RCT. Second, it is obvious that the two matches with balance prioritization improve on the first match and on the balance from the RCT.

Finally, in Table 10.4 we include outcome estimates. The RCT estimate shows a positive, but modest effect of the intervention that is not statistically significant. The estimate from the unmatched sample, without any statistical adjustment, is negative and significant at the 0.10 level. This is not surprising, since the treated group was targeted for enrollment in the RCT. This result is consistent with a pattern where the treated group generally has poorer reading performance than many schools in the district. We find that for Match 1, the bias between the COS and RCT estimates is reduced but not substantially. However, the estimates from the two matches based on balance prioritization are much closer to the RCT estimates. See Keele et al. [3] for a more complete outcome analysis.

10.5 Discussion

Multilevel matched designs may be viewed as an alternative to specialized regression models frequently used to analyze multilevel data. For example, fixed-effects regression models are commonly employed with multilevel data. A fixed-effects model includes an intercept term for each of the clusters in the data. Certain similarities between the methods are present. For instance, within-cluster

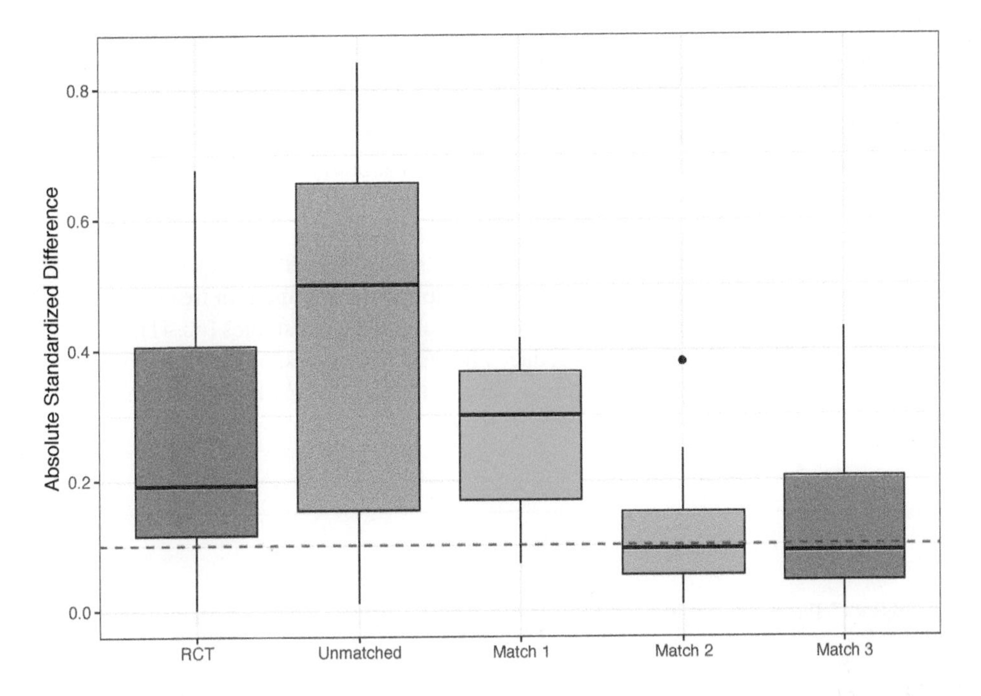

FIGURE 10.1
Distribution of Standardized Differences for RCT, Unmatched, and 3 Matches.

TABLE 10.4
Estimates of the effect of Achieve 3000 Intervention on Reading Test Scores. Outcome is standardized reading score. Results are from a mixed model with random intercept. Match 1 includes no balance refinements. Match 2 prioritizes balance on prior reading test scores. Match 3 prioritizes overall balance.

	Point Estimate	95% CI	p-value
RCT	0.048	[−0.13, 0.23]	0.598
Unmatched	−0.13	[−0.28, 0.02]	0.088
Match 1	−0.084	[−0.16, −0.013]	0.023
Match 2	0.01	[−0.068, 0.089]	0.785
Match 3	−0.0053	[−0.1, 0.093]	0.909

matching in cases with individualistic assignment only compares treated and control units within the same cluster, much as the regression coefficient for treatment in a fixed-effects regression is a partial coefficient for the adjusted regression in which each cluster receives its own intercept. However, multilevel matching offers several advantages. Most importantly, constructing a matched design in a multilevel setting involves thinking primarily about the distribution of treatment assignment rather than the distribution of the outcome variable, and whenever treatment assignment is easier to model or more plausibly random than outcomes, matching may provide substantial benefits relative to outcome regression. Constructing a matched design in multilevel data also requires careful consideration of whether treatment is at the individual or the cluster level and provides clear guidance for inference in either case, unlike outcome regression where distinctions between treatment mechanisms and handling of clustering in inference can become murky [39]. Matched designs also avoid some of the parametric assumptions associated with linear modeling, offer transparent balance summaries describing their success in adjusting for measured variables, and offer support for randomization inference and associated methods of sensitivity analysis. We find them to be a compelling design option in a wide variety of practical settings in the biomedical and social sciences.

One frontier worthy of exploration is the use of multilevel matching in two-stage observational studies. Two-stage studies involve treatment assignments at both levels; for instance, a school may decide to participate in an experimental treatment, and then individual teachers within that school may decide on an individual basis whether to participate. Such two-stage studies essentially involve two separate treatment processes. Another version of this setting arises frequently in clustered observational studies where individuals may fail to comply with treatment in treated clusters. While weighting methods have been introduced for two-stage observational studies [40, 41], and randomization inference methods similar to those introduced for matched estimators here have been introduced for two-stage randomized trials [42–44], multilevel matched designs for this setting remain to be introduced.

References

[1] Lindsay C Page, Matthew Lenard, and Luke Keele. The design of clustered observational studies in education. *AERA Open*, 6(3):1–14, July-Sept 2020.

[2] Guido W Imbens. Matching methods in practice: Three examples. *Journal of Human Resources*, 50(2):373–419, 2015.

[3] Luke J Keele, Matthew Lenard, and Lindsay Page. Matching methods for clustered observational studies in education. *Journal of Educational Effectiveness*, 14 (3), 696–725.

[4] Donald B Rubin. Comment: Which ifs have causal answers. *Journal of the American statistical association*, 81(396):961–962, 1986.

[5] David Roxbee Cox. *Planning of experiments.* Wiley, New York, 1958.

[6] Paul R Rosenbaum. *Observational Studies.* Springer, New York, NY, 2002.

[7] Junyeop Kim and Michael Seltzer. Causal inference in multilevel settings in which selection processes vary across schools. cse technical report 708. Technical report, National Center for Research on Evaluation, Standards, and Student Testing (CRESST), Center for Student Evaluation (CSE)., 2007.

[8] Bruno Arpino and Fabrizia Mealli. The specification of the propensity score in multilevel observational studies. *Computational Statistics & Data Analysis*, 55(4):1770–1780, 2011.

[9] Felix J Thoemmes and Stephen G West. The use of propensity scores for nonrandomized designs with clustered data. *Multivariate Behavioral Research*, 46(3):514–543, 2011.

[10] Jee-Seon Kim and Peter M Steiner. Multilevel propensity score methods for estimating causal effects: A latent class modeling strategy. In *Quantitative psychology research*, pages 293–306. Springer, Switzerland, 2015.

[11] Youjin Lee, Trang Q Nguyen, and Elizabeth A Stuart. Partially pooled propensity score models for average treatment effect estimation with multilevel data. *arXiv preprint arXiv:1910.05600*, 2019.

[12] José R Zubizarreta, Mark Neuman, Jeffrey H Silber, and Paul R Rosenbaum. Contrasting evidence within and between institutions that provide treatment in an observational study of

alternate forms of anesthesia. *Journal of the American Statistical Association*, 107(499):901–915, 2012.

[13] Guanglei Hong and Stephen W Raudenbush. Evaluating kindergarten retention policy: A case study of causal inference for multilevel observational data. *Journal of the American Statistical Association*, 101(475):901–910, 2006.

[14] Samuel D. Pimentel, Lindsay C Page, Matthew Lenard, and Luke J. Keele. Optimal multilevel matching using network flows: An application to a summer reading intervention. *Annals of Applied Statistics*, 12(3):1479–1505, 2018.

[15] Paul R Rosenbaum. Optimal matching for observational studies. *Journal of the American Statistical Association*, 84(408):1024–1032, 1989.

[16] Ben B. Hansen and Stephanie Olsen Klopfer. Optimal full matching and related designs via network flows. *Journal of Computational and Graphical Statistics*, 15(3):609–627, 2006.

[17] José R Zubizarreta. Using mixed integer programming for matching in an observational study of kidney failure after surgery. *Journal of the American Statistical Association*, 107(500):1360–1371, 2012.

[18] Samuel D Pimentel, Rachel R Kelz, Jeffrey H Silber, and Paul R Rosenbaum. Large, sparse optimal matching with refined covariate balance in an observational study of the health outcomes produced by new surgeons. *Journal of the American Statistical Association*, 110(510):515–527, 2015.

[19] Magdalena Bennett, Juan Pablo Vielma, and José R Zubizarreta. Building representative matched samples with multi-valued treatments in large observational studies. *Journal of Computational and Graphical Statistics*, 29(4):744–757, 2020.

[20] Luke J. Keele and José Zubizarreta. Optimal multilevel matching in clustered observational studies: A case study of the effectiveness of private schools under a large-scale voucher system. *Journal of the American Statistical Association*, 112(518):547–560, 2017.

[21] José R Zubizarreta, Ricardo D Paredes, Paul R Rosenbaum, et al. Matching for balance, pairing for heterogeneity in an observational study of the effectiveness of for-profit and not-for-profit high schools in chile. *The Annals of Applied Statistics*, 8(1):204–231, 2014.

[22] Samuel D Pimentel and Rachel R Kelz. Optimal tradeoffs in matched designs comparing us-trained and internationally trained surgeons. *Journal of the American Statistical Association*, 115(532):1675–1688, 2020.

[23] Paul R Rosenbaum and Donald B Rubin. Constructing a control group using multivariate matched sampling methods that incorporate the propensity score. *The American Statistician*, 39(1):33–38, 1985.

[24] Paul R. Rosenbaum. Optimal Matching of an Optimally Chosen Subset in Observational Studies. *Journal of Computational and Graphical Statistics*, 21(1):57–71, 2012.

[25] José R Zubizarreta, Caroline E Reinke, Rachel R Kelz, Jeffrey H Silber, and Paul R Rosenbaum. Matching for several sparse nominal variables in a case-control study of readmission following surgery. *The American Statistician*, 65(4):229–238, 2011.

[26] Kewei Ming and Paul R Rosenbaum. Substantial gains in bias reduction from matching with a variable number of controls. *Biometrics*, 56(1):118–124, 2000.

[27] Ben B Hansen. Full matching in an observational study of coaching for the sat. *Journal of the American Statistical Association*, 99(467):609–618, 2004.

[28] Samuel D Pimentel, Frank Yoon, and Luke Keele. Variable-ratio matching with fine balance in a study of the peer health exchange. *Statistics in medicine*, 34(30):4070–4082, 2015.

[29] Ronald A Fisher. *The Design of Experiments*. Oliver and Boyd, Edinburgh, 1935.

[30] Ben B Hansen, Paul R Rosenbaum, and Dylan S Small. Clustered treatment assignments and sensitivity to unmeasured biases in observational studies. *Journal of the American Statistical Association*, 109(505):133–144, 2014.

[31] Richard J Murnane and John B Willett. *Methods matter: Improving causal inference in educational and social science research.* Oxford University Press, Oxford, UK, 2010.

[32] Stephen W Raudenbush. Statistical analysis and optimal design for cluster randomized trials. *Psychological Methods*, 2(2):173, 1997.

[33] Alberto Abadie and Guido W Imbens. Bias-corrected matching estimators for average treatment effects. *Journal of Business & Economic Statistics*, 29(1):1–11, 2011.

[34] Elizebeth A Stuart. Matching methods for causal inference: A review and a look forward. *Statistical Science*, 25(1):1–21, 2010.

[35] Daniel E Ho, Kosuke Imai, Gary King, and Elizabeth A Stuart. Matching as nonparametric preprocessing for reducing model dependence in parametric causal inference. *Political analysis*, 15(3):199–236, 2007.

[36] Guido W. Imbens and Donald B. Rubin. *Causal Inference For Statistics, Social, and Biomedical Sciences: An Introduction*. Cambridge University Press, Cambridge, UK, 2015.

[37] Vivian C Wong, Peter M Steiner, and Kylie L Anglin. What can be learned from empirical evaluations of nonexperimental methods? *Evaluation review*, 42(2):147–175, 2018.

[38] Paul R. Rosenbaum. *Design of Observational Studies*. Springer-Verlag, New York, 2010.

[39] Alberto Abadie, Susan Athey, Guido W Imbens, and Jeffrey Wooldridge. When should you adjust standard errors for clustering? Technical report, National Bureau of Economic Research, 2017.

[40] Brian G Barkley, Michael G Hudgens, John D Clemens, Mohammad Ali, Michael E Emch, et al. Causal inference from observational studies with clustered interference, with application to a cholera vaccine study. *Annals of Applied Statistics*, 14(3):1432–1448, 2020.

[41] Lan Liu, Michael G Hudgens, Bradley Saul, John D Clemens, Mohammad Ali, and Michael E Emch. Doubly robust estimation in observational studies with partial interference. *Stat*, 8(1):e214, 2019.

[42] Hyunseung Kang and Luke Keele. Spillover effects in cluster randomized trials with noncompliance. *arXiv preprint arXiv:1808.06418*, 2018.

[43] Hyunseung Kang and Luke Keele. Estimation methods for cluster randomized trials with noncompliance: A study of a biometric smartcard payment system in india. *arXiv preprint arXiv:1805.03744*, 2018.

[44] GW Basse, A Feller, and P Toulis. Randomization tests of causal effects under interference. *Biometrika*, 106(2):487–494, 2019.

11

Effect Modification in Observational Studies

Kwonsang Lee, Jesse Y. Hsu

CONTENTS

11.1 Introduction

In an observational study of treatment effects, subjects are not assigned at random to treatment or control, so they may differ visibly with respect to measured pretreatment covariates, $\mathbf{x}$, and may also differ with respect to a covariate not measured, u. Visible differences in $\mathbf{x}$ are removed by adjustments, such as matching, but there is invariably concern that adjustments failed to compare comparable individuals, that differing outcomes in treated and control groups reflect neither a treatment effect nor chance but rather a systematic bias from failure to control some unmeasured covariate, u. A sensitivity analysis asks: What would u have to be like to materially and substantively alter the conclusions of an analysis that presumes adjustments for the observed $\mathbf{x}$ suffice to eliminate bias?

Once one can measure sensitivity to bias, it is natural to ask: What aspects of design and analysis affect sensitivity to bias? An aid to answering this question is the power of a sensitivity analysis and a number, the design sensitivity, that characterizes the power in large samples [1]. Some test statistics tend to exaggerate the reported sensitivity to unmeasured biases [2], whereas some design elements tend to make studies less sensitive to bias [1,3]. Generally, larger effects are less sensitive than smaller ones. This last point suggests that effect modification – that is, an interaction between a pretreatment covariate and the magnitude of a treatment effect – might matter for sensitivity to

DOI: 10.1201/9781003102670-11

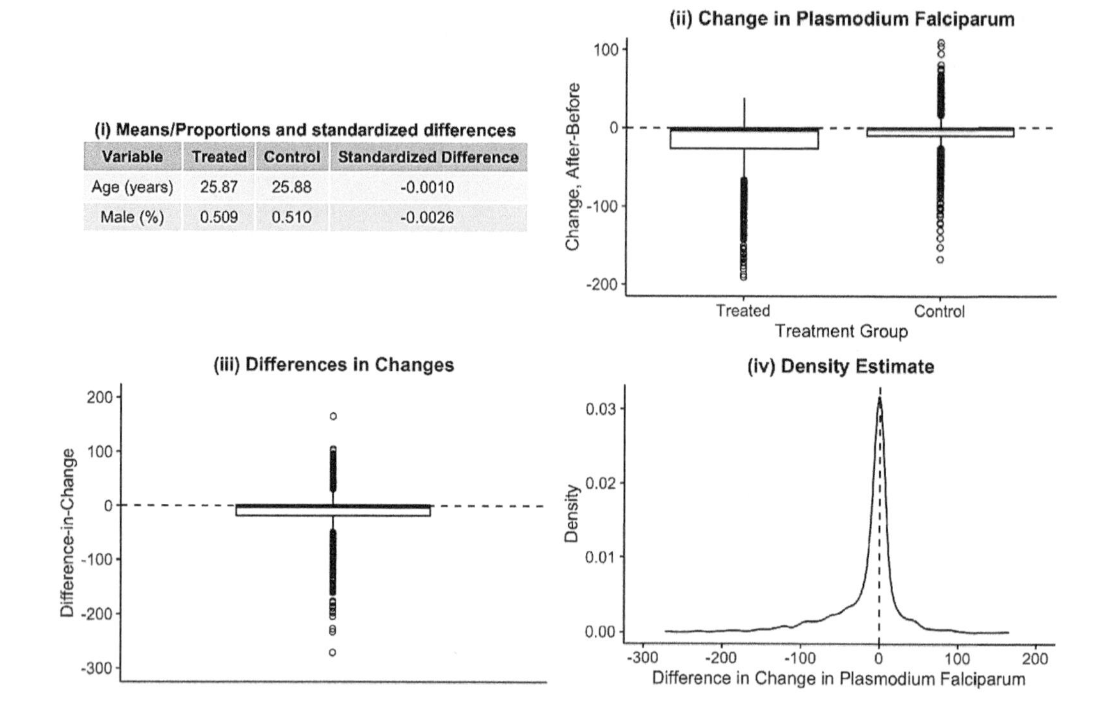

(i) Means/Proportions and standardized differences

Variable	Treated	Control	Standardized Difference
Age (years)	25.87	25.88	-0.0010
Male (%)	0.509	0.510	-0.0026

FIGURE 11.1
Age, sex, and parasite density in 1560 treated/control pairs matched for age and sex. After matching, the distribution of age and sex are similar, whereas the after-minus-before changes in parasite density exhibit a greater decline in the treated group. The treated-minus-control pair differences in changes in parasite density are typically negative, though many are near zero, with a long thick negative tail to their density.

unmeasured biases. Unfortunately, such an interaction may be uncertain or unexpected. How should one conduct a sensitivity analysis in the absence of *a priori* knowledge of where the effect will turn out to be large or small? Before developing the technical aspects, it is helpful to consider a motivating example.

11.1.1 Motivating example: Malaria in West Africa

Working with the government of Nigeria, the World Health Organization contrasted several strategies to control malaria in the Garki project [4]. We will look at one of these, namely, spraying with an insecticide, propoxur, together with mass administration of a drug, sulfalene-pyrimethamine, at high frequency. Matching for an observed covariate $\mathbf{x}$ consisting of age and gender, we paired 1560 treated subjects with 1560 untreated controls, making $I = 1560$ matched pairs. As is typically true in statistical applications of matching, there are $1560 + 1560 = 3120$ distinct individuals in the 1560 matched pairs – that is, no one is used twice. Also, the matching used only age, gender, and assigned treatment and so is "on the basis of $\mathbf{x}$ alone" [5]; therefore, if individuals were independent prior to matching, then outcomes in distinct pairs are conditionally independent given $\mathbf{x}$, treatment assignment, and the pairing.

The outcome is a measure of the frequency of *plasmodium falciparum* in blood samples, that is, the frequency of a protozoan parasite that causes malaria. We evaluate the effect of the malaria treatment (insecticide & drug) on the malaria parasite frequency with the 1560 matched pairs.

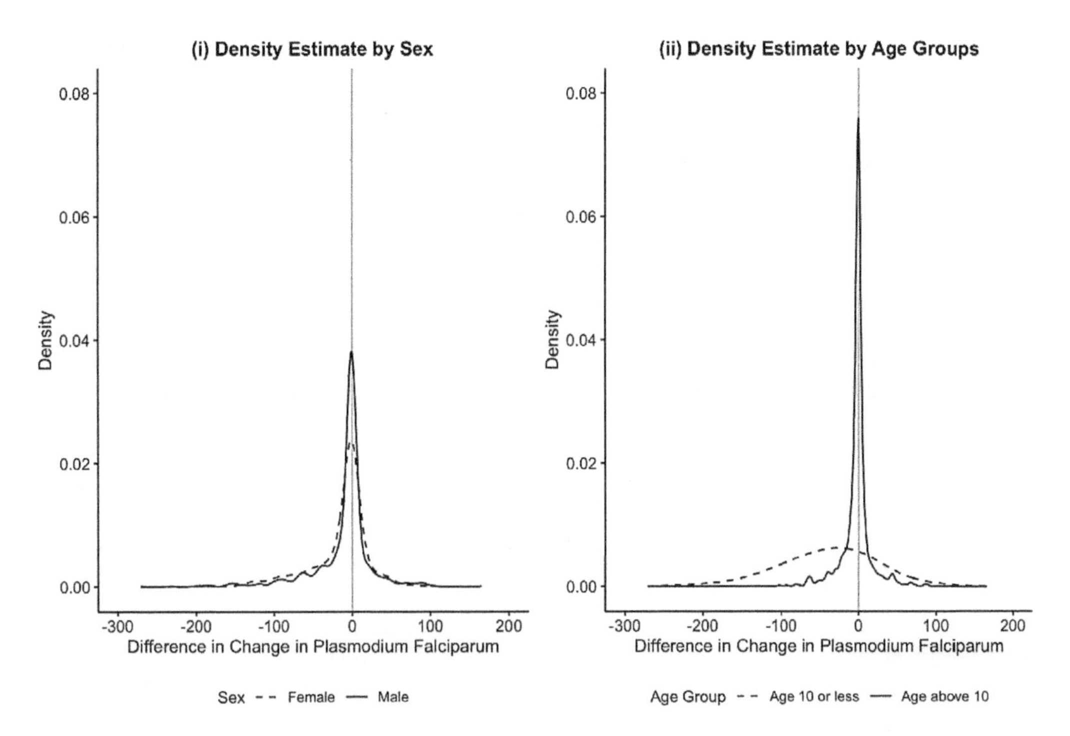

FIGURE 11.2
Densities of the treated-minus-control differences in changes in parasite density separately for pairs of (i) children older than 10 years and children 10 years old or younger; and (ii) female and male children.

Figure 11.1 displays (i) the close match for age and sex, (ii) after-minus-before changes in parasite frequency in treated and control groups, ignoring the matching, (iii) the matched pair treated-minus-control difference in after-minus-before changes in parasite frequencies, and (iv) a density estimate for this difference in changes. Density estimates use the default settings in R but with double the default bandwidth. Although declines in parasite frequency are more common in the treated group, many differences in changes are close to zero.

Figure 11.2 splits the 1560 pairs into (i) two age groups, 447 pairs of young children aged 10 or less, and 1113 pairs of children older than 10 years; and (ii) two sex groups, 766 pairs of female children, and 794 pairs of male children. The impression from Figure 11.2 is that the treatment was of much greater benefit to young children than to older ones but no different among female and male children.

Table 11.1 displays four sensitivity analyses and reports the upper bound on the one sided P-values testing no treatment effect using Wilcoxon's signed rank test. The sensitivity parameter, Γ, represents odds that in a matched pair, one unit has Γ times higher odds of being treated because of unmeasured confounding variable. In column I, all 1560 matched pairs differences were used. Using Wilcoxon's statistic, we would judge the results to be sensitive to a bias of magnitude $\Gamma = 2$, because the upper bound on the one sided P-value testing no treatment effect exceeds the conventional 0.05 level. Columns II-III of Table 11.1 repeat the sensitivity analyses separately for younger and older children, respectively. Despite the reduced sample size, the 447 pairs of young children exhibit an association with treatment that is far less sensitive to unmeasured bias than the full sample of 1560 pairs, with the test statistic being insensitive at $\Gamma = 3.8$.

TABLE 11.1
Various sensitivity analyses for the treated-minus-control difference in after-minus-before changes in parasite frequency in blood samples.

Analysis	Aggregate	Subgroup Specific		Truncated Product
Label	I	II	III	IV
Sample Size	1560	447	1113	1560
Γ	All	Young	Old	Combined
1.0	**0.000**	**0.000**	**0.000**	**0.000**
1.4	**0.000**	**0.000**	**0.048**	**0.000**
1.5	**0.000**	**0.000**	0.243	**0.000**
1.9	**0.011**	**0.000**	0.996	**0.000**
2.0	0.072	**0.000**	1.000	**0.000**
3.7	1.000	**0.022**	1.000	**0.045**
3.8	1.000	**0.034**	1.000	0.067
3.9	1.000	0.050	1.000	1.000
4.0	1.000	0.071	1.000	1.000

11.2 Review of Effect Modification in Observational Studies

In observational studies, it is known that certain patterns of treatment effects are more resistant than others to being explained away as the consequence of unmeasured biases in treatment assignment; see, for instance, [1,6,7].

Effect modification occurs when the size of a treatment effect or its stability varies with the level of a pretreatment covariate: the effect modifier. Effect modification affects the sensitivity of ostensible treatment effects to unmeasured biases. Other things being equal, larger or more stable treatment effects are insensitive to larger unmeasured biases; see [1,8]. As a consequence, discovering effect modification when it is present is an important aspect of appraising the evidence that distinguishes treatment effects from potential unmeasured biases, which is a concern in every observational study. In particular, Hsu et al. [9,10] discussed sensitivity analysis in observational studies with potential effect modification. Many authors have discussed effect modification from different perspectives, placing less emphasis on its role in confirmatory analyses that distinguish treatment effects from unmeasured biases in observational studies [11–17].

11.2.1 Notation and review: experiments and observational studies

There are I pairs, $i = 1, \ldots, I$, of two subjects, $j = 1, 2$, one treated with $W_{ij} = 1$, the other control with $W_{ij} = 0$, so $W_{i1} + W_{i2} = 1$ for each i. By design, matched pairs are disjoint: no individual appears in more than one matched pair. Subjects in each pair are matched for observed covariates $\mathbf{x}_{ij}$, so $\mathbf{x}_{i1} = \mathbf{x}_{i2} = \mathbf{x}_i$, say, for each i, but they may differ in terms of a covariate u_{ij} that was not measured. Each subject has two potential responses, Y_{1ij} if treated, $Y_{0,ij}$ if control, exhibiting response $Y_{ij} = W_{ij}Y_{1ij} + (1 - W_{ij})Y_{0ij}$, so the effect that is caused by the treatment, $Y_{1ij} - Y_{0ij}$ is not seen from any subject; see [18, 19]. Fisher's null hypothesis H_0 of no treatment effect asserts that $Y_{1ij} = Y_{0ij}$ for all i and j [20]. Simple algebra shows that the treated-minus-control pair difference in observed responses is $PD_i = (W_{i1} - W_{i2})(Y_{i1} - Y_{i2})$ which equals $(W_{i1} - W_{i2})(Y_{0,i1} - Y_{0,i2}) = \pm(Y_{0,i1} - Y_{0,i2})$ if Fisher's hypothesis H_0 is true. Write $\mathcal{F} = \{(Y_{1ij}, Y_{0ij}, \mathbf{x}_{ij}, u_{ij}), i = 1, \ldots, I, j = 1, 2\}$ for the potential responses and covariates, and write $\mathcal{W}$ for the event that $W_{i1} + W_{i2} = 1$ for each i. Let $q_i \geq 0$ be a function of $|PD_i|$ such that $q_i = 0$ if $|PD_i| = 0$, and let $\text{sgn}(y) = 1$ if $y > 0$ and $\text{sgn}(y) = 0$ if $y \leq 0$, so that under H_0 in a paired randomized experiment, the test statistic $T = \sum \text{sgn}\,(PD_i)\; q_i$ is the sum of I independent random variables, taking the value 0 with probability 1 if $q_i = 0$ and otherwise taking the values q_i and 0 with equal probabilities 1/2. For instance, if q_i is the rank of $|PD_i|$, then this yields the familiar null distribution of Wilcoxon's signed rank statistic, and many other statistics may be expressed in this form, including the permutational t-statistic [21], tests based on order statistics [22, 23], M-statistics [24], and various U-statistics [22, 25, 26].

In a randomized experiment, $W_{i1} = 1 - W_{i2}$ is determined by I independent flips of a fair coin and $\mathcal{W}$ is the set containing the 2^I possible values $\mathbf{w}$ of $\mathbf{W} = (W_{11}, W_{12}, \ldots, W_{I2})$, so $\mathbf{w} \in \mathcal{W}$ if $W_{ij} \in \{0, 1\}$ and $\pi_i = \Pr(W_{i1} = 1 \mid \mathcal{F}, \mathcal{W}) = \frac{1}{2}$ for each i, and this becomes the basis for randomization inferences, for instance for tests of Fisher's null hypothesis or for confidence intervals or point estimates formed by inverting hypothesis tests. A randomization inference derives the null distribution given $(\mathcal{F}, \mathcal{W})$ of a test statistic as its permutation distribution by using the fact that the 2^I possible values of $\mathbf{W}$ each have probability 2^{-I} in a randomized paired experiment; see [20, 27, 28].

11.2.2 Sensitivity analysis

A simple model for sensitivity analysis in observational studies assumes that, in the population prior to matching for x, treatment assignments are independent and two individuals, ij and $i'j'$, with the same observed covariates, $x_{ij} = x_{i'j'}$, may differ in their odds of treatment by at most a factor of $\Gamma \geq 1$,

$$\frac{1}{\Gamma} \leq \frac{\Pr\left(W_{ij} = 1 \mid \mathcal{F}\right) \Pr\left(W_{i'j'} = 0 \mid \mathcal{F}\right)}{\Pr\left(W_{i'j'} = 1 \mid \mathcal{F}\right) \Pr\left(W_{ij} = 0 \mid \mathcal{F}\right)} \leq \Gamma \text{ whenever } x_{ij} = x_{i'j'}; \tag{11.1}$$

then the distribution of $\mathbf{W}$ is returned to $\mathcal{W}$ by conditioning on $\mathbf{W} \in \mathcal{W}$. Write $\mathcal{U} = [0, 1]^{2I}$ for the $2I$-dimensional unit cube. It is easy to check that $1/(1+\Gamma) \leq \pi_i \leq \Gamma/(1+\Gamma)$ and conditioning on $\mathbf{W} \in \mathcal{W}$ are the same as assuming

$$\begin{aligned} \Pr\left(\mathbf{W} = \mathbf{w} \mid \mathcal{F}, \mathcal{W}\right) &= \prod_{i=1}^{I} \frac{\exp\left\{\gamma\left(w_{i1}u_{i1} + w_{i2}u_{i2}\right)\right\}}{\exp\left(\gamma u_{i1}\right) + \exp\left(\gamma u_{i2}\right)} \\ &= \frac{\exp\left(\gamma \mathbf{w}^T \mathbf{u}\right)}{\sum_{\mathbf{b} \in \mathcal{W}} \exp\left(\gamma \mathbf{b}^T \mathbf{u}\right)} \text{ with } \mathbf{u} \in \mathcal{U} \end{aligned} \tag{11.2}$$

where $\gamma = \log(\Gamma)$. Let $\overline{\overline{T}}$ be the sum of I independent random variables taking the value 0 with probability 1 if $q_i = 0$ and otherwise the value q_i with probability $\Gamma/\left(1+\Gamma\right)$ and the value 0 with probability $1/\left(1+\Gamma\right)$, and define $\overline{T}$ similarly but with $\Gamma/\left(1+\Gamma\right)$ and $1/\left(1+\Gamma\right)$ interchanged. It is straightforward to show that under (11.2), if H_0 is true, then

$$\begin{aligned} &\Pr\left(\overline{T} \geq v \mid \mathcal{F}, \mathcal{W}\right) \leq \Pr\left(T \geq v \mid \mathcal{F}, \mathcal{W}\right) \\ &\qquad \leq \Pr\left(\overline{\overline{T}} \geq v \mid \mathcal{F}, \mathcal{W}\right) \text{ for all } \mathbf{u} \in \mathcal{U}, \end{aligned} \tag{11.3}$$

and, as $I \to \infty$, the upper bound $\Pr\left(\overline{\overline{T}} \geq v \mid \mathcal{F}, \mathcal{W}\right)$ in (11.3) may be approximated by

$$\Pr\left(\overline{\overline{T}} \geq v \mid \mathcal{F}, \mathcal{W}\right) \doteq 1 - \Phi\left[\frac{v - \left\{\Gamma/\left(1+\Gamma\right)\right\} \sum q_i}{\sqrt{\left\{\Gamma/\left(1+\Gamma\right)^2\right\} \sum q_i^2}}\right]; \tag{11.4}$$

see, for instance, [29, 30] for the case of matched binary responses. For instance, a sensitivity analysis may report the range of possible P-values or point estimates that are consistent with the data and a bias of at most Γ for several values of Γ.

11.2.3 Design sensitivity

If after matching for observed covariates, an observational study were free of bias from unmeasured covariates u in the sense that $\Pr(\mathbf{W} = \mathbf{w} \mid \mathcal{F}, \mathcal{W}) = 2^{-I}$, and if the association between treatment W and response Y was the consequence of a treatment effect, not bias, then there would be no way to know this from the data. Call the situation just described, with an effect and no unmeasured bias, the "favorable situation."An investigator cannot know if she is in the favorable situation, and the best she can hope to say is that the conclusions are insensitive to small and moderate biases as measured by Γ. The power of a sensitivity analysis is the probability that she will be able to say this when she is indeed in the favorable situation. That is, for a specific Γ, the power of an α-level sensitivity analysis is the probability that the upper bound on the P-value in (11.4) is less than or equal to α, this probability being computed in the favorable situation.

In the favorable situation, there is typically a value $\widetilde{\Gamma}$ called the design sensitivity such that, as the sample size increases, $I \to \infty$, the upper bound on the P-value in (11.4) tends to zero when the analysis is performed with $\Gamma < \widetilde{\Gamma}$ and it tends to 1 when the analysis is performed with $\Gamma > \widetilde{\Gamma}$. Somewhat more precisely, if the Y_i are independent and identically distributed observations from some distribution, and if H_0 is rejected for a specific $\Gamma \geq 1$ when the upper bound on the P-value in (11.4) is $\leq \alpha$, conventionally $\alpha = 0.05$, then the probability of rejection or the power of the sensitivity analysis is tending to 1 for $\Gamma < \widetilde{\Gamma}$ and to 0 for $\Gamma > \widetilde{\Gamma}$ as $I \to \infty$; see [1,3]. For example, if $Y_i = \tau + \varepsilon_i$ where the ε_i are sampled from the standard Normal distribution and $\tau = 1/2$, then $\widetilde{\Gamma} = 3.2$ for Wilcoxon's signed rank statistic, whereas if the ε_i are sampled from the t-distribution on three degrees of freedom and $\tau = 1$, the corresponding design sensitivities are $\widetilde{\Gamma} = 6.0$ for Wilcoxon's statistic; see [25].

11.3 Effect Modification in a Few Nonoverlapping Prespecified Groups

What is a good strategy for conducting a sensitivity analysis when the treatment effect varies across two or a few pretreatment and prespecified subgroups? This section describes the use of the truncated product method to combine sensitivity analyses across the subgroups; for instance, Column IV in Table 11.1.

11.3.1 Combining independent P-values using their truncated product

In the context of the prespecified groups, the I pairs have been partitioned into L nonoverlapping groups of pairs based on a pretreatment covariate controlled by matching, and H_ℓ asserts that there is no treatment effect in group ℓ. The conjunction $H_\wedge = H_1 \wedge H_2 \wedge \cdots \wedge H_L$ asserts that all L hypotheses are true. Let P_ℓ, $\ell = 1, \ldots, L$, be valid, statistically independent P-values testing hypotheses H_ℓ, $\ell = 1, \ldots, L$, respectively, so $\Pr(P_\ell \leq \alpha) \leq \alpha$ for all $\alpha \in [0, 1]$ if H_ℓ is true. Zaykin et al. proposed testing $H_\wedge$ using a truncated product of P-values, $P_\wedge = \prod_{\ell=1}^{L} P_\ell^{\chi(P_\ell \leq \widetilde{\alpha})}$, where $\chi(E) = 1$ if event E occurs and $\chi(E) = 0$ otherwise, so $P_\wedge$ is the product of the P-values that are less than or equal to $\widetilde{\alpha}$ [31]. As an example, consider the case of $L = 2$ hypotheses with $\widetilde{\alpha} = 0.05$, so only P-values ≤ 0.05 are included in $P_\wedge$. The Bonferroni inequality would reject at level 0.05 with $L = 2$ hypotheses if $\min(P_1, P_2) \leq 0.05/2 = 0.025$; however, the method of Zaykin et al. gives P-value 0.05 if $P_\wedge = 0.025$, so $P_\wedge$ rejects whenever the Bonferroni inequality rejects and would also reject if $P_1 = P_2 = 0.05$.

Zaykin et al. obtain the distribution of $P_\wedge$ by a calculus argument, but it may alternatively but equivalently be written as a binomial mixture of gamma distributions. Let $F_k(\cdot)$ be the cumulative gamma distribution with shape parameter k and scale parameter 1, so that, in particular, $F_k(w) = 0$

for $w < 0$, and the sum of k independent exponential random variables has distribution $F_k(\cdot)$. If the P_ℓ are independent uniform random variables, then for $0 < w \leq 1$

$$\Pr(P_\wedge \leq w) = \sum_{k=1}^{L} \binom{L}{k} \widetilde{\alpha}^k (1-\widetilde{\alpha})^{L-k} \left[1 - F_k\left\{-\log\left(\frac{w}{\widetilde{\alpha}^k}\right)\right\}\right] \tag{11.5}$$

or in R,

$$\Pr(P_\wedge \leq w) =$$
$$\texttt{sum(dbinom(1:L, L,}\,\widetilde{\alpha}\texttt{)} * \texttt{(1} - \texttt{pgamma(}-\texttt{log(w/(}\widetilde{\alpha}\texttt{\^{}(1:L))), 1:L)))}. \tag{11.6}$$

In a sensitivity analysis, suppose that the I pairs are divided to L groups and let P_ℓ be a P-value testing H_ℓ using the pairs in group ℓ and computed from (11.2) for a specific unknown $\mathbf{u}$ and $\gamma = \log(\Gamma)$, and let $\overline{\overline{P}}_{\Gamma\ell}$ be the corresponding upper bound in (11.3). If the bias is at most Γ, then (11.3) implies $P_\ell \leq \overline{\overline{P}}_{\Gamma\ell}$ and $\Pr\left(\overline{\overline{P}}_{\Gamma\ell} \leq \alpha \mid \mathcal{F}, \mathcal{W}\right) \leq \Pr(P_\ell \leq \alpha \mid \mathcal{F}, \mathcal{W}) \leq \alpha$ for $\alpha \in [0, 1]$, $\ell = 1, \ldots, L$. Because the truncated product is a monotone increasing function, it follows that $\overline{\overline{P}}_{\Gamma\wedge} = \prod_{\ell=1}^{L} \left(\overline{\overline{P}}_{\Gamma\ell}\right)^{\chi\left(\overline{\overline{P}}_{\Gamma\ell} \leq \widetilde{\alpha}\right)}$ is an upper bound for $P_\wedge = \prod_{\ell=1}^{L} P_\ell^{\chi(P_\ell \leq \widetilde{\alpha})}$. Combining these two facts, if the bias is at most Γ then $\Pr\left(\overline{\overline{P}}_{\Gamma\wedge} \leq w \mid \mathcal{F}, \mathcal{W}\right) \leq \Pr(P_\wedge \leq w \mid \mathcal{F}, \mathcal{W})$ where $\Pr(P_\wedge \leq w \mid \mathcal{F}, \mathcal{W})$ is at most (11.5). If we calculate w such (11.5) equals α, conventionally $\alpha = 0.05$, and if we reject H_0 when $\overline{\overline{P}}_{\Gamma\wedge} \leq w$, then we will falsely reject H_0 with probability at most α if the bias is at most Γ. Column IV of Table 11.1 performs these calculations for the malaria data using $\widetilde{\alpha} = 0.05$. In Table 11.1, $\overline{\overline{P}}_{\Gamma\ell}$, $\ell = 1, 2$ are computed for young and old pairs, and these are combined into the truncated product $\overline{\overline{P}}_{\Gamma\wedge}$, whose P-value is determined from (11.5). The results in column IV of Table 11.1 testing H_0 using $\overline{\overline{P}}_{\Gamma\wedge}$ are much less sensitive to bias than the results in column I using all of the pairs in a single analysis. To emphasize, combining two independent sensitivity analyses yields less sensitivity to unmeasured bias than a single sensitivity analysis that uses all of the data, and this occurred because the treatment effect appears to be much larger for children aged 10 or less. Indeed, the sensitivity Γ for $\overline{\overline{P}}_{\Gamma\wedge}$ is only slightly worse than knowing *a priori* that attention should focus on the young pairs in Table 11.1.

Rejecting $H_0 = H_\wedge$ suggests there is an effect in at least one subgroup ℓ, but it does not provide an inference about specific subgroups. Of course, it would be interesting to know which subgroups are affected. Closed testing was proposed by Marcus et al. as a general method for converting a test of a global null hypothesis into a multiple inference procedure for subhypotheses [32]. Using the Wilcoxon's signed rank statistic in Table 11.1 with $\Gamma = 1$, closed testing rejects no effect $H_0 = H_1 \wedge H_2$ and then rejects both H_1 and H_2. Using the Wilcoxon's signed rank statistic in Table 11.1 with $\Gamma = 3$, closed testing rejects no effect $H_0 = H_1 \wedge H_2$ and then rejects H_1 but not H_2. In words, there is some evidence of a treatment effect for both those under and over ten years of age, the evidence about the young children being insensitive to a large bias of $\Gamma = 3$, while the evidence for older individuals is sensitive to some biases smaller than $\Gamma = 1.4$.

11.4 Discovering Effect Modification in Matched Observational Studies

In the previous section, we discussed a way to discover effect modification when we have non-overlapping pre-specified groups. However, in reality, it could be difficult to specify these groups before analysis. Without any prior domain knowledge, there will be no guidance and we have to specify groups from data. Such an adaptive method usually requires special attention.

In this section, we introduce two methods, CART and Submax, for discovering effect modification. The CART method is useful when there is large effect modification. By using part of outcome information, promising groups can be discovered and they can be treated as pre-specified groups. For moderate effect modification, the Submax method can be alternatively considered. Instead of discovering promising groups, each variable is examined whether it is an effect modifier or not.

11.4.1 CART method

Using the regression trees [33], as implemented in `R` with default settings in the `rpart` package [34], we regressed the ranks of $|PD_i|$ on $\mathbf{x}$ among I matched pairs matching for $\mathbf{x}$. Once a regression tree identifies a few promising subgroups, we can test the null hypothesis of no treatment effect by pooling evidence from the subgroups. We call this method the *CART* method. When applying the CART method, we use the data twice: once to build promising groups, and again to test the null hypothesis H_0 of no treatment effect and examine the sensitivity of that test to unmeasured biases. With subgroups discovered by an analysis of the data, a different assignment of treatments would have produced different subgroups, hence different null hypotheses, and perhaps even a different number of null hypotheses. Propositions 1 and 2 in [10] make assertions about the level of certain tests or testing procedures. Specifically, Proposition 1 says that whenever a group of pairs is entirely unaffected, a test with nominal level α will falsely reject with probability at most α, despite the fact that the groups were constructed using the data. Proposition 2 says that when closed testing is applied with component testing having nominal level α, the familywise error rate is strongly controlled at α: the chance of falsely rejecting at least one true hypothesis is at most α.

In §11.1.1, the matching controlled for age and gender. Working with these ranks of absolute differences in the outcome, a measure of frequency of a protozoan parasite that causes malaria, the regression tree algorithm ignored gender and split age into four bins with three cuts at 7.5 years, 17.5 years, and 32.25 years; see top portion of Figure 11.3. Beginning with the youngest, the bins contained 340 individuals under age 7.5, 243 between 7.5 and 17.5, 413 between 17.5 and 32.25, and 564 at least 32.5 years old, where $1560 = 340 + 243 + 413 + 564$. The ranks, of course, ranged from 1 to 1560, but the mean ranks were 1241 below age 7.5, 992 between ages 7.5 and 17.5, 659 between 17.5 and 32.5, and 501 above 32.5. This partition turned out to be fairly good advice.

Using the truncated product method to combine sensitivity analyses from the four subgroups in Table 11.2, the Wilcoxon's signed rank statistic in Table 11.2 with $\Gamma = 1$, the closed testing procedure rejects no effect $H_0 = H_1 \wedge H_2 \wedge H_3 \wedge H_4$ (column V), and then rejects all H_1, H_2, H_3 and H_4 (columns I–IV). Using the Wilcoxon's signed-rank statistic in Table 11.2 with $\Gamma = 3.6$, closed testing rejects no effect $H_0 = H_1 \wedge H_2 \wedge H_3 \wedge H_4$ (column V), and then rejects H_1 (column I) but not H_2, H_3 and H_4 (columns II–IV). In words, there is some evidence of a treatment effect for all subgroups of age, the evidence about the children with age <7.5 years being insensitive to a large bias of $\Gamma = 3.6$, while the evidence for older individuals is sensitive to some biases smaller than $\Gamma = 1.9$ for age 7.5–17.5 years, $\Gamma = 1.3$ for age 17.5–32.5 years and $\Gamma = 1.1$ for age $\geq$32.5 years.

If one confines attention to the 340 children under age 7.5 with the largest mean ranks of $|PD_i|$, then the sensitivity analysis for the Wilcoxon's statistic yields an upper bound (11.4) on the one-sided P-value of 0.045 at $\Gamma = 3.7$; Column III in Table 11.3. This result is similar to that in Column IV in Table 11.1 which cut at 10 years and used 447 pairs. If the split is made at 17.5 years, using $340 + 243 = 583$ pairs, the results are more sensitive to bias, with an upper bound on the one-sided P-value of 0.085 at $\Gamma = 3.2$; Column VI in Table 11.3.

There are several difficulties with the approach just described. First, large values of $|PD_i|$ at certain values of $\mathbf{x}_i$ are compatible with either a large typical effect or with greater instability at this $\mathbf{x}_i$, and we cannot distinguish these before looking at PD_i and $\mathbf{W}$, which we cannot do without affecting the level of the test. Second, combining a few leaves of a small tree may produce higher expected ranks of $|PD_i|$, but it also lowers the sample size, and one cannot shop around for the most favorable of several analyses without paying a price for multiple testing. Instead of performing one

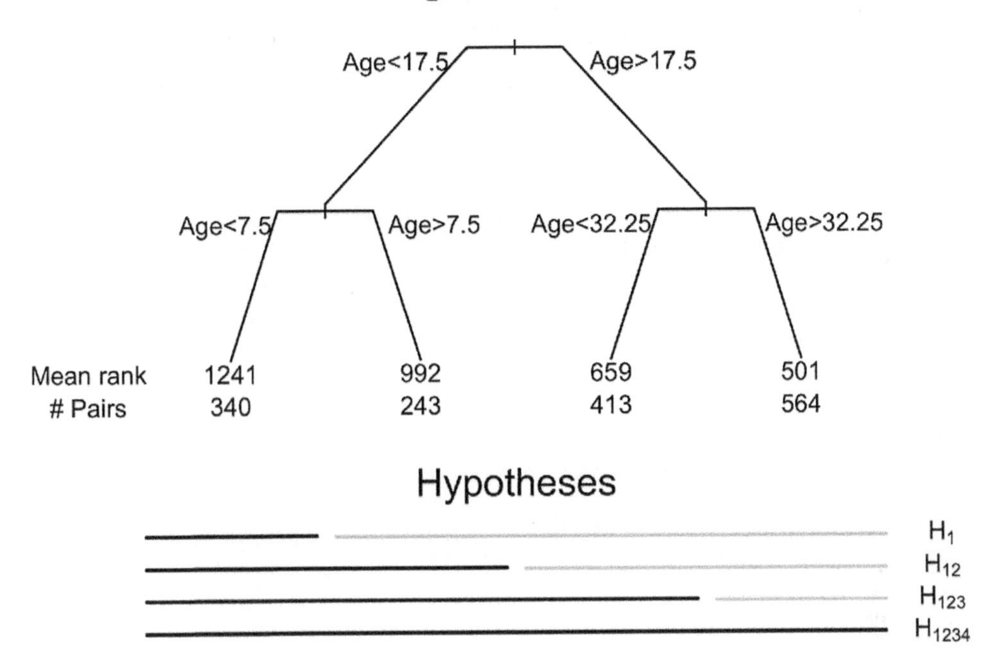

FIGURE 11.3
Top: The regression tree formed four groups of matched pairs fitting the ranks of $|PD_i|$ using age and gender. Bottom: Four hypotheses were tested simultaneously: (i) H_1 no effect for age <7.5; (ii) H_{12} no effect for age <17.5; H_{123} no effect for age <32.5; $H_{1234} = H_0$ no effect for all ages.

TABLE 11.2
Sensitivity analysis for the treated-minus-control difference in after-minus-before changes in parasite frequency in blood samples among subgroups identified by the regression tree.

Analysis	Subgroup Specific 1				
Label	I	II	III	IV	V
Sample Size	340	243	413	564	1560
Γ	Age <7.5	Age 7.5–17.5	Age 17.5–32.5	Age ≥32.5	Combined*
1.0	**0.000**	**0.000**	**0.000**	**0.000**	**0.000**
1.1	**0.000**	**0.000**	**0.001**	**0.007**	**0.000**
1.2	**0.000**	**0.000**	**0.006**	0.055	**0.000**
1.3	**0.000**	**0.000**	**0.032**	0.213	**0.000**
1.4	**0.000**	**0.000**	0.112	0.478	**0.000**
1.9	**0.000**	**0.035**	0.918	0.999	**0.000**
2.0	**0.000**	0.068	0.966	1.000	**0.000**
3.6	**0.009**	0.981	1.000	1.000	**0.045**
3.7	**0.014**	0.988	1.000	1.000	0.061
4.0	**0.040**	0.997	1.000	1.000	0.152
4.1	0.054	0.998	1.000	1.000	1.000

* Combined using the truncated product method at $\widetilde{\alpha} = 0.05$

TABLE 11.3
Sensitivity analysis for the treated-minus-control difference in after-minus-before changes in parasite frequency in blood samples among subgroups identified by the regression tree.

Analysis	Subgroup Specific 1			Subgroup Specific 2		
Label	I	II	III	IV	V	VI
Sample Size	340	1220	1560	583	977	1560
Γ	Age <7.5	Age ≥7.5	Combined*	Age <17.5	Age ≥17.5	Combined*
1.0	**0.000**	**0.000**	**0.000**	**0.000**	**0.000**	**0.000**
1.5	**0.000**	**0.028**	**0.000**	**0.000**	0.517	**0.000**
1.6	**0.000**	0.167	**0.000**	**0.000**	0.814	**0.000**
3.1	**0.001**	1.000	**0.002**	**0.024**	1.000	**0.048**
3.2	**0.001**	1.000	**0.004**	0.043	1.000	0.085
3.8	**0.020**	1.000	**0.041**	0.395	1.000	1.000
3.9	**0.029**	1.000	0.057	0.481	1.000	1.000
4.1	0.054	1.000	1.000	0.646	1.000	1.000

* Combined using the truncated product method at $\widetilde{\alpha} = 0.05$

test of H_0, while being uncertain as to which one test to perform, we perform four tests of one H_0, adjusting for multiple testing using the technique in [35]. The four tests concern hypotheses H_1, H_{12}, H_{123}, and H_{1234} in the bottom portion of Figure 11.4, where H_{1234} is the same as Fisher's hypothesis H_0 of no effect. Because these are four tests of one null hypothesis, all computed from the same data, the tests are highly correlated, and a correction for multiple testing that takes the high correlation into account is a small correction; see §4.3 in [9] for more details.

Though multiple testing corrections can be applied to any tree to control the familywise error rate, it is more important to choose an appropriate tree at the beginning. A CART tree can be selected by some predetermined criterion or cross-validation. However, there is no theoretical support that such trees perform well. Instead, domain knowledge can be helpful for choosing a tree. It can give some guidance whether we need to grow a tree further or trim it. For instance, in the study of surgical mortality at hospitals with superior nursing [36], the 130 surgical procedures were considered to examine effect modification. As a tree grows, CART will create several clusters of the procedures. When irrelevant surgical procedures are in the same cluster, we need to divide this cluster further, by growing a tree, to ensure each cluster contains relevant surgical procedures. Conversely, some clusters can be fragmented, so we need to integrate them into one larger cluster by trimming it. See §3 in [36] for more details. To control for false discovery rate when searching for effect modification in observational studies, Karmakar et al. [37] developed an approach for a collection of adaptive hypotheses identified from the data on matched-pairs design.

11.4.2 Submax method

The previous method to effect modification constructs a few promising subgroups from several measured covariates using CART. A limitation of this approach is that it is hard to study the power of such a technique except by simulation, because the CART step does not lend itself to such an evaluation. In the current section we discuss a different approach – the submax method–for which a theoretical evaluation is possible. The submax method has formulas for power and design sensitivity, and permits statements about Bahadur efficiency. The new submax method achieves the largest– i.e., best–of the design sensitivities for the subgroups, and the highest Bahadur efficiency of the subgroups; moreover, both the power formula and a simulation confirm that the asymptotic results are a reasonable guide to performance in samples of practical size.

A simple setting of the Submax method uses p binary covariates. The main purpose is to examine whether each covariate is an effect modifier or not while minimizing the power loss due to multiple testing. According to levels of each covariate x, the entire population can be divided into two smaller subgroups. If one of the subgroups has a larger effect size than the other, it is likely to detect effect modification, which usually leads to a smaller P-value. Two tests are defined for each covariates, and thus $2p$ tests can be defined for all covariates. It is still possible that there is no effect modification at all. To account for this situation, we include one test for the entire population. In total, there are $2p+1$ tests for testing H_0. As p increases, the number of the tests increase, which makes more difficult to find effect modification when there are too many uninformative covariates. To mitigate such issue, we create tests that are associated with each other. Association can be naturally generated when designing test statistics.

Let's revisit the malaria example. We have $p = 2$ binary variables, age and sex. Therefore, we consider $2p + 1 = 5$ comparisons: (1) all, (2) young, (3) old, (4) female, and (5) male. The test for young is associated with the test for female since the two groups share young female. The group of young female can be understood as two-way interaction. However, tests for some pairs are not associated with each other. To construct the association, we consider two-way interaction groups are disjoint. The tests for these groups are used as building blocks for constructing comparisons. Let (T_1, T_2, T_3, T_4) be the test statistics for young female, young male, old female and old male. Then, each comparison can be constructed by these four statistics. For instance, the test statistic for young is defined as $S_2 = T_1 + T_2$. Similarly, the statistic S_4 for female is defined as $S_4 = T_1 + T_3$. The statistic S_1 for all is defined as $S_1 = T_1 + T_2 + T_3 + T_4$. In brief, we test the hypothesis of no treatment effect at all (using S_1), plus the four hypotheses of no effect in $2p = 4$ overlapping subgroups.

Before evaluating the comparisons $S_k, k = 1, \ldots, K$ in general, we describe sensitivity analysis with parameter $\Gamma \geq 1$. To compute the upper bound on the P-value for overlapping S_k, we compute the upper P-value bound for non-overlapping T_g first. An approximation to the bound is obtained by considering a G-dimensional multivariate normal distribution. Subject to (11.1) for a given $\Gamma \geq 1$, find the maximum expectation, $\mu_{\Gamma g}$, of T_g. Also, among all treatment assignment probabilities that satisfy (11.1) and that achieve the maximum expectation $\mu_{\Gamma g}$, find the maximum variance, $\nu_{\Gamma g}$, of T_g; see [38] for detailed discussion. If treatment assignments were governed by the probabilities satisfying (11.1) that yield $\mu_{\Gamma g}$ and $\nu_{\Gamma g}$, then, under H_0 and mild conditions on the q_{gij}, the joint distribution of the G statistics $(T_g - \mu_{\Gamma g})/\nu_{\Gamma g}^{1/2}$, converges to a G-dimensional Normal distribution with expectation $\mathbf{0}$ and covariance matrix $\mathbf{I}$ as $\min(I_g) \to \infty$. Write $\boldsymbol{\mu}_\Gamma = (\mu_{\Gamma 1}, \ldots \mu_{\Gamma G})^T$ and $\mathbf{V}_\Gamma$ for the $G \times G$ diagonal matrix with gth diagonal element $\nu_{\Gamma g}$.

For p binary covariates, we consider $G = 2^p$ disjoint p-way interaction groups. Let $\mathbf{C}$ be the $K \times G$ matrix whose K rows are the $\mathbf{c}_k^T = (c_{1k}, \ldots, c_{Gk})$, $k = 1, \ldots, K$. Thus, the statistic S_k is represented as a linear combination of T_g using a 0-1 vector c_k, $S_k = \sum_{g=1}^{G} c_{gk} T_g$. In the malaria example, $c_1^T = (1, 1, 1, 1)$ for all and $c_2^T = (1, 1, 0, 0)$ for young. The $\mathbf{C}$ matrix in this example is

$$\mathbf{C} = \begin{bmatrix} 1 & 1 & 1 & 1 \\ 1 & 1 & 0 & 0 \\ 0 & 0 & 1 & 1 \\ 1 & 0 & 1 & 0 \\ 0 & 1 & 0 & 1 \end{bmatrix}$$

Define $\boldsymbol{\theta}_\Gamma = \mathbf{C}\boldsymbol{\mu}_\Gamma$ and $\boldsymbol{\Sigma}_\Gamma = \mathbf{C}\mathbf{V}_\Gamma\mathbf{C}^T$, noting that $\boldsymbol{\Sigma}_\Gamma$ is not typically diagonal. Write $\theta_{\Gamma k}$ for the kth coordinate of $\boldsymbol{\theta}_\Gamma$ and $\sigma_{\Gamma k}^2$ for the kth diagonal element of $\boldsymbol{\Sigma}_\Gamma$. Define $D_{\Gamma k} = (S_k - \theta_{\Gamma k})/\sigma_{\Gamma k}$ and $\mathbf{D}_\Gamma = (D_{\Gamma 1}, \ldots, D_{\Gamma K})^T$. Finally, write $\boldsymbol{\rho}_\Gamma$ for the $K \times K$ correlation matrix formed by dividing the element of $\boldsymbol{\Sigma}_\Gamma$ in row k and column k' by $\sigma_{\Gamma k}\,\sigma_{\Gamma k'}$. Subject to (11.1) under H_0, at the treatment assignment probabilities that yield the $\mu_{\Gamma g}$ and $\nu_{\Gamma g}$, the distribution of $\mathbf{D}_\Gamma$ is converging to a Normal distribution, $N_K(\mathbf{0}, \boldsymbol{\rho}_\Gamma)$, with expectation $\mathbf{0}$ and covariance matrix $\boldsymbol{\rho}_\Gamma$ as $\min(I_g) \to \infty$. Using this null distribution, the null hypothesis H_0 is tested using $D_{\Gamma\max} = \max_{1 \leq k \leq K} D_{\Gamma k}$. The α critical

TABLE 11.4
Submax: Five standardized deviates $D_{\Gamma k}, k = 1, \ldots, 5$ for Wilcoxon's test and their maximum $D_{\Gamma\max}$ are shown. The critical value $\kappa_{\Gamma,\alpha} = 2.19$ for $\Gamma \geq 1, \alpha = 0.05$. Deviates larger than 2.19 are in bold.

Analysis	Aggregate	Subgroup Specific				Max
Sample Size	1560	447	1113	766	794	
Group	All	Young	Old	Female	Male	
Γ	$D_{\Gamma 1}$	$D_{\Gamma 2}$	$D_{\Gamma 3}$	$D_{\Gamma 4}$	$D_{\Gamma 5}$	$D_{\Gamma\max}$
1.0	**9.33**	**12.30**	**6.44**	**6.10**	**7.12**	**12.30**
1.5	**2.91**	**8.83**	0.71	1.62	**2.52**	**8.83**
1.6	1.91	**8.32**	−0.20	0.91	1.80	**8.32**
1.7	0.97	**7.84**	−1.05	0.25	1.13	**7.84**
2.0	−1.56	**6.61**	−3.33	−1.51	−0.67	**6.61**
3.0	−7.93	**3.69**	−9.16	−5.99	−5.20	**3.69**
3.7	−11.34	**2.25**	−12.30	−8.40	−7.62	**2.25**
3.8	−11.78	2.07	−12.71	−8.71	−7.94	2.07
3.9	−12.21	1.89	−13.11	−9.01	−8.24	1.89
4.0	−12.63	1.72	−13.50	−9.31	−8.54	1.72
4.1	−13.05	1.55	−13.89	−9.60	−8.84	1.55

value $\kappa_{\Gamma,\alpha}$ for $D_{\Gamma\max}$ solves

$$1 - \alpha = \Pr\left(D_{\Gamma\max} < \kappa_{\Gamma,\alpha}\right) = \Pr\left(\frac{S_k - \theta_{\Gamma k}}{\sigma_{\Gamma k}} < \kappa_{\Gamma,\alpha},\ k = 1, \ldots, K\right) \tag{11.7}$$

under H_0. In general, $\kappa_{\Gamma,\alpha}$ depends upon both Γ and α. The multivariate Normal approximation to $\kappa_{\Gamma,\alpha}$ is obtained using the `qmvnorm` function in the `mvtnorm` package in R, as applied to the $N_K(\mathbf{0}, \boldsymbol{\rho}_\Gamma)$ distribution; see [39]. Notice that this approximation to $\kappa_{\Gamma,\alpha}$ depends upon Γ only through $\boldsymbol{\rho}_\Gamma$, which in turn depends upon Γ only through $\nu_{\Gamma g}$. The critical value $\kappa_{\Gamma,\alpha}$ for $D_{\Gamma\max}$ is larger than $\Phi^{-1}(1-\alpha)$ because the largest of K statistics $D_{\Gamma k}$ has been selected, and it reflects the correlations $\boldsymbol{\rho}_\Gamma$ among the coordinates of $\mathbf{D}_\Gamma$.

Table 11.4 performs the test for the malaria data using the Wilcoxon signed rank test statistic. The row of Table 11.4 for $\Gamma = 1$ consists of Normal approximations to randomization tests, while the rows with $\Gamma > 1$ examine sensitivity to bias from nonrandom treatment assignment. For $\Gamma = 1$, the test statistic $D_{\Gamma\max} = 12.30 \geq \kappa_{\Gamma,\alpha} = 2.19$. Thus, Fisher's hypothesis of no treatment effect would be rejected at level $\alpha = 0.05$ if the data had come from a randomized experiment with $\Gamma = 1$. The maximum statistic is based on 447 young pairs, $D_{\Gamma\max} = D_{\Gamma 2}$, but $D_{\Gamma k} \geq \kappa_{\Gamma,\alpha} = 2.19$ for every subgroup, $k = 1, \ldots, 5$. At $\Gamma = 1.6$, the deviates except for $D_{\Gamma 2}$ no longer exceed 2.19. The deviate $D_{\Gamma 2}$ far exceeds 2.19, which leads to strong evidence of effect modification even when the random treatment assignment is violated. The maximum deviate $D_{\Gamma\max}$ is attained as $D_{\Gamma 2}$ for any value of Γ and exceeds $\kappa_{\Gamma,\alpha}$ up to $\Gamma = 3.7$. This indicates that the strongest evidence, the least sensitive evidence, of a treatment effect on malaria parasites is for young. Though we pay a price for multiple testing, the conclusion is not different from the CART method.

Unlike the CART method, when applying the Submax method, the number of comparisons to test linearly increases as the number of covariates increases. There are 2^p interaction groups, but only $2p + 1$ comparisons are used to test. Consider a simple, balanced case under H_0 and outcomes are continuously distributed and hence untied with probability one. Also, assume the same number of matched pairs in each group, $I_1 = \cdots = I_G = \bar{I}$. The Wilcoxon signed rank statistic T_g is computed from the $\bar{I}$ pairs in each group g. In this case, $\mu_{\Gamma g} = \{\Gamma/(1+\Gamma)\}\bar{I}(\bar{I}+1)/2$ and $\nu_{\Gamma g} = \{\Gamma/(1+\Gamma)^2\}\,\bar{I}(\bar{I}+1)(2\bar{I}+1)/6$; see [40]. Because of the symmetry, the correlation matrix

TABLE 11.5
The critical constant $\kappa_{\Gamma,\alpha}$ in a simple, balanced case with Wilcoxon's signed rank statistics T_g and $I_1 = \cdots = I_G = \bar{I}$.

p	$K = 2p+1$	$\kappa_{\Gamma,\alpha}$	Bonferroni
0	1	1.64	1.64
1	3	2.03	2.13
2	5	2.20	2.33
3	7	2.32	2.45
4	9	2.40	2.54
7	15	2.55	2.71
12	25	2.70	2.88

ρ_Γ of $D_{\Gamma k}$ has the simple form. $D_{\Gamma 1}$ has correlation $1/\sqrt{2} = 0.707$ with $D_{\Gamma k}$ for $k \geq 2$, the two consecutive comparisons for the two categories of the same binary variable are uncorrelated, and all other comparisons have correlation 0.5. In this simple and balanced case, Table 11.5 shows the critical constant $\kappa_\alpha = \kappa_{\Gamma=1,\alpha}$ for various p values. The key point in Table 11.5 is that κ_α increases very slowly for larger values of p. The Submax method for 7 potential effect modifiers has the critical value $\kappa_\alpha = 2.54$ that is almost the same as the critical value of the Bonferroni method for screening 4 potential effect modifiers. Similarly, the Submax method for 12 effect modifiers and the Bonferroni method for 7 effect modifiers require the similar level of the critical value.

The main goal is to discover effect modification by testing the global null hypothesis H_0, specifically Fisher's hypothesis of no treatment effect in the study at all as a whole. However, we may be interested in testing the hypothesis, say H_k, that asserts there is no effect in subgroup S_k, say no effect in the subgroup of young. We would like to simultaneously test all K hypotheses $H_k, k = 1, \ldots, K$ while controlling for a familywise error rate α. Simultaneous inference can be done by using the closed testing method [32]. The precise meaning of Table 11.4 can be further examined with the closed testing method; see [41] for more details.

11.4.3 Comparison between CART and submax methods

The performance of the Submax method was evaluated in the previous section using Table 11.4. This table can be also used to compare the Submax method with the CART method. Although the statistic $S_2 = T_1 + T_2$ for young is differently defined from the statistic used for the CART method discussed in § 11.4.1, a rough comparison is possible. The CART method used $|PD_i|$ without the sign, and discovered that age is a strong potential effect modifier, but no evidence about sex. Given this discovery, the CART method does not need to consider $D_{\Gamma 4}, D_{\Gamma 5}$. Also, $D_{\Gamma 1}$ is not necessary since CART made the age split. Thus, only the two deviates $D_{\Gamma 2}, D_{\Gamma 3}$ can be considered, and the critical value via the Bonferroni adjustment is 1.96 for an one-sided alternative. From Table 11.4, we can see that $D_{\Gamma 2} > 1.96$ until $\Gamma = 3.8$, and then conclude that the evidence remains valid up to $\Gamma = 3.8$. This new summarized Γ value is slightly greater than the value $\Gamma = 3.7$ obtained from the Submax method. The Submax method performs slightly worse than the CART method does since it pays a price for checking all the covariates. The price is not too high in this example. In more general situations, given that CART successfully discovers effect modifiers, the CART performs better, but this superiority is not significant as shown in Table 11.5.

However, the CART method does not always discover true effect modifiers. A strong effect modifier induces large effect modification. Its effect on $|PD_i|$ is so strong that CART hardly misses it. For instance, the malaria example is the case of large effect modification. It is possible that some covariate is an important effect modifier, but not strong enough to be discovered by CART. In such cases, the CART method can fail to discover a meaningful effect modification structure and

Submax may have more power to detect effect modification. More discussion about moderate effect modification can be found in [41].

The Submax and CART methods can be combined in several ways. For instance, an investigator may combine a few potential effect modifiers with a few subgroups suggested by CART and apply the Submax method to the considered subgroups. Alternatively, some potential effect modifiers can be chosen as the covariates consisting of the CART tree output, and then we run the Submax method for the chosen covariates only. In the malaria example, there is evidence through CART that age is an effect modifier, but no evidence about sex. We may be interested in checking whether there is effect modification by sex within the young group; that is, comparison between young female and young male. Instead of considering four two-way interaction groups, we can consider three groups (i.e., young female, young male, old) only. The Submax method can be applied with an appropriate **C** matrix.

11.5 Discovering Effect Modification Using Sample-Splitting: Denovo

The previously proposed methods contain two different tasks: (1) discovering effect modifiers or groups with heterogeneous effects and (2) confirming/evaluating effect modification via hypothesis testing. The performance of each task can be improved with sample-splitting. A sample-splitting approach divides the total sample into two smaller samples: discovery sample and inference sample. The two divided samples are separately used for corresponding tasks. The *Denovo* method can improve the effect modification discovery.

11.5.1 Discovery step

The magnitude of treatment effect can vary across levels of covariates. When the magnitude is large or moderate, we are easily able to recognize it by using the methods discussed above. However, effect modification can be small. It is possible that there is no theoretical gain from discovering small effect modification compared to the price for multiple testing correction.

Even though effect modification is small, we are still willing to discover it for purposes of better understanding of treatment effects. Discovering small effect modification may be not useful to assert, for example, the claim that the conclusion is insensitive to unmeasured bias. However, it can be very useful to know about which subpopulations benefit the most from a treatment (or are the most vulnerable to a harmful exposure). Information about effect modification further suggests policy implication. As an example, during the COVID-19 pandemic, vaccines have been limited, so we need to decide who needs to be vaccinated first. If we have information about who benefits most from vaccinated, decision making or setting up logistics can be much easier.

In addition to discovering small effect modification, the Denovo method can be useful to discover a more complex effect modification structure that cannot be discovered by the Submax method. If the effect of a treatment is confined to a specific two-way interaction group, strategies of checking covariates one by one may fail to discover this group. For example, assume that the young male group has the effect of the treatment, but other groups (young female, old male, old female) do not. In this case, sex is an effect modifier *within* the young group, but may be not an effect modifier globally. The Submax method is likely to miss this group when age is not a strong effect modifier. Also, the CART method is likely to miss it.

To discover the true effect modification structure, sample-splitting can be a good strategy. Of course, we have to set aside a part of the total sample and spend it for discovery. Only the rest of the sample can be used for making inference, which makes it seem that sample-splitting leads to reduction in the practical sample size for analysis. However, a strategy of splitting work can be more

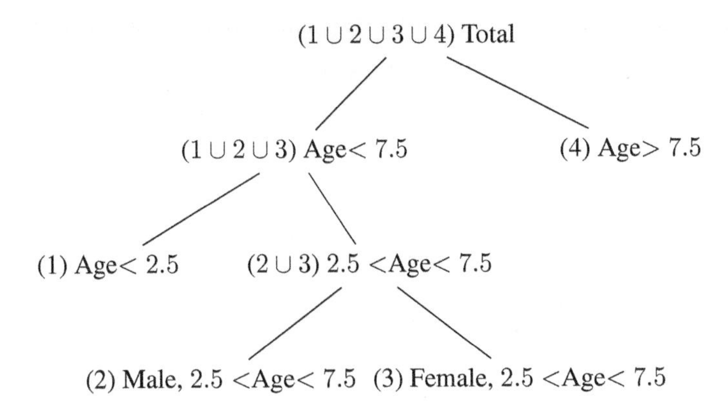

FIGURE 11.4
The discovered tree using the discovery sample of $0.25n = 390$.

efficient and thus leads to a causal conclusion more robust to unmeasured bias. We will show it with the malaria example in the following section.

The discovery step is straightforward. We randomly split the total (matched) sample into two subsamples: the discovery and inference subsamples. We suggest to use $0.25n$ for the discovery subsample size. While the inference subsample is set aside, the discovery subsample is used to fit CART. In this step, we use the sign, actually all the information about outcomes. Some subgroups as the terminal nodes of the CART output can be discovered, and they are examined by using the remaining inference subsample. Any methods that provide disjoint subgroups can be applied in this discovery step. After trying several methods, an investigator can choose one (or a combined structure) since the discovery subsample is not used for inference. Thus, this subsample can be considered as a test sample in machine learning literature, and a method to minimize the test error can be chosen.

There is a main issue in applying the sample-splitting strategy. How do we divide the total sample? The optimal ratio between the discovery and inference samples is critical in the overall performance of discovery and inference. However, there is no theory discovered yet. Through several simulation studies, we find that a (25%, 75%) splitting ratio has the best performance. The best performance can be defined in several ways. During the discovery step, the primary interest is to find the subgroups with heterogeneous treatment effects. We want to avoid missing true heterogeneous subgroups. However, high sensitivity tests have low specificity. Falsely discovered subgroups (that are thought to have heterogeneous effect, but actually do not have) can reduce the power of a test during the inference step. The (25%, 75%) splitting ratio has a great balance between discovering true heterogeneous subgroups and statistical power. In addition to the splitting-ratio, randomness in splitting can be problematic to provide a consistent result. When a different subsample is considered, a different CART output is discovered. Subsampling techniques are helpful to find a stable effect modification structure. More discussions about the optimal ratios and randomness in sample-splitting are in [42].

Let us look at the malaria example again. The Denovo method is applied to the malaria example with the splitting ratio (25%, 75%). A random sample of $0.25n = 390$ is chosen from the original matched pair sample. Figure 11.4 shows the tree output by regressing 390 treated-minus-control paired differences PD_i on age and sex. There are three splits in the tree and four terminal nodes (groups). The four disjoint groups are (1) Age< 2.5, (2) Male & 2.5 <Age< 7.5, (3) Female & 2.5 <Age< 7.5, and (4) Age> 7.5. The tree is quite different from what we evaluated in the previous section. The cutoff values for age are 2.5 and 7.5 instead of 10 used in the Submax methods. Also, We can confirm that the two data-driven methods show the same cutoff value 7.5. The second and third subgroups are constructed by splitting twice and more specific than the previously considered subgroups. During the discovery step, these specific groups are suspected to have large effects. In the following subsection, we will see how to confirm effect modification with the tree like Figure 11.4.

11.5.2 Inference step

Suppose that there is a given tree that is a partitioning of the total sample. Assume that there are G terminal nodes and each subgroup g represents a terminal node in the partition. To utilize the structure of trees, we trace back how the partition is built. When growing a tree, a certain internal node is chosen and forced to split into two subsequent nodes. Since there are G terminal nodes, the number of all nodes is $2G-1$. Excluding the initial node, we consider $G-2$ internal nodes and G terminal nodes. Write $\mathbf{\Pi} = (\Pi_1, \ldots, \Pi_{2G-2})^\top$ for these $2G-2$ nodes in total. To illustrate $\mathbf{\Pi}$, consider Figure 11.4. The internal node of $2.5 < \text{Age} < 7.5$ represents the union of (2) male with $2.5 < \text{Age} < 7.5$ and (3) female with $2.5 < \text{Age} < 7.5$. We simply denote this internal node as $2 \cup 3$. In the malaria example, the tree can be represented by $\mathbf{\Pi} = (1, 2, 3, 4, 2 \cup 3, 1 \cup 2 \cup 3)^\top$.

With G terminal nodes, $G-2$ internal nodes are considered for discovering effect modification. It may seem counter-intuitive to consider internal nodes since more comparisons implies paying more of a price for multiple testing. However, the inclusion of the internal nodes has several beneficial aspects. First, some of the terminal nodes may have a small number of matched pairs, which leads to a lack of power for detecting effect modification. Including the internal nodes can compensate for this lack of power. Second, when $\mathbf{\Pi}$ is much more complicated than the true structure, considering only the terminal nodes is misleading. This is important especially when $\mathbf{\Pi}$ is not given and has to be estimated. Overfitting a tree leads to an unnecessarily complex structure, but including internal nodes can correct this problem.

The inference step is the same as the Submax procedure except for constructing the $\mathbf{C}$ matrix. Define the test statistics $T_g, g = 1, \ldots, G$ for G groups. The comparison vector of $(2G-2)$ test statistics is constructed from T_g by using the $(2G-2) \times G$ conversion matrix $\mathbf{C}$. The matrix $\mathbf{C}$ creates the $(2G-2)$ correlated test statistics $S_k, k = 1, \ldots, 2G-2$ from mutually independent statistics T_g. In the malaria example, the $\mathbf{C}$ matrix can be constructed as

$$\mathbf{C} = \begin{bmatrix} 1 & 0 & 0 & 0 \\ 0 & 1 & 0 & 0 \\ 0 & 0 & 1 & 0 \\ 0 & 0 & 0 & 1 \\ 0 & 1 & 1 & 0 \\ 1 & 1 & 1 & 0 \end{bmatrix}.$$

The last two rows represent the internal nodes.

After defining the $\mathbf{C}$ matrix, we can define the deviates $D_{\Gamma k}, k = 1, \ldots, 2G-2$. The maximum deviate $D_{\Gamma\max}$ can be considered to test the global null hypothesis H_0 of no treatment effect at all. For $\Gamma \geq 1$, consider the following testing procedure,

$$\text{Reject } H_0 \text{ if } D_{\Gamma\max} = \max_{1 \leq k \leq 2G-2} \frac{S_k - \theta_{\Gamma k}}{\sigma_{\Gamma k}} \geq \kappa_{\Gamma,\alpha}, \tag{11.8}$$

where $\kappa_{\Gamma,\alpha}$ is the critical value obtained from a multivariate Normal distribution. In general, $\kappa_{\Gamma,\alpha}$ varies with Γ, but for the Wilcoxon's statistic, $\kappa_{\Gamma,\alpha}$ is constant for any Γ.

Now, we are ready to confirm effect modification as a form of the tree in Figure 11.4. The inference subsample of size 1170 is used to test the global null hypothesis H_0. Table 11.6 shows the deviates for the considered comparisons. The critical value $\kappa_{\Gamma,\alpha} = 2.57$ increased quite compared to the Submax method's $\kappa_{\Gamma,\alpha} = 2.19$. The Submax method actively utilizes the correlation between the subgroups, but the Denovo method uses the correlation induced by the tree structure. Among the disjoint subgroups $\{D_{\Gamma 1}, D_{\Gamma 2}, D_{\Gamma 3}, D_{\Gamma 4}\}$, $D_{\Gamma 2}$ has the largest deviate indicating that the

TABLE 11.6
Denovo: Six standardized deviates $D_{\Gamma k}, k = 1, \ldots, 6$ for Wilcoxon's test and their maximum $D_{\Gamma\max}$ are shown. The critical value $\kappa_{\Gamma,\alpha} = 2.57$ for $\alpha = 0.05$. Deviates larger than 2.57 are in bold.

Analysis	Subgroups						Max
Group	1	2	3	4	$2 \cup 3$	$1 \cup 2 \cup 3$	
Sample Size	86	96	71	917	167	253	
Γ	$D_{\Gamma 1}$	$D_{\Gamma 2}$	$D_{\Gamma 3}$	$D_{\Gamma 4}$	$D_{\Gamma 5}$	$D_{\Gamma 6}$	$D_{\Gamma\max}$
1.0	**4.76**	**7.48**	**3.43**	**6.79**	**8.15**	**9.40**	**9.40**
2.0	2.20	**4.93**	1.05	−2.03	**4.72**	**5.12**	**5.12**
3.0	0.85	**3.73**	−0.26	−7.23	**3.00**	**2.93**	**3.73**
3.2	0.64	**3.55**	−0.47	−8.08	**2.74**	**2.60**	**3.55**
3.3	0.54	**3.47**	−0.57	−8.49	**2.62**	2.44	**3.47**
3.4	0.44	**3.39**	−0.67	−8.89	2.50	2.29	**3.39**
3.6	0.26	**3.24**	−0.86	−9.65	2.27	2.00	**3.24**
3.7	0.17	**3.17**	−0.95	−10.03	2.16	1.86	**3.17**
3.8	0.08	**3.10**	−1.03	−10.39	2.06	1.72	**3.10**
3.9	0.00	**3.03**	−1.12	−10.74	1.96	1.59	**3.03**
4.0	−0.09	**2.97**	−1.20	−11.09	1.86	1.46	**2.97**
4.7	−0.61	**2.57**	−1.74	−13.35	1.24	0.65	**2.57**
4.8	−0.68	2.52	−1.81	−13.65	1.16	0.54	2.52

group of male aged 3-7 is the least sensitive to unmeasured bias. Also, the large deviates of $D_{\Gamma 5}$ and $D_{\Gamma 6}$ are due to $D_{\Gamma 2}$. Interestingly, the sample size for $D_{\Gamma 2}$ is 96 that is much smaller than the young group's sample size 447 in Table 11.4. However, from the sensitive analysis, we find the conclusion is unchanged until $\Gamma = 4.7$ that is largely improved from $\Gamma = 3.7$ of the Submax method. We spent only 75% of the sample to test H_0, but the sensitive analysis results are significantly improved. In addition, we can check again that age is an important effect modifier and gather another important information that sex is an effect modifier within 2.5 <Age< 7.5.

We can imagine several situations to understand Table 11.6. First, we can consider a case when there is only one split by the age cutoff value 7. In this case, we have two subgroups, $1 \cup 2 \cup 3$ and 4. Using the Bonferroni correction, the critical value is 1.96. Thus, the sensitivity parameter can be summarized as $\Gamma = 3.6$. This Γ value is not too different from the Submax method's value $\Gamma = 3.7$. If we further divide $1 \cup 2 \cup 3$ into 1 and $2 \cup 3$, the comparisons are $D_{\Gamma 1}, D_{\Gamma 4}, D_{\Gamma 5}$. The Γ value is then 3.7 using a new critical value 2.12, but not significantly improved. The above two situations imply that discovering $D_{\Gamma 2}$ is critically important to improving the sensitive analysis result. In addition, one more analysis is possible. If we use only the four terminal nodes (i.e., $\{D_{\Gamma 1}, D_{\Gamma 2}, D_{\Gamma 3}, D_{\Gamma 4}\}$) of the tree without the initial nodes, the critical value gets smaller as 2.24. The sensitivity analysis can be further improved to $\Gamma = 4.8$.

As shown above, spending a part of the sample to discover more specific subgroups is a good strategy as they can be easily missed by the other methods. However, there is one thing to be careful about. A discovered tree often needs to be adjusted for the purpose of the study. Too specific subgroups can be harmful if the discovered structure is needlessly more complicated than the true structure. An easy remedy is to include internal nodes to avoid unfortunate situations. An investigator can choose an appropriate tree by trimming or growing the discovered tree. For instance, for public policy, discovering too complicated subgroups is undesirable. On the contrary, for personalized medicine, discovering too simple subgroups is not informative. Selecting a suitable size of a tree can be critical for the inference step. See [42] for more discussion about an appropriate tree.

11.6 Discussion

In this chapter, we provided an overview of studying effect modification in matched observational studies. By discovering effect modification, we expect to report much firmer causal conclusions in subgroups with larger effects. That is, we expect the design sensitivity and the power of the sensitivity analysis to be larger, so we expect to report findings that are insensitive to larger unmeasured biases in these subgroups. Such a discovery is important in three ways. First, the finding about the affected subgroup is typically important in its own right as a description of that subgroup. Second, if there is no evidence of an effect in the complementary subgroup, then that may be news as well. Third, if a sensitivity analysis convinces us that the treatment does indeed cause effects in one subgroup, then this fact demonstrates the treatment does sometimes cause effects, and it makes it somewhat more plausible that smaller and more sensitive effects in other subgroups are causal and not spurious. This is analogous to the situation in which we discover that heavy smoking causes lots of lung cancer, and are then more easily convinced that secondhand smoke causes some lung cancer, even though the latter effect is much smaller and more sensitive to unmeasured bias.

The methods we introduced have been applied to matched observational study data. In the malaria example, two covariates, age and gender, were exactly matched among the 1560 matched pairs. In studying effect modification, it is convenient to have treatment-control pairs matched to have the same values of the effect modifiers. It is often difficult to match exactly for many covariates $\mathbf{x}$, but not so difficult to match for a few covariates and simply balance the rest. However, before we examined the matched pairs, we did not know which covariates would be suggested as possible effect modifiers, so we did not know which covariates should be exactly matched and which other covariates could merely be balanced. The strength-k matching provides a solution to this task. Strength-k balance means that every subset of k covariates is exactly balanced. In strength-k matching, all covariates should be exactly matched in the maximum number of pairs. One can always rematch the inexact pairs to be exact for k or fewer covariates. Also, this rematching does not alter the fact that all rematched pairs constitute a strength-k match of all covariates, because that property is unaffected by who is matched to whom. For more details, see §6 in [10].

The CART and Denovo methods used a tree structure to study effect modification. The size of the tree is important for later analysis to justify a large sample approximation. However, as the sample size grows, it is likely to obtain a larger tree with more specific subgroups. When discovering a tree, its size is usually selected with cross-validation, so is not fixed. Theoretical results about how cross-validation affects the performance of the methods are not known yet. To ensure the validity of the proposed methods, the maximum depth of a tree can be specified initially. The depth may depend on the sample size and research questions of interest. The R package for the Denovo method is available at `https://github.com/kwonsang`.

References

[1] Paul R Rosenbaum. Design sensitivity in observational studies. *Biometrika*, 91(1):153–164, 2004.

[2] Paul R Rosenbaum. Design sensitivity and efficiency in observational studies. *Journal of the American Statistical Association*, 105(490):692–702, 2010.

[3] Paul R Rosenbaum, PR Rosenbaum, and Briskman. *Design of observational studies*, volume 10. Springer, 2010.

[4] Louis Molineaux, Gabriele Gramiccia, World Health Organization, et al. *The Garki project: research on the epidemiology and control of malaria in the Sudan savanna of West Africa.* World Health Organization, 1980.

[5] Paul R Rosenbaum and Donald B Rubin. Constructing a control group using multivariate matched sampling methods that incorporate the propensity score. *The American Statistician*, 39(1):33–38, 1985.

[6] Elizabeth A Stuart and David B Hanna. Commentary: Should epidemiologists be more sensitive to design sensitivity? *Epidemiology*, 24(1):88–89, 2013.

[7] José R Zubizarreta, Magdalena Cerdá, and Paul R Rosenbaum. Effect of the 2010 chilean earthquake on posttraumatic stress reducing sensitivity to unmeasured bias through study design. *Epidemiology (Cambridge, Mass.)*, 24(1):79, 2013.

[8] Paul R Rosenbaum. Heterogeneity and causality: Unit heterogeneity and design sensitivity in observational studies. *The American Statistician*, 59(2):147–152, 2005.

[9] Jesse Y Hsu, Dylan S Small, and Paul R Rosenbaum. Effect modification and design sensitivity in observational studies. *Journal of the American Statistical Association*, 108(501):135–148, 2013.

[10] Jesse Y Hsu, José R Zubizarreta, Dylan S Small, and Paul R Rosenbaum. Strong control of the familywise error rate in observational studies that discover effect modification by exploratory methods. *Biometrika*, 102(4):767–782, 2015.

[11] Susan Athey and Guido Imbens. Recursive partitioning for heterogeneous causal effects. *Proceedings of the National Academy of Sciences*, 113(27):7353–7360, 2016.

[12] Andrew Chesher. Testing for neglected heterogeneity. *Econometrica: Journal of the Econometric Society*, 52(4):865–872, 1984.

[13] Richard K Crump, V Joseph Hotz, Guido W Imbens, and Oscar A Mitnik. Nonparametric tests for treatment effect heterogeneity. *The Review of Economics and Statistics*, 90(3):389–405, 2008.

[14] Peng Ding, Avi Feller, and Luke Miratrix. Randomization inference for treatment effect variation. *Journal of the Royal Statistical Society: Series B: Statistical Methodology*, pages 655–671, 2016.

[15] Steven F Lehrer, R Vincent Pohl, and Kyungchul Song. Targeting policies: Multiple testing and distributional treatment effects. Technical report, National Bureau of Economic Research, 2016.

[16] Xun Lu and Habert White. Testing for treatment dependence of effects of a continuous treatment. *Econometric Theory*, 31(5):1016–1053, 2015.

[17] Stefan Wager and Susan Athey. Estimation and inference of heterogeneous treatment effects using random forests. *Journal of the American Statistical Association*, 113(523):1228–1242, 2018.

[18] Donald B Rubin. Estimating causal effects of treatments in randomized and nonrandomized studies. *Journal of Educational Psychology*, 66(5):688, 1974.

[19] Jerzy Splawa-Neyman, Dorota M Dabrowska, and TP Speed. On the application of probability theory to agricultural experiments essay on principles. section 9. *Statistical Science*, 5(4):465–472, 1990.

[20] Ronald Aylmer Fisher. Statistical methods for research workers. In *Breakthroughs in statistics*, pages 66–70. Springer, 1992.

[21] B Li Welch. On the z-test in randomized blocks and latin squares. *Biometrika*, 29(1/2):21–52, 1937.

[22] BM Brown. Symmetric quantile averages and related estimators. *Biometrika*, 68(1):235–242, 1981.

[23] Gottfried E Noether. Some simple distribution-free confidence intervals for the center of a symmetric distribution. *Journal of the American Statistical Association*, 68(343):716–719, 1973.

[24] JS Maritz. A note on exact robust confidence intervals for location. *Biometrika*, 66(1):163–170, 1979.

[25] Paul R Rosenbaum. A new u-statistic with superior design sensitivity in matched observational studies. *Biometrics*, 67(3):1017–1027, 2011.

[26] W Robert Stephenson. A general class of one-sample nonparametric test statistics based on subsamples. *Journal of the American Statistical Association*, 76(376):960–966, 1981.

[27] Erich L Lehmann and Joseph P Romano. Unbiasedness: Applications to normal distributions. In *Testing statistical hypotheses*, pages 150–211. Springer, 2005.

[28] Paul R Rosenbaum. Randomized experiments. In *Observational studies*, pages 19–70. Springer, 2002.

[29] Paul R Rosenbaum. Sensitivity to hidden bias. In *Observational studies*, pages 105–170. Springer, 2002.

[30] Paul R Rosenbaum. Attributing effects to treatment in matched observational studies. *Journal of the American statistical Association*, 97(457):183–192, 2002.

[31] Dmitri V Zaykin, Lev A Zhivotovsky, Peter H Westfall, and Bruce S Weir. Truncated product method for combining p-values. *Genetic Epidemiology: The Official Publication of the International Genetic Epidemiology Society*, 22(2):170–185, 2002.

[32] Ruth Marcus, Peritz Eric, and K Ruben Gabriel. On closed testing procedures with special reference to ordered analysis of variance. *Biometrika*, 63(3):655–660, 1976.

[33] Leo Breiman, Jerome H Friedman, Richard A Olshen, and Charles J Stone. *Classification and regression trees*. Routledge, 2017.

[34] Terry M Therneau, Elizabeth J Atkinson, et al. An introduction to recursive partitioning using the rpart routines. Technical report, Technical report Mayo Foundation, 1997.

[35] Paul R Rosenbaum. Testing one hypothesis twice in observational studies. *Biometrika*, 99(4):763–774, 2012.

[36] Kwonsang Lee, Dylan S Small, Jesse Y Hsu, Jeffrey H Silber, and Paul R Rosenbaum. Discovering effect modification in an observational study of surgical mortality at hospitals with superior nursing. *Journal of the Royal Statistical Society: Series A (Statistics in Society)*, 181(2):535–546, 2018.

[37] Bikram Karmakar, Ruth Heller, and Dylan S Small. False discovery rate control for effect modification in observational studies. *Electronic Journal of Statistics*, 12(2):3232–3253, 2018.

[38] Joseph L Gastwirth, Abba M Krieger, and Paul R Rosenbaum. Asymptotic separability in sensitivity analysis. *Journal of the Royal Statistical Society: Series B (Statistical Methodology)*, 62(3):545–555, 2000.

[39] Alan Genz and Frank Bretz. *Computation of multivariate normal and t probabilities*, volume 195. Springer Science & Business Media, 2009.

[40] Paul R Rosenbaum. *Observational studies*. Springer, 2002.

[41] Kwonsang Lee, Dylan S Small, and Paul R Rosenbaum. A powerful approach to the study of moderate effect modification in observational studies. *Biometrics*, 74(4):1161–1170, 2018.

[42] Kwonsang Lee, Dylan S Small, and Francesca Dominici. Discovering heterogeneous exposure effects using randomization inference in air pollution studies. *Journal of the American Statistical Association*, 116(534):569–580, 2021.

12

Optimal Nonbipartite Matching

Robert A. Greevy, Jr. and Bo Lu

CONTENTS

Matching is a powerful design tool to remove measured confounding in both observational studies and experiments. Conventional matching designs focus on two-group setup, namely, treated and control groups, which is also known as bipartite matching. In more complex real-world scenarios, multiple treatment groups (more than two) are not uncommon, either in the form of different dose levels or as several unordered intervention arms. This chapter reviews the methodology for creating matched pairs in the presence of multiple groups, or in situations without clear grouping, which is referred to as nonbipartite matching. Such a design may be used to match with doses of treatment, or with multiple control groups, or with various time points in a longitudinal study, or as an aid to strengthen the instrumental variable analysis.

12.1 Introduction

We usually think of matching for settings where there are two distinct groups, an exposed group and an unexposed group. Smokers versus non-smokers is a classic example. We call these settings "bipartite," referring to two separate parts. However, exposures are often more continuous in their nature. Consider studying smokers with the exposure being measured as a self-reported average cigarettes per day smoked. Here someone smoking 10 cigarettes is a heavy smoker compared to someone smoking 1 cigarette per day, but they are a light smoker compared to someone smoking 20 cigarettes per day. We could think of the levels of cigarettes per day as creating many groups or there

DOI: 10.1201/9781003102670-12

being one big group, smokers, with varying levels of exposure. Studies that are not with two distinct groups, being either one big group or many groups, are referred to as "nonbipartite" (sometimes hyphenated "non-bipartite"). This chapter discusses nonbipartite matching methods for such settings.

Matching is a popular tool for causal inference utilizing data with non-randomized treatment assignment or exposure status. Matching forms groups that are similar in the distributions of their observed covariates. For covariates to be confounders, they must be associated with both the outcome and the exposure. By balancing the distributions of covariates between groups, matching breaks the covariates' associations with the exposure, thus controlling for confounding. The covariates remain predictive of the outcome, but they no longer induce biased inference when comparing the outcome between the matched groups [1].

Matching can be an appealing alternative to model-based analyses for several reasons. Matching is akin to experimental trials in that the burden of controlling for confounding is placed on the study's design, rather than its analysis. This can facilitate a simple comparison of outcomes between groups, similar to the *unadjusted* analysis of a randomized trial. The study design can allow using methods with no parametric assumptions. Following Fisher's suggestion of randomization being the reasoned basis for causal inference, randomization-based testing and estimation strategies can be readily implemented in a matched design [2]. Matching methods that create nonoverlapping sets, e.g. matching without replacement, preserve an independence structure that facilitates using conventional estimates of standard errors [3]. Additionally, matching does not preclude the use of model-based analyses. Model-based analyses may still be performed on a matched cohort, as they are in some randomized trials. Through balancing the covariate distributions, matching makes these *adjusted analyses* more robust [4].

The control for confounding is only as good as the covariates being balanced. Imagine a study of smoking and cardiovascular disease that did not include sex as a covariate when men are known to be heavier smokers and to have greater cardiovascular risk. The application of matching methods places clear focus on which covariates are being balanced and on the balance being achieved. The approach encourages the integration of existing knowledge when designing and evaluating the study. The quality of covariate balance may be assessed and iteratively improved without creating an opportunity to cherry-pick the study's results [5]. Unlike checking model fit and examining residuals, which are generally performed with the analysis results unblinded, checking the performance of the matching can and should be done fully blinded to the analysis results, i.e., prior to any analysis of the outcome.

Finally, many matching methods select the cohort where the data is most capable of providing meaningful information. Imagine a study of the impact of smoking among school aged children in first through twelfth grade. A matched analysis would put little to no emphasis on the youngest children, among whom smoking is extremely rare. Finding the cohort most suited for analysis may not be trivial because of the interplay between the exposure and the covariates. Consider a common dichotomization for smoking adults: <10 cigarettes per day is low and ≥ 10 is high. This would not be a good choice for smoking children where smoking intensity is strongly correlated with age and other covariates. What threshold is appropriate for a "heavy" smoker may vary across regions of the covariate space. As will be described below, a nonbipartite matched design would seek out the high and low smoking levels for each region of the multidimensional covariate space.

This chapter will first review optimal matching in the bipartite case briefly and then extend to the nonbipartite case with statistical software availabilities. A small numerical example is presented to fix the idea. Three examples from the literature are reviewed with more details to illustrate different ways of using the optimal nonbipartite matching design.

12.2 Optimal Nonbipartite Matching

This section reviews the commonly used bipartite matching design and provides an overview of the optimal nonbipartiate matching design. A small numerical example is used to illustrate the algorithm.

12.2.1 Bipartite matching

To lay the foundation of the methods, it is helpful to begin with the familiar bipartite case. Consider matching smokers to nonsmokers. Each person exists as a point in a multidimensional covariate space. If there are enough covariates, especially continuous covariates, no two people will occupy the exact same point in that space. But some points will be closer than others. A good match connects two points that are sufficiently close in this covariate space. What constitutes closeness must be defined by the researchers with a measurment of "distance." For example, the researchers could calculate the Euclidean distance between any two points in the covariate space. Continuous covariates could be specified as is or transformed to a meaningful scale, e.g., $log(income)$. Categorical covariates could be specified by $k-1$ indicator variables, where k is the number of categories. A problem with this distance measure is the scale of a variable will have a big impact on how much weight that variable has on the distance. For example, the impact of differences in age would be much greater when measured in months instead of years. One commonly used solution is to use the Mahalanobis distance, which accounts for the standard deviation of the variables as well as their correlation structure. It has its own limitations in that rare categorical variables may acquire undue weight. The reweighted Mahalanobis distance allows researchers to adjust the relative weights of the variables to reflect their clinical importance [6]. In causal analysis with observational data, another popular distance metric is through propensity score, which is a scalar summary of all relevant high dimensional covariates [7].

The problem of forming good matches may be imagined as a graphing problem, and thus, take advantage of developments in graph theory. In the language of graph theory, our subjects are "nodes," and a match between two subjects is an edge connecting two nodes. A graph is a set of edges satisfying certain conditions. In mathematical notation, $G = (V, E)$, where $V = \{v_i, i = 1, \ldots, n\}$ denotes the set of nodes and $E = \{[v_i, v_j], v_i, v_j \in V\}$ denotes the set of edges. In a bipartite matching, the nodes are divided *a priori* into two groups, e.g. smokers and non-smokers. An edge may not be formed between two nodes in the same group, constraining the set of possible edges, E. The edge between $[v_i, v_j]$ is associated with a nonnegative distance, d_{ij}, – a smaller distance meaning greater similarity in the nodes' covariate values. The sum of the distances of the edges in a matched sample reflects the overall quality of the matching. The graph with the smallest possible sum of distances, under given constraints, is the optimally matched set. Formally, it is to find a set of edges such that no node can appear in two different edges (M for matched sample):

$$x_{ij} = \begin{cases} 1, & \text{if } [v_i, v_j] \in M \\ 0, & \text{if } [v_i, v_j] \notin M \end{cases}$$

which solves the restricted minimization problem

$$min \sum_{[v_i, v_j] \in E} d_{ij} \cdot x_{ij}$$

For bipartite matching, subjects from the same group are not allowed to match to each other. Thus the distances between those subjects are set to infinity, or in practical terms, a very large positive number. By taking certain edges out of the matching, it effectively reduces the search space for the optimal solution, hence leads to a shorter running time. But it also limits the possibility of otherwise feasible matches and makes the design less flexible. For example, in a cohort of smokers with different number of cigarettes smoked per day (CPD), a simple two-group split may classify 10 CPD or less as light smoking and more than 10 CPD as heavy smoking. Then, matching someone smoking 15 CPD to another person smoking 10 CPD is allowable, but matching between persons with 20 CPD and 15 CPD is not possible, even though both scenarios have a dose difference of 5 units. Nonbipartite matching may be more appropriate in situations involving various doses of exposure, as it does not require grouping.

12.2.2 Nonbipartite matching

A nonbipartite matching removes the condition that one group must match to another group. A set of $2I$ nodes will have a matched graph with I edges when I is positive integer. An optimal nonbipartite match divides $2I$ nodes into I nonoverlapping pairs to minimize the sum of the I distances within the I pairs. This is also known as complete match in graph theory as every single node in the graph is matched. In the nonbipartite matching, typically one node is allowed to match to any other node (as long as their distance is not infinity) since there is no group structure to forbid certain matches. Therefore, the bipartite matching can be viewed as a special case of the nonbipartite matching with restriction on the distance matrix. The minimum distance nonbipartite matching problem is a combinatorial optimization problem that can be solved with a polynomial time algorithm [8]. Specifically, the number of arithmetic operations needed to find an optimal nonbipartite matching of N subjects is $O(N^3)$, which is comparable to solving an optimal bipartite matching problem of the same size.

The original nonbipartite matching algorithm solves the complete match problem. In reality, the study sample size may be an odd number or we prefer to match a subset of the full sample to improve matching quality. A practical strategy of using sink was introduced in the matching literature to create a matched sample of any desirable size with the original complete match algorithm [9, 10]. Specifically, suppose the study sample size is K and the researchers would like to create a matched sample with size of $2I$ ($K \geq 2I$). Since there are $K - 2I$ units not included in the final match, we can introduce a set of sink units with the size of $K - 2I$, namely, $s_1, \cdots, s_{K-2I}$. The distance between sinks and the original units are all set to 0, and the distance between sinks are set to infinity, i.e. $d_{\{k,s_i\}} = 0$ and $d_{\{s_i,s_j\}} = \infty$ for $k = 1, \cdots, K, i, j = 1, ..., K - 2I$. The infinity distance between sinks prevents matching between any two sink units, so they will all be matched to the original units. With $K - 2I$ original units being taken away by sinks, only $2I$ original units are left and matched. In the next subsection, a small numerical example is used to illustrate the nonbipariate matching and the strategy of using sinks.

To implement the optimal nonbipartite matching, Lu et al. provides a R package named *nbpMatching* (https://cran.r-project.org/web/packages/ nbpMatching/index.html) [11]. It is based on a Fortran implementation of Derigs' algorithm [12]. An implementation in C is also available [13]. Zubizarreta's R package *designmatch* implements the optimal nonbipariate matching using mixed integer programming (https://cran.r-project.org/web/packages/design match/index.html).

12.2.3 A small example

This small numerical example is based on a setup from Figure 2 in Lu et al. [11], where there are six units to be matched. Table 12.1 presents the distance matrix between these six units, with some infinity distance prohibiting matching between those pairs. The distance between the unit and itself is zero, but matching to itself is not allowed. The optimal result for this sample is $\{[1, 6], [2, 4], [3, 5]\}$ and the total distance is $10 + 10 + 10 = 30$. Since the data is not split into two groups, the bipartite matching algorithm does not apply and the optimal nonbipartite matching algorithm must be used.

If researchers want to improve the matching quality, sinks may be used to take some harder-to-match units away. Table 12.2 presents the updated distance matrix with two sink units added (with asterisks). There are eight units in total and sinks have zero distance with the original units and

TABLE 12.1
A 6 × 6 distance matrix for nonbipartite matching with six units.

ID	1	2	3	4	5	6
1	0	1	∞	2	∞	10
2	1	0	2	10	100	∞
3	∞	2	0	∞	10	∞
4	2	10	∞	0	30	100
5	∞	100	10	30	0	100
6	10	∞	∞	100	100	0

TABLE 12.2
A 8 × 8 distance matrix for nonbipartite matching with sinks.

ID	1	2	3	4	5	6	7*	8*
1	0	1	∞	2	∞	10	0	0
2	1	0	2	10	100	∞	0	0
3	∞	2	0	∞	10	∞	0	0
4	2	10	∞	0	30	100	0	0
5	∞	100	10	30	0	100	0	0
6	10	∞	∞	100	100	0	0	0
7*	0	0	0	0	0	0	0	∞
8*	0	0	0	0	0	0	∞	0

infinity distance among themselves. Because the distance from original units to sinks is 0, the total distance equals the total distance between paired original units. The algorithm minimizes that total distance, so it makes the optimal choice of two units to remove. As a result, four pairs are created under the optimal nonbipartite match and they are $\{[1,4],[2,3],[5,7],[6,8]\}$ with a total distance of $2+2+0+0=4$. The two sinks take away two original units, so the final match is $\{[1,4],[2,3]\}$. This is the optimal set of four original units if two units have to be removed. The matching quality is much improved with a maximal individual edge distance of 2 versus 10 from the previous match.

12.3 Illustrative Examples

In this section, we showcase the utility of the optimal nonbipartite matching on improving statistical inference by reviewing four examples. The first example directly utilizes the nonbipartite matching to infer media campaign effects when there are multiple ordinal dose groups. The second example uses the nonbipartite matching to construct an exact test comparing two multivariate distributions based on adjacency. The third example applies this algorithm to strengthen the instrumental variable approach for healthcare program evaluation, which is also known as near-far matching in Baiocchi et al. [14]. The last part is a brief review on applying the optimal nonbiparite matching algorithm to conduct multivariate matching before randomization, which provides the practitioners with a powerful tool to better balance many variables at the same time, hence to improve the experiment design.

12.3.1 The impact of the office of national drug control policy media campaign

Evaluating media campaigns or marketing strategies is challenging due to the fact that not only the exposure to media is self-selected, but there is also no natural control group if the large scale campaign is carried out across the country. An example of such study is the nationwide media campaign launched by the United States Office of National Drug Control Policy (ONDCP) to reduce illegal drug use by young Americans. Since almost all teenagers with access to TV, radio, newspaper, etc., were exposed to the media campaign at some degree, a sensible evaluation approach is to compare teens who received different exposures to the campaign, but who were similar in terms of baseline characteristics. It is referred to as a matching-with-doses design, which forms pairs with very different doses of treatment in such a way that the final contrast groups with high versus low dose have similar distributions of observed covariates. Then it resembles a randomized study with high/low dose assignment, barring any unmeasured confounding.

To evaluate the impact of this antidrug media campaign, Lu et al. classified 521 teenagers, from a pilot dataset of this study, into five ordinal dose groups based on how often they reported seeing anitdrug commercials (1 denoting least exposure and 5 for most exposure) [10]. The following ordinal logit model was used to estimate the five-level dose distribution:

$$\log \frac{Pr(Z_k \geq d)}{Pr(Z_k < d)} = \theta_k + \beta^T x_k, \quad \text{for } d = 2, 3, 4, 5,$$

where Z_k is the dose level. Note that the distribution of doses depended on the observed covariates only through the propensity score component, $e(x_k) = \beta^T x_k$. So the maximum likelihood estimtate $\hat{\beta}^T x_k$ was used in the matching to balance covariate distributions. If a dose-response relationship existed, subjects with similar covariates but very different dose exposures were more likely to show significant effects. Therefore, in matching with doses, the goal was not only to balance the observed covariates, but also to produce pairs with very different doses. The following distance metric was constructed to combine both covariates and dose information:

$$\Delta(x_i, x_j) = \frac{(\hat{\beta}^T x_i - \hat{\beta}^T x_j)^2 + \epsilon}{(Z_i - Z_j)^2},$$

where $\epsilon > 0$ was a sufficiently small positive number that would not affect the optimal matching result. It served two purposes: first, it ensured the distance between two subjects with the same dose equal ∞, regardless of covariate values; second, for subjects with identical covariates, the distance became smaller as the dose discrepancy increased.

Since there are multiple dose groups, the conventional bipartite matching algorithm does not work. The optimal nonbipartite matching is used to create pairs with similar covariates but different dose levels. One technical issue is that, when the sample size is odd, a sink needs to be used to make the total size an even number to facilitate matching. The sink unit has zero distance to all 521 original subjects and infinity distance to itself. Following the procedures described in Lu et al. [10], 260 teen pairs were formed and one subject was discarded because he was matched to the sink. All 22 covariates were balanced after matching, as no significant differences were observed between high- and low-dose groups. Table 2 from the original paper (refer to the original paper for table content) presents the number of pairs for each high-low dose combination, e.g., there are 14 pairs in which the high dose teen had a dose level of 5 and the low dose teen had a dose level of 1. About 40% of the pairs had dose difference of one, and another 40% of the pairs had dose difference of two.

Lu and colleagues analyzed the pilot data based on 251 pairs (those who had complete information on outcomes) to illustrate the methodology, and also acknowledged that this data might be too limited in scope and too unrepresentative of the whole sample. There were four questions about intention for drug use over the next year, two about marijuana and two about inhalants. All four questions used the same four point scale: (1) I definitely will not, (2) I probably will not, (3) I probably will, and (4) I definitely will, which was coded as 1-4 indicating an increasing likelihood of drug use. For each question, they calculated the difference of response between the two subjects in each matched pair. If the high dose teen reported "I definitely will not" or 1 and the low dose teen reported "I definitely will" or 4, then the difference would be $1 - 4 = -3$. Generally, negative values signified greater intentions to use drugs by the low-dose teen in a pair, and positive values signified the opposite. The Wilcoxon signed rank test was applied to each of the four questions, and none showed significant results. Therefore, there was no evidence that dose level was associated with stated drug use intention. They also considered a strengthened test by combining the results from the four questions and adding the corresponding signed rank statistics [15]. This combined test also showed no significance.

12.3.2 The cross-match test

It is an important statistical problem to compare two groups of data at the distribution level. For example, if the treated group has a distribution of $F(\cdot)$ and the control group has a distribution of $G(\cdot)$, the null hypothesis is phrased as: $H_0 : F(\cdot) = G(\cdot)$. If the response is a scalar with continuous values, this yields the conventional, unidimensional two-sample comparison problem. The popular choice of exact distribution-free tests for such one-dimensional problem includes the Wilcoxon rank sum test, the Kolmogorov-Smirnov (KS) test, etc. But when the response is multidimensional with no clear way of ordering them, the problem becomes quite challenging.

Rosenbaum used the optimal nonbipartite matching to construct an exact distribution-free test for comparing whether the outcomes from two populations followed the same distribution [16]. His motivating example compared brain activities during language tasks in subjects with and without arteriovenous abnormalities. The brain activities were measured by functional magnetic resonance imaging (FMRI) and there were multiple continuous measures (one for each task) for each subject. The measures were on the relative activation in left and right hemispheres. Since there was no easy way to come up with a composite index to rank these measures, Rosenbaum proposed a new cross-match test based on interpoint distances.

To implement the test, observations from both groups were pooled together and the grouping information was tentatively ignored. The optimal nonbipartite matching was used to create matched pairs among all observations based on the Mahalanobis distance (without grouping). After matching, the original grouping information was revealed. Matched pairs were classified into three categories based on their compositions, namely, pairs consisting of treated subjects only, pairs consisting of control subjects only, and pairs consisting of one treated and one control subject. If the two groups had the same distribution, one would expect to see a good mix of the three types of pairs, as the treated and control groups mingled well with each other. A substantially lower number in the third category would provide evidence against the null hypothesis that the two groups had exactly the same distribution. With I matched pairs ($N = 2I$ subjects), denote A_k be the number of pairs with exactly k treated subjects, $k = 0, 1, 2$. The null hypothesis of no difference was tested by using A_1. Specifically, the null distribution of A_1 was given by

$$Pr(A_1 = a_1) = \frac{2^{a_1} I!}{\binom{N}{n} a_0! a_1! a_2!}$$

where $a_k \geq 0$ was the number of pairs with type A_k and n was the number of treated subjects, where $a_1 + 2a_2 = n$. Rosenbaum also derived the normal approximation version of the exact test and compared the test with the well-known KS test in a simulation study, which showed both tests had similar type-I errors but the cross-match test had higher powers. It was shown that the cross-match test was consistent for comparing any two discrete distributions with finitely many mass points. It could also be extended to comparing continuous distributions given that the distributions can be approximated well by discrete distributions with finitely many mass points.

12.3.3 Strengthening instrumental variable analyses, near-far matching

An instrument is a random influence towards acceptance of a treatment that affects outcomes only through treatment acceptance. In settings in which treatment assignment is deliberate, there may still exist some essentially random influences to treatment acceptance. Such random component of treatment assignment can be extracted to obtain unbiased causal effect estimation. An instrument is weak if such random influences barely affect treatment assignment, or strong if they are critical in determining the treatment. Weak instruments are more susceptible to bias when IV analysis assumptions are violated [17].

Baiocchi et al. tried to address a causal question regarding regionalization of intensive care for premature infants, using an IV approach. Hospitals varied in their ability to care for premature infants, so regionalization of care suggested that high-risk mothers deliver at hospitals with greater capabilities [18]. Under such system, within a region, mothers were sorted into hospitals of varied capability based on the risks faced by the newborn, rather than affiliation or proximity. In practice, it was not clear whether delivering high risk infants at more capable hospitals could actually reduce mortality, as sorting by risk might be too inaccurate to affect health. Since a randomized study was deemed infeasible, Baiocchi and colleagues decided to exploit the IV approach by considering proximity as the instrument. A high-risk mother was more likely to deliver at a high-level hospital if it was close to home. The travel time to a high-level hospital was likely to affect the outcome only through the way it might alter whether the mother delivered at that hospital. If this was true, proximity would be an instrument for care at high-level hospitals. The mother's risk, however, might be related to geography, largely through socioeconomic factors that vary with geography, which would consequently violate the assumption of the IV approach. To tackle this challenge, the authors adopted the nonbipartite matching design to build a stronger instrument which was more robust to potential violation of IV assumptions.

Baiocchi and colleagues used a Pennsylvania dataset consisting of nearly 200,000 premature births in years 1995–2005. The hospitals were classified as high-level and low-level. Travel time was calculated as the time from the centroid of mother's zip code to the closest low-level and high-level hospitals. The degree of encouragement to deliver at a low-level hospital was the difference in travel times, high-minus-low, which was termed as excess travel time. Distance strongly encouraged the mother to deliver at a low-level hospital if the excess travel time was positive and large. To pair similar mothers together while allowing as large excess travel time as possible, the discrepancy metric (using the term "discrepancy" to avoid confusion with geographical distance to hospitals) was defined with two key components. The first component measured the discrepancies between mothers' characteristics and the second component was about excess travel time. The construction of discrepancy matrix took several steps, as the authors tried to balance many covariates, including year of birth, gestational age, socioeconomic status, numbers of congenital disorders, etc. (detailed description could be found in their paper) To explore the potential impact of weak versus strong instrument, they also considered two levels of excess travel time discrepancies. A substantial penalty was added to the discrepancy between any pair of babies whose excess travel time differed in absolute value by at most Λ, with $\Lambda = 0$ in the first match and $\Lambda = 25$ minutes in the second match.

When $\Lambda = 0$, it allowed any pair of babies to match as long as their excess travel times were not the same. When $\Lambda = 25$, it had a more stringent requirement that only pairs with 25 minutes or more excess travel time difference could be matched. The first match was likely to yield a weaker instrument with much larger sample size since it had minimum requirement on the instrument. The second match would yield a stronger instrument with much smaller sample size as the two babies in any pair needed to have excess travel times at least 25 minutes apart. With excess travel time being the "treatment," there was no clear cut for two treatment groups. Instead, babies had various doses of "treatment," so the optimal nonbipartite matching was utilized to form pairs. Note that sinks were used to take away unmatchable babies in the second match. The post-matching balance checking revealed that the first match (99,174 pairs of babies) produced pairs with very good balance but small mean excess travel time difference, which implied a weak encouragement to deliver at low-level hospitals. The second match showed bigger mean excess travel time difference, but the covariate balance was not acceptable. Therefore, they considered a third match called Half-25, in which half of the babies were required a difference in excess travel time of 25 minutes. Half-25 match (49,587 pairs of babies) turned out to work really well with very good covariate balance and much bigger mean excess travel time difference, which indicated a stronger instrument. Table 2 from the original paper (refer to the original paper for table content) showed the magnitude of encouragement, use of low-level hospital, and mortality in two matched comparisons. The stronger instrument match produced more encouragement as the travel time difference was bigger than the weaker instrument match. The encouragement seemed to work in the sense that more mothers delivered at the low-level hospital in the far group where the excess travel time to high-level hospital was large. But it did not seem to produce much effect on infant mortality.

As discussed by Angrist et al., a meaningful causal effect estimator would be the effect ratio, which was the ratio of two average treatment effect – the effect of distance on mortality over the effect of distance on where mother delivered [19]. It was interpreted as the average increase in mortality caused by delivering at the low-level hospitals among mothers who would deliver at a low-level hospital if and only if no high-level hospital was close by. Baiocchi and colleagues developed the inferential procedure for this effect ratio estimator in matched observational studies and applied it to the perinatal care study. The point estimates were quite similar under both weaker and stronger instruments (0.0092 vs. 0.009) and both were significant. But the length of 95% confidence interval of the stronger instrument was about half of that for weaker instrument, which made the estimate more robust. Overall, the evidence suggested a nearly one percent increase in mortality due to the use of low-level hospitals, provided that there were no unmeasured confounders and the IV assumption held.

12.3.4 Matching before randomization

Blocking or pairing before randomization is a fundamental principle of experimental design. But for most practices, this principle is only applied to one or two blocking variables. The optimal nonbipartite matching algorithm makes it easy to conduct multivariate matching prior to randomization to improve the balance for many covariates at the same time. Greevy and colleagues presented an algorithm for its implementation and a case-study of its performance [20]. Suppose there are $2n$ subjects to be randomized in an experiment. The naive complete randomization may not produce well balanced covariates between treated and control groups, especially for some covariates with low prevalence. The optimal nonbipartite matching divides the $2n$ subjects into n pairs of two subjects to minimize the sum of the n distances within pairs, then the randomization is applied within each pair. Such matching before randomization ensures the comparability of two treatment arms, avoiding the disasters occasionally experienced in unmatched randomization. To evaluate its empirical performance, Greevy and colleagues conducted simulations using 132 patients with 14 covariates from an actual unmatched randomized experiment. Under repeated randomizations with and without matching, they found that every one of the 14 covariates was substantially better balanced when randomization was performed within matched pairs. Even after covariance adjustment for chance imbalances in the 14 covariates, matched randomizations provided more accurate estimates than unmatched randomizations, the increase in accuracy being equivalent to a 7% increase in sample size, on average. In randomization tests of no treatment effect, matched randomizations using the signed rank test also showed higher power than unmatched randomizations using the rank sum test.

12.4 Discussion

Nonbiparite matching, also known as matching without groups [21], extends the conventional matching between two groups to a more flexible structure, hence provides many additional options for matched designs in both observational studies and experiments. In randomized studies, when all subjects are set to be randomized at the same time, nonbipartite matching can be used to conduct matching before randomization to improve covariate balance, which will result in efficiency gain in the analysis [20]. With observational data, the nonbipartite matching can form pairs between multiple ordinal dose groups, multiple unordered groups, or within a dataset without specific grouping [22,23]. Such design can also be used in longitudinal data, when the treatment assignment may depend on time-varying covariates [24]. This is referred to as risk-set matching , which will be covered in detail in chapter 9.

A technical note is that the nonbiparite matching still forms matched pairs, even though there are multiple treatment groups. To make a direct comparison across multiple groups, polygon-shaped matched sets with one subject from each treatment group may be more desirable. This is referred to as poly-matching in the literature [25], where the optimal solution is an NP-hard problem. Approximate optimal solutions based on either bipartite or nonbipartite matching algorithms are available [26–28].

12.5 Acknowledgement

This work was partially supported by grant DMS-2015552 from National Science Foundation. This work was also supported, in part, by the National Center for Advancing Translational Sciences of the National Institutes of Health under Grant Number UL1TR002733. The content is solely the responsibility of the authors and does not necessarily represent the official views of National Science Foundation and National Institutes of Health.

References

[1] W.C. Cochran and S.P. Chambers. The planning of observational studies of human populations. *Journal of the Royal Statistical Society, Series A*, 128:234–255, 1965.

[2] R.A. Fisher. *The Design of Experiments*. Edinburgh: Oliver & Boyd, 1935.

[3] B.B. Hansen and S.O. Klopfer. Optimal full matching and related designs via network flows. *Journal of Computational and Graphical Statistics*, 15:609–627, 2006.

[4] D. Ho, K. Imai, G. King, and Stuart E. Matching as nonparametric preprocessing for reducing model dependence in parametric causal inference. *Potilical Analysis*, 15:199–236, 2007.

[5] E.A. Stuart and D.B. Rubin. Best practices in quasi–experimental designs: matching methods for causal inference. In Osborne, J. (Ed.), *Best Practices in Quantitative Methods*, 155–176, SAGE Publications, Inc., 2008.

[6] R.A. Greevy Jr, C.G. Grijalva, C.L. Roumie, C. Beck, A.M. Hung, H.J. Murff, X. Liu, and M.R. Griffin. A non-bipartite propensity score analysis of the effects of teacher–student relationships on adolescent problem and prosocial behavior. *Journal of Youth and Adolescence*, 46:1661–1687, 2012.

[7] P.R. Rosenbaum and D.B. Rubin. The central role of propensity score in observational studies for causal effects. *Biometrika*, 70:41–55, 1983.

[8] C.H. Papadimitriou and K. Steiglitz. *Combinatorial Optimization: Algorithms and Complexity*. Englewood Cliffs: Prentice Hall, 1982.

[9] K. Ming and P.R. Rosenbaum. A note on optimal matching with variable controls using the assignment algorithm. *Journal of Computational and Graphical Statistics*, 10:455–463, 2001.

[10] B. Lu, E. Zanutto, R. Hornik, and P.R. Rosenbaum. Matching with doses in an observational study of a media campaign against drug abuse. *Journal of the American Statistical Association*, 96(456):1245–1253, December 2001.

[11] B. Lu, R. Greevy, X. Xu, and C. Beck. Optimal nonbipartite matching and its statistical applications. *The American Statistician*, 65:21–30, 2011.

[12] U. Derigs. Solving non-bipartite matching problems vis shortest path techniques. *Annals of Operations Reserach*, 13:225–261, 1988.

[13] W. Cook and A. Rohe. Computing minimum-weight perfect matchings. *INFORMS Journal on Computing*, 11:138–148, 1999.

[14] M. Baiocchi, D. Small, L. Yang, D. Polsky, and P. Groeneveld. Near/far matching: a study design approach to instrumental variables. *Health Services and Outcomes Research Methodology*, 12:237–253, 2012.

[15] P.R. Rosenbaum. Signed rank statistics for coherent predictions. *Biometrics*, 53:556–566, 1997.

[16] P.R. Rosenbaum. An exact distribution-free test comparing two multivariate distributions based on adjacency. *Journal of the Royal Statistical Society, Series B*, 67:515–530, 2005.

[17] J. Bound, Jaeger D.A., and Baker R.M. Problems with instrumental variables estimation when the correlation between the instruments and the endogenous explanatory variable is weak. *Journal of the American Statistical Association*, 90:443–450, 1995.

[18] M. Baiocchi, D. Small, S. Lorch, and P.R. Rosenbaum. Building a stronger instrument in an observational study of perinatal care for premature infants. *Journal of the American Statistical Association*, 105:1285–1296, 2010.

[19] J.D. Angrist, Imbens G.W., and Rubin D.B. Identification of causal effects using instrumental variables. *Journal of the American Statistical Association*, 91:444–455, 1996.

[20] R. Greevy, B. Lu, J. Silber, and P.R. Rosenbaum. Optimal multivariate matching before randomization: An algorithm and a case-study. *Biostatistics*, 5:263–275, 2004.

[21] P.R. Rosenbaum. *Design of Observational Studies*. New York: Springer, 2010.

[22] B. Lu and P.R. Rosenbaum. Optimal pair matching with two control groups. *Journal of Computational and Graphical Statistics*, 13:422–434, 2004.

[23] I. Obsuth, A. L. Murray, T. Malti, P. Sulger, D. Ribeaud, and M. Eisner. A non-bipartite propensity score analysis of the effects of teacher–student relationships on adolescent problem and prosocial behavior. *Journal of Youth and Adolescence*, 46:1661–1687, 2017.

[24] B. Lu. Propensity score matching with time-dependent covariates. *Biometrics*, 61:721–728, 2005.

[25] G. Nattino, C. Song, and B. Lu. Poly-matching algorithm in observational studies with multiple treatment groups. *Computational Statistics & Data Analysis*, 167:107364, 2022.

[26] B. Karmakar, D. Small, and P.R. Rosenbaum. Using approximation algorithms to build evidence factors and related designs for observational studies. *Journal of Computational and Graphical Statistics*, 28:698–709, 2019.

[27] M. Bennett, Vielma J.P., and Zubizarreta J.R. Building representative matched samples with multi-valued treatments in large observational studies. *Journal of Computational and Graphical Statistics*, 29:744–757, 2020.

[28] G. Nattino, B. Lu, J. Shi, S. Lemeshow, and H. Xiang. Triplet matching for estimating causal effects with three treatment arms: A comparative study of mortality by trauma center level. *Journal of the American Statistical Association*, 116:44–53, 2021.

13

Matching Methods for Large Observational Studies

Ruoqi Yu

CONTENTS

13.1 Introduction: Challenges and Opportunities for Large Datasets

As technologies have rapidly developed in recent decades, data sets have grown in size and become more accessible for analysis, e.g., electronic health records, medical claims data, educational databases, and social media data. For instance, using data from US Medicare, Kelz et al. [1] formed 18,200 matched pairs of two patients to compare surgical outcomes treated by new and experienced surgeons. Silber et al. [2] used Pediatric Health Information System (PHIS) data to build 23,582

DOI: 10.1201/9781003102670-13

matched pairs of two children, comparing the surgical outcomes of children on Medicaid and other health insurance.

The increasing sample size has posed tremendous challenges to matching in observational studies in recent decades. Commonly used methods to construct matched observational studies include propensity scores [3, 4], prognostic scores, fine balance and related techniques [4–6], minimum cost flow algorithms that minimize the total distance within matched sets [7, 8], and mixed-integer programs that directly target the covariate balance [9]. These techniques have satisfactory performance for small or moderate data sets with thousands of people but confront computational difficulties for larger data sets consisting of millions of individuals. How can we build matched samples for large-scale observational studies efficiently without detracting the appealing balance properties?

A simple and common practice divides a large sample with millions of people into smaller groups by matching exactly on a few discrete or rounded covariates; then, each subgroup consisting of thousands of individuals is considered separately for matching. This approach is reasonable and practical, but it has some unattractive aspects. First, exact matching in groups gives overriding importance to the covariates that define the groups, which may have no scientific basis. Due to this hard constraint of prioritization, other covariates of equal importance may not have the chance of achieving adequate balance. In addition, cross-group matches can sometimes be closer than within-group matches. Categorizing continuous covariates to form exact-match groups forbids close matches that cross the category boundaries, risking more significant gaps inside categories. Furthermore, without carefully dividing the whole sample, some groups may still be too large to match, whereas some other groups need to be merged before the group is large enough for matching. As a result, creating groups of practical size can be subjective to investigators who construct the match, leading to unreproducible analysis. More importantly, matching everyone at once brings substantial statistical advantages that other matching techniques can be applied on a larger scale, improving their effectiveness. For instance, a matching technique called "fine balance" tries to make groups comparable by counter-balancing without constraining individuals in each matched set to have the same covariate values [5]. Splitting people into groups unnecessarily limits the performance of fine balance, as the more people considered for matching simultaneously, the more opportunities fine balance has to work well [12]. Can we overcome these limitations?

This chapter is organized as follows. Section 13.2 focuses on size-scalable matching techniques for the common case of optimal pair matching. Section 13.3 is devoted to incorporating other matching techniques when the sample size is large. In Section 13.4, we briefly discuss several different matching designs – matching with multiple controls, matching without pairing, full matching, and coarsened exact matching.

13.2 Optimal Pair Matching

13.2.1 Challenges of working with dense graphs

Suppose there are T treated units, $\mathcal{T} = \{\tau_1, \dots, \tau_T\}$, and C potential controls, $\mathcal{C} = \{\gamma_1, \dots, \gamma_C\}$, where $\mathcal{T} \cap \mathcal{C} = \emptyset$, and $T \leq C$. Let $|\cdot|$ denote the number of elements of a finite set, then $|\mathcal{T}| = T$ and $|\mathcal{C}| = C$. In pair matching without replacement, we would like to match every treated unit $\tau \in \mathcal{T}$ to a similar control $\gamma \in \mathcal{C}$ such that each potential control is used at most once. Optimal matching, first proposed by Rosenbaum [8], aims to build matched sets with similar units by minimizing total distances among all matched sets. Finding an optimal match is equivalent to solving a minimum cost network flow problem, whose computational complexity depends on the sample size and the number of candidate pairs under consideration [10].

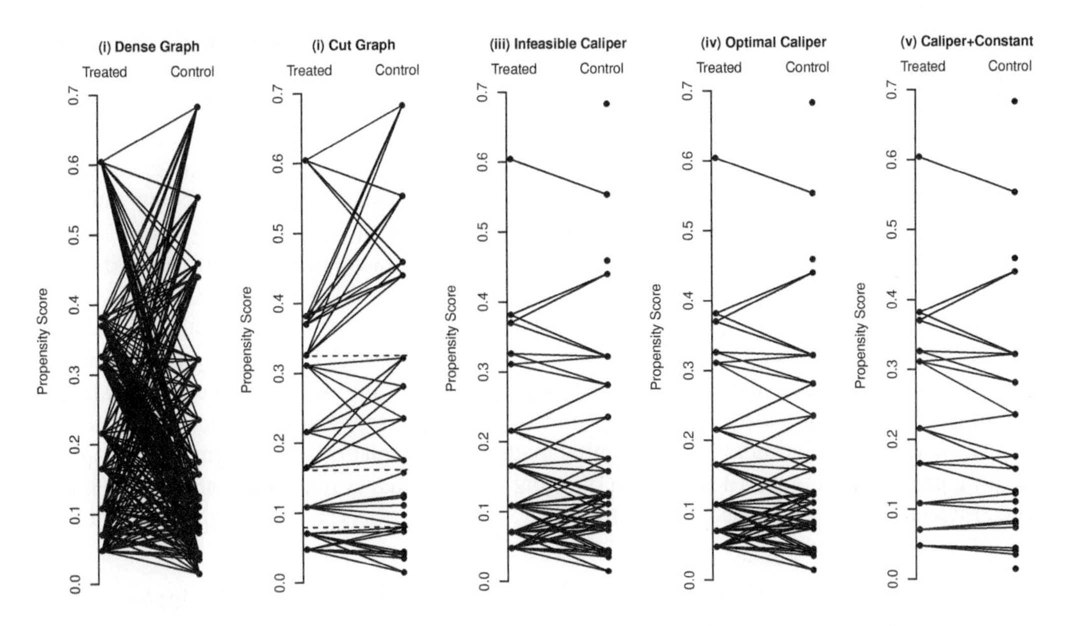

FIGURE 13.1
Five bipartite graphs with different numbers of edges, where the vertical axis is the propensity score. Graph (i) is a dense graph containing all possible pairings. Graph (ii) cuts the graph into four parts based on propensity score quartiles. Graph (iii) has a caliper that is too small, so pair matching is not feasible. Graph (iv) uses the smallest caliper that permits a feasible pair matching. Graph (v) has the smallest caliper and the smallest upper bound on the number of closest candidate controls for treated units.

When the treatment W is binary, the main part of the network is a bipartite graph consisting of two sets of nodes corresponding to the treated and control groups. Each edge in the bipartite graph connects nodes in different sets, representing the binary decision variable: Should this treated unit be matched to this connected control unit or someone else? Each edge has a cost attached to it, which is a distance measuring the similarity between any treated-control pair in terms of observed covariates. The choice of distance may involve the propensity score, a Mahalanobis distance or its variation, and other considerations. Traditional optimal matching considers all possible candidate treated-control pairs, therefore deals with a dense graph with edges connecting each treated node to all control nodes. As a result, it takes $O(C^3)$ steps to find an optimal matching [10], which can be computationally intensive for large observational studies.

More concretely, consider a toy example in Figure 13.1, omitting for the moment incorporating other matching techniques. The example uses a random sample of the public data from the 2005-2006 National Health and Nutrition Examination Survey (NHANES), consisting of 10 daily smokers as the treated group and 20 nonsmokers as potential controls. This data can be obtained from the `R` package `bigmatch`, following the command "`set.seed(79); nhs<-nh0506[sample(1:(dim(nh0506)[1]),30),]`", using `R` version 4.1.0. For a much larger real application using data from US Medicaid, see a recent paper by Yu et al. [11].

In Figure 13.1, nodes represent 30 people in the toy example, and edges represent candidate matches. Each panel of Figure 13.1 is a so-called bipartite graph – nodes are divided into two parts for $T = 10$ treated and $C = 20$ controls, and each edge connects a treated node to a control node. Figure 13.1(i) contains all possible $200 = 10 \times 20$ edges, so it is a complete and dense bipartite graph.

In other words, the edge set $\mathcal{B}$ is the direct product, $\mathcal{B} = \mathcal{T} \times \mathcal{C}$, so $|\mathcal{B}| = T \times C$. In the simplest case of pair matching, we would like to pick 10 edges that do not share a node in Figure 13.1(i). The problem is not trivial because two treated nodes may want the same potential control, so pairing each treated node to the closest control may not produce a feasible match. In contrast, Figure 13.1(i) entails an optimization problem with $200 = 10 \times 20$ binary decision variables subject to constraints requiring 10 non-overlapping treated-control pairs. Following this idea, optimal matching with 1000 treated units and 2000 potential controls utilizes Figure 13.1(i) containing $2 \times 10^6 = 1000 \times 2000$ edges, a reasonable size in practice. If we had a small administrative database with 30,000 treated units to 60,000 controls, Figure 13.1(i) would have $1.8 \times 10^9 = 30,000 \times 60,000$ edges. In this case, optimal matching using Figure 13.1(i) would not be practical with current computational tools.

It is easy to see that many of the $200 = 10 \times 20$ possible pairings have poor quality, therefore, do not deserve serious consideration. For instance, we do not want to match the upper left treated node to the lower right control node because their propensity scores are far apart. Motivated by this observation, one solution to make matching methods size-scalable is to reformulate the optimal matching problem in Figure 13.1(i) into a different problem that is reasonable and can be solved more efficiently. How can we accomplish this?

13.2.2 A practical solution with unattractive limitations: Dividing the large population into smaller groups

Recall from the previous section that the computational difficulty of optimal matching $O(C^3)$ grows much faster than linearly with the number of candidate pairs under consideration, i.e., the number of edges in the bipartite graph. Therefore, it would be much easier to solve many small problems with fewer edges than one single large problem with more edges. Figure 13.1(ii) demonstrates a toy version of the common practice in large data sets motivated by this fact. Specifically, the whole population of 30 people is split into four groups based on propensity score quartiles – essentially the horizontal dashed lines in Figure 13.1(ii). All edges crossing a dashed horizontal line are removed, and a treated subject is only connected to controls in the same group. As a result, there are 70 edges or decision variables in Figure 13.1(ii), rather than 200 in Figure 13.1(i). In addition, Figure 13.1(ii) reduces the original optimal matching problem to four unrelated small problems that can be solved independently one by one. For instance, there are four treated units and four controls in the top quartile of Figure 13.1(ii). We would like to select one of the $4! = 24$ possible pairings to minimize the total distance within the four pairs. Suppose we divide 30,000 treated individuals and 60,000 potential controls into 30 groups of 1000 treated units and 2000 controls. Then a graph analogous to Figure 13.1(ii) consists of 30 separate subproblems each with $2 \times 10^6 = 1000 \times 2000$ edges or decision variables. Although one would need to solve 30 problems rather than one, each problem could be solved in a reasonable amount of time.

The idea of splitting into subpopulations based on observed covariates is practical and reasonable, but it can restrict the possible matches in undesirable ways. Some of the limitations are already visible in the toy example. First, the top group in Figure 13.1(ii) has four treated units and four potential controls, which forces us to use all four controls, only leaving matching who to whom open. As a result, matching does not reduce the bias in this group. In addition, no control in the top group is close to the bottom treated unit with a propensity score of 0.325. However, this treated unit must be matched to one of the four controls as only four controls are available in this group. Although this treated unit is very close to a control unit barely on the opposite side of the group boundary and the second group has enough potential controls, the quartile dividers do not permit the cross-bin matches. A worse situation would occur if one of the four controls were not in the top group. Then it would leave four treated units and three controls in this bin, so matching all 10 treated units to 10 distinct controls would be impossible with the group dividers in Figure 13.1(ii). Although pair matching of all treated units might be feasible in Figure 13.1(i), but subdividing the population might make it infeasible. In addition, splitting into bins can limit the effectiveness of other matching

TABLE 13.1
Counts for Hispanic or not in four groups split based on propensity quartiles.

Propensity Score	$[0, 0.080]$		$(0.080, 0.161]$		$(0.161, 0.325]$		$(0.325, 1]$	
Hispanic	Yes	No	Yes	No	Yes	No	Yes	No
Treated	1	1	1	0	0	3	0	4
Control	5	1	0	6	1	3	0	4

techniques. Consider fine balance of the indicator for being Hispanic as an example. As shown in Table 13.1, this toy dataset has two Hispanic units and eight non-Hispanic units in the treated group, plus six Hispanic units and 14 non-Hispanic units in the control group. Therefore, it is possible to select two Hispanic and eight non-Hispanic units from the 20 potential controls to finely balance the marginal distribution of the indicator of being Hispanic. However, in the group with propensity score in $(0.080, 0.161]$, the only treated unit is Hispanic, whereas all six controls are not. As a result, fine balance of the Hispanic indicator is impossible in this quartile. Splitting into groups unnecessarily limits what fine balance can do [12].

13.2.3 A new approach: sparsifying the network

13.2.3.1 Optimal caliper

Caliper matching [12] is a common matching technique to improve the balance of a covariate. For instance, we could impose a caliper of $x \pm 2$ on age x if we would like to match a treated unit to any control who is either up to two years older or two years younger. Rosenbaum and Rubin [3] advocated a combined matching strategy of using the Mahalanobis distance within a propensity score caliper, which guarantees a close match on the propensity score and seeks a close match on other covariates if several such matches exist. In general, a caliper is a function $\kappa : \mathbf{R} \longmapsto \mathbf{R}^2$ mapping x to $\kappa_1(x) \leq \kappa_2(x)$ such that a treated subject with covariate value x may be matched to any control with covariate values in the interval $[\kappa_1(x), \kappa_2(x)]$. A common choice is a symmetric caliper that $[\kappa_1(x), \kappa_2(x)] = [x - w, x + w]$ for some fixed $w \geq 0$, such as the two-year caliper on age. Alternatively, we could use different calipers for subjects with various characteristics. For example, we might prefer a very short caliper for very young children and a longer caliper for people in middle age since the difference in body and brain development between a 1-year old and a 3-year old is much larger than that between a 32-year old and a 34-year old. Another popular choice is to use an asymmetric caliper [13] to offset the bias in the study. For instance, if treated subjects are typically older than controls, we might prefer an asymmetric caliper $[x - 1, x + 3]$ to counteract the tendency to pairing a younger control to an older treated individual. We focus on the symmetric caliper in the current section and discuss other variations of calipers in Section 13.3.

A caliper can eliminate some edges from the dense graph in Figure 13.1(i) but need not produce disconnected components as in Figure 13.1(ii). Narrower calipers remove more edges or decision variables to accelerate the computation but risk the feasibility of matching if the caliper is too small. Figures 13.1(iii) and (iv) impose symmetric propensity score calipers of 0.0763 and 0.0764, respectively, in Figure 13.1(i). Despite the tiny discrepancy (less than 0.001) between these two calipers, Figures 13.1(iii) and 13.1(iv) differ from a crucial perspective. The caliper of 0.0763 is too small that no pair matching of distinct individuals exists since the four treated units with propensity scores in the range $(0.3, 0.4)$ compete for three controls. Although the caliper in Figure 13.1(iv) is only a little larger, it permits a feasible matching. Yu et al. [11] defined the optimal caliper of the form $\pm w$ as the smallest caliper $w \geq 0$ such that pair matching is feasible. Then the optimal caliper w in Figure 13.1(i) belongs to the short interval $[0.0763, 0.0764]$, where the upper bound of 0.0764 is feasible. By applying a caliper of 0.0764, Figure 13.1(iv) only has 54 edges connecting treated units and controls whose propensity scores are close.

Figure 13.1(iv) is more attractive than Figure 13.1(ii) because, unlike Figure 13.1(ii), there are no boundaries in Figure 13.1(iv) to prevent us from matching similar individuals. By applying an optimal caliper in a large dense bipartite graph, Figure 13.1(iv) removes the maximum number of edges that a caliper of the form $\pm w$ can possibly remove without creating infeasibility issues. We then use this new, sparser graph to minimize the total distance within matched pairs, constrained by the optimal caliper plus additional matching constraints. This revised problem can be solved more efficiently than the original optimal matching problem due to the reduced number of decision variables.

Suppose each individual has a real-valued score, $\rho : \mathcal{T} \cup \mathcal{C} \to \mathbf{R}$. How can we find an accurate estimate of the optimal caliper for $\rho(\cdot)$? Yu et al. [11] proposed a new technique to quickly find a short interval containing the optimal caliper by utilizing an iterative use of a variant of Glover's algorithm [14] for matching in a convex bipartite graph. In a doubly convex graph defined below, Lipski and Preparata [15] used a doubly-ended queue to efficiently implement Glover's algorithm with complexity $O(C)$, which runs much faster than minimum distance matching in either a dense or sparse graph. Specifically, for a bipartite graph $(\mathcal{T} \cup \mathcal{C},\ \mathcal{B})$ with nodes $\mathcal{T} \cup \mathcal{C}$ and edges $\mathcal{B} \subseteq \mathcal{T} \times \mathcal{C}$, we call it convex if it is possible to order the nodes of $\mathcal{T} \cup \mathcal{C}$ so that for each treated node $\tau \in \mathcal{T}$, all control nodes $\gamma \in \mathcal{C}$ connected to τ are consecutive. A convex bipartite graph $(\mathcal{T} \cup \mathcal{C},\ \mathcal{B})$ is doubly convex if it is also convex with the roles of $\mathcal{T}$ and $\mathcal{C}$ reversed.

To find the optimal caliper, first sort the nodes of $\mathcal{T}$ and $\mathcal{C}$ separately by the score $\rho(\cdot)$. Then any caliper $\varkappa > 0$ generates a doubly convex graph $(\mathcal{T} \cup \mathcal{C},\ \mathcal{B})$, with the edge set $\mathcal{B}$ only including an edge (τ, γ) if and only if $|\rho(\tau) - \rho(\gamma)| \leq \varkappa$. To determine the feasibility of a caliper $\varkappa > 0$, we apply Glover's algorithm in the corresponding doubly convex graph $(\mathcal{T} \cup \mathcal{C},\ \mathcal{B})$ to study whether pair matching is feasible in $\mathcal{B}$. Although Glover's algorithm can provide more information, this is all we need. More importantly, the time needed to find the optimal caliper is negligible compared with solving the minimum cost flow problem since both sorting and Glover's algorithm are much faster than finding an optimal match. We determine the optimal caliper $\varkappa$ by binary search with initial choices $\varkappa_{\min} = 0$ and $\varkappa_{\max} = \max_{\iota \in \mathcal{T} \cup \mathcal{C}} \rho(\iota) - \min_{\iota \in \mathcal{T} \cup \mathcal{C}} \rho(\iota)$, and a tolerance $\epsilon > 0$. Let $\text{glover}(\varkappa) = 1$ if pair matching is feasible in the bipartite graph $(\mathcal{T} \cup \mathcal{C},\ \mathcal{B})$ with caliper $\varkappa$ on the score $\rho(\cdot)$; otherwise, let $\text{glover}(\varkappa) = 0$. Then the optimal caliper can be estimated with error ϵ using the following algorithm:

1. If $\text{glover}(\varkappa_{\min}) = 1$, stop; pair matching is feasible with the smallest caliper $\varkappa = \varkappa_{\min} = 0$.

2. If $\text{glover}(\varkappa_{\max}) = 0$, stop; pair matching is infeasible for every $\varkappa$.

3. If $\varkappa_{\max} - \varkappa_{\min} < \epsilon$, stop; caliper $\varkappa_{\max}$ is feasible and within ϵ of the optimal caliper.

4. Define $\overline{\varkappa} = (\varkappa_{\max} + \varkappa_{\min})/2$. If $\text{glover}(\overline{\varkappa}) = 1$, set $\varkappa_{\max} \leftarrow \overline{\varkappa}$; if $\text{glover}(\overline{\varkappa}) = 0$, set $\varkappa_{\min} \leftarrow \overline{\varkappa}$. Go to step 3.

For the toy example in Figure 13.1, a binary search suggests $[\varkappa_{\min}, \varkappa_{\max}] = [0.0763, 0.0764]$. That is, pair matching is infeasible with caliper 0.0763 in Figure 13.1(iii) but feasible with caliper 0.0764 in Figure 13.1(iv). If $\rho(\cdot)$ is the propensity score, then $\varkappa_{\max} - \varkappa_{\min} \leq 1$, and the interval $[\varkappa_{\min}, \varkappa_{\max}]$ has length at most 2^{-I} after I iterations of steps 3 and 4. For instance, after $I = 10$ iterations, the length of the interval $[\varkappa_{\min}, \varkappa_{\max}]$ is no more than $2^{-10} = 0.000977 < 0.001$.

13.2.3.2 Optimal number of nearest neighbors inside a caliper

A limitation of Figure 13.1(iv) is that the treated units where the propensity score distributions in the treated and control groups overlap extensively still have many edges or decision variables. The larger the sample size, the more severe the issue is. Any subgraph of Figure 13.1(iv) maintains the optimal caliper of 0.0764, but not every subgraph would permit a feasible pair match; for instance, Figure 13.1(iii) is an infeasible subgraph of Figure 13.1(iv). How could we find a subgraph of Figure 13.1(iv) that removes more edges, maintains feasibility, and retains close candidate pairs?

Suppose that we would like to retain at most the ν nearest neighbors of each treated unit in Figure 13.1(iv). Formally speaking, for each treated subject $\tau \in \mathcal{T}$, sort $|\rho(\tau) - \rho(\gamma)|$ into increasing order, and define $o_v(\tau)$ to be the νth of the C sorted values of $|\rho(\tau) - \rho(\gamma)|$. We then define the bipartite graph $(\mathcal{T} \cup \mathcal{C}, \mathcal{B})$ so that it includes edge (τ, γ) in $\mathcal{B}$ if and only if $|\rho(\tau) - \rho(\gamma)| \leq \min\{\varkappa, o_v(\tau)\}$. How small can ν be to remain a feasible pair matching within an optimal caliper in Figure 13.1(iv)? It is clear from Figure 13.1(iii) that $\nu = 2$ is too small, because the four treated units with propensity scores in the range $(0.3, 0.4)$ are competing for three controls. A pair match is possible in Figure 13.1(v), indicating $\nu = 3$ is feasible. With the optimal caliper, $\varkappa$, estimated as above, we can determine the minimum feasible value of ν by a second iterative use of Glover's algorithm. That is, the first application of Glover's algorithm determines the optimal caliper, and then the second application determines the smallest feasible ν among subgraphs of the optimal caliper graph. By combining these two techniques, we can significantly reduce the number of edges in the bipartite graph. As demonstrated in Figure 13.1(v), some treated subjects are connected to less than $\nu = 3$ controls. For instance, the treated subject with the largest propensity score has only one neighbor, not $\nu = 3$ neighbors, because the caliper has sensibly eliminated distant controls as neighbors. As a result, Figure 13.1(v) has 25 edges. In a more realistic example with 30,000 treated units and 60,000 potential controls, the dense graph would have $1.8 \times 10^9 = 30,000 \times 60,000$ edges or decision variables. With $\nu = 100$, there would be $3 \times 10^6 = 30,000 \times 100$ edges or decision variables, comparable to a dense bipartite graph with one twentieth as many nodes or people.

Knowing the minimum feasible ν does not require investigators to use this minimal ν. Instead, it informs an important message that matching within an optimal caliper will remain feasible if attention is restricted to at most ν nearest neighbors. An intermediate graph with a slightly larger ν would offer more candidate pairs for matching, hence may lead to a smaller Mahalanobis distance on covariates other than the propensity score or other desired properties such as covariate balance. So considerations besides the caliper on the propensity score would have a substantial influence on the final match. For instance, a feasible graph satisfying the optimal caliper and with at most $\nu = 4$ nearest neighbors is an intermediate graph between Figure 13.1(iv) and Figure 13.1(v), with 5 more edges than Figure 13.1(v). If $(\mathcal{T} \cup \mathcal{C}, \mathcal{B})$ were complete with $\mathcal{B} = \mathcal{T} \times \mathcal{C}$, as in Figure 13.1(i), Theorem 9.13 of [10] gives a bound of $O\left(C^3\right)$ for optimal matching. On the other hand, even with growing values of ν, we have a time bound of $O\left\{C^2 \log(C)\right\}$ providing $\nu = O\{\log(C)\}$, which is much smaller than $O(C^3)$.

13.2.3.3 Practical considerations

These methods are implemented in an `R` package `bigmatch` available on `CRAN`. There are many minor but valuable variations on this theme. See §13.3 and Yu et al. [11] for more discussion. Essentially, the `bigmatch` package does two things.

First, the package creates a sparser but feasible graph for matching by producing a graph like Figure 13.1(v) rather than Figure 13.1(i); however, the actual graph is vastly larger in every sense than Figure 13.1(v). This step has two tasks: (i) pick a caliper on the propensity score (or some other score) yielding Figure 13.1(iv); then (ii) pick a limit ν on the number of nearest neighbors, moving from Figure 13.1(iv) to Figure 13.1(v). Although `bigmatch` removes most of the edges, it carefully avoids removing too many edges to ensure that matching is still possible. To avoid a large optimal caliper resulted from outliers among the scores, $\rho(\tau)$, $\tau \in \mathcal{T}$, investigators can replace the scores by their ranks. The `optcal` function in the `bigmatch` package returns the optimal caliper along with an interval indicating the precision of estimating the caliper. You need not use this caliper – you may use a larger one – but a smaller caliper does not permit a feasible pair matching. The `optconstant` function takes a caliper you specify and determines the minimum feasible value of ν, the upper bound on the number of nearest neighbors. You can use this minimum feasible ν or use a larger one, but no pair matching exists if you use a smaller ν. With the lower limits on the caliper and nearest neighbors, you are now ready for the next step.

Second, in a graph like Figure 13.1(v), the `bigmatch` package offers a suite of standard techniques for optimal matching in observational studies, such as propensity score calipers, near-fine balance, minimizing a robust covariate distance, exact matching, near-exact matching. For a review of these standard methods, see Part II of Rosenbaum's book [16]. Although there are various nonstandard implementations of these standard methods to avoid computing or storing information in Figure 13.1(i) that plays no role in Figure 13.1(v), these methods work in their usual way from the user's point of view. In this step, with a caliper and nearest neighbors restriction that are at least as large as the minimums determined in step one and your other matching requirements, the `nfmatch` function computes the optimal match subject to your specifications.

The minimum feasible caliper and constraint on the number of nearest neighbors yield a sparse graph and perhaps the fastest computation in step two. However, speed is one consideration among others. Setting the caliper and nearest neighbor restriction to be higher than their minimum feasible values gives `nfmatch` more options of searching for a close, balanced match, perhaps producing a better match in terms of covariate balance. It is reasonable to construct and compare a few matched samples and pick the most satisfactory one for the outcome analysis.

13.3 Incorporating Other Matching Techniques

13.3.1 Exact matching for a nominal covariate

Several important matching techniques can be adapted for use in sparsified networks. Figure 13.2 adds two features to the bipartite graphs in Figure 13.1 that improve covariate balance in matching. The first feature, exact matching for a nominal covariate, is discussed in the current section, and the second feature, near-fine balance, is discussed in the next section.

An exact matching constraint on a nominal variable $\xi(\cdot)$ requires that a treated unit can only be matched to any control with the same value of $\xi(\cdot)$. This constraint is feasible if there are at least as many controls as treated subjects for all possible values of $\xi(\cdot)$. We can implement the exact matching constraint in the network structure by removing the edges connecting the treated units and controls with different $\xi(\cdot)$ values. For instance, in the toy example in Figure 13.1, an exact match for gender is possible because there are three treated men and nine control men, and seven treated women and 11 control women. Figure 13.2 removes edges in Figure 13.1(i) connecting a treated man to a control woman or a treated woman to a control man, to achieve this requirement.

With fewer edges in the initial graph, the optimal caliper on the propensity score increases to 0.2133 in Figure 13.2 from 0.0764 in Figure 13.1(iv). Although the tightest feasible constraint on the number of nearest neighbors ν within the optimal caliper stays the same as $\nu = 3$ in Figure 13.1(v), the minimum feasible ν may have increased with other data sets due to this additional requirement. In Figure 13.2, all treated individuals are connected to no more than $\nu = 3$ controls with the same gender and differing on the propensity score by at most 0.2133. Notably, the caliper and ν are the smallest choices possible to permit a feasible pair matching.

Despite the additional requirement of exact matching on a nominal variable $\xi(\cdot)$, we can still apply Glover's algorithm to find the optimal caliper and the optimal number of nearest neighbors, with some modifications. We need to first sort all the individuals based on the exact matching nominal variable $\xi(\cdot)$, and within each exactly matched level, by the score $\rho(\cdot)$. For each caliper $\varkappa$, we form the edge set $\mathcal{B}$ by including $(\tau, \gamma) \in \mathcal{B}$ if and only if $\xi(\tau) = \xi(\gamma)$ and $|\rho(\tau) - \rho(\gamma)| \leq \varkappa$ for a fixed number $\varkappa > 0$, which produces a doubly convex bipartite graph. As a result, we can determine the smallest $\varkappa$ such that paring matching is feasible in $\mathcal{B}$ efficiently with Glover's method. A similar analogy applies to the minimal feasible number of nearest neighbors.

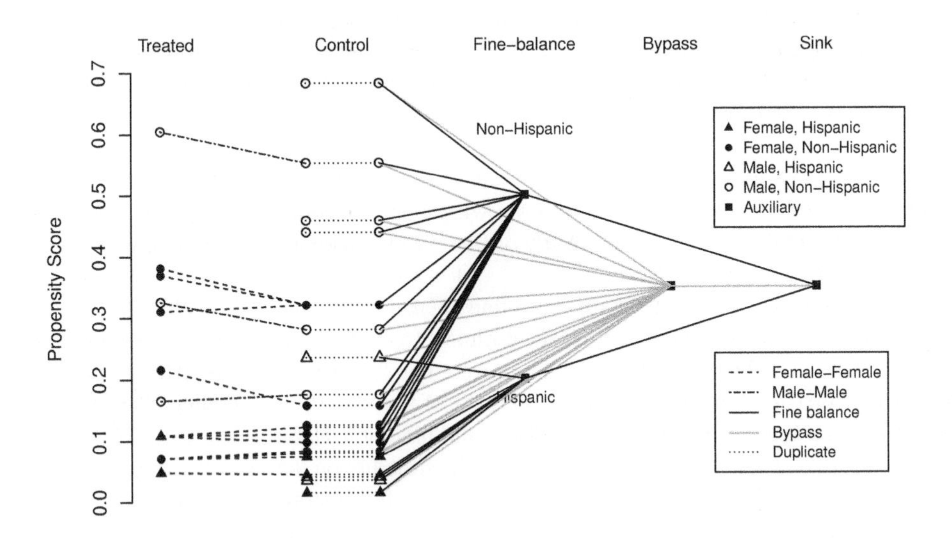

FIGURE 13.2
A bipartite graph matching exactly for gender, with near-fine balance of Hispanic or not. The optimal caliper is now 0.2133, and the minimum number of nearest neighbors is $\nu = 3$ with this caliper. The duplicate edges connect γ to γ', with capacity 1 to ensure that each control may be matched at most once. The solid grey edges retain feasibility through a penalized bypass, β, of the fine balance constraints.

The algorithm above finds a uniform caliper and constraint on the nearest neighbors for all exact matching categories. A straightforward but valuable variation is using different calipers and numbers of nearest neighbors for each category to account for the various natures of exact matching groups, e.g., hospitals or disease types. This trick can reduce more candidate pairs than a uniform choice which is equivalent to the most conservative option that is feasible among all categories.

Another generalization deals with the case when the nominal variable has a lexical order, e.g., the International Classification of Diseases (ICD-10) code for diagnosis or procedure and zip code. The hierarchical relationship helps control the degree of exact matching if it is infeasible match exactly on the original variable. For instance, if exact matching with the 5-digit zip code is not feasible for a particular area, we could match the first four digits or less in that area. On the other hand, for regions with more controls than treated subjects, we will keep using the 5-digit zip code. Generally speaking, investigators need to conduct several contingency tables to determine the degree of exact matching for each category (the optimal depth for matching exactly on a variable with lexical order) and then find the corresponding optimal caliper and constraint on the number of nearest neighbors.

13.3.2 Fine balance and related techniques

Another commonly used matching technique to balance a nominal variable is fine balance, first proposed by [5], which requires the matched treated and control groups to have the same marginal distribution without restricting who is matched to whom. For instance, the treated group in Figures 13.1 and 13.2 contains two Hispanics and eight non-Hispanics. Fine balance forces the control group to include two Hispanics and eight non-Hispanics, ignoring whether Hispanics are matched to Hispanics or non-Hispanics. Fine balance is not always feasible. If there were fewer than two Hispanics or eight non-Hispanics in the control group, fine balance would be infeasible. Let Υ be the number of possible values of the fine balance nominal covariate; so $\Upsilon = 2$ in this toy example. We can determine whether fine balance is feasible in Figure 13.1(i) by constructing a $2 \times \Upsilon$ contingency table recording the treatment by the nominal covariate. However, Figure 13.2 imposes additional constraints – the exact matching for gender, the caliper on the propensity score, and the limit on the number of nearest neighbors. Even if the control group did include two Hispanics and eight non-Hispanics, fine balance might be infeasible with these additional constraints. Therefore, no simple tabulation can determine the feasibility of fine balance with a reduced number of candidate pairs.

Near-fine balance means the deviation from fine balance is as small as possible [6]. The formal definition of near-fine balance requires the total absolute differences in frequencies between the treated and control groups over the categories to be minimized. For instance, if fine balance for the indicator of Hispanics is infeasible, the next minimal difference of $2 = |2-1| + |8-9| = |2-3| + |8-7|$ can be achieved by including either one Hispanic and nine non-Hispanics or three Hispanics and seven non-Hispanics in the matched control group. The structure on the right in Figure 13.2 imposes a near-fine balance constraint for Hispanic or not as a soft constraint. Here, a soft constraint is implemented by altering the objective function – the total covariate distance within matched pairs – to penalize violations of fine balance through the grey edges. The minimum cost flow algorithm tries to avoid penalized violations of fine balance, but tolerates the minimum number of violations needed to ensure feasibility. Utilizing a soft constraint is necessary when dealing with large observational studies because the calipers and nearest neighbors constraints produce a reduced bipartite graph.

Pimentel et al. [4] generalized the concept of near-fine balance, introducing a tree-structured hierarchy of near-fine balance constraints, allowing the user to express a preference for certain kinds of deviations from fine balance over other deviations. This method is handy when trying to balance the joint distribution of several nominal variables with different priorities. The proposed idea of refined balance can be applied in conjunction with the optimal caliper and restriction on nearest neighbors described in §13.2 by adding multiple layers of near-fine balance nodes in the auxiliary structure on the right in Figure 13.2.

13.3.3 Directional and non-directional penalties

To improve the quality of a matched sample, Yu and Rosenbaum [13] changed the matching requirements with directional and non-directional penalties when diagnostics have suggested the initial match is not adequate. As further motivation, we extend §13.4 of [13] and apply symmetric and asymmetric calipers for a Normally distributed covariate to reduce the bias in §13.3.3.1. The general strategy of adjusting distances asymmetrically to favor pairs against the direction of the bias in optimal matching is discussed in §13.3.3.2. Essentially, we can connect directional penalties with a standard idea in integer programming – the Lagrangian relaxation of the constrained matching problem which penalizes the objective function aiming to enforce the balance constraints [17]. Notably, directional and non-directional penalties are compatible with the size-scalable matching methods discussed in §13.2, as they simply adjust covariate distances of the remaining candidate pairs. The methods are implemented in an `R` package `DiPs`, which is publicly available from `CRAN`.

13.3.3.1 A simple illustration: Symmetric and asymmetric calipers for matching on a single Normal covariate

In 1973, Cochran and Rubin [12, §2.3] discussed an intuitive but wise technique of caliper matching to ensure the covariate similarity of matched pairs. For instance, caliper matching for age with a non-negative caliper $\eta \geq 0$ requires a treated subject to be matched to a control whose age differs no more than η years from the treated subject. With $\eta = 4$, a 16-year old treated unit can only

TABLE 13.2
Bias after caliper matching with a symmetric/asymmetric caliper $x_t - x_c \in [-\eta_1, \eta_2]$ on a single Normally distributed covariate. The initial bias before matching is μ. The smallest absolute bias after caliper matching for each μ is labeled in **bold**. This table is an extension of Table 3 of [13] by considering more caliper and initial bias choices.

Caliper		Initial bias μ				
$-\eta_1$	η_2	−0.25	0	0.25	0.5	1
−0.8	1.2	0.08	0.15	0.21	0.28	0.40
−0.9	1.1	**0.01**	0.07	0.14	0.20	0.33
−1.0	1.0	−0.06	**−0.00**	0.06	0.13	0.25
−1.1	0.9	−0.14	−0.07	**−0.01**	0.05	0.17
−1.2	0.8	−0.21	−0.15	−0.08	**−0.02**	0.09
−1.3	0.7	−0.29	−0.22	−0.16	−0.10	**0.02**
−1.4	0.6	−0.36	−0.30	−0.23	−0.17	−0.06
−1.5	0.5	−0.44	−0.37	−0.31	−0.25	−0.14

be matched to any control who is 12-20 years old. Such a caliper of the form $[-\eta, \eta]$ is called a symmetric caliper. However, if treated subjects are typically older than controls, we are more likely to include matched pairs with an older treated unit and a younger control than the opposite within the symmetric caliper. As a result, the residual imbalances after symmetric caliper matching are likely to remain tilted in the original direction. In this situation, an asymmetric caliper of the same length might be preferable, requiring $x_t - x_c \in [-\eta_1, \eta_2]$ where $\eta_1 > \eta_2 \geq 0$. Such a caliper of the form $[-\eta_1, \eta_2]$ with $\eta_1 \neq \eta_2$ is called an asymmetric caliper.

To demonstrate the merits of asymmetric calipers, consider the simple case of matching two individuals on one Normally distributed covariate, as discussed in [13]. Specifically, suppose the treated subject has covariate $x_t \sim N(\mu, 1)$ and the control has covariate $x_c \sim N(0, 1)$. Then the bias before matching is $\mathrm{E}(x_t - x_c) = \mu$. How much bias can caliper matching reduce? Suppose that we first sample x_t, then independently sample x_c conditional on the caliper constraint $x_t - x_c \in [-\eta_1, \eta_2]$. We use numerical integration to calculate the bias after caliper matching, namely

$$\mathbf{E}\{\mathbf{E}(x_t - x_c \mid x_t, x_t - x_c \in [-\eta_1, \eta_2])\}.$$

Table 13.2 is an extension of Table 3 of [13] and summarizes the bias after matching for various choices of initial biases μ and calipers $[-\eta_1, \eta_2]$. The symmetric caliper $[-1, 1]$ works very well when the initial bias is zero, achieving the smallest residual bias of almost zero. However, when the initial bias is non-zero, the asymmetric caliper with the same length can achieve a much smaller bias than the symmetric one. For instance, from the last column of Table 13.2, we can observe that with a relatively large initial bias $\mu = 1$, the symmetric caliper reduces the bias to 0.25, but the asymmetric caliper $[-1.3, 0.7]$ removes 98% of the initial bias and leaves a residual bias of 0.02.

Table 13.2 provides several insightful information. First, the best asymmetric caliper often works better than the symmetric caliper with the same length in reducing biases. In addition, the optimal degree of asymmetry among calipers with the same length depends on the direction and scale of the initial bias μ. More asymmetry is needed to offset a larger initial bias. Using a caliper with an inappropriate asymmetry may worsen the situation, increasing the bias rather than decreasing it.

With the merits of asymmetric calipers, a natural question arises: can we use an asymmetric caliper to remove edges in the network and reduce the computational difficulties in large observational studies? The good news is yes. But how can we proceed? A simple way is to find the optimal symmetric caliper and then extend the caliper in one direction to offset the initial bias. With an extended caliper covering the original optimal caliper, no feasibility issue occurs. For instance, if we go back to the toy example in Figure 13.1, we have the optimal symmetric caliper for the propensity score is $[-0.0764, 0.0764]$. Since the propensity score in the treated group is typically larger than that in the control group, it is natural to use an asymmetric caliper such as $[-0.0764 \times 2, 0.0764]$ to achieve a smaller bias in the propensity score. A more formal approach is directly applying Glover's algorithm to find the optimal asymmetric caliper with a specified asymmetry. For instance, what is the best $\eta > 0$ in the calipers of the form $[-2\eta, \eta]$? Glover's algorithm is still applicable in this case since a caliper can produce a doubly convex bipartite graph no matter its asymmetry. The investigators may want to try several choices of asymmetry and select the one that gives the best covariate balance.

13.3.3.2 Optimal matching with asymmetric adjustments to distances

Including asymmetric adjustments in caliper matching is one example of asymmetrically adjusting the distances to obtain a better match, where an appropriate choice of asymmetry is critical to reducing the bias maximally. Similar ideas can be generalized to other asymmetric distance adjustments. Suppose we have K functions of observed covariates x_k, $k = 1, \ldots, K$, for each individual. For covenience, we allow several functions refer to the same covariate; for instance, x_k for $k = 1, 2, 3$ might all be functions of age but in different forms. This strategy is helpful because balancing the means of functions of covariates can balance variances, covariances, and quantiles. Write x_{k,τ_t} and x_{k,γ_c} for the observed value of x_k for treated subject τ_t and control γ_c, respectively. Let d_{ktc} denote a comparison of x_{k,τ_t} and x_{k,γ_c} using some distance measure, and let a_{tc} denote a binary indicator on whether γ_c is matched to τ_t. To balance x_k in the treated and control groups, consider the average d_{ktc} for matched individuals (i.e., $a_{tc} = 1$),

$$\overline{d}_k = \frac{1}{T} \sum_{t=1}^{T} \sum_{c=1}^{C} a_{tc} d_{ktc}.$$

The smaller $\overline{d}_k$ is, the closer the matched treated and control groups are. How can we choose the comparison d_{ktc}? In this section, we describe several easy-to-use measures discussed in [13].

The difference $d_{ktc} = x_{k,\tau_t} - x_{k,\gamma_c}$ is an intuitive and natural comparison of x_{k,τ_t} and x_{k,γ_c}. However, the corresponding $\overline{d}_k$ – the average difference of x_k between the matched treated and control groups – has a limitation: it permits the existence of large positive/negative differences, allowing them to cancel with each other to give a small average. To avoid significant differences in both signs, a better choice of d_{ktc} might be the absolute difference $|x_{k,\tau_t} - x_{k,\gamma_c}|$, which can be decomposed into two hockey-stick measures of the form

$$d'_{ktc} = \max(0,\, x_{k,\tau_t} - x_{k,\gamma_c}) \geq 0, \quad d''_{ktc} = \max(0,\, x_{k,\gamma_c} - x_{k,\tau_t}) \geq 0.$$

Then we have $|x_{k,\tau_t} - x_{k,\gamma_c}| = d'_{ktc} + d''_{ktc}$. We can control the average absolute difference by focusing on the two averages $\overline{d}'_k$ and $\overline{d}''_k$, the corresponding averages $\overline{d}_k$ obtained by replacing d_{ktc} with d'_{ktc} and d''_{ktc}, respectively. More importantly, we can tilt against the direction of bias by adjusting the weights in their weighted sum $\lambda_1 \overline{d}'_k + \lambda_2 \overline{d}''_k$.

Another popular choice of d_{ktc} is $\text{sign}(x_{k,\tau_t} - x_{k,\gamma_c})$, where

$$\text{sign}(x_{k,\tau_t} - x_{k,\gamma_c}) = \begin{cases} 1 & \text{if } x_{k,\tau_t} - x_{k,\gamma_c} > 0 \\ 0 & \text{if } x_{k,\tau_t} - x_{k,\gamma_c} = 0 \\ -1 & \text{if } x_{k,\tau_t} - x_{k,\gamma_c} < 0 \end{cases}.$$

Then the corresponding $\overline{d}_k = 0$ if the median of $x_{k,\tau_t} - x_{k,\gamma_c}$ for all matched pairs is zero. In addition, we would prefer obtaining a zero difference to canceling out differences with opposite signs. To control the median difference with this preferred property, we apply a similar trick as before and decompose $\text{sign}(x_{k,\tau_t} - x_{k,\gamma_c})$ into two parts,

$$d'_{ktc} = 1 \text{ if } x_{k,\tau_t} - x_{k,\gamma_c} > 0, \;\; d'_{ktc} = 0 \text{ otherwise;}$$

$$d''_{ktc} = 1 \text{ if } x_{k,\gamma_c} - x_{k,\tau_t} > 0, \quad d''_{ktc} = 0 \text{ otherwise.}$$

Then controlling the two averages $\overline{d}'_k$ and $\overline{d}''_k$ can reduce the median difference and the number of non-zero differences. Again, we can assign $\overline{d}'_k$ and $\overline{d}''_k$ difference weights to tilt against the direction of bias.

The asymmetric calipers discussed in §13.3.3.1 can also be implemented under this framework by choosing

$$d_{ktc} = \begin{cases} 1 \text{ if } x_{k,\tau_t} - x_{k,\gamma_c} < -\eta_1 \\ 0 \text{ if } -\eta_1 \leq x_{k,\tau_t} - x_{k,\gamma_c} \leq \eta_2 \\ 1 \text{ if } x_{k,\tau_t} - x_{k,\gamma_c} > \eta_2 \end{cases},$$

where $-\eta_1 \leq 0 \leq \eta_2$. If all matched pairs satisfy the caliper constraint on x_k, i.e., $x_{k,\tau_t} - x_{k,\gamma_c} \in [-\eta_1, \eta_2]$, the corresponding $\overline{d}_k = 0$,

One common concern about optimal matching is that the matched sample may include a few pairs with large distances δ_{tc} since optimal matching minimizes the total distance among all matched pairs $\sum_{t=1}^{T} \sum_{c=1}^{C} a_{tc}\delta_{tc}$ without directly focusing on the single within-pair distances. To control the maximum distance in the matched set, Rosenbaum [18] discussed an efficient algorithm to find the smallest possible threshold ψ for the maximum paired distance that permits a feasible match. In practice, it is more straightforward and usually adequate to pick a moderate but somewhat arbitrary value for ψ in most applications. Suppose $\psi > 0$ is a moderate threshold for the within-pair distances. By choosing

$$d_{ktc} = \begin{cases} 1 & \text{if } \delta_{tc} > \psi \\ 0 & \text{otherwise} \end{cases},$$

controlling $\overline{d}_k$ can reduce the number of distances exceeding ψ.

After choosing K comparisons d_{ktc}, $k = 1, \ldots, K$, suppose that we would like to have $\overline{d}_k \leq \epsilon_k$, $k = 1, \ldots, K$ for fixed $\epsilon_k \geq 0$. Recall in §13.2, the original matching problem without these new balance constraints $\overline{d}_k \leq \epsilon_k$ can be solved by a relatively efficient algorithm in polynomial time [10, Theorem 11.2]. In contrast, the new problem with additional constraints $\overline{d}_k \leq \epsilon_k$ is a general integer program and can be much more challenging to solve. Specifically, the general integer programming problem is NP-complete and cannot be solved in polynomial time [19, Theorem 18.1]. Lagrangian relaxation, proposed by Fisher [17], is one of the approximation techniques to solve the new constrained problem. It involves solving several easy problems in the original form with adjusted objective function, so it is computationally tractable. In light of Lagrangian relaxation of the linear imbalance summary requirements, Yu and Rosenbaum [13] proposed the following iterative method:

1. Choose a set of directional penalties $\lambda_1 \geq 0, \ldots, \lambda_K \geq 0$.
2. Define new distances $\delta^*_{tc} = \delta_{tc} + \sum_{k=1}^{K} \lambda_k \left(d_{ktc} - \epsilon_k\right)$.
3. Solve the conventional matching problem with revised distances δ^*_{tc}.

Each iteration can be accomplished efficiently, in $O\left(C^3\right)$ steps, if we consider all possible treated-control pairs. Combining these techniques with the methods discussed in §13.2 to remove some candidate pairs, we can solve this revised problem in $O\left\{C^2 \log\left(C\right)\right\}$ steps. We can repeat these three steps several times, adjusting the directional penalties/Lagrangian multipliers λ_k at each iteration to enforce the new balance constraints $\overline{d}_k \leq \epsilon_k$, $k = 1, \ldots, K$. A rule of thumb is to increase λ_k if the balance constraints are not satisfied and decrease λ_k otherwise.

Using large penalties is common in many matching techniques to improve covariate balance. Investigators can implement balance requirements as soft constraints to meet the desired conditions as closely as possible. For instance, a large penalty can be applied to enforce a propensity score caliper without generating infeasibility issues or to balance the marginal distribution of a nominal variable with minimum deviations. On the other hand, directional penalties work against the direction of bias. An over-adjustment can reverse the direction of the bias but not adequately reduce the bias. Therefore, investigators must tune directional penalties carefully to achieve satisfactory results. In practice, a few adjustments to directional penalties can efficiently improve the quality of a matched sample, as shown in the real data application in [13, §3].

13.4 Other Matching Designs

13.4.1 Extension to multiple controls

Another matching design of interest when the control group is much larger than the treated group is matching with multiple controls, which allows each treated unit to be matched to a fixed number of controls. A conceptually simple way to match with $\kappa \geq 2$ controls is to duplicate each treated subject κ times so that $\mathcal{T} = \{\tau_1, \ldots, \tau_T\}$ is replaced by $\mathcal{T}^* = \{\tau_1, \tau_{12}, \ldots, \tau_{1\kappa}, \ldots, \tau_T, \tau_{T2}, \ldots, \tau_{T\kappa}\}$, and then apply the method for pair matching described in §13.2. Doing so does little harm in determining the optimal caliper and optimal restriction on the number of nearest neighbors due to the linear computational complexity of Glover's algorithm on a doubly convex bipartite graph. However, to construct a match with $\kappa \geq 2$ controls by solving a minimum cost flow problem, duplication is unwise because it entails storing the same edge several times. A more efficient implementation is to update the supply of each treated node to κ in the network setup for the minimum cost flow problem.

This idea is analogous for matching with a variable number of controls. In other words, treated subjects may be matched to different numbers of controls. Yoon [20] introduced the concept of entire number to decide how many controls should be matched to each treated subject τ_t based on the propensity score. After determining these numbers κ_t, everything will be similar to matching with a fixed number of controls. The only difference is we now duplicate κ_t times for treated subject τ_t. Then we can apply Glover's algorithm to determine the optimal caliper and smallest feasible constraint on the number of nearest neighbors. Again, while using minimum cost flow algorithms to find the match, we should change the supply of each treated node τ_t to κ_t instead of a fixed number κ to save computer memory and reduce computational intensity.

13.4.2 Full matching

13.4.2.1 Sparsifying the network for full matching

Unlike matching with multiple controls, another class of matching methods deals with a broader class of matches, permitting both many-to-one and one-to-many matched sets [21]. Full matching can achieve a smaller total distance within matched sets as it minimizes the total distance over a larger class of matches. As a result, it can usually achieve better covariate balance than matching with a fixed number of controls. However, the unrestricted form of full matching might produce highly skewed matching ratios, which can harm the efficiency of treatment effect estimation [22] and the simplicity of matched analysis [3]. To overcome this limitation, investigators can impose structural restrictions to full matching, adding upper limits on a and b for which $a : 1$ matches and $1 : b$ matches are permitted [7, 23]. With or without structural restrictions, full matching experiences the same computation difficulties for large datasets as pair matching.

Fredrickson et al. [24] adapted the methods discussed in §13.2 to full matching and found an interesting but different result. The key idea is to determine the feasibility of full matching, which is much easier than determining the feasibility of pair matching. Specifically, Corollary 4 in [21] gives a sufficient condition for full matching to be feasible: all treated subjects $\tau_t \in \mathcal{T}$ have at least one candidate control $\gamma_c \in \mathcal{C}$ to be matched to, i.e., each treated node must be connected to at least one

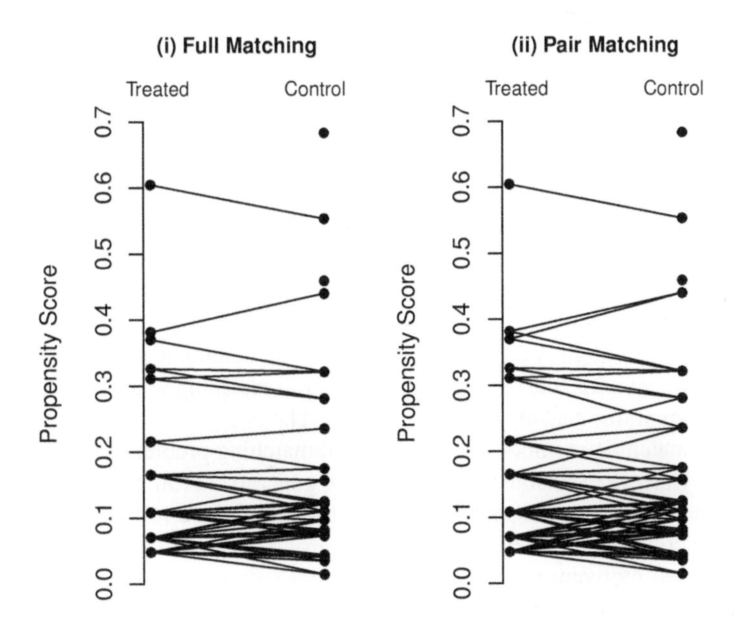

FIGURE 13.3
Two bipartite graphs: (i) optimal caliper for unrestricted full matching (ii) optimal caliper for pair matching.

control node in the bipartite graph analogous to Figure 13.1. Thus, the optimal caliper for unrestricted full matching is the maximum over the treated group of minimum distances to all potential controls, $\varkappa_f = \max_{\tau_t \in \mathcal{T}} \min_{\gamma_c \in \mathcal{C}} d_{tc}$, as discussed in [24]. Notably, finding this caliper does not impose convexity on the graph structure or other conditions on the distances d_{tc}, so the choice of d_{tc} can be flexible, e.g., the Mahalanobis distance based on all observed covariates or the distance based on a single score. If calculating a distance is takes constant time, finding the optimal caliper for full matching takes $O(C^2)$ steps. In the situation where $d_{tc} = |\rho(\tau_t) - \rho(\gamma_c)|$ for some real-valued score $\rho(\cdot)$, which is the same scenario considered by Yu et al. [11], Fredrickson et al. [24] proposed an efficient algorithm that determines the optimal caliper $\varkappa_f$ in $O\left\{C \log(C)\right\}$ operations.

Applying narrow calipers can reduce the computational burden for full matching in large observational studies but may create highly imbalanced structures including matched sets with many treated subjects sharing the same control or matching one treated subject to many controls. To overcome the drawbacks of highly skewed matching ratios described above, one option is to consider full matching with symmetric restrictions: matching in ratios of $h : 1$ up to $1 : h$, for some positive integer h. This is equivalent to pair matching when $h = 1$ and unrestricted full matching when $h = \infty$. Note that the optimal caliper for unrestricted full matching $\varkappa_f$ provides a lower bound on the optimal caliper for any restricted match, such as pair matching or restricted full matching, but may not permit a feasible restricted match. Therefore, to find the optimal caliper for restricted full matching, one can compute $\varkappa_f$ and then check if the caliper is also feasible for full matching with the additional structural constraints. If not, we need to increase the caliper size until a feasible restricted full matching exists. As illustrated in Figure 13.3, with our toy dataset in §13.2, the optimal caliper for propensity score with unrestricted full matching is 0.0579. In this example, this caliper is also feasible for restricted full matching with ratios of $2 : 1$ up to $1 : 2$. On the other hand, the full matching caliper is too narrow for pair matching, for which the smallest feasible caliper is 0.0764.

13.4.2.2 Generalized full matching

To allow full matching in more complex settings, e.g., when there are several treatment conditions or the sample size is large, Savje et al. [25] considered a generalized version of full matching, which reduced to the original full matching in a special case. Specifically, suppose we have L treatment arms indexed by $1, \ldots, L$. A feasible generalized full matching with constraint $H = (h_1, ..., h_L, q)$ subclassifies all individuals into disjoint matched sets such that (i) each matched set has at least q units in total, and (ii) each matched set consists of at least h_l units under treatment l. For instance, for a study with three treatment conditions, the constraint $H = (3, 3, 2, 10)$ restricts each matched set of a feasible matching to contain at least 10 units in total, with at least 3 units under the first and second treatment arms and at least 2 units under the third condition. Although the definition of generalized full matching is slightly different from the original definition of full matching in [21], for studies with two treatment arms, the optimal generalized full matching with constraint $H = (1, 1, 2)$ is equivalent to a original full matching [21, Proposition 1].

Similar to other matching methods, generalized full matching problems are formulated as optimization problems. However, the objective function for generalized full matching departs from the conventional full matching in two aspects [25]. First, the generalized full matching focuses on the bottleneck objective function of the maximum within-group distance rather than the total distance as in [22]. The primary motivation for this change is that an approximation algorithm can find a near-optimal solution to the new optimization problem efficiently. In addition to the computational considerations, minimizing the maximum distance can avoid match sets of poor quality produced by targeting the total distance as the objective, a similar motivation as imposing an upper bound for the within-pair distances in §13.3.3.2. The second deviation is that generalized full matching considers the distance across different treatment arms as in the existing literature and the distances between individuals under the same treatment condition. The new objective can capture the important within-group differences for studies with more than two treatment arms.

The optimal generalized full matching problem is NP-hard [26]. To achieve computational tractability, Savje et al. [25] extended the blocking algorithm proposed by [26] and introduced a near-optimal approximation algorithm for generalized full matching, which is computationally efficient for large observational studies (with computational complexity of $O\left\{N^2 \log(N)\right\}$ if there are N individuals in the study). A match produced by this approximation algorithm is guaranteed to achieve an objective value within four times of the optimal objective value, ensuring that the quality of the match is reasonable.

For simplicity, we consider the generalized full matching in the conventional setting with a binary treatment, i.e., with constraint $H = (1, 1, 2)$. The near-optimal generalized full match can be constructed with the following algorithm; see Section 4.3 of [25] for the general version of the algorithm. The methods are implemented in an `R` package `quickmatch` available on `CRAN`.

1. Construct a graph G that contains a node for each unit and three types of edges: (i) self-loop for each node, (ii) an edge from each treated node to its closest control node, (iii) an edge from each control node to its closest treated node. For each node i, denote its closest neighborhood as $N[i] = \{j : G \text{ has edge } i \to j\}$.

2. Find a set of units S, called seeds, such that their closed neighborhoods in G are non-overlapping and maximal (i.e., adding any additional vertex to S would create some overlap).

3. Assign a label to each seed $s \in S$. Then assign the same label to all nodes in the closest neighborhood of s, $N[s]$.

4. For each node i without a label, select one of its labeled neighbors in $N[i]$, say i'. Then assign the label of i' to i.

5. Form the matched groups using nodes sharing the same label.

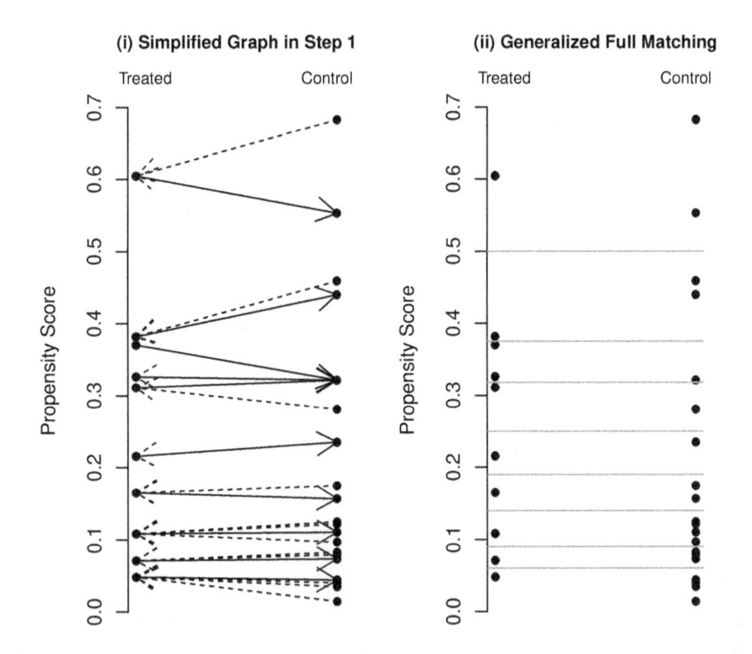

FIGURE 13.4
(i) A simplified version of graph G constructed in step 1 without the self-loops for each node. Solid arrows are from treated units to controls, and dashed arrows are from controls to treated units. (ii) The constructed full matching by the approximation algorithm. Gray lines define the matched sets.

Figure 13.4 demonstrates the approximation algorithm with the toy dataset used in the previous sections. Figure 13.4(i) is a simplified version of graph G constructed in step 1 without the self-loops for each node. In the second step, nine seeds are identified, which leads to 9 matched sets in Figure 13.4(ii).

13.4.3 Matching without pairing

In contrast to characterizing the matching problem as a network flow problem, another class of matching methods solves integer programs directly by using commercial solvers like CPLEX and Gurobi [9, 27]. These methods usually use a quadratic-sized formulation with $T \times C$ binary decision variables corresponding to the indicators of who is matched to whom, analogous to edges in a dense graph in Figure 13.1(i). An advantage of this class of methods is that investigators can impose various balance requirements as constraints of the integer programs. However, with more flexible constraints, the problem becomes a general integer program, which is NP-complete [19]. In addition, the quadratic-sized formulation with large-scale observational studies may be too large to store in computer memory.

To reduce the computational intensity, Bennett et al. [28] considered reducing the size of the formulation by ignoring the pairing information. As a result, the new formulation has only C decision variables representing whether each control is included in the matched control group. For instance, in the toy example in Section 13.2, there are 20 binary decision variables in the reduced form instead of $10 \times 20 = 200$ binary decision variables in the original formulation. This linear-sized formulation is more effective than the conventional quadratic-sized formulation because it allows us to handle considerably larger datasets in a reasonable amount of time. Since this new formulation ignores the pairing information, we cannot minimize the total covariate distance among all matched pairs.

Instead, the objective function can target the discrepancy of distributions in the matched treated and control groups, e.g., the Kolmogorov-Smirnov statistics of marginal distributions and the difference in covariate means and quantiles. With an appropriately chosen objective function suitable for both formulations, Bennett et al. [28] demonstrated the equivalency of these two formulations in terms of the optimal objective value and strength of formulation (i.e., the gap between the optimal value of the original integer program and its linear program relaxation). In particular, when there are two covariates or nested covariates, the problem is equivalent to its linear program relaxation, hence can be solved in polynomial time. This reduced formulation is computationally intractable in theory without any special structures, but it usually works well in practice.

This new formulation is implemented in the `designmatch` package. After matching for balance, the selected controls can be re-matched to reduce outcome heterogeneity and sensitivity to hidden biases [29].

13.4.4 Coarsened exact matching

Coarsened exact matching (CEM) proposed by [30] is a matching algorithm that works as a subclassification method. Basically, one simply coarsens each variable to a reasonable degree of clustering and then performs an exact match on these coarsened variables. Strata including only treated subjects or only controls are excluded from the matched data. This method has several nice statistical properties, as described in [30]. In addition, it is very fast even in large samples since forming the matched sets with CEM only involves tabulations with computational complexity $O(N)$ if there are N individuals in the study.

The initial step of coarsening data is critical to the quality of CEM but is subjective to users. If the coarsened bins are too narrow, the matched treated and control groups will be very similar, but the sample size of the final matched data may be very small. Using very wide coarsened bins discards few observations, but the residual differences within the large strata can bias the outcome analysis. The number of observations may still be too small after choosing coarsening level as the largest that the investigators find reasonable. This is likely to happen in the presence of a single continuous covariate with little coarsening or higher-order interactions of discrete covariates. In this case, Iacus et al. [30] offered a "progressive coarsening" procedure to help users decide the coarsening levels by demonstrating how the sample size of the match increases as loosening the coarsening level for each variable. If the remaining sample is still too small after revising the coarsening step, the only options include collecting more data, incorporating other methods to reduce the bias, or tolerating the significant uncertainty for the inferences [31]. This problem becomes less severe as the sample size of the observational study increases. These methods are implemented in `R` package `cem` on `CRAN`.

13.5 Summary

We have outlined several recently proposed methods for finding different matching designs with large observational studies, illustrated them with a toy example of 30 individuals from NHANES data, and mentioned the corresponding software. These methods improve the computational efficiency in different ways. The first class of techniques focuses on excluding the candidate matches that are far apart and do not deserved to be included in the matched data. Alternatively, matched samples can be constructed using approximation algorithms or in two steps by first deciding whom to include in the match then pairing within the selected samples. Another type of matching method directly uses tabulation on the coarsened covariates.

13.6 Acknowledgments

Part of this chapter is adapted from the author's joint paper with Paul Rosenbaum and Jeffrey Silber, "Matching methods for observational studies derived from large administrative databases," which appeared with discussion in *Statistical Science* in 2020. The author is grateful to the Institute of Mathematical Statistics for their copyright policy which permits use by authors of publications in IMS journals.

References

[1] Rachel R Kelz, Morgan M Sellers, Bijan A Niknam, James E Sharpe, Paul R Rosenbaum, Alexander S Hill, Hong Zhou, Lauren L Hochman, Karl Y Bilimoria, Kamal Itani, et al. A national comparison of operative outcomes of new and experienced surgeons. *Annals of Surgery*, 273(2):280–288, 2021.

[2] Jeffrey H Silber, Paul R Rosenbaum, Wei Wang, Shawna R Calhoun, Joseph G Reiter, Orit Even-Shoshan, and William J Greeley. Practice style variation in medicaid and non-medicaid children with complex chronic conditions undergoing surgery. *Annals of Surgery*, 267(2):392–400, 2018.

[3] Paul R Rosenbaum and Donald B Rubin. Constructing a control group using multivariate matched sampling methods that incorporate the propensity score. *American Statistician*, 39(1):33–38, 1985.

[4] Samuel D Pimentel, Rachel R Kelz, Jeffrey H Silber, and Paul R Rosenbaum. Large, sparse optimal matching with refined covariate balance in an observational study of the health outcomes produced by new surgeons. *Journal of the American Statistical Association*, 110(510):515–527, 2015.

[5] Paul R Rosenbaum, Richard N Ross, and Jeffrey H Silber. Minimum distance matched sampling with fine balance in an observational study of treatment for ovarian cancer. *Journal of the American Statistical Association*, 102(477):75–83, 2007.

[6] Dan Yang, Dylan S Small, Jeffrey H Silber, and Paul R Rosenbaum. Optimal matching with minimal deviation from fine balance in a study of obesity and surgical outcomes. *Biometrics*, 68(2):628–636, 2012.

[7] Ben B Hansen and Stephanie Olsen Klopfer. Optimal full matching and related designs via network flows. *Journal of Computational and Graphical Statistics*, 15(3):609–627, 2006.

[8] Paul R Rosenbaum. Optimal matching for observational studies. *Journal of the American Statistical Association*, 84(408):1024–1032, 1989.

[9] José R Zubizarreta. Using mixed integer programming for matching in an observational study of kidney failure after surgery. *Journal of the American Statistical Association*, 107(500):1360–1371, 2012.

[10] Bernhard H Korte and Jens Vygen. *Combinatorial Optimization*. New York: Springer, 2012.

[11] Ruoqi Yu, Jeffrey H Silber, and Paul R Rosenbaum. Matching methods for observational studies derived from large administrative databases. *Statistical Science*, 35(3):338–355, 2020.

[12] William G Cochran and Donald B Rubin. Controlling bias in observational studies: A review. *Sankhyā: The Indian Journal of Statistics, Series A*, 35(4):417–446, 1973.

[13] Ruoqi Yu and Paul R Rosenbaum. Directional penalties for optimal matching in observational studies. *Biometrics*, 75(4):1380–1390, 2019.

[14] Fred Glover. Maximum matching in a convex bipartite graph. *Naval Research Logistics Quarterly*, 14(3):313–316, 1967.

[15] Witold Lipski and Franco P Preparata. Efficient algorithms for finding maximum matchings in convex bipartite graphs and related problems. *Acta Informatica*, 15(4):329–346, 1981.

[16] Paul R Rosenbaum. *Design of Observational Studies*. New York: Springer, 2010.

[17] Marshall L Fisher. The lagrangian relaxation method for solving integer programming problems. *Management Science*, 27(1):1–18, 1981.

[18] Paul R Rosenbaum. Imposing minimax and quantile constraints on optimal matching in observational studies. *Journal of Computational and Graphical Statistics*, 26(1):66–78, 2017.

[19] Alexander Schrijver. *Theory of Linear and Integer Programming*. New York, John Wiley & Sons, 1998.

[20] Frank B Yoon. *New methods for the design and analysis of observational studies*. PhD thesis, University of Pennsylvania, 2009.

[21] Paul R Rosenbaum. A characterization of optimal designs for observational studies. *Journal of the Royal Statistical Society: Series B (Methodological)*, 53(3):597–610, 1991.

[22] Ben B Hansen. Full matching in an observational study of coaching for the sat. *Journal of the American Statistical Association*, 99(467):609–618, 2004.

[23] Xing Sam Gu and Paul R Rosenbaum. Comparison of multivariate matching methods: Structures, distances, and algorithms. *Journal of Computational and Graphical Statistics*, 2(4):405–420, 1993.

[24] Mark M Fredrickson, Josh Errickson, and Ben B Hansen. Comment: Matching methods for observational studies derived from large administrative databases. *Statistical Science*, 35(3):361–366, 2020.

[25] Fredrik Sävje, Michael J Higgins, and Jasjeet S Sekhon. Generalized full matching. *Political Analysis*, 29(4):423–447, 2021.

[26] Michael J Higgins, Fredrik Sävje, and Jasjeet S Sekhon. Improving massive experiments with threshold blocking. *Proceedings of the National Academy of Sciences*, 113(27):7369–7376, 2016.

[27] Cinar Kilcioglu and José R Zubizarreta. Maximizing the information content of a balanced matched sample in a study of the economic performance of green buildings. *Annals of Applied Statistics*, 10(4):1997–2020, 2016.

[28] Magdalena Bennett, Juan Pablo Vielma, and José R Zubizarreta. Building representative matched samples with multi-valued treatments in large observational studies. *Journal of Computational and Graphical Statistics*, 29(4):744–757, 2020.

[29] José R Zubizarreta, Ricardo D Paredes, and Paul R Rosenbaum. Matching for balance, pairing for heterogeneity in an observational study of the effectiveness of for-profit and not-for-profit high schools in chile. *Annals of Applied Statistics*, 8(1):204–231, 2014.

[30] Stefano M Iacus, Gary King, and Giuseppe Porro. Causal inference without balance checking: Coarsened exact matching. *Political Analysis*, 20(1):1–24, 2012.

[31] Stefano M Iacus, Gary King, and Giuseppe Porro. cem: Software for coarsened exact matching. *Journal of Statistical Software*, page Forthcoming, 30(9): 1–27, 2009.

[32] Paul R Rosenbaum and Donald B Rubin. The central role of the propensity score in observational studies for causal effects. *Biometrika*, 70(1), 41–55, 1983.

[33] Ben B Hansen. The prognostic analogue of the propensity score. *Biometrika*, 95(2), 481–488, 2008.

[34] Ruoqi Yu. How well can fine balance work for covariate balancing. *Biometrics*, `https://doi.org/10.1111/biom.13771`, 2022.

Part III

Weighting

14

Overlap Weighting

Fan Li

CONTENTS

Weighting is one of the fundamental methods in causal inference. The main idea of weighting is to re-weight each unit to create a weighted population, namely, the target population, where the treatment and control groups are comparable in baseline characteristics. The dominant weighting scheme in causal inference has been inverse probability weighting (IPW). IPW is a propensity score weighting method and targets the average treatment effect (ATE) estimand. However, IPW has a well-known limitation: it is subject to excessive variance due to extreme weights when covariates are poorly overlapped between treatment groups. There has been an increasingly large literature on new weighting methods in causal inference [1–4]. The majority of these methods target at the ATE estimand and devise new ways to reduce the large variance in estimating ATE when covariates are poorly overlapped. In this chapter, we describe the general framework of balancing weights, which generalizes propensity score weighting beyond IPW. This framework allows analysts to flexibly specify a target population first and then estimate the corresponding treatment effect. In particular, we focus on a special case of balancing weights, the overlap weight (OW), which possesses desirable theoretical and empirical properties, and scientifically meaningful interpretation [5–8].

14.1 Causal Estimands on a Target Population

Suppose we have a sample of N units. Each unit i ($i = 1, 2, \ldots, N$) has a binary treatment indicator Z_i, with $Z_i = 0$ being control and $Z_i = 1$ being treated, and a vector of p covariates

DOI: 10.1201/9781003102670-14

$X_i = (X_{1i}, \cdots, X_{pi})$. Assuming the Stable Unit Treatment Value Assumption (SUTVA) [9], for each unit i, there are a pair of potential outcomes $\{Y_{1i}, Y_{0i}\}$ mapped to the treatment and control status, of which only the one corresponding to the observed treatment is observed, denoted by $Y_i = Z_i Y_{1i} + (1 - Z_i) Y_{0i}$; the other potential outcome is counterfactual. The propensity score is the probability of a unit being assigned to the treatment group given the covariates [10]: $e(x) = \Pr(Z_i = 1 | X_i = x)$.

Causal effects are contrasts of potential outcomes of the same units. We define the conditional average treatment effect (CATE) at covariate value x as

$$\tau(x) = E(Y_{1i} - Y_{0i} \mid X_i = x) = \mu_1(x) - \mu_0(x), \tag{14.1}$$

where $\mu_z(x) = E(Y_{zi} \mid X_i = x)$ is the conditional expectation of the potential outcomes under treatment level z ($z = 0, 1$) over a population. Here we focus on comparing potential outcomes averaged over a distribution of the covariates, namely, a *target population*. One can formulate a causal estimand on any pre-specified target population as follows. Assume the observed sample is drawn from a population with probability density of covariates $f(x)$. Let $g(x)$ denote the covariate density of the target population, which may be different from $f(x)$. We call the ratio $h(x) = g(x)/f(x)$ the *tilting function*, which re-weights the distribution of the observed sample to represent the target population. Then we can represent the average treatment effect on the target population g by a weighted average treatment effect (WATE) :

$$\tau^h = E_g(Y_{1i} - Y_{0i}) = \frac{E[h(x)(\mu_1(x) - \mu_0(x))]}{E[h(x)]}. \tag{14.2}$$

In the formulation (14.2), when $h(x) = 1$, the target population is the same as the population where the study sample is drawn from, and the WATE reduces to the standard ATE estimand $\tau^{\text{ATE}} = E(Y_{1i} - Y_{0i})$. ATE has been the focus of a large majority of causal inference literature until recently. However, the automatic focus on ATE is not always warranted in practice. First, often the available sample does not represent a natural population of scientific interest. For example, patients usually have to meet some criteria in order to be included in a clinical trial and thus may not resemble of the general patient population. Also, in observational comparative effectiveness studies, the study sample is sometimes a so-called "convenience" sample, e.g., patients drawn from a few selected hospitals or clusters, who may be substantially different from the general patient population. In such cases, scientific interpretation of the ATE estimated from the available sample is opaque. Second, the ATE corresponds to the effect of hypothetically switching every unit in the study population from one treatment to the other, a scenario rarely conceivable in medical studies because physicians are well aware that the treatment might be harmful to patients with certain characteristics. Third, researchers commonly exclude some units from the final analysis, e.g. unmatched units or units with extremely large weights. The remaining sample may only represent a subpopulation of the original targeted population and this subpopulation can vary substantially depending on the specific algorithm. In summary, the target population is often different from the population that is represented by the available sample. Therefore, it is important to explore alternative target populations and estimands.

14.2 Balancing Weights

Li, Morgan, and Zaslavsky [5] proposed the *balancing weights* framework to allow analysts to estimate the WATE on any target population. Specifically, let $f_z(x) = \Pr(X = x \mid Z = z)$ be the density of X in the $Z = z$ group. Then it is straightforward to show

$$f_1(x) \propto f(x)e(x), \quad \text{and} \quad f_0(x) \propto f(x)(1 - e(x)),$$

TABLE 14.1
Examples of balancing weights and corresponding target population under different tilting function h.

Target Population	$h(x)$	Weight (w_1, w_0)
Overall	1	$\left(\frac{1}{e(x)}, \frac{1}{1-e(x)}\right)$
Treated	$e(x)$	$\left(1, \frac{e(x)}{1-e(x)}\right)$
Control	$1-e(x)$	$\left(\frac{1-e(x)}{e(x)}, 1\right)$
Overlap	$e(x)(1-e(x))$	$(1-e(x), e(x))$
Trimmed	$\mathbf{1}(\alpha < e(x) < 1-\alpha)$	$\left(\frac{\mathbf{1}(\alpha<e(x)<1-\alpha)}{e(x)}, \frac{\mathbf{1}(\alpha<e(x)<1-\alpha)}{1-e(x)}\right)$
Matching	$\min\{e(x), 1-e(x)\}$	$\left(\frac{\min\{e(x),1-e(x)\}}{e(x)}, \frac{\min\{e(x),1-e(x)\}}{1-e(x)}\right)$

where $\propto$ means proportional to. For a given a $g(x)$ or equivalently tilting function $h(x) = g(x)f(x)$, to estimate τ^h, we can weight $f_z(x)$ to equal $g(x)$ using the following weights (proportional up to a normalizing constant):

$$\begin{cases} w_1(x) \propto \frac{f(x)h(x)}{f(x)e(x)} = \frac{h(x)}{e(x)}, & \text{for } Z = 1, \\ w_0(x) \propto \frac{f(x)h(x)}{f(x)(1-e(x))} = \frac{h(x)}{1-e(x)}, & \text{for } Z = 0. \end{cases} \tag{14.3}$$

The class of weights defined in (14.3) are called *balancing weights* because they balance the weighted distributions of the covariates between comparison groups, both towards the target population:

$$f_1(x)w_1(x) = f_0(x)w_0(x) = f(x)h(x) = g(x). \tag{14.4}$$

Balancing weights include several widely used propensity score weighting schemes as special cases. Choice of the tilting function h determines the target population, estimand, and weights. Statistical, scientific, and policy considerations all may come into play in choosing h in a specific application. For example, as discussed before, when $h(x) = 1$, the corresponding target population $f(x)$ is the overall (combining treated and control) population that is represented by the study sample, the weights (w_1, w_0) are the IPW $\{1/e(x), 1/(1-e(x))\}$ [11], and the estimand is the ATE for the overall population. When $h(x) = e(x)$, the target population is the treated population, the weights are the ATT weight $\{1, e(x)/(1-e(x))\}$ [12], and the estimand is the average treatment effect for the treated (ATT), $\tau^{\text{ATT}} = E(Y_{1i} - Y_{0i} \mid Z_i = 1)$. When $h(x) = 1 - e(x)$, the target population is the control population, the weights are $\{(1-e(x))/e(x), 1\}$, and the estimand is the average treatment effect for the control (ATC), $\tau^{\text{ATC}} = E(Y_{1i} - Y_{0i} \mid Z_i = 0)$. When h is an indicator function, one can define the ATE on subpopulations with specific baseline characteristics, e.g. a subpopulation with a certain age or gender. Other examples of balancing weight include the *matching weight* [13], corresponding to $h(x) = \min\{e(x), 1-e(x)\}$, and the *entropy weight* [14], corresponding to $h(x) = -(e(x)\log(e(x)) + (1-e(x))\log(1-e(x))\}$.

An important special case of balancing weight is propensity score trimming, which focus on a target population with adequate covariate overlap. In particular, Crump et al. [1] recommended to use $h(x) = \mathbf{1}(\alpha < e(x) < 1-\alpha)$, where $\mathbf{1}(\cdot)$ is an indicator function, with a pre-specified threshold $\alpha \in (0, 1/2)$ to limit the analysis to the subpopulation with adequate covariate overlap. Commonly used trimming threshold is 0.01, 0.05 or 0.1. A drawback of the trimming method is that the results may be sensitive to the choice of the threshold. The above examples of balancing weights are summarized in Table 14.1. In the next sections, we will focus on the special case of *overlap weight* [5].

Once the target population and estimand is decided, the central task is to estimate the corresponding WATE estimand. Following the convention in the literature, we maintain two standard assumptions throughout this chapter: (A1) *unconfoundedness*: $\Pr(Z_i \mid Y_{1i}, Y_{0i}, X_i) = \Pr(Z_i \mid X_i)$, implying that there is no unmeasured confounder; (A2) *overlap (or positivity)*: $0 < \Pr(Z_i = 1 \mid X_i) < 1$, implying that every unit's probability of being assigned to each treatment condition is bounded away from 0 and 1. Under these two assumptions, a consistent moment estimator for the WATE estimand (14.2) with any tilting function h is a Hájek estimator:

$$\hat{\tau}_g = \frac{\sum_i w_1(X_i) Z_i Y_i}{\sum_i w_1(X_i) Z_i} - \frac{\sum_i w_0(X_i)(1 - Z_i) Y_i}{\sum_i w_0(X_i)(1 - Z_i)}. \tag{14.5}$$

In observational studies, propensity scores are usually unknown and must be first estimated from the data, e.g. from a logistic regression, and the weights w are calculated from plugging the estimated propensity scores into formula (14.3). In theory, any type of balancing weights guarantee balance in the overall multivariable distribution of covariates between treatment groups. But this does not imply that using balancing weights with estimated propensity scores would balance the distribution of each individual covariate. In practice, balance of individual covariate before and after weighting is routinely checked by the metric of absolute standardized difference (ASD) [15]:

$$\text{ASD} = \left| \frac{\sum_{i=1}^N w_1(X_i) Z_i X_{ik}}{\sum_{i=1}^N w_1(X_i) Z_i} - \frac{\sum_{i=1}^N w_0(X_i)(1 - Z_i) X_{ik}}{\sum_{i=1}^N w_0(X_i)(1 - Z_i)} \right| \Big/ \sqrt{\frac{s_1^2 + s_0^2}{2}}, \tag{14.6}$$

where $k (= 1, ..., p)$ is a specific covariate, or the target population standardized difference (PSD) [16], $\max\{\text{PSD}_0, \text{PSD}_1\}$, where

$$\text{PSD}_z = \left| \frac{\sum_{i=1}^N w_z(X_i) \mathbf{1}\{Z_i = z\} X_{ik}}{\sum_{i=1}^N w_z(X_i) \mathbf{1}\{Z_i = z\}} - \frac{\sum_{i=1}^N h(X_i) X_{ik}}{\sum_{i=1}^N h(X_i)} \right| \Big/ \sqrt{\frac{s_1^2 + s_0^2}{2}}, \tag{14.7}$$

where s_z^2 is the unweighted variance of the covariate in group z. Setting $w_0 = w_1 = 1$ corresponds to the unweighted mean differences. As we shall show in a real application in Section 14.6.2, imbalance of individual covariate after propensity score weighting is common.

14.3 Overlap Weighting

The *overlap weight (OW)* is a type of balancing weight corresponding to the tilting function $h(x) = e(x)(1 - e(x))$,

$$\begin{cases} w_1(X_i) \propto 1 - e(X_i), & \text{for } Z_i = 1, \\ w_0(X_i) \propto e(X_i), & \text{for } Z_i = 0. \end{cases} \tag{14.8}$$

OW weights each unit by its probability of being assigned to the opposite group. The tilting function $h(x)$ maximizes when the propensity score is 1/2, and gradually reduces when the propensity score moves toward 0 or 1. Unlike IPW, the overlap weight is bounded and thus is less sensitive to extreme weights. Compared to the aforementioned practice of trimming extreme weights, OW continuously down-weights the units in the tail of the propensity score distribution and thus avoids arbitrary choosing a cutoff point for trimming. In implementation, OW must be normalized so that the sum of the weights within a treatment group is 1, i.e., all weights in one group need to be divided by the sum of weights of that group. Normalization is automatically done in the Hájek estimator (14.5).

OW was first proposed by Alan Zaslavsky, who applied it to a series of medical studies, starting in [17]. The concept of OW has also been independently discovered in various contexts [18, 19]. Li et al. [5] coined the name "overlap weights" and established two important theoretical properties, as described below.

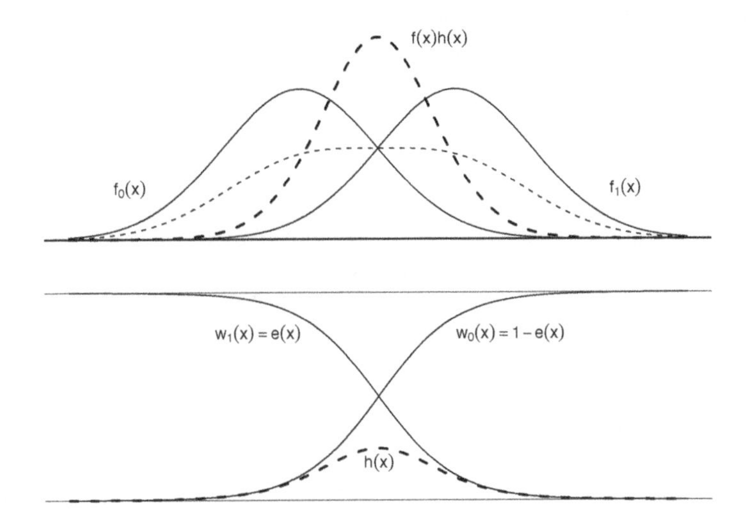

FIGURE 14.1
Overlap weights for two normally-distributed groups with different means. In the upper panel, the left and right solid lines, the thin and thick dashed lines represent the density of the covariate in the control, treated, combined ($h(x) = 1$), and overlap weighted ($h(x) = e(x)(1 - e(x))$) populations, respectively. In the lower panel, the two solid lines represent $w_0(x)$, $w_1(x)$ and the dashed line represents $h(x) = e(x)(1 - e(x))$.

Property 1 (Minimum variance). *If the variance of the potential outcomes is homoscedastic between the two groups, i.e.,* $Var(Y_{1i} \mid X_i) = Var(Y_{0i} \mid X_i)$*, then the tilting function* $h(x) = e(x)(1 - e(x))$ *gives the smallest asymptotic variance of the Hájek estimator* $\hat{\tau}_g$ *in* (14.5) *among all* $h(x)$.

OW is constructed directly from the minimum variance property, which implies that, among all balancing weights, OW leads to a target population on which the average treatment effect can be estimated with the most precision. Though statistically derived, OW corresponds to a scientifically meaningful target population, which consists of units whose characteristics could appear with substantial probability in either treatment group. This target population is called the overlap population and the corresponding estimand is the *average treatment effect on the overlap population (ATO)*. In medicine, the overlap population is known as the population in *clinical equipoise*, that is, patients for whom clinical consensus is ambiguous or divided, and thus research on these patients may be most needed. This is the population eligible to be enrolled in randomized clinical trials, where the treatment effect can be estimated with the most interval validity. In this sense, OW extracts the information on the subpopulation in an observational study that is the closest to a randomized trial [7]. In policy, these might be the units whose treatment assignment would be most responsive to a policy shift as new information is obtained. Matching and IPW with trimming methods target at similar populations with adequate overlap between treatments, but in a less transparent fashion. Figure 14.1 illustrates OW using a simple example of two univariate normal populations with equal size and variance. The upper panel shows the target population density $g(x)$, which maximizes where the treated and control groups most overlap. The lower panel shows that the OW tilting function $h(x) = e(x)(1 - e(x))$ peaks where the propensity score is 1/2, and places more emphasis on units with propensity scores close to 1/2, who could be in either group, relative to those with propensity scores close to 0 or 1. In practice, the overlap population as well as any target population belongs to the balancing weights family can be characterized by a descriptive table on the weighted averages of each covariate. A example is given in Section 14.6.2.

As a technical note, the minimal variance property is proved under (i) true propensity scores, and (ii) a homoscedasticity assumption. Nonetheless, it has been consistently shown in simulations and real applications that neither is crucial in practice [6, 20], and OW provides the most efficient causal estimates under limited overlap comparing to IPW and trimming [21, 22]. Because the analytical expression of ATO involves the true propensity score, one may question whether OW would be more sensitive to misspecification of the propensity score model than IPW. Recent work has demonstrated the opposite [14]. A possible explanation is that OW targets at the subpopulation with the most overlap, whose propensity scores are close to 0.5. As such, the estimated propensity scores, even if from a misspecified model, unlikely deviate much from the true scores. This phenomenon exemplifies the dictum "design trumps analysis" for objective causal inference [23] in the sense that OW attempts to solve the lack of overlap problem from a design rather than analysis perspective by shifting the target population.

The second property of OW concerns finite-sample balance of individual covariates.

Property 2 (Exact balance). *When the propensity scores are estimated by maximum likelihood under a logistic regression model,* logit $e(X_i) = \beta_0 + X_i\beta'$, *overlap weights lead to exact balance in the means of any included covariate between treatment and control groups. That is,*

$$\frac{\sum_i X_{ik} Z_i(1-\hat{e}(X_i))}{\sum_i Z_i(1-\hat{e}(X_i))} = \frac{\sum_i X_{ik}(1-Z_i)\hat{e}(X_i)}{\sum_i (1-Z_i)\hat{e}(X_i)}, \quad \text{for } k = 1, \ldots, p, \tag{14.9}$$

where $\hat{e}_i = \{1 + \exp[-(\hat{\beta}_0 + x_i\hat{\beta}')]\}^{-1}$ *and* $\hat{\beta} = (\hat{\beta}_1, ..., \hat{\beta}_p)$ *is the MLE for the regression coefficients.*

The exact balance property is a finite-sample result, not relying on asymptotic arguments. And it holds for any included predictor covariate and their functions – including high order and interaction terms – in the logistic propensity model. Technically, the exact balance property is unique to the logistic link and binary treatments. However, even if the propensity score is estimated via other models such as probit or machine learning models [24], or via multinomial models in the case of multiple treatments, the resulting OW generally leads to superior balance – in terms of both mean and multivariate measure such as Mahalanobis distance – compared to other balancing weights. This is as expected because overlap weight focuses on the population with most covariate overlap between treatment groups.

A direct corollary of the exact balance property is that if the postulated logistic propensity score model includes any interaction term with a binary covariate, then the resulting OW leads to exact balance in the means in the subgroups defined by that binary covariate. Building on this result, Yang et al. [25, 26] proposed to first use LASSO [27] to select important covariate–subgroup interactions in the propensity score model and then use OW to estimate the causal effects within pre-specified subgroups. Such an approach achieves covariate balance (and thus reduces bias) compared to IPW or matching in estimating subgroup causal effects.

Variance of the Hájek OW estimator (14.5) can be estimated by an empirical sandwich estimator [6], which also accounts for the uncertainty of estimating the propensity scores [28]. The variance can also be estimated using bootstrap, where one re-estimates the propensity scores and re-calculates the causal estimate in each of the bootstrap samples.

An increasingly popular estimator in the weighting literature is the augmented weighting estimator, which combines propensity score weighting and outcome model to improve robustness and efficiency [29]. A general form of the augmented weighting estimator with an arbitrary tilting function h and its corresponding balancing weight w is:

$$\begin{aligned}\hat{\tau}^{h,aug} \;=\;& \frac{\sum_i w_1(X_i)Z_i\{Y_i - \hat{\mu}_1(X_i)\}}{\sum_i w_1(X_i)Z_i} - \frac{\sum_i w_0(X_i)(1-Z_i)\{Y_i - \hat{\mu}_0(X_i)\}}{\sum_i w_0(X_i)(1-Z_i)} \\ &+ \frac{\sum_i h(X_i)\{\hat{\mu}_1(X_i) - \hat{\mu}_0(X_i)\}}{\sum_i h(X_i)},\end{aligned} \tag{14.10}$$

where $\hat{\mu}_z(X_i) = \widehat{E}(Y_i|X_i, Z_i = z)$ is the predicted outcome of unit i from an outcome model. With IPW, the augmented weighting estimator (14.10) is called the doubly–robust estimator [28, 29] because it is consistent if either the propensity score model or the outcome model, but not necessarily both, is correctly specified. When both models are correctly specified, the augmented IPW estimator $\hat{\tau}^{h,aug}$ is more efficient than the Hájek estimator $\hat{\tau}^h$. With OW, because the definition of the ATO estimand involves the true propensity score, $\hat{\tau}^{h,aug}$ is not consistent if the propensity model is misspecified. Nonetheless, $\hat{\tau}^{h,aug}$ is singly–robust in the sense that it is consistent for ATO if the propensity model is correctly specified, regardless of the outcome model specification. Interestingly, simulations and empirical applications consistently suggest that with OW, the Hájek estimator and the augmented estimator are very similar in terms of bias and variance. This is not surprising because it is well known that most methods lead to similar treatment effect estimates in randomized experiments. OW mimics the design of a randomized experiment; therefore, similar to a randomized study, outcome augmentation adds little value to the OW estimate of causal effect in an observational study.

14.4 Extensions of Overlap Weighting

In recent years, OW has been extended to many important settings [6, 30], expanding its applicability in practice. This subsection reviews some of the extensions.

14.4.1 Multiple treatments

Causal inference studies often involve more than two treatments. Li and Li [16] extend the framework of balancing weights to multiple treatments. Though the conceptual and technical extension is straightforward, multiple treatments require new notations and formulae, which are briefly reviewed below. Suppose we have J ($J \geq 3$) treatment groups, and let $Z_i \in \{1, \ldots, J\}$ stand for the treatment received by unit i. Further define $D_{ij} = \mathbf{1}\{Z_i = j\}$ as a set of multinomial indicators, satisfying $\sum_{i=1}^{J} D_{ij} = 1$ for all j. We denote the potential outcome for unit i under treatment j as $Y_{j,i}$, of which only the one corresponding to the actual treatment, $Y_i = Y_{Z_i,i}$, is observed. The generalized propensity score is the probability of receiving a potential treatment j given X_i [31]: $e_j(X_i) = \Pr(Z_i = j|X_i)$, with the constraint that $\sum_{j=1}^{J} e_j(X_i) = 1$.

Similar to the case of binary treatments, let $\mu_j(x) = E(Y_{j,i} \mid X_i = x)$ be the conditional expectation of the potential outcome in treatment j. Assume the observed sample is drawn from a population with probability density of covariates $f(x)$. Any target population, denoted by its covariate density $g(x)$, has a one-to-one map to a tilting function $h(x) = g(x)/f(x)$. Then, the average potential outcome for the jth treatment among the target population is

$$\mu_j^h = E_g(Y_{j,i}) = \frac{E\{h(x)\mu_j(x)\}}{E\{h(x)\}}. \tag{14.11}$$

Causal estimands can then be constructed in a general manner as contrasts based on μ_j^h. For example, the most commonly seen estimands in multiple treatments are the pairwise average treatment effects between groups j and j': $\tau_{j,j'}^h = \mu_j^h - \mu_{j'}^h$. This definition can be generalized to arbitrary linear contrasts. Denote $\boldsymbol{a} = (a_i, \cdots, a_J)$ as a contrast vector of length J. A general class of additive estimands is $\tau^h(\boldsymbol{a}) = \sum_{j=1}^{J} a_j \mu_j^h$. Specific choices for $\boldsymbol{a}$ with nominal and ordinal treatments can be found in [16].

It is straightforward to generalize the balancing weights framework in Section 14.2 to the case of multiple treatments. Let $f_j(x) = f(x \mid Z_i = j)$ be the density of the covariates in the observed jth group, we have $f_j(X) \propto f(x)e_j(x)$. For any pre-specified tilting function $h(x)$ and equivalently the target population $g(x) = f(x)h(x)$, the balancing weight for treatment j ($j = 1, ..., J$) is:

$$w_j(x) \propto \frac{f(x)h(x)}{f(x)e_j(x)} = \frac{h(x)}{e_j(x)}. \tag{14.12}$$

The above weights balance the weighted distributions of the covariates across the J comparison groups, $f_j(x)w_j(x) = f(x)h(x)$, for all j.

Similar as the case of binary treatments, to estimate μ_j^h, we need to maintain two standard assumptions: (A3) *weak unconfoundedness*: $Y_{j,i} \perp \mathbf{1}\{Z_i = j\} \mid X_i$, for all j, and (A4) *overlap*: the generalized propensity score is bounded away from 0 and 1: $0 < e_j(x) < 1$, for all j and x. Under these assumptions, a consistent estimator for μ_j^h with any tilting function $h(x)$ is a Hájek estimator:

$$\hat{\mu}_j^h = \frac{\sum_{i=1}^N w_j(X_i)D_{ij}Y_i}{\sum_{i=1}^N w_j(X_i)D_{ij}}. \tag{14.13}$$

The corresponding target estimand of each weighting scheme is its pairwise – between each pair of treatments – counterpart in binary treatments.

Under homoscedasticity, the function $h(x) = \{\sum_j 1/e_j(x)\}^{-1}$, i.e., the harmonic mean of the generalized propensity scores of all groups, minimizes the sum of the asymptotic variance of the Hájek estimator $\hat{\mu}_j^h$ of all groups among all choices of h [16]. Consequently, one can define the corresponding OW for treatment group j as:

$$w_j(x) \propto \frac{1/e_j(x)}{\sum_{k=1}^J 1/e_k(x)}, \tag{14.14}$$

which reduces to the OW in binary treatments (14.8) when $J = 2$.

The maximum of the harmonic mean function h is attained when $e_j(X) = 1/J$ for all j, that is, when the units have the same propensity to each of the treatments. Figure 14.2 demonstrates a ternary plot of this tilting function when $J = 3$. It is evident that this tilting function gives the most relative weight to the covariate regions in which none of the propensities are close to zero, and down-weighs the region where there is lack of overlap in at least one dimension. Therefore, we can interpret the corresponding target population to be the subpopulation with the most overlap in covariates among all groups.

14.4.2 Time-to-even outcomes

Survival, or more generally time-to-event outcomes are common in comparative effectiveness research and require unique handling because they are usually incompletely observed due to right-censoring. A popular approach to draw causal inference with survival outcomes is to combine standard survival estimators with propensity score methods. For example, one can construct the Kaplan-Meier estimator on a propensity weighted sample to estimate the counterfactual survival curves [32, 33]. Another way is to fit a survival outcome model, such as the Cox proportional hazards model [34] to a propensity score weighted sample to estimate the causal hazard ratio [35] or the counterfactual survival curves [36]. The existing literature on propensity score methods with survival outcomes mostly focuses on IPW, but the aforementioned drawbacks of IPW carry over to the survival outcome setting. Below we show how to extend the balancing weights framework, particularly OW, to incorporate survival outcomes and address these drawbacks.

To accommodate survival outcomes, we first introduce some new notations and causal estimands. To simplify the discussion, we proceed under binary treatments. Under SUTVA, each unit has a

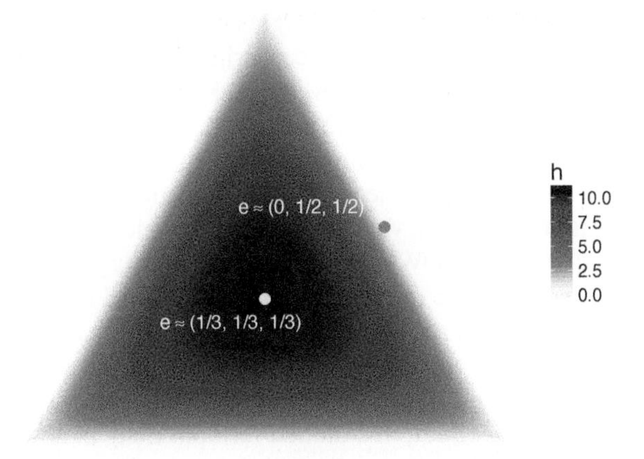

FIGURE 14.2
Ternary plot of the optimal tilting h – the harmonic mean – as a function of the generalized propensity score vector with $J = 3$ treatments. Each point in the triangular plane represents a unit with certain values of the generalized propensity scores. The value of each generalized propensity score is proportional to the orthogonal distance from that point to each edge. Overlap weighting scheme emphasizes the centroid region with good overlap, e.g., units with $e(x) \approx (1/3, 1/3, 1/3)$, and smoothly down-weighs the edges, e.g., units with $e(x) \approx (0, 1/2, 1/2)$.

potential survival time $T_i(z)$ and a potential censoring time $C_i(z)$ under treatment z. let T_i and C_i denote the observed survival time and censoring time for each unit, respectively. The survival time might be right censored when $C_i \leq T_i$ so we observe $V_i = \min(T_i, C_i)$ and the indicator whether the subject failed within the study period $\delta_i = \mathbf{1}_{T_i \leq C_i}$. There are several causal estimands with survival outcomes [37]. Below we focus on a class of WATE defined by the counterfactual survival functions, denoted by $S^z(t \mid X) = \Pr(T(z) \geq t \mid X)$ for $z = 0, 1$. As before, we use a tilting function $h(x)$ to represent a target population $g(x) = h(x)f(x)$. The survival probability causal effect (SPCE), or causal risk difference, on target population g is

$$\tau_h^{SPCE}(t) = E_g\{S^1(t) - S^0(t)\} = \frac{E[h(x)\{S^1(t \mid x) - S^0(t \mid x)\}]}{E\{h(x)\}}. \quad (14.15)$$

In order to identify the SPCE , besides the standard unconfoundedness and overlap assumptions, we additionally need an assumption on the censoring mechanism: (A5) *covariate dependent censoring*. $T_i(z) \perp C_i(z) \mid \{X_i, Z_i\}$, which states that the censoring time is independent of survival time given the covariates in each group. The key to drawing causal inference with survival outcomes is to address the selection bias associated with the right censoring of survival outcomes. The censoring process can be represented by the survival distribution of the potential censoring time, denoted by $K_c^z(t, X) = \Pr(C \geq t \mid X, Z = z)$ for $z = 0, 1$. The censoring score $K_c^z(t, X)$ is usually unknown and must be estimated from the observed data, e.g., from a Cox model or a parametric Weibull model. Cheng et al. [20] proposed an estimator for SPCE that combines balancing weights and inverse probability of censoring weights:

$$\begin{aligned}\hat{\tau}_h^{SPCE}(t) \quad &= \quad \left(1 - \frac{\sum_i w(X_i) Z_i \mathbf{1}(V_i \leq t)\delta_i / \widehat{K}_c^1(V_i, X_i)}{\sum_i w(X_i) Z_i}\right) \\ &\quad - \left(1 - \frac{\sum_i w(X_i)(1 - Z_i)\mathbf{1}(V_i \leq t)\delta_i / \widehat{K}_c^0(V_i, X_i)}{\sum_i w(X_i)(1 - Z_i)}\right) \quad (14.16)\end{aligned}$$

where $w(X_i)$ is the balancing weight corresponding to the tilting function h and $\delta_i / \widehat{K}_c^z(V_i, X_i)$ is the estimated inverse probability of censoring weights applied only to the non-censored observations throughout the study follow-up. When the censoring process does not depend on covariates, estimator (14.16) reduces to a propensity score weighted Kaplan-Meier estimator [32,38]. Cheng et al. [20] proved that under Assumptions A1, A2 and A5, estimator (14.16) is point-wise consistent of $\tau_h^{SPCE}(t)$ for any time $0 \leq t \leq t_{max}$ with any titling function h. Moreover, under certain homoscedasticity conditions, OW achieves the smallest asymptotic point-wise variance for estimating $\tau_h^{SPCE}(t)$ among the class of balancing weight. Extensive simulations in [20] show that, similar to the case of non-censored outcomes, OW consistently outperforms IPW and trimming methods in terms of bias, variance, and coverage in estimating the SPCE, and the advantage increases as the degree of covariate overlap between groups decreases.

An alternative approach for propensity score weighting with survival outcome is through the jackknife pseudo-observations [39,40], which can handle several causal estimands, including SPCE and the restricted average causal effect, in a unified fashion. Zeng et al. [8] proved the minimum variance property of OW also holds with the pseudo-observation-based propensity score weighting estimator.

14.4.3 Covariate adjustment in randomized experiments

We have so far discussed OW in the context of observational studies. Below we will show the use of OW and more generally propensity score weighting as a covariate adjustment method in randomized experiments [26,41–43].

Randomization ensures balance of both measured and unmeasured confounders in expectation; however, chance imbalance of baseline covariates is common in a single randomized experiment, especially when the sample size is small. For covariates that are predictive of the outcome, adjusting for their imbalance in the analysis – known as covariate adjustment – can improve the statistical power of the treatment effect estimates. Covariate adjustment is traditionally conducted via outcome regression, i.e., fit an analysis of covariance (ANCOVA) model of the outcome on the treatment, covariates, and possibly their interactions and take the coefficient of the treatment variable as the estimated treatment effect [44]. Even if the outcome regression model is misspecified, the point estimate remains consistent [45] but the standard errors may increase.

Propensity score weighting for covariate adjustment in randomized experiments bypasses the outcome model. Instead, it fits a *working* propensity score model and then estimates the treatment effect by a moment estimator (14.2), i.e., the weighted difference of the outcome between the two treatment arms according to a chosen weighting scheme. This procedure is exactly the same as that in an observational study; however, it differs in the goal, estimands, theory, and operating characteristics. First, in randomized experiments (for stratified experiments, assuming we are within the stratum), the true propensity score for all units is the randomization probability r, known and fixed for any covariate value x: $e(x) = \Pr(Z_i = 1 \mid X_i) = \Pr(Z_i = 1) = r$. Therefore, as long as $h(x)$ is a function of the true propensity score, different tilting functions lead to the same target population g, and the WATE estimand in (14.2) reduces to ATE: $\tau^h = \tau$. This is distinct from observational studies, where the true propensity scores usually vary between units, and consequently different tilting functions correspond to different target populations and estimands. Second, the estimated propensity score is not an estimate of the randomization probability, rather it is simply a numerical one-dimensional summary of the covariates and thus is called a *working* score. Here the goal of propensity score weighting is not to remove confounding and reduce bias, instead it is to reduce variance in estimating the treatment effect.

Comparing to ANCOVA, propensity score weighting for covariate adjustment in randomized experiments avoids specifying an outcome model and thus the potential loss of efficiency when the model is misspecified. Operationally, propensity score weighting can straightforward handle non-continuous outcomes, whereas ANCOVA often fails to converge in cases of rare binary or ordinal outcomes. Zeng et al. [43] established that, as long as the tilting function is a smooth function of the true propensity score, its corresponding balancing weight is semiparametric efficient and asymptotically equivalent to the ANCOVA estimator [44]. This includes OW and IPW as special cases. Moreover, owing to its exact balance property, OW has a unique advantage of completely removing chance imbalance when the propensity score is estimated by a logistic regression, which improves the face validity of a randomized experiment as well as efficiency. Through extensive simulations, Zeng et al. [43] demonstrated that OW consistently outperforms IPW in finite samples and improves the efficiency over ANCOVA when the outcome model is incorrectly specified.

14.5 Implementation and Software

Propensity score weighting is popular in applied research because its implementation and programming is simple, arguably much simpler than matching methods. Below is an outline of conducting propensity score weighting analysis, which will be illustrated in a real data example later.

1. Estimate the propensity scores from a model, e.g., logistic model or machine learning methods. Using that model, calculate the weight of each unit according to a selected weighting scheme.

2. Check the overall overlap and balance of the covariates between treatment groups, which is usually visualized by overlaid histograms of the estimated propensity scores of each group. Display ASDs or PSDs before and after weighting in a baseline characteristics table (often known as "Table 1" in medical research articles; an example is given in Table 14.3) and visualized via a Love plot (an example is given in Figure 14.5). A rule of thumb for determining adequate balance is when ASD or PSD of all covariates is controlled within 0.1 [15]. In the same table, also present the weighted average of each covariate: $\sum_{i=1}^{N} h(X_i)X_i / \sum_{i=1}^{N} h(X_i)$, which characterizes the corresponding target population.

3. Estimate the treatment effect by the Hájek estimator (14.5), or the augmented estimator (14.10), in which case one needs to specify an additional outcome model.

In propensity score analysis of observational studies, an important step is to choose specification of the propensity score model. In theory true propensity scores lead to balance of any observed covariate [10]. Therefore, a common practice is to use the balance of the weighted covariates between treatment group to determine if a propensity score model is adequately specified. This is often conducted in an iterative fashion [46]: (1) first specify a propensity score model and then check the resulting covariate balance, e.g., using the ASD metric; (2) when imbalance of some covariates is detected, add interaction or higher order terms in the propensity score model and re-fit and re-check the resulting balance; (3) repeat this process until most important covariates are balanced. Such a procedure, however, focus on the problem of "how to balance," namely, given a set of covariates, how to make the treatment groups have similar covariate distribution.

A seeming drawback of OW is that it does not allow such a check-and-refit procedure because it automatically achieves exact or nearly exact mean balance of any covariate entering a propensity score model. To understand why this is not a true drawback, it is important to differentiate two tasks in causal inference: "what to balance," namely, identifying the covariates that should be balanced in a causal study to reduce bias, versus "how to balance," namely, given a set of covariates, how to make the treatment groups have similar covariate distribution. Most weighting and matching methods, combined with the check-and-refit procedure, address the problem of "how to balance" instead of "what to balance." Because the ultimate goal of causal inference is to estimate treatment effects, finding "what to balance" is arguably more crucial. One can obtain perfect balance of a covariate via statistical methods, but if that covariate is not a confounder, i.e., covariates that are predictive of both outcome and treatment assignment, then such balancing does not improve causal inference. In a sense, OW already liberates analysts from the "how to balance" problem owing to the exact balance property, and thus allows analysts to focus on finding "what to balance." In practice, we recommend to first identify the set of confounders from substantive knowledge, e.g. based on discussion with domain scientists and then specify a propensity score model that includes all these confounders.

OW has been implemented in several R packages, including `PSW` [47] and `WeightIt` [48]. In particular, the most comprehensive package for propensity score weight is the R package `PSweight` developed by Zhou et al. [49], which supports (i) several balancing weights, including OW, IPW, matching weights, entropy weights, trimming, (ii) binary and multiple treatments, (iii) Hájek and augmented estimators, (iv) ratio estimands for binary and count outcomes, (v) time-to-event outcomes. `PSweight` also provides diagnostic tables and graphs for visualizing target population and covariate balance.

14.6 Illustration

This section provides two illustrative examples of propensity score weighting analysis and compares several balancing weights.

14.6.1 A simulated example

We use a simple simulated example to illustrate the operating characteristics of OW, IPW, and ATT weights under limited overlap. To highlight the main message, we consider a scenario with a large shift in covariate between two groups and homogeneous treatment effect. Each group has 1000 units, and each unit has a single covariate X_i: for units in the control group ($Z_i = 0$), X_i is generated from $\mathrm{N}(0, 1)$, and for units in the treated group ($Z_i = 1$), X_i is generated from $\mathrm{N}(2, 1)$. The potential outcomes Y_{zi} are generated from $Y_{zi} = X_i + \tau z + \epsilon_i$, with $\epsilon_i \sim \mathrm{N}(0, 1)$. This outcome model implies a homogeneous true treatment effect $\tau = 1$ across all units, and thus the true effect on any subpopulation is also 1.

We estimate the propensity scores from a logistic regression with a main effect of X. Figure 14.3 displays the distributions of the weighted covariate distributions as well as the estimated propensity scores within each group under OW, IPW, ATT weights. Distributions of the estimated propensity scores indicate severe lack of overlap, as expected from the large shift of the covariates between the two groups. OW leads to nearly identical weighted covariate distributions between the two groups. IPW attempts to match the covariate distribution within each group to the overall covariate distribution, but the severe lack of overlap renders the units with propensity scores close to 0 or 1 to be up-weighted dramatically. Consequently, the IPW weighted covariate distribution in the two groups deviates from each other in the tail region. A similar pattern is observed with the ATT weights in the control group, as the lack of overlap leads to extreme weights in the control units with the highest X values. ASD of X is reported in Table 14.2. While each weighting scheme improves the balance in covariate means, OW achieves exact balance (ASD equals 0), whereas both IPW and ATT weights still lead to noticeable imbalance, with ASD being 0.08 and 0.03, respectively.

We then estimate treatment effects using the Hájek estimator. Table 14.2 presents the estimates based on OW, IPW and ATT weights, along with the estimated standard error, obtained from 1000 bootstrap simulations. For comparison, we also report the ordinary least square (OLS) estimate from the true outcome model, which serves as the optimal benchmark. We can see that the unweighted estimate, that is, the raw difference in the means between the groups, is extremely biased, with relative bias being about 300%. OW leads to the smallest bias and variance in estimating τ among the three weighting methods, matching the OLS estimate from the true model, without requiring

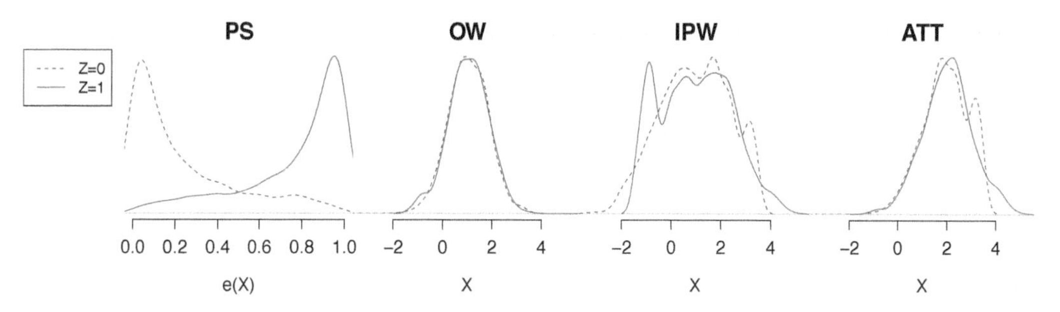

FIGURE 14.3
Distribution of propensity scores, and weighted covariate distributions by OW, IPW, and ATT weights, respectively, within each treatment group, of the simulated example in Section 14.6.1.

TABLE 14.2
Absolute standardized difference (ASD), estimated treatment effect, and corresponding standard error from OW, IPW and ATT weights in the simulated example in Section 14.6.1. The true treatment effect is $\tau = 1$.

	Unweighted	OW	IPW	ATT	OLS
ASD	1.59	0	0.08	0.03	-
$\hat{\tau}_h$	2.95	1.00	0.79	1.17	1.00
SE($\hat{\tau}_h$)	0.05	0.02	0.43	0.11	0.02

correctly specifying the outcome model. IPW and ATT weights both yield a relative bias of about 20%, and a standard error over 20 and 5 times of that of OW, respectively. This simple example illustrates that, although in theory IPW and ATT weights lead to covariate balance and easy-to-interpret target populations, in practice both methods can lead to substantial imbalance and inflated bias and variance when there is limited covariate overlap between treatments. In contrast, OW provides optimal covariate balance and unbiased point estimate with low variance regardless of the degree of overlap or outcome generating process. Though our simulations focus on a simple case of homogeneous treatment effects, these patterns are consistently observed in several comprehensive simulation studies with heterogeneous treatment effects [6, 8, 14, 20, 50].

14.6.2 The National Child Development Survey data

The National Child Development Survey (NCDS) is a longitudinal study of children born in the United Kingdom (UK) in 1958. NCDS collected information such as educational attainment, familial backgrounds, and socioeconomic and health well-being on 17,415 individuals. The data is publicly available online. For this simple illustration, our goal is to estimate the causal effect of education attainment on wages from the observational NCDS data. We followed [51] to pre-process the data and obtain a subset of 3,642 males employed in 1991 with complete educational attainment and wage information. We use imputation by chained equations [52] to impute missing covariates and obtain a single imputed data set for the analysis. The outcome variable is log transformed gross hourly wage in Pounds. The treatment variable is educational attainment, and we created a binary indicator of whether one had attained any academic qualification, with 2399 and 1,243 individuals who attained and had not attained any academic qualification, respectively. We consider twelve pre-treatment covariates. The variable `white` indicates whether an individual identified himself as white race; `scht` indicates the school type they attended at age 16; `qmab` and `qmab2` are math test scores at age 7 and 11; `qvab` and `qvab2` are two reading test scores at age 7 and 11; `sib_u` stands for the number of siblings; `agepa` and `agema` are the ages of parents in year 1974; in the same year,

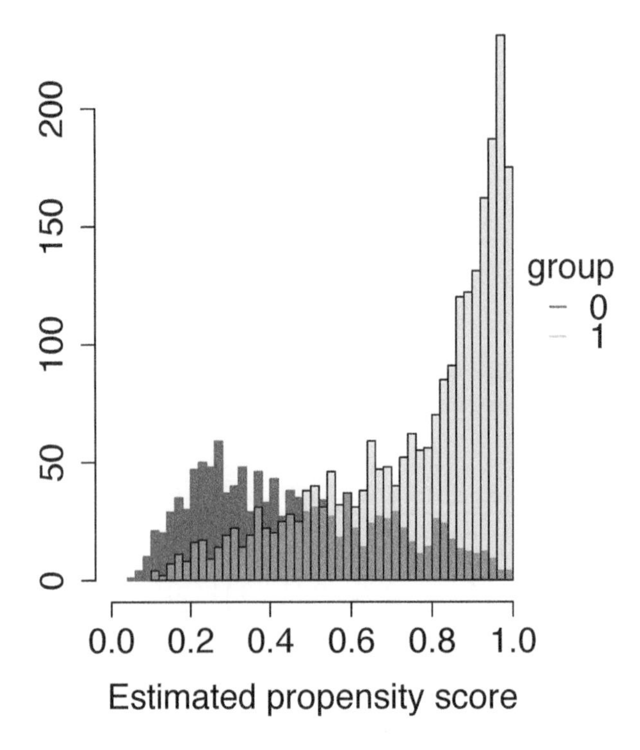

FIGURE 14.4
Histogram of the estimated propensity scores in the NCDS study.

the employment status of mother `maemp` was also collected; `paed_u` and `maed_u` are the years of education of the parents.

We first estimate the propensity scores via a logistic regression model with main effects of all covariates and several interactions suggested by [51]. Figure 14.4 presents the histogram of the estimated propensity scores. The histogram suggests that though overall distributions of the covariates are imbalanced between the two groups, the range of the propensity scores between the groups is well overlapped. Figure 14.5 presents the Love plot of the PSD metric of each covariate before and after three weighting schemes: OW, IPW, ATT. Both graphs are produced by the `plot.SumStat()` function in the R `PSweight` package. Clearly, the unweighted mean differences are substantially larger than the commonly used balance threshold 0.1, while propensity score weighting in general improves covariate balance. Among the three weighting schemes, OW and IPW have controlled the maximum PSD for each covariate to be below 0.1; as guaranteed by theory, OW provides exact balance, with the maximum PSD for each covariate being zero. Table 14.3 presents the unadjusted mean by treatment status and weighted overall mean by each weighting scheme of a subset of covariates. We can see different weighting methods result in marked difference in several covariates. We recommend to provide such summary tables in real applications because they concretely characterize the target population corresponding to any chosen weighting method.

Based on the estimated propensity scores, we can estimate the treatment effect on different target populations using their corresponding weighting scheme by the Hájek estimator. For example, using OW, we estimated the ATO to be 0.179 with standard error of 0.016; using IPW, we estimated the

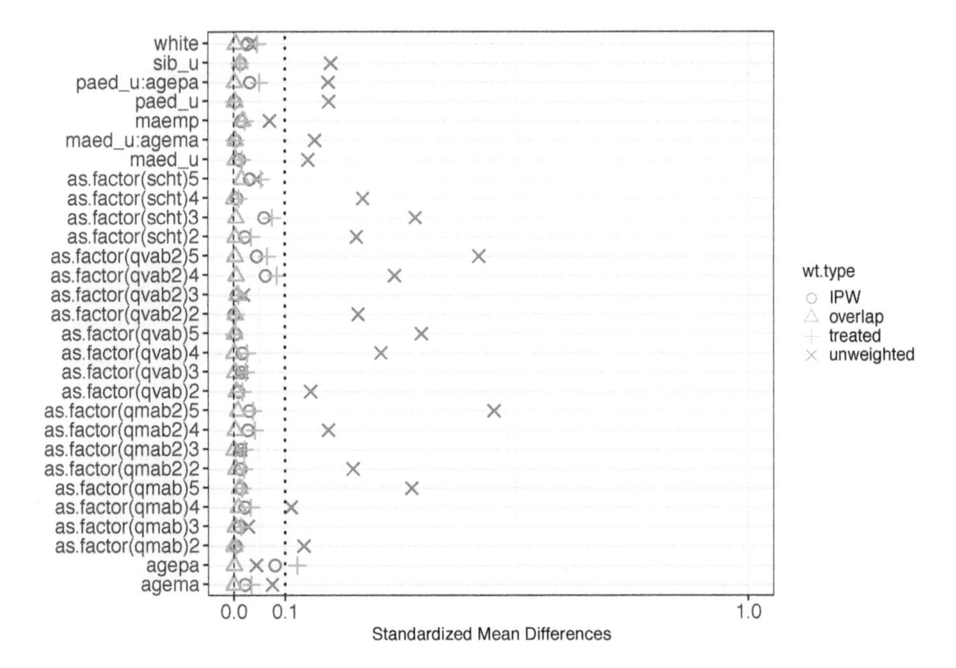

FIGURE 14.5
Love plot of the maximum PSD metric of each covariate in the NCDS study.

TABLE 14.3
Unadjusted mean by group and weighted overall mean (by IPW, ATT, and OW, respectively) of selected covariates in the NCDS study. Note: School type, math, and reading scores are categorical variables, each with multiple categories. Here we only present the summary of one category for each variable to illustrate.

	Unadjusted			Adjusted overall		
	No edu	Any edu	Overall	IPW	ATT	OW
Race (White)	.96	.97	.97	.97	.98	.97
Mother employed	.48	.53	.51	.51	.52	.51
School type	.30	.15	.20	.18	.16	.26
Math score at age 7	.23	.15	.18	.18	.15	.22
Math score at age 11	.28	.14	.19	.19	.14	.26
Reading score at age 7	.26	.17	.20	.19	.16	.26
Reading score at age 11	.30	.15	.20	.20	.15	.29
Father age in 1974	46.32	46.74	46.60	46.90	47.24	46.47
Mother age in 1974	43.19	43.83	43.61	43.67	43.96	43.45
Father education (yrs)	8.12	9.15	8.80	8.82	9.18	8.44
Mother education (yrs)	8.20	8.96	8.70	8.69	8.94	8.46
Number of siblings	2.09	1.56	1.74	1.75	1.56	1.86

ATE for the overall population to be 0.193 with standard error of 0.021. This suggests that obtaining any academic attainment would statistically significantly increase the hourly wage. In this study, the point estimates from different weighting schemes are similar, though OW leads to the smallest variance. This is as expected given that the covariates between the two groups are well overlapped. When there is severe lack of overlap or/and heterogeneous treatment effects, different weighting schemes are expected to lead to much more variable treatment effect estimates; for such an example, see [6]. More details, including R code, of the NCDS example can be found in [49].

14.7 Discussion

In this chapter we have discussed the general class of propensity-score-based balancing weights and highlighted a special member of the class, overlap weight (OW). OW shifts the traditional goalpost from finding an "optimal" estimate of ATE on the overall population to finding an "optimal" subpopulation on which the ATE can be estimated with the most precision and internal validity. The target population of OW is the subpopulation with substantial chance of being assigned to either treatment conditions, namely, the population in "clinical equipoise" in medical research. OW possesses scientifically meaningful interpretation and desirable theoretical properties, including minimum variance and exact mean balance, and has been repeatedly shown in simulations and real applications to outperform alternative weighting methods. For these reasons, OW has been increasingly popular in real world applications, including several high profile COVID-19 studies, e.g., [7, 53].

The development of OW and more broadly balancing weights is motivated by a key yet often overlooked question in causal inference: "what population is the ATE defined on?" In practice, we apply a statistical method to a sample, but usually we want to interpret the results on a target population where the sample is drawn from. If the study sample is representative of a scientifically meaningful population, then ATE can be interpreted as the treatment effect on that population. But in many real world situations, the sample does not represent a natural population; consequently, the ATE estimated from such a sample has an opaque interpretation.

A central message from the balancing weight framework is that, instead of fixating on a specific weighting method, analysts should first specify a target population and then estimate the treatment effect on that population from the observed sample using the corresponding weighting scheme. The singular focus on ATE is neither justified nor necessary. To better understand this point, it is worthwhile to compare surveys where IPW originates from and causal inference. The starting point of survey sampling is *design*, comprised of an explicitly specified target population and corresponding survey weights or sampling probabilities. A well-designed survey rarely has extreme weights. In contrast, the starting point of causal inference with observational studies is *data*, and one uses the observed data to attempt to re-construct the design. Because observational data are not sampled based on a pre-specified design, lack of overlap and consequently extreme weights are common. Automatic focus on ATE (and equivalently IPW) implicitly assumes that the sample is representative of a well-defined target population, which is often not the case in practice, and thus often obscures the design element in observational studies. Furthermore, in comparative effectiveness studies, there is a trade-off between interval validity and external validity [54], which measures how generalizable the results are to a different population. OW achieves optimal internal validity at the cost of external validity. In contrast, other weighting scheme such as IPW may be more suitable if external validity is the priority.

OW can be extended in several directions. First, continuous treatments [55] are common in practice, and the goal is usually to estimate a dose–response relationship. Ensuring adequate overlap is challenging in continuous treatments because of the small sample size in each level of treatment. Operationally, we can view a continuous treatment as the limit of multiple ordered treatments, and thus it is straightforward to extend the OW method in Section 14.4.1 to continuous treatments. However, conceptually the meaning of overlap is ambiguous with continuous treatments and requires more formal investigation. Second, OW has so far been developed mostly in cross-sectional settings. Many observational studies involve longitudinal treatments, where lack of overlap becomes more severe as time evolves. We could frame longitudinal treatments as multiple treatments where each treatment path is a treatment level. However, to calculate OW, one need to know all the time-varying covariates, including the counterfactual intermediate outcomes under each treatment path, which are not observed for each unit. One possibility is to specify an outcome model to impute the counterfactual intermediate outcomes, but this would counter the simplicity of OW. Wallace and Moodie [19] independently discovered OW in the context of identifying dynamic treatment regimes, but that method does not apply to estimate marginal effects of longitudinal treatments. Overall, extension of OW to longitudinal treatments remains to be a desirable but challenging open topic. Third, a recent trend in biopharmaceutical development is to augment clinical trials with real-world evidence. In particular, in clinical trials of rare or severe diseases, there has been an increasing need to construct external or synthetic controls using historical data [56]. Propensity score matching is often used to match historical and concurrent data, but sometimes results in discarding a large number of unmatched units, which is undesirable when the concurrent trial has a small sample size. OW can bypass such a problem and construct external controls that are the most similar to the concurrent trial. Furthermore, one may combine OW with the Bayesian dynamic pooling method [57] to adjust for both measured and unmeasured confounding between historical and concurrent data.

References

[1] Richard Crump, Joseph Hotz, Guido Imbens, and Oscar Mitnik. Dealing with limited overlap in estimation of average treatment effects. *Biometrika*, 96(1):187–199, 2009.

[2] Kosuke Imai and Marc Ratkovic. Covariate balancing propensity score. *Journal of the Royal Statistical Society: Series B (Statistical Methodology)*, 76(1):243–263, 2014.

[3] José R. Zubizarreta. Stable weights that balance covariates for estimation with incomplete outcome data. *Journal of the American Statistical Association*, 110(511):910–922, 2015.

[4] Qingyuan Zhao. Covariate balancing propensity score by tailored loss functions. *The Annals of Statistics*, 47(2):965–993, 2019.

[5] Fan Li, Kari Lock Morgan, and Alan M. Zaslavsky. Balancing Covariates via Propensity Score Weighting. *Journal of the American Statistical Association*, 113(521):390–400, 2018.

[6] Fan Li, Laine E Thomas, and Fan Li. Addressing extreme propensity scores via the overlap weights. *American journal of Epidemiology*, 188(1):250–257, 2019.

[7] Laine Thomas, Fan Li, and Michael Pencina. Overlap weighting: A propensity score method that mimics attributes of a randomized clinical trial. *Journal of the American Medical Association*, 323(23):2417–2418, 2020.

[8] Shuxi Zeng, Fan Li, Liangyuan Hu, and Fan Li. Propensity score weighting analysis for survival outcomes using pseudo observations. *Statistica Sinica*, 2021. doi:10.5705/ss.202021.0175

[9] Donald B Rubin. Randomization analysis of experimental data: The Fisher randomization test comment. *Journal of the American Statistical Association*, 75(371):591–593, 1980.

[10] Paul R. Rosenbaum and Donald B. Rubin. The central role of the propensity score in observational studies for causal effects. *Biometrika*, 70(1):41–55, 1983.

[11] Daniel G Horvitz and Donovan J Thompson. A generalization of sampling without replacement from a finite universe. *Journal of the American Statistical Association*, 47(260):663–685, 1952.

[12] Keisuke Hirano and Guido W Imbens. Estimation of causal effects using propensity score weighting: An application to data on right heart catheterization. *Health Services and Outcomes Research Methodology*, 2:259–278, 2001.

[13] Liang Li and Tom Greene. A weighting analogue to pair matching in propensity score analysis. *International Journal of Biostatistics*, 9(2):1–20, 2013.

[14] Yunji Zhou, Roland A Matsouaka, and Laine Thomas. Propensity core weighting under limited overlap and model misspecification. *Statistical Methods in Medical Research*, 29(12):3721–3756, 2020.

[15] Peter C Austin and Elizabeth A Stuart. Moving towards best practice when using inverse probability of treatment weighting (iptw) using the propensity score to estimate causal treatment effects in observational studies. *Statistics in Medicine*, 34(28):3661–3679, 2015.

[16] Fan Li and Fan Li. Propensity score weighting for causal inference with multiple treatments. *The Annals of Applied Statistics*, 13(4):2389–2415, 2019.

[17] Eric C Schneider, Paul D Cleary, Alan M Zaslavsky, and Arnold M Epstein. Racial disparity in influenza vaccination: Does managed care narrow the gap between African Americans and whites? *JAMA*, 286(12):1455–1460, 2001.

[18] Richard Crump, Joseph Hotz, Guido Imbens, and Oscar Mitnik. Moving the goalposts: Addressing limited overlap in the estimation of average treatment effects by changing the estimand. Technical Report 330, National Bureau of Economic Research, Cambridge, MA, September 2006.

[19] Michael P Wallace and Erica EM Moodie. Doubly-robust dynamic treatment regimen estimation via weighted least squares. *Biometrics*, 71(3):636–644, 2015.

[20] Cao Cheng, Fan Li, Laine E Thomas, and Fan Li. Addressing extreme propensity scores in estimating counterfactual survival functions via the overlap weights. *American Journal of Epidemiology*, 191(6), 1140–1151, 2022.

[21] Huzhang Mao, Liang Li, and Tom Greene. Propensity score weighting analysis and treatment effect discovery. *Statistical Methods in Medical Research*, 28(8):2439–2454, 2019.

[22] Kazuki Yoshida, Sonia Hernández-Díaz, Daniel H. Solomon, John W. Jackson, Joshua J. Gagne, Robert J. Glynn, and Jessica M. Franklin. Matching weights to simultaneously compare three treatment groups comparison to three-way matching. *Epidemiology*, 28(3):387–395, 2017.

[23] Donald B Rubin. For objective causal inference, design trumps analysis. *The Annals of Applied Statistics*, 2(3):808–840, 2008.

[24] Brian K Lee, Justin Lessler, and Elizabeth A Stuart. Improving propensity score weighting using machine learning. *Statistics in Medicine*, 29(3):337–346, 2010.

[25] Siyun Yang, Elizabeth Lorenzi, Georgia Papadogeorgou, Daniel M Wojdyla, Fan Li, and Laine E Thomas. Propensity score weighting for causal subgroup analysis. *Statistics in Medicine*, 40:4294–4309, 2021.

[26] Siyun Yang, Fan Li, Laine E Thomas, and Fan Li. Covariate adjustment in subgroup analyses of randomized clinical trials: A propensity score approach. *Clinical Trials*, 2021 Jul 16, 2021.

[27] Robert Tibshirani. Regression shrinkage and selection via the lasso. *Journal of the Royal Statistical Society: Series B (Methodological)*, 58(1):267–288, 1996.

[28] Jared K Lunceford and Marie Davidian. Stratification and weighting via the propensity score in estimation of causal treatment effects: a comparative study. *Statistics in Medicine*, 23(19):2937–2960, 2004.

[29] Heejung Bang and James M Robins. Doubly robust estimation in missing data and causal inference models. *Biometrics*, 61(4):962–973, 2005.

[30] Laine Thomas, Fan Li, and Michael Pencina. Using propensity score methods to create target populations in observational clinical research. *Journal of the American Medical Association*, 323(5):466–467, 2020.

[31] Guido W Imbens. The role of the propensity score in estimating dose-response functions. *Biometrika*, 87(3):706–710, 2000.

[32] James M Robins and Dianne M Finkelstein. Correcting for noncompliance and dependent censoring in an aids clinical trial with inverse probability of censoring weighted (ipcw) log-rank tests. *Biometrics*, 56(3):779–788, 2000.

[33] Alan E Hubbard, Mark J Van Der Laan, and James M Robins. Nonparametric locally efficient estimation of the treatment specific survival distribution with right censored data and covariates in observational studies. In *Statistical Models in Epidemiology, the Environment, and Clinical Trials*, pages 135–177. Springer, 2000.

[34] David R Cox. Regression models and life-tables. *Journal of the Royal Statistical Society: Series B (Methodological)*, 34(2):187–202, 1972.

[35] Peter C Austin and Elizabeth A Stuart. The performance of inverse probability of treatment weighting and full matching on the propensity score in the presence of model misspecification when estimating the effect of treatment on survival outcomes. *Statistical Methods in Medical Research*, 26(4):1654–1670, 2017.

[36] Stephen R Cole and Miguel A Hernán. Adjusted survival curves with inverse probability weights. *Computer Methods and Programs in Biomedicine*, 75(1):45–49, 2004.

[37] Huzhang Mao, Liang Li, Wei Yang, and Yu Shen. On the propensity score weighting analysis with survival outcome: Estimands, estimation, and inference. *Statistics in Medicine*, 37(26):3745–3763, 2018.

[38] Glen A Satten and Somnath Datta. The Kaplan–Meier estimator as an inverse-probability-of-censoring weighted average. *The American Statistician*, 55(3):207–210, 2001.

[39] Per K Andersen and Maja Pohar P. Pseudo-observations in survival analysis. *Statistical Methods in Medical Research*, 19(1):71–99, 2010.

[40] Per K Andersen, Elisavet Syriopoulou, and Erik T Parner. Causal inference in survival analysis using pseudo-observations. *Statistics in Medicine*, 36(17):2669–2681, 2017.

[41] Elizabeth J Williamson, Andrew Forbes, and Ian R White. Variance reduction in randomised trials by inverse probability weighting using the propensity score. *Statistics in Medicine*, 33(5):721–737, 2014.

[42] Changyu Shen, Xiaochun Li, and Lingling Li. Inverse probability weighting for covariate adjustment in randomized studies. *Statistics in Medicine*, 33(4):555–568, 2014.

[43] Shuxi Zeng, Fan Li, Rui Wang, and Fan Li. Propensity score weighting for covariate adjustment in randomized clinical trials. *Statistics in Medicine*, 40(4):842–858, 2021.

[44] Anastasios A Tsiatis, Marie Davidian, Min Zhang, and Xiaomin Lu. Covariate adjustment for two-sample treatment comparisons in randomized clinical trials: A principled yet flexible approach. *Statistics in Medicine*, 27(23):4658–4677, 2008.

[45] Winston Lin. Agnostic notes on regression adjustments to experimental data: Reexamining Freedman's critique. *The Annals of Applied Statistics*, 7(1):295–318, 2013.

[46] Guido W Imbens and Donald B Rubin. *Causal Inference in Statistics, Social, and Biomedical Sciences*. Cambridge University Press, 2015.

[47] Huzhang Mao and Liang Li. *PSW: Propensity Score Weighting Methods for Dichotomous Treatments*, 2018. R package version 1.1-3.

[48] Noah Greifer. *WeightIt: Weighting for Covariate Balance in Observational Studies*, 2020. R package version 0.10.2.

[49] Tianhui Zhou, Guangyu Tong, Fan Li, Laine Thomas, and Fan Li. *PSweight: Propensity Score Weighting for Causal Inference*, 2020. R package version 1.1.8.

[50] Serge Assaad, Shuxi Zeng, Chenyang Tao, Shounak Datta, Nikhil Mehta, Ricardo Henao, Fan Li, and Lawrence Carin Duke. Counterfactual representation learning with balancing weights. In *International Conference on Artificial Intelligence and Statistics*, pages 1972–1980. PMLR, 2021.

[51] Erich Battistin and Barbara Sianesi. Misclassified treatment status and treatment effects: an application to returns to education in the united kingdom. *Review of Economics and Statistics*, 93(2):495–509, 2011.

[52] Stef Van Buuren and Karin Groothuis-Oudshoorn. mice: Multivariate imputation by chained equations in R. *Journal of Statistical Software*, 45(1):1–67, 2011.

[53] Neil Mehta, Ankur Kalra, Amy S Nowacki, Scott Anjewierden, Zheyi Han, Pavan Bhat, Andres E Carmona-Rubio, Miriam Jacob, Gary W Procop, Susan Harrington, et al. Association of use of angiotensin-converting enzyme inhibitors and angiotensin ii receptor blockers with testing positive for coronavirus disease 2019 (COVID-19). *JAMA Cardiology*, 5(9):1020–1026, 2020.

[54] Elizabeth A Stuart, Stephen R Cole, Catherine P Bradshaw, and Philip J Leaf. The use of propensity scores to assess the generalizability of results from randomized trials. *Journal of the Royal Statistical Society: Series A (Statistics in Society)*, 174(2):369–386, 2011.

[55] Keisuke Hirano and Guido W Imbens. The propensity score with continuous treatments. *Applied Bayesian Modeling and Causal Inference from Incomplete-data Perspectives*, 226164:73–84, 2004.

[56] Jessica Lim, Rosalind Walley, Jiacheng Yuan, Jeen Liu, Abhishek Dabral, Nicky Best, Andrew Grieve, Lisa Hampson, Josephine Wolfram, Phil Woodward, et al. Minimizing patient burden through the use of historical subject-level data in innovative confirmatory clinical trials: review of methods and opportunities. *Therapeutic Innovation & regulatory Science*, 52(5):546–559, 2018.

[57] Chenguang Wang, Heng Li, Wei-Chen Chen, Nelson Lu, Ram Tiwari, Yunling Xu, and Lilly Q Yue. Propensity score-integrated power prior approach for incorporating real-world evidence in single-arm clinical studies. *Journal of Biopharmaceutical Statistics*, 29(5):731–748, 2019.

15

Covariate Balancing Propensity Score

Kosuke Imai and Yang Ning

Harvard University and Cornell University

CONTENTS

15.1 Introduction

Covariate balancing propensity score (CBPS) was originally proposed by [1] as a general methodology to improve the estimation of propensity score for causal inference. Propensity score, which is defined as the conditional probability of treatment assignment given observed covariates [2], plays an essential role as part of covariate adjustment methods including matching, weighting, and regression modeling. If the propensity score is correctly estimated, these covariate adjustment methods make the distribution of covariates equal between the treatment and control group, allowing researchers to draw valid causal inference under the standard unconfoundedness assumption. Yet, in practice, propensity score is unknown and must be estimated from data. This can lead to the misspecification of propensity score model and ultimately bias in causal effect estimation.

The basic idea of CBPS is simple. It estimates the propensity score such that covariates are balanced between the treatment and control groups after being inversely weighted by the estimated propensity score. In applied research, analysts often check covariate balance after making matching or weighting adjustment based on the estimated propensity score in an attempt to diagnose the potential misspecification of propensity score model. CBPS exploits this idea but directly optimizes covariate balance when estimating the propensity score. This allows researchers to avoid repeating the process of estimating the propensity score and checking the resulting covariate balance.

Suppose that we have a random sample of n observations from a population $\mathcal{P}$. Let T_i represent the binary (for now) treatment variable, which is equal to 1 if unit i receives the treatment and is equal to 0 if the unit belongs to the control group. We use $\boldsymbol{X}_i$ to denote the set of observed pre-treatment covariates. Then, the propensity score is defined as,

$$\pi(\boldsymbol{x}) = \Pr(T_i = 1 \mid \boldsymbol{X}_i = \boldsymbol{x}), \tag{15.1}$$

for $\boldsymbol{x} \in \mathcal{X}$ where $\mathcal{X}$ is the support of covariates $\boldsymbol{X}_i$. Following [2], we assume that the propensity

DOI: 10.1201/9781003102670-15

score is bounded away from 0 and 1,

$$0 < \pi(\boldsymbol{x}) < 1 \quad \text{for any } \boldsymbol{x} \in \mathcal{X}. \tag{15.2}$$

Now, consider the maximum likelihood estimation of the propensity score, based on a parametric model $\pi_{\boldsymbol{\beta}}(\boldsymbol{x})$ where $\boldsymbol{\beta}$ is a finite-dimensional parameter. A commonly used model includes a logistic regression. We can show that this maximum likelihood estimation implicitly balances the first derivative of the propensity score with respect to the model parameters. Formally, the score condition implies,

$$\frac{1}{n}\sum_{i=1}^{n}\left\{\frac{T_i\pi'_{\boldsymbol{\beta}}(\boldsymbol{X}_i)}{\pi_{\boldsymbol{\beta}}(\boldsymbol{X}_i)} - \frac{(1-T_i)\pi'_{\boldsymbol{\beta}}(\boldsymbol{X}_i)}{1-\pi_{\boldsymbol{\beta}}(\boldsymbol{X}_i)}\right\} = 0 \tag{15.3}$$

where $\pi'_{\boldsymbol{\beta}}(\boldsymbol{X}_i) = \partial\pi_{\boldsymbol{\beta}}(\boldsymbol{X}_i)/\partial\boldsymbol{\beta}$. Equation (15.3) represents the sample analogue of the population constraint that balances, by weighting each observation according to the inverse of its propensity score, the covariates between the treatment and control groups through the derivative $\pi'_{\boldsymbol{\beta}}(\boldsymbol{X}_i)$.

If the propensity score is correctly estimated, however, we should be able to balance *any* function of observed covariates $f(\cdot)$. This observation leads to the following general definition of covariate balancing condition,

$$\mathbb{E}\left\{\left(\frac{T_i}{\pi_{\boldsymbol{\beta}}(\boldsymbol{X}_i)} - \frac{1-T_i}{1-\pi_{\boldsymbol{\beta}}(\boldsymbol{X}_i)}\right) f(\boldsymbol{X}_i)\right\} = 0, \tag{15.4}$$

where a typical choice of $f(\cdot)$ is a lower order polynomial function. [1] suggests the use of the generalized method of moments estimator [3] to incorporate this additional covariate balancing condition as well as the original score condition.

There are several alternative formulations of covariate balancing conditions. First, Equation (15.4) weights the treatment and control groups such that their covariate distributions match with that of the target population. This means that we can have two separate covariate balancing conditions,

$$\mathbb{E}\left\{\left(\frac{T_i}{\pi_{\boldsymbol{\beta}}(\boldsymbol{X}_i)} - 1\right) f(\boldsymbol{X}_i)\right\} = 0, \tag{15.5}$$

$$\mathbb{E}\left\{\left(1 - \frac{1-T_i}{1-\pi_{\boldsymbol{\beta}}(\boldsymbol{X}_i)}\right) f(\boldsymbol{X}_i)\right\} = 0. \tag{15.6}$$

In addition we may be interested in adjusting the covariate distribution of the control group so that it matches with that of the treatment group. Such an adjustment is useful when estimating the average treatment effect for the treated (ATT). In this case the covariate balancing condition becomes,

$$\mathbb{E}\left\{\left(T_i - \frac{(1-T_i)\pi(\boldsymbol{X}_i)}{1-\pi_{\boldsymbol{\beta}}(\boldsymbol{X}_i)}\right) f(\boldsymbol{X}_i)\right\} = 0. \tag{15.7}$$

15.2 Extensions

15.2.1 A continuous treatment

It is possible to extend the idea of covariate balancing conditions to other settings. [4] considers a continuous treatment. They first standardize both treatment and covariates using the following transformation,

$$\widetilde{T}_i = s_T^{-1/2}(T - \overline{T}), \quad \widetilde{\boldsymbol{X}}_i = \boldsymbol{S}_{\boldsymbol{X}}^{-1/2}(\boldsymbol{X} - \overline{\boldsymbol{X}}) \tag{15.8}$$

where $\overline{T}$ and s_T are the sample mean and variance of the treatment, respectively, and $\overline{\boldsymbol{X}}$ and $\boldsymbol{S_X}$ are the sample mean and variance of the covariates, respectively. Then, the covariate balancing condition is given by,

$$\mathbb{E}\left(\frac{p(\widetilde{T}_i)}{p(\widetilde{T}_i \mid \widetilde{\boldsymbol{X}}_i)}\widetilde{T}_i\widetilde{\boldsymbol{X}}_i\right) = 0 \tag{15.9}$$

where $p(\widetilde{T}_i \mid \widetilde{\boldsymbol{X}}_i)$ is the generalized propensity score [5,6] and $p(\widetilde{T}_i)$ is the normalizing weight used in the marginal structural models Fong et al [4,7] develops both parametric and nonparametric estimation approaches (generalized method of moments and empirical likelihood estimation, respectively) and term the resulting estimator CBGPS (covariate balancing generalized propensity score). We refer readers to the original article for the details of these estimation strategies.

15.2.2 A dynamic treatment

Another extension is the application of the CBPS methodology to longitudinal data settings. The propensity score has been used extensively as part of marginal structural models [7]. A major practical problem is that the weights used in marginal structural models are based on the product of estimated propensity score at each time period. This means that the estimation of causal effects is often sensitive to the misspecification of propensity scores especially when the number of time periods is moderate or large.

[8] derives the covariate balancing conditions under the longitudinal settings. Consider the simplest case with two time periods, $J = 2$. There are four possible treatment combinations, where $(T_{i1}, T_{i2}) \in \mathcal{T} = \{(0,0),(0,1),(1,0),(1,1)\}$ with T_{ij} representing the treatment condition for unit $i = 1, 2, \ldots, n$ at time $j = 1, 2$. Suppose that we observe covariates at both time periods, i.e., $\boldsymbol{X}_{i1}, \boldsymbol{X}_{i2}$ where $\boldsymbol{X}_{ij}$ is causally prior to T_{ij} for $j = 1, 2$. Finally, the potential outcome is observed after all the other variables are realized and is denoted by $Y_i(t_1, t_2)$ with $(t_1, t_2) \in \mathcal{T}$. The observed outcome then is given by $Y_i = Y_i(T_{i1}, T_{i2})$.

There are two propensity scores under this setting – one for each time period. At the second time period, the covariate balancing condition tries to balance the second period covariates $\boldsymbol{X}_{i2}$ given the baseline covariates $\boldsymbol{X}_{i1}$ and the treatment assignment at the first time period T_{i1}. This is given by,

$$\mathbb{E}\{X_{i2}(t_1)\} = \mathbb{E}[\mathbf{1}\{T_{i1} = t_1, T_{i2} = t_2\}w_i(\bar{t}_2, \overline{\boldsymbol{X}}_{i2}(t_1))\boldsymbol{X}_{i2}(t_1)] \tag{15.10}$$

for $(t_1, t_2) \in \mathcal{T}$ where $\boldsymbol{X}_{i2}(t_1)$ represents the potential values of the second period covariates when the first time period treatment assignment is t_1, $\bar{t}_2 = (t_1, t_2)$ is the treatment history, and $\overline{\boldsymbol{X}}_{i2}(t_1) = (\boldsymbol{X}_{i1}, \boldsymbol{X}_{i2}(t_1))$ is the covariate history. The marginal structural model weight is given as the product of propensity scores in the denominator with the normalizing factor in the numerator,

$$w_i(\bar{t}_2, \overline{\boldsymbol{X}}_{i2}(t_1)) = \frac{\Pr(T_{i1} = t_1)}{\Pr(T_{i1} = t_1 \mid \boldsymbol{X}_{i1})} \times \frac{\Pr(T_{i2} = t_2 \mid T_{i1} = t_1)}{\Pr(T_{i2} = t_2 \mid T_{i1} = t_1, \boldsymbol{X}_{i1}, \boldsymbol{X}_{i2}(t_1))}. \tag{15.11}$$

[8] shows that these covariate balancing conditions can be written compactly as the following orthogonal moment conditions using the observed data notation,

$$\mathbb{E}\{(-1)^{T_{i2}}w_iX_{i2}\} = \mathbb{E}\{(-1)^{T_{i1}+T_{i2}}w_iX_{i2}\} = 0 \tag{15.12}$$

where $w_i = w_i(\overline{T}_{i2}, \overline{\boldsymbol{X}}_{i2})$ is the observed weight with $\overline{T}_{i2} = (T_{i1}, T_{i2})$ and $\overline{\boldsymbol{X}}_{i2} = (\boldsymbol{X}_{i1}, \boldsymbol{X}_{i2})$.

For the baseline covariates, we simply balance them across all four treatment combinations, i.e.,

$$\mathbb{E}(X_{i1}) = \mathbb{E}[\mathbf{1}\{T_{i1} = t_1, T_{i2} = t_2\}w_i(\bar{t}_2, \overline{\boldsymbol{X}}_{i2}(t_1))\boldsymbol{X}_{i1}]. \tag{15.13}$$

Again, this leads to the following orthogonal covariate balancing conditions,

$$\mathbb{E}\{(-1)^{T_{i1}}w_iX_{i1}\} = \mathbb{E}\{(-1)^{T_{i2}}w_iX_{i1}\} = \mathbb{E}\{(-1)^{T_{i1}+T_{i2}}w_iX_{i2}\} = 0. \tag{15.14}$$

[8] generalizes this consruction of covariate balancing conditions to an arbitrary number of time periods and estimates the propensity scores using the generalized method of moments. They provide empirical evidence that the CBPS substantially improves the robustness of marginal structural models.

While these papers have documented promising empirical performance of the CBPS methodology in various settings, it is important to examine its properties as well. For the remaining of the chapter, we first review the recent theoretical developments of the CBPS methodology. In particular we focus on the optimal choice of the covariate balancing conditions. We then discuss how the CBPS methodology and theory can be extended to the settings with high-dimensional covariates. Finally, we briefly discuss the software implementation of the CBPS and suggest future research.

15.3 Theory

In the above formulation, the choice of covariate balancing conditions in Equation (15.4), i.e., $f(\cdot)$, is left unspecified. Although in practice researchers often choose simple functions such as $f(\mathbf{x}) = \mathbf{x}$ and $f(\mathbf{x}) = \mathbf{x}^2$, it is important to study the optimal choice. Here, we briefly review the theoretical results of [9]. We first derive the optimal choice of the covariate balancing conditions under a locally misspecified propensity score model. From this result, we then construct an optimal CBPS estimator (oCBPS) for the average treatment effect (ATE) and establish the double robustness property of this estimator.

We begin by introducing the following notation,

$$\mathbb{E}(Y_i(0) \mid \boldsymbol{X}_i) = K(\boldsymbol{X}_i) \quad \text{and} \quad \mathbb{E}(Y_i(1) \mid \boldsymbol{X}_i) = K(\boldsymbol{X}_i) + L(\boldsymbol{X}_i), \tag{15.15}$$

where $K(\cdot)$ and $L(\cdot)$ represent the conditional mean of the potential outcome under the control condition and the conditional average treatment effect (CATE), respectively. Then, the ATE can be written as,

$$\mu = \mathbb{E}(Y_i(1) - Y_i(0)) = \mathbb{E}(L(\boldsymbol{X}_i)). \tag{15.16}$$

Let $\hat{\boldsymbol{\beta}}$ denote the estimator of $\boldsymbol{\beta}$ by solving the covariate balancing condition as an estimating equation (15.4). The Horvitz-Thompson estimator for the ATE is given by,

$$\hat{\mu}_{\hat{\boldsymbol{\beta}}} = \frac{1}{n}\sum_{i=1}^{n}\left(\frac{T_i Y_i}{\pi_{\hat{\boldsymbol{\beta}}}(\boldsymbol{X_i})} - \frac{(1-T_i)Y_i}{1-\pi_{\hat{\boldsymbol{\beta}}}(\boldsymbol{X_i})}\right). \tag{15.17}$$

To study how the bias and variance of $\hat{\mu}_{\hat{\boldsymbol{\beta}}}$ depend on the function $f(\cdot)$, we focus on the local misspecification of propensity score model using the general framework of [10]. Specifically, we assume that the true propensity score $\pi(\boldsymbol{X_i})$ is related to the working model $\pi_{\boldsymbol{\beta}}(\boldsymbol{X_i})$ through the exponential tilt for some $\boldsymbol{\beta}^*$,

$$\pi(\boldsymbol{X}_i) = \pi_{\boldsymbol{\beta}^*}(\boldsymbol{X}_i)\exp(\xi \cdot u(\boldsymbol{X}_i; \boldsymbol{\beta}^*)), \tag{15.18}$$

where $u(\boldsymbol{X}_i; \boldsymbol{\beta}^*)$ is a function determining the direction of misspecification and $\xi \in \mathbb{R}$ represents the magnitude of misspecification with $\xi = 0$ implying correct specification. We further assume $\xi = o(1)$ as $n \to \infty$ so that the true propensity score $\pi(\boldsymbol{X}_i)$ is in a local neighborhood of the working model $\pi_{\boldsymbol{\beta}^*}(\boldsymbol{X}_i)$. This approach is designed to account for the uncertainty of the specification of the propensity score model encountered in practical settings.

Under the model given in Equation (15.18) and some mild regularity conditions, we can show,

$$n^{1/2}(\hat{\mu}_{\hat{\boldsymbol{\beta}}} - \mu) \xrightarrow{d} N(B(f), \sigma^2(f)), \tag{15.19}$$

where $B(f)$ and $\sigma^2(f)$ are the asymptotic bias and variance of $\hat{\mu}_{\hat{\boldsymbol{\beta}}}$ which depends on the choice of the function $f(\cdot)$. The explicit form of $B(f)$ and $\sigma^2(f)$ is given in [9].

Since the ultimate goal of the CBPS methodology is to estimate the treatment effects (i.e., ATE in this section), we define the optimal choice of the function $f(\cdot)$ as the one that minimizes the asymptotic mean squared error (AMSE) of $\hat{\mu}_{\hat{\boldsymbol{\beta}}}$, i.e., $B^2(f) + \sigma^2(f)$ using the result given in Equation (15.19). A key theoretical result of [9] is that such optimal choice of $f(\cdot)$ exists and is characterized by the following condition: $f(\cdot)$ is optimal if there exists a vector $\boldsymbol{\alpha}$ such that

$$\boldsymbol{\alpha}^\top f(\boldsymbol{X}_i) = \pi_{\boldsymbol{\beta}^*}(\boldsymbol{X}_i)\mathbb{E}(Y_i(0) \mid \boldsymbol{X}_i) + (1 - \pi_{\boldsymbol{\beta}^*}(\boldsymbol{X}_i))\mathbb{E}(Y_i(1) \mid \boldsymbol{X}_i). \tag{15.20}$$

This result implies that the optimal choice of $f(\cdot)$ is not unique. Recall that $f = (f_1, ..., f_m)$ where m is the number of functions used to balance covariates. If we choose $f_1(\boldsymbol{X}_i) = \pi_{\boldsymbol{\beta}^*}(\boldsymbol{X}_i)\mathbb{E}(Y_i(0) \mid \boldsymbol{X}_i) + (1 - \pi_{\boldsymbol{\beta}^*}(\boldsymbol{X}_i))\mathbb{E}(Y_i(1) \mid \boldsymbol{X}_i)$, for example, then Equation (15.20) is always satisfied regardless of how the other $m - 1$ functions are specified. The lack of uniqueness means that we may need to use an initial estimator for $\boldsymbol{\beta}$ based on, for example, maximum likelihood. In addition, we also need to estimate the conditional mean functions, $\mathbb{E}(Y_i(0) \mid \boldsymbol{X}_i)$ and $\mathbb{E}(Y_i(1) \mid \boldsymbol{X}_i)$, using some parametric or nonparametric methods.

To overcome this problem, [9] introduced the optimal CBPS estimator (oCBPS) that does not require any initial estimator. Assume that we have a class of pre-specified functions $h(\cdot) \in \mathbb{R}^{m_1}$ and $g(\cdot) \in \mathbb{R}^{m_2}$. The oCBPS estimator $\hat{\boldsymbol{\beta}}_O$ is defined as the solution to the following estimating equations,

$$\frac{1}{n}\sum_{i=1}^{n}\left(\frac{T_i}{\pi_{\boldsymbol{\beta}}(\boldsymbol{X}_i)} - \frac{1 - T_i}{1 - \pi_{\boldsymbol{\beta}}(\boldsymbol{X}_i)}\right) h(\boldsymbol{X}_i) = 0, \tag{15.21}$$

and

$$\frac{1}{n}\sum_{i=1}^{n}\left(\frac{T_i}{\pi_{\boldsymbol{\beta}}(\boldsymbol{X}_i)} - 1\right) g(\boldsymbol{X}_i) = 0. \tag{15.22}$$

For simplicity, we only focus on the exact identified case with $m_1 + m_2 = p$ where p is the dimensionality of $\boldsymbol{\beta}$. To see how Equations (15.21) and (15.22) are related to the optimality condition in Equation (15.20), we rewrite Equation (15.22) as,

$$\frac{1}{n}\sum_{i=1}^{n}\left(\frac{T_i}{\pi_{\boldsymbol{\beta}}(\boldsymbol{X}_i)} - \frac{1 - T_i}{1 - \pi_{\boldsymbol{\beta}}(\boldsymbol{X}_i)}\right) g(\boldsymbol{X}_i)(1 - \pi_{\boldsymbol{\beta}}(\boldsymbol{X}_i)) = 0. \tag{15.23}$$

Thus, the oCBPS estimator is a special case of the general CBPS estimator given in Equation 15.4 with $f(\cdot) = [h(\cdot), g(\cdot)(1 - \pi_{\boldsymbol{\beta}})]$. If the functions $K(\cdot)$ and $L(\cdot)$ lie in the linear space spanned by $h(\cdot)$ and $g(\cdot)$ respectively (i.e., $K(\cdot) = \boldsymbol{\alpha}_1^\top h(\cdot)$ and $L(\cdot) = \boldsymbol{\alpha}_2^\top g(\cdot)$ for some vectors $\boldsymbol{\alpha}_1$ and $\boldsymbol{\alpha}_2$), then the optimality condition in Equation (15.20) holds. To see this, note

$$\begin{aligned}
&\boldsymbol{\alpha}_1^\top h(\boldsymbol{X}_i) + \boldsymbol{\alpha}_2^\top g(\boldsymbol{X}_i)(1 - \pi_{\boldsymbol{\beta}}(\boldsymbol{X}_i)) \\
&= K(\boldsymbol{X}_i) + L(\boldsymbol{X}_i)(1 - \pi_{\boldsymbol{\beta}}(\boldsymbol{X}_i)) \\
&= \pi_{\boldsymbol{\beta}^*}(\boldsymbol{X}_i)\mathbb{E}(Y_i(0) \mid \boldsymbol{X}_i) + (1 - \pi_{\boldsymbol{\beta}^*}(\boldsymbol{X}_i))\mathbb{E}(Y_i(1) \mid \boldsymbol{X}_i).
\end{aligned}$$

With the oCBPS estimator $\hat{\boldsymbol{\beta}}_O$, [9] proved the following theorem that establishes the double robustness and locally semiparametric efficiency of the ATE estimator $\hat{\mu}_{\hat{\boldsymbol{\beta}}_O}$. We reproduce the result here.

Theorem 15.1. *Under mild regularity conditions given in [9], the estimator $\hat{\mu}_{\hat{\boldsymbol{\beta}}_O}$ is doubly robust in the sense that $\hat{\mu}_{\hat{\boldsymbol{\beta}}_O} \xrightarrow{p} \mu$ if either of the following conditions holds:*

1. The propensity score model is correctly specified:

$$\mathbb{P}(T_i = 1 \mid \boldsymbol{X}_i) = \pi_{\boldsymbol{\beta}}(\boldsymbol{X}_i).$$

2. There exist $\boldsymbol{\alpha}_1, \boldsymbol{\alpha}_2$ *such that* $K(\cdot) = \boldsymbol{\alpha}_1^\top h(\cdot)$ *and* $L(\cdot) = \boldsymbol{\alpha}_2^\top g(\cdot)$.
In addition, if both conditions hold, then the estimator $\hat{\mu}_{\hat{\boldsymbol{\beta}}_O}$ *satisfies*

$$n^{1/2}(\hat{\mu}_{\hat{\boldsymbol{\beta}}_O} - \mu) \xrightarrow{d} N(0, V_{opt}),$$

where

$$V_{opt} = \mathbb{E}\left[\frac{\mathbb{V}(Y_i(1) \mid \boldsymbol{X}_i)}{\pi(\boldsymbol{X}_i)} + \frac{\mathbb{V}(Y_i(0) \mid \boldsymbol{X}_i)}{1 - \pi(\boldsymbol{X}_i)} + \{L(\boldsymbol{X}_i) - \mu\}^2\right]$$

is the semiparametric variance bound. Thus, $\hat{\mu}_{\hat{\boldsymbol{\beta}}_O}$ *is a locally semiparametric efficient estimator [11].*

In addition, the estimator $\hat{\mu}_{\hat{\boldsymbol{\beta}}_O}$ is shown to enjoy better high order asymptotic properties than the standard doubly robust estimator, such as the augmented inverse probability weighting estimator (AIPW). The oCBPS method can be also extended to the nonparametric regression setting. We refer the interesting readers to [9] for additional theoretical results.

15.4 High-Dimensional CBPS

In many observational studies, researches may collect a large number of covariates $\boldsymbol{X}$, whose dimension p can be comparable or even larger than the sample size n. To deal with such high-dimensional covariates, one common approach is to impose regularization terms, such as Lasso [12] or SCAD [13], in the maximum likelihood estimation of the propensity score and outcome models. However, due to the use of the regularization terms, this approach can lead to a biased estimate of the average treatment effect.

Ning et al [14] generalized the CBPS methodology to the estimation of the ATE with high-diemsional covariates. In this section, we briefly review their methodology and theory. For simplicity, we focus on the estimation of $\mu_1^* = \mathbb{E}(Y(1))$. Assume the following working models for the treatment and outcome variables,

$$\mathbb{P}(T_i = 1 \mid \boldsymbol{X}_i) = \pi(\boldsymbol{X}_i^\top \boldsymbol{\beta}), \ \mathbb{E}\{Y_i(1) \mid \boldsymbol{X}_i\} = \boldsymbol{X}_i^\top \boldsymbol{\alpha},$$

where $\pi(\cdot)$ is a known function and $\boldsymbol{\beta}$ and $\boldsymbol{\alpha}$ are p-dimensional unknown parameters. When the goal is to estimate $\mathbb{E}(Y(1))$ and the outcome model is linear in $\boldsymbol{X}$, the theory in Section 15.3 yields that the optimal covariate balancing equation has the form

$$\frac{1}{n}\sum_{i=1}^{n}\left(\frac{T_i}{\pi(\boldsymbol{X}_i^\top \boldsymbol{\beta})} - 1\right)\boldsymbol{X}_i = 0. \tag{15.24}$$

However, when $p > n$, the above covariate balancing equation does not have a unique solution and therefore the CBPS is not well defined.

The main idea in [14] is to estimate the propensity score $\pi(\boldsymbol{X}_i^\top \boldsymbol{\beta})$ such that the following covariate balancing condition is satisfied,

$$\sum_{i=1}^{n}\left(\frac{T_i}{\pi(\boldsymbol{X}_i^\top \boldsymbol{\beta})} - 1\right)\boldsymbol{\alpha}^{*\top}\boldsymbol{X}_i = 0, \tag{15.25}$$

where $\boldsymbol{\alpha}^*$ is the true value of $\boldsymbol{\alpha}$ in the outcome regression when it is correctly specified. Otherwise, $\boldsymbol{\alpha}^*$ corresponds to some least false value of $\boldsymbol{\alpha}$ under model misspecification, see [14] for the precise definition. We refer to Equation (15.25) as the weak covariate balancing equation.

It is easy to see that the propensity score, which satisfies Equation (15.24), also satisfies the weak covariate balancing equation, but not vice versa. It turns out that it is sufficient to find an estimate of the propensity score, that approximately satisfies the weak covariate balancing equation, to remove the bias from the estimation of the propensity score model. The algorithm proposed by [14] is summarized in Algorithm 1.

Algorithm 1 High-dimensional CBPS estimator

Require: Dataset: $(Y_i, T_i, X_i), 1 \leq i \leq n$ and tuning parameters $\lambda, \lambda' \geq 0$.

Step 1: Estimate the propensity score model by the penalized quasi-likelihood estimator

$$\hat{\boldsymbol{\beta}} = \arg\min_{\boldsymbol{\beta} \in \mathbb{R}^d} -\frac{1}{n}\sum_{i=1}^{n}\int_0^{\boldsymbol{\beta}^\top \boldsymbol{X}_i}\left(\frac{T_i}{\pi(u)} - 1\right)du + \lambda\|\boldsymbol{\beta}\|_1,$$

where the Lasso penalty can be replaced with the non-convex penalty [13].

Step 2: Estimate the outcome model by the weighted regression

$$\tilde{\boldsymbol{\alpha}} = \arg\min_{\boldsymbol{\alpha} \in \mathbb{R}^d} \frac{1}{n}\sum_{i=1}^{n}\frac{T_i\pi'(\hat{\boldsymbol{\beta}}^\top \boldsymbol{X}_i)}{\pi^2(\hat{\boldsymbol{\beta}}^\top \boldsymbol{X}_i)}(Y_i - \boldsymbol{\alpha}^\top \boldsymbol{X}_i)^2 + \lambda'\|\boldsymbol{\alpha}\|_1,$$

where $\pi'(\cdot)$ is the derivative of $\pi(\cdot)$.

Step 3: Let $\tilde{S} = \{j : |\tilde{\alpha}_j| > 0\}$ denote the support of $\tilde{\boldsymbol{\alpha}}$ and $\boldsymbol{X}_{\tilde{S}}$ denote the corresponding entries of $\boldsymbol{X}$. We calibrate the estimator $\hat{\boldsymbol{\beta}}_{\tilde{S}}$ in Step 1 in order to balance $X_{\tilde{S}}$. Specifically, we define $\tilde{\gamma}$ as the solution to the following equation

$$\sum_{i=1}^{n}\left\{\frac{T_i}{\pi(\boldsymbol{\gamma}^\top \boldsymbol{X}_{i\tilde{S}} + \hat{\boldsymbol{\beta}}_{\tilde{S}^c}^\top \boldsymbol{X}_{i\tilde{S}^c})} - 1\right\}\boldsymbol{X}_{i\tilde{S}} = 0. \tag{15.26}$$

We denote the calibrated PS estimator by $\tilde{\boldsymbol{\beta}} = (\tilde{\gamma}, \hat{\boldsymbol{\beta}}_{\tilde{S}^c})$ and $\tilde{\pi}_i = \pi(\tilde{\boldsymbol{\beta}}^\top \boldsymbol{X}_i)$.

return the estimator $\hat{\mu}_1 = \frac{1}{n}\sum_{i=1}^{n} T_i Y_i/\tilde{\pi}_i$.

Algorithm 1 has three steps. In Step 1, we obtain an initial estimate of the propensity score by maximizing a penalized quasi-likelihood function. The proposed quasi-likelihood function is obtained by integrating Equation (15.24). However, this initial estimator may have a large bias because many covariates are not balanced. To remove this bias, we further select a set of covariates that are predictive of the outcome variable based on a weighted regression in Step 2. The weight may depend on the initial estimator $\hat{\boldsymbol{\beta}}$ and is chosen to reduce the bias of the estimator under model misspecification. Finally, in Step 3, we calibrate the propensity score by balancing the selected covariates $\boldsymbol{X}_{\tilde{S}}$.

While in the last step only the selected covariates are balanced, it approximately satisfies the weak covariate balancing equation. To see this, consider the following heuristic argument,

$$\sum_{i=1}^{n}\left(\frac{T_i}{\tilde{\pi}_i} - 1\right)\boldsymbol{\alpha}^{*\top}\boldsymbol{X}_i \approx \sum_{i=1}^{n}\left(\frac{T_i}{\tilde{\pi}_i} - 1\right)\tilde{\boldsymbol{\alpha}}^\top \boldsymbol{X}_i = \sum_{i=1}^{n}\left(\frac{T_i}{\tilde{\pi}_i} - 1\right)\tilde{\boldsymbol{\alpha}}_{\tilde{S}}^\top \boldsymbol{X}_{i\tilde{S}} = 0,$$

where the approximation holds if the estimator $\tilde{\boldsymbol{\alpha}}$ is close to $\boldsymbol{\alpha}^*$, the first equality follows from $\tilde{\boldsymbol{\alpha}}_{\tilde{S}^c} = 0$ and the second equality holds due to the covariate balancing condition given in Equation (15.26). Finally, we estimate $\mathbb{E}(Y(1))$ by the Horvitz-Thompson estimator.

The asymptotic properties of the estimator $\hat{\mu}_1$ have been studied thoroughly in [14]. In the following, we only reproduce part of their main results. Recall that $\boldsymbol{\alpha}^*$ denotes the true value of $\boldsymbol{\alpha}$ in the outcome regression when it is correctly specified or the least false value under model misspecification. We define $\boldsymbol{\beta}^*$ in the propensity score in a similar manner. Let $s_1 = \|\boldsymbol{\alpha}^*\|_0$ and $s_2 = \|\boldsymbol{\beta}^*\|_0$ denote the number of nonzero entries in $\boldsymbol{\alpha}^*$ and $\boldsymbol{\beta}^*$. Then, we have the following theorem.

Theorem 15.2. *Under the regularity conditions given in [14], by choosing tuning parameters* $\lambda \asymp \lambda' \asymp \{\log(p \vee n)/n\}^{1/2}$, *if* $(s_1 + s_2)\log p = o(n)$ *as* $s_1, s_2, p, n \to \infty$, *then the estimator* $\hat{\mu}_1$ *has the following asymptotic linear expansion*

$$\hat{\mu}_1 - \mu_1^* = \frac{1}{n}\sum_{i=1}^{n}\left[\frac{T_i}{\pi(\boldsymbol{X}_i^\top\boldsymbol{\alpha}^*)}\{Y_i(1) - \boldsymbol{\alpha}^{*\top}\boldsymbol{X}_i\} + \boldsymbol{\alpha}^{*\top}\boldsymbol{X}_i - \mu_1^*\right] + o_p(1),$$

as long as either the propensity score or outcome regression is correctly specified.

The asymptotic normality of $\hat{\mu}_1$ follows directly from this theorem. It is worthwhile to mention that, with $s_1, s_2 = O(1)$, Theorem 15.2 holds even if the dimension p grows exponentially fast with the sample size n under the condition $p = o(\exp(n))$. In addition, the asymptotic linear expansion of $\hat{\mu}_1$ holds if either the propensity score or outcome regression is correctly specified. This double-robustness implies that one can construct valid confidence intervals for μ_1^* in high-dimensional setting even if either the propensity score model or outcome regression model is misspecified (but not both).

The high-dimensional CBPS method is closely related to some recent proposals in causal inference [15–21] and more generally high-dimensional inference for regression models [22–25]. We refer to [14] for further discussion.

15.5 Software Implementation

The CBPS methodology mentioned above can be implemented through an easy-to-use open-source R package, CBPS: Covariate Balancing Propensity Score, available for download at `https://CRAN.R-project.org/package=CBPS` [26].

15.6 Future Research

This chapter reviewed the development of the covariate balancing propensity score (CBPS) methodology, which was originally published in 2014. Since then, various extensions and theoretical developments have appeared. Both the propensity score and covariate balance play a fundamental role in causal inference with observational data. The CBPS combines these concepts into a single framework based on a simple idea that one should estimate the propensity score by both balancing covariates and predicting treatment assignment. Future research should consider the development of better estimation algorithms as well as extensions to other settings including multiple (or even high-dimensional) treatments.

References

[1] Kosuke Imai and Marc Ratkovic. Covariate balancing propensity score. *Journal of the Royal Statistical Society, Series B (Statistical Methodology)*, 76(1):243–263, January 2014.

[2] Paul R. Rosenbaum and Donald B. Rubin. The central role of the propensity score in observational studies for causal effects. *Biometrika*, 70(1):41–55, 1983.

[3] Lars Peter Hansen. Large sample properties of generalized method of moments estimators. *Econometrica*, 50(4):1029–1054, July 1982.

[4] Christian Fong, Chad Hazlett, and Kosuke Imai. Covariate balancing propensity score for a continuous treatment: Application to the efficacy of political advertisements. *Annals of Applied Statistics*, 12(1):156–177, 2018.

[5] Keisuke Hirano and Guido W. Imbens. The propensity score with continuous treatments. In *Applied Bayesian Modeling and Causal Inference from Incomplete-Data Perspectives: An Essential Journey with Donald Rubin's Statistical Family*, chapter 7. Wiley, 2004.

[6] Kosuke Imai and David A. van Dyk. Causal inference with general treatment regimes: Generalizing the propensity score. *Journal of the American Statistical Association*, 99(467):854–866, September 2004.

[7] James M. Robins, Miguel Ángel Hernán, and Babette Brumback. Marginal structural models and causal inference in epidemiology. *Epidemiology*, 11(5):550–560, September 2000.

[8] Kosuke Imai and Marc Ratkovic. Robust estimation of inverse probability weights for marginal structural models. *Journal of the American Statistical Association*, 110(511):1013–1023, September 2015.

[9] Jianqing Fan, Kosuke Imai, Inbeom Lee, Han Liu, Yang Ning, and Xiaolin Yang. Optimal Covariate Balancing Conditions in Propensity Score Estimation. *Journal of Business & Economic Statistics*, 41(1): 97–110, 2023.

[10] John Copas and Shinto Eguchi. Local model uncertainty and incomplete-data bias. *Journal of the Royal Statistical Society, Series B (Methodological)*, 67(4):459–513, 2005.

[11] James M Robins, Andrea Rotnitzky, and Lue Ping Zhao. Estimation of regression coefficients when some regressors are not always observed. *Journal of the American Statistical Association*, 89(427):846–866, 1994.

[12] Robert Tibshirani. Regression shrinkage and selection via the lasso. *Journal of the Royal Statistical Society: Series B (Methodological)*, 58(1):267–288, 1996.

[13] Jianqing Fan and Runze Li. Variable selection via nonconcave penalized likelihood and its oracle properties. *Journal of the American statistical Association*, 96(456):1348–1360, 2001.

[14] Yang Ning, Peng Sida, and Kosuke Imai. Robust estimation of causal effects via a high-dimensional covariate balancing propensity score. *Biometrika*, 107(3):533–554, 2020.

[15] Max H Farrell. Robust inference on average treatment effects with possibly more covariates than observations. *Journal of Econometrics*, 189(1):1–23, 2015.

[16] Victor Chernozhukov, Denis Chetverikov, Mert Demirer, Esther Duflo, Christian Hansen, and Whitney Newey. Double machine learning for treatment and causal parameters. *arXiv preprint arXiv:1608.00060, 2016*, 2016.

[17] Susan Athey, Guido W Imbens, and Stefan Wager. Approximate residual balancing: debiased inference of average treatment effects in high dimensions. *Journal of the Royal Statistical Society: Series B (Statistical Methodology)*, 80(4):597–623, 2018.

[18] Zhiqiang Tan. Model-assisted inference for treatment effects using regularized calibrated estimation with high-dimensional data. *The Annals of Statistics*, 48(2):811–837, 2020.

[19] Oliver Dukes, Vahe Avagyan, and Stijn Vansteelandt. Doubly robust tests of exposure effects under high-dimensional confounding. *Biometrics*, 76(4):1190–1200, 2020.

[20] Jelena Bradic, Stefan Wager, and Yinchu Zhu. Sparsity double robust inference of average treatment effects. *arXiv preprint arXiv:1905.00744*, 2019.

[21] Ezequiel Smucler, Andrea Rotnitzky, and James M Robins. A unifying approach for doubly-robust ℓ_1 regularized estimation of causal contrasts. *arXiv preprint arXiv:1904.03737*, 2019.

[22] Yang Ning and Han Liu. A general theory of hypothesis tests and confidence regions for sparse high dimensional models. *The Annals of Statistics*, 45(1):158–195, 2017.

[23] Cun-Hui Zhang and Stephanie S Zhang. Confidence intervals for low dimensional parameters in high dimensional linear models. *Journal of the Royal Statistical Society: Series B (Statistical Methodology)*, 76(1):217–242, 2014.

[24] Adel Javanmard and Andrea Montanari. Confidence intervals and hypothesis testing for high-dimensional regression. *The Journal of Machine Learning Research*, 15(1):2869–2909, 2014.

[25] Matey Neykov, Yang Ning, Jun S Liu, and Han Liu. A unified theory of confidence regions and testing for high-dimensional estimating equations. *Statistical Science*, 33(3):427–443, 2018.

[26] Christian Fong, Marc Ratkovic, and Kosuke Imai. CBPS: R package for covariate balancing propensity score. available at the Comprehensive R Archive Network (CRAN), 2021. `https://CRAN.R-project.org/package=CBPS`.

16

Balancing Weights for Causal Inference

Eric R. Cohn, Eli Ben-Michael, Avi Feller, and José R. Zubizarreta

CONTENTS

16.1 Introduction

Covariate balance is central to both randomized experiments and observational studies. In randomized experiments, covariate balance is expected by design: when treatment is randomized, covariates are balanced in expectation, and observed differences in outcomes between the treatment groups can be granted a causal interpretation. In observational studies, where the treatment assignment is not controlled by the investigator, such balance is not guaranteed, and, subject to certain assumptions, one can adjust the data to achieve balance and obtain valid causal inferences. In fact, if investigators believe that treatment assignment is determined only by observed covariates, then balance is sufficient to remove confounding and establish causation from association.

DOI: 10.1201/9781003102670-16

Weighting is a popular method for adjusting observational data to achieve covariate balance that has found use in a range of disciplines such as economics (e.g., [1]), education (e.g., [2]), epidemiology (e.g., [3]), medicine (e.g., [4]), and public policy (e.g., [5]). In many cases the weights are a function of the propensity score, the conditional probability of treatment given the observed covariates [6]. Weighting by the true inverse propensity score guarantees covariate balance in expectation. In practice, however, the propensity score must be estimated. In the modeling approach to weighting, researchers estimate the propensity score directly, such as via logistic regression, and then plug in these estimates to obtain unit-level weights (see, e.g., [7]). An important risk of this approach is that, if the propensity score model is wrong, the balancing property of the resulting weights no longer holds, and treatment effect estimates may be biased [8]. Additionally, even if the model is correct, the resulting weights can fail to balance covariates in a given sample. This is often addressed by re-estimating the weights and balance checking anew in an iterative, sometimes ad hoc fashion [9].

Alternative weighting methods, which we term *balancing weights*, directly target the balancing property of the inverse propensity score weights by estimating weights that balance covariates in the sample at hand [10]. Such weights are found by solving an optimization problem that minimizes covariate imbalance across treatment groups while directly incorporating other concerns related to the ad hoc checks described earlier for modeling approaches (e.g., minimizing dispersion, constraining the weights to be non-negative, etc.). A variety of balancing approaches have appeared in the literature (e.g., [11–15]).

In this chapter, we introduce the balancing approach to weighting for covariate balance and causal inference. In Section 16.2 we begin by providing a framework for causal inference in observational studies, including typical assumptions necessary for the identification of average treatment effects. In Section 16.3 we then motivate the task of finding weights that balance covariates and unify a variety of methods from the literature. In Section 16.4 we discuss several implementation and design choices for finding balancing weights in practice and discuss the trade-offs of these choices using an example from the canonical LaLonde data [16, 17]. In Section 16.5 we discuss how to estimate effects after weighting, also using this applied example. In Section 16.6 we conclude with a discussion and future directions.

16.2 Framework

16.2.1 Notations, estimand, and assumptions

Our setting is an observational study with n independent and identically-distributed triplets (X_i, W_i, Y_i) for $i = 1, ..., n$, where $X_i \in \mathbb{R}^d$ are observed covariates, $W_i \in \{0, 1\}$ is a binary treatment indicator, and $Y_i \in \mathbb{R}$ is an outcome of interest. There are $n_1 = \sum_{i=1}^{n} W_i$ treated units and $n_0 = n - n_1$ control units. We operate under the potential outcomes framework for causal inference [18, 19] and impose the Stable Unit Treatment Value Assumption (SUTVA), which requires only one version of treatment and no interference between units [20]. Under this setup, each unit has two potential outcomes $Y_i(0)$ and $Y_i(1)$, but only one outcome is observed: $Y_i = (1 - W_i)Y_i(0) + W_iY_i(1)$. Since the study units are a random sample from the distribution of (X, W, Y), and we are generally interested in population-level estimands here, we often drop the subscript i.

In this chapter we focus on estimating the Average Treatment Effect on the Treated (ATT), defined as:

$$\text{ATT} := \mathbb{E}[Y(1) - Y(0)|W = 1] = \mathbb{E}[Y(1)|W = 1] - \mathbb{E}[Y(0)|W = 1].$$

Other causal estimands may be of interest to investigators, and we can easily extend the key ideas here to those quantities; see, for example, [21].

We directly observe $Y(1)$ for units assigned to treatment; therefore $\mathbb{E}[Y(1)|W = 1] = \mathbb{E}[Y|W = 1]$, and $n_1^{-1}\sum_{i=1}^n W_i Y_i$ is an unbiased estimator of this potential outcome mean. The challenge then is to estimate $\mu_0 := \mathbb{E}[Y(0)|W = 1]$. In this chapter we focus on identifying this quantity assuming strong ignorability.

Assumption 5 (Strong ignorability).

1. **Ignorability**. $W \perp\!\!\!\perp (Y(0), Y(1))|X$
2. **Overlap**. $e(x) > 0$, where $e(x) := \mathbb{E}[W|X = x]$ is the propensity score[1]

16.2.2 The central role of inverse propensity score weights

Under Assumption 5, we can use inverse propensity weights (IPW) to identify the unobservable control potential outcome mean by re-weighting the treatment group.

$$\mathbb{E}\left[(1 - W)\gamma^{\text{IPW}} Y\right] = \mathbb{E}\left[\frac{(1 - W)e(X)}{(1 - e(X))} Y\right] = \mathbb{E}[WY(0)]. \tag{16.1}$$

Importantly, the IPW weights satisfy:

$$\mathbb{E}[(1 - W)\gamma^{\text{IPW}}(X)f(X)] = \mathbb{E}[Wf(X)] \tag{16.2}$$

for all bounded functions $f(\cdot)$. That is, the weights balance all bounded functions of the covariates between the control group and the target treated group. In fact, due to the Law of Large Numbers, these weights can be expected (with probability tending to 1) to balance such functions in a given sample: that is $n^{-1}\sum_{i=1}^n (1 - W_i)\gamma^{\text{IPW}}(X_i)f(X_i) \approx n^{-1}\sum_{i=1}^n W_i f(X_i)$. Targeting this property of the weights, rather than targeting the propensity score itself, motivates the balancing approach to weighting that we consider below.

16.2.3 Bounding the error

To make progress on finding weights in practice, we first characterize the error of a general weighting estimator. In particular a general weighted estimator for $\mathbb{E}[WY(0)]$ takes the form:

$$\widehat{\mathbb{E}}[WY(0)] = \widehat{\mu}_0 = \frac{1}{n}\sum_{i=1}^n (1 - W_i)\widehat{\gamma}_i Y_i \tag{16.3}$$

for estimated weights $\widehat{\gamma}_i$. The error of this estimator has the following decomposition:

$$\begin{aligned}\widehat{\mu}_0 - \mu_0 = {} & \underbrace{\frac{1}{n}\sum_{i=1}^n (1 - W_i)\widehat{\gamma}_i m(X_i, 0) - \frac{1}{n}\sum_{i=1}^n W_i m(X_i, 0)}_{\text{bias from imbalance}} + \underbrace{\frac{1}{n}\sum_{i=1}^n (1 - W_i)\widehat{\gamma}_i \epsilon_i}_{\text{noise}} \\ & + \underbrace{\frac{1}{n}\sum_{i=1}^n W_i m(X_i, 0) - \mu_0}_{\text{sampling variation}}\end{aligned} \tag{16.4}$$

where $m(X_i, 0) := \mathbb{E}[Y_i|X = X_i, W_i = 0]$ denotes the outcome model and $\epsilon_i := Y_i - m(X_i, 0)$ denotes the residual. The second and third terms have expectation 0, and so the bias of our weighted

[1] Other (sometimes stronger) overlap assumptions exist for different results. See, e.g., "sufficient overlap" in [21]

estimator $\widehat{\mu}_0$ of μ_0 is driven solely by the first term: that is, the imbalance in the conditional mean function $m(X, 0)$ between the treated and control groups.

The question becomes, then, how to achieve balance on this typically unknown function $m(X, 0)$. One way of doing this is to consider a model class $\mathcal{M}$ for $m(X, 0)$ (e.g., the class of models linear in X) and the maximal imbalance over functions in that class, $\text{imbalance}_{\mathcal{M}}(\widehat{\gamma})$.

$$\begin{aligned}
\left|\text{design-conditional bias}\right| &:= \left|\mathbb{E}\left[\widehat{\mu}_0 - \mu_0 \middle| X_1, ..., X_n, W_1, ..., W_n\right]\right| \\
&\leq \text{imbalance}_{\mathcal{M}}(\widehat{\gamma}) \\
&:= \max_{m \in \mathcal{M}} \left| \frac{1}{n}\sum_{i=1}^{n}(1 - W_i)\widehat{\gamma}_i m(X_i, 0) - \frac{1}{n}\sum_{i=1}^{n} W_i m(X_i, 0)\right|.
\end{aligned}$$

This reformulates the task of minimizing imbalance in $m(X, 0)$ in terms of imbalance over the candidate functions $m \in \mathcal{M}$. Balance over all $m \in \mathcal{M}$ is in turn sufficient (though not necessary) for controlling the bias in Equation 16.4. This objective therefore allows investigators to avoid the difficult task of specifying the particular form of the outcome model m and to instead consider the potentially much broader class to which m belongs. Furthermore, as discussed in the previous section, because the IPW γ^{IPW} satisfy the sample balance property (in probability), weighting by the IPW should (with probability tending to 1) lead to an unbiased estimate of μ_0. Thus, seeking weights that balance a particular class of functions of X is motivated both by the balancing property of the inverse propensity score weights and by the error bound above.

16.3 Two Approaches to Estimating the Inverse Propensity Score Weights

Section 16.1 introduced two approaches to weighting for causal inference. In this next section we briefly discuss the traditional modeling approach and some of its drawbacks. We then discuss the balancing approach, some design choices that come along with it, and the formal connection between the two approaches.

16.3.1 The modeling approach

A common approach to weighting involves estimating a model for the propensity score $e(x)$ and then plugging in the estimated propensity scores to create the weights for the control group [22]:

$$\widehat{\gamma}_i^{\text{IPW}}(X_i) = \frac{\widehat{e}(X_i)}{1 - \widehat{e}(X_i)}$$

A variety of methods exist for estimating the propensity score $e(X)$. While the use of parametric models such as logistic regression remain the most common, flexible machine learning approaches, such as boosting [23], have become increasingly popular [24]. Modeling approaches are useful for their well-established statistical properties [25], which allow for the calculation of confidence intervals by using, e.g., an estimator for the asymptotic variance or resampling methods. Additionally, if the model for the propensity score is correct (that is, produces consistent estimates of $e(X)$), then the estimated weights are consistent for the true weights, yielding an asymptotically unbiased estimator $\widehat{\mu}_0$ of μ_0.

In practice, however, the propensity score model is unknown and must be posited based on subject matter knowledge or determined in a data-driven manner. Errors in the propensity score model can magnify errors in the estimated weights due to inversion and/or multiplication. Further, the estimated weights, even if correctly specified, may not ensure balance in a given sample, since this balance is only guaranteed in expectation. To confirm correct specification of the propensity score model, a variety of best-practice post hoc checks are available [26], including assessing covariate balance after weighting and examining the distribution of the weights (e.g., mean, median, minimum, and maximum). Many heuristics and statistical tests exist to aid in these assessments in practice. If the modeling approach underperforms on one of these diagnostics, the investigator may specify a new model [27], truncate extreme weights [28], modify the estimand [29], or make other adjustments.

16.3.2 The balancing approach

An alternative approach for IPW weights instead directly targets the balancing property (Equation 16.2) by seeking weights that balance covariates in the sample at hand. This approach is motivated both by the balancing property and the error decomposition discussed in Section 16.2. As we discuss in Section 16.3.4, the resulting balancing weights can also be motivated as a method of moments estimator for the true IPW weights.

Typically, the balancing approach posits an optimization problem that jointly minimizes imbalance and the dispersion of the weights. More formally, balancing approaches seek weights that solve the following general optimization problem:

$$\begin{aligned} \text{minimize}_{\gamma} \quad & h_{\lambda}\left(\text{imbalance}_{\mathcal{M}}(\gamma)\right) + \sum_{i:W_i=0} f(\gamma_i) \\ \text{subject to} \quad & [\text{constraints}], \end{aligned} \tag{16.5}$$

where $\text{imbalance}_{\mathcal{M}}(\gamma)$ is the "imbalance metric," e.g., the maximal imbalance over $\mathcal{M}$; $h_{\lambda}(\cdot)$ is the "imbalance penalty," e.g., $h_{\lambda}(x) = \lambda^{-1}x^2$; and $f(\cdot)$ is the "dispersion penalty," e.g., $f(x) = x^2$. The additional optional constraints target other desirable qualities of resulting weighted estimates. For example, an average-to-one constraint, i.e., $n^{-1}\sum_{i=1}^{n}(1 - W_i)\gamma_i = 1$, forces the weighted estimate $\widehat{\mu}_0$ to be translation invariant; that is, if we estimate μ_0 when $m = f(x)$, then we estimate $\mu_0 + t$ when $m = f(x) + t$ for $t \in \mathbb{R}$.[2] Coupling this with a nonnegativity constraint, i.e., $\gamma_i \geq 0$, for all $i : W_i = 0$, restricts any weighted estimates to be an interpolation (rather than an extrapolation) of the observed data. Together these constraints restrict the weights to the n_0-simplex, and so together are sometimes called a simplex constraint. We discuss the role of these constraints further in Section 16.4.4.

Many existing approaches can be written in the form of (16.5) for particular choices of h_{λ} and f, including the stable balancing weights [12], the lasso minimum distance estimator [13], the regularized calibrated propensity score estimation [14], and entropy balancing [11]. Even the linear regression of Y on X among the control units can be written in this form, and the weights implied by this regression are a particular form of exact balancing weights in which the weights can be negative [30, see].

16.3.3 Balance and model classes

Arguably, the most important design choice in the balancing approach is the choice of the model class $\mathcal{M}$ over which imbalance is minimized. Also, the choice of $\mathcal{M}$ affects the particular form of the optimization problem in Equation 16.5 due to the relationship between $\mathcal{M}$ and the imbalance metric.

Perhaps the simplest $\mathcal{M}$ is the class of linear models in the covariates X. That is, investigators might posit that $m \in \mathcal{M} = \{\beta^T X\}$ for some $\beta \in \mathbb{R}^p$. However, without a constraint on the coefficients β, the maximal imbalance over $\mathcal{M}$ is unbounded unless the weights exactly balance the covariates. This can lead to weights with possibly high variance, and may yield no solution if

[2] In fact, the average-to-one constraint is a form of balancing constraint in and of itself by allowing intercepts into the model.

additional constraints are included (e.g., the simplex constraint). This can be avoided by constraining the norm of the coefficients β, such as $||\beta||_1 \leq 1$ or $||\beta||_2 \leq 1$. These constraints correspond to particular imbalance metrics in Equation 16.5. Specifically, for $\mathcal{M} = \{\beta^T X : ||\beta||_p \leq 1\}$, the imbalance metric is equal to:

$$\text{imbalance}_{\mathcal{M}}(\gamma) = \left\| \frac{1}{n}\sum_{i=1}^{n}(1 - W_i)\gamma_i X_i - \frac{1}{n}\sum_{i=1}^{n} W_i X_i \right\|_q \qquad 1/p + 1/q = 1 \tag{16.6}$$

Thus, the one-norm constraint corresponds to an infinity-norm balancing metric, i.e., Equation 16.5 controls the maximal imbalance over the covariates; and the two-norm constraint corresponds to a two-norm balancing metric, i.e., Equation 16.5 controls the Euclidean distance between the covariate mean vectors.

Linear models are often too restrictive in practice. In this case investigators might posit that the outcome model is instead linear in a finite collection of transformations of the covariates $\phi_1(X), \ldots, \phi_K(X)$, such as the covariates, their squares, and their two-way interactions. In this case, $\mathcal{M} = \{\beta^T \phi(X) : ||\beta||_p \leq 1\}$ for $\beta \in \mathbb{R}^K$, and the imbalance metric appears as in (16.6), albeit with $X_i = (X_{1i}, \ldots, X_{pi})^T$ replaced by $\phi(X_i) = (\phi_1(X_i), \ldots, \phi_K(X_i)^T)$.

When all covariates are binary, the outcome model m always lies in a model class that is linear in a finite number of covariate functions: namely, all possible interactions of covariates. This is because the implied model class $\mathcal{M}$ includes all possible functions of the covariates. However, this approach can become untenable when the number of covariates increases, and is often not necessary in practice if the investigator believes that only interactions up to a certain depth (e.g., only up to two-way interactions) determine the outcome model (see, e.g., [31]). In other cases, the researcher may attempt to balance all interactions but place less weight (or none at all) on higher-order interactions, for example, by finding weights that exactly balance main terms and lower-order interactions and approximately balance higher-order interactions [32].

A nonparametric approach when some covariates are continuous is more difficult, as functions of continuous covariates can be nonlinear or otherwise highly complex. Thus, nonparametric models involving even one continuous covariate require an infinite-dimensional basis (though a finite basis can often be a good enough approximation). Common bases are the Hermite polynomials $1, x_1, x_2, x_1^2 - 1, x_2^2 - 1, x_1 x_2, \ldots$ (written here for two covariates) or sines and cosines of different frequencies. In general the model class $\mathcal{M}$ may be expressed in terms of basis functions $\phi_1(x), \phi_2(x), \ldots$ and corresponding $\lambda_1, \lambda_2, \ldots$:

$$\mathcal{M} = \left\{ m(x) = \sum_j \beta_j \phi_j(x) : \sum_j \beta_j / \lambda_j^2 \leq 1 \right\} \tag{16.7}$$

With an infinite number of basis functions to balance, we cannot expect to find weights that can balance all of them equally well. However, the relative importance of balancing each basis function can be measured by λ_j. This is reflected in the corresponding imbalance function induced by this model class,

$$\text{imbalance}^2_{\mathcal{M}}(\gamma) = \sum_j \lambda_j^2 \left(\frac{1}{n}\sum_{i=1}^{n} \phi_j(X_i) - \frac{1}{n}\sum_{i=1}^{n} W_i \gamma(X_i) \phi_j(X_i) \right)^2. \tag{16.8}$$

This measure weights the imbalance in basis function j by the value of λ_j, corresponding to, e.g., placing higher priority on lower order terms in a polynomial expansion or lower frequency terms in a Fourier expansion.

Taking another view, we note that Equation 16.7 is the unit ball of a reproducing kernel Hilbert space (RKHS), which suggests examining the balancing problem from the perspective of RKHS theory and focusing on the properties of the functions rather than their representations under a particular basis expansion. Balancing weights for RKHS models are helpful because they allow for optimization problems involving infinite components (e.g., in Equation 16.7) to be reduced to finite-dimensional problems via the "kernel trick," in which the inner products from (16.5) are replaced with kernel evaluations. For kernel $k_j(X_i) = k(X_i, X_j)$, this leads to weights that minimize

$$\frac{1}{n^2}\sum_{j=1}^{n}\sum_{i=1}^{n}\{W_iW_jk_j(X_i) - 2(1-W_i)W_jk_j(X_i)\gamma_i + (1-W_i)(1-W_j)k_j(X_i)\gamma_i\gamma_j\} \quad (16.9)$$

This objective upweights control units that are similar to treated units (second term) while recognizing that control units for which there are many others that are similar need not have too large a weight (third term). Different choices of kernels correspond to different assumptions on the RKHS $\mathcal{M}$. See [21] for technical details and [33], [34], [35], [36], and [37] for more complete discussions of RKHS models.

Another approach to defining the model class $\mathcal{M}$ begins with the Kolmogorov-Smirnov (KS) statistic, which quantifies the distance between two cumulative distribution functions (cdfs). Weights that minimize the maximal KS statistic (i.e., to control imbalance on the entire distribution of continuous covariate) correspond to an additive model over the covariates with the summed componentwise total variation as the imbalance metric: $\mathcal{M} = \{\sum_{j=1}^{p} f_j(x_j) : \sum_{j=1}^{p} ||f_j||_{TV} \leq 1\}$, where $||f_j||_{TV} = \sum_j \int |f_j'(x)|dx$.

16.3.4 The primal-dual connection

While the balancing and modeling approaches may seem at odds, in fact, the two are related via the Lagrangian dual of the optimization problem in (16.5). Implicitly, this problem fits a generalized linear model for the inverse propensity score. The choices for the imbalance and dispersion penalties determine the particular form and penalization for this model. To make our discussion concrete, consider the balancing problem (16.5) with squared imbalance and dispersion penalties $h_\lambda(x) = \lambda^{-1}x^2$ and $f(x) = x^2$, without constraints. Then, the weights that solve (16.5) can be equivalently characterized as the solution to the following optimization problem:

$$\min_{\gamma} \underbrace{\frac{1}{n}\sum_{i=1}^{n}(1-W_i)\left(\gamma^{\text{IPW}}(X_i) - \gamma(X_i)\right)^2}_{\text{MSE for } \gamma^{\text{IPW}}}$$
$$- \underbrace{\left(\frac{1}{n}\sum_{i=1}^{n}W_i\gamma(X_i) - \frac{1}{n}\sum_{i=1}^{n}(1-W_i)\gamma^{\text{IPW}}(X_i)\gamma(X_i)\right)}_{\text{mean zero "noise"}} + \frac{\lambda}{2}||\gamma||^2_{\mathcal{M}} \quad (16.10)$$

The dual form in (16.10) can be thought of as a regularized regression for the treatment odds γ^{IPW}. In expectation, it finds weights that minimize the mean square error for the true treatment odds, among the control population, with a mean zero "noise" term in finite samples. The model class $\mathcal{M}$ governs the type of regularization on this model through its gauge norm $||\gamma||_{\mathcal{M}}$.[3] For example, if we choose the class of models linear in transformed covariates $\phi(x)$ with a p norm constraint in Equation (16.6), then the weights are also linear with $\gamma(x) = \eta \cdot \phi(x)$, and the model is regularized via the p norm: $||\gamma||_{\mathcal{M}} = ||\eta||_p$. In this way we can understand the balancing problem for different model classes as estimating γ^{IPW} under different forms of regularization, including, e.g., sparsity penalties [14, 38] and kernel ridge regression [34].

[3]The gauge of the model $\mathcal{M}$ is $||\gamma||_{\mathcal{M}} = \inf\{\alpha > 0 : \gamma \in \alpha\mathcal{M}\}$.

16.3.5 Connections to regression

A common approach to estimating treatment effects in observational studies involves least squares linear regression. One approach (the "conventional" approach) uses the estimated coefficient on W from a linear regression of Y on (X, W) as an estimator for the ATE, and another approach (the "fully interacted" approach) fits linear regressions of Y on X separately in the treated and untreated groups and uses the difference in imputed outcomes from these models as an estimate of the ATE. In fact, both of these are implicitly balancing approaches. Their implied weights exactly balance the covariate means in the treated and control groups, albeit with respect to different covariate profiles. The conventional regression weights balance the mean of covariate X_p in each group to a weighted combination of the treated and control sample means for X_p, where the combining weights are proportional to the relative variance of X_p in each group. The fully interacted regression weights balance the mean of covariate X_p in each group to its overall sample mean. Indeed, the weights can be expressed as solutions to a particular convex optimization problem, as in Equation (16.5) (see Section 6 of Chattopadhyay and Zubizarreta, [8]).

16.4 Computing the Weights in Practice

In this section, we discuss the design choices and trade-offs that arise for the balancing approach in practice. These choices include

1. the model class $\mathcal{M}$ to which the outcome model m belongs,
2. the imbalance and dispersion penalties (h_λ and f in Equation 16.5, respectively),
3. the additional constraints to specify in Equation 16.5,

among other concerns. We argue that choosing the model class $\mathcal{M}$ is by far the most important of these. Our discussion centers on the classic example from [16]. Specifically, our goal is to estimate the effect of a job training program on reported earnings in the year 1978. Following ([39], [17]), data come from the National Supported Work Demonstration (NSW) randomized trial, with the control group from this trial replaced with controls drawn from the Population Survey of Income Dynamics (PSID). The resulting data set is therefore a constructed observational study in which we can compare estimates using observational causal inference methods to those from the NSW trial.

16.4.1 Choosing what to balance

In Section 16.3.3 we discussed the connection between assumptions about the outcome model $m(X, 0)$, the bias of the weighting estimator, and the functions of covariates for which balance is sought. The LaLonde data have three continuous covariates (age, reported earnings in 1974, and reported earnings in 1975) and four binary covariates (indicators of Black race, Hispanic ethnicity, marriage, and not having a high school degree). Comparing treatment group means gives an initial picture of covariate balance in the data. In Table 16.1 we see that the treatment groups differ greatly on nearly all observed covariates. On average, control individuals are older, less likely to be Black or Hispanic, more likely to be married and a high school graduate, and have higher reported past earnings.

Investigating balance with respect to covariate interactions reveals further imbalances; for example, the imbalance in reported earnings in 1975 is greater in the non-Black subgroup than in the Black subgroup. Table 16.2 shows the means for the continuous covariates among the subgroups defined by values on two of the binary covariates.

TABLE 16.1
Covariate means for the LaLonde data.

Covariate	Control Group Mean	Treatment Group Mean
Age (years)	34.9	25.8
Black (%)	25.1	84.3
Hispanic (%)	3.3	5.9
Married (%)	86.6	18.9
No high school degree (%)	30.5	70.8
Reported earnings, 1974 (thousands of $)	19.4	2.1
Reported earnings, 1975 (thousands of $)	19.1	1.5

TABLE 16.2
Continuous covariate means for the LaLonde data, by subgroup.

Subgroup	Covariate	Control Group Mean	Treated Group Mean
Black	Age (years)	34.2	26.0
	Reported earnings, 1974 (thousands of $)	14.6	2.2
	Reported earnings, 1975 (thousands of $)	14.0	1.5
Non-Black	Age (years)	35.1	24.9
	Reported earnings, 1974 (thousands of $)	21.1	1.8
	Reported earnings, 1975 (thousands of $)	20.8	1.8
High school degree	Age (years)	33.1	26.9
	Reported earnings, 1974 (thousands of $)	21.5	3.4
	Reported earnings, 1975 (thousands of $)	21.3	1.4
No high school degree	Age (years)	38.8	25.4
	Reported earnings, 1974 (thousands of $)	14.5	1.5
	Reported earnings, 1975 (thousands of $)	14.0	1.6

In this section, we consider the implications of different choices for the outcome model class $\mathcal{M}$, which determines what functions of covariates to balance and/or the balance metric. We use the form of the problem in Equation 16.5, with the imbalance penalty $h_\lambda(x) = \lambda^{-1}x^2$ (λ is defined below), the dispersion penalty $f(x) = x^2$, and including the simplex constraint to rule out extrapolation. We standardize all covariates to have a mean 0 and standard deviation 1. Under this setup, we consider the performance of the balancing approach for three designs corresponding to three model classes $\mathcal{M}$: (1) linear in the covariates with a one-norm constraint on the coefficients ("marginal design"), (2) linear in the covariates and up to their three way interactions with a one-norm constraint on the coefficients ("interaction design"), and (3) an RKHS equipped with the RBF kernel[4] ("RKHS design"). For the first two, this corresponds to balancing (1) main terms and (2) main terms and up to three-way interactions (using Hermite polynomials), and (3) corresponds to the use of the RBF kernel in the kernel trick (Equation 16.9). For the marginal and interaction designs, $\lambda = 1 \times 10^{-3}$, and for the RKHS design, $\lambda = 1 \times 10^{-1}$.

[4]Specifically, we used the kernel $k(x, y) = \exp(-\sigma||x - y||^2)$ with $\sigma = 0.01$. This corresponds to the assumption that $\mathcal{M}$ is an RKHS with kernel k given above. In order for the weights that minimize (16.9), with k an RBF kernel, to reduce bias, the functions in $\mathcal{M}$ must be continuous, square integrable, and satisfy a technical regularity condition (see [40]). However, these restrictions are relatively weak; the associated feature space of this RKHS is infinite-dimensional, and the RBF kernel is exceedingly popular among kernel regression methods.

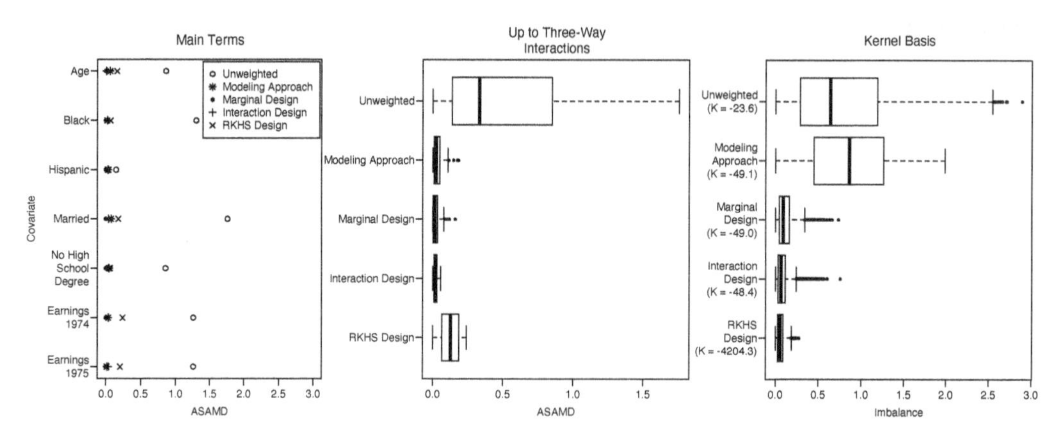

FIGURE 16.1
ASAMDs after weighting to balance various functions of the covariates.

In Figure 16.1, we evaluate how the different balancing designs (including a traditional modeling approach for the weights, where we model the propensity score using a logistic regression of the treatment on main terms) produce covariate balance in the LaLonde data using the average standardized absolute (weighted) mean differences (ASAMDs) for several functions of the covariates of interest (e.g., means). The leftmost plot shows the balance for main terms, where we see that all designs achieve relatively good balance on main terms compared to before weighting, particularly the marginal and interaction designs, which directly target minimizing imbalance on these functions. In the middle plot, the distributions of ASAMDs for up to three-way interactions are plotted before and after weighting, showing improvements for all designs over the unweighted data. Again, the method that directly targets these covariate functions for balance (i.e., the interaction design) shows the best performance, though the marginal design does almost as well, leaving a few higher order terms relatively unbalanced. The rightmost plot shows the distributions of ASAMDs for 5000 random features generated as in [41] from the RBF kernel, and the RKHS design shows the best performance, as expected. We also include the value of the objective function in (16.9) for each design, where, as expected, RKHS has the smallest value.

While some heuristics exist (e.g., ASAMDs < 0.1) for assessing covariate balance, it is important to evaluate balance in conjunction with considering the sample size, as there is a trade-off between the two. We discuss this in Section 16.4.3.

16.4.2 Maximizing the effective sample size

In addition, the designs can be compared on their effective sample sizes (ESS) for the control group. The effective sample size is calculated using the formula due to [42]: $\left(\sum_{i=1}^{n}(1-W_i)\widehat{\gamma}_i\right)^2 / \sum_{i=1}^{n}(1-W_i)\widehat{\gamma}_i^2$. Without weights, the ESS reduces to the sample size; with weights it measures the amount of data that will contribute to weighted analyses, measured in units of sample size. The ESS provides a measure of the variance of the weights in the sense that higher variance of the weights corresponds to lower ESS (see, e.g., [8]). Table 16.3 presents the control group ESSs for the three designs and for the unweighted data. All designs result in a highly reduced effective sample due to the high imbalances before weighting. Additionally, as the design becomes more complex (in the sense of balancing more basis functions), the ESS decreases, as is typical.

TABLE 16.3
Effective sample sizes (ESS) for the various weighting designs.

Weighting Design	Control Group Effective Sample Size
Unweighted	2490.0
Marginal	59.8
Interaction	41.4
RKHS	38.1

For any covariate adjustment method, there is a balance-dispersion trade-off, where better balance on more covariate functions typically leads to more highly dispersed weights. We discuss this more thoroughly in the next section.

16.4.3 Choosing the objective functions: the balance-dispersion trade-off

The objective function in Equation 16.5 includes tuning parameters to control how much minimizing imbalance is prioritized over minimizing dispersion. This relates directly to the bias-variance trade-off inherent in any estimation method, where balance relates to the bias of the estimator and dispersion is related to its variance: roughly, better balance leads to lower bias, and less dispersion leads to lower variance [43]. This trade-off is governed by the tuning parameter λ in Equation 16.5. In this section we fix the rest of the design choices in Equation (16.5) (namely, we set the imbalance penalty $h_\lambda(x) = \lambda^{-1}x^2$ and the dispersion penalty $f(x) = x^2$ and use the interaction design) and investigate the effect of varying λ on the resulting weights.

The black line in the lefthand panel of Figure 16.2 plots imbalance (in the form of maximum ASAMD) versus dispersion (in the form of ESS) when balance is sought on up to three-way interactions of covariates in the LaLonde data. The points represent the maximum ASAMD and ESS for a particular value of λ, which was varied from $\lambda = 1 \times 10^{-3}$ to $\lambda = 1 \times 10^{3}$ on a logarithmic scale. We see that, as balance improves (i.e., maximum ASAMD decreases), the ESS decreases. Investigators can use a plot like this to judge how much balance must be sacrificed in order to achieve any particular ESS, and design decisions can be made accordingly. For example, in the plot below, we see that to achieve relatively good balance (e.g., ASAMD < 0.1) for all covariate functions, we must sacrifice considerable effective sample size – effective sample size drops to ≈ 65, roughly 3 percent of the total sample.

16.4.4 Extrapolating and interpolating

In the previous sections we sought weights that balance covariates subject to a simplex constraint (i.e., average-to-one and nonnegativity constraints on the weights). This restricts any weighted estimates to be an interpolation of the observed data: that is, it ensures sample-boundedness, which restricts the estimate to lie between the minimum and maximum observed values [44]. Investigators typically seek to avoid extrapolation, which makes the validity of estimates more reliant on the modeling assumptions (i.e., assumptions about the outcome model class $\mathcal{M}$). However, relaxing the simplex constraint to a simple average-to-one (i.e., normalization) constraint can result in weights that better trade-off balance and dispersion. In the lefthand panel of Figure 16.2, we see that relaxing the non-negativity constraint leads to better balance for a given effective sample size: that is, the trade-off between balance and the dispersion of the weights is less extreme when the non-negativity constraint is relaxed, particularly when the effective sample size is small. We can also measure how far the resulting weights divided by n are from the simplex (i.e., the region $\{w_i : \sum_{i=1}^n (1 - W_i)w_i = 1, w_i > 0,\ \forall i : W_i = 0\}$). In Figure 16.3 we see that as the effective sample size increases, the weights are closer to the simplex, implying less extrapolation when

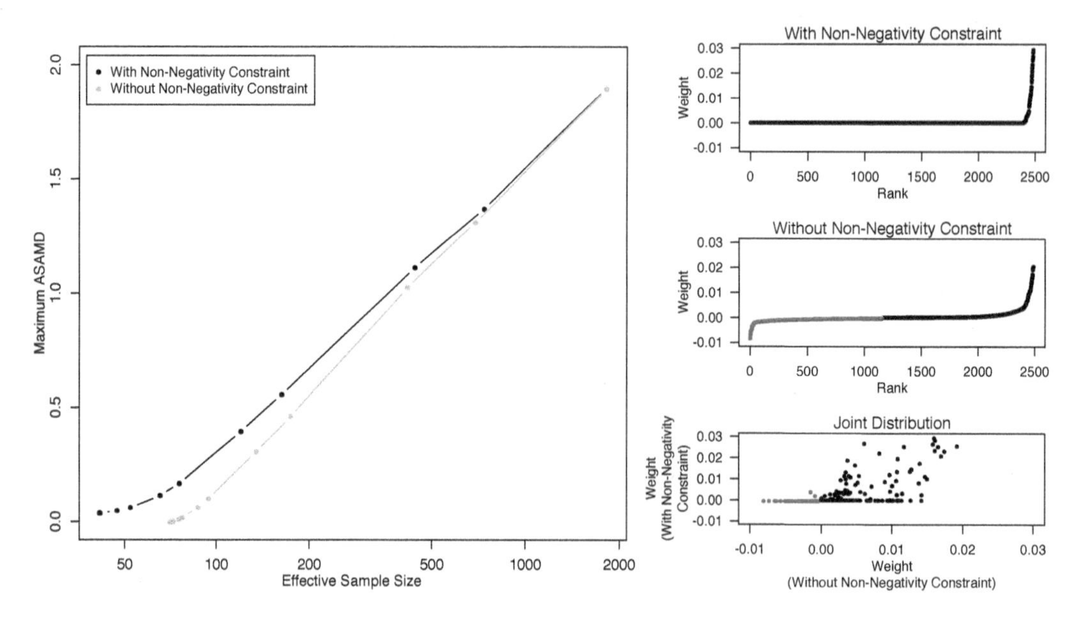

FIGURE 16.2
Balance-dispersion trade-off and plots of the weights for the LaLonde data, with and without the non-negativity constraint.

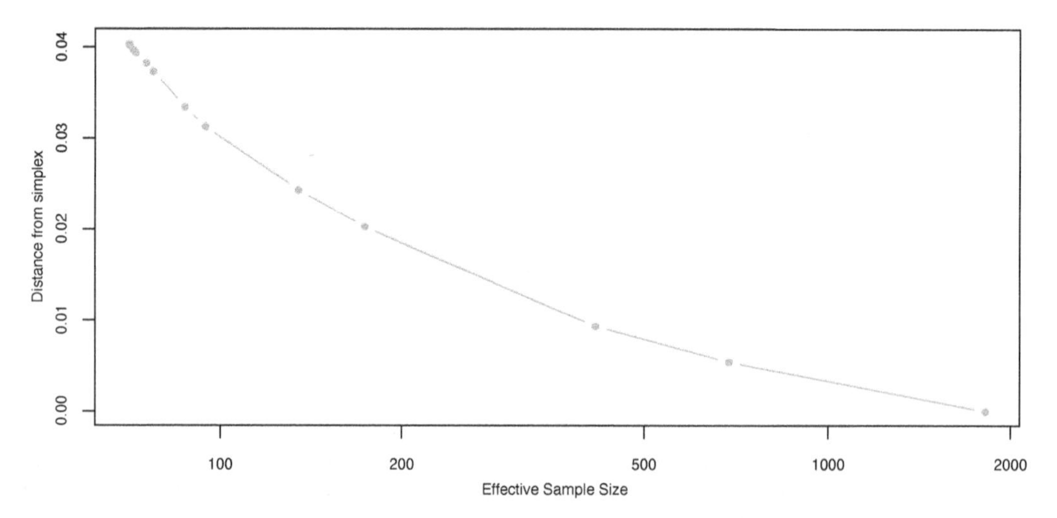

FIGURE 16.3
Distance from the simplex versus effective sample size, without the non-negativity constraint.

computing weighted estimators. Considering this suggests a trade-off between extrapolation and balance, where limiting extrapolation comes at the expense of balance.

We can also assess extrapolation by examining the extent to which non-negativity is violated when the simplex constraint is relaxed. The righthand panels of Figure 16.2 show how the unit-specific weights change before and after removing the non-negativity constraint, where negative weights are highlighted in red. The top-right and middle-right panels of Figure 16.2 plot the weights with and without the non-negativity constraint by order of their rank within that set. Whereas generally, the negative weights are smaller in magnitude than the non-negative weights, there are some units with

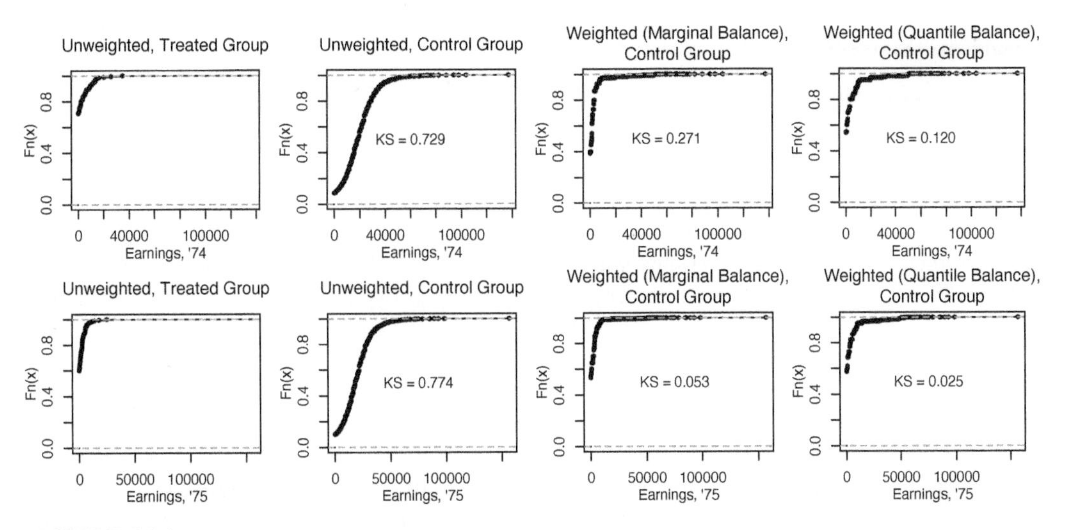

FIGURE 16.4
Empirical pdfs of reported earnings covariates for various balancing approaches.

extreme negative weights, suggesting a high degree of extrapolation is admitted for this design. The bottom-right panel shows the joint distribution of the unit-specific weights, with and without the non-negativity constraint. This panel shows that, for the most part, the units that receive negative weights receive very small weights (nearly zero in most cases) when non-negativity is enforced: that is, there is a subset of units such that, with negativity allowed, they contribute non-trivially to weighted estimates, but with negativity disallowed, they contribute almost nothing.

16.4.5 Additional options for balancing in practice

Investigators can target balance on entire distributions of covariates (including, possibly, joint distributions) by balancing their quantiles. They can also combine approaches, for example, by balancing quantiles of more important continuous covariates and the means of less important ones, as is commonly done with matching methods [45]. We demonstrate this approach using the LaLonde example, where up to three-way interactions of covariates are balanced along with the quantiles of two continuous covariates: reported earnings in 1974 and reported earnings in 1975. This approximates balancing the KS statistics for these two continuous covariates, as in the limit (i.e., as the number of quantiles approaches infinity) it minimizes their KS statistics.

In Figure 16.4 we compare this approach with the approach of just balancing the interactions in terms of the empirical cdfs of the two earnings covariates. The top row of charts shows the empirical cdfs for reported earnings in 1974 and the bottom row shows those for reported earnings in 1975. The left-most column reports the unweighted empirical cdfs for the treated group: i.e., the distribution we might want the weighted control group to resemble. The second from the left-most column shows the unweighted control group empirical cdfs – which have much more mass at the upper end of the distribution (i.e., high earnings) than the treated group does. The second from the right-most column shows the empirical cdfs after weighting the control group to balance only the three-way interactions of all covariates, and the right-most column shows the empirical pdfs after balancing the quantiles of the reported earnings covariates in addition. We also report the KS statistics for the various control group empirical cdfs. We see that only balancing the main terms and up to three-way interaction terms results in more highly similar empirical cdfs even when this is not explicitly targeted, though the KS statistic is lowest when balancing the quantiles directly. This better balance is traded off with a lower ESS, which drops from 41.4 to 31.0.

16.5 Estimating Effects

16.5.1 The role of augmentation

In practice, weighting approaches alone can still lead to meaningful covariate imbalance. The size of the balancing problem implied by $\mathcal{M}$ may be too large to yield a solution for a given sample; for example, the number of covariates p may be too large to achieve exact balance on everything (especially when $p \gg n$ and/or for continuous covariates, as explained in Section 16.3.3). This is the "curse of dimensionality," and in many cases we must settle for weights that only approximately balance. In that case researchers often augment weighted estimates with an estimate of the outcome model to remove bias due to remaining imbalance. Augmented estimators can be thought of as an interpolation between a pure outcome model-based estimator (which often involves negative weights, i.e., extrapolation) and a pure balancing weights-based estimator (which protects from extrapolation). This technique is common with propensity score approaches, where it is typically connected to the notion of double-robustness: the estimator of the potential outcome mean is consistent if either the propensity score model for the weights or the outcome model is correctly-specified [46]. With $\widehat{m}(X, 0)$ a model for the outcome $\mathbb{E}[Y|X, W = 0]$, the augmented weighting estimator takes the form:

$$\widehat{\mu}_0^{aug} = \underbrace{\frac{1}{n}\sum_{i=1}^{n}(1 - W_i)\widehat{\gamma}_i Y_i}_{\text{weighting estimator}} + \underbrace{\frac{1}{n}\sum_{i=1}^{n}(1 - W_i)(\widehat{m}(X_i, 0) - \widehat{\gamma}_i\widehat{m}(X_i, 0))}_{\text{bias correction via imputation}}. \tag{16.11}$$

Numerous methods exist for estimating $\widehat{m}(X, 0)$, including the linear regression of Y on X among those with $W = 0$ or machine learning methods. While the traditional approach to forming the estimator in (16.11) involves a modeling approach for the weights (i.e., as in augmented inverse propensity weighting (AIPW)), recent proposals have combined the bias correction in (16.11) with a balancing weights approach. These methods include the approximate residual balancing approach of [47], the augmented minimax linear estimation approach of [48], and the automatic double machine learning approach of [49].

Like the error decomposition for a general weighted estimator in Equation 16.4, we can decompose the error from an augmented estimator as follows,

$$\begin{aligned}\widehat{\mu}_0^{aug} - \mu_0 = &\underbrace{\frac{1}{n}\sum_{i=1}^{n}(1 - W_i)\widehat{\gamma}_i\delta m(X_i, 0) - \frac{1}{n}\sum_{i=1}^{n} W_i\delta m(X_i, 0)}_{\text{bias from imbalance}} \\ &+ \underbrace{\frac{1}{n}\sum_{i=1}^{n}(1 - W_i)\widehat{\gamma}_i\epsilon_i}_{\text{noise}} + \underbrace{\frac{1}{n}\sum_{i=1}^{n} W_i m(X_i, 0) - \mu_0}_{\text{sampling variation}}\end{aligned} \tag{16.12}$$

where $\delta = \widehat{m}(\cdot, 0) - m(\cdot, 0)$ is the regression error. Intuitively, the imbalance with augmentation should be smaller than without because the error term δm should be smaller than m. Similar to our approach in Section 16.2.3, where we bound the expectation of the error by the maximal imbalance in models m over the model class $\mathcal{M}$, we can bound the expectation of the error (16.12) by the maximal imbalance over a model class for δm.

In the next section, we briefly discuss the asymptotic properties of the estimators $\widehat{\mu}_0$ and $\widehat{\mu}_0^{aug}$.

16.5.2 Asymptotic properties

We begin by considering the asymptotic properties of $\widehat{\mu}_0 - \tilde{\mu}_0$, where $\tilde{\mu}_0 = n^{-1}\sum_{i=1}^n W_i m(X_i, 0)$ is the sample analog of μ_0. By focusing on the sample analog, we drop the sampling variation term from the decompositions in Equations (16.4) and (16.12), which is mean-zero, and any estimate of $\tilde{\mu}_0$ will also be an estimate of μ_0. Thus, the remaining terms from the decomposition in (16.4) are the bias from imbalance and the noise. The latter term is a mean-zero sum of independent and identically–distributed (iid) random variables, and so basic asymptotic theory applies and it will converge to a mean-zero Normal distribution. If the first term (the imbalance) goes to 0 in probability at a rate faster than the variance of the noise term (which is $n^{1/2}$), then the bias will be small relative to the variance and $\hat{\mu}_0$ will be an asymptotically normal estimator of μ_0. In short, we need the bias to be $o_p(n^{-1/2})$.

Ben-Michael et al. [21] argues that, without augmentation, we cannot guarantee the imbalance is $o_p(n^{-1/2})$ for a general enough model class $\mathcal{M}$ (namely, a Donsker class; see also [48], Remark 4). Further, even if we correctly specify $\mathcal{M}$, the best we may hope for – without further assumptions – is that the imbalance is $O_p(n^{-1/2})$.

This approach of attaining sufficiently small imbalance does work, however, with augmentation, where we consider balance over a shrinking model class rather than a fixed one. In this case we consider the imbalance in δm, which we re-index with the sample size n to denote its dependence on the sample size. That is, we consider the model class $\mathcal{M}_n = \{\delta m_n\}$, where $\delta m_n = \widehat{m}_n - m$. If we further assume that the regression model is cross-fit (as is increasingly standard practice; see, e.g., [50], [51], [52]) and consistent, then we have asymptotic normality. Namely,

Theorem 16.1. *Let $\widehat{m}(0, X)^{(-i)}$ be a consistent cross-fit estimator of $m(0, X)$. Then*

$$\frac{\widehat{\mu}_0^{aug} - \tilde{\mu}_0}{\widehat{V}} \to N(0,1) \;\; where \;\; \widehat{V} = \frac{1}{n^2}\sum_{i=1}^{n}(1 - W_i)\widehat{\gamma}(X_i)^2\{Y_i - \widehat{m}(0, X_i)^{(-i)}\}^2 \tag{16.13}$$

if $\sqrt{n^{-1}\sum_{i=1}^n (1 - W_i)\widehat{\gamma}(X_i)^2} \to 1$, imbalance in δm is $o_p(n^{-1/2})$, each weight's square is small relative to the sum of them all, and the noise sequence $\epsilon_1, ..., \epsilon_n$ is uniformly square integrable conditional on the design.

Ben-Michael et al. [21] includes technical details and derivations for this theorem, which uses results from [53]. Theorem 16.1 thus provides an estimator for the asymptotic variance that allows for simple computation of confidence intervals (e.g., at level α via $\widehat{\mu}_0^{aug} \pm z_{1-\alpha/2}\sqrt{\widehat{V}/n}$). Bootstrapping techniques are also available for variance and/or confidence interval estimation.

16.5.3 Estimates

We return to our goal of estimating the control potential outcome mean for the treated μ_0 using the observational control units from the LaLonde data. Now that the design is complete (insofar as the weights have been constructed), we may proceed to the outcome analysis. For the sake of illustration, we use the weights that balance up to three-way interactions. We construct estimates with and without augmentation using the constructions in (16.3) and (16.12), respectively. In this example, we have access to the control group from the randomized trial – which should provide an unbiased estimate of μ_0—for comparison, and we use this to benchmark the observational results. Figure 16.5 provides these results, along with 95% confidence intervals constructed via the bootstrap.

Figure 16.5 shows that, before weighting, the observational estimate is dramatically different from the trial estimate (i.e., the ground truth). Both the weighted and augmented estimates are much closer to the truth, and their confidence intervals contain it. The augmented estimate is slightly closer to the truth than the non-augmented estimate, and its confidence interval is shorter.

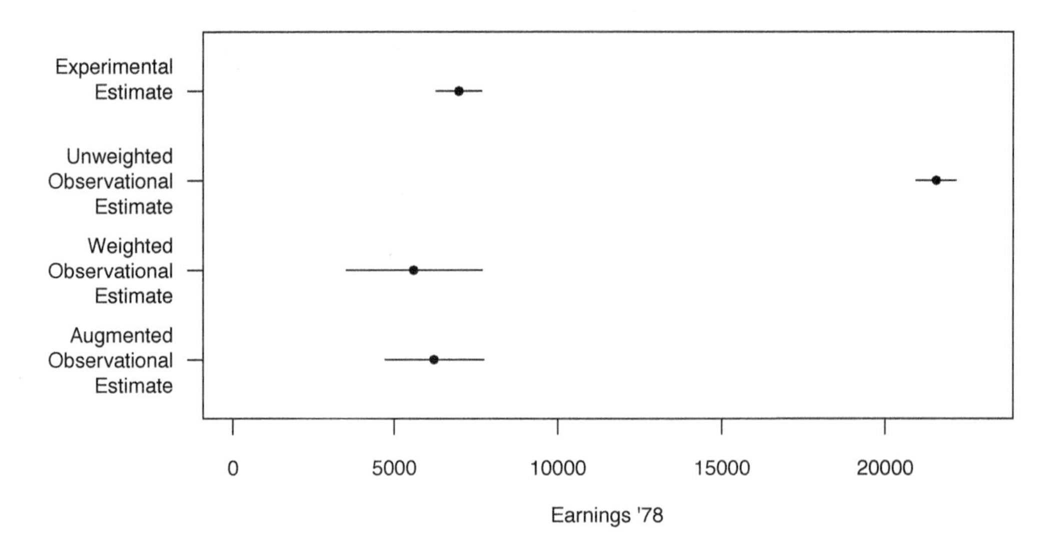

FIGURE 16.5
Estimates of $\mathbb{E}[Y(0)|W=1]$.

16.6 Concluding Remarks

In this chapter we provide an introduction to the balancing approach to weighting for causal inference, which seeks weights that directly balance covariates and/or functions of covariates to produce unbiased estimates of treatment effects from observational data. This approach differs from the traditional modeling approach to weighting, which posits a model for the propensity score and then applies a function to the estimated probabilities (e.g., the inverse) to produce the weights. The two approaches are related via the dual form presented in Section 16.3.4, showing that the balancing approach implicitly fits a penalized generalized linear model for the propensity score weights. The choice of the covariate functions to balance has a direct relation to assumptions about the potential outcome model(s), as described in Section 16.3. We discuss various design choices for the balancing approach in practice in Section 16.4 using the canonical LaLonde data ([16], [39], [17]), which we hope can guide readers as they work through these considerations in their own data.

This chapter is intended as an introductory and accessible guide to the balancing approach, and we invite readers to consult more detailed discussions of some of the issues discussed in, for example, [8, 10, 21], as well as the other literature referenced throughout. In particular, [21] discusses the balancing approach for estimands besides the ATT, such as the Average Treatment Effect (ATE).

References

[1] Stefan Tübbicke. Entropy balancing for continuous treatments. *Journal of econometric methods*, 11(1):71–89, 2021.

[2] C.A. Stone and Y. Tang. Comparing propensity score methods in balancing covariates and recovering impact in small sample educational program evaluations. *Practical Assessment, Research and Evaluation*, 18, 11 2013.

[3] Mark Lunt, Daniel Solomon, Kenneth Rothman, Robert Glynn, Kimme Hyrich, Deborah P. M Symmons, and Til Stürmer. Different Methods of balancing covariates leading to different effect estimates in the presence of effect modification," *American Journal of Epidemiology*, 169(7):909–917, 2009.

[4] Jessica M Franklin, Jeremy A Rassen, Diana Ackermann, Dorothee B Bartels, and Sebastian Schneeweiss. Metrics for covariate balance in cohort studies of causal effects. *Statistics in Medicine*, 33(10):1685–1699, 2014.

[5] Beth Ann Griffin, Greg Ridgeway, Andrew R Morral, Lane F Burgette, Craig Martin, Daniel Almirall, Rajeev Ramchand, Lisa H Jaycox, and Daniel F McCaffrey. Toolkit for weighting and analysis of nonequivalent groups (twang), 2014.

[6] Paul R Rosenbaum and Donald B Rubin. The central role of the propensity score in observational studies for causal effects. *Biometrika*, 70(1):41–55, 1983.

[7] Daniel Daniel Westreich, Justin Justin Lessler, and Michele Jonsson Michele Jonsson Funk. Propensity score estimation: machine learning and classification methods as alternatives to logistic regression. *Journal of clinical epidemiology*, 63(8):826–833, 2010.

[8] Ambarish Chattopadhyay, Christopher H Hase, and José R Zubizarreta. Balancing vs modeling approaches to weighting in practice. *Statistics in Medicine*, 39(24):3227–3254, 2020.

[9] Elizabeth A Stuart. Matching methods for causal inference: A review and a look forward. *Statistical science*, 25(1):1–21, 2010.

[10] David A Hirshberg and José R Zubizarreta. On two approaches to weighting in causal inference. *Epidemiology (Cambridge, Mass.)*, 28(6):812–816, 2017.

[11] Jens Hainmueller. Entropy balancing for causal effects: A multivariate reweighting method to produce balanced samples in observational studies. *Political Analysis*, 20(1):25–46, 2012.

[12] José R Zubizarreta. Stable weights that balance covariates for estimation with incomplete outcome data. *Journal of the American Statistical Association*, 110(511):910–922, 2015.

[13] Victor Chernozhukov, Whitney K Newey, and Rahul Singh. Automatic debiased machine learning of causal and structural effects. *arXiv preprint*, 2018.

[14] Z Tan. Regularized calibrated estimation of propensity scores with model misspecification and high-dimensional data. *Biometrika*, 107(1):137–158, 2020.

[15] Kosuke Imai and Marc Ratkovic. Covariate balancing propensity score. *Journal of the Royal Statistical Society. Series B, Statistical Methodology*, 76(1):243–263, 2014.

[16] Robert J LaLonde. Evaluating the econometric evaluations of training programs with experimental data. *The American Economic Review*, 76(4):604–620, 1986.

[17] Rajeev H Dehejia and Sadek Wahba. Causal effects in nonexperimental studies: Reevaluating the evaluation of training programs. *Journal of the American Statistical Association*, 94(448):1053–1062, 1999.

[18] J. Neyman. On the application of probability theory to agricultural experiments. *Statistical Science*, 5(5):463–480, 1923.

[19] Donald B Rubin. Estimating causal effects of treatments in randomized and nonrandomized studies. *Journal of Educational Psychology*, 66(5):688, 1974.

[20] Donald B Rubin. Randomization analysis of experimental data: the fisher randomization test comment. *Journal of the American Statistical Association*, 75(371):591–593, 1980.

[21] Eli Ben-Michael, Avi Feller, David A Hirshberg, and Jose R Zubizarreta. The balancing act for causal inference. *arXiv preprint*, 2021.

[22] Jamie M Robins, A Rotnitzky, and L P Zhao. Estimation of regression coefficients when some regressors are not always observed. *Journal of the American Statistical Association*, 89(427):846–866, 1994.

[23] Daniel F McCaffrey, Greg Ridgeway, and Andrew R Morral. Propensity score estimation with boosted regression for evaluating causal effects in observational studies. *Psychological Methods*, 9(4):403, 2004.

[24] Brian K Lee, Justin Lessler, and Elizabeth A Stuart. Improving propensity score weighting using machine learning. *Statistics in Medicine*, 29(3):337–346, 2010.

[25] Keisuke Hirano, Guido W Imbens, and Geert Ridder. Efficient estimation of average treatment effects using the estimated propensity score. *Econometrica*, 71(4):1161–1189, 2003.

[26] Peter C Austin and Elizabeth A Stuart. Moving towards best practice when using inverse probability of treatment weighting (iptw) using the propensity score to estimate causal treatment effects in observational studies. *Statistics in Medicine*, 34(28):3661–3679, 2015.

[27] Til Sturmer, Michael Webster-Clark, Jennifer L Lund, Richard Wyss, Alan R Ellis, Mark Lunt, Kenneth J Rothman, and Robert J Glynn. Propensity score weighting and trimming strategies for reducing variance and bias of treatment effect estimates: A simulation study. *American Journal of Epidemiology*, 190(8):1659–1670, 2021.

[28] Valerie S Harder, Elizabeth A Stuart, and James C Anthony. Propensity score techniques and the assessment of measured covariate balance to test causal associations in psychological research. *Psychological Methods*, 15(3):234–249, 2010.

[29] Fan Li and Laine E Thomas. Addressing extreme propensity scores via the overlap weights. *American Journal of Epidemiology*, 188(1):250–257, 2019.

[30] Ambarish Chattopadhyay and Jose R. Zubizarreta (2022). On the implied weights of linear regression for causal inference. Biometrika, 2022, forthcoming.

[31] Yair Ghitza and Andrew Gelman. Deep interactions with MRP: Election turnout and voting patterns among small electoral subgroups. *American Journal of Political Science*, 57(3):762–776, 2013.

[32] Eli Ben-Michael, Avi Feller, and Erin Hartman. Multilevel calibration weighting for survey data, 2021.

[33] Chad Hazlett. Kernel balancing: A flexible non-parametric weighting procedure for estimating causal effects. *Statistica Sinica*, 30(3):1155–1189, 2020.

[34] David A Hirshberg, Arian Maleki, and Jose R Zubizarreta. Minimax linear estimation of the retargeted mean. *arXiv preprint*, 2019.

[35] Nathan Kallus. Generalized optimal matching methods for causal inference. *Journal of Machine Learning Research*, 21, 2020.

[36] Rahul Singh, Liyuan Xu, and Arthur Gretton. Generalized kernel ridge regression for nonparametric structural functions and semiparametric treatment effects, 2021.

[37] Raymond K W Wong and Kwun Chuen Gary Chan. Kernel-based covariate functional balancing for observational studies. *Biometrika*, 105(1):199–213, 2018.

[38] Yixin Wang and Jose R Zubizarreta. Minimal dispersion approximately balancing weights: Asymptotic properties and practical considerations. *Biometrika*, 107(1):93–105, 2020.

[39] Rajeev H Dehejia and Sadek M Wahba. *Propensity score matching methods for non-experimental causal studies*, volume no. 6829. of *NBER working paper series*. National Bureau of Economic Research, Cambridge, MA, 1998.

[40] Ha Quang Minh. Some Properties of Gaussian reproducing Kernel Hilbert spaces and their implications for function approximation and learning theory. *Constructive Approximation*, 32(2):307–338, 2009.

[41] Ali Rahimi and Benjamin Recht. Random features for large-scale kernel machines. In J. Platt, D. Koller, Y. Singer, and S. Roweis, editors, *Advances in Neural Information Processing Systems*, volume 20. Curran Associates, Inc., 2007.

[42] Leslie Kish. Survey sampling. *John Wiley and Sons, New York*, 26, 1965.

[43] Gary King, Christopher Lucas, and Richard A Nielsen. The balance-sample size frontier in matching methods for causal inference. *American Journal of Political Science*, 61(2):473–489, 2017.

[44] James Robins, Mariela Sued, Quanhong Lei-Gomez, and Andrea Rotnitzky. Comment: Performance of double-robust estimators when "inverse probability" weights are highly variable. *Statistical Science*, 22(4), Nov 2007.

[45] Paul R Rosenbaum. Modern algorithms for matching in observational studies. 7(1):143–176, 2020.

[46] James M Robins, Andrea Rotnitzky, and Lue Ping Zhao. Estimation of regression coefficients when some regressors are not always observed. *Journal of the American Statistical Association*, 89(427):846–866, 1994.

[47] Susan Athey, Guido W Imbens, and Stefan Wager. Approximate residual balancing: debiased inference of average treatment effects in high dimensions. *Journal of the Royal Statistical Society. Series B, Statistical Methodology*, 80(4):597–623, 2018.

[48] David A. Hirshberg and Stefan Wager. Augmented minimax linear estimation, 2020.

[49] Victor Chernozhukov, Whitney K Newey, and Rahul Singh. Automatic debiased machine learning of causal and structural effects, 2021.

[50] Victor Chernozhukov, Denis Chetverikov, Mert Demirer, Esther Duflo, Christian Hansen, Whitney Newey, and James Robins. Double/debiased machine learning for treatment and structural parameters. *The Econometrics Journal*, 21(1):C1–C68, 2018.

[51] A Schick. On asymptotically efficient estimation in semiparametric models. *The Annals of statistics*, 14(3):1139–1151, 1986.

[52] Wenjing Zheng and Mark J van der Laan. Cross-validated targeted minimum-loss-based estimation. In *Targeted Learning*, Springer Series in Statistics, pages 459–474. Springer New York, New York, NY, 2011.

[53] C. C Heyde and P Hall. *Martingale limit theory and its application*. Academic Press, 2014.

17

Assessing Principal Causal Effects Using Principal Score Methods

Alessandra Mattei, Laura Forastiere, Fabrizia Mealli

CONTENTS

17.1 Introduction

Decisions in many branches of sciences, including social sciences, medical care, health policy and economics, depend critically on appropriate comparisons of competing treatments, interventions and policies. Causal inference is used to extract information about such comparisons. A main statistical framework for causal inference is the potential outcomes approach [1–6], also known as the Rubin Causal Model [7], where a causal effect is defined as the contrast of potential outcomes under different treatment conditions for a common set of units. See Imbens and Rubin [8] for a textbook discussion.

In several studies, inference on causal effects is made particularly challenging by the presence of post-treatment variables (e.g., mediators, treatment compliance, intermediate endpoints), potentially affected by the treatment and also associated with the response. Adjusting treatment comparisons for post-treatment variables requires special care in that post-treatment variables are often confounded by the unit's characteristics that are also related to the response. Therefore, the observed values of an

DOI: 10.1201/9781003102670-17

intermediate variable generally encode both the treatment condition as well as the characteristics of the unit, and thus, naive methods, conditioning on the observed values of that intermediate variable, do not lead to properly defined causal effects, unless strong assumptions are imposed.

Within the potential outcomes approach to causal inference, principal stratification [9] is a principled approach for addressing causal inference problems, where causal estimands are defined in terms of post-treatment variables. A principal stratification with respect to a post-treatment variable is a partition of units into latent classes, named *principal strata*, defined by the joint potential values of that post-treatment variable under each of the treatments being compared. Because principal strata are not affected by treatment assignment, they can be viewed as a pre-treatment covariate, and thus, conditioning on principal strata leads to well-defined causal effects, named principal causal effects. Principal causal effects are local causal effects in the sense that they are causal effects for sub-populations defined by specific (union of) principal strata.

In the causal inference literature there is a widespread agreement on the important role of principal stratification to deal with post-treatment complications in both experimental and observational studies. Principal stratification has been successfully applied to deal with the following issues: non-compliance [10, 11]; censoring by death, where the primary outcome is not only unobserved but also undefined for subjects who die [12–14]; loss to follow-up or missing outcome data [15]; surrogate endpoints [9, 16]; and mediation [17–22], although its role in these settings is still controversial [23].

Early contributions on principal stratification mainly concentrate on studies with a binary treatment and a binary post-treatment variable, and we mainly focus on this setting throughout the chapter. Nevertheless, the literature has rapidly evolved offering a growing number of studies extending principal stratification to deal with the presence of intermediate variables that are multivariate, such as in studies suffering from combinations of post-treatment complications [12, 13, 24], categorical [25], continuous [26–28] or time-dependent [29], and to studies with longitudinal treatments [30] and multilevel treatments [31].

Principal stratification allows dealing with confounded post-treatment intermediate variables by providing meaningful definitions of causal effects conditional on principal strata. However, it also raises serious inferential challenges due to the latent nature of the principal strata. In general, principal stratum membership is not observed and the observed values of the treatment and intermediate variable cannot uniquely identify principal strata. Instead, groups of units sharing the same observed values of the treatment and post-treatment variable generally comprise mixtures of principal strata. For instance, in randomized studies with non-compliance, after randomization, some subjects assigned to the active treatment will not take it, but effectively take the control/standard treatment. Similarly, some subjects assigned to the control condition will actually take the active treatment. The actual treatment received is a confounded intermediate variable, in the sense that subjects receiving and subjects not receiving the active treatment may systematically differ in characteristics associated with the potential outcomes. The principal stratification with respect to the compliance behavior partitions subjects into four types defined by the joint potential values of the treatment that would be received under assignment to the control and the active treatment. Without additional assumptions, above and beyond randomization, the compliance status is not observed for any subject, e.g., the group of subjects assigned to the control condition who are observed to effectively not take treatment comprises a mixture of subjects who would always comply with their assignment, the compliers, by taking the active treatment if assigned to treatment and taking the control if assigned control; and subjects who would never take the active treatment, the never-takers, irrespective of their assignment.

Some common assumptions for disentangling these mixtures, which relate also to the econometric Instrumental Variables framework (IV), are the monotonicity assumption, which allows the effect of the treatment on the intermediate variable to be only in one direction, essentially limiting the number of principal strata [10, 11, 17], and exclusion restrictions, which rule out any effect of treatment on the outcome for units in principal strata where the intermediate variable is unaffected by the treatment [10]. Monotonicity and exclusion restrictions have been often invoked in randomized studies with non-compliance to non-parametrically identify and estimate the average causal effect for compliers

(CACE) [10]. More generally, under monotonicity and exclusion restrictions we can uniquely disentangle the observed distribution of outcomes into a mixture of conditional outcome distributions associated to each of the latent strata [32]. Depending on the empirical setting, monotonicity and exclusion restrictions may be questionable. For instance, in randomized experiments with non-compliance, exclusion restrictions appear plausible in blind or double-blind experiments, but less so in open-label ones [11, 24]. Outside the non-compliance/instrumental variables setting, assumptions such as the exclusion restrictions cannot be invoked, because they would rule out a priori the effects of interest, that is, causal effects for subpopulations of units where the intermediate variable is not affected by the treatment. For example, in studies with outcomes censored by death, such as Quality of Life, which is censored by death for subjects who die before the end of the follow-up, focus is on causal effects for always survivors, subjects who would live no matter how treated, that is, subjects for which the post-treatment survival indicator is unaffected by the treatment. In mediation analysis causal effects for units for which the mediator is unaffected by the treatment provide information on the direct effect of the treatment. Exclusion restrictions would imply that causal effects for always survivors in studies with outcomes censored by death and direct effects in mediation studies are zero.

In the literature various sets of assumptions have been introduced and investigated, including alternative structural assumptions, such as assumptions of ranked average score/stochastic dominance [24, 33], distributional assumptions, such as the assumption of normal outcome distributions within principal strata [13, 14] and assumptions that either involve the existence of an instrument for the intermediate variable [15, 19] or rely on additional covariates or secondary outcomes [34–39]. A fully model-based likelihood or Bayesian approach to principal stratification [11–14, 17, 24, 26, 28, 40], can be used, carefully thinking about the plausibility of parametric assumptions [41]. In addition, the problem of unbounded likelihood function may arise and jeopardize inference [34, 42] and Bayesian analyses may be sensitive to priors.

Without exclusion restrictions and/or parametric assumptions, principal causal effects can be generally only partially identified [19, 33, 43, 44], but large sample bounds are typically too wide for practical use, although there exist several strategies for sharpening them [33, 37, 38, 45–48].

A line of the literature proposes an alternative approach to deal with identification issues in principal stratification analysis dealing with confounded post-treatment variables: it invokes principal ignorability assumptions, which are conditional independence assumptions between (some) principal strata and (some) potential outcomes for the primary outcome given covariates. Principal ignorability assumptions imply that, given the observed covariates, there is no unmeasured confounding between certain principal strata and the outcome under specific treatment conditions. As a consequence, the distribution of potential outcomes is exchangeable across different principal strata and valid inference on principal casual effects can be conducted by comparing treated and control units belonging to different principal strata but with the same distributions of background characteristics.

As opposed to the general principal stratification framework, where outcome information is required to disentangle mixtures of principal strata, under principal ignorability assumptions studies with intermediate variables should be designed and analyzed without using information on the primary outcome to create subgroups of units belonging to different principal strata with similar distributions of the background variables. The idea is not new in causal inference; it borrows from the literature on the design and analysis of regular observational studies, i.e., non-experimental studies where the assignment mechanism is strongly ignorable [49]. Indeed, it is widely recognized that for objective causal inference, regular observational studies should be carefully designed to approximate randomized experiments using only background information (without any access to outcome data). This is done by creating subgroups of treated and control units with similar distributions of background variables [50, 51]. Among others, the covariate-adjustment method based on the propensity score, defined as the treatment probability given covariates, is a popular method to identify and adjust for these groups of individuals [49].

In a similar spirit, the literature on principal ignorability has introduced the *principal score* [52], defined as the conditional probability of belonging to a latent principal stratum given covariates.

In the same way the propensity score is a key tool in the design and analysis of observational causal studies with regular treatment assignment mechanisms, the principal score plays a central role in principal stratification analysis under principal ignorability assumptions. Principal scores are generally unknown and their identification may be challenging: even in randomized experiments with two treatments and a single binary intermediate variable, each observed group defined by the assigned treatment and the observed value of the intermediate variable comprises a mixture of two principal strata, and thus, principal scores are not identified without additional assumptions. Monotonicity assumptions are usually invoked, ruling out the existence of some principal strata and unmasking principal stratum membership for some units. Therefore, together with the ignorability of treatment assignment, which implies that principal strata have the same distribution in both treatment arms (at least conditional on covariates), monotonicity assumptions allow one to identify and estimate principal scores for each unit.

Early principal stratification analyses based on principal ignorability assumptions and principal scores date back to Follmann [53], who refers to the conditional principal stratum membership probability given the covariates as a *compliance score*, and to Hill et al. [52], who introduced the term *principal score*. Both papers focus on the analysis of randomized experiments with one-sided non-compliance, where a strong monotonicity assumption, implying that there exist only two latent principal strata, holds by design, and thus, the principal score is identified and can be easily estimated. In subsequent years various researchers have contributed to the literature on principal stratification analysis based on principal scores, under the assumptions of principal ignorability and monotonicity. Applied contributions providing prominent examples of how principal score methods work in practice include Crèpon et al. [54], Schochet et al. [55] and Zhai et al. [56]. Key theoretical and conceptual contributions include Jo and Stuart [57], Jo et al. [58] and Stuart and Jo [59], who further investigate the performance of principal score methods in randomized experiments with one-sided non-compliance (see also [60] for a simulation study), and Ding and Lu [61], Feller et al. [62], Joffe et al. [63], and Jiang et al. [64], who consider the use of principal scores in a more general setup, where there may exist more than two principal strata. Specifically, Joffe et al. [63] suggest the use of principal scores to identify and estimate general principal causal effects; Ding and Lu [61] provide theoretical justification for principal stratification analysis based on principal scores; Feller et al. [62] offer an excellent review of methods based on propensity scores and introduce new sets of assumptions under which principal causal effects can be identified and estimated mixing and matching principal ignorability assumptions and exclusion restrictions; and Jiang et al. [64] develop multiply robust estimators for principal causal effects, showing that they are consistent if two of the treatment, intermediate variable, and outcome models are correctly specified, and that they are locally efficient if all three models are correctly specified.

The chapter will draw from this literature, offering a comprehensive overview of principal score methods based on principal ignorability and monotonicity assumptions. We will also discuss the importance of conducting sensitivity analysis to assess the robustness of the results with respect to violations of principal ignorability and monotonicity assumptions in principal stratification analysis based on principal scores. Indeed, principal ignorability and monotonicity assumptions are fundamentally untestable and, in some cases, they are not easily justifiable according to subject-matter knowledge. We will review the literature on sensitivity analysis with respect to principal ignorability and monotonicity assumptions [61, 62, 64] .

Throughout the chapter we will illustrate and discuss the key concepts using two public health randomized experiments. The first study is a randomized experiment aiming to assess causal effects of Brest-Self Examination (BSE) teaching methods [24, 65]. The study was conducted between January 1988 and December 1990 at the Oncologic Center of the Faenza Health District in Italy. We will refer to this study as the *Faenza Study*. In the Faenza study, a sample of women who responded to a pretest BSE questionnaire was randomly assigned to either a standard leaflet-based information program (control group) or to a new in-person teaching program (intervention group). Women in the standard treatment group were mailed an information leaflet describing how to perform BSE correctly. In addition to receiving the mailed information, women assigned to the new intervention group were invited to the Faenza Oncologic Center to attend a training course on BSE techniques held by specialized medical staff. Unfortunately, some of the women assigned to the treatment group did not attend the course, but only received the standard information leaflet. On the contrary, women assigned to the control group could not access the in-person course, implying a one-sided non-compliance setting with only two compliance types (never-takers and compliers). The outcome of interest is post-treatment BSE practice a year after the beginning of the study as reported in a self-administered questionnaire. The post-treatment questionnaire also collected information on quality of self-exam execution. BSE quality is a secondary outcome censored by "death," with the censoring event defined by BSE practice: BSE quality is not defined for women who do not practice BSE. We will use the Faeza study to discuss principal stratification analysis based on principal score in settings of one-sided non-compliance and in settings with outcome truncated by death. We will consider the two complications one at a time, defining separate principal scores for each post-treatment variable. In practice we should deal with them simultaneously [24, 65].

The second study is the flu shot encouragement experiment conducted between 1978 and 1980 at a general medicine clinic in Indiana (USA). In this study, physicians were randomly selected to receive a letter encouraging them to inoculate patients at risk for flu under U.S. Public Health Service Criteria. The treatment of interest is the actual flu shot, and the outcome is an indicator for flu-related hospital visits during the subsequent winter. The study suffers from two-sided non-compliance: some of the study participants, whose physician received the encouragement letter, did not actually receive the vaccine; and some of the study participants, whose physician did not receive the encouragement letter, actually received the flu shot. The study was previously analyzed by Hirano et al. [11] using a model-based Bayesian approach to inference and by Ding and Lu [61] using principal score methods.

The chapter is organized as follows. In Section 21.2, we first introduce the notation and the basic framework of principal stratification. Then, we review some identifying (structural) assumptions – ignorability of the treatment assignment mechanism, monotonicity, exclusion restrictions and principal ignorability assumptions – and discuss their plausibility in different settings. In Section 17.4 we define principal scores, discuss their properties and provide some details on the estimator under monotonicity. In Section 17.5 we review identification and estimation strategies of principal causal effects based on weighting on principal score under monotonicity and principal ignorability assumptions. In Section 17.6 we deal with the specification of the principal score model. In Section 17.7 we review and discuss methods to conducting sensitivity analysis with respect to principal ignorability and monotonicity. We offer some discussion on possible extensions of principal stratification analysis based on principal scores under principal ignorability in Section 17.8 together with some concluding remarks.

17.2 Potential Outcomes and Principal Stratification

Consider a causal study involving a group of n units, indexed by $i = 1, \dots, n$. For each unit i, we observe a vector of pre-treatment covariates, X_i, the assigned treatment, W_i, the outcome of primary interest, Y_i, and an additional post-treatment variable, S_i, which may not be a main endpoint, but it is deemed to be useful in understanding the underlying causal mechanism and in explaining treatment effect heterogeneity. Oftentimes, S_i is an intermediate variable, lying between the treatment and the outcome, but it may also be a variable occurring after the realization of the main endpoint [66]. Without loss of generality, throughout the chapter, we mainly refer to S_i as an intermediate variable, where intermediate means that S_i is not the primary endpoint.

Throughout the chapter, we will consider studies where the treatment variable is binary, so each unit i can be assigned and/or exposed to the active/new treatment $W_i = 1$ or the control/standard treatment, $W_i = 0$. In addition we will mainly focus on studies with a single binary post-treatment variable.

We define causal effects using the potential outcomes framework under the Stable Unit Treatment Value Assumption (SUTVA) [5]. SUTVA rules out hidden versions of treatment and interference between units, that is, a unit's potential outcome does not depend on the treatment and intermediate variables of others, and a unit's potential value of the intermediate variable does not depend on the treatment of others. SUTVA is a critical assumption, which may be questionable, especially in studies where the outcomes are transmissible (e.g., infectious diseases or behaviors) and where units are grouped in clusters (e.g., clustered encouragement designs) or interact on a network. In the principal stratification literature multi-level and hierarchical models have been proposed to account for clustering and relax SUTVA in clustered encouragement designs [31, 67–70]. Here, for ease of presentation, we maintain SUTVA throughout. Under SUTVA, for each unit i, there are two associated potential outcomes for the intermediate and the primary outcome: let S_{0i} and Y_{0i} be the potential values of the intermediate variable and outcome if unit i were assigned to the control treatment, $W_i = 0$, and let S_{1i} and Y_{1i} be the potential values of the intermediate variable and outcome if unit i were assigned to the active treatment, $W_i = 1$.

A causal effect of treatment assignment, W, on the outcome Y, is defined to be a comparison between treatment and control potential outcomes, Y_{1i} versus Y_{0i}, on a common set of units, G, that is, a comparison of the ordered sets $\{Y_{0i}, i \in G\}$ and $\{Y_{1i}, i \in G\}$. For instance, we can be interested in the difference between the means of Y_{1i} and Y_{0i} for all units, $ACE = E[Y_{1i} - Y_{0i}]$, or for subgroups of units defined by values of covariates, $ACE_x = E[Y_{1i} - Y_{0i} \mid X_i = x]$.

Principal stratification uses the joint potential values of the intermediate variable to define a stratification of the population into principal strata. Formally, the *basic principal stratification* with respect to a post-treatment variable S is the partition of subjects into sets such that all subjects in the same set have the same vector $U_i = (S_{0i}, S_{1i})$.

In studies with two treatments, the basic principal stratification with respect to a binary intermediate variable, S, partitions subjects into four latent groups: $\mathcal{P}_0 = \{(0,0), (0,1), (1,0), (1,1)\}$, which we label as $\{00, 01, 10, 11\}$, respectively. A principal stratification $\mathcal{P}$ with respect to the post-treatment variable S is then a partition of units into sets being the union of sets in the basic principal stratification. A common principal stratification is the basic principal stratification itself: $\mathcal{P} = \mathcal{P}_0 = \{\{00\}, \{01\}, \{10\}, \{11\}\}$. Other examples of principal stratification are $\mathcal{P} = \{\{00, 11\}, \{01, 10\}\}$ and $\mathcal{P} = \{\{00, 10, 11\}, \{01\}\}$. We denote by U_i an indicator of the principal stratum membership, with $U_i \in \mathcal{P}$. A principal causal effect with respect to a principal stratification $\mathcal{P}$ is defined as a comparison of potential outcomes under standard versus new treatment within a principal stratum, ζ, in that principal stratification $\mathcal{P}$: Y_{0i} versus Y_{1i}, for all i with $U_i = u$, for all $u \in \mathcal{P}$ [9]. Because the principal stratum variable, U_i, is unaffected by the treatment, any principal causal effect is a well-defined local causal effect. Here we mainly focus on average principal causal effects with respect to the basic principal stratification:

$$ACE_u = E\left[Y_{1i} - Y_{0i} \mid U_i = u\right] \qquad u \in \{00, 01, 10, 11\} \tag{17.1}$$

By definition, principal causal effects provide information on treatment effect heterogeneity. Depending on the context, principal strata may have a specific scientific meaning and principal causal effects may be of intrinsic interest *per se*. Furthermore, we could be interested in further investigating the heterogeneity of treatment effect with respect to observed characteristics. To this end, let $ACE_{u|x}$ denote the conditional average principal causal effect for the subpopulation of units of type $U_i = u$ with the same values of the covariates, x: $ACE_{u|x} = E\left[Y_{1i} - Y_{0i} \mid U_i = u, X_i = x\right]$.

In randomized experiments with non-compliance, as our running examples, S is the actual treatment received and principal stratification classifies units into four groups according to their compliance status: $U_i = 00$ for subjects who would never take the treatment, irrespective of their assignment; $U_i = 01$ for subjects who always comply with their assignment; $U_i = 10$ for subjects who would always do the opposite of their assignment; and $U_i = 11$ for subjects who would always take the treatment, irrespective of their assignment. In the literature these principal strata are usually referred to as never-takers, compliers, defiers and always-takers, respectively [10]. In these settings, defiers are often ruled out invoking monotonicity of the compliance status. In randomized experiments with non-compliance, in addition to the intention-to-treat effect of the assignment, which is identified thanks to randomization, the interest is usually on the effect of the actual treatment receipt. Because the actual treatment is not randomized, we must rely on alternative identifying assumptions. If we could assume ignorability of the treatment conditional on covariates, we would be able to identify the treatment effect by comparing treated and untreated units within strata defined by the covariates. However, in non-compliance settings, the treatment receipt is usually confounded. Principal stratification focuses on well-defined causal effects within principal strata, ACE_u, and provides alternative identifying assumptions. The estimand of interest is usually the average causal effect for compliers, ACE_{01}, also named complier average causal effect (CACE). Because under monotonicity compliers are the only units that can be observed under both the active and control treatment according to the protocol, the ACE_{01} can provide information on the effect of the actual treatment. Furthermore, regardless of the identifying assumptions, principal causal effects can provide useful information on treatment effect heterogeneity.

In studies where the primary outcome is censored by death, in the sense that it is not only unobserved but also undefined for subjects who die, S is the survival status with $S = 1$ for survival and $S = 0$ for dead. For instance, in a clinical study where focus is on assessing causal effects of a new versus standard drug treatment on quality of life one year after assignment, quality of life is censored by death for patients with $S = 0$, that is, quality of life is nod defined for patients who die before reaching the end-point of one-year post-randomization survival. As opposed to the case where outcomes are missing, i.e., not observed, for patients who experience the censoring event, i.e., death, quality of life is not defined, without any hidden value masked by death. In these settings, the basic principal stratification classifies units into the following four groups: always survivors – subjects who would live no matter how treated, $U_i = 11$; never survivors – subjects who would die no matter how treated, $U_i = 00$; compliant survivors – subjects who would live if treated but would die if not treated, $U_i = 01$; and defiant survivors – subjects who would die if treated but would live otherwise, $U_i = 10$. Because a well-defined real value for the outcome under both the active treatment and the control treatment exists only for the latent stratum of always survivors, ACE_{11} is the only well-defined and scientifically meaningful principal causal effect in this context [71]. This causal estimand of interest, ACE_{11}, is often referred to as the survivor average causal effect (SACE). As opposed to the non-compliance setting where the primary causal estimand of interest might not be a principal causal effect, in settings affected by censoring by death the SACE is the only well-defined causal estimand of intrinsic interest. The problem of censoring by death arises not only in studies where the censoring event is death and the primary outcome (e.g., quality of life) is not defined for those who die. For instance, in the Faenza study, when focus is on causal effects on BSE quality, S is the indicator for BSE practice and the survivor average causal effect on BSE quality, ACE_{11}, is the average causal effect of the assignment on BSE quality for women who would practice BSE irrespective of their assignment to receive the information leaflet only or to also attend the in-person training [24, 65]. Censoring events can also be found in other fields. For instance, in labor economics, where focus is on the causal effect of a job-training program on wages, wages are censored by unemployment, in the sense that they are not defined for people who are unemployed [12–14]. In educational studies where interest is in assessing the effect of a special educational intervention in high school on final test scores, test score is censored by death for students who do not finish high school [33].

In studies with loss at follow-up or missing outcome data, the intermediate variable S is the response indicator and units can be classified into principal strata according to their response behavior, defined by the joint values of the potential response indicator under each treatment condition [15].

In studies where S is a surrogate or a mediator, precious information can be obtained by separately looking at casual effects for principal strata where the intermediate variable is unaffected by the treatment, ACE_{00} and ACE_{11}, and casual effects for principal strata where the intermediate variable

TABLE 17.1
Observed data pattern and principal strata for the observed groups.

Observed Group $O_{w,s}$	Actual Treatment Assigned W_i	Observed Intermediate Variable S_i	Possible Latent Principal Strata U_i	
$O_{0,0}$	0	0	00	01
$O_{0,1}$	0	1	10	11
$O_{1,0}$	1	0	00	10
$O_{1,1}$	1	1	01	11

is affected by the treatment, ACE_{10} and ACE_{01}. In this literature ACE_{00} and ACE_{11} and ACE_{10} and ACE_{01} are, respectively, named *dissociative effects* and *associative effects*, because they measure effects on the outcome that are dissociative and associative with effects on the intermediate variable [9]. In problems of surrogate endpoints, a good surrogate should satisfy the property of causal necessity, which implies that $ACE_{00} = 0$ and $ACE_{11} = 0$ [9], and the property of causal sufficiency, which implies that $ACE_{10} \neq 0$ and $ACE_{01} \neq 0$ [16]. In mediation studies, evidence on the direct effect of the treatment on the primary outcome is provided by dissociative effects, ACE_{00} and ACE_{11}; in principal strata where $S_{0i} \neq S_{1i}$, causal effects are associative, and thus, they combine direct effects and indirect effects through the mediator [20, 23].

17.3 Structural Assumptions

The fundamental problem of causal inference [4,7] is that for each unit only one potential outcome can be observed for each post-treatment variable, the potential outcome corresponding to the treatment actually assigned. Formally, the observed intermediate variable and outcome are $S_i = W_i S_{1i} + (1 - W_i)S_{0i}$ and $Y_i = W_i Y_{1i} + (1 - W_i)Y_{0i}$. Therefore, principal stratum membership, U_i, is generally missing for each unit. According to the observed values of the assigned treatment and the intermediate variable, units can be partitioned into four groups: $O_{0,0} = \{i : W_i = 0, S_i = 0\}$, $O_{0,1} = \{i : W_i = 0, S_i = 1\}$, $O_{1,0} = \{i : W_i = 1, S_i = 0\}$, $O_{1,1} = \{i : W_i = 1, S_i = 1\}$, each of which is a mixture of two latent principal strata as shown in Table 17.1. For instance, the observed group $O_{0,0}$, comprising units who are observed to be assigned to control and to have a value of the intermediate variable, S, equal to zero, are a mixture of the 00 principal stratum and the 01 principal stratum, both with $S_{0i} = 0$. Similarly, a subject in the observed group $O_{1,1}$, i.e., assigned to treatment and observed with a value of the intermediate variable equal to 1, exhibits a potential value $S_{1i} = 1$, and, thus, may be a subject of type 01 or 11. Analogous explanations apply to other observed groups in the Table 17.1.

The latent nature of principal strata and the missingness of principal stratum membership make inference on principal causal effects a challenging task. Several assumptions have been proposed to facilitate their identification and estimation. We now review some of these structural assumptions and discuss their plausibility.

17.3.1 Ignorable treatment assignment mechanism

Throughout the chapter we mainly assume that treatment assignment is independent of all the potential outcomes and covariates:

Assumption 6. *(Ignorability of treatment assignment).*

$$W_i \perp\!\!\!\perp (X_i, S_{0i}, S_{1i}, Y_{0i}, Y_{1i})$$

Assumption 6 amounts to state that there is no unmeasured confounding between the treatment assignment and intermediate variable and between the treatment assignment and outcome.

Assumption 6 holds by design in randomized experiments where covariates do not enter the assignment mechanism, such as completely randomized experiments. Both the Faenza study and the flu shot study are randomized experiments where randomization is performed without taking into account the values of the pretreatment variables, and thus, Assumption 6 holds by design.

Assumption 6 implies that the distribution of the principal stratification variable, U_i, is the same in both treatment arms, both unconditionally and conditionally on covariates.

In stratified randomized experiments the treatment assignment mechanism is ignorable by design conditionally on the design variables: $W_i \perp\!\!\!\perp (S_{0i}, S_{1i}, Y_{0i}, Y_{1i}) \mid X_i$. In observational studies ignorability of the treatment assignment mechanism conditionally on the observed covariates is a critical assumption, and it strongly relies on subject matter knowledge, which allows one to rule out the existence of relevant unmeasured confounders that affect the treatment as well as the outcome and intermediate variable.

When the treatment assignment is ignorable only conditional on covariates, it is the conditional distribution of the principal stratification variable, U_i, given covariates that is the same in both treatment arms.

17.3.2 Monotonicity assumptions

Monotonicity assumptions on the intermediate variable are often considered.

Assumption 7. *(Strong monotonicity). $S_{0i} = 0$ for all $i = 1, \ldots, n$*

Strong monotonicity rules out the presence of units of type $U_i = 11$ and $U_i = 10$, namely, always-takers and defiers in randomized experiments with non-compliance. In randomized studies with one-sided non-compliance, strong monotonicity holds by design. For instance, in the Faenza study, non-compliance is one-sided, because women assigned to the control group cannot access the BSE training course. Thus, strong monotonicity holds, implying that there exist only two principal strata of women: compliers and never-takers.

Assumption 8. *(Monotonicity). $S_{1i} \geq S_{0i}$ for all $i = 1, \ldots, n$*

Monotonicity rules out the presence of units of type $U_i = 10$. Its plausibility depends on the substantive empirical setting and cannot be in general tested using the observed data. It has, however, testable implications: under monotonicity the observed proportion of units with $S_i = 1$ in the treatment group should not be lower than that in the control group. In the flu shot experiment, non-compliance is two-sided and so, in principle, all the four compliance types may exist. Nevertheless, we can reasonably expect that physicians receiving the encouragement letter are unlikely to decide not to give the flu shot to patients to whom they would have given it in the absence of the letter. Therefore, in the flu shot study we can assume a non-negative effect of the encouragement on treatment receipt, which allows us to invoke the assumption of monotonicity of non-compliance (Assumption 8). This will in turn rule out the existence of defiers ($U_i = 10$), patients who would receive the vaccine if their physician did not receive the letter, but would not receive the vaccine if their physician did receive the letter.

Table 17.2 shows the relationship between observed groups and principal strata under strong monotonicity (upper panel) and monotonicity (bottom panel). Table 17.2 makes it clear that under strong monotonicity (Assumption 7) we observe the principal stratum membership for some units. In particular, the observed two groups of treated units $O_{1,0} = \{i : W_i = 1, S_i = 0\}$ and $O_{1,1} =$

TABLE 17.2
Observed data pattern and principal strata for the observed groups under monotonicity assumptions.

Strong Monotonicity

Observed Group $O_{w,s}$	Actual Treatment Assigned W_i	Observed Intermediate Variable S_i	Possible Latent Principal Strata U_i
$O_{0,0}$	0	0	00 01
$O_{1,0}$	1	0	00
$O_{1,1}$	1	1	01

Monotonicity

Observed Group $O_{w,s}$	Actual Treatment Assigned W_i	Observed Intermediate Variable S_i	Possible Latent Principal Strata U_i
$O_{0,0}$	0	0	00 01
$O_{0,1}$	0	1	11
$O_{1,0}$	1	0	00
$O_{1,1}$	1	1	01 11

$\{i : W_i = 1, S_i = 1\}$ comprise only units belonging to the principal stratum 00 and 01, respectively. The control group, that is, the observed group $O(0, 0) = \{i : W_i = 0, S_i = 0\}$, still comprises a mixture of the two latent principal strata, $U_i = 00$ and $U_i = 01$. Under monotonicity (Assumption 8), the observed group $O_{0,1} = \{i : W_i = 0, S_i = 1\}$ comprises only units belonging to the principal stratum 11, and the observed group $O_{1,0} = \{i : W_i = 1, S_i = 0\}$ comprises only units belonging to the principal stratum 00. The observed groups $O_{0,0} = \{i : W_i = 0, S_i = 0\}$ and $O_{1,1} = \{i : W_i = 1, S_i = 1\}$ comprise a mixture of the two types of units: units of types $U_i = 00$ and $U_i = 01$, and units of type $U_i = 01$ and $U_i = 11$, respectively.

17.3.3 Exclusion restrictions

Two additional assumptions that are often invoked, especially in the analysis of randomized experiments with non-compliance, are exclusion restrictions for units with $U_i = 00$ and units with $U_i = 11$. The first exclusion restriction assumes that within the subpopulation of units with $U_i = 00$, the two potential outcomes Y_{0i} and Y_{1i} are the same:

Assumption 9. *(Exclusion restriction for units of type* 00*).* $Y_{0i} = Y_{1i}$ *for all* i *with* $U_i = 00$

The second exclusion restriction assumes that within the subpopulation of units with $U_i = 11$, the two potential outcomes Y_{0i} and Y_{1i} are the same:

Assumption 10. *(Exclusion restriction for units of type* 11*).* $Y_{0i} = Y_{1i}$ *for all* i *with* $U_i = 11$

Assumptions 9 and 10 rule out any effect of treatment on the outcome for the two types of units for whom there is no effect of treatment on the intermediate variable. They respectively imply that $ACE_{00} = 0$ and $ACE_{11} = 0$, formalizing the notion that any effect of the treatment on the outcome should be mediated by an effect of treatment on the intermediate variable. In randomized experiment with non-compliance, exclusion restrictions appear plausible in blind or double-blind experiments, where the outcome may be affected only by the treatment received. In these settings under monotonicity and exclusion restrictions, the average causal effects for compliers, ACE_{01}, is equal to the ratio of the average causal effects on Y and S: $ACE_{01} = E[Y_{1i} - Y_{0i}]/E[S_{1i} - S_{0i}]$ [10]:

Proposition 17.1. *Suppose that ignorability of treatment assignment (Assumption 6), (strong) monotonicity (Assumption 7 or Assumption 8), and exclusion restrictions (Assumptions 9 and 10) hold, then*

$$ACE_{01} = \frac{E[Y_{1i} - Y_{0i}]}{E[S_{1i} - S_{0i}]}$$

Exclusion restriction assumptions may be questionable in open-label randomized experiments, as the Faenza study and the flu shot encouragement study, where treatment assignment may have a direct effect on the outcome. In the Faenza study the exclusion restriction for never-takers, women who would not participate in the BSE training course irrespective of their treatment assignment, might be violated if the decision of some women to not attend the BSE training course induces them not to perform BSE and take instead other screening actions (e.g., mammography, breast ultrasonography).

In the flu shot study the exclusion restriction for always-takers, patients who would take the flu shot regardless of their physician's receipt of the encouragement letter, may be arguable if always-takers comprise weaker patients, and thus, the letter prompted physicians to take other medical actions for this type of patients, e.g., by advising them about ways to avoid exposure or by providing them other medical treatment. The exclusion restriction for never-takers, which implies that never-takers are completely unaffected by their physicians' receipt of the letter, may be a reasonable assumption if never-takers and their physicians do not regard the risk of flu as high enough to warrant inoculation, and thus, they are not subject to other medical actions either [11]. On the other hand, also the exclusion restriction for never-takers may be arguable, if never-takers' physicians were particularly meticulous, and thus, the letter encouraged them to take other precautions for never-takers, given that they will not get the flu shot.

Outside the non-compliance setting, ACE_{00} and ACE_{11} may be effects of primary interest. For instance, in problems of censoring by death, surrogacy and mediation, assessing whether ACE_{00} and/or ACE_{11} is zero is the actual scientific question of interest. Therefore, in these settings, exclusion restrictions would rule out a priori the effects of interest, and thus, cannot be invoked.

17.3.4 Principal ignorability assumptions

As an alternative, a line of research relies on principal ignorability assumptions. Although there exist various definitions of principal ignorability, here we mainly define principal ignorability assumptions as conditional statistical independence assumptions.

Assumption 11. *(Strong principal ignorability).*

$$Y_{0i} \perp\!\!\!\perp U_i \mid X_i \tag{17.2}$$

and

$$Y_{1i} \perp\!\!\!\perp U_i \mid X_i \tag{17.3}$$

Strong principal ignorability requires that each potential outcome, Y_{0i} and Y_{1i}, is independent of the bi-variate variable $U_i = (S_{0i}, S_{1i})$ conditionally on the covariates. Specifically, it assumes conditional independence of Y_{0i} and U_i (Equation (17.2)) and of Y_{1i} and U_i (Equation (17.3)) given the covariates, X_i. Therefore, conditionally on X_i the marginal distributions of the control and treatment potential outcomes are the same across principal strata, and thus, $ACE_x = ACE_{00|x} = ACE_{10|x} = ACE_{01|x} = ACE_{11|x}$.

In non-compliance and mediation settings, strong principal ignorability implies that the causal effect of the assignment, conditional on covariates, is the same for all principal strata, regardless of whether the intermediate variable is affected by the assignment. This could either preclude any impact from the intermediate variable or it could mean that the direct impact of the assignment is heterogenous and can compensate the lack of the effect of the intermediate variable for never-takers (and always-takers). For instance, in a randomized evaluation of a job training program with non-compliance never-takers could receive an equivalent program to compliers [62]. However, in general, this assumption is quite difficult to justify in practice, especially in non-compliance and mediation settings.

Assumption 12. *(Weak principal ignorability).*

$$Y_{0i} \perp\!\!\!\perp S_{1i} \mid S_{0i}, X_i \tag{17.4}$$

and

$$Y_{1i} \perp\!\!\!\perp S_{0i} \mid S_{1i}, X_i \tag{17.5}$$

Weak principal ignorability requires that the potential outcome Y_{wi} is independent of the marginal potential outcome for the intermediate variable under the opposite treatment, i.e., $S_{(1-w)i}$ given S_{wi} and covariates, $w = 0, 1$. Specifically, weak principal ignorability assumes that conditional on the covariates, X_i, Y_{0i} is independent of S_{1i} given S_{0i} (Equation (17.4)), and Y_{1i} is independent of S_{0i} given S_{1i} (Equation (17.5)). Equation (17.4) implies that conditionally on X_i, the distributions of the control potential outcomes, Y_{0i}, are the same across strata $U_i = 00$ and $U_i = 01$ (with $S_{0i} = 0$) and across strata $U_i = 10$ and $U_i = 11$ (with $S_{0i} = 1$). Similarly, Equation (17.5) implies that, conditionally on X_i, the distributions of the treatment potential outcomes, Y_{1i}, are the same across strata $U_i = 00$ and $U_i = 10$ (with $S_{1i} = 0$) and across strata $U_i = 01$ and $U_i = 11$ (with $S_{1i} = 1$). Assumption 12 (weak principal ignorablity) is strictly weaker than Assumption 11 (strong principal ignorablity), because it applies to a subset of units in each treatment arm. Principal ignorability assumptions, implying that treatment and/or control potential outcomes have the same distribution across some principal strata, are types of homogeneity assumptions, similar to homogeneity assumptions recently introduced in causal mediation analysis [27, 67].

Principal ignorability assumptions require that the set of the observed pre-treatment variables, X_i, includes all the confounders between the intermediate variable, S_i, and the outcome, Y_i. [1]

Under strong monotonicity, $U_i = (S_{0i}, S_{1i}) = (0, S_{1i})$, and thus, principal stratum membership is defined by S_{1i} only. In this setting, strong and weak principal ignorability (Assumptions 11 and 12) are equivalent: they both require that conditionally on the covariates, the distributions of the control and treatment potential outcomes for units of type $U_i = 01$ and units of type $U_i = 00$ are the same: $Y_{0i} \mid U_i = 00, X_i \sim Y_{0i} \mid U_i = 01, X_i$ and $Y_{1i} \mid U_i = 00, X_i \sim Y_{1i} \mid U_i = 01, X_i$. In the Faenza study where strong monotonicity holds by design, Equations (17.2) and (17.4) imply that for women assigned to control, that is, women who only receive mailed information on how to perform BSE correctly, the distribution of their outcome, BSE practice, given the covariates, does not depend on whether they would have attended the BSE training course if offered. Although this assumption is very strong, it may be plausible in the Faenza study, given that women assigned to the control treatment have no access to the training course, and thus, the observed training course participation S_i takes on the same value, zero, for all of them, irrespective if they are compliers or never-takers. Equations (17.3) and (17.5) imply that for women invited to attend the BSE training course, their likelihood of practising BSE is unrelated to whether they actually participate in the course, given the covariates. This assumption is difficult to justify.

It is worth noting that, under strong monotonicity, Equations (17.3) and (17.5) are superfluous for identification: Equation (17.5) naively holds because $S_{0i} = 0$ for all units $i = 1, \dots, n$ under strong monotonicity; and Equation (17.3) is not required because we directly observe the principal stratum membership for units assigned to treatment, we can directly estimate the outcome distributions under treatment for units of type $U_i = 00$ and $U_i = 01$.

---

[1] Ignorability of the intermediate variable in the form of sequential ignorability also rules out unmeasured confounders between the intermediate variable and the outcome. For an in-depth comparison between sequential and principal ignorability assumptions, see [18].

Under monotonicity, there exist three principal strata: 00, 01, 11, and under strong principal ignorability they all have the same conditional distributions of the control and treatment potential outcomes given the covariates: $Y_{0i} \mid U_i = 01, X_i \sim Y_{0i} \mid U_i = 00, X_i \sim Y_{0i} \mid U_i = 11, X_i$ and $Y_{1i} \mid U_i = 01, X_i \sim Y_{1i} \mid U_i = 00, X_i \sim Y_{1i} \mid U_i = 11, X_i$. In the flu shot study Assumption 11 states that, given covariates, whether a patient actually receives the influenza vaccine is unrelated to that patient's potential hospitalization status under treatment and under control. Equivalently, given covariates, whether a patient actually receive the influenza vaccine is unrelated to the effect of the receipt of the encouragement letter by his/her physician on hospitalization. Strong principal ignorability is a quite strong assumption and may be implausible in practice. However, under monotonicity, it has testable implications because we observe the outcome under treatment for units of type $U_i = 00$ (never-takers) and the outcome under control for units of type $U_i = 11$ (always-takers).

Under monotonicity, the key implications of weak principal ignorability are that, conditionally on X_i, (i) the distributions of the control potential outcomes, Y_{0i}, are the same across strata $U_i = 00$ and $U_i = 01$ (with $S_{0i} = 0$): $Y_{0i} \mid U_i = 01, X_i \sim Y_{0i} \mid U_i = 00, X_i$; and (ii) the distributions of the treatment potential outcomes are the same across strata $U_i = 01$ and $U_i = 11$ (with $S_{1i} = 1$): $Y_{1i} \mid U_i = 01, X_i \sim Y_{1i} \mid U_i = 11, X_i$.

In the flu shot study, Equation (17.4) states that, for patients whose physician does not receive the encouragement letter (patients assigned to control) and who are not inoculated ($S_{0i} = 0$), their hospitalization status (Y_{0i}) is unrelated to whether they would have been inoculated if their physician had received the encouragement letter (S_{1i}), given covariates. Similarly, Equation (17.5) states that, for patients whose physician receives the encouragement letter (patients assigned to treatment) and who receive a flu shot ($S_{1i} = 1$), their hospitalization status (Y_{1i}) is unrelated to whether they would have received a flu shot if their physician had not received the encouragement letter (S_{0i}), given covariates.

It is worth noting that if focus is on average principal causal effects, we can use weaker versions of the principal ignorability assumptions, requiring mean independence rather than statistical independence:

Assumption 13. *(Strong principal ignorability - Mean independence).*

$$E[Y_{0i} \mid U_i, X_i] = E[Y_{0i} \mid X_i] \tag{17.6}$$

and

$$E[Y_{1i} \mid U_i, X_i] = E[Y_{1i} \mid X_i] \tag{17.7}$$

Assumption 14. *(Weak principal ignorability - Mean independence).*

$$E[Y_{0i} \mid S_{1i}, S_{0i}, X_i] = E[Y_{0i} \mid S_{0i}, X_i] \tag{17.8}$$

and

$$E[Y_{1i} \mid S_{1i}, S_{0i}, X_i] = E[Y_{1i} \mid S_{1i}, X_i] \tag{17.9}$$

Although most of the existing results on principal stratification analysis based on principal scores hold under these weaker versions of strong and weak principal ignorability (Assumptions 13 and 14) [61, 62, 64], we prefer to define principal ignorability assumptions in terms of statistical independence (Assumptions 11 and 12).

In some studies exclusion restriction may be plausible for at least one of the two types of units $U_i = 00$ and $U_i = 11$ for which treatment assignment has no effect on the intermediate variable. In these settings we can combine exclusion restriction and principal ignorability assumptions [62]. For instance, in the flu shot study, the exclusion restriction for always-takers (patients of type $U_i = 11$) appears arguable, but the exclusion restriction for never-takers (patients of type $U_i = 00$) is relatively uncontroversial. Therefore, we can assume that (i) never-takers are completely unaffected by their physicians' receipt of the letter (Assumption 9 – exclusion restriction for never-takers) and that (ii) the conditional distributions given covariates of the treatment outcomes are the same for always-takers and compliers (Assumption 12 – weak principal ignorability – Equation (17.5) for patients with $S_{1i} = 1$):

Assumption 15. *(Exclusion restriction for units of type $U_i = 00$ and weak principal ignorability).*

(i) Exclusion restriction for units of type 00*:*

$$Y_{0i} = Y_{1i} \qquad \textit{for all } i : U_i = 00$$

(ii) Weak principal ignorability for units with $S_{1i} = 1$*:*

$$Y_{1i} \perp\!\!\!\perp S_{0i} \mid S_{1i} = 1, X_i$$

Similarly, in some studies it might be reasonable to make the following assumption:

Assumption 16. *(Exclusion restriction for units of type $U_i = 11$ and weak principal ignorability).*

(i) Exclusion restriction for units of type 11*:*

$$Y_{0i} = Y_{1i} \qquad \textit{for all } i : U_i = 11$$

(ii) Weak principal ignorability for units with $S_{0i} = 0$*:*

$$Y_{0i} \perp\!\!\!\perp S_{1i} \mid S_{0i} = 0, X_i$$

17.4 Principal Scores

Suppose that ignorability of treatment assignment (Assumption 6) and strong principal ignorability (Assumption 11) hold. Then,

$$\begin{aligned} ACE_{u|x} &= E[Y_{1i} - Y_{0i} \mid U_i = u, X_i = x] \\ &= E[Y_{1i} - Y_{0i} \mid X_i = x] \\ &= E[Y_{1i} \mid W_i = 1, X_i = x] - E[Y_{0i} \mid W_i = 0, X_i = x] \\ &= E[Y_i \mid W_i = 1, X_i = x] - E[Y_i \mid W_i = 0, X_i = x], \end{aligned}$$

where the second equality follows from strong principal ignorability and the third equality follows from ignorability of treatment assignment.

Similarly, suppose that ignorability of treatment assignment (Assumption 6) and weak principal ignorability (Assumption 12) hold. Let $u = s_0 s_1$, $s_0, s_1 \in \{0, 1\}$. Then,

$$\begin{aligned} ACE_{u|x} &= E[Y_{1i} - Y_{0i} \mid U_i = u, X_i = x] \\ &= E[Y_{1i} \mid S_{0i} = s_0, S_{1i} = s_1, X_i = x] - E[Y_{0i} \mid S_{0i} = s_0, S_{1i} = s_1, X_i = x] \\ &= E[Y_{1i} \mid S_{1i} = s_1, X_i = x] - E[Y_{0i} \mid S_{0i} = s_0, X_i = x] \\ &= E[Y_{1i} \mid S_{1i} = s_1, W_i = 1, X_i = x] - E[Y_{0i} \mid S_{0i} = s_0, W_i = 0, X_i = x] \\ &= E[Y_i \mid S_i = s_1, W_i = 1, X_i = x] - E[Y_i \mid S_i = s_0, W_i = 0, X_i = x], \end{aligned}$$

where the third equality follows from weak principal ignorability and the fourth equality follows from ignorability of treatment assignment. Therefore under ignorability of treatment assignment and strong/weak principal ignorability, we can non-parametrically identify conditional principal causal effects, $ACE_{u|x}$.

To further identify principal causal effects, ACE_u, an additional assumption, such as monotonicity, is needed to identify the conditional distribution of the covariates given the principal stratum variable. Thus, under ignorability of treatment assignment, strong/weak principal ignorability, and monotonicity, the principal causal effects ACE_u are non-parametrically identified by:

$$\begin{aligned} ACE_u &= E[Y_{1i} - Y_{0i} \mid U_i = u] \\ &= E\left[E[Y_{1i} - Y_{0i} \mid U_i = u, X_i = x] \mid U_i = u\right] \\ &= E\left[ACE_{u|x} \mid U_i = u\right] \end{aligned}$$

where the outer expectation is over the conditional distribution of the covariates given the principal stratum variable.

The non-parametric identification of principal causal effects under strong principal ignorability implies that we can estimate them by estimating the mean difference of observed outcomes between treated and untreated units adjusting for observed covariates. Under weak principal ignorability we also need to adjust for the observed value of the intermediate variable. Working within cells defined by the covariates, although feasible in principle, may be difficult in practice with a large number of pre-treatment variables, possibly including continuous covariates. Borrowing from the propensity score literature, we can face the curse of dimensionality using the principal score.

17.4.1 The principal score: definition and properties

The principal score is defined as the conditional probability of being in a latent principal strata given pre-treatment characteristics:

$$e_u(x) = \Pr\left(U_i = u \mid X_i = x\right) \qquad u = 00, 10, 01, 11 \tag{17.10}$$

The principal score has two key properties [57, 61, 62]:

1. *Balancing of pre-treatment variables across principal strata given the principal score.*

   $$\mathbb{I}\{U_i = u\} \perp\!\!\!\perp X_i \mid e_u(X_i),$$

   where $\mathbb{I}\{\cdot\}$ is the indicator function.

   The balancing property states that the principal score is a balancing score in the sense that within groups of units with the same value of the principal score for a principal stratum u, $e_u(x)$, the probability that $U_i = u$ does not depend on the value of the covariates, X_i:

   $$\Pr\left(U_i = u \mid X_i, e_u(X_i)\right) = \Pr\left(U_i = u \mid e_u(X_i)\right).$$

   In other words, within cells with the same value of $e_u(x)$, the distribution of covariates is the same for those belonging to the principal stratum u and for those belonging to other principal strata.

2. *Principal ignorability given the principal score.* If principal ignorability holds conditional on the covariates, it holds conditional on the principal score. Formally,

   (a) *Strong principal ignorability given the principal score.* If

   $$Y_{0i} \perp\!\!\!\perp \mathbb{I}\{U_i = u\} \mid X_i \quad \textit{and} \quad Y_{1i} \perp\!\!\!\perp \mathbb{I}\{U_i = u\} \mid X_i$$

for all u, then

$$Y_{0i} \perp\!\!\!\perp \mathbb{I}\{U_i = u\} \mid e_u(X_i) \quad \textit{and} \quad Y_{1i} \perp\!\!\!\perp \mathbb{I}\{U_i = u\} \mid e_u(X_i)$$

for all u.

(b) *Weak principal ignorability given the principal score.* If

$$Y_{0i} \perp\!\!\!\perp \mathbb{I}\{S_{1i} = s_1\} \mid S_{0i} = s_0, X_i$$

and

$$Y_{1i} \perp\!\!\!\perp \mathbb{I}\{S_{0i} = s_0\} \mid S_{1i} = s_1, X_i$$

for all s_0, s_1, then

$$Y_{0i} \perp\!\!\!\perp \mathbb{I}\{S_{1i} = s_1\} \mid S_{0i} = s_0, e_u(X_i)$$

and

$$Y_{1i} \perp\!\!\!\perp \mathbb{I}\{S_{0i} = s_0\} \mid S_{1i} = s_1, e_u(X_i)$$

for $u = s_0 s_1$, for all s_0, s_1.

These properties of the principal score imply that under ignorability of treatment assignment (Assumption 6) and strong principal ignorability (Assumption 11)

$$\begin{aligned} ACE_{u|x} &= E[Y_{1i} \mid W_i = 1, e_u(x)] - E[Y_{0i} \mid W_i = 0, e_u(x)] \\ &= E[Y_i \mid W_i = 1, e_u(x)] - E[Y_i \mid W_i = 0, e_u(x)], \end{aligned}$$

and under ignorability of treatment assignment (Assumption 6) and weak principal ignorability (Assumption 12), for $u = (s_0, s_1)$, $s_0, s_1 \in \{0, 1\}$,

$$\begin{aligned} ACE_{u|x} &= E[Y_{1i} \mid W_i = 1, S_{1i} = s_1, e_u(x)] - E[Y_{0i} \mid W_i = 0, S_{0i} = s_0, e_u(x)] \\ &= E[Y_i \mid W_i = 1, S_i = s_1, e_u(x)] - E[Y_i \mid W_i = 0, S_i = s_0, e_u(x)]. \end{aligned}$$

17.4.2 Estimating the principal score

Unfortunately, we do not know the principal scores, because the observed data are mixtures of two latent principal strata. Even under ignorability of treatment assignment, without monotonicity, we can only partially identify them. On the other hand, (strong) monotonicity allows us to point identify the principal scores.

Suppose that ignorability of treatment assignment (Assumption 6) holds, so that, for $u = 00, 10, 01, 11$,

$$\begin{aligned} e_u(x) &= \Pr(U_i = u \mid X_i = x) \\ &= \Pr(U_i = u \mid X_i = x, W_i = 0) \\ &= \Pr(U_i = u \mid X_i = x, W_i = 1). \end{aligned}$$

This result implies that the principal stratum variable, U_i, has the same conditional distribution given the covariates in both treatment arms. Therefore, if we could observe the principal stratum membership for units in a treatment arm, we could easily infer the conditional distribution of U_i for all units. Let $p_w(x) = \Pr(S_i = 1 \mid W_i = w, X_i = x)$, $w = 0, 1$, denote the conditional probability of $S_i = 1$ under treatment status w given covariates, X. Under strong monotonicity (Assumption 7), $U_i = 00$ or $U_i = 01$. Because within the treatment group, $S_i = 1$ if and only if $U_i = 01$ and $S_i = 0$ if and only if $U_i = 00$, and ignorability of treatment assignment guarantees that U_i has the same distribution in both treatment arms conditional on covariates, we have that $e_{01}(x) = \Pr(U_i =$

$01 \mid X_i = x, W_i = 1) = p_1(x)$ and $e_{00}(x) = \Pr(U_i = 00 \mid X_i = x, W_i = 1) = 1 - p_1(x)$. Under monotonicity (Assumption 8), $e_{00}(x) = \Pr(U_i = 00 \mid X_i = x, W_i = 1) = 1 - p_1(x)$ and $e_{11}(x) = \Pr(U_i = 11 \mid X_i = x, W_i = 0) = p_0(x)$, and $e_{01}(x) = 1 - e_{00}(x) - e_{11}(x) = p_1(x) - p_0(x)$. Therefore, under (strong) monotonicity, in principle, we could non-parametrically estimate principal scores. However, using a fully non-parametric approach is not feasible in many realistic settings, including high-dimensional or continuous covariates, so that some form of smoothing is essential. Regression models for binary or categorical variables can be used for the principal score $e_u(x)$. Let α denote the parameter vector of the regression model for the principal stratum variable, U_i, and let $e_u(x;\alpha) = \Pr(U_i = u \mid X_i = x; \alpha)$ be the principal score corresponding to the specified regression model under parameter vector α.

Under strong monotonicity, U_i takes only two values: 00 and 01, and the principal stratum membership is known in the treatment arm depending on the value of S_i. Therefore, we can simply use a regression model for binary variables (e.g., a logistic regression model) of S_i on X_i using only data from the treatment group, and then, estimate $e_u(x)$ for the whole sample, including the control group, using the predicted probabilities: $e_{01}(x) = p_1(x)$ and $e_{00}(x) = 1 - p_1(x)$.

Under monotonicity (Assumption 8), U_i takes three values, 00, 01, and 11. In the literature there exist alternative approaches for obtaining model-based estimates of the principal score in this setting. Specifically, marginal and joint methods have been proposed. Using the fact that under monotonicity we can directly observe units of type $U_i = 00$ under treatment and units of type $U_i = 11$ under control, the marginal method specifies two marginal regression models for binary variables, namely, a regression model of S_i on X_i in the control group and a regression model of S_i on X_i in the treatment group. Then, for each unit, i, we can estimate $e_{00}(X_i)$ and $e_{11}(X_i)$ via the predicted probabilities from the regression of S_i on X_i in the treatment and the control group, respectively. Let $\widehat{e}_{00}(X_i)$ and $\widehat{e}_{11}(X_i)$ denote the estimates of $e_{00}(X_i)$ and $e_{11}(X_i)$, respectively. Finally, $e_{01}(X_i)$ is estimated using the complement to one of the estimates of $e_{00}(X_i)$ and $e_{11}(X_i)$: $\widehat{e}_{01}(X_i) = 1 - \widehat{e}_{00}(X_i) - \widehat{e}_{11}(X_i)$.

A potential drawback of the marginal method is that it could lead to estimates of $e_{01}(x)$ that are outside $[0, 1]$. The joint method deals with this issue, by treating the principal stratum variable, U_i, as missing data and jointly modeling the principal scores by using a model for categorical variables. There are four possible patterns of missing and observed data in U_i corresponding to the four possible observed groups, $O_{w,s} = \{i : W_i = w, S_i = s\}$, $w = 0, 1$, $s = 0, 1$. Therefore, the likelihood function is

$$\begin{aligned}\mathcal{L}(\alpha) \quad \propto \quad & \prod_{i:W_i=0,S_i=0} \left[e_{00}(X_i;\alpha) + e_{01}(X_i;\alpha)\right] \times \prod_{i:W_i=0,S_i=1} e_{11}(X_i;\alpha) \times \\ & \prod_{i:W_i=1,S_i=0} e_{00}(X_i;\alpha) \times \prod_{i:W_i=1,S_i=1} \left[e_{01}(X_i;\alpha) + e_{11}(X_i;\alpha)\right].\end{aligned}$$

The likelihood function has a complex form involving mixtures of distributions, which challenge inference. However, we can obtain an estimate of the model parameter vector, α, and thus, of the principal scores, $e_u(X_i;\alpha)$, by using missing data methods, such as the EM algorithm [72] (see [61] for details on EM estimation of principal scores) or the data augmentation (DA) algorithm [73]. It is worth noting that the joint method allows us to estimate the principal scores even without monotonicity. Alternative smoothing techniques can be used, including nonparametric regression methods and machine learning methods.

17.4.3 Choosing the specification of the principal score model

A critical issue in the model-based estimate of the principal score concerns the choice of the specification of the principal score model. The choice of the parametric model should be motivated by substantive knowledge. In addition, the balancing property of the principal score can help. The goal is to obtain estimates of the principal score that best balance the covariates between treated and control units within each principal stratum. This criterion is somewhat vague, and we elaborate on its implementation later (see Section 17.6). From this perspective, we can use a two-part procedure for specifying the principal score model. First we specify an initial model, motivated by substantive knowledge. Second, we assess the statistical adequacy of an estimate of that initial model, by checking whether the covariates are balanced for treated and control units within each principal stratum. In principle, we can iterate back and forth between these two stages, specification of the model and assessment of that model, each time refining the specification of the model, by, e.g., adding higher polynomial and interaction terms of the covariates. It is worth noting that estimation of the principal scores only requires information on the covariates, X_i, the treatment assignment variable, W_i, and the intermediate variable, S_i: it does not use the outcome data. As in the design of observational studies with regular assignment mechanism [8,50,51], this outcome-free strategy guarantees against the temptation to search for favorable outcome models and effect estimates.

17.5 Estimating Principal Causal Effects Using Principal Scores

In the literature there exist various strategies for estimating principal causal effects under principal ignorability, including regression [63], matching [52,57], and weighting [34,57,59,62,64]. We focus here on weighting on principal score methods. Simulation studies have shown that in finite samples weighting performs slightly better than matching [57,60].

Let $e_u = \Pr(U_i = u)$, $u = 00, 10, 01, 11$, denote the marginal probabilities of the principal stratum variable, U_i. It is worth noting that

$$\begin{aligned} E\left[Y_{wi} \mid U_i = u\right] &= E\left[E\left[Y_{wi} \mid X_i = x, U_i = u\right] \mid U_i = u\right] \\ &= \int E\left[Y_{wi} \mid X_i = x, U_i = u\right] f_{X_i \mid U_i}(x \mid u)\, dx \\ &= \int E\left[Y_{wi} \mid X_i = x, U_i = u\right] \frac{f_{X_i}(x) e_u(x)}{e_u}\, dx \\ &= E\left[E\left[Y_{wi} \mid X_i = x, U_i = u\right] \frac{e_u(x)}{e_u}\right] \end{aligned}$$

where the first/second equality follows from the law of iterated expectations, the third equality follows form the Bayes's formula, and the outer expectation in the last equality is over the distribution of the covariates.

The following propositions summarize the main results on weighting under Assumption 6, which holds by design in randomized studies where randomization does not depend on covariates. Proposition 17.2 shows identification of average principal causal effects under strong monotonicity. For instance, it may be used to draw inference on principal causal effects in randomized experiments with one-sided non-compliance, such as the Faenza study. Recall that under strong monotonicity, strong principal ignorability (Assumption 11) and weak principal ignorability (Assumption 12) are equivalent.

Proposition 17.2. *Suppose that ignorability of treatment assignment (Assumption 6) and strong monotonicity (Assumption 7) hold.*

If strong/weak principal ignorability (Assumption 11/12) holds, then

$$\begin{aligned} ACE_{01} &= E\left[\omega_{01}(X_i) \cdot Y_i \mid W_i = 1\right] - E\left[\omega_{01}(X_i) \cdot Y_i \mid W_i = 0\right] \\ ACE_{00} &= E\left[\omega_{00}(X_i) \cdot Y_i \mid W_i = 1\right] - E\left[\omega_{00}(X_i) \cdot Y_i \mid W_i = 0\right] \end{aligned} \tag{17.11}$$

If only Equation (17.2)/(17.4) *of strong/weak principal ignorability (Assumption 11/12) holds, then*

$$\begin{aligned} ACE_{01} &= E[Y_i \mid W_i = 1, S_i = 1] - E[\omega_{01}(X_i) \cdot Y_i \mid W_i = 0] \\ ACE_{00} &= E[Y_i \mid W_i = 1, S_i = 0] - E[\omega_{00}(X_i) \cdot Y_i \mid W_i = 0] \end{aligned} \tag{17.12}$$

where

$$\omega_{01}(x) = \frac{e_{01}(x)}{e_{01}} \quad \text{and} \quad \omega_{00}(x) = \frac{e_{00}(x)}{e_{00}} = \frac{1 - e_{01}(x)}{1 - e_{01}}$$

Proposition 17.3 shows identification of average principal causal effects under monotonicity, which implies that there exist three latent principal strata. We may use results in Proposition 17.3 to analyze the flu shot data [61].

Proposition 17.3. *Suppose that ignorability of treatment assignment (Assumption 6) and monotonicity (Assumption 7) hold.*

If strong principal ignorability (Assumption 11) holds, then

$$ACE_u = E[\omega_u(X_i) \cdot Y_i \mid W_i = 1] - E[\omega_u(X_i) \cdot Y_i \mid W_i = 0] \tag{17.13}$$

where $w_u(x) = \dfrac{e_u(x)}{e_u}$, *for* $u = 00, 01, 11$.

If weak principal ignorability (Assumption 12) holds, then

$$\begin{aligned} ACE_{00} &= E[Y_i \mid W_i = 1, S_i = 0] - E[\omega_{0,00}(X_i) \cdot Y_i \mid W_i = 0, S_i = 0] \\ ACE_{01} &= E[\omega_{1,01}(X_i) \cdot Y_i \mid W_i = 1, S_i = 1] - E[\omega_{0,01}(X_i) \cdot Y_i \mid W_i = 0, S_i = 0] \\ ACE_{11} &= E[\omega_{1,11}(X_i) \cdot Y_i \mid W_i = 1, S_i = 1] - E[Y_i \mid W_i = 0, S_i = 1] \end{aligned} \tag{17.14}$$

where

$$\begin{aligned} \omega_{0,01}(x) &= \frac{e_{01}(x)}{e_{00}(x) + e_{01}(x)} \Big/ \frac{e_{01}}{e_{00} + e_{01}} \\ \omega_{1,01}(x) &= \frac{e_{01}(x)}{e_{01}(x) + e_{11}(x)} \Big/ \frac{e_{01}}{e_{01} + e_{11}} \\ \omega_{0,00}(x) &= \frac{e_{00}(x)}{e_{00}(x) + e_{01}(x)} \Big/ \frac{e_{00}}{e_{00} + e_{01}} \\ \omega_{1,11}(x) &= \frac{e_{11}(x)}{e_{01}(x) + e_{11}(x)} \Big/ \frac{e_{11}}{e_{01} + e_{11}} \end{aligned}$$

It is worth noting that, under strong principal ignorability, Equations (17.11) and (17.14) also identify principal causal effects under strong monotonicity and monotonicity, respectively. Therefore, under strong principal ignorability, there are two possible approaches to the identification of the principal causal effects, which could be compared to see if they provide consistent estimtates.

Proposition 17.4 shows identification of average principal causal effects under monotonicity and a combination of exclusion restrictions and principal ignorability assumptions. Proposition 17.4 offers an alternative approach to the analysis of the flu shot encouragement study, where exclusion restriction for always-takers ($U_i = 11$) is controversial but exclusion restriction for never-takers ($U_i = 00$) is deemed to be reasonable.

Proposition 17.4. *Suppose that ignorability of treatment assignment (Assumption 6) and monotonicity (Assumption 7) hold.*

If Assumption 15 holds, then $ACE_{00} = 0$ *with*

$$E[Y_{1i} \mid U_i = 00] = E[Y_{0i} \mid U_i = 00] = E\left[Y_i \mid W_i = 1, S_i = 0\right];$$

$$\begin{aligned} ACE_{01} &= E\left[\omega_{1,01}(X_i) \cdot Y_i \mid W_i = 1, S_i = 1\right] - \\ &\left\{ \frac{e_{00} + e_{01}}{e_{01}} E\left[Y_i \mid W_i = 0, S_i = 0\right] - \frac{e_{00}}{e_{01}} E\left[Y_i \mid W_i = 1, S_i = 0\right] \right\} \end{aligned}$$

and

$$ACE_{11} = E\left[\omega_{1,11}(X_i) \cdot Y_i \mid W_i = 1, S_i = 1\right] - E\left[Y_i \mid W_i = 0, S_i = 1\right]$$

If Assumption 16 holds, then $ACE_{11} = 0$ *with*

$$E[Y_{1i} \mid U_i = 11] = E[Y_{0i} \mid U_i = 11] = E\left[Y_i \mid W_i = 0, S_i = 1\right];$$

$$\begin{aligned} ACE_{01} &= \left\{ \frac{e_{11} + e_{01}}{e_{01}} E\left[Y_i \mid W_i = 1, S_i = 1\right] - \frac{e_{11}}{e_{01}} E\left[Y_i \mid W_i = 0, S_i = 1\right] \right\} - \\ &E\left[\omega_{0,01}(X_i) \cdot Y_i \mid W_i = 0, S_i = 0\right] \end{aligned}$$

and

$$ACE_{00} = E\left[Y_i \mid W_i = 1, S_i = 0\right] - E\left[\omega_{0,00}(X_i) \cdot Y_i \mid W_i = 0, S_i = 0\right].$$

The practical use of the above propositions requires the derivation of appropriate estimators for the principal scores, $e_u(x)$, the principal strata proportions, e_u, and the expectations of the outcome over subpopulations involved. We can use simple moment-based estimators of the quantities in Propositions 17.2, 17.3 and 17.4. In particular, we replace the principal scores $e_u(x)$ and e_u with their consistent estimators, $\widehat{e}_u(X_i)$ and $\widehat{e}_u = \sum_{i=1}^n \widehat{e}_u(X_i)/n$.

Define $n_w = \sum_{i=1}^n \mathbb{I}\{W_i = w\}$ and $n_{ws} = \sum_{i=1}^n \mathbb{I}\{W_i = w\} \cdot \mathbb{I}\{S_i = s\}$, $w = 0, 1$, $s = 0, 1$. Then, moment-based estimators of the expectations of the outcome over subpopulations involved in Propositions 17.2, 17.3 and 17.4 are

$$\widehat{E}\left[Y_i \mid W_i = w, S_i = s\right] = \frac{\sum_{i=1}^n Y_i \cdot \mathbb{I}\{W_i = w\} \cdot \mathbb{I}\{S_i = s\}}{n_{ws}}$$

and

$$\widehat{E}\left[w_u(X_i) \cdot Y_i \mid W_i = w\right] = \frac{1}{n_w} \sum_{i=1}^n \widehat{w}_u(X_i) \cdot Y_i \cdot \mathbb{I}\{W_i = w\}$$

$$\widehat{E}\left[w_{1,u}(X_i) \cdot Y_i \mid W_i = 1, S_i = 1\right] = \frac{1}{n_{11}} \sum_{i=1}^n \widehat{w}_{1,u}(X_i) \cdot Y_i \cdot \mathbb{I}\{W_i = 1\} \cdot \mathbb{I}\{S_i = 1\} \qquad u = 01, 11$$

$$\widehat{E}\left[w_{0,u}(X_i) \cdot Y_i \mid W_i = 0, S_i = 0\right] = \frac{1}{n_{00}} \sum_{i=1}^n \widehat{w}_{0,u}(X_i) \cdot Y_i \cdot \mathbb{I}\{W_i = 0\} \cdot \mathbb{I}\{S_i = 0\} \qquad u = 00, 01.$$

We could also normalize the weights to unity within groups using the following estimators of the weighted averages of the outcomes:

$$\begin{aligned}
&\widehat{E}\left[w_u(X_i) \cdot Y_i \mid W_i = w\right] \\
&\quad = \frac{\sum_{i=1}^n \widehat{w}_u(X_i) \cdot Y_i \cdot \mathbb{I}\{W_i = w\}}{\sum_{i=1}^n \widehat{w}_u(X_i) \cdot \mathbb{I}\{W_i = w\}} \\
&\quad = \frac{\sum_{i=1}^n \widehat{e}_u(X_i) \cdot Y_i \cdot \mathbb{I}\{W_i = w\}}{\sum_{i=1}^n \widehat{e}_u(X_i) \cdot \mathbb{I}\{W_i = w\}}
\end{aligned}$$

$$\begin{aligned}
&\widehat{E}\left[w_{1,u}(X_i) \cdot Y_i \mid W_i = 1, S_i = 1\right] \\
&\quad = \frac{\sum_{i=1}^n \widehat{w}_{1,u}(X_i) \cdot Y_i \cdot \mathbb{I}\{W_i = 1\} \cdot \mathbb{I}\{S_i = 1\}}{\sum_{i=1}^n \widehat{w}_{1,u}(X_i) \cdot \mathbb{I}\{W_i = 1\} \cdot \mathbb{I}\{S_i = 1\}} \\
&\quad = \frac{\sum_{i=1}^n \frac{\widehat{e}_u(X_i)}{\widehat{e}_{01}(X_i)+\widehat{e}_{11}(X_i)} \cdot Y_i \cdot \mathbb{I}\{W_i = 1\} \cdot \mathbb{I}\{S_i = 1\}}{\sum_{i=1}^n \frac{\widehat{e}_u(X_i)}{\widehat{e}_{01}(X_i)+\widehat{e}_{11}(X_i)} \cdot \mathbb{I}\{W_i = 1\} \cdot \mathbb{I}\{S_i = 1\}} \qquad u = 01, 11
\end{aligned}$$

$$\begin{aligned}
&\widehat{E}\left[w_{0,u}(X_i) \cdot Y_i \mid W_i = 0, S_i = 0\right] \\
&\quad = \frac{\sum_{i=1}^n \widehat{w}_{0,u}(X_i) \cdot Y_i \cdot \mathbb{I}\{W_i = 0\} \cdot \mathbb{I}\{S_i = 0\}}{\sum_{i=1}^n \widehat{w}_{0,u}(X_i) \cdot \mathbb{I}\{W_i = 0\} \cdot \mathbb{I}\{S_i = 0\}} \\
&\quad = \frac{\sum_{i=1}^n \frac{\widehat{e}_u(X_i)}{\widehat{e}_{00}(X_i)+\widehat{e}_{01}(X_i)} \cdot Y_i \cdot \mathbb{I}\{W_i = 0\} \cdot \mathbb{I}\{S_i = 0\}}{\sum_{i=1}^n \frac{\widehat{e}_u(X_i)}{\widehat{e}_{00}(X_i)+\widehat{e}_{01}(X_i)} \cdot \mathbb{I}\{W_i = 0\} \cdot \mathbb{I}\{S_i = 0\}} \qquad u = 00, 01
\end{aligned}$$

For inference, we can use nonparametric bootstrap, which allows us to naturally account for both uncertainty from principal score estimation and sampling uncertainty. Bootstrap inference is based on the following procedure:

1. Fit the principal score model and estimate the principal causal effects on the original data set;
2. Generate B bootstrap data sets by sampling n observations with replacement from the original data set;
3. Fit the principal score model and estimate the principal causal effects on each of the B bootstrap data sets.

Bootstrap standard errors and $(1-\alpha)$ confidence intervals for the principal causal effects can be obtained by calculating the standard deviation and the $(\alpha/2)^{th}$ and $(1-\alpha/2)^{th}$ percentiles of the bootstrap distribution.

It is worth noting that inference depends on the identifying assumptions. Generally, stronger identifying assumptions (e.g., strong principal ignorability versus weak principal ignorability) increase precision, by leading to estimators with smaller variance. Nevertheless, estimators assuming stronger assumptions may be biased if the underlying assumptions do not hold (see [62] for some theoretical justifications on this issue).

In the above estimation procedure for principal causal effects, pre-treatment variables are only used to predict principal stratum membership and construct the weights. Pre-treatment variables may contain precious information about both the principal strata and the outcome distributions, and thus, adjusting for covariates may help improving statistical efficiency. To this end, weighted regression models can be used to estimate principal causal effects [57, 59]. Alternative covariate-adjusted estimators can be derived using the following equality

$$\begin{aligned}
ACE_u = {} & E[Y_{1i} - \beta_{1,u}' X_i \mid U_i = u] - E[Y_{0i} - \beta_{0,u}' X_i \mid U_i = u] \\
& + (\beta_{1,u} - \beta_{0,u})' E[X_i \mid U_i = u],
\end{aligned}$$

which holds for all u and all fixed vectors, $\beta_{w,u}$, $u = 00, 10, 01, 11$; $w = 0, 1$. Specifically, we can apply Propositions 17.2 and 17.3 treating the residuals $Y_{wi} - \beta'_{w,u} X_i$ as new potential outcomes [61]:

Corollary 1. *Suppose that ignorability of treatment assignment (Assumption 6) and strong monotonicity (Assumption 7) hold.*

If strong/weak principal ignorability (Assumption 11/12) holds, then

$$\begin{aligned} E[Y_{wi} - \beta'_{w,u} X_i \mid U_i = u] &= E\left[\omega_u(X_i) \cdot (Y_i - \beta'_{w,u} X_i) \mid W_i = w\right] \quad w = 0, 1 \\ E[X_i \mid U_i = u] &= E\left[\omega_u(X_i) \cdot X_i \mid W_i = 0\right] \\ &= E\left[\omega_u(X_i) \cdot X_i \mid W_i = 1\right] \end{aligned}$$

If only Equation (17.2)/(17.4) *of strong/weak principal ignorability (Assumption 11/12) holds, then*

$$\begin{aligned} E[Y_{1i} - \beta'_{1,01} X_i \mid U_i = 01] &= E\left[Y_i - \beta'_{1,01} X_i \mid W_i = 1, S_i = 1\right] \\ E[Y_{0i} - \beta'_{0,01} X_i \mid U_i = 01] &= E\left[\omega_{01}(X_i) \cdot (Y_i - \beta'_{0,01} X_i) \mid W_i = 0\right] \\ E[X_i \mid U_i = 01] &= E\left[X_i \mid W_i = 1, S_i = 1\right] \\ &= E\left[\omega_{01}(X_i) \cdot X_i \mid W_i = 0\right] \\ E[Y_{1i} - \beta'_{1,00} X_i \mid U_i = 00] &= E\left[Y_i - \beta'_{1,00} X_i \mid W_i = 1, S_i = 0\right] \\ E[Y_{0i} - \beta'_{0,00} X_i \mid U_i = 00] &= E\left[\omega_{00}(X_i) \cdot (Y_i - \beta'_{0,01} X_i) \mid W_i = 0\right] \\ E[X_i \mid U_i = 00] &= E\left[X_i \mid W_i = 1, S_i = 0\right] \\ &= E\left[\omega_{00}(X_i) \cdot X_i \mid W_i = 0\right] \end{aligned}$$

Corollary 2. *Suppose that ignorability of treatment assignment (Assumption 6) and monotonicity (Assumption 7) hold.*

If strong principal ignorability (Assumption 11) holds, then

$$\begin{aligned} E\left[Y_{1i} - \beta'_{w,u} X_i \mid W_i = w\right] &= E\left[\omega_u(X_i) \cdot (Y_i - \beta'_{w,u} X_i) \mid W_i = w\right] \quad w = 0, 1 \\ E[X_i \mid U_i = u] &= E\left[\omega_u(X_i) \cdot X_i \mid W_i = 0\right] \\ &= E\left[\omega_u(X_i) \cdot X_i \mid W_i = 1\right] \end{aligned}$$

$u = 00, 01, 11.$

If weak principal ignorability (Assumption 12) holds, then

$$\begin{aligned} E\left[Y_{1i} - \beta'_{1,00} X_i \mid U_i = 00\right] &= E\left[Y_i - \beta'_{1,00} X_i \mid W_i = 1, S_i = 0\right] \\ E\left[Y_{0i} - \beta'_{0,00} X_i \mid U_i = 00\right] &= E\left[\omega_{0,00}(X_i) \cdot (Y_i - \beta'_{0,00} X_i) \mid W_i = 0, S_i = 0\right] \\ E[X_i \mid U_i = 00] &= E\left[X_i \mid W_i = 1, S_i = 0\right] \\ &= E\left[\omega_{0,00}(X_i) \cdot X_i \mid W_i = 0, S_i = 0\right] \\ E\left[Y_{1i} - \beta'_{1,01} X_i \mid U_i = 01\right] &= E\left[\omega_{1,01}(X_i) \cdot (Y_i - \beta'_{1,01} X_i) \mid W_i = 1, S_i = 1\right] \\ E\left[Y_{0i} - \beta'_{0,01} X_i \mid U_i = 01\right] &= E\left[\omega_{0,01}(X_i) \cdot (Y_i - \beta'_{0,01} X_i) \mid W_i = 0, S_i = 0\right] \\ E[X_i \mid U_i = 01] &= E\left[\omega_{1,01}(X_i) \cdot X_i \mid W_i = 1, S_i = 1\right] \\ &= E\left[\omega_{0,01}(X_i) \cdot X_i \mid W_i = 0, S_i = 0\right] \\ E\left[Y_{1i} - \beta'_{1,11} X_i \mid U_i = 11\right] &= E\left[\omega_{1,11}(X_i) \cdot (Y_i - \beta'_{1,11} X_i) \mid W_i = 1, S_i = 1\right] \\ E\left[Y_{0i} - \beta'_{0,11} X_i \mid U_i = 11\right] &= E\left[Y_i - \beta'_{0,11} X_i \mid W_i = 0, S_i = 1\right] \\ E[X_i \mid U_i = 11] &= E\left[\omega_{1,11}(X_i) \cdot X_i \mid W_i = 1, S_i = 1\right] \\ &= E\left[X_i \mid W_i = 0, S_i = 1\right] \end{aligned}$$

For any fixed vectors $\beta_{w,u}$, covariate-adjustment estimators for principal strata effects can be derived using the empirical analog of the expectations in corollaries 1 and 2 with $e_u(x)$ and e_u again replaced by their consistent estimators, $\widehat{e}_u(X_i)$ and $\widehat{e}_u = \sum_{i=1}^n \widehat{e}_u(X_i)/n$. For instance, under Assumptions 6, 7 and 12, a possible covariate-adjustment estimator for ACE_{01} is

$$\widehat{ACE}_{01}^{adj} = \frac{\sum_{i=1}^n \omega_{1,01}(X_i) \cdot (Y_i - \beta_{1,01}' X_i) \cdot \mathbb{I}\{W_i = 1\} \cdot \mathbb{I}\{S_i = 1\}}{n_{11}} -$$
$$\frac{\sum_{i=1}^n \omega_{0,01}(X_i) \cdot (Y_i - \beta_{0,01}' X_i) \cdot \mathbb{I}\{W_i = 0\} \cdot \mathbb{I}\{S_i = 0\}}{n_{00}} +$$
$$\frac{1}{n_{00} + n_{11}} (\beta_{1,01} - \beta_{0,01})' \left[\sum_{i=1}^n \omega_{1,01}(X_i) \cdot X_i \cdot \mathbb{I}\{W_i = 1\} \cdot \mathbb{I}\{S_i = 1\} + \right.$$
$$\left. \sum_{i=1}^n \omega_{1,01}(X_i) \cdot X_i \cdot \mathbb{I}\{W_i = 0\} \cdot \mathbb{I}\{S_i = 0\} \right]$$

The vectors $\beta_{w,u}$ can be chosen as linear regression coefficients of potential outcomes on the space spanned by the covariates for units of type $U_i = u$:

$$\beta_{w,u} = E\left[X'X \mid U_i = u\right]^{-1} E\left[XY_w \mid U_i = u\right]$$

where X is a matrix with i^{th} row equals to X_i and Y_w is a $n-$dimensional vector with i^{th} element equal to Y_{wi}. Each component of the above least squares formula is identifiable under ignorability of treatment assignment, (strong) monotonicity, and strong/weak principal ignorability [61]. Simulation studies suggest that this type of covariate-adjusted estimators perform better than the regression method proposed by Jo and Stuart [57]: they are robust to mispecification of the outcome models and have smaller variance [61].

Recently, Jiang et al. [64] have generalized Proposition 17.3 assuming ignorability of treatment assignment conditionally on covariates: $W_i \perp\!\!\!\perp (S_{0i}, S_{1i}, Y_{0i}, Y_{1i}) \mid X_i$. Under ignorability of treatment assignment given covariates, monotonicity and weak principal ignorability (Assumptions 6, 8, 12), the identification formulas in Equation (17.14) still hold with an additional weighting term based on the inverse of the treatment probability (see Theorem 1 in [64]). Theorem 1 in [64] also introduces two additional sets of identification formulas, providing overall three nonparametric identification formulas for each principal causal effect. These three identification formulas lead also to develop three alternative estimators. Moreover, Jiang et al. [64] show that appropriately combining these estimators through either the efficient influence functions in the semi-parametric theory or the model-assisted estimators in the survey sampling theory, we can obtain new estimators for the principal causal effects, which are triply robust, in the sense that they are consistent if two of the treatment, intermediate variables, and outcome models are correctly specified, and they are locally efficient if all three models are correctly specified. Simplified versions of these results hold under ignorability of treatment assignment as defined in Assumption 6 and/or strong monotonicity (Assumptions 7). See [64] for details.

17.6 Assessing Principal Score Fit

Assessing the adequacy of the specification for the principal scores could rely on the balancing property of the true principal score: $\mathbb{I}\{U_i = u\} \perp\!\!\!\perp X_i \mid e_u(x)$. This is challenging because we cannot compare the covariate distributions across principal strata that are not directly observed. However, we can use the fact that, along with randomization, this balancing property of the principal score allows us to estimate the expectation of any stratum-specific function of the covariates, $h(x)$, via a principal score weighted average. Specifically, Ding and Lu [61] show that if the treatment assignment mechanism satisfies Assumption 6, such that, for all u,

$$E[h(X_i) \mid U_i = u] = [h(X_i) \mid U_i = u, W_i = 1] = E[h(X_i) \mid U_i = u, W_i = 0]$$

then, under strong monotonicity (Assumption 7), we have the following balancing conditions:

$$\begin{aligned} & E[h(X_i) \mid U_i = 01, W_i = 1] \\ & \quad = \quad E[h(X_i) \mid W_i = 1, S_i = 1] = E[\omega_{01} \cdot h(X_i) \mid W_i = 0] = \\ & E[h(X_i) \mid U_i = 01, W_i = 0] \end{aligned}$$

$$\begin{aligned} & E[h(X_i) \mid U_i = 00, W_i = 1] \\ & \quad = \quad E[h(X_i) \mid W_i = 1, S_i = 0] = E[\omega_{00} \cdot h(X_i) \mid W_i = 0] = \\ & E[h(X_i) \mid U_i = 00, W_i = 0] \end{aligned}$$

Similarly, under monotonicity (Assumption 8), we have

$$\begin{aligned} & E[h(X_i) \mid U_i = 01, W_i = 1] \\ & \quad = \quad E[\omega_{1,01} \cdot h(X_i) \mid W_i = 1, S_i = 1] = E[\omega_{0,01} \cdot h(X_i) \mid W_i = 0, S_i = 0] = \\ & E[h(X_i) \mid U_i = 01, W_i = 0] \end{aligned}$$

$$\begin{aligned} & E[h(X_i) \mid U_i = 00, W_i = 1] \\ & \quad = \quad E[h(X_i) \mid W_i = 1, S_i = 0] = E[\omega_{0,00} \cdot h(X_i) \mid W_i = 0, S_i = 0] = \\ & E[h(X_i) \mid U_i = 00, W_i = 0] \end{aligned}$$

$$\begin{aligned} & E[h(X_i) \mid U_i = 11, W_i = 1] \\ & \quad = \quad E[\omega_{1,11} \cdot h(X_i) \mid W_i = 1, S_i = 1] = E[h(X_i) \mid W_i = 0, S_i = 1] = \\ & E[h(X_i) \mid U_i = 11, W_i = 0] \end{aligned}$$

In practice, we replace the true principal scores with the estimated principal scores and investigate whether, at least approximately, the previous equalities hold. If the above balancing conditions are violated, we need to specify more flexible models for the principal scores to account for the residual dependence of the principal stratum variable, U_i, on the covariates, X_i.

Jiang et al. [64] generalize these balancing conditions to studies where the treatment assignment mechanism is ignorable conditional on the observed covariates and introduce additional sets of balancing conditions using different sets of weights.

We can assess covariate balance using various summary statistics of the covariate distributions by treatment status within principal strata. For instance, borrowing from the literature on the role of the propensity score in the design of a regular observational study [8], we can measure the difference between the covariate distributions focusing on locations or dispersion using the normalized differences within principal strata, as a measure of the difference in location:

$$\Delta_u = \frac{E[X_i \mid U_i = u, W_i = 1] - E[X_i \mid U_i = u, W_i = 0]}{\sqrt{(V[X_i \mid U_i = u, W_i = 1] + V[X_i \mid U_i = u, W_i = 0])/2}},$$

and the logarithm of the ratio of standard deviations within each principal stratum as a measure of the difference in dispersion:

$$\Gamma_u = \log \frac{\sqrt{V[X_i \mid U_i = u, W_i = 1]}}{\sqrt{V[X_i \mid U_i = u, W_i = 0]}}.$$

These quantities can be easily estimated either directly from the observed data or via weighting using the estimated principal score according to the above balancing conditions (or to the version in [64]). For instance, under monotonicity, the mean and the variance of each covariate for units of type $U_i = 01$ can be estimated as follows:

$$\widehat{E}[X_i \mid U_i = 01, W_i = w] = \frac{\sum_{i=1}^{n} \widehat{\omega}_{w,01}(X_i) \cdot X_i \cdot \mathbb{I}\{W_i = w\}}{\sum_{i=1}^{n} \widehat{\omega}_{w,01}(X_i) \cdot \mathbb{I}\{W_i = w\}}$$

and

$$\widehat{V}[X_i \mid U_i = 01, W_i = w] = \frac{\sum_{i=1}^{n} \widehat{\omega}_{w,01}(X_i) \cdot \left(X_i - \widehat{E}[X_i \mid U_i = 01, W_i = w]\right)^2 \cdot \mathbb{I}\{W_i = w\}}{\sum_{i=1}^{n} \widehat{\omega}_{w,01}(X_i) \cdot \mathbb{I}\{W_i = w\} - 1}.$$

17.7 Sensitivity Analysis

Principal stratification analysis based on principal score methods critically depends on the assumptions of monotonicity and principal ignorability. These critical assumptions are fundamentally not testable and may not be easily justified using subject-matter knowledge. Therefore, it is important to consider ways to assess the plausibility of these assumptions from the data at hand.

Comparing results obtained under alternative sets of identifying assumptions may provide some insights on their validity [57, 62]. Suppose that (strong) monotonicity holds. Results under exclusion restriction assumptions may provide some evidence on the plausibility of principal ignorability assumptions, at least in some contexts [57, 59]. Under principal ignorability, principal score methods allow us to identify and estimate principal causal effects for units of type $U_i = 00$ and/or units of type $U_i = 11$, and thus, inform us on the plausibility of exclusion restriction assumptions: if estimates of principal causal effects for units of type $U_i = 00/U_i = 11$ obtained under principal ignorability are not meaningfully different from zero, the data provide no evidence against the exclusion restriction for units of type $U_i = 00/U_i = 11$. On the other hand, comparing results obtained under strong principal ignorability and weak principal ignorability provides empirical evidence on the plausibility of strong principal ignorability: if estimates of the average treatment outcome for units of type $U_i = 00$ and/or estimates of the average control outcome for units of type $U_i = 11$ are largely unchanged under strong principal ignorability and weak principal ignorability, inferences based on strong principal ignorability are more defensible. Nevertheless, in practice, it may be generally preferable to invoke weak principal ignorability, which is strictly weaker than strong principal ignorability, even if it may lead to a loss in precision [62].

More principled approaches to sensitivity analysis have been recently proposed by Ding and Lu [61] and Feller et al. [62]. Below we briefly review these approaches.

17.7.1 Sensitivity analysis for principal ignorability

Principal ignorability assumptions require that all the variables affecting both the principal stratum and the potential outcomes are observed and pre-treatment. Therefore, generally, the higher the number of covariates we observe, the more plausible principal ignorability assumptions are. Nevertheless, principal ignorability assumptions are not directly testable from the observed data, and we can never rule out the possibility that there exist unmeasured covariates confounding the relationship between the principal stratum variable and the outcome variable. In the Faenza study, variables like knowledge of screening tests and breast cancer risk perceptions and attitudes can be viewed as important counfounders, associated with both the compliance behavior and BSE practice. Therefore, principal ignorability assumptions may be untenable whenever the observed variables do not properly account for them. Although the Faenza study includes information on pre-treatment BSE practice and knowledge of breast pathophysiology, which may be considered as proxies of breast cancer risk perceptions and attitudes and knowledge of screening tests, a sensitivity analysis with respect to principal ignorability assumptions is still valuable, making inference based on those assumptions more defensible. In the flu shot encouragement study, Ding and Lu [61] find that results based on principal ignorability assumptions are consistent with those obtained by Hirano et al. [11] using a model-based Bayesian approach to inference, suggesting that the exclusion restriction may be plausible for never-takers, but does not hold for always-takers. The coherence between the two analyses supports the plausibility of principal ignorability, but does not prove it, and thus, performing a sensitivity analysis for principal ignorability is recommendable.

A sensitivity analysis relaxes the principal ignorability assumptions without replacing them with additional assumptions, and thus, it leads to ranges of plausible values for principal causal effects, with the width of these ranges corresponding to the extent to which we allow the principal ignorability assumptions to be violated.

Feller et al. [62] propose a partial identification-based approach to sensitivity analysis, where estimates of principal causal effects under principal ignorability assumptions are compared with their corresponding nonparametric bounds. We can interpret estimates of principal causal effects based on principal ignorability assumptions as possible guesses of the true principal causal effects within the bounds, so that estimates of principal causal effects based on principal ignorability outside of the corresponding bounds provide strong evidence against principal ignorability. Unfortunately, Feller et al. [62] do not provide any technical detail on the implementation of the partial identification approach, describing it as a valuable direction for future work.

A formal framework to assess the sensitivity of the deviations from the principal ignorability assumptions has been developed by Ding and Lu [61]. Specifically, they focus on violations of the weaker version of the (weak) principal ignorability assumption (Assumption 14). Assumption 14 implies that (i) under strong monotonicity, the conditional means of the control potential outcomes are the same for principal strata 01 and 00 given covariates, $E[Y_i(0) \mid U_i = 01, X_i] = E[Y_i(0) \mid U_i = 00, X_i]$; and (ii) under monotonicity, the conditional means of the control potential outcomes are the same for principal strata 01 and 00 given covariates, and the conditional means of the treatment potential outcomes are the same for principal strata 01 and 11 given covariates, $E[Y_i(0) \mid U_i = 01, X_i] = E[Y_i(0) \mid U_i = 00, X_i]$ and $E[Y_i(1) \mid U_i = 01, X_i] = E[Y_i(1) \mid U_i = 11, X_i]$. Given that Assumption 14 is weaker than Assumption 12, results derived under Assumption 12 in Propositions 17.2 and 17.3 still hold under the weaker version centered on conditional means, and clearly, if Assumption 14 does not hold, Assumption 12 does not hold either.

The sensitivity analysis proposed by Ding and Lu [61] is based on the following proposition:

Proposition 17.5. *Suppose that ignorability of treatment assignment (Assumption 6) holds.*

Under strong monotonicity (Assumption 7), define

$$\epsilon = \frac{E\left[Y_{0i} \mid U_i = 01\right]}{E\left[Y_{0i} \mid U_i = 00\right]}.$$

Then, for a fixed value of ϵ, *we have*

$$\begin{aligned} ACE_{01} &= E\left[Y_i \mid W_i = 1, S_i = 1\right] - E\left[\omega_{01}^{\epsilon}(X_i) \cdot Y_i \mid W_i = 0\right] \\ ACE_{00} &= E\left[Y_i \mid W_i = 1, S_i = 0\right] - E\left[\omega_{00}^{\epsilon}(X_i) \cdot Y_i \mid W_i = 0\right] \end{aligned} \tag{17.15}$$

where

$$\omega_{01}^{\epsilon}(x) = \frac{\epsilon \cdot e_{01}(x)}{\left[\epsilon \cdot e_{01}(x) + e_{00}(x)\right] \cdot e_{01}} \quad \text{and} \quad \omega_{00}^{\epsilon}(x) = \frac{e_{00}(x)}{\left[\epsilon \cdot e_{01}(x) + e_{00}(x)\right] \cdot e_{00}}$$

Under monotonicity (Assumption 8), define

$$\epsilon_1 = \frac{E\left[Y_{1i} \mid U_i = 01\right]}{E\left[Y_{1i} \mid U_i = 11\right]} \quad \text{and} \quad \epsilon_0 = \frac{E\left[Y_{0i} \mid U_i = 01\right]}{E\left[Y_{0i} \mid U_i = 00\right]}.$$

Then, for fixed values of (ϵ_0, ϵ_1), *we have*

$$\begin{aligned} ACE_{00} &= E\left[Y_i \mid W_i = 1, S_i = 0\right] - E\left[\omega_{0,00}^{\epsilon_0}(X_i) \cdot Y_i \mid W_i = 0, S_i = 0\right] \\ ACE_{01} &= E\left[\omega_{1,01}^{\epsilon_1}(X_i) \cdot Y_i \mid W_i = 1, S_i = 1\right] - E\left[\omega_{0,01}^{\epsilon_0}(X_i) \cdot Y_i \mid W_i = 0, S_i = 0\right] \\ ACE_{11} &= E\left[\omega_{1,11}^{\epsilon_1}(X_i) \cdot Y_i \mid W_i = 1, S_i = 1\right] - E\left[Y_i \mid W_i = 0, S_i = 1\right] \end{aligned} \tag{17.16}$$

where

$$\begin{aligned} \omega_{0,01}(x) &= \frac{\epsilon_0 \cdot e_{01}(x)}{e_{00}(x) + \epsilon_0 \cdot e_{01}(x)} \Big/ \frac{e_{01}}{e_{00} + e_{01}} \\ \omega_{1,01}(x) &= \frac{\epsilon_1 \cdot e_{01}(x)}{\epsilon_1 \cdot e_{01}(x) + e_{11}(x)} \Big/ \frac{e_{01}}{e_{01} + e_{11}} \\ \omega_{0,00}(x) &= \frac{e_{00}(x)}{e_{00}(x) + \epsilon_0 \cdot e_{01}(x)} \Big/ \frac{e_{00}}{e_{00} + e_{01}} \\ \omega_{1,11}(x) &= \frac{e_{11}(x)}{\epsilon_1 \cdot e_{01}(x) + e_{11}(x)} \Big/ \frac{e_{11}}{e_{01} + e_{11}} \end{aligned}$$

The sensitivity parameters ϵ and (ϵ_0, ϵ_1) capture deviations from the principal ignorability assumption 14. The principal score for units of type $U_i = 01$, $e_{01}(x)$, is over-weighted by the sensitivity parameter ϵ under strong monotonicity and by the sensitivity parameters ϵ_0 and ϵ_1 in the treatment group and control control group, respectively, under monotonicity. For $\epsilon = 1$ and $(\epsilon_0, \epsilon_1) = (1, 1)$, no extra weight is applied and the same identification results hold as those under (weak) principal ignorability (Assumption 12) shown in Propositions 17.2 and 17.3, respectively. The further away the values of the sensitivity parameters are from 1, the stronger the deviation from principal ignorability is. Given a set of reasonable values for the sensitivity parameters, ϵ or (ϵ_0, ϵ_1), we can calculate a lower and upper bound for the average principal causal effects over that set, and assess whether inferences based on principal ignorability assumptions are defensible. Note that Proposition 17.5 implicitly assumes that the sensitivity parameters ϵ and (ϵ_0, ϵ_1) do not depend on the covariates.

The choice of the sensitivity parameters deserves some discussion. Suppose that strong monotonicity holds. The sensitivity parameter ϵ compares the average outcomes under control for units of type $U_i = 01$ and units of type $U_i = 00$ with the same values of the covariates. If the outcome Y is binary, it is the relative risk of U_i on the control potential outcome, Y_{0i}, given covariates, X_i. This interpretation suggests that we can select the range of ϵ according to subject-matter knowledge on the relationship between the control potential outcome and the principal stratum variable. For instance, in the Faenza study, where strong monotonicity holds by design, it might be reasonable to believe that on average the never-takers ($U_i = 00$) are women who feel that preforming BSE correctly requires some experience, and thus, we can assume a deviation from homogeneous outcomes under principal ignorability in the direction corresponding to $\epsilon < 1$. Similarly, under monotonicity, the sensitivity parameters ϵ_0 and ϵ_1 can be selected on the basis of background knowledge. For instance, in the flu shot study, background knowledge suggests that the never-takers may comprise the healthiest patients and the always-takers may comprise the weakest patients, and thus, a plausible choise of the sensitivity parameters is within the range $\epsilon_0 > 1$ and $\epsilon_1 < 1$ [11,61]. Ding and Lu [61] perform the sensitivity analysis with respect to principal ignorability for the flu shot study, finding that the point and interval estimates vary with the sensitivity parameters (ϵ_0, ϵ_1), but the final conclusions do not change substantially, and thus, inferences based on principal ignorability are credible.

Interestingly, Proposition 17.5 implies testable conditions of principal ignorability and exclusion restrictions. For instance, under strong monotonicity, if $ACE_{00} = 0$ and $\epsilon = 1$, then

$$E\left[Y_i \mid W_i = 1, S_i = 0\right] = E\left[\omega_{00}(X_i) \cdot Y_i \mid W_i = 0, S_i = 0\right].$$

If the observed data show evidence against this condition, then we must reject either $ACE_{00} = 0$ or $\epsilon = 1$. Therefore, we can test $ACE_{00} = 0$, assuming $\epsilon = 1$, and we can test $\epsilon = 1$, assuming $ACE_{00} = 0$. It is worth noting that Proposition 17.5 implies testable conditions for compatibility of principal ignorability and exclusion restrictions. Suppose that we are willing to impose principal ignorability, then we can test $ACE_{00} = 0$. If the test leads to reject the null hypothesis that $ACE_{00} = 0$, then we may reject the exclusion restriction assumption. Nevertheless, we may also have doubts about principal ignorability.

Similarly, Proposition 17.5 implies testable conditions for compatibility of principal ignorability and exclusion restrictions under monotonicity:

$$\begin{aligned}
&\text{If } ACE_{00} = 0 \text{ and } \epsilon_0 = 1 \text{ then}\\
&\qquad E\left[Y_i \mid W_i = 1, S_i = 0\right] = E\left[\omega_{0,00}(X_i) \cdot Y_i \mid W_i = 0, S_i = 0\right];\\
&\text{If } ACE_{11} = 0 \text{ and } \epsilon_1 = 1 \text{ then}\\
&\qquad E\left[Y_i \mid W_i = 0, S_i = 1\right] = E\left[\omega_{1,11}(X_i) \cdot Y_i \mid W_i = 1, S_i = 1\right].
\end{aligned}$$

Jiang et al. [64] extend Proposition 17.5 to studies where the assignment mechanism is (or is assumed) ignorable conditionally on covariates, by also allowing the sensitivity parameters to depend on the covariates, and derive the semi-parametric efficiency theory for sensitivity analysis.

17.7.2 Sensitivity analysis for monotonicity

In principal stratification analysis based on principal score methods, monotonicity is a core assumption. Monotonicity of the intermediate variable is often plausible by design and/or background knowledge in randomized studies with non-compliance, but it is often debatable outside the non-compliance setting. For instance, in the Faenza study, when focus is on assessing causal effects on treatment assignment on BSE quality, monotonicity of BSE practice may be questionable: for some type of women that are not willing to attend the in-person course, the assignment to the course might have a negative effect on BSE practice, because they might think that they are not able to perform it correctly [24].

Without monotonicity, all the four principal strata are present, and thus, we cannot even identify their proportions without further assumptions. Ding and Lu [61] propose an approach to conduct sensitivity analysis with respect to monotonicity. This approach uses as sensitivity parameter the ratio between the conditional probabilities of strata $U_i = 10$ and $U_i = 01$ given the covariates,

$$\xi = \frac{\Pr(U_i = 10 \mid X_i)}{\Pr(U_i = 01 \mid X_i)}.$$

Note that ξ is also the ratio between the marginal probabilities of strata $U_i = 10$ and $U_i = 01$: $\xi = \Pr(U_i = 10)/\Pr(U_i = 01)$.

The sensitivity parameter takes on values in $[0, +\infty)$: when $\xi = 0$ monotonicity holds; when $\xi > 0$ we have a deviation from monotonicity, with zero average causal effect on the intermediate variable S for $\xi = 1$, and positive and negative average causal effect on the intermediate variable S for $0 < \xi < 1$ and $\xi > 1$, respectively.

Let $p_w = \Pr(S_i = 1 \mid W_i = w)$, $w = 0, 1$, denote the marginal probability of S under treatment status w. Following Ding and Lu [61], without loss of generality, we shall assume that $p_1 - p_0 \geq 0$ and $0 \leq \xi \leq 1$. Suppose that the treatment assignment is ignorable according to Assumption 6. Then, for a fixed value of ξ, we can identify the principal scores by

$$e_{01}(x) = \frac{p_1(x) - p_0(x)}{1-\xi} \qquad e_{00}(x) = 1 - p_0(x) - \frac{p_1(x) - p_0(x)}{1-\xi}$$

$$e_{11}(x) = p_1(x) - \frac{p_1(x) - p_0(x)}{1-\xi} \qquad e_{10}(x) = \frac{\xi \cdot (p_1(x) - p_0(x))}{1-\xi};$$

and the principal strata proportions by

$$e_{01} = \frac{p_1 - p_0}{1-\xi} \qquad e_{00} = 1 - p_0 - \frac{p_1 - p_0}{1-\xi}$$

$$e_{11} = p_1 - \frac{p_1 - p_0}{1-\xi} \qquad e_{10} = \frac{\xi \cdot (p_1 - p_0)}{1-\xi}.$$

These identifying equations for the principal strata proportions imply bounds on sensitivity parameter ξ. Formally, we have that

$$0 \leq \xi \leq 1 - \frac{p_1 - p_0}{\min\{p_1, 1 - p_0\}} \leq 1.$$

Therefore, the observed data provide an upper bound for the sensitivity parameter ξ when the average causal effect on the intermediate variable, S, is non-negative, simplifying the non-trival task of selecting values for the sensitivity parameter, ξ: we need to conduct sensitivity analysis varying ξ within the empirical version of the above bounds.

Moreover, under weak principal ignorability (Assumption 12), for a fixed value of ξ, we have that

$$ACE_u = E\left[\omega_{1,u}(X_i) \cdot Y_i \mid W_i = 1, S_i = s_1\right] - E\left[\omega_{0,u}(X_i) \cdot Y_i \mid W_i = 0, S_i = s_0\right]$$

for $u = s_0 s_1 \in \{00, 10, 01, 11\}$ and

$$\omega_{0,01}(x) = \frac{e_{01}(x)}{e_{00}(x) + e_{01}(x)} \bigg/ \frac{e_{01}}{e_{00} + e_{01}} \qquad \omega_{1,01}(x) = \frac{e_{01}(x)}{e_{01}(x) + e_{11}(x)} \bigg/ \frac{e_{01}}{e_{01} + e_{11}}$$

$$\omega_{0,00}(x) = \frac{e_{00}(x)}{e_{00}(x) + e_{01}(x)} \bigg/ \frac{e_{00}}{e_{00} + e_{01}} \qquad \omega_{1,00}(x) = \frac{e_{00}(x)}{e_{00}(x) + e_{10}(x)} \bigg/ \frac{e_{00}}{e_{00} + e_{10}}$$

$$\omega_{0,11}(x) = \frac{e_{11}(x)}{e_{10}(x) + e_{11}(x)} \bigg/ \frac{e_{11}}{e_{10} + e_{11}} \qquad \omega_{1,11}(x) = \frac{e_{11}(x)}{e_{01}(x) + e_{11}(x)} \bigg/ \frac{e_{11}}{e_{01} + e_{11}}$$

$$\omega_{0,10}(x) = \frac{e_{10}(x)}{e_{10}(x) + e_{11}(x)} \bigg/ \frac{e_{10}}{e_{10} + e_{11}} \qquad \omega_{1,10}(x) = \frac{e_{10}(x)}{e_{00}(x) + e_{10}(x)} \bigg/ \frac{e_{10}}{e_{00} + e_{10}}.$$

17.8 Discussion

The chapter has introduced and discussed principal stratification analysis under different versions of principal ignorability using principal scores.

As the propensity score in observational studies, the principal score plays a similar role in settings with post-assignment variables, in that it may be used to design a principal stratification analysis and to develop estimation strategies of principal causal effects. Through several examples, we have tried also to convey the idea that, despite this similarity, principal stratification under principal ignorability is more critical for several reasons.

Principal ignorability is an assumption about conditional independence of the outcome of interest and the principal strata, which we do not completely observe. This makes it more difficult to identify covariates that allow to break the potential dependence. The design plays a crucial role here: complications or analysis with intermediate variables should be anticipated in protocols and covariates that are predictive of both the intermediate and the outcome should be collected in order to make principal ignorability plausible (see, also, Griffin et al. [74] for some discussion on the role of covariates in principal stratification analysis).

The fact that principal strata are only partially observed renders the estimation of the principal score in general more complicated than estimating a propensity score and requires some form of monotonicity. This is one reason why the extension of principal score analysis to more complex settings such as multivalued intermediate variables, sequential treatment, and so on may not be so straightforward, although conceptually feasible. In general we would need to extend monotonicity or find other plausible assumptions to reduce the number of principal strata (see for example [30]).

In addition, when the principal strata are more than two, as with binary treatment, binary intermediate variable and monotonicity, the estimation procedures need to be modified accordingly. In particular, the principal score of one principal strata must be rescaled by the probability of one of the potential values of the intermediate variable, which is given by the sum of multiple principal scores [61]. For instance, compare the weights involved in Proposition 17.2 and in Proposition 17.3. In particular, consider, e.g., the expected control potential outcome for units of type $U_i = 01$, $E[Y_{0i} \mid U_i = 01]$. When there exit only two principal strata, 00 and 01, $E[Y_{0i} \mid U_i = 01]$ is obtained as a weighted average of the outcomes for units with $W_i = 0$ and $S_i = 0$ with weights given by $\omega_{01}(x) = e_{01}(x)/e_{01} = P(U_i = 01 \mid X_i = x)/P(U_i = 01)$ (see Proposition 17.2). When there exit three principal strata, 00, 01, and 11, we need to take into account that units assigned to the control treatment (with $W_i = 0$) for which $S_i = 0$, are units of type $U_i = 00$ or $U_i = 01$; they cannot be units of type $U_i = 11$. Therefore their outcomes need to be weighted using the weights $\omega_{0,01}(x)$, defined by rescaling the ratio $e_{01}(x)/e_{01} = P(U_i = 01 \mid X_i = x)/P(U_i = 01)$ by $(e_{00}(x) + e_{01}(x)) \,/\, (e_{00} + e_{01}) = (P(U_i = 00 \mid X_i = x) + P(U_i = 01 \mid X_i = x)) \,/\, (P(U_i = 00) + P(U_i = 01))$ (see Proposition 17.2).

Likewise, with multivalued intermediate variables in principle we can derive theoretical results under principal ignorability by rescaling the weights in a similar way. However, a continuous intermediate variable results in infinitely many principal strata, and, thus, requires more structural assumptions and more advanced statistical inferential tools to estimate the principal scores and outcome distributions conditional on continuous variables.

Furthermore, theoretical results can also be derived for the case of multivalued treatment when principal strata are defined by the potential values of the intermediate variable under each treatment level and principal causal effects are defined as pairwise comparisons within principal strata or combinations of principal strata.

We discussed primarily point estimators using the principal score and provided some guidelines to assess some crucial underlying assumptions. However, principal stratification analysis can be very naturally conducted under a Bayesian framework [4, 17, 27, 31, 36, 67]. Bayesian inference does not necessarily require neither principal ignorability nor monotonicity because it does not require full identification. From a Bayesian perspective, the posterior distribution of the parameters of interest is derived by updating a prior distribution to a posterior distribution via a likelihood, irrespective of whether the parameters are fully or partially identified [17, 24, 36, 67, 75]. This is achieved at the cost of having to specify parametric models for the principal strata and the outcomes, although Bayesian nonparametric tools can be developed and research is still active in this area [28].

The Bayesian approach also offers an easy way to conduct sensitivity analysis to deviations from both principal ignorability and monotonicity, by checking how the posterior distribution of causal parameters change under specific deviations from the assumptions.

Bayesian inference can indeed be conducted also under principal ignorability, and it would offer also a way of multiply imputing the missing intermediate outcomes. Under principal ignorability, the posterior variability of the parameters of interest will be smaller, but the posterior distributions might lead to misleading conclusions if the underlying principal ignorability assumption does not hold.

When principal ignorability assumptions appear untenable, as an alternative to the Bayesian approach, a partial identification approach can be used, deriving nonparametric bounds on principal causal effects. Unfortunately, unadjusted bounds are often too wide to be informative. The causal inference literature has dealt with this issue, developing various strategies for shrinking bounds on principal causal effects, based on pre-treatment covariates and/or secondary outcomes [33,37,38,45–48].

Until very recently principal stratification analysis under principal ignorability assumptions has been mostly conducted under the ignorability assumption that the treatment assignment mechanism is independent of both potential outcomes and covariates, and some type of monotonicity assumption. Nowadays there is ongoing work aimed to extend principal stratification analysis under principal ignorability assumptions: Jiang et al. [64] develop multiply robust estimators for principal causal effects under principal ignorability, which can be used also to analyze block-randomized experiments and observational studies, where the assignment mechanism is assumed to be ignorable conditional on pre-treatment covariates; Han et al. [76] propose a new estimation technique based on a stochastic monotonicity, which is weaker than the deterministic monotonicity usually invoked.

We believe that further extensions of principal stratification analysis under principal ignorability assumptions might be worthwhile, providing a fertile ground for future work.

References

[1] Jerzy S Neyman. On the application of probability theory to agricultural experiments. essay on principles. section 9.(tlanslated and edited by dm dabrowska and tp speed, statistical science (1990), 5, 465-480). *Annals of Agricultural Sciences*, 10:1–51, 1923.

[2] Donald B Rubin. Estimating causal effects of treatments in randomized and nonrandomized studies. *Journal of Educational Psychology*, 66(5):688, 1974.

[3] Donald B Rubin. Assignment to treatment group on the basis of a covariate. *Journal of Educational Statistics*, 2(1):1–26, 1977.

[4] Donald B Rubin. Bayesian inference for causal effects: The role of randomization. *The Annals of Statistics*, 6(1):34–58, 1978.

[5] Donald Rubin. Discussion of "randomization analysis of experimental data in the fisher randomization test" by d. basu. *Journal of the American Statistical Association*, 75:591–593, 1980.

[6] Donald B Rubin. Comment: Neyman (1923) and causal inference in experiments and observational studies. *Statistical Science*, 5(4):472–480, 1990.

[7] Paul W Holland. Statistics and causal inference. *Journal of the American Statistical Association*, 81(396):945–960, 1986.

[8] Guido W Imbens and Donald B Rubin. *Causal inference in statistics, social, and biomedical sciences.* Cambridge University Press, 2015.

[9] Constantine E Frangakis and Donald B Rubin. Principal stratification in causal inference. *Biometrics*, 58(1):21–29, 2002.

[10] Joshua D Angrist, Guido W Imbens, and Donald B Rubin. Identification of causal effects using instrumental variables. *Journal of the American statistical Association*, 91(434):444–455, 1996.

[11] Keisuke Hirano, Guido W Imbens, Donald B Rubin, and Xiao-Hua Zhou. Assessing the effect of an influenza vaccine in an encouragement design. *Biostatistics*, 1(1):69–88, 2000.

[12] Michela Bia, Alessandra Mattei, and Andrea Mercatanti. Assessing causal effects in a longitudinal observational study with "truncated" outcomes due to unemployment and nonignorable missing data. *Journal of Business & Economic Statistics*, pages 1–12, 2021.

[13] Paolo Frumento, Fabrizia Mealli, Barbara Pacini, and Donald B Rubin. Evaluating the effect of training on wages in the presence of noncompliance, nonemployment, and missing outcome data. *Journal of the American Statistical Association*, 107(498):450–466, 2012.

[14] JL Zhang, DB Rubin, and F Mealli. Likelihood-based analysis of causal effects via principal stratification: new approach to evaluating job training programs. *Journal of the American Statistical Association*, 104:166–176, 2009.

[15] Alessandra Mattei, Fabrizia Mealli, and Barbara Pacini. Identification of causal effects in the presence of nonignorable missing outcome values. *Biometrics*, 70(2):278–288, 2014.

[16] Peter B Gilbert and Michael G Hudgens. Evaluating candidate principal surrogate endpoints. *Biometrics*, 64(4):1146–1154, 2008.

[17] Michela Baccini, Alessandra Mattei, and Fabrizia Mealli. Bayesian inference for causal mechanisms with application to a randomized study for postoperative pain control. *Biostatistics*, 18(4):605–617, 2017.

[18] Laura Forastiere, Alessandra Mattei, and Peng Ding. Principal ignorability in mediation analysis: through and beyond sequential ignorability. *Biometrika*, 105(4):979–986, 2018.

[19] Alessandra Mattei and Fabrizia Mealli. Augmented designs to assess principal strata direct effects. *Journal of the Royal Statistical Society: Series B (Statistical Methodology)*, 73(5):729–752, 2011.

[20] Fabrizia Mealli and Donald B Rubin. Assumptions allowing the estimation of direct causal effects. *Journal of Econometrics*, 112(1):79–79, 2003.

[21] Donald B Rubin. Direct and indirect causal effects via potential outcomes. *Scandinavian Journal of Statistics*, 31(2):161–170, 2004.

[22] Thomas R Ten Have and Marshall M Joffe. A review of causal estimation of effects in mediation analyses. *Statistical Methods in Medical Research*, 21(1):77–107, 2012.

[23] Fabrizia Mealli and Alessandra Mattei. A refreshing account of principal stratification. *The International Journal of Biostatistics*, 8(1), 2012. doi: 10.1515/1557-4679.1380. PMID: 22611592.

[24] Alessandra Mattei and Fabrizia Mealli. Application of the principal stratification approach to the faenza randomized experiment on breast self-examination. *Biometrics*, 63(2):437–446, 2007.

[25] Avi Feller, Todd Grindal, Luke Miratrix, and Lindsay C Page. Compared to what? variation in the impacts of early childhood education by alternative care type. *The Annals of Applied Statistics*, 10(3):1245–1285, 2016.

[26] Hui Jin and Donald B Rubin. Principal stratification for causal inference with extended partial compliance. *Journal of the American Statistical Association*, 103(481):101–111, 2008.

[27] Chanmin Kim, Michael J Daniels, Joseph W Hogan, Christine Choirat, and Corwin M Zigler. Bayesian methods for multiple mediators: Relating principal stratification and causal mediation in the analysis of power plant emission controls. *The Annals of Applied Statistics*, 13(3):1927, 2019.

[28] Scott L Schwartz, Fan Li, and Fabrizia Mealli. A bayesian semiparametric approach to intermediate variables in causal inference. *Journal of the American Statistical Association*, 106(496):1331–1344, 2011.

[29] Alessandra Mattei, Fabrizia Mealli, and Peng Ding. Assessing causal effects in the presence of treatment switching through principal stratification. *arXiv preprint arXiv:2002.11989*, 2020.

[30] Constantine E Frangakis, Ronald S Brookmeyer, Ravi Varadhan, Mahboobeh Safaeian, David Vlahov, and Steffanie A Strathdee. Methodology for evaluating a partially controlled longitudinal treatment using principal stratification, with application to a needle exchange program. *Journal of the American Statistical Association*, 99(465):239–249, 2004.

[31] Laura Forastiere, Patrizia Lattarulo, Marco Mariani, Fabrizia Mealli, and Laura Razzolini. Exploring encouragement, treatment, and spillover effects using principal stratification, with application to a field experiment on teens' museum attendance. *Journal of Business & Economic Statistics*, 39(1):244–258, 2021.

[32] Guido W Imbens and Donald B Rubin. Estimating outcome distributions for compliers in instrumental variables models. *The Review of Economic Studies*, 64(4):555–574, 1997.

[33] Junni L Zhang and Donald B Rubin. Estimation of causal effects via principal stratification when some outcomes are truncated by "death". *Journal of Educational and Behavioral Statistics*, 28(4):353–368, 2003.

[34] Peng Ding, Zhi Geng, Wei Yan, and Xiao-Hua Zhou. Identifiability and estimation of causal effects by principal stratification with outcomes truncated by death. *Journal of the American Statistical Association*, 106(496):1578–1591, 2011.

[35] Zhichao Jiang, Peng Ding, and Zhi Geng. Principal causal effect identification and surrogate end point evaluation by multiple trials. *Journal of the Royal Statistical Society: Series B: Statistical Methodology*, pages 829–848, 2016.

[36] Alessandra Mattei, Fan Li, and Fabrizia Mealli. Exploiting multiple outcomes in bayesian principal stratification analysis with application to the evaluation of a job training program. *The Annals of Applied Statistics*, 7(4):2336–2360, 2013.

[37] Fabrizia Mealli and Barbara Pacini. Using secondary outcomes to sharpen inference in randomized experiments with noncompliance. *Journal of the American Statistical Association*, 108(503):1120–1131, 2013.

[38] Fabrizia Mealli, Barbara Pacini, and Elena Stanghellini. Identification of principal causal effects using additional outcomes in concentration graphs. *Journal of Educational and Behavioral Statistics*, 41(5):463–480, 2016.

[39] Fan Yang and Dylan S Small. Using post-outcome measurement information in censoring-by-death problems. *Journal of the Royal Statistical Society: Series B (Statistical Methodology)*, 78(1):299–318, 2016.

[40] Corwin M Zigler and Thomas R Belin. A bayesian approach to improved estimation of causal effect predictiveness for a principal surrogate endpoint. *Biometrics*, 68(3):922–932, 2012.

[41] Avi Feller, Evan Greif, Luke Miratrix, and Natesh Pillai. Principal stratification in the twilight zone: Weakly separated components in finite mixture models. *arXiv preprint arXiv:1602.06595*, 2016.

[42] Paolo Frumento, Fabrizia Mealli, Barbara Pacini, and Donald B Rubin. The fragility of standard inferential approaches in principal stratification models relative to direct likelihood approaches. *Statistical Analysis and Data Mining: The ASA Data Science Journal*, 9(1):58–70, 2016.

[43] Kosuke Imai. Sharp bounds on the causal effects in randomized experiments with "truncation-by-death". *Statistics & Probability Letters*, 78(2):144–149, 2008.

[44] Junni L Zhang, Donald B Rubin, and Fabrizia Mealli. Evaluating the effects of job training programs on wages through principal stratification. In *Modelling and Evaluating Treatment Effects in Econometrics*. Emerald Group Publishing Limited, 2008.

[45] Jing Cheng and Dylan S Small. Bounds on causal effects in three-arm trials with non-compliance. *Journal of the Royal Statistical Society: Series B (Statistical Methodology)*, 68(5):815–836, 2006.

[46] Leonardo Grilli and Fabrizia Mealli. Nonparametric bounds on the causal effect of university studies on job opportunities using principal stratification. *Journal of Educational and Behavioral Statistics*, 33(1):111–130, 2008.

[47] David S Lee. Training, wages, and sample selection: Estimating sharp bounds on treatment effects. *The Review of Economic Studies*, 76(3):1071–1102, 2009.

[48] Dustin M Long and Michael G Hudgens. Sharpening bounds on principal effects with covariates. *Biometrics*, 69(4):812–819, 2013.

[49] Paul R Rosenbaum and Donald B Rubin. The central role of the propensity score in observational studies for causal effects. *Biometrika*, 70(1):41–55, 1983.

[50] Donald B Rubin. The design versus the analysis of observational studies for causal effects: parallels with the design of randomized trials. *Statistics in Medicine*, 26(1):20–36, 2007.

[51] Donald B Rubin. For objective causal inference, design trumps analysis. *The Annals of Applied Statistics*, 2(3):808–840, 2008.

[52] Jennifer Hill, Jane Waldfogel, and Jeanne Brooks-Gunn. Differential effects of high-quality child care. *Journal of Policy Analysis and Management: The Journal of the Association for Public Policy Analysis and Management*, 21(4):601–627, 2002.

[53] Dean A Follmann. On the effect of treatment among would-be treatment compliers: An analysis of the multiple risk factor intervention trial. *Journal of the American Statistical Association*, 95(452):1101–1109, 2000.

[54] Bruno Crépon, Florencia Devoto, Esther Duflo, and William Parienté. Estimating the impact of microcredit on those who take it up: Evidence from a randomized experiment in morocco. *American Economic Journal: Applied Economics*, 7(1):123–50, 2015.

[55] Peter Z Schochet and John Burghardt. Using propensity scoring to estimate program-related subgroup impacts in experimental program evaluations. *Evaluation Review*, 31(2):95–120, 2007.

[56] Fuhua Zhai, Jeanne Brooks-Gunn, and Jane Waldfogel. Head start's impact is contingent on alternative type of care in comparison group. *Developmental Psychology*, 50(12):2572, 2014.

[57] Booil Jo and Elizabeth A Stuart. On the use of propensity scores in principal causal effect estimation. *Statistics in Medicine*, 28(23):2857–2875, 2009.

[58] Booil Jo, Elizabeth A Stuart, David P MacKinnon, and Amiram D Vinokur. The use of propensity scores in mediation analysis. *Multivariate Behavioral Research*, 46(3):425–452, 2011.

[59] Elizabeth A Stuart and Booil Jo. Assessing the sensitivity of methods for estimating principal causal effects. *Statistical Methods in Medical Research*, 24(6):657–674, 2015.

[60] Raphaël Porcher, Clémence Leyrat, Gabriel Baron, Bruno Giraudeau, and Isabelle Boutron. Performance of principal scores to estimate the marginal compliers causal effect of an intervention. *Statistics in Medicine*, 35(5):752–767, 2016.

[61] Peng Ding and Jiannan Lu. Principal stratification analysis using principal scores. *Journal of the Royal Statistical Society: Series B (Statistical Methodology)*, 79(3):757–777, 2017.

[62] Avi Feller, Fabrizia Mealli, and Luke Miratrix. Principal score methods: Assumptions, extensions, and practical considerations. *Journal of Educational and Behavioral Statistics*, 42(6):726–758, 2017.

[63] Marshall M Joffe, Dylan S Small, and Chi-Yuan Hsu. Defining and estimating intervention effects for groups that will develop an auxiliary outcome. *Statistical Science*, 22(1):74–97, 2007.

[64] Zhichao Jiang, Shu Yang, and Peng Ding. Multiply robust estimation of causal effects under principal ignorability. *arXiv preprint arXiv:2012.01615*, 2020.

[65] Fabrizia Mealli, Guido W Imbens, Salvatore Ferro, and Annibale Biggeri. Analyzing a randomized trial on breast self-examination with noncompliance and missing outcomes. *Biostatistics*, 5(2):207–222, 2004.

[66] Leah Comment, Fabrizia Mealli, Sebastien Haneuse, and Corwin Zigler. Survivor average causal effects for continuous time: A principal stratification approach to causal inference with semicompeting risks. *arXiv preprint arXiv:1902.09304*, 2019.

[67] Laura Forastiere, Fabrizia Mealli, and Tyler J VanderWeele. Identification and estimation of causal mechanisms in clustered encouragement designs: Disentangling bed nets using bayesian principal stratification. *Journal of the American Statistical Association*, 111(514):510–525, 2016.

[68] Constantine E Frangakis, Donald B Rubin, and Xiao-Hua Zhou. Clustered encouragement designs with individual noncompliance: Bayesian inference with randomization, and application to advance directive forms. *Biostatistics*, 3(2):147–164, 2002.

[69] Booil Jo, Tihomir Asparouhov, and Bengt O Muthén. Intention-to-treat analysis in cluster randomized trials with noncompliance. *Statistics in Medicine*, 27(27):5565–5577, 2008.

[70] Booil Jo, Tihomir Asparouhov, Bengt O Muthén, Nicholas S Ialongo, and C Hendricks Brown. Cluster randomized trials with treatment noncompliance. *Psychological Methods*, 13(1):1, 2008.

[71] Donald B Rubin. Causal inference through potential outcomes and principal stratification: application to studies with" censoring" due to death. *Statistical Science*, 21(3):299–309, 2006.

[72] Arthur P Dempster, Nan M Laird, and Donald B Rubin. Maximum likelihood from incomplete data via the em algorithm. *Journal of the Royal Statistical Society: Series B (Methodological)*, 39(1):1–22, 1977.

[73] Martin A Tanner and Wing Hung Wong. The calculation of posterior distributions by data augmentation. *Journal of the American Statistical Association*, 82(398):528–540, 1987.

[74] Beth Ann Griffin, Daniel F McCaffery, and Andrew R Morral. An application of principal stratification to control for institutionalization at follow-up in studies of substance abuse treatment programs. *The Annals of Applied Statistics*, 2(3):1034, 2008.

[75] Paul Gustafson. Bayesian inference for partially identified models. *The International Journal of Biostatistics*, 6(2), Article 17, 2010. doi: 10.2202/1557-4679.1206. PMID: 21972432.

[76] Shasha Han, Larry Han, and Jose R. Zubizarreta. Principal resampling for causal inference. *Mimeo*, 2021.

18

Incremental Causal Effects: An Introduction and Review

Matteo Bonvini, Alec McClean

MB and AM contributed equally.

Zach Branson, Edward H. Kennedy

CONTENTS

In this chapter, we review the class of causal effects based on incremental propensity score interventions proposed by [1]. The aim of incremental propensity score interventions is to estimate the effect of increasing or decreasing subjects' odds of receiving treatment; this differs from the average treatment effect, where the aim is to estimate the effect of everyone deterministically receiving versus not receiving treatment. We first present incremental causal effects for the case when there is a single

DOI: 10.1201/9781003102670-18

binary treatment, such that it can be compared to average treatment effects and thus shed light on key concepts. In particular, a benefit of incremental effects is that positivity – a common assumption in causal inference – is not needed to identify causal effects. Then we discuss the more general case where treatment is measured at multiple time points, where positivity is more likely to be violated and thus incremental effects can be especially useful. Throughout, we motivate incremental effects with real-world applications, present nonparametric estimators for these effects, and discuss their efficiency properties, while also briefly reviewing the role of influence functions in functional estimation. Finally, we show how to interpret and analyze results using these estimators in practice and discuss extensions and future directions.

18.1 Introduction

Understanding the effect of a variable A, the treatment, on another variable Y, the outcome, involves measuring how the distribution of Y changes when the distribution of A is manipulated in some way. By "manipulating a distribution," we mean that we imagine a world where we can change the distribution of A to some other distribution of our choice, which we refer to as the "intervention distribution." The intervention distribution defines the causal effect of A on Y, i.e., the causal estimand. For instance, suppose A is binary and is completely randomized with some probability c_0; one may then ask how the average outcome Y would change if the randomization probability were set to some other constant c_1.

In order to answer questions about causal effects, we adopt the potential outcomes framework ([2]). We suppose that each subject in the population is linked to a number of "potential outcomes," denoted by Y^a, that equal the outcome Y that would have been observed if the subject had received treatment $A = a$. In practice, only one potential outcome is observed for each subject, because one cannot go back in time and assign a different treatment value to the same subject. As a result, we must make several assumptions to identify and estimate causal effects. One common assumption is "positivity," which says that each subject in the population must have a non-zero chance of receiving each treatment level. Positivity is necessary to identify the average treatment effect, which is the most common causal estimand in the literature; however, positivity may be very unrealistic in practice, particularly in many time-point analyses where the number of possible treatment regimes scales exponentially with the number of time-points.

In this chapter we introduce and review the class of "incremental causal effects" proposed in [1]. Incremental causal effects are based on an intervention distribution that shifts the odds of receiving a binary treatment by a user-specified amount. Crucially, incremental causal effects are well-defined and can be efficiently estimated even when positivity is violated. Furthermore, they represent an intuitive way to summarize the effect of the treatment on the outcome, even in high-dimensional, time-varying settings. In Section 18.2 we first review how to identify and estimate the average treatment effect, as well as review the notion of static interventions, dynamic interventions, and stochastic interventions. A limitation of the average treatment effect is that it only considers a static and deterministic intervention, where all subjects are either assigned to treatment or control, which may be improbable if positivity is violated or nearly violated. On the other hand, dynamic and stochastic interventions allow the treatment to vary across subjects, thereby lessening dependence on positivity. Incremental causal effects assume a particular stochastic intervention, which we discuss in depth for binary treatments in Section 18.3 before discussing time-varying treatments in Section 18.4. We then demonstrate how to use incremental effects in practice in Section 18.5. We conclude with a discussion of extensions and future directions for incremental causal effects in Section 18.6.

18.2 Preliminaries: Potential Outcomes, the Average Treatment Effect, and Types of Interventions

In this section to introduce and focus ideas, we consider studies with a single binary exposure of interest. Longitudinal studies with time-varying exposures are presented in Section 18.4. Here, for each of n subjects, we observe a binary treatment A, a vector of covariates X, and an outcome Y, each indexed by $i = 1, \ldots, n$. The goal is to estimate the causal effect of the treatment A on the outcome Y, with the complication that subjects self-select into treatment in a way that may depend on the covariates X. A fundamental quantity in this setting is the probability of receiving treatment $\pi(X) \equiv \mathbb{P}(A = 1 \mid X)$, known as the propensity score. We also let $\mathbb{1}(\mathcal{E})$ be a binary indicator for whether event $\mathcal{E}$ occurs.

When the treatment is binary and there is no interference, each subject has only two potential outcomes, Y^1 and Y^0. As a running example, we will consider estimating the effect of behavioral health services on probationers' likelihood of re-arrest. Recidivism is a colossal societal issue in the United States. Millions of people are on probation in the U.S. and around half of the people who leave prison are re-arrested within one year [3,4]. Further, outcomes may be even worse for people with mental illness or substance use disorders [5,6]. Because of this, some researchers have posited that attending behavioral health services (e.g., talk therapy) during probation may help probationers avoid criminal behavior and being re-arrested. In this example the covariate information X for each probationer includes demographic information such as age, gender, and race, the treatment A denotes whether a probationer attended a behavioral health service over, say, the first six months of probation, and the outcome Y denotes whether they were re-arrested within a year. The goal in this example is to assess if attending services lowers one's chance of being re-arrested. In Section 18.4 we consider the more complex case where the covariates X, treatment A, and outcome Y can be time-varying.

First we will consider estimating the average treatment effect for this example. The average treatment effect implies a so-called *static deterministic* intervention that assumes two extreme counterfactual scenarios, where either every subject receives treatment ($A = 1$) or every subject receives control ($A = 0$), which may not be realistic for many applications (including the above example from criminology). Then, we will consider estimating incremental effects, which instead posit a stochastic dynamic intervention that interpolates counterfactual scenarios in between the "everyone receives treatment" and "everyone receives control" scenarios.

18.2.1 Average treatment effects

We first review the *average treatment effect* (ATE), because it is by far the most common causal estimand of interest. The ATE is defined as:

$$\text{ATE} \equiv \mathbb{E}(Y^1 - Y^0). \tag{18.1}$$

The ATE compares two quantities: $\mathbb{E}(Y^1)$, the mean outcome when all subjects receive treatment; and $\mathbb{E}(Y^0)$, the mean outcome when all subjects receive control. Within our example, these are the recidivism rate when everyone attends behavioral health services and the recidivism rate when no one attends behavioral health services, respectively.

The fundamental problem of causal inference ([7]) is that each subject receives either treatment or control – never both – and thus we only observe one of the two potential outcomes for each subject. In other words, the difference $Y^1 - Y^0$ is only partially observed for each subject. Thus, assumptions must be made in order to identify and thereby estimate the ATE. It is common to make the following three assumptions to estimate the ATE:

Assumption 17. *(Consistency). $Y = Y^a$ if $A = a$.*

Assumption 18. *(Exchangeability).* $A \perp\!\!\!\perp \{Y\}^a \mid X$ *for* $a = 1$ *and* $a = 0$.

Assumption 19. *(Positivity). There exists* $\epsilon > 0$ *such that* $\mathbb{P}(\epsilon \leq \pi(X) \leq 1 - \epsilon) = 1$.

Consistency says that if an individual takes treatment a, we observe their potential outcome under that treatment; consistency allows us to write the observed outcomes as $Y = AY^1 + (1 - A)Y^0$ and would be violated if, for example, there were interference between subjects, such that one subject's treatment affected another subject's outcome. Meanwhile, exchangeability says that treatment is effectively randomized within covariate strata, in the sense that treatment is independent of subjects' potential outcomes – as in a randomized experiment – after conditioning on covariates. This assumption is also called "no unmeasured confounding," which means that there are no variables beyond X that induce dependence between treatment and the potential outcomes. Finally, positivity says that the propensity score is bounded away from zero and one for all subjects. With these three assumptions, we have

$$\mathbb{E}(Y^a) = \mathbb{E}\{\mathbb{E}(Y^a \mid X)\} = \mathbb{E}\{\mathbb{E}(Y^a \mid X, A = a)\} = \mathbb{E}\{\mathbb{E}(Y \mid X, A = a)\} = \mathbb{E}\left\{\frac{Y\mathbb{1}(A = a)}{\pi(X)}\right\},$$

i.e., the mean outcome if all were assigned $A = a$ reduces to the regression function $\mathbb{E}(Y \mid X, A = a)$, averaged over the distribution of X. The average treatment effect is thus ATE $= \mathbb{E}\{\mathbb{E}(Y \mid X, A = 1) - \mathbb{E}(Y \mid X, A = 0)\}$.

18.2.2 When positivity is violated

We will give special attention to positivity, for several reasons; it is often violated in practice ([8,9]), and it is also the only causal assumption that is, in principle, testable. It is important to remember that the ATE characterizes what would happen in two extreme counterfactual scenarios: one counterfactual world where every subject receives treatment, versus another where every subject receives control. Although we never observe either of these two extreme scenarios, they nonetheless must be plausible in order for the ATE to be a sensible estimand to target. In other words, it must be plausible that every subject could receive treatment or control – i.e., positivity must hold. Positivity would be violated if there are subjects for whom some treatment level is essentially impossible to receive. For example, in our running criminology example, some probationers might be required to attend behavioral health services as a condition of their probation, and thus their probability of attending treatment is one; conversely some probationers may be extremely unlikely to go to treatment. For both these groups, positivity would not hold. In this case, the ATE in (18.1) may not be a sensible estimand to target, since there is essentially zero chance of observing Y^0 or Y^1 for some subjects.

In some settings researchers may know *a priori* that positivity does not hold for a particular application, while in others they may be concerned that positivity does not hold after conducting preliminary exploratory data analysis (e.g., if estimated propensity scores for some subjects are close to zero or one). Furthermore, even if positivity technically holds in a population, near-positivity violations in a sample can have disastrous effects on estimators for the ATE and related estimands. For example, it is well-known that inverse propensity score weighting and doubly robust estimators can suffer from high variance when propensity scores are extreme ([10, 11]). This does not mean that these estimators are unsatisfactory – when propensity scores are close to zero or one, at least one of the two extreme scenarios the ATE considers ("everyone receives treatment") or ("everyone receives control") is unlikely, and high variance is thus an appropriate signal of the uncertainty we have about those scenarios. Simple regression-based estimators may mask this high variance, but really only through extrapolation, which merely trades high variance for high bias. In fact, it can be shown ([12]) that, under exchangeability, no regular estimator of the ATE can have asymptotic variance smaller than

$$\mathbb{E}\Big(\frac{\text{Var}(Y^1 \mid X)}{\pi(X)} + \frac{\text{Var}(Y^0 \mid X)}{1-\pi(X)} + \big[\mathbb{E}(Y^1 - Y^0 \mid X) - \mathbb{E}\{\mathbb{E}(Y^1 - Y^0 \mid X)\}\big]^2\Big).$$

This is called the semiparametric efficiency bound, which is the equivalent notion to the Cramer-Rao lower bound in parametric models. See [13] for a review on this topic and the precise definition of regular estimators. Because the efficiency bound involves the reciprocal of the propensity score, it is clear that near-positivity violations have a deleterious effect on the precision with which the ATE can be estimated.

In the presence of positivity violations where there are substantial limits on how well one can estimate the ATE in (18.1), an alternative option is to target different estimands. For example, matching methods restrict analyses to a subsample that exhibits covariate balance between treatment and control, and thus positivity can be more plausible for that subsample ([14, 15]). In this case, the targeted estimand would be the ATE *for the matched subsample* instead of the ATE in the entire population. Similarly, propensity score trimming aims to remove subjects for whom propensity scores are outside the interval $[\epsilon, 1-\epsilon]$, for some user-specified $\epsilon > 0$ ([16]). If the true propensity scores were known, this trimming would ensure positivity holds by design, and the targeted estimand is the ATE *for the trimmed subsample* instead of the ATE. A benefit of these approaches is that standard ATE estimators can still be used, simply within a subsample. However, complications arise because the estimand then depends on the sample: The subsample ATE is only defined after a matching algorithm is chosen or propensity scores are estimated. Thus, interpretability may be a concern, because the causal effect is only defined for the sample at-hand instead of the broader population of interest. A way to overcome this shortcoming is to define the estimand based on the trimmed true propensity scores, i.e. $\mathbb{E}\{Y^1 - Y^0 \mid \pi(X) \in [\epsilon, 1-\epsilon]\}$. However, estimating the quantity $\mathbb{1}\{\pi(X) \in [\epsilon, 1-\epsilon]\}$ is challenging in flexible, nonparametric models because it is a non-smooth transformation of the data-generating distribution; as such, $\sqrt{n}$-consistent estimators do not generally exist without imposing further assumptions.

The estimands discussed so far – the ATE in (18.1) and subsample ATEs implied by matching or propensity score trimming – all represent *static interventions*, where the treatment A is set to fixed values across a population of interest. In what follows, we will consider alternative estimands that researchers can target in the face of positivity violations. These estimands concern effects of *dynamic* (non-static) interventions, where, in counterfactual worlds, the treatment can be allowed to vary across subjects, instead of being set to fixed values. Indeed, the static interventions discussed so far – where every subject receives treatment or every subject receives control – may be impossible to implement for many applications, whereas non-static interventions may be closer to what we would expect to be possible in practice.

18.2.3 Dynamic interventions

Dynamic interventions allow the intervention to depend on subjects' covariate information; thus, the intervention is allowed to vary across subjects, depending on their covariate values, albeit still in a deterministic way. Dynamic interventions often arise in medical situations, where treatment is only given to people with severe symptoms, and thus positivity only holds for a subset of the population. In this sense dynamic interventions are quite similar to the interventions implied by matching or propensity score trimming, discussed at the end of the previous subsection. In particular matching or propensity score trimming considers a *static* intervention that only occurs within a subpopulation (which is often defined in terms of covariates or propensity scores), whereas dynamic interventions consider an intervention that occurs across the whole population, but the treatment for each subject will depend on their covariate values.

As an example, let's say we want to measure the effect of providing behavioral health services only for probationers for whom positivity holds, i.e., for probationers for whom $\mathbb{P}(A = 1 \mid X) \in [\epsilon, 1-\epsilon]$. In other words we would like to answer the question, "What is the causal effect of providing behavioral health services to probationers who can plausibly choose whether or not to attend services?" We could address this question by considering the following dynamic intervention:

$$d_a(X) = \begin{cases} a & \text{if } \mathbb{P}(A = 1|X) \in [\epsilon, 1-\epsilon] \\ 1 & \text{if } \mathbb{P}(A = 1|X) > 1-\epsilon \\ 0 & \text{if } \mathbb{P}(A = 1|X) < \epsilon \end{cases}, \quad a \in \{0, 1\}$$

A similar dynamic intervention was discussed in [17]. Under this intervention, we set subjects to treatment a when positivity holds; otherwise, treatment is fixed at $a = 1$ for subjects that will almost certainly receive treatment and $a = 0$ for subjects that will almost certainly receive control. In this case the causal estimand is $\mathbb{E}\left\{Y^{d_1(X)} - Y^{d_0(X)}\right\}$, where $\mathbb{E}\left\{Y^{d_a(X)}\right\}$ denotes the average outcome when A is set according to $d_a(X)$ across subjects. Note that $\mathbb{E}\left\{Y^{d_1(X)}\right\}$ differs from $\mathbb{E}(Y^1)$ in the ATE (18.1), to the extent that there are subjects for whom $\mathbb{P}(A = 1 \mid X) < \epsilon$, in which case some subjects receive control when estimating $\mathbb{E}\left\{Y^{d_1(X)}\right\}$ but not $\mathbb{E}(Y^1)$. In this case, $\mathbb{E}\left\{Y^{d_1(X)} - Y^{d_0(X)}\right\}$ is equivalent to the ATE within a propensity score trimmed subsample, but the hypothetical intervention is employed across the entire population.

More generally, a dynamic intervention implies causal effects on the subpopulation whose treatment is affected by that dynamic intervention. Because practitioners may realistically only be able to intervene on a subset of the population, causal effects implied by dynamic interventions may be relevant for many applications. That said, dynamic interventions assume that we can deterministically assign treatment or control to all subjects as a function of their covariates. In reality treatment assignment may still be stochastic when an intervention is employed. In this case the intervention is considered a stochastic intervention.

18.2.4 Stochastic interventions

A *stochastic intervention* assigns treatment randomly based on a probability distribution. As a simple example, consider a Bernoulli trial, where we assign each subject to treatment with probability p, and otherwise we assign them to control. In this case the probability distribution of treatment for each subject is $Q(a) = p^a(1-p)^{1-a}$ for $a \in \{0, 1\}$, and we might consider the causal effect of changing the parameter p. Under this intervention, although we control subjects' probability of receiving treatment, the actual realization of treatment is stochastic, unlike the static and dynamic interventions discussed so far. Stochastic interventions can be considered generalizations of static and dynamic interventions, and the static intervention implied by the ATE in (18.1) is a special case of the Bernoulli trial example, which contrasts the average outcome when $p = 1$ versus when $p = 0$.

Stochastic interventions are perhaps closest to what can be realistically implemented in many applications. For example, in our criminology example, it's difficult to imagine a world where we could deterministically force probationers to attend or not attend behavioral health services; however, we may well be able to affect probationers' likelihood of attending services (e.g., by providing public transportation stipends, thereby improving the accessibility of services). In this case we may be interested in measuring the effect of making services more accessible, which can be viewed as a proxy for the effect of increasing p in the Bernoulli trial example. As we will see in Section 18.3, incremental effects assume a particular stochastic intervention that is similar to the Bernoulli trial example and approximates many interventions that we may expect to see in practice. Thus, before discussing incremental effects specifically, we outline some fundamental ideas for stochastic interventions more generally.

Let $Q(a \mid x)$ denote the probability distribution of treatment for a stochastic intervention, and let $\mathbb{E}(Y^Q)$ denote the average outcome under this intervention. The quantity $Q(a \mid x)$ is also known as the *intervention distribution*, because it is a distribution specified by an intervention. This intervention can depend on the covariates x, and in this case, it is a *dynamic* stochastic intervention. Or, we can have $Q(a \mid x) = Q(a)$ for all x, which would be a static (but potentially stochastic) intervention. The quantity $\mathbb{E}(Y^Q)$ can then be written as a weighted average of the various potential outcomes Y^a, with weights given by $Q(a \mid x)$:

$$\mathbb{E}(Y^Q) = \int_{\mathcal{A}} \int_{\mathcal{X}} \mathbb{E}(Y^a \mid X = x)\, dQ(a \mid x)\, d\mathbb{P}(x) \tag{18.2}$$

Causal effects under stochastic interventions are often framed as contrasting different $\mathbb{E}(Y^Q)$ for different $Q(a)$ distributions. For example, as stated earlier, the ATE in (18.1) can be viewed as contrasting $\mathbb{E}(Y^Q)$ for two different point-mass distributions, where every subject receives treatment or every subject receives control.

Under consistency and exchangeability (Assumptions 17 and 18), we can identify (18.2) as

$$\mathbb{E}(Y^Q) = \int_{\mathcal{A}} \int_{\mathcal{X}} \mathbb{E}(Y \mid X = x, A = a) dQ(a \mid x) d\mathbb{P}(x) \tag{18.3}$$

where $\mathcal{A}$ denotes the set of all possible treatment values, and $\mathcal{X}$ are all possible covariate values. Note that the notation $dQ(a \mid x)$ acknowledges that the intervention distribution may depend on covariates. In the binary treatment case, this reduces to

$$\mathbb{E}(Y^Q) = \int_{\mathcal{X}} \Big\{ \mathbb{E}(Y \mid X = x, A = 0) Q(A = 0 \mid x) + \mathbb{E}(Y \mid X = x, A = 1) Q(A = 1 \mid x) \Big\} d\mathbb{P}(x).$$

Notably, positivity (Assumption 19) may not be needed to identify (18.2), depending on the definition of Q. We now turn to incremental propensity score interventions, which are stochastic interventions defining an intuitive causal estimand that does not rely on positivity for identification.

18.3 Incremental Propensity Score Interventions

An *incremental propensity score intervention*, first introduced in [1], is a stochastic dynamic intervention in which the odds of receiving treatment under the observed distribution $\mathbb{P}$ are multiplied by δ. In other words this causal effect asks what would happen if everyone's odds of receiving treatment were multiplied by δ. That is, letting Q denote the intervention distribution:

$$\delta = \frac{Q(A = 1 \mid X)/(1 - Q(A = 1 \mid X)}{\mathbb{P}(A = 1 \mid X)/(1 - \mathbb{P}(A = 1 \mid X)} = \frac{\text{odds}_Q(A = 1 \mid X)}{\text{odds}_{\mathbb{P}}(A = 1 \mid X)}$$

This implies that, for $0 < \delta < \infty$, the probability of receiving treatment under Q is

$$Q(A = 1 \mid x) \equiv q(x; \delta, \pi) = \frac{\delta \pi(x)}{\delta \pi(x) + 1 - \pi(x)} \tag{18.4}$$

where $\pi(x) = \mathbb{P}(A = 1 \mid X = x)$. In our recidivism example this intervention would shift each probationer's probability of attending a healthcare service. The incremental parameter δ is user-specified and controls the direction and magnitude of the propensity score shift. It tells us how much the intervention changes the odds of receiving treatment.

For example, if $\delta = 1.5$, then the intervention increases the odds of treatment by 50% for everyone. If $\delta = 1$, then we are left with the observation propensity scores, and $q(x; \delta, \pi) = \pi(x)$. As δ increases from 1 towards ∞, the shifted propensity scores approach 1, and as δ decreases from 1 towards 0, the shifted propensity scores approach 0. There are other interventions one might consider, but shifting the odds of treatment is a simple intervention that gives an intuitive explanation for the parameter δ.

Remark 18.1. *It is possible to let δ depend on X, thereby allowing the intervention distribution Q to modify the treatment process differently based on subjects' covariates. This would lead to more nuanced definitions of treatment effects, potentially at the cost of losing straightforward interpretation of the estimated effects. In fact, taking δ to be constant is not an assumption; it just defines the particular causal estimand that is targeted for inference.*

Incremental propensity score interventions allow us to avoid the tricky issues with positivity that were discussed in Section 18.2. There are two groups of people for whom the positivity assumption is violated: people who never attend treatment ($\pi = 0$) and people who always attend treatment ($\pi = 1$). Incremental interventions avoid assuming positivity for these groups because the intervention leaves their propensity score unchanged: It has the useful property that $\pi = 0 \implies q = 0$ and $\pi = 1 \implies q = 1$.

We cannot know *a priori* whether positivity is violated, so this intervention allows us to compute effects on our data without worrying whether positivity holds. Thus, this intervention differs from the dynamic intervention in Section 18.2, because we do not make our intervention depend on the propensity score of each individual in our sample. Practically, this means we could still include in our sample people who always attend or never attend treatment; e.g., in our running recidivism example, we could still include individuals who must attend treatment as part of their probation, and the causal effect is still well-defined.

Remark 18.2. *The reader may wonder why we do not consider simpler interventions, such as $q(x; \delta, \pi) = \pi + \delta$ or $q(x; \delta, \pi) = \pi \cdot \delta$. One reason is that these interventions require the range of δ to depend on the distribution $\mathbb{P}$, because otherwise $q(x; \delta, \pi)$ may fall outside the unit interval. In contrast, for any δ, the incremental propensity score intervention constrains Q so that $0 \leq q(x; \delta, \pi) \leq 1$.*

Remark 18.3. *If positivity holds, incremental interventions contain the ATE as a special case. If $\pi(x)$ is bounded away from zero and one almost surely, then $\mathbb{E}(Y^Q)$ tends to $\mathbb{E}(Y^1)$ as $\delta \to \infty$ and to $\mathbb{E}(Y^0)$ as $\delta \to 0$. Thus, both $\mathbb{E}(Y^1)$ and $\mathbb{E}(Y^0)$ can be approximated by taking δ to be very large or very small, and the ATE can be approximated by taking their contrast.*

18.3.1 Identification

Under consistency and exchangeability (Assumptions 17 and 18), we can plug the incremental intervention distribution $Q(a \mid x) = q(x; \delta, \pi)^a \{1 - q(x; \delta, \pi)\}^{1-a}$ into equation (18.3) to derive an identification expression for $\psi(\delta) = \mathbb{E}\{Y^{Q(\delta)}\}$, the expected outcome if the treatment distribution is intervened upon and set to $Q(a \mid x)$.

Theorem 18.1. Under Assumptions 17 and 18, and if $\delta \in (0, \infty)$, the incremental effect $\psi(\delta) = \mathbb{E}\{Y^{Q(\delta)}\}$ for the propensity score intervention as defined in equation (18.4) satisfies

$$\psi(\delta) = \mathbb{E}[\frac{\delta\pi(X)\mu(X,1) + \{1 - \pi(X)\}\mu(X,0)}{\delta\pi(X) + \{1 - \pi(X)\}}] = \mathbb{E}[\frac{Y(\delta A + 1 - A)}{\delta\pi(X) + \{1 - \pi(X)\}}] \tag{18.5}$$

where $\mu(x, a) = \mathbb{E}(Y \mid X = x, A = a)$.

Theorem 18.1 offers two ways to link the incremental effect $\psi(\delta)$ to the data generating distribution $\mathbb{P}$. The first involves both the outcome regressions $\mu(x,a)$ and the propensity score $\pi(x)$, and the second just the propensity score. The former is a weighted average of the regression functions $\mu(x,1)$ and $\mu(x,0)$, where the weight on $\mu(x,a)$ is given by the fluctuated intervention propensity score $Q(A=1 \mid x)$, while the latter is a weighted average of the observed outcomes, where the weights are related to the intervention propensity score and depend on the observed treatment.

Remark 18.4. *From the identification expression in equation* (18.5), *one can notice that even if there are subjects for which* $\mathbb{P}(A=a \mid X)=0$, *so that* $\mu(X,a)=\mathbb{E}(Y \mid X, A=a)$ *is not defined because of conditioning on a zero-probability event,* $\psi(\delta)$ *is still well-defined, because those subjects receive zero weight when the expectation over the covariates distribution is computed.*

18.3.2 Estimation

Theorem 18.1 provides two formulas to link the causal effect $\psi(\delta)$ to the data generating distribution $\mathbb{P}$. The next step is to estimate $\psi(\delta)$ relying on these identification results. The first estimator, which we call the "outcome-based estimator," includes estimates for both μ and π:

$$\widehat{\psi}(\delta)=\frac{1}{n}\sum_{i=1}^{n}\left[\frac{\delta\widehat{\pi}(X_i)\widehat{\mu}(X_i,1)+\{1-\widehat{\pi}(X_i)\}\widehat{\mu}(X_i,0)}{\delta\widehat{\pi}(X_i)+\{1-\widehat{\pi}(X_i)\}}\right]$$

The second estimator motivated by the identification result is the inverse-probability-weighted (IPW) estimator:

$$\widehat{\psi}(\delta)=\frac{1}{n}\sum_{i=1}^{n}\left[\frac{Y_i(\delta A_i+1-A_i)}{\delta\widehat{\pi}(X_i)+\{1-\widehat{\pi}(X_i)\}}\right]$$

Both estimators are generally referred to as "plug-in" estimators, since they take estimates for $\widehat{\mu}$ or $\widehat{\pi}$ and plug them directly into the identification results. If we assume we can estimate μ and π with correctly specified parametric models, then both estimators inherit parametric rates of convergence and can be used to construct valid confidence intervals.

However, in practice, parametric models are likely to be misspecified. Thus, we may prefer to estimate the nuisance regression functions μ and π with nonparametric models. However, if nonparametric models are used for either of the plug-in estimators without any correction, the estimator are generally sub-optimal, because they inherit the large bias incurred in estimating the regression functions in large classes. For example, the best possible convergence rate in mean-square-error of an estimator of a regression function that belongs to a Hölder class of order α, essentially a function that is α-times differentiable, scales at $n^{-2\alpha/(2\alpha+d)}$, where n is the sample size and $d \geq 1$ is the dimension of the covariates ([18]). This rate is slower than n^{-1} for any α and d. Because plug-in estimators with no further corrections that use agnostic nuisance function estimators typically inherit this "slow rate," they lose n^{-1} efficiency if the nuisance functions are not estimated parametrically.

Remark 18.5. *We remind readers that the issues regarding plug-in estimators outlined in the previous paragraph are not isolated to incremental interventions; they apply generally to plug-in estimators of functionals. So, for example, these problems also apply to the outcome-based and IPW estimators for the ATE.*

Semiparametric efficiency theory provides a principled way to construct estimators that make a more efficient use of flexibly estimated nuisance functions ([19–23]).[1] Such estimators are based

[1] For a gentle introduction to the use of influence functions in functional estimation, we refer to [24] and the tutorial at `http://www.ehkennedy.com/uploads/5/8/4/5/58450265/unc_2019_cirg.pdf`

on *influence functions* and are designed for parameters that are "smooth" transformations of the distribution $\mathbb{P}$. The parameter $\psi(\delta)$ is an example of a smooth parameter. The precise definition of smoothness in this context can be found in Chapter 25 of [20]. Informally, however, we can note that $\psi(\delta)$ only involves differentiable functions of π and μ, thereby suggesting that it is a smooth parameter.

If we let $\Phi(\mathbb{P}) \in \mathbb{R}$ denote some smooth target parameter, we can view $\Phi(\mathbb{P})$ as a functional acting on the space of distribution functions. One key feature of smooth functionals is that they satisfy a functional analog of a Taylor expansion, sometimes referred to as the *von Mises expansion.* Given two distributions $\mathbb{P}_1$ and $\mathbb{P}_2$, the von Mises expansion of $\Phi(\mathbb{P})$ is

$$\Phi(\mathbb{P}_1) - \Phi(\mathbb{P}_2) = \int \phi(z; \mathbb{P}_1)\{d\mathbb{P}_1(z) - d\mathbb{P}_2(z)\} + R(\mathbb{P}_1, \mathbb{P}_2)$$

where $\phi(z; \mathbb{P})$ is the (mean-zero) influence function and R is a second-order remainder term. For the purpose of estimating $\Phi(\mathbb{P})$, the above expansion is useful when $\mathbb{P}_2 = \mathbb{P}$ denotes the true data-generating distribution and $\mathbb{P}_1 = \widehat{\mathbb{P}}$ is its empirical estimate. Thus, the key step in constructing estimators based on influence functions is to express the first-order bias of plug-in estimators as an expectation with respect to $\mathbb{P}$ of a particular, fixed function of the observations and the nuisance parameters, referred to as the influence function. Because the first-order bias is expressed as an expectation with respect to the data-generating $\mathbb{P}$, it can be estimated with error of order n^{-1} simply by replacing $\mathbb{P}$ by the empirical distribution. This estimate of the first-order bias can be subtracted from the plug-in estimator, so that the resulting estimator will have second-order bias without any increase in variance. This provides an explicit recipe for constructing "one-step" bias-corrected estimators of the form

$$\widehat{\Phi}(\mathbb{P}) = \Phi(\widehat{\mathbb{P}}) + \frac{1}{n}\sum_{i=1}^{n} \phi(Z_i; \widehat{\mathbb{P}})$$

If $\phi(Z; \mathbb{P})$ takes the form $\varphi(Z; \mathbb{P}) - \Phi(\mathbb{P})$, then the estimator simplifies to $\widehat{\Phi}(\mathbb{P}) = n^{-1}\sum_{i=1}^{n} \varphi(Z_i; \widehat{\mathbb{P}})$. Remarkably, there are other ways of doing asymptotically equivalent bias corrections, such as targeted maximum likelihood ([25]).

The functional $\psi(\delta)$ satisfies the von Mises expansion for

$$\begin{aligned} \phi(Z; \mathbb{P}) &= \frac{\delta\pi(X)\phi_1(Z) + \{1 - \pi(X)\}\phi_0(Z)}{\delta\pi(X) + \{1 - \pi(X)\}} + \frac{\delta\gamma(X)\{A - \pi(X)\}}{\{\delta\pi(X) + 1 - \pi(X)\}^2} - \psi(\delta) \\ &\equiv \varphi(Z; \mathbb{P}) - \psi(\delta) \end{aligned}$$

where $\gamma(X) = \mu(X, 1) - \mu(X, 0)$ and

$$\phi_a(Z) = \frac{\mathbb{1}(A = a)}{\mathbb{P}(A = a \mid X)}\{Y - \mu(X, a)\} + \mu(X, a).$$

To highlight that $\varphi(Z; \mathbb{P})$ depends on $\mathbb{P}$ through π and μ and that it also depends on δ, we will write $\varphi(Z; \mathbb{P}) \equiv \varphi(Z; \delta, \pi, \mu)$. The influence-function-based one-step estimator of $\psi(\delta)$ is therefore

$$\widehat{\psi}(\delta) = \frac{1}{n}\sum_{i=1}^{n} \varphi(Z_i; \delta, \widehat{\pi}, \widehat{\mu}) \tag{18.6}$$

Remark 18.6. *The function* $\phi_a(Z)$ *is the un-centered influence function for the parameter* $\mathbb{E}\{\mu(X, a)\}$*, and is thus part of the influence function for the ATE under exchangeability (Assumption 2), which can be expressed as* $\mathbb{E}\{\mu(X, 1)\} - \mathbb{E}\{\mu(X, 0)\}$*. Thus, one may view the first* *term in the expression for $\widehat{\psi}(\delta)$ roughly as a weighted mean of the influence functions for the ATE. The second term in the expression arises from the need to estimate $\pi(X)$.*

It can be advantageous to construct $\widehat{\pi}$ and $\widehat{\mu}$ on a sample that is independent from that used to average $\varphi(Z;\delta,\widehat{\pi},\widehat{\mu})$, such that the observations Z_i in (18.6) are not used to estimate π and μ. The role of the two samples can then be swapped so that two estimates of $\psi(\delta)$ are computed and their average is taken to be the final estimate. This technique, known as cross-fitting, has a long history in statistics ([26–30]) and is useful to preserve convergence to a Gaussian distribution while avoiding restrictions on the class of nuisance functions (π and μ) and their estimators; see also Remark 18.8.

The full estimator of $\psi(\delta)$ that incorporates sample splitting is detailed in Algorithm 1:

Algorithm 1 Split the data into K folds (e.g., 5), where fold $k \in \{1,\ldots,K\}$ has n_k observations. For each fold k:

1. Build models $\widehat{\pi}_{-k}(X)$, $\widehat{\mu}_{1,-k}(X)$ and $\widehat{\mu}_{0,-k}(X)$ using the observations **not** contained in fold k.
2. For each observation j in fold k, calculate its un-centered influence function $\varphi(Z_j;\delta,\widehat{\pi}_{-k}(X_j),\widehat{\mu}_{-k}(X_j))$ using the models $\widehat{\pi}_{-k}$ and $\widehat{\mu}_{-k}$ calculated in the Step 1.
3. Calculate an estimate for $\psi(\delta)$ within fold k by averaging the estimates of the un-centered influence function:

$$\widehat{\psi}_k(\delta) = \frac{1}{n_k}\sum_{j\in k}\varphi\{Z_j;\ \delta,\widehat{\pi}_{-k}(X_j),\widehat{\mu}_{-k}(X_j)\}$$

Calculate the final estimate $\widehat{\psi}(\delta)$ as the average of the K estimates from each fold:

$$\widehat{\psi} = \frac{1}{K}\sum_{k=1}^{K}\widehat{\psi}_k(\delta)$$

An implementation of the estimator described in Algorithm 1 can be found in the `R` package `npcausal` as the function `ipsi`. The package can be installed from Github at `https://github.com/ehkennedy/npcausal`.

18.3.3 Properties of the estimator

18.3.3.1 Pointwise inference

Consider the estimator described in Algorithm 1; for each fold k, π and μ are estimated using all units except those in fold k. The units outside fold k are independent of the units in fold k, and the units in fold k are used to calculate the sample average. This means that, under mild regularity conditions, the estimators from each fold have variances of order n^{-1} conditional on the training sample, as they are sample averages of independent observations. Conditioning on the training sample, the bias can be upper bounded by a multiple of

$$B \equiv \int\{\pi(x)-\widehat{\pi}(x)\}^2 d\mathbb{P}(x) + \int\{\pi(x)-\widehat{\pi}(x)\}\max_a\{\mu_a(x)-\widehat{\mu}_a(x)\}d\mathbb{P}(x) \tag{18.7}$$

This bias term is second-order. Thus, the bias can be $o_{\mathbb{P}}(n^{-1/2})$, and of smaller order than the standard error, even if π and μ are estimated at slower rates. If the bias is asymptotically negligible

($o_{\mathbb{P}}(n^{-1/2})$), then

$$\frac{\sqrt{n}\{\widehat{\psi}(\delta)-\psi(\delta)\}}{\sigma(\delta)}=\frac{\sqrt{n}\{\widetilde{\psi}(\delta)-\psi(\delta)\}}{\sigma(\delta)}+o_{\mathbb{P}}(1)\rightsquigarrow N(0,1),$$

where $\sigma^2(\delta)=\text{Var}\{\varphi(Z;\delta,\pi,\mu)\}$ and $\widetilde{\psi}(\delta)=n^{-1}\sum_i\varphi(Z_i;\delta,\pi,\mu)$. Given a consistent estimator of the variance $\widehat{\sigma}^2(\delta)$, by Slutsky's theorem,

$$\frac{\sqrt{n}\{\widehat{\psi}(\delta)-\psi(\delta)\}}{\widehat{\sigma}(\delta)}\rightsquigarrow N(0,1). \tag{18.8}$$

The bias term in (18.7) can be $o_{\mathbb{P}}(n^{-1/2})$ even when the nuisance functions are estimated nonparametrically with flexible machine learning methods. By the Cauchy-Schwarz inequality,

$$|B|\leq\|\pi-\widehat{\pi}\|^2+\|\pi-\widehat{\pi}\|\max_a\|\mu_a-\widehat{\mu}_a\|,$$

where $\|f\|^2=\int f^2(z)d\mathbb{P}(z)$. Therefore, it is sufficient that the product of the integrated MSEs for estimating π and μ_a converge to zero faster than $n^{-1/2}$ to ensure that the bias is asymptotically negligible. This can happen, for instance, if both π and μ_a are estimated at faster than $n^{-1/4}$ rates, which is possible in nonparametric models under structural conditions such as sparsity or smoothness.

The convergence statement in (18.8) provides an asymptotic Wald-type $(1-\alpha)$-confidence interval for $\psi(\delta)$,

$$\widehat{\psi}(\delta)\pm z_{1-\alpha/2}\cdot\frac{\widehat{\sigma}(\delta)}{\sqrt{n}}, \tag{18.9}$$

where z_τ is the τ-quantile of a standard normal. This confidence interval enables us to conduct valid inference for $\psi(\delta)$ for a particular value of δ.

Remark 18.7. *If π and μ_a both lie a Hölder class of order α, then the best estimators in the class have MSEs of order $n^{-2\alpha/(2\alpha+d)}$, where d is the dimension of the covariates ([18]). Therefore, the bias term would be of order $o_{\mathbb{P}}(n^{-1/2})$ and asymptotically negligible whenever $d<2\alpha$. Similarly, if μ_a follows a s-sparse linear model and π a s-sparse logistic model, [31] shows that $s^2(\log d)^{3+2\delta}=o(n)$, for some $\delta>0$, is sufficient for the bias to be asymptotically negligible.*

Remark 18.8. *To see why estimating the nuisance functions on an independent sample may help, consider the following expansion for $\widehat{\psi}(\delta)$ defined in* (18.6)*:*

$$\begin{aligned}\widehat{\psi}(\delta)-\psi(\delta)=&(\mathbb{P}_n-\mathbb{P})\{\varphi(Z;\delta,\widehat{\pi},\widehat{\mu})-\varphi(Z;\delta,\pi,\mu)\}+(\mathbb{P}_n-\mathbb{P})\{\varphi(Z;\delta,\pi,\mu)\}\\&+\mathbb{P}\{\varphi(Z;\delta,\widehat{\pi},\widehat{\mu})-\varphi(Z;\delta,\pi,\mu)\}\end{aligned}$$

where we used the shorthand notation $\mathbb{P}_n\{g(Z)\}=n^{-1}\sum_{i=1}^n g(Z_i)$ and $\mathbb{P}\{g(Z)\}=\int g(z)d\mathbb{P}(z)$. The second term will be of order $O_{\mathbb{P}}(n^{-1/2})$ by the Central Limit Theorem, whereas the third term can be shown to be second order (in fact, upper bounded by a multiple of B in (18.7)*) by virtue of $\varphi(Z;\delta,\pi,\mu)$ being the first-order influence function.*

Cross-fitting helps with controlling the first term. If $g(Z_i)\mathbb{1}\{g\}(Z_j)$, then $(\mathbb{P}_n-\mathbb{P})\{g(Z)\}=O_{\mathbb{P}}(\|g(Z)\|/\sqrt{n})$, where $\|g(Z)\|^2=\int g^2(z)d\mathbb{P}(z)$. If $\widehat{\pi}$ and $\widehat{\mu}$ are computed on a separate training sample, then, given that separate sample, the first term is

$$(\mathbb{P}_n-\mathbb{P})\{\varphi(Z;\delta,\widehat{\pi},\widehat{\mu})-\varphi(Z;\delta,\pi,\mu)\}=O_{\mathbb{P}}\left\{\frac{\|\varphi(Z;\delta,\widehat{\pi},\widehat{\mu})-\varphi(Z;\delta,\pi,\mu)\|}{\sqrt{n}}\right\}$$

Therefore, because of cross-fitting, the first-term is $o_{\mathbb{P}}(n^{-1/2})$ as long as $\|\varphi(Z;\delta,\widehat{\pi},\widehat{\mu})-\varphi(Z;\delta,\pi,\mu)\|=o_{\mathbb{P}}(1)$. This is a very mild consistency requirement. In the absence of cross-fitting, *the first term would not be a centered average of independent observations. It would still be* $o_{\mathbb{P}}(n^{-1/2})$ *if* $\|\varphi(Z;\delta,\widehat{\pi},\widehat{\mu})-\varphi(Z;\delta,\pi,\mu)\| = o_{\mathbb{P}}(1)$ *and the process* $\{\sqrt{n}(\mathbb{P}_n-\mathbb{P})\{\varphi(Z;\delta,\overline{\pi},\overline{\mu})\}:\overline{\pi}\in\mathcal{P},\overline{\mu}\in\mathcal{M}\}$ *is stochastically equicontinuous, where* $\mathcal{P}$ *and* $\mathcal{M}$ *are the classes where* $\widehat{\pi}$ *and* $\widehat{\mu}$*, together with their limits, live, respectively. This would require* $\mathcal{P}$ *and* $\mathcal{M}$ *to satisfy specific "Donsker-type" conditions. For example,* $\mathcal{P}$ *and* $\mathcal{M}$ *could not be overly complex in the sense of having infinite uniform entropy integrals. Therefore, cross-fitting allows us to avoid these Donsker-type conditions on* $\mathcal{P}$ *and* $\mathcal{M}$ *and still have the first term in the decomposition above be* $o_{\mathbb{P}}(n^{-1/2})$*. For more discussion, please see Lemma 1 in [32] and Chapter 2 in [33].*

18.3.3.2 Uniform inference

The Wald-type confidence interval in equation (18.9) is valid for a particular value of δ. For example, if we only cared about what might happen if we multiply the odds of attending treatment by two, then we can use the estimator in Algorithm 1 with $\delta = 2$ and the confidence interval in (18.9). However, in almost any applied analysis, we care about comparing different levels of shifts δ. In the most basic example, we would care at least about $\widehat{\psi}(\delta=1)$ and $\widehat{\psi}(\delta\neq 1)$ so we can compare the observational data to a counterfactual intervention. In general, it is useful to vary $\delta\in\mathcal{D}\subseteq(0,\infty)$ and trace a curve $\widehat{\psi}(\delta)$ as a function of δ. In this section we outline the convergence of $\widehat{\psi}(\delta)$ to $\psi(\delta)$ in the function space $\ell^\infty(\mathcal{D})$, i.e., the space of all bounded functions on $\mathcal{D}$.

If $\mathcal{D}\subseteq(0,\infty)$, then the function $\delta\mapsto\varphi(Z;\delta,\pi,\mu)$ belongs to the class of Lipschitz functions, which is sufficiently small to satisfy Donsker conditions. As such, the quantity $\sqrt{n}\{\widetilde{\psi}(\delta)-\psi(\delta)\}/\sigma(\delta)$ converges to a mean-zero Gaussian process in $\ell^\infty(\mathcal{D})$:

$$\frac{\sqrt{n}\{\widetilde{\psi}(\delta)-\psi(\delta)\}}{\sigma(\delta)}\rightsquigarrow G(\delta)$$

where $G(\cdot)$ is a mean-zero Gaussian process with covariance

$$\mathbb{E}\{G(\delta_1)G(\delta_2)\}=\mathbb{E}[\{\frac{\varphi(Z;\delta_1,\pi,\mu)-\psi(\delta_1)}{\sigma(\delta_1)}\}\{\frac{\varphi(Z;\delta_2,\pi,\mu)-\psi(\delta_2)}{\sigma(\delta_2)}\}]$$

This means that, under essentially the same conditions that guarantee asymptotic normality at a fixed value of δ, it is also the case that any finite linear combination $\psi(\delta_1),\psi(\delta_2),\ldots,\psi(\delta_m)$ is asymptotically distributed as a multivariate Gaussian.

Establishing sufficient conditions to achieve convergence to a Gaussian process allows us to conduct uniform inference across many δ's; i.e., we can conduct inference for $\psi(\delta)$ for many δ's without issues of multiple testing. In particular, we can construct confidence bands around the curve $\widehat{\psi}(\delta)$ that covers the true curve with a desired probability across all δ. The bands can be constructed as $\widehat{\psi}(\delta)\pm\widehat{c}_\alpha\widehat{\sigma}(\delta)$, where $\widehat{c}_\alpha$ is an estimate of the $(1-\alpha)$-quantile of $\sup_{\delta\in\mathcal{D}}\sqrt{n}|\widehat{\psi}(\delta)-\psi(\delta)|/\widehat{\sigma}(\delta)$. We can estimate the supremum of Gaussian processes using the multiplier bootstrap, which is computationally efficient. We refer the reader to [1] for full details.

With the uniform confidence interval, we can also conduct a test of no treatment effect. If the treatment has no effect on the outcome, then $Y\mathbb{1}\{A\}\mid X$ and $\psi(\delta)=\mathbb{E}(Y)$ under exchangeability (Assumption 2). Given the uniform confidence band, the null hypothesis of no incremental intervention effect,

$$H_0:\psi(\delta)=\mathbb{E}(Y)\text{ for all }\delta\in\mathcal{D},$$

can be tested by checking whether a $(1-\alpha)$ band contains a straight horizontal line over $\mathcal{D}$. That is, we reject H_0 at level α if we cannot run a straight horizontal line through the confidence band. We can also compute a p-value as

$$\widehat{p} = \sup\left[\alpha : \inf_{\delta \in \mathcal{D}} \left\{\widehat{\psi}(\delta) + \widehat{c}_{\alpha}\widehat{\sigma}(\delta)/\sqrt{n}\right\} \geq \sup_{\delta \in \mathcal{D}}\{\widehat{\psi}(\delta) - \widehat{c}_{\alpha}\widehat{\sigma}(\delta)/\sqrt{n}\}\right].$$

Geometrically, the p-value corresponds to the α at which we can no longer run a straight horizontal line through our confidence band. At $\alpha = 0$ we have an infinitely wide confidence band and we always fail to the reject H_0. Increasing α corresponds to a tightening confidence band. In Section 18.5 we give a visual example of how to conduct this test. But first, we discuss incremental propensity score interventions when treatment is time-varying.

18.4 Time-varying Treatments

18.4.1 Notation

In this section we extend the prior results from the one time point case to the time-varying setup. In the time-varying setup we consider n samples $(Z_1, ..., Z_n)$ IID from some distribution $\mathbb{P}$ over T time-points such that

$$Z = (X_1, A_1, X_2, A_2, ..., X_T, A_T, Y)$$

where X_t are time-varying covariates, A_t is time-varying treatment, and Y is the outcome. In many studies only a few covariates are time-varying and most are measured at baseline; all baseline covariates are included in X_1. Additionally, in some studies there are time-varying outcome variables Y_t; in that case, we can include the time-varying outcomes as part of the time-varying covariates X_{t+1}. We consider potential outcomes $Y^{\overline{a}_T}$, which is the outcome we would have observed if the treatment regime had been $\overline{A}_T = \overline{a}_T$, where $\overline{A}_t$ is the history of A_t from period 1 until t. We also define $H_t = (\overline{A}_{t-1}, \overline{X}_t)$ as the full history up until treatment in period t.

First, we make two time-varying assumptions that are analogous to Assumptions 17 and 18:

Assumption 20. *(Consistency).* $Y = Y^{\overline{a}_T}$ *if* $\overline{A}_T = \overline{a}_T$.

Assumption 21. *(Sequential Exchangeability).* $A_t \mathbb{1} \{Y\}^{\overline{a}_T} \mid H_t$.

Usually, positivity, is also assumed. As we saw in Section 18.3, positivity will not be required for incremental interventions, but we will state it here:

Assumption 22. *(Positivity). There exists* $\epsilon > 0$ *such that* $\mathbb{P}(\epsilon < \mathbb{P}(\overline{A}_T = \overline{a}_T \mid X) < 1 - \epsilon) = 1$ *for any treatment regime* $\overline{a}_T$.

Consistency says that the observed outcome under a treatment sequence equals the potential outcome under that treatment sequence. Sequential exchangeability says the outcome *at each time point* is effectively randomized within covariate history strata. Positivity says that everyone has a non-zero chance of taking each treatment regime. We discussed each assumption in detail in Section 18.3.1 and showed how these assumptions might be violated.

18.4.2 Marginal structural models

In time-varying causal inference, or generally when the treatment can take many values, a tension arises between the number of possible interventions and how well we can estimate effects at each intervention. For example, a binary treatment administered at T time points leads to 2^T possible treatment regimes and as many possible effects to estimate. Without further restrictions, the number of parameters and the sample size required to estimate every effect grow exponentially with T.

A popular approach to reduce the number of parameters is to assume that the expected potential outcomes under different interventions vary smoothly. For instance, one can specify a model of the form

$$\mathbb{E}(Y^{\overline{a}_T}) = m(\overline{a}_T;\beta),$$

where $m(\overline{a}_T;\beta)$ is a function known up to a finite-dimensional parameter $\beta \in \mathbb{R}^p$. This model, termed a *marginal structural model* (MSM), imposes a parametric structure on the relationship between the potential outcomes and the treatment regime. A simple example allows the potential outcomes only to depend on the cumulative treatment attendance:

$$m(\overline{a}_T;\beta) = \beta_0 + \beta_1\left(\sum_{t=1}^{T} a_t\right).$$

A popular approach to estimate β is through the moment condition

$$\mathbb{E}[h(\overline{A}_T)W(Z)\{Y - m(\overline{A}_T;\beta)\}] = 0 \tag{18.10}$$

where

$$W(Z) = \frac{1}{\prod_{t=1}^{T}\mathbb{P}(A_t \mid H_t)}$$

and h is some arbitrary function of the treatment. This moment condition suggests an inverse-propensity weighted estimator, where we estimate $\widehat{\beta}$ by solving the empirical analog of (18.10):

$$\frac{1}{n}\sum_{i=1}^{n}\left\{h(\overline{A}_{T,i})\widehat{W}(Z_i)\{Y_i - m(\overline{A}_{T,i};\widehat{\beta})\}\right\} = 0.$$

We can also estimate β using a doubly robust version of the moment condition ([34]).

MSMs are a major advance toward performing sound causal inference in time-varying settings. However, there are two important issues they cannot easily resolve. First, while specifying a model for $\mathbb{E}(Y^{\overline{a}_T})$ is a less stringent requirement than, say, specifying a model for the outcome given treatment and covariates, it can still lead to biased estimates if $m(\overline{a}_T;\beta)$ is not correctly specified. Second, identifying and estimating β still relies on positivity, which is unlikely to be satisfied when there are many time points or treatment values. In fact, even if positivity holds by design as it would in an experiment, we are unlikely to observe all treatment regimes in a given experiment simply because the number of possible treatment regimes grows exponentially with the number of time points. This poses a computational challenge even when positivity holds, because the product of densities in the denominator of (18.10) may be very small, resulting in an unstable estimate of β.

There are ways to mitigate the two drawbacks outlined above, but neither issue can be completely fixed. First, one does not necessarily need to assume the MSM is correctly specified. Instead, one can use the "working model" approach and estimate a projection of $\mathbb{E}(Y^{\overline{a}_T})$ onto the MSM $m(\overline{a}_T;\beta)$. In this approach one estimates β as the parameter that yields the best approximation of the causal effect in the function class described by $m(\overline{a}_T;\beta)$ ([35]). This approach is beneficial because it allows for model-free inference: we can construct valid confidence intervals for the projection parameter β regardless of whether the MSM is correctly specified. However, we are still only estimating a projection; so, this approach can be of limited practical relevance if the model is grossly misspecified and the projection onto the model has very little bearing on reality.

Meanwhile, to mitigate in-sample positivity violations, a popular approach is to "stabilize" the propensity scores by replacing $W(Z)$ in (18.10) with

$$\widetilde{W}(Z) = \prod_{t=1}^{T} \frac{\mathbb{P}(A_t \mid \overline{A}_{t-1})}{\mathbb{P}(A_t \mid H_t)}.$$

See [36] and [37] for discussions of the advantages and disadvantages of using the stabilized weights $\widetilde{W}(Z)$ instead of $W(Z)$. Another approach is propensity score trimming, which we also discussed in the one time point case. However, as we discussed in Section 18.2, these ad-hoc approaches are not guaranteed to solve positivity violations.

Remark 18.9. *MSMs struggle with positivity violations in the same way traditional ATE estimators struggle with Positivity violations in the one time point case. In Section 18.2 we outlined how these positivity issues lead naturally to dynamic and stochastic interventions. A similar logic applies here, although perhaps more urgently, because positivity violations occur in time-varying analyses more frequently.*

Remark 18.10. *MSMs have also been criticized because, even if the model is correctly specified and positivity is not violated, the causal effect of the treatment may be hard to interpret and visualize since it may be implicitly encoded in a complicated model $m(\overline{a}_T;\beta)$. While this is a solvable problem, we mention it so the reader might compare the difficulty of interpreting a complicated model $m(\overline{a}_T;\beta)$ with the ease of understanding the delta curve graphs in Section 18.5.*

18.4.3 Time-varying incremental effects

A *time-varying incremental propensity score intervention* takes the same form as the one time-point intervention. So, at time t, we multiply the odds of receiving treatment by δ. Thus, we have a stochastic dynamic intervention $Q_t(A_t \mid H_t)$ where

$$Q_t(A_t = 1 \mid H_t) \equiv q_t(H_t;\delta,\pi_t) = \frac{\delta\pi_t(H_t)}{\delta\pi_t(H_t) + 1 - \pi_t(H_t)} \text{ for } 0 < \delta < \infty. \quad (18.11)$$

for $\pi_t(H_t) = \mathbb{P}(A_t = 1 \mid H_t)$. This is the same intervention as in equation (18.4), just with time subscripts added and conditioning on all the past up to treatment in time t (i.e., H_t). There are two main differences between the intervention (18.11) and the incremental intervention described in Section 18.3:

1. This intervention happens over every time period from $t = 1$ to T, the end of the study.

2. This intervention requires the time-varying analogs of consistency and exchangeability, Assumptions 20 and 21.

Despite these differences, most of the machinery developed in Section 18.3 applies here. The intuition about what happens when we shift $\delta \to 0$ or $\delta \to \infty$ is the same. Unfortunately, the results and proofs look much more imposing at first glance, but that is due to the time-varying nature of the data, not any change in the incremental intervention approach.

The incremental approach is quite different from the MSM approach. The incremental intervention is a stochastic dynamic intervention that shifts propensity scores in each time period, whereas MSMs describe a static intervention for what would happen if everyone took treatment $\overline{a}_T$. Consequently, the incremental intervention framework does not require us to assume positivity (Assumption 22) or a parametric model $m(\overline{A}_T;\beta)$, whereas MSMs require both.

Remark 18.11. *The time-varying incremental intervention can actually allow δ to vary over t, but we omit this generalization for ease of exposition and interpretability. In other words, in equation* (18.11) *we could use δ_t instead of δ and allow δ_t to vary with t. Whether allowing δ to vary with time is useful largely depends on the context. In some applications it may be enough to study interventions that keep δ constant across time. On the other hand, one can imagine interventions whose impact varies over time; for example, some encouragement mechanism might have an effect that decays toward zero with time. Either way, the theory and methodology presented here would remain valid.*

18.4.4 Identification

The following theorem is a generalization of Theorem 18.1; it shows that the mean counterfactual outcome under the incremental intervention $\{q_t(H_t; \delta, \pi_t)\}_{t=1}^{T}$ is identified under Assumptions 20 and 21.

Theorem 18.2. (Theorem 1, [1])
Under Assumptions 20 and 21, for $\delta \in (0, \infty)$, the incremental propensity score effect $\psi(\delta)$ satisfies

$$\psi(\delta) = \sum_{\overline{a}_T \in \mathcal{A}^T} \int_{\mathcal{X}} \mu(h_T, a_T) \prod_{t=1}^{T} \frac{a_t \delta \pi_t(h_t) + (1 - a_t)\{1 - \pi_t(h_t)\}}{\delta \pi_t(h_t) + 1 - \pi_t(h_t)} d\mathbb{P}(x_t \mid h_{t-1}, a_{t-1}) \tag{18.12}$$

and

$$\psi(\delta) = \mathbb{E}\{Y \prod_{t=1}^{T} \frac{\delta A_t + 1 - A_t}{\delta \pi_t(H_t) + 1 - \pi_t(H_t)}\} \tag{18.13}$$

where $\mathcal{X} = \mathcal{X}_1 \times \cdots \times \mathcal{X}_T$ and $\mu(h_T, a_T) = \mathbb{E}[Y \mid H_T = h_T, A_T = a_t]$.

To see why the identification formula in equation (18.12) arises, notice that with generic interventions $dQ(a_t \mid h_t)$ we can extend equation (18.3) to multiple time points with

$$\mathbb{E}(Y^Q) = \int_{\mathcal{A}_1 \times \cdots \times \mathcal{A}_T} \int_{\mathcal{X}_1 \times \cdots \times \mathcal{X}_T} \mu(h_T, a_T) \prod_{t=1}^{T} dQ(a_t \mid h_t) d\mathbb{P}(x_t \mid h_{t-1}, a_{t-1}),$$

which follows by Robins' g-formula ([38]) and replacing the general treatment process with a generic stochastic intervention $dQ(a_t \mid h_t)$. Then, we can replace the generic $dQ(a_t \mid h_t)$ with $a_t q_t(H_t; \delta, \pi_t) + (1 - a_t)\{1 - q_t(H_t; \delta, \pi_t)\}$ and $\int_{\mathcal{A}_1 \times \cdots \times \mathcal{A}_T}$ with $\sum_{\overline{a}_T \in \mathcal{A}_T}$. The full proof is shown in the appendix of [1]. Just like in the $T = 1$ case, $\psi(\delta)$ is well-defined even if $\pi_t(h_t) = 0$ or 1 for some h_t.

18.4.5 Estimation

In Section 18.3, we presented two "plug-in" estimators, briefly reviewed the theory of influence functions, and presented an influence-function-based estimator that allows for nonparametric nuisance function estimation. In the multiple time point case, the parameter $\psi(\delta)$ is still smooth enough to possess an influence function. Thus, the discussion in Section 18.3 also applies here, although the notation becomes more involved.

Again, there are two plug-in estimators. The outcome-based g-computation style estimator is motivated by equation (18.12) in Theorem 18.2 and can be constructed with

$$\psi(\delta) = \sum_{\overline{a}_T \in \mathcal{A}^T} \int_{\mathcal{X}} \widehat{\mu}(h_T, a_T) \prod_{t=1}^{T} \frac{a_t \delta \widehat{\pi}_t(h_t) + (1 - a_t)\{1 - \widehat{\pi}_t(h_t)\}}{\delta \widehat{\pi}_t(h_t) + 1 - \widehat{\pi}_t(h_t)} d\widehat{\mathbb{P}}(x_t \mid h_{t-1}, a_{t-1})$$

where $\mathcal{X} = \mathcal{X}_1 \times \cdots \times \mathcal{X}_T$. Similarly, the IPW estimator can be constructed with:

$$\widehat{\psi}(\delta) = \frac{1}{n}\sum_{i=1}^{n}\left\{Y_i \prod_{t=1}^{T} \frac{(\delta A_{t,i} + 1 - A_{t,i})}{\delta\widehat{\pi}_t(H_{t,i}) + 1 - \widehat{\pi}_t(H_{t,i})}\right\}.$$

As before, both estimators inherit the convergence rates of the nuisance function estimators for $\widehat{\pi}_t$ and $\widehat{\mu}$, and typically attain parametric rates of convergence only with correctly specified restrictive models for **every** nuisance function. As with the single time-point case, we may be skeptical as to whether specifying correct parametric models is possible, and this motivates an influence-function-based approach. Theorem 2 in [1] derives the influence function for $\psi(\delta)$.

Theorem 18.3. (Theorem 2, [1]) *The un-centered efficient influence function for $\psi(\delta)$ in a nonparametric model with unknown propensity scores is given by*

$$\begin{aligned}
\varphi(Z;\eta,\delta) \equiv \sum_{t=1}^{T} & \left[\frac{A_t\{1-\pi_t(H_t)\} - (1-A_t)\delta\pi(H_t)}{\delta/(1-\delta)}\right] \\
& \times \left[\frac{\delta\pi_t(H_t)m_t(H_t,1) + \{1-\pi_t(H_t)\}m_t(H_t,0)}{\delta\pi_t(H_t) + 1 - \pi_t(H_t)}\right] \\
& \times \left\{\prod_{s=1}^{t}\frac{\delta A_s + 1 - A_s}{\delta\pi_s(H_s) + 1 - \pi_s(H_s)}\right\} + \prod_{t=1}^{T}\frac{(\delta A_t + 1 - A_t)Y}{\delta\pi_t(H_t) + 1 - \pi_t(H_t)}
\end{aligned}$$

where $\eta = \{m_1, \ldots, m_T, \pi_1, \ldots, \pi_T\}$, $m_T(h_T, a_T) = \mu(h_T, a_T)$ and for $t = 0, \ldots, T-1$:

$$\begin{aligned}
m_t(h_t, a_t) = \int_{\mathcal{R}_t} \mu(h_T, a_T) & \prod_{s=t+1}^{T} [a_s q_s(H_s;\delta,\pi_s) \\
& + (1-a_s)\{1 - q_s(H_s;\delta,\pi_s)\}] d\mathbb{P}(x_s \mid h_{s-1}, a_{s-1})
\end{aligned}$$

with $\mathcal{R}_t = (\mathcal{H}_T \times \mathcal{A}_T) \setminus \mathcal{H}_t$.

The following influence-function-based estimator is a natural consequence of Theorem 18.3:

$$\widehat{\psi}(\delta) = \frac{1}{n}\sum_{i=1}^{n}\varphi(Z_i;\widehat{\eta},\delta)$$

Similarly to the $T = 1$ case, this estimator is optimal with nonparametric models under certain smoothness or sparsity constraints. Again, it is advantageous to construct this estimator using cross-fitting, since it allows fast convergence rates of $\widehat{\psi}(\delta)$ without imposing Donsker-type conditions on the estimators of η. The detailed algorithm is provided as Algorithm 1 of [1]. For intuition, the reader can imagine the algorithm as an extension of Algorithm 1 in this chapter, where in step 1 we estimate all the nuisance functions in Theorem 18.3. However, unlike Algorithm 1, we can estimate m_t recursively backwards though time, and this is outlined in Remark 2 and Algorithm 1 from [1]. The `ipsi` function in the `R` package `npcausal` can be used to calculate incremental effects in time-varying settings.

18.4.6 Inference

As in the $T = 1$ case, we provide both pointwise and uniform inference results. The theory is essentially the same as in the one time-point case, but the conditions to achieve convergence to a Gaussian distribution or a Gaussian process need to be adjusted to handle multiple time points.

Theorem 3 in [1] states that, under mild regularity conditions, if

$$\left(\sup_{\delta\in\mathcal{D}} \|\widehat{m}_{t,\delta} - m_{t,\delta}\| + \|\widehat{\pi}_t - \pi_t\|\right) \|\widehat{\pi}_s - \pi_s\| = o_{\mathbb{P}}(n^{-1/2}) \quad \text{for } s \le t \le T$$

then

$$\frac{\sqrt{n}\{\widehat{\psi}(\delta) - \psi(\delta)\}}{\widehat{\sigma}(\delta)} \rightsquigarrow \mathbb{G}(\delta)$$

in $l^\infty(\mathcal{D})$, where $\widehat{\sigma}(\delta)$ is a consistent estimate of $\text{Var}\{\varphi(Z;\eta,\delta)\}$ and $\mathbb{G}(\cdot)$ is a mean-zero Gaussian process with covariance $\mathbb{E}\{\mathbb{G}(\delta_1)\mathbb{G}(\delta_2)\} = \mathbb{E}\{\widetilde{\varphi}(Z;\eta,\delta_1)\widetilde{\varphi}(Z;\eta,\delta_2)\}$ and $\widetilde{\varphi}(z;\eta,\delta) = \{\varphi(z;\eta,\delta) - \psi(\delta)\}/\sigma(\delta)$. In particular, for a given fixed value of δ, we have

$$\frac{\sqrt{n}\{\widehat{\psi}(\delta) - \psi(\delta)\}}{\widehat{\sigma}(\delta)} \rightsquigarrow N(0,1).$$

The main assumption underlying this result is that the product of the L_2 errors for estimating m_t and π_t is of smaller order than $n^{-1/2}$. This is essentially the same small-bias condition discussed in Section 18.3.3.1 for the $T = 1$ case, adjusted to handle multiple time points. This requirement is rather mild, because m_t and π_t can be estimated at slower-than-$\sqrt{n}$ rates, say $n^{-1/4}$ rates, without affecting the efficiency of the estimator. As discussed in Remark 18.7, $n^{-1/4}$ rates are attainable in nonparametric models under smoothness or sparsity constraints.

Finally, the convergence statements above allow for straightforward pointwise and uniform inference as discussed in Sections 18.3.3.1 and 18.3.3.2. In particular, we can construct a Wald-type confidence interval

$$\widehat{\psi}(\delta) \pm z_{1-\alpha/2}\frac{\widehat{\sigma}(\delta)}{\sqrt{n}}$$

at each δ using the algorithm outlined above. We can also create uniformly valid confidence bands covering $\delta \mapsto \psi(\delta)$ via the multiplier bootstrap and use the bands to test for no treatment effect, as in Section 18.3.3.2. We refer to Sections 4.3 and 4.4 in [1] for additional technical details.

18.5 Example Analysis

In this section we show an example curve that one could obtain to describe incremental effects. One could obtain Figure 18.1 (Figure 3 from [1]) by running the `ipsi` function in the `npcausal` package. [1] reanalyzed a dataset on the effects of incarceration on marriage prevalence. For this figure, they used 10 years of data (2001-2010) for 4781 individuals from the National Longitudinal Survey of Youth. The dataset contains baseline covariates like demographic information and delinquency indicators, as well as many time-varying variables such as employment and earnings. They defined the outcome Y as whether someone was married at the end of the study (i.e., in 2010), and defined the treatment A_t as whether someone was incarcerated in a given year.

Figure 18.1 shows their estimated $\psi(\delta)$ curve. The x-axis is the odds multiplier δ, which ranges from 0.2 to 5. Because δ is a multiplier, this range is a natural choice ($0.2 = \frac{1}{5}$). The y-axis shows the estimated marriage prevalence $\widehat{\psi}(\delta)$ as a proportion; so, 0.3 corresponds to 30% of subjects being married in 2010. The black line is the point estimate of $\psi(\delta)$ ranging across δ; as discussed previously, because $\psi(\delta)$ is smooth in δ, we can interpolate between the point estimates and plot a

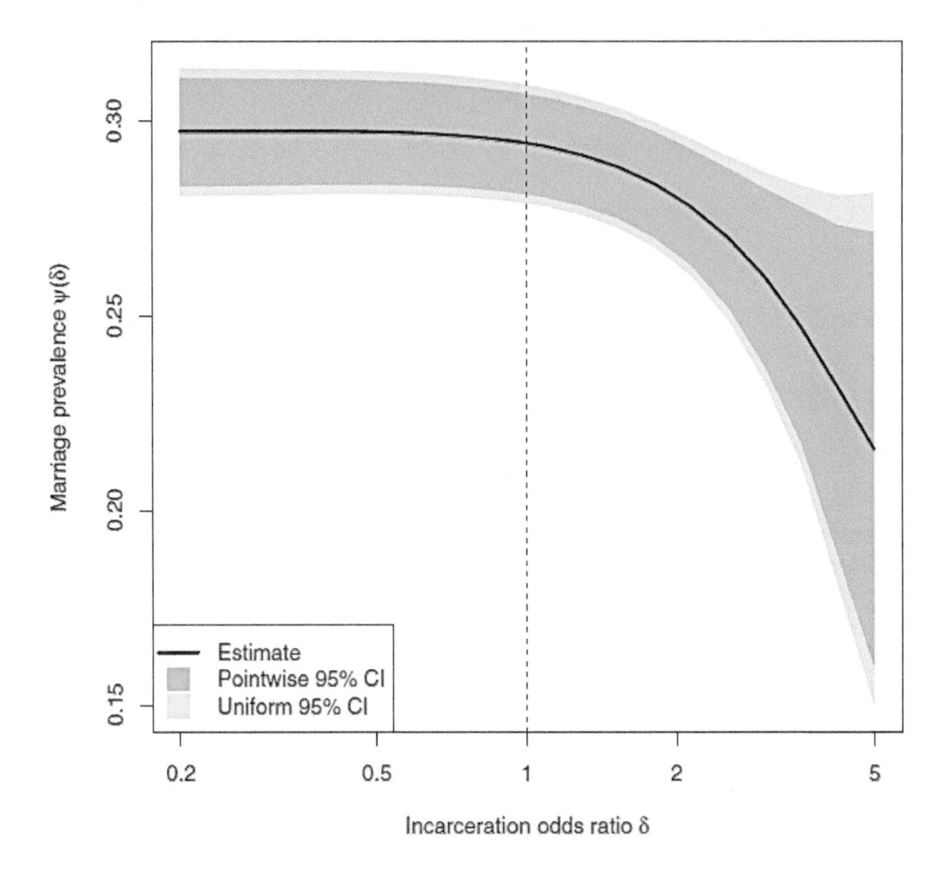

FIGURE 18.1
Estimated marriage prevalence 10 years post-baseline, if the incarceration odds were multiplied by factor δ, with pointwise and uniform 95% confidence bands.

smooth line. The darker blue confidence band is a pointwise 95% confidence interval that would give us valid inference at a single δ value. The lighter blue confidence band is the uniform 95% confidence interval that allows us to perform inference across the whole range of δ values.

In this example we see that incarceration has a strong effect on marriage rates: Estimated marriage rates decrease with higher odds of incarceration ($\delta > 1$) but only increase slightly with lower odds of incarceration ($\delta < 1$). To be more specific, at observational levels of incarceration (i.e., leaving the odds of incarceration unchanged) they estimated a marriage rate of $\psi(1) = 0.294$, or 29.4%. If the odds of treatment were doubled, they estimated marriage rates would decrease to 0.281, and if they were quadrupled they would decrease even further to 0.236. Conversely, if the odds of treatment were halved, they estimated marriage rates would only increase to 0.297; the estimated marriage rates are the same if the odds are quartered.

Finally, we can use the uniform confidence interval in Figure 18.1 to test for no effect across the range $\delta \in [0.2, 5]$, and we reject the null hypothesis at the $\alpha = 0.05$ level with a p-value of 0.049. By eye, we can roughly see that it is just about impossible to run a horizontal line through the uniform 95% confidence interval. But it is very close, and the p-value is 0.049, only just below 0.05.

18.6 Extensions & Future Directions

18.6.1 Censoring & dropout

In time-varying settings, it is common that subjects are censored or dropout so that we no longer observe them after a certain time point. This complicates causal inference, in that time-varying covariates, treatment, and outcomes are not observed for all subjects across all time points in the study. The incremental propensity score approach can be extended to account for censoring and dropout ([39]). As with almost all analyses with censoring, we must assume that the data is "missing at random" (MAR) or "missing completely at random" (MCAR) if we wish to make progress. The MCAR assumption is much stronger and thus less likely to be true, so researchers aim to use the weaker MAR assumption. In our time-varying setup, this corresponds to assuming that, given the past data, subjects' values of future data are independent of whether or not they are censored. This assumption allows for identification and estimation of the estimand $\psi(\delta)$. The estimator is very similar to what we outlined in Section 18.4, but includes an inverse weighting term for the probability of censoring. This is common in time-varying causal analyses with missing data (e.g., [40]), and takes the same form as inverse propensity weighting to account for missing potential outcomes.

18.6.2 Future directions

In this chapter we have outlined the simplest approach to estimate incremental effects without considering any major complications. As we discussed in the previous section, this approach has been extended to account for time-varying censoring ([39]). There are many natural extensions that have already been developed for standard estimands that would make sense in the incremental effects paradigm, including continuous treatment effects, conditional treatment effects, instrumental variables analyses, and sensitivity (to unmeasured confounding) analyses. Additionally, the incremental effects approach generates a few of its own unique problems. First, as we highlighted in Remarks 18.1 and 18.11, we can allow δ to vary with covariates and time. Second, we might hope to characterize the stochastic treatment intervention distribution in more detail than just the parameter δ. For example, we may be interested in estimating the average number of treatments, the first treatment, or the modal number of treatments under the incremental propensity score intervention. In this case we would need to develop an estimator for the distribution of treatment attendance over time for various δ.

18.7 Discussion

This note serves as an introduction to and a review of [1], which describes a class of causal effects based on incremental propensity score interventions. Such effects can be identified even if positivity is violated and thus can be useful in time-varying settings when the number of possible treatment sequences is large and positivity is less likely to hold. We have compared this class of effects to other effects commonly estimated in practice, such as average treatment effects or coefficients in MSM, and highlighted the scenarios where it can be most useful. Along the way, we have reviewed the estimation and inferential procedure proposed in [1] and briefly reviewed the underlying semiparametric efficiency theory.

References

[1] Edward H Kennedy. Nonparametric causal effects based on incremental propensity score interventions. *Journal of the American Statistical Association*, 114(526):645–656, 2019.

[2] Donald B Rubin. Estimating causal effects of treatments in randomized and nonrandomized studies. *Journal of Educational Psychology*, 66(5):688, 1974.

[3] BJS. Bureau of justice statistics web page. https://www.bjs.gov/index.cfm?ty=kfdetail&iid=487, 2021.

[4] BJS. Bureau of justice statistics 2018 update on prisoner recidivism: A 9-year follow-up period (2005-2014). https://www.bjs.gov/content/pub/pdf/18upr9yfup0514_sum.pdf, 2018.

[5] HJ. Steadman, FC. Osher, PC. Robbins, B. Case, and S. Samuels. Prevalence of serious mental illness among jail inmates. *Psychiatric Services*, 60(6):761–765, 2009.

[6] Jennifer L. Skeem, Sarah Manchak, and Jillian K. Peterson. Correctional policy for offenders with mental illness: Creating a new paradigm for recidivism reduction. *Law and Human Behavior*, 35(2):110–126, 2011.

[7] Paul W Holland. Statistics and causal inference. *Journal of the American Statistical Association*, 81(396):945–960, 1986.

[8] Daniel Westreich and Stephen R Cole. Invited commentary: Positivity in practice. *American Journal of Epidemiology*, 171(6):674–677, 2010.

[9] Maya L Petersen, Kristin E Porter, Susan Gruber, Yue Wang, and Mark J Van Der Laan. Diagnosing and responding to violations in the positivity assumption. *Statistical Methods in Medical Research*, 21(1):31–54, 2012.

[10] Leslie Kish. Weighting for unequal pi. *Journal of Official Statistics*, pages 183–200, 1992.

[11] Matias Busso, John DiNardo, and Justin McCrary. New evidence on the finite sample properties of propensity score reweighting and matching estimators. *Review of Economics and Statistics*, 96(5):885–897, 2014.

[12] Jinyong Hahn. On the role of the propensity score in efficient semiparametric estimation of average treatment effects. *Econometrica*, 66(2):315–331, 1998.

[13] Whitney K Newey. Semiparametric efficiency bounds. *Journal of Applied Econometrics*, 5(2):99–135, 1990.

[14] Daniel E Ho, Kosuke Imai, Gary King, and Elizabeth A Stuart. Matching as nonparametric preprocessing for reducing model dependence in parametric causal inference. *Political Analysis*, 15(3):199–236, 2007.

[15] Elizabeth A Stuart. Matching methods for causal inference: A review and a look forward. *Statistical science: a review journal of the Institute of Mathematical Statistics*, 25(1):1, 2010.

[16] Richard K Crump, V Joseph Hotz, Guido W Imbens, and Oscar A Mitnik. Dealing with limited overlap in estimation of average treatment effects. *Biometrika*, 96(1):187–199, 2009.

[17] Kelly L Moore, Romain Neugebauer, Mark J van der Laan, and Ira B Tager. Causal inference in epidemiological studies with strong confounding. *Statistics in Medicine*, 31(13):1380–1404, 2012.

[18] Alexandre B Tsybakov. *Introduction to Nonparametric Estimation*. New York: Springer, 2009.

[19] Peter J Bickel, Chris AJ Klaassen, Ya'acov Ritov, and Jon A Wellner. *Efficient and Adaptive Estimation for Semiparametric Models*. Baltimore: Johns Hopkins University Press, 1993.

[20] Aad W van der Vaart. *Asymptotic Statistics*. Cambridge: Cambridge University Press, 2000.

[21] Edward H Kennedy. Semiparametric theory and empirical processes in causal inference. *In: Statistical Causal Inferences and Their Applications in Public Health Research*, pages 141–167, Springer, 2016.

[22] James M Robins, Lingling Li, Eric Tchetgen Tchetgen, and Aad W van der Vaart. Quadratic semiparametric von mises calculus. *Metrika*, 69(2-3):227–247, 2009.

[23] Anastasios A Tsiatis. *Semiparametric Theory and Missing Data*. New York: Springer, 2006.

[24] Aaron Fisher and Edward H Kennedy. Visually communicating and teaching intuition for influence functions. *The American Statistician*, 75(2):162–172, 2021.

[25] Mark J van der Laan and Daniel B Rubin. Targeted maximum likelihood learning. *UC Berkeley Division of Biostatistics Working Paper Series*, 212, 2006.

[26] Chris AJ Klaassen. Consistent estimation of the influence function of locally asymptotically linear estimators. *The Annals of Statistics*, 15(4):1548–1562, 1987.

[27] Peter J Bickel and Yaacov Ritov. Estimating integrated squared density derivatives: Sharp best order of convergence estimates. *Sankhyā*, pages 381–393, 1988.

[28] James M Robins, Lingling Li, Eric J Tchetgen Tchetgen, and Aad W van der Vaart. Higher order influence functions and minimax estimation of nonlinear functionals. *Probability and Statistics: Essays in Honor of David A. Freedman*, pages 335–421, 2008.

[29] Wenjing Zheng and Mark J van der Laan. Asymptotic theory for cross-validated targeted maximum likelihood estimation. *UC Berkeley Division of Biostatistics Working Paper Series*, Paper 273:1–58, 2010.

[30] Victor Chernozhukov, Denis Chetverikov, Mert Demirer, Esther Duflo, Christian Hansen, Whitney Newey, and James M Robins. Double/debiased machine learning for treatment and structural parameters. *The Econometrics Journal*, 21(1):C1–C68, 2018.

[31] Max H Farrell. Robust inference on average treatment effects with possibly more covariates than observations. *Journal of Econometrics*, 189(1):1–23, 2015.

[32] Edward H Kennedy, S Balakrishnan, and M G'Sell. Sharp instruments for classifying compliers and generalizing causal effects. *The Annals of Statistics*, 48(4):2008–2030, 2020.

[33] Aad W van der Vaart and Jon A Wellner. *Weak Convergence and Empirical Processes*. Springer, 1996.

[34] Heejung Bang and James M Robins. Doubly robust estimation in missing data and causal inference models. *Biometrics*, 61(4):962–973, 2005.

[35] Romain Neugebauer and Mark J van der Laan. Nonparametric causal effects based on marginal structural models. *Journal of Statistical Planning and Inference*, 137(2):419–434, 2007.

[36] Stephen R Cole and Miguel A Hernán. Constructing inverse probability weights for marginal structural models. *American Journal of Epidemiology*, 168(6):656–664, 2008.

[37] Denis Talbot, Juli Atherton, Amanda M Rossi, Simon L Bacon, and Genevieve Lefebvre. A cautionary note concerning the use of stabilized weights in marginal structural models. *Statistics in Medicine*, 34(5):812–823, 2015.

[38] James M Robins. A new approach to causal inference in mortality studies with a sustained exposure period - application to control of the healthy worker survivor effect. *Mathematical Modelling*, 7(9-12):1393–1512, 1986.

[39] Kwangho Kim, Edward H Kennedy, and Ashley I Naimi. Incremental intervention effects in studies with many timepoints, repeated outcomes, and dropout. *arXiv preprint arXiv:1907.04004*, 2019.

[40] James M Robins, Miguel Angel Hernan, and Babette Brumback. Marginal structural models and causal inference in epidemiology. *Epidemiology*, 11(5):550–560, 2000.

19

Weighting Estimators for Causal Mediation

Donna L. Coffman
University of South Carolina

Megan S. Schuler
RAND Corporation

Trang Q. Nguyen
Johns Hopkins University

Daniel F. McCaffrey
ETS

CONTENTS

DOI: 10.1201/9781003102670-19

19.1 Introduction

The focus of this chapter is weighting estimators for causal mediation analysis – specifically, we focus on natural direct and indirect effects and interventional direct and indirect effects. Using potential outcomes notation, we define specific estimands and the needed identifying assumptions. We then describe estimation methods for the estimands. We detail relevant R packages that implement these weighting methods and provide sample code, applied to our case study example. More detail about the R packages is given at the end of the chapter in Section 19.7.

19.1.1 Overview of causal mediation

Mediation analysis, by definition, involves questions about causal effects and mechanisms. An important scientific goal in many fields of research is determining to what extent the total effect of an exposure on an outcome is mediated by an intermediate variable on the causal pathway between the exposure and outcome. A graph that illustrates a simple mediation model is shown in Figure 19.1 where $Y \equiv$ outcome, $A \equiv$ exposure, $C \equiv$ pre-exposure covariates, and $M \equiv$ mediator. Note that we use "exposure" broadly to refer to a non-randomized or randomized condition, treatment, or intervention.

The **total effect** of A on Y includes two possible causal paths from A to Y: the path $A \to M \to Y$ is the **indirect effect** of A on Y through M and the path $A \to Y$ is the **direct effect** of A on Y that does not go through M. Direct and indirect effects are of scientific interest because they provide a framework to quantify and characterize the mechanism by which an exposure affects a given outcome.

19.1.2 Introduction to case study

Our motivating example applies mediation analysis to health disparities research. Our specific focus is to examine potential mediating pathways that explain substance use disparities among sexual minority (e.g., lesbian/gay or bisexual) women, using data from the National Survey of Drug Use and Health (NSDUH). Specifically, lesbian/gay and bisexual (LGB) women report higher rates of smoking and alcohol use than heterosexual women. We conceptualize sexual minority status as the exposure of interest (note that these groups reflect the measurement of sexual identity in the NSDUH;

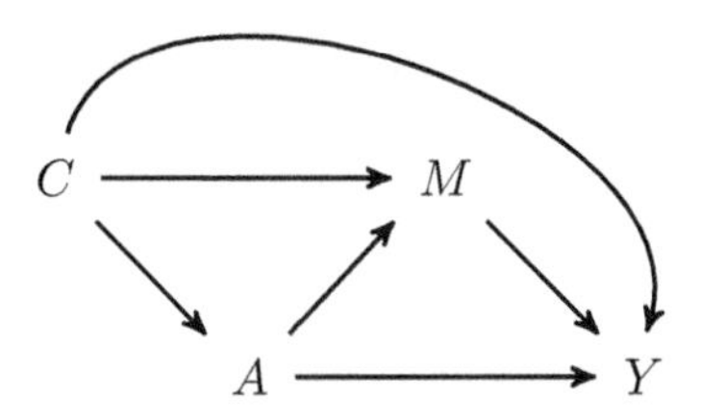

FIGURE 19.1
A simple mediation model.

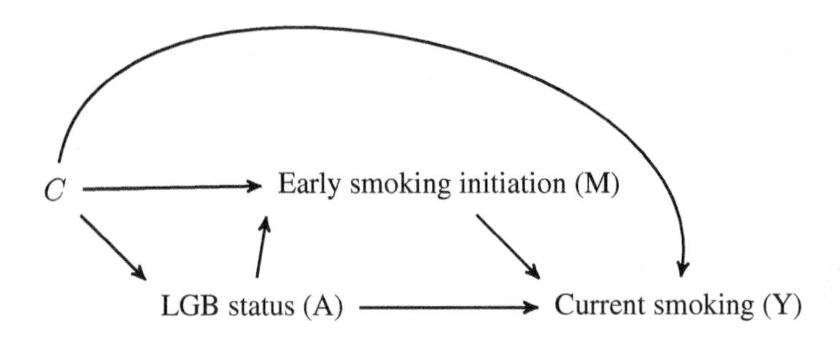

FIGURE 19.2
Mediation model for case study.

individuals may identify as a broader range of sexual identities), in that it gives rise to experiences of "minority stress," namely, excess social stressors experienced by individuals in a marginalized social group (e.g., LGB individuals). Manifestations of minority stress may include experiences of stigma, discrimination, bullying, and family rejection, among others. Substance use among LGB individuals has been theorized to reflect, in part, a coping strategy to minority stress experiences. In our example, the particular outcome of interest is current smoking among LGB women, which we know to be disproportionately higher than among heterosexual women [1]. We apply mediation analysis to elucidate potential causal pathways that may give rise to these elevated rates of smoking. Our hypothesized mediator is early smoking initiation (i.e., prior to age 15); that is, we hypothesize that LGB women are more likely to begin smoking at an earlier age than heterosexual women, potentially in response to minority stressors. In turn, early smoking initiation, which is a strong risk factor for developing nicotine dependence, may contribute to higher rates of smoking among LGB women. In summary, the exposure is defined as sexual minority status (1 = LGB women, 0 = heterosexual women), the mediator is early smoking initiation (1 = early initiation, 0 = no early initiation), and the outcome is current smoking in adulthood (1 = yes, 0 = no). Baseline covariates include age, race/ethnicity, education level, household income, employment status, marital status, and urban vs. rural residence. Figure 19.2 depicts the mediation model for our motivating example.

19.1.3 Potential outcomes notation

Consider the case in which A is a binary indicator of the exposure, indicating the exposed condition ($A = 1$) or the comparison condition ($A = 0$). There are two potential outcomes for each study participant corresponding to each exposure level a: the outcome had they received the exposure, denoted Y_1, and the outcome had they received the comparison condition, denoted Y_0. These two potential outcomes, Y_1 and Y_0, exist for all individuals in the population regardless of whether the individual received the exposure or comparison condition. However, we can observe only one of these outcomes for each participant depending on which exposure condition the individual actually receives.

The mediator is an "intermediate" outcome of the exposure and itself has potential values. For each exposure level a, there is a corresponding potential mediator value, denoted M_a. Also, there is a corresponding potential outcome that reflects the outcome value that would arise under the specific exposure level a and the specific potential mediator value M_a – this potential outcome is denoted $Y_{(a,M_a)}$. Causal definitions of direct and indirect effects require extending the potential outcomes framework such that there is a potential outcome for each treatment and mediator pair. For the case of a binary exposure A, there are four potential outcomes for an individual, formed by crossing both exposure values with both potential mediator values: $Y_{(1,M_1)}$, $Y_{(0,M_0)}$, $Y_{(1,M_0)}$, and $Y_{(0,M_1)}$. Only $Y_{(1,M_1)}$ and $Y_{(0,M_0)}$, which correspond to the individual receiving $A = 1$ or $A = 0$, respectively, can be observed in practice. The other two potential outcomes are hypothetical quantities (i.e., the mediator value is manipulated to take on the value it would have under the other exposure condition); these are necessary to define the causal estimands of interest, as we detail later.

Our use of the above notation implicitly makes several assumptions, often collectively referred to as the stable unit treatment value assumption [2]. First, we assume that an individual's potential outcomes are not influenced by any other individual's treatment status and that there are not multiple versions of the treatment. These assumptions ensure that the potential outcomes are well-defined. Additionally, we invoke the consistency and composition assumptions [2]. The consistency assumption states that the outcome observed for an individual is identical to (i.e., consistent with) the potential outcome that corresponds to their observed exposure value; similarly, their observed mediator value is the potential mediator value that corresponds to their observed exposure value, that is $M = M_a$ and $Y = Y_a$ if $A = a$. For example, if an individual's sexual identity is LGB ($A = 1$), then their observed mediator value M equals M_1 and their observed outcome, Y, equals Y_1. Similarly, if an individual's sexual identity is heterosexual ($A = 0$), then their observed mediator value M equals M_0 and their observed outcome, Y, equals Y_0. The consistency assumption, extended to the joint exposure-mediator, is that if observed exposure $A = a$ and observed mediator $M = m$, then the observed outcome, Y, equals $Y_{(a,m)}$. Finally, the composition assumption pertains to the nested counterfactuals and states that $Y_a = Y_{(a,M_a)}$, that is intervening to set $A = a$, the potential outcome Y_a equals the nested potential outcome $Y_{(a,M_a)}$, when intervening to set $A = a$ and $M = M_a$, the value that the mediator would obtain if A had been a. For example, using both the consistency and composition assumptions, if an individual's sexual identity is LGB ($A = 1$), then their observed outcome Y equals Y_1, which equals $Y_{(1,M_1)}$ and similarly, if an individual's sexual identity is heterosexual ($A = 0$), then their observed outcome Y equals Y_0, which equals $Y_{(0,M_0)}$.

19.2 Natural (In)direct Effects

Mediation is inherently about **causal** mechanisms, and causal effects are defined as the difference between two potential outcomes for an individual. We begin by introducing the potential outcomes needed to define the natural direct and indirect effects.

19.2.1 Definitions

When we cross the possible exposure values and potential mediator values, there are four potential outcome values:

- $Y_{(1,M_1)}$, the potential outcome under the exposed condition and the mediator value corresponding to the exposed condition.
- $Y_{(0,M_0)}$, the potential outcome under the unexposed condition and the mediator value corresponding to the unexposed condition.
- $Y_{(1,M_0)}$, the potential outcome under the exposed condition and the mediator value corresponding to the *unexposed* condition.
- $Y_{(0,M_1)}$, the potential outcome under the unexposed condition and the mediator value corresponding to the *exposed* condition.

As discussed previously, the latter two potential outcomes, $Y_{(1,M_0)}$ and $Y_{(0,M_1)}$, are often referred to as cross-world counterfactuals or cross-world potential outcomes. They are never observed for any individual, yet allow us to more precisely define causal estimands for direct and indirect effects. We begin by defining the total effect (TE) of A on Y:

$$TE = Y_{(a,M_a)} - Y_{(a',M_{a'})} = Y_a - Y_{a'} \tag{19.1}$$

where a and a' are two different levels of the exposure (e.g., $1 =$ LGB and $0 =$ heterosexual).

The natural direct effect (NDE) and natural indirect effect (NIE), which sum to produce the TE, are defined as follows:

$$NDE_{a'} = Y_{(a,M_{a'})} - Y_{(a',M_{a'})} \tag{19.2}$$

$$NIE_a = Y_{(a,M_a)} - Y_{(a,M_{a'})} \tag{19.3}$$

Note that the NDE and NIE definitions rely on the hypothetical (unobservable) potential outcomes. Consider the following decomposition of TE in the case of a binary exposure ($a = 1$ and $a' = 0$):

$$\begin{aligned} TE &= \overbrace{Y_1 - Y_0}^{\text{total effect}} = Y_{(1,M_1)} - Y_{(0,M_0)} \\ &= \overbrace{Y_{(1,M_1)} - Y_{(1,M_0)}}^{\text{natural indirect effect}} + \overbrace{Y_{(1,M_0)} - Y_{(0,M_0)}}^{\text{natural direct effect}} \\ &= NIE_1 + NDE_0 \end{aligned} \tag{19.4}$$

This decomposition is obtained by adding and subtracting $Y_{(1,M_0)}$, the potential outcome we would observe in a world where the exposure $A = 1$ and M is artificially manipulated to take the value it would naturally have under the condition $A = 0$. We can similarly define an alternative decomposition for NDE_1 and NIE_0, by adding and subtracting $Y_{(0,M_1)}$ as follows:

$$\begin{aligned} TE &= \overbrace{Y_1 - Y_0}^{\text{total effect}} = Y_{(1,M_1)} - Y_{(0,M_0)} \\ &= \overbrace{Y_{(1,M_1)} - Y_{(0,M_1)}}^{\text{natural direct effect}} + \overbrace{Y_{(0,M_1)} - Y_{(0,M_0)}}^{\text{natural indirect effect}} \\ &= NDE_1 + NIE_0 \end{aligned} \tag{19.5}$$

The subscripts for NDE denote the condition to which the mediator is held constant, whereas the subscripts for NIE denote the condition to which the exposure is held constant. Each decomposition includes an NIE and an NDE corresponding to opposite subscripts. NDE_0 is sometimes called "pure" direct effect and the NIE_1 is sometimes referred to as the "total" indirect effect. The NDE_1 is sometimes referred to as the "total" direct effect and the NIE_0 is sometimes referred to as the "pure" indirect effect [3]. Although we do not use them here, these labels are sometimes used in software output.

In the context of our motivating example, the NDE_0 term, $Y_{(1,M_0)} - Y_{(0,M_0)}$, compares adult smoking status corresponding to LGB versus heterosexual status, holding early smoking initiation status to the value that would be obtained if heterosexual. The individual NDE_0 will be non-null only if LGB status has an effect on adult smoking status when early smoking initiation status is held fixed – namely, if LGB status has a **direct** effect on the outcome, not through the mediator. The population version of this effect is $NDE_0 = E\left(Y_{(1,M_0)} - Y_{(0,M_0)}\right)$.

The NIE_1 term $Y_{(1,M_1)} - Y_{(1,M_0)}$ compares adult smoking status under the early smoking initiation status that would arise with and without the exposure condition (i.e., LGB status), for those in the exposure group (i.e., LGB women). The individual NIE_1 will be non-null only if LGB status has an **indirect** effect on adult smoking status via early smoking initiation among LGB women. The population version of this effect is $NIE_1 = E\left(Y_{(1,M_1)} - Y_{(1,M_0)}\right)$. Throughout the remainder of the chapter, $NDE_{a'}$, NIE_a, and TE all refer to the population versions of the effects.

The above effect definitions, given as marginal mean differences, can also be defined on other scales (e.g., odds ratio, risk ratio) and as conditional on covariates. In particular, our case study involves a binary outcome, thus effect definitions on the ratio scale are relevant. On the ratio scale (either odds or risk ratios), the $NDE_{a'}$ and NIE_a are defined as follows:

$$\begin{aligned} NDE_{a'} &= E(Y_{(a,M_{a'})})/E(Y_{(a',M_{a'})}) \\ NIE_a &= E(Y_{(a,M_a)})/E(Y_{(a,M_{a'})}) \\ TE &= NDE_{a'} \times NIE_a \end{aligned}$$

In addition, the effects may be defined conditionally on covariates. For example, on the ratio scale,

$$\begin{aligned} NDE_{a'|c} &= E(Y_{(a,M_{a'})}|c)/E(Y_{(a',M_{a'})}|c) \\ NIE_{a|c} &= E(Y_{(a,M_a)}|c)/E(Y_{(a,M_{a'})}|c) \\ TE_c &= NDE_{a'|c} \times NIE_{a|c} \end{aligned}$$

19.2.2 Identifying assumptions

In order to identify the natural (in)direct effects, we must impose assumptions that link the potential outcomes to our actual observed data. The positivity assumption requires that all individuals have some positive probability of receiving each level of the exposure given baseline covariates (i.e., $p(A = a|C = c) > 0$ for $a = 0, 1$ and all c). That is, within all levels of C, there are both exposed (i.e., $A = 1$) and unexposed (i.e., $A = 0$) individuals.

In addition, the positivity assumption requires that, within all levels of C, the range of mediator values in the exposed condition covers that of the unexposed condition when using the decomposition in Equation 19.4. If using the decomposition in Equation 19.5, the positivity assumption requires that, within levels of C, the range of mediator values in the unexposed condition covers that of the exposed condition [4]. If the positivity assumption does not hold, individuals for whom it does not hold can be removed from the sample (note that a causal effect is not meaningful for individuals for whom positivity does not hold).

Additionally, sequential ignorability refers to a set of assumptions regarding confounding. The nonparametric assumptions typically made for identification of natural (in)direct effects conditioning on pre-exposure or baseline variables, C, are the following, for $a, a' = 0, 1$:

1. Conditional on baseline confounders, C, there are no unobserved confounders of the effect of A on Y, $Y_{am} \perp A|C$.

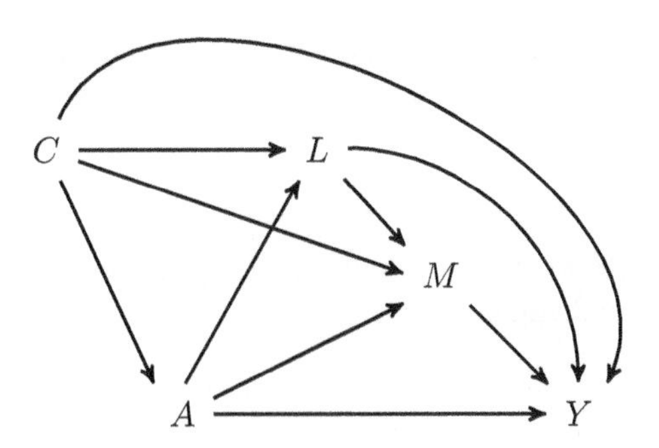

FIGURE 19.3
Mediation model in which Assumption 4 is violated. $L \equiv$ post-exposure confounders.

2. Conditional on C, there are no unobserved confounders of the effect of A on M, $M_a \perp A|C$.

3. Conditional on A and C, there are no unobserved confounders of the effect of M on Y, $Y_{am} \perp M|A, C$.

4. There are no confounders (observed or unobserved) of the effect of M on Y that are affected by A, $Y_{am} \perp M_{a'}|C$.

If individuals are randomly assigned to levels of exposure, then Assumptions 1 and 2 should hold. However, Assumptions 3 and 4 may not hold even when exposure is randomized. See VanderWeele (2015) [5] or Nguyen et al. (2021) [4] for further discussion of these identifying assumptions.

Assumption 4, often referred to as the cross-world independence assumption [6], is violated when there exist post-treatment confounders, denoted L, that may have been influenced by the exposure but are also confounders of the mediator and outcome, as shown in Figure 19.3. Note that if they are not influenced by the exposure they do not violate Assumption 4. However, in practice, it is often difficult to discern whether they have been affected by the exposure (as they occur after the exposure). These variables may be viewed as additional mediators, and many of the weighting schemes introduced in the next section apply to multiple mediators considered en bloc. For simplicity though, we begin by introducing the weighting methods for a single mediator. In Section 19.4 we discuss weighting estimators for multiple mediators.

19.3 Estimation of Natural (In)direct Effects for a Single Mediator

We begin by presenting weighting estimators for a single mediator for the natural direct and indirect effects. The basic idea is to obtain estimates of $E\left(Y_{(1,M_1)}\right)$, $E\left(Y_{(0,M_0)}\right)$, $E\left(Y_{(1,M_0)}\right)$, and $E\left(Y_{(0,M_1)}\right)$. These estimates are then plugged into Equations 19.4 or 19.5 to obtain estimates of the natural (in)direct effects. For $E\left(Y_{(1,M_1)}\right)$ and $E\left(Y_{(0,M_0)}\right)$, we can use the usual inverse propensity weights and compute the average of the observed outcome in the exposure group weighted by $1/p(A=1|C)$ and in the unexposed group weighted by $1/(1-p(A=1|C))$, where $p(A=1|C)$ is estimated. Alternatively, these weights may be stabilized by replacing the numerator with marginal probabilities thereby reducing the variability of the weights. In this case, the exposure group is weighted by $p(A=1)/p(A=1|C)$ and the unexposed group is weighted by $(1-p(A=1)/(1-p(A=1|C))$.

Next, we need to obtain estimates of $E\left(Y_{(1,M_0)}\right)$ (for the decomposition in Equation 19.4) and/or $E\left(Y_{(0,M_1)}\right)$ (for the decomposition in Equation 19.5). It is the estimation of these cross-world potential outcomes that complicates mediation analysis. Several approaches have been proposed for estimating these cross-world potential outcomes, which we now present in detail.

19.3.1 Ratio of mediator probability weighting (RMPW)

Hong (2010) [7] was the first to define weights to estimate cross-world potential outcomes $E\left(Y_{(a,M_{a'})}\right)$. The weight, for individual i in the group that experiences exposure level a, is

$$w_{i(a,M_{a'})} = \frac{p(M_{a'} = M_i|C = C_i)}{p(M_a = M_i|C = C_i)} \frac{1}{p(A = a|C = C_i)}, \tag{19.6}$$

which is equal to the following formula (which is used for estimation)

$$w_{i(a,M_{a'})} = \overbrace{\frac{p(M = M_i|A = a', C = C_i)}{p(M = M_i|A = a, C = C_i)}}^{\text{mediator probability ratio weight}} \overbrace{\frac{1}{p(A = a|C = C_i)}}^{\text{IPW (for A=a)}}. \tag{19.7}$$

The first term in this weight is a ratio of the probabilities that the potential mediator $M_{a'}$ (in the numerator) and the potential mediator M_a (in the denominator) take on the person's observed mediator value M_i, given their covariate values C_i. The point of a probability ratio weight is to morph one distribution so that it resembles another. Consider a subpopulation where the individuals share the same C values. Within this subpopulation we are using data from individuals with actual exposure a to infer the mean of $Y_{(a,M_{a'})}$, but their mediator values are M_a, not $M_{a'}$. Probability ratio weighting shifts the mediator distribution so it resembles the desired distribution, that of $M_{a'}$, effectively correcting the mismatch so that the weighted outcome distribution resembles that of the cross-world potential outcome (within levels of C).

The second term is the usual inverse probability weight (IPW) that equates the exposed and unexposed groups for the distribution of C and shifts that distribution to mimic to the distribution of C in the full sample. The stabilized version of the inverse probability weight, $\frac{p(A=a)}{p(A=a|C=C_i)}$, can also be used.

When estimating $E\left(Y_{(1,M_0)}\right)$ (for the decomposition in Equation 19.4), the weight is:

$$w_{i(1,M_0)} = \frac{p(M = M_i|A = 0, C = C_i)}{p(M = M_i|A = 1, C = C_i)} \frac{1}{p(A = 1|C = C_i)}. \tag{19.8}$$

Similarly, for estimating $E\left(Y_{(0,M_1)}\right)$ (for the decomposition in Equation 19.5), the weight is:

$$w_{i(0,M_1)} = \frac{p(M = M_i|A = 1, C = C_i)}{p(M = M_i|A = 0, C = C_i)} \frac{1}{p(A = 0|C = C_i)}. \tag{19.9}$$

In practice, these inverse probability weights can be estimated using logistic regression for a binary exposure. For a binary mediator variable, each of the mediator probabilities can be estimated using logistic regression. For a non-binary mediator, estimation of the first term (now a ratio of probability densities) becomes more complicated.

The weighted outcome model is fit by duplicating the exposed individuals when estimating $E\left(Y_{(1,M_0)}\right)$ for the decomposition in Equation 19.4. Let D be an indicator variable that equals 1 for the duplicates and 0 otherwise. Thus, the analytic data set is composed of the original exposed and unexposed individuals for which $D = 0$ and duplicated exposed individuals in which $D = 1$. Weights are assigned as follows. If $A = 0$ and $D = 0$, then the weight for individual i, w_i, is

TABLE 19.1
Estimates from rmpw.

	Estimate	Std.Error	p-val
Natural Direct Effect (NDE)	0.128	0.008	<0.001
Natural Indirect Effect (NIE)	0.033	0.003	<0.001

$p(A=0)/p(A=0|C)$. If $A=1$ and $D=1$, then $w_i = p(A=1)/p(A=1|C)$, that is, the usual inverse propensity weights used to obtain estimates of $E\left(Y_{(1,M_1)}\right)$ and $E\left(Y_{(0,M_0)}\right)$. If $A=1$ and $D=0$, then $w_i = w_{i(1,M_0)}$. In assigning these weights, the outcome model can be specified as follows to simultaneously estimate the natural (in)direct effects:

$$\mathrm{E}(Y) = \beta_0 + \beta_1 A + \beta_2 D$$

where β_1 represents an estimate of the NDE_0 and β_2 represents an estimate of the NIE_1. This outcome model was later termed the "natural effect model" [8].

19.3.1.1 Applied example

As detailed in Section 19.7, RMPW can be implemented in R using the rmpw package [9] or the medflex package [10]. Below we present results from implementing RMPW on our case study data (using the rmpw package).

Table 19.1 presents estimates on the marginal risk difference scale. Note that by default, the rmpw package returns a single estimate for the NDE and the NIE, thereby assuming NDE_1=NDE_0 and NIE_1=NIE_0 or in other words, that there is no interaction between the exposure and the mediator. See Section 19.7 for details on changing this default option. The total effect can be obtained by summing the NDE and NIE estimate – in this case, the total effect estimate is $0.128+0.033=0.161$. While this TE represents a marginal effect, this is very similar to the crude difference in smoking rates observed in the NSDUH data of 16% (31% among LGB women vs. 15% among heterosexual women). The NIE estimate is statistically significant, indicating that early smoking initiation represents a significant pathway regarding adult smoking status, with early initiation accounting for approximately a 3.3% increase in magnitude in adult smoking rates. Comparing the ratio of the NIE and TE (i.e., the proportion mediated on the risk difference scale [5]), this suggests that the early initiation pathway accounts for approximately 20% of the observed disparities in adulthood. Additionally, a statistically significant NDE estimate indicates that LGB identity also has a significant effect on adult smoking that is not attributed to the early smoking initiation pathway.

19.3.2 Huber (2014)

Under the previously stated assumptions of consistency, positivity, and sequential ignorability (i.e., C is strictly pre-exposure, not affected by A), Huber [11] used the following manipulation (i.e., Bayes Rule)

$$p(M|A,C) = \frac{p(A|M,C)p(M|C)}{p(A|C)}$$

to obtain an easier set of weights to estimate. The formula below (which is equivalent to Hong's weights [7]) shows the weight for individuals i in the group that experienced exposure level a.

$$w_{i(a,M_{a'})} = \overbrace{\frac{p(A=a'|M_i,C_i)}{p(A=a|M_i,C_i)}}^{\text{Odds Weight}} \overbrace{\frac{1}{p(A=a'|C_i)}}^{\text{IPW (for A=a')}} \tag{19.10}$$

TABLE 19.2
Estimates from `causalweight`.

	Estimate	Std.Error	p-val
TE	0.156	0.009	0.000
NDE_0	0.122	0.009	0.000
NIE_1	0.034	0.003	0.000
NDE_1	0.117	0.010	0.000
NIE_0	0.039	0.002	0.000

TE = total effect; NDE = natural direct effect; NIE = natural indirect effect

It is the combination of an odds weight that morphs the $A = a$ group to mimic the joint (C, M) distribution of the $A = a'$ group, and an inverse probability weight that shifts the C distribution to that of the full sample. Note that the denominator of the IPW term here is different from that in Hong's formula; it is the conditional probability of being in the $A = a'$ (not $A = a$) group.

Equation 19.10 implies that for the decomposition in Equation 19.4, the cross-world weights for treated units are

$$w_{i(1,M_0)} = \frac{p(A = 0 \mid M_i, C_i)}{p(A = 1 \mid M_i, C_i)} \frac{1}{p(A = 0 \mid C_i)},$$

and for the other decomposition, the cross-world weights for control units are

$$w_{i(0,M_1)} = \frac{p(A = 1 \mid M_i, C_i)}{p(A = 0 \mid M_i, C_i)} \frac{1}{p(A = 1 \mid C_i)}.$$

Huber [11] used logistic or probit regression to estimate these weights. More recently, Generalized Boosted Modeling (GBM) has been implemented in the `R` package `twangMediation` for estimating the weights [12]. GBM is a nonparametric machine learning algorithm that has been shown to outperform logistic regression [13]. Once the weights have been estimated, they are used to compute a weighted average of the observed outcomes. Specifically, to estimate $E\left(Y_{(a,M_a)}\right)$ and $E\left(Y_{(a',M_{a'})}\right)$, we use the usual inverse propensity weights and compute the average of the observed outcomes in the exposure group weighted by $\frac{p(A=1)}{p(A=1|C)}$ and in the unexposed group weighted by $\frac{1-p(A=1)}{1-p(A=1|C)}$, where $p(A = 1|C)$ is estimated based on a model specified for the exposure. For estimating $E\left(Y_{(a,M_{a'})}\right)$ and $E\left(Y_{(a',M_a)}\right)$, we use the weights in Equation 19.10. Using these estimates of the counterfactuals, the natural (in)direct effects can be computed using the definitions in Section 19.2.1.

19.3.2.1 Applied example

As detailed in Section 19.7, Huber's approach can be implemented in `R` using the `causalweight` package (which estimates weights using logistic or probit regression) [14], the `mediationClarity` package (which currently uses logistic regression) [15], or the `twangMediation` package (which uses GBM or logistic regression) [12]. Below we present results from implementing Huber's approach on our case study data using `causalweight` and `twangMediation`.

Table 19.2 presents results estimated with logistic regression using the `causalweight` package. The TE estimate of 0.156 represents a difference in magnitude of 15.6% in adult smoking rates between LGB and heterosexual women. Both NIE estimates are statistically significant, indicating that for both LGB and heterosexual women, early smoking initiation represents a significant pathway regarding adult smoking status (with early initiation accounting for approximately a 3–4% increase

TABLE 19.3
Estimates from `twangMediation`.

	Estimate	Std.Error	CI.min	CI.max
TE	0.123	0.009	0.105	0.141
NDE_0	0.097	0.009	0.080	0.115
NIE_1	0.025	0.003	0.020	0.031
NDE_1	0.094	0.009	0.076	0.112
NIE_0	0.029	0.001	0.026	0.031

in magnitude in adult smoking rates). Additionally, both NDE estimates are significant, indicating that LGB identity also has a significant effect on adult smoking that is not attributed to the early smoking initiation pathway.

Table 19.3 presents results estimated with GBM using the `twangMediation` package. The TE estimate of 0.123 represents a difference in magnitude of 12.3% in adult smoking rates between LGB and heterosexual women. Both NIE estimates are statistically significant, indicating that for both LGB and heterosexual women, early smoking initiation represents a significant pathway regarding adult smoking status (with early initiation accounting for approximately a 2–3% increase in magnitude in adult smoking rates). Additionally, both NDE estimates are significant, indicating that LGB identity also has a significant effect on adult smoking that is not attributed to the early smoking initiation pathway. Differences in estimates between the `twangMediation` and `causalweight` packages can be attributed to differences between GBM and logistic estimation of weights.

19.3.3 Nguyen et al. (2022)

Nguyen et al. (2022) [16] identified a third way to estimate the cross-world weights, based on the equivalence between the two weight formulas given by Equations 19.7 and 19.10 with a third expression:

$$w_{i(a,M_{a'})} = \frac{1}{p(A=a)} \frac{p(C=C_i, M=M_i \mid A=a')\frac{p(A=a')}{p(A=a'|C=C_i)}}{p(C=C_i, M=M_i \mid A=a)}. \tag{19.11}$$

Here the first term can be ignored as it is a constant (dropping it results in stabilized weights). The second term can be interpreted as the ratio of two densities. The denominator is the joint density of (C, M) for units in the $A = a$ condition. The numerator is the joint density of (C, M) for units in the $A = a'$ condition but where these units have been weighted to mimic the covariate distribution of the population. This means the cross-world weights essentially morph the $A = a$ subsample to mimic the (C, M) distribution in *the $A = a'$ subsample that has been weighted to mimic the population C distribution*. We do not use Equation 19.11 as a formula, but rely on this insight to develop a simple method for estimating the weights. It involves two steps:

1. Estimate weights for the $A = a'$ subsample to mimic the full sample C distribution.
2. Estimate weights for the $A = a$ subsample to mimic the (C, M) distribution of the weighted $A = a'$ sample from the previous step. These weights are the cross-world weights.

In practice these steps can be implemented via probability models (using IPW weights from the propensity score model in the first step and using odds weights from a weighted model for A given C, M in the second step) or via direct estimation of weights for optimal balance.

Conceptually, the connection among the three weighting estimation methods above is that the objective of the cross-world weights (for estimating $E(Y_{(a,M_{a'})})$ is to weight the $A = a$ subsample so that it mimics the full sample (population) C distribution and the M given C distribution in

TABLE 19.4
Estimates obtained using Nguyen et al. (2021) weights.

	Estimate	Std.Error	t-value	p-val
NDE_0	0.097	0.005	19.034	<0.000
NIE_1	0.027	0.005	5.322	<0.000

individuals with $A = a'$. The latter identifies the distribution of $M_{a'}$, so intuitively this achieves a swapping out of the M_a distribution for the $M_{a'}$ distribution. The three weighting estimation methods all achieve this but in different ways. Hong's (2010) [17] method directly achieves this by combining an inverse probability weight that achieves the target C distribution and a density ratio weight that achieves the target M given C distribution. Huber's (2014) [11] formula achieves this in a zig-zag manner where the $A = a$ subsample is first weighted by an odds weight to mimic the (C, M) distribution of the $A = a'$ subsample, and then is weighted by an inverse probability weight to morph the C distribution to that of the full sample. The third method identified by Nguyen et al. (2022) [16], like the first method, is direct weighting. The first step simply serves to set up the right target for use in the weighting of the $A = a$ subsample. This works because the weighted $A = a'$ subsample obtained in the first step has both elements of the target: the same C distribution as in the full sample (due to the inverse probability weighting), and the right M given C distribution (as these are individuals with $A = a'$).

In terms of ease of use, the three methods require fitting different models (i.e., estimating different functions), so each may be more convenient in different settings. The last two methods do not require modeling mediator densities, so are easier to use generally (because densities are hard to estimate) and especially when the mediator is non-binary or in the case of multiple mediators. The third method is the only one that weights the source ($A = a$ subsample) to mimic a target in one step rather than relying on multiplying two weights. Therefore, methods that directly estimate weights (i.e., methods that do not necessarily require fitting probability models) could be used to morph one sample to mimic another.

19.3.3.1 Applied example

As detailed in Section 1.7, Nguyen's approach can be implemented in R using the `mediationClarity` package [15]. Table 19.4 presents the results from our case study using this weighting estimation approach based on probability models. The estimated effects presented are on the marginal risk difference scale. Summing the NDE_0 and NIE_1 estimates, we obtain a TE estimate of 0.124, representing a difference of 12.4 percentage points in adult smoking prevalence between LGB and heterosexual women (standardized to the full population covariate distribution). The NIE_1 estimate is significant, indicating that early smoking initiation represents a significant pathway regarding adult smoking status (accounting for a 2.7 percentage points increase in smoking prevalence). Also, the NDE_0 estimate is significant, indicating that LGB identity also has a significant effect on adult smoking that is not attributed to the early smoking initiation pathway.

19.3.4 Albert (2012)

Albert (2012) proposed a weighting based approach that also does not specify a model for the mediator. This approach has an advantage if the mediator is non-binary. To implement this approach, a model for the outcome and a model for the exposure given pre-exposure covariates are specified. Suppose that the reference level of the exposure is $A = a'$. We need estimates for the counterfactuals, $E\left(Y_{(a,M_a)}\right)$, $E\left(Y_{(a',M_{a'})}\right)$, and $E(Y_{(a,M_{a'})})$. For $E\left(Y_{(a,M_a)}\right)$ and $E\left(Y_{(a',M_{a'})}\right)$, we can use the usual inverse propensity weights and compute the average of the observed outcome in the exposure group weighted by $\frac{p(A=1)}{p(A=1|C)}$ and in the unexposed group weighted by $\frac{1-p(A=1)}{1-p(A=1|C)}$. For these weights,

TABLE 19.5
Estimates from `CMAverse` for the Albert (2012) approach.

	Estimate	Std.error	CI.min	CI.max	p-val
TE	1.940	0.079	1.775	2.106	<0.000
NDE_0	1.718	0.059	1.616	1.831	<0.000
NDE_1	1.680	0.068	1.545	1.816	<0.000
NIE_0	1.155	0.017	1.125	1.193	<0.000
NIE_1	1.129	0.028	1.078	1.190	<0.000
% Mediated	0.236	0.040	0.162	0.314	<0.000

$p(A = 1|C)$ is estimated based on a model specified for the exposure. $E(Y_{(a,M_{a'})})$ is estimated by obtaining predicted values from the outcome model, $E(Y|A = a, M = m, C = c)$. Specifically, for each individual with $A = a'$, obtain a predicted value for the outcome if the individual had been in exposure $A = a$ instead of $A = a'$ (using their observed values of the mediator and covariates). A weighted average of these predicted values for individuals with $A = a'$ is computed using the weights $\frac{1-p(A=1)}{1-p(A=1|C)}$ and provides an estimate of $E(Y_{(a,M_{a'})})$.

19.3.4.1 Applied example

As detailed in Section 19.7, Albert's weighting method can be implemented in R using the `CMAverse` package [18]. Below we present results from implementing Albert's approach on our case study data. As shown in Table 19.5, the TE estimate is 1.94 on the OR scale, indicating that LGB women have nearly twice the odds of adult smoking as heterosexual women. Both NIE estimates are statistically significant, indicating that LGB status increases the odds of adult smoking by 1.1–1.2 times through the pathway of early smoking initiation. Based on the NDE estimates, we conclude that LGB status increases the odds of adult smoking by 1.7–1.8 times through alternative pathways excluding early initiation. Additionally, the proportion mediated estimate indicates that early initiation accounts for approximately 24% of the total disparities in adult smoking. Under the rare outcome assumption, and effects given on the OR scale, the proportion mediated on a risk difference scale [5] for the decomposition in Equation 19.4 is calculated as follows:

$$\% \text{ Mediated} = \frac{NDE_0(NIE_1 - 1)}{TE - 1}.$$

19.3.5 Inverse odds ratio weighting (IORW)

The Inverse Odds Ratio Weighting (IORW) method [19] involves the use of an odds ratio-based weight that relates the exposure and the mediator. Specifically, this weight is calculated by specifying a model that regresses the exposure variable on the mediator and pre-exposure covariates. The weight is then used in an outcome model that does not include the mediator (i.e., only includes the exposure variable and pre-exposure covariates) to estimate the NDE, which is given by the coefficient for the exposure variable and is conditional on covariates included in this model. Given the aforementioned identification assumptions, the NDE may be subtracted from the TE to obtain an estimate of the NIE in the case of a continuous outcome. In the case of a binary, count, or survival outcome, the NIE may be obtained by dividing the TE estimate by the NDE estimate. The outcome model for estimating the TE is the same as that for estimating the NDE, except that it is not weighted. The TE estimate is given by the coefficient for the exposure variable, which is conditional on the covariates included in this model. Thus, all of these effects – TE, NDE, and NIE – are conditional on the covariates included in the outcome models.

TABLE 19.6
Estimates from CMAverse using IORW.

	Estimate	Std.error	CI.min	CI.max	p-val
TE	1.988	0.068	1.846	2.104	<0.000
NDE_0	1.743	0.071	1.608	1.884	<0.000
NIE_1	1.141	0.033	1.077	1.211	<0.000
% Mediated	0.249	0.051	0.148	0.352	<0.000

This approach takes advantage of the invariance of the OR (i.e., the OR for two variables will be the same regardless of which of the two variables is the independent variable and which is the dependent variable). The weighting deactivates the indirect effects through the mediator by rendering the exposure and mediator independent. Because this approach does not rely on a model for the mediator, it is very flexible. For example, it easily accommodates mediators of any variable type (i.e., continuous, categorical).

If the exposure variable is binary, then a logistic regression model can be specified in which the mediator(s) and pre-exposure covariates are predictors of the exposure. The weight is then computed as the inverse of the predicted OR from this model for the individuals in the exposure group, as follows:

$$w_i = \frac{p(A=0|M_i, C_i)}{p(A=1|M_i, C_i)}.$$

Individuals in the control group are given a weight of one. Alternatively, more stable weights for the exposure individuals may be obtained by using the inverse of the predicted odds from this model, which is referred to as inverse odds weighting [19].

This approach is flexible in that the exposure variable need not be binary. For example, if there are three levels for the exposure variable, then a multinomial logistic regression, ordinal logistic regression, or other appropriate model could be used. Again, individuals in the control (i.e., unexposed) group are given a weight of one. Individuals in the other groups are given weights of the form:

$$w_i = \frac{p(A=0|M_i, C_i)}{p(A=a|M_i, C_i)}.$$

In addition the model for the exposure variable can be flexible (e.g., could include polynomial terms for the mediator(s)). Similarly, the outcome model is flexible in that it may be any generalized linear model with a nonlinear link function, a Cox proportional hazard model in the case of a survival outcome, or a quantile regression.

19.3.5.1 Applied example

As detailed in Section 19.7, the IORW method can be implemented using the CMAverse package [18] in R. When implementing IORW, the CMAverse package generates conditional estimates on the OR scale. Table 19.6 presents results from implementing the IORW method on our case study data.

On the OR scale, the TE estimate is 1.99, indicating that LGB women have nearly twice the odds of adult smoking as heterosexual women. The NIE_1 estimate is statistically significant, indicating that LGB status increases the odds of adult smoking by 1.14 times through the pathway of early smoking initiation. The proportion mediated estimate indicates that early initiation accounts for approximately 25% of the total disparities in adult smoking. Additionally, the statistically significant NDE_0 estimate indicates that LGB status increases the odds of adult smoking by 1.74 times through alternative pathways excluding early initiation.

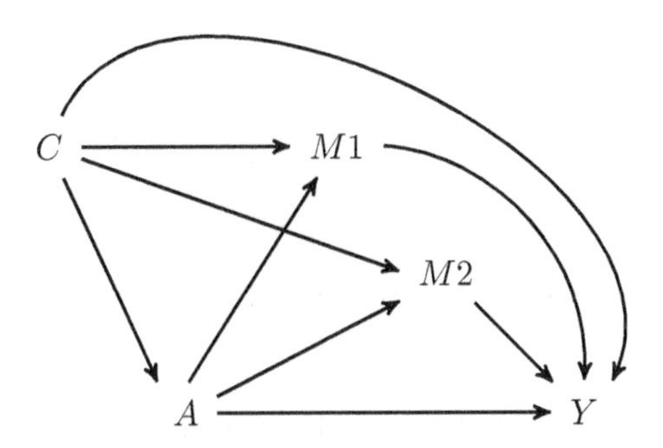

FIGURE 19.4
Multiple mediation model in which mediators are independent.

19.4 Natural (In)direct Effects for Multiple Mediators

There are different ways in which multiple mediators may be hypothesized to mediate the effect of an exposure on an outcome. First, the mediators may themselves be causally ordered or not. Multiple (non-causally ordered) mediators may be hypothesized when an intervention is designed to target multiple mediators that are then hypothesized to affect the outcome. Second, the mediators may be correlated or assumed to be conditionally independent given C, as in Figure 19.4. Third, interactions between the mediators may or may not be present. The weighting approaches presented in the previous section apply to what is commonly referred to as the joint natural (in)direct effects [20], that is, the NIE and NDE estimands are calculated to reflect mediation jointly through all mediators, rather than separate path-specific mediation effects (e.g., [21]). Thus, we will limit our presentation to these effects.

For example, considering our case study, in addition to the original mediator, early cigarette initiation (denoted as $M1$), we now include another mediator, early alcohol initiation (denoted as $M2$), and the outcome is now nicotine and/or alcohol dependence in adulthood. We may be interested in $A \rightarrow M1 \rightarrow Y$, $A \rightarrow M2 \rightarrow Y$, or the direct effect of A on Y that is not due to either $M1$ or $M2$. If a researcher considers the mediators one at a time, then for example, in estimating the effect of $A \rightarrow M1 \rightarrow Y$, the effect through $M2$ will be incorporated into the direct effect. More often though, researchers will be interested in how a treatment or exposure affects the outcome through both mediators, including their interaction (if any).

19.4.1 Notation and assumptions

Considering non-causally ordered mediators, the notation and assumptions for a single mediator extend to multiple mediators. For example, the observed outcome, Y, is equal to $Y_{(a,M1_a,M2_a)}$ when intervening to set $A = a$. The previous ignorability assumptions are extended as follows:

1. Conditional on baseline confounders, C, there are no unobserved confounders of the effect of A on Y: $Y_{a,m1,m2} \perp A|C$.

2. Conditional on C, there are no unobserved confounders of the effect of A on $M1$ or $M2$: $M1_a, M2_a \perp A|C$.

3. Conditional on A and C, there are no unobserved confounders of the effect of $M1$ or $M2$ on Y: $Y_{a,m1,m2} \perp M1, M2|A, C$.

4. There are no confounders (observed or unobserved) of the effect of $M1$ or $M2$ on Y that is affected by A: $Y_{a,m1,m2} \perp M1_{a'}, M2_{a''}|C$ and $M1_{a'} \perp M2_{a''}|C$.

19.4.2 Definitions

The joint natural (in)direct effects of $M1, M2$ are defined as [20]

$$\begin{aligned} NDE_{a'} &= Y_{(a,M1_{a'},M2_{a'})} - Y_{(a',M1_{a'},M2_{a'})} \\ NIE_{a} &= Y_{(a,M1_{a},M2_{a})} - Y_{(a,M1_{a'},M2_{a'})} \end{aligned}$$

As before,

$$\begin{aligned} TE &= NDE_1 + NIE_0 \\ &= NDE_0 + NIE_1 \end{aligned}$$

The joint NIE is the effect of A on Y that is mediated through $M1$ and/or $M2$, and the joint NDE is the effect that is not mediated through $M1$ or $M2$.

19.4.3 Estimation

Below, we detail multiple estimation approaches that can be used with multiple (independent) mediators.

19.4.3.1 RMPW

The RMPW method may be extended to multiple mediators [22] for the special case where all the mediators are independent of one another conditional on baseline covariates and exposure, as in Figure 19.4. This condition can and should be checked using data. For example, with continuous mediators, one can model the mediators given exposure and covariates, and check within each exposure condition whether the residuals are uncorrelated. In this special case, to obtain an estimate of $E(Y_{(a,M1_{a'},M2_{a''})})$, the following weights are applied to individuals i in the $A = a$ group, and the outcome is averaged using these weights in this group:

$$\frac{p(M1 = M1_i|A = a', C = C_i)}{p(M1 = M1_i|A = a, C = C_i)} \frac{p(M2 = M2_i|A = a'', C = C_i)}{p(M2 = M2_i|A = a, C = C_i)} \frac{1}{p(A = a|C = C_i)}.$$

This approach assumes correct models for the exposure and mediator but not the outcome. Because these weights rely on models for the mediators, the weights may become unstable if any of the mediators are continuous. Multiple mediators using RMPW are not currently implemented in the `rmpw` or `medflex` R packages.

19.4.3.2 Huber (2014)

Huber's [11] weight formula applies regardless of whether there is one or more mediators. The weights are obtained via models for the exposure, rather than the mediators, which is particularly advantageous with non-binary or multiple mediators. To obtain an estimate of $E(Y_{(a,M1_{a'},M2_{a'})})$,

TABLE 19.7
Multiple mediator setting: Estimates from `causalweight`.

	Estimate	Std.Error	p-val
TE	0.110	0.008	0.000
NDE_0	0.079	0.008	0.000
NIE_1	0.031	0.003	0.000
NDE_1	0.077	0.008	0.000
NIE_0	0.033	0.002	0.000

the following weights may be used to obtain a weighted average of Y:

$$\frac{p(A = a'|M1_i, M2_i, C_i)}{p(A = a|M1_i, M2_i, C_i)} \frac{1}{p(A = a'|C_i)}. \tag{19.12}$$

The `causalweight` package implements Huber's weighting approach, estimated using logistic or probit regression, the `mediationClarity` package implements it using logistic regression, and the `twangMediation` package implements it using GBM or logistic regression. We demonstrate this approach with the `causalweight` and `twangMediation` packages, using an extension of our prior LGB disparities analyses examining the effect of a single mediator, early smoking initiation, on adult smoking status. We now consider another mediator, early alcohol initiation, `alc15`, in addition to early smoking initiation, `cig15`. The outcome is an indicator for whether an individual meets criteria for either alcohol or nicotine dependence, `alc_cig_depend`, in adulthood.

Recall that `causalweight` provides marginal risk difference estimates. As shown in Table 19.7, the TE estimate of 0.110 represents a difference in magnitude of 11.0% in the prevalence of alcohol/nicotine dependence between LGB and heterosexual women (standardized to the full population covariate distribution). This indicates a significant disparity. Both NIE estimates are significant, indicating that early initiation of alcohol and smoking jointly represents a significant mediating pathway to adult dependence status (with early initiation accounting for approximately a 3% increase in magnitude in adult dependence rates). Examining the ratio of NIE to TE, we conclude that early initiation accounts for approximately 30% of the adult disparity in alcohol/nicotine dependence. Additionally, both NDE estimates are significant, indicating that LGB identity also has a significant effect on adult alcohol/nicotine dependence that is not attributed to early initiation of alcohol or smoking.

Table 19.8 presents the estimates from using `twangMediation` to implement Huber's approach. Recall that `twangMediation` provides marginal risk difference estimates. As shown in Table 19.8, the TE estimate of 0.084 represents a difference in magnitude of 8.4% in the rates of alcohol/nicotine dependence between LGB and heterosexual women, indicating a significant disparity. Both NIE estimates are significant, indicating that early initiation of alcohol and smoking jointly represent a significant mediating pathway to adult dependence status (with early initiation accounting for approximately a 3% increase in magnitude in adult dependence rates). Examining the ratio of NIE to TE, we conclude that early initiation accounts for approximately 30% of the adult disparity in alcohol/nicotine dependence. Additionally, both NDE estimates are significant, indicating that LGB identity also has a significant effect on adult alcohol/nicotine dependence that is not attributed to early initiation of alcohol or smoking.

19.4.3.3 Nguyen et al. (2022)

The weighting method in Nguyen et al. (2022) [16] is agnostic to whether the mediator is a single variable or multiple variables. The estimated results from the `mediationClarity` R package for NDE_0 and NIE_1 are shown in Table 19.9. Results are shown on the marginal risk difference

TABLE 19.8
Multiple mediator setting: Estimates from `twangMediation`.

	Estimate	Std.error	CI.min	CI.max
TE	0.084	0.008	0.068	0.099
NDE_0	0.059	0.008	0.044	0.074
NIE_1	0.025	0.003	0.018	0.032
NDE_1	0.056	0.008	0.041	0.072
NIE_0	0.027	0.001	0.025	0.030

TABLE 19.9
Multiple mediator setting: Estimates using the Nguyen et al. [16] weights.

	Estimate	Std.Error	t-value	p-val
NDE_0	0.060	0.004	13.336	<0.000
NIE_1	0.027	0.004	6.029	<0.000

scale. Summing across the NDE and NIE estimates, we obtain a TE estimate of 0.087 representing a difference in magnitude of 8.7% in the rates of alcohol/nicotine dependence between LGB and heterosexual women. The NIE estimate is significant, indicating that early initiation of alcohol and smoking jointly represent a significant mediating pathway to adult dependence status (with early initiation accounting for approximately a 2.7% increase in magnitude in adult dependence rates). Additionally, both NDE estimates are significant, indicating that LGB identity also has a significant effect on adult alcohol/nicotine dependence that is not attributed to early initiation of alcohol or smoking.

19.4.3.4 Albert (2012)

VanderWeele and Vansteelandt [20] extended Albert's (2012) [23] approach to multiple mediators. This approach can be used in the presence of mediator – mediator interactions as well. To obtain an estimate of $E(Y_{(a,M1_a,M2_a)})$, the following weights are used to obtain a weighted average of Y for individuals with $A = a$,

$$\frac{p(A = a)}{p(A = a|C_i)}$$

and for individuals with $A = a'$, the weights

$$\frac{p(A = a')}{p(A = a'|C_i)}$$

are used to obtain a weighted average of Y as an estimate of $E(Y_{(a',M1_{a'},M2_{a'})})$. Finally, an estimate of $E(Y_{(a,M1_{a'},M2_{a'})})$ is obtained using an outcome model, $E(Y|A = a, M1, M2, C)$, and then for each individual with $A = a'$, obtain a predicted value for the outcome if the individual had been in exposure $A = a$ instead of $A = a'$. Each individual's observed values of the mediators and covariates are used in obtaining the predicted values. A weighted average of these predicted values for individuals with $A = a'$ is computed using the weights in Equation 19.4.3.4 and provides an estimate of $E(Y_{(a,M1_{a'},M2_{a'})})$. As is the case for Albert's (2012) [23] approach, this extension to multiple mediators does not require a model for the mediators. Thus, it is advantageous when the analyst is more comfortable specifying models for the exposure and outcome.

VanderWeele & Vansteelandt's [20] weighting approach for multiple mediators is implemented in the `CMAverse` R package. Table 19.10 presents the estimates from our case study. On the OR scale, the TE estimate is 1.826, indicating that LGB women have approximately 1.8 times the odds of adult

TABLE 19.10
Multiple mediator setting: Estimates using VanderWeele & Vansteelandt's [20] weighting approach.

	Estimate	Std.error	CI.min	CI.max	p-val
TE	1.826	0.077	1.680	1.982	<0.000
NDE_0	1.472	0.055	1.375	1.578	<0.000
NIE_1	1.241	0.029	1.188	1.297	<0.000
NDE_1	1.503	0.062	1.393	1.623	<0.000
NIE_0	1.215	0.018	1.178	1.247	<0.000
CDE	1.512	0.061	1.408	1.628	<0.000
% Mediated	0.429	0.038	0.365	0.489	<0.000

TABLE 19.11
Multiple mediator setting: IORW estimates.

	Estimate	Std.error	CI.min	CI.max	p-val
TE	1.772	0.070	1.641	1.924	<0.000
NDE_0	1.573	0.080	1.442	1.758	<0.000
NIE_1	1.126	0.035	1.064	1.192	<0.000
% Mediated	0.258	0.065	0.128	0.372	<0.000

alcohol/nicotine dependence as heterosexual women. Both NIE estimates are statistically significant, indicating that LGB status increases the odds of adult dependence 1.2 times through the pathways of early alcohol and smoking initiation. Results indicate that early alcohol/smoking initiation accounts for approximately 43% of the total disparities in adult alcohol/nicotine dependence. Additionally, the NDE estimates indicate that LGB status increases the odds of adult dependence by approximately 1.5 times through alternative pathways excluding early initiation.

19.4.3.5 IORW

IORW is easily extended to multiple mediators [24]. As in the case of a single mediator, this approach does not specify a model for the mediators. It can accommodate multiple mediators of mixed variable types, and interactions between the mediators and exposure without the need to specify them.

The IORW approach for multiple mediators is implemented in the `CMAverse` package in R. Table 19.11 presents the estimates from our case study. The TE estimate indicates that LGB women have approximately 1.8 times the odds of adult alcohol/nicotine dependence as heterosexual women. The NIE_1 estimate is statistically significant, indicating that LGB status increases the odds of adult dependence by 1.13 times through the pathways of early alcohol and smoking initiation, which accounts for approximately 26% of the total disparities in adult alcohol/nicotine dependence. The statistically significant NDE estimate indicates that LGB status increases the odds of adult alcohol/nicotine dependence by 1.57 times through alternative pathways excluding early alcohol and smoking initiation.

19.5 Interventional (In)direct Effects

Due to the likelihood of time-varying confounders of the mediator and outcome that have themselves been influenced by the exposure (see Figure 19.3), an alternative set of direct and indirect effects have been defined that are known as interventional effects [25]. These effects are defined in terms of interventions at the population level rather than the individual level and have also been called stochastic interventions or randomized intervention analogs. The advantage is that they are identified without invoking cross-world independence assumptions but the disadvantage is that the direct and indirect effects do not necessarily sum to the total effect.

19.5.1 Definitions

We use the same potential outcomes notation defined in Section 20.1. In addition, we let $G_{a|C}$ denote a value of the mediator randomly drawn from the population distribution of those with exposure level a conditional on baseline confounders, C. This value is used in place of M_0 or M_1 in the definitions of the natural direct and indirect effects given previously. The interventional effects differ from the natural effects in that the mediator is fixed to a value randomly drawn from the mediator distribution of the exposed or the unexposed rather than the individual's particular mediator value in the presence or absence of exposure. Specifically, the interventional direct and indirect effects, defined by VanderWeele et al (2014) [25], are

$$\begin{aligned} IDE_{a'} &= Y_{(a,G_{a'|C})} - Y_{(a',G_{a'|C})} \\ IIE_a &= Y_{(a,G_{a|C})} - Y_{(a,G_{a'|C})} \end{aligned}$$

These effects sum to the overall effect,

$$\begin{aligned} OE &= (Y_{(a,G_{a'|C})} - Y_{(a',G_{a'|C})}) + (Y_{(a,G_{a|C})} - Y_{(a,G_{a'|C})}) \\ &= Y_{(a,G_{a|C})} - Y_{(a',G_{a'|C})} \end{aligned}$$

referred to as such because it may or may not equal the total effect [25].

Other than this definition of interventional (in)direct effects, there are also other possible definitions. One of us has argued that the definition of interventional effects should be informed by the specific research question in each application. See a thorough and accessible discussion of this topic in Nguyen et al. (2021) [26]. For simplicity we will restrict current attention to the interventional (in)direct effects defined above.

19.5.2 Identifying assumptions

The assumptions needed to identify the interventional (in)direct effects are as follows:

1. Conditional on baseline confounders, C, there are no unobserved confounders of the effect of A on Y, $Y_{am} \perp A|C$.

2. Conditional on C, there are no unobserved confounders of the effect of A on M, $M_a \perp A|C$.

3. Conditional on A, C, and L, there are no unobserved confounders of the effect of M on Y, $Y_{am} \perp M|A, C, L$.

   where L is a post-exposure confounder as shown in Figure 19.3.

19.5.3 Estimation of interventional (in)direct effects

To estimate $IDE_{a'}$ and IIE_a by weighting, we need weights that target the three hypothetical intervention conditions $(a', G_{a'|C})$, $(a, G_{a'|C})$ and $(a, G_{a|C})$. As before, for concreteness, think of $a' = 0, a = 1$. The weights that target $(a', G_{a'|C})$ are defined for units in the (actual) $A = a'$

TABLE 19.12
Interventional effect estimates.

	Estimate	SE	t-value	p-value
IDE_0	0.137	0.005	25.829	<2e-16
IIE_1	−0.002	0.005	−0.293	0.769

condition using

$$w_{a'a'} = \frac{1}{p(A = a' \mid C_i)} \frac{p(M = M_i \mid A = a', C_i)}{p(M = M_i \mid A = a', C_i, L_i)}.$$

The weights that target $(a, G_{a'|C})$ are defined for units in the $A = a$ condition using the formula

$$w_{aa'} = \frac{1}{p(A = a \mid C_i)} \frac{p(M = M_i \mid A = a', C_i)}{p(M = M_i \mid A = a, C_i, L_i)}.$$

The weights that target $(a, G_{a|C})$ are defined also for units in the $A = a$ condition using the formula

$$w_{aa} = \frac{1}{p(A = a \mid C_i)} \frac{p(M = M_i \mid A = a, C_i)}{p(M = M_i \mid A = a, C_i, L_i)}.$$

Each of these three formulas includes two terms. The first is an inverse-propensity weight that helps mimic the C distribution of the full sample. The second term is a ratio of mediator densities (or probabilities), where the denominator reflects the actual mediator distribution which depends on L, and the numerator reflects the target mediator distribution ($G_{a'|C}$, $G_{a'|C}$ and $G_{a|C}$) which does not condition on L. In the absence of L, these weights reduce to the same weights for natural (in)direct effects: w_{00} and w_{11} become the inverse-propensity weights, and $w_{aa'}$ becomes the $w_{(a,M_{a'})}$ defined in (19.7).

For R code to implement these weights in the simple case where the mediator is binary, see section 19.7.7.

19.5.3.1 Applied example

For the sake of illustration, here we take `distress` (a binary variable indicating whether the participant had elevated psychological distress in the past year) as the mediator M, and `cig15` as the intermediate confounder L. The implementation of the weights for estimating interventional (in)direct effects are shown with R code in section 19.7.7. The estimated effects are presented in Table 19.12.

The results indicate that the IIE_1 is not statistically significant. Thus, there is not a statistically significant pathway through distress. There is, however, a statistically significant IDE_0, indicating that LGB status has a significant effect on adult nicotine dependence through other pathways (i.e., not attributable to distress). The sum of the IDE_0 and IIE_1 is 0.135 and is referred to as the overall effect.

19.6 Assessing Balance Across Groups after Weighting

Causal mediation analysis involves the comparison of groups with observed differences in their exposure and mediator status. The key assumption of causal mediation analysis is that conditional on observed covariates, those comparisons are unconfounded and can yield consistent estimates of the various causal estimands described earlier in this chapter. These ignorability assumptions cannot be tested. However, regardless of whether these assumptions hold, if there are differences across groups in the observed covariates that are related to either the outcome or the mediator, then they can introduce bias into estimates unless adjusted for through modeling or weighting. Thus, at minimum, for estimation methods that use weighting, the weighted distributions of the covariates should match across comparison groups and methods that use modeling need to check the adequacy of the model for the covariates.

Various methods estimate the total, direct, and indirect effects as differences of weighted means, $\sum_i w_{a,M_{a'},i}Y_i / \sum_i w_{a,M_{a'},i}$, for $a = 0, 1$ and $a' = 0, 1$. For instance, Huber's method [11] estimates NIE_1 as $\sum_i w_{1,M_1,i}Y_i / \sum_i w_{1,M_1,i} - \sum_i w_{1,M_0,i}Y_i / \sum_i w_{1,M_0,i}$, where the sample is restricted to the treatment group and other corresponding differences for the other effect estimates. Any differences in the distributions of the covariates when weighted by each of these sets of weights could result in bias in the TE, NIE, and NDE estimates. Moreover, if the weights were known and not estimated, then $E\left[w_{(a,M'_a)}C_j \mid A = a\right] = E\left[C_j\right]$ for all the covariates C_j. Thus, if the estimated weights are adequate, then the weighted means of all of the covariates should be approximately equal.

Checking that the weighted means and distributions of the covariates are approximately the same provides a check on the potential for bias and a check on the adequacy of the estimates of the conditional probability functions used in estimating the weights (e.g., $P(A = a \mid C)$ or $P(A = a \mid C, M)$ for $a = 0, 1$). The `twangMediation` package, for instance, compares the weighted means of all covariates used in the models for each of the pairwise differences used to estimate the five effects, TE, NDE_0, NIE_1, NDE_1, and NIE_0. For each pair of weights, the Kolmogorov–Smirnov (K-S) statistic is used to compare the weighted distributions for each covariate.

Balance of the covariates is not sufficient for removing possible bias in the estimated effects. The cross-world weighted distribution of the mediator for the treatment group must match the mediator distribution of the control group weighted to match the population, i.e., the IPW distribution for the control group. Otherwise, the estimate of the cross-world mean $E(Y_{(1,M_0)})$ and therefore, the NIE_1 and NDE_0 could be biased. Similarly, for estimating $E(Y_{(0,M_1)})$ and NIE_0 and NDE_1, the cross-world weighted distribution of the mediator for the control group must match the IPW distribution of the mediator in the treatment group. Hence, the comparability of these weighted distributions should also be checked. To our knowledge, `twangMediation` is the only package that provides these checks. For example, the `twangMediation` package provides a graphical comparison of the density of the cross-world weighted mediator distributions for the treatment (control) group with the density of the IPW distribution of the mediator in the control (treatment) group. It also provides the standardized mean difference in the two weighted distributions and the K-S statistic as a measure of the similarity of the distributions. If the densities are visually similar and the standardized mean difference and the K-S statistics are small, then the distributions are similar, which supports the unbiasedness of the effect estimates.

Tables 19.13, 19.14, and 19.15 present the output from `twangMediation` for the checks described above. Table 19.13 is the *unweighted* balance table corresponding to the TE estimand. The `tx.mn` and `ct.mn` columns represent the mean covariate values for the treatment and control groups, respectively. The `std.eff.sz` column shows the standardized mean differences between the treatment and control groups; values near 0 indicate very similar covariate values across groups. As shown in Table 19.13, the treatment and control groups exhibit sizeable differences on several covariates at baseline, prior to weighting. Table 19.14 is the *weighted* balance table corresponding to the TE estimand. We can see that weighting has significantly improved balance, with standardized mean differences uniformly near 0.

Table 19.15 is the weighted balance table corresponding to the NIE_1 estimand. Note that the `tx.mn` and `ct.mn` columns represent the mean covariate values for the treatment group weighted by w_{11} weights and the *treatment* group weighted by w_{10} weights, respectively. Examining the

TABLE 19.13
Balance Table for Unweighted Covariate Differences for the TE estimand.

	tx.mn	tx.sd	ct.mn	ct.sd	std.eff.sz
age:1	0.555	0.497	0.319	0.466	0.497
age:2	0.239	0.427	0.229	0.42	0.026
age:3	0.17	0.376	0.312	0.463	−0.311
age:4	0.035	0.185	0.14	0.347	−0.313
race:1	0.568	0.495	0.588	0.492	−0.042
race:2	0.149	0.356	0.133	0.34	0.045
race:3	0.172	0.377	0.18	0.384	−0.02
race:4	0.112	0.315	0.098	0.298	0.044
educ:1	0.121	0.326	0.106	0.308	0.048
educ:2	0.297	0.457	0.224	0.417	0.173
educ:3	0.383	0.486	0.363	0.481	0.042
educ:4	0.199	0.399	0.307	0.461	−0.237
income:1	0.295	0.456	0.201	0.401	0.23
income:2	0.351	0.477	0.302	0.459	0.108
income:3	0.125	0.331	0.154	0.361	−0.082
income:4	0.229	0.42	0.343	0.475	−0.242
employ:1	0.441	0.497	0.507	0.5	−0.132
employ:2	0.216	0.411	0.192	0.394	0.059
employ:3	0.098	0.297	0.05	0.219	0.206
employ:4	0.193	0.395	0.215	0.411	−0.052
employ:5	0.052	0.222	0.035	0.185	0.087

standardized mean differences, we conclude that weighting has achieved very good covariate balance. Similar tables are also provided for NIE_0, NDE_1, and NDE_1.

19.7 R Packages and Code to Implement Causal Mediation Weighting

We now provide a more detailed overview of the following R packages that implement weighting methods for causal mediation: `rmpw`, `medflex`, `causalweight`, `twangMediation`, `CMAverse`, and `mediationClarity`. We will also provide code (not part of a package) to implement weighting-based estimation of interventional (in)direct effects in the presence of a post-treatment confounder for the simple case where the mediator is a binary variable.

19.7.1 The `rmpw` package

The `rmpw` package [9] implements RMPW [17]. Estimation is performed in a single step using the function `rmpw()`. Variable names are specified for the treatment, mediator, and outcome. Additionally, separate vectors of covariates are specified for the propensity score model (using `propensity_x`) and the outcome model (`outcome_x`). Note that `rmpw` can only accommodate a single mediator.

TABLE 19.14
Balance table for weighted covariate differences for the TE estimand.

	tx.mn	tx.sd	ct.mn	ct.sd	std.eff.sz
age:1	0.349	0.477	0.343	0.475	0.013
age:2	0.23	0.421	0.23	0.421	0.001
age:3	0.296	0.456	0.298	0.457	−0.004
age:4	0.125	0.331	0.13	0.336	−0.014
race:1	0.589	0.492	0.586	0.492	0.005
race:2	0.134	0.341	0.135	0.342	−0.002
race:3	0.177	0.381	0.179	0.383	−0.006
race:4	0.1	0.3	0.1	0.3	0.001
educ:1	0.104	0.305	0.108	0.31	−0.012
educ:2	0.23	0.421	0.232	0.422	−0.003
educ:3	0.367	0.482	0.365	0.481	0.004
educ:4	0.299	0.458	0.296	0.456	0.007
income:1	0.213	0.409	0.211	0.408	0.005
income:2	0.311	0.463	0.307	0.461	0.009
income:3	0.149	0.356	0.151	0.358	−0.006
income:4	0.327	0.469	0.331	0.471	−0.009
employ:1	0.504	0.5	0.501	0.5	0.008
employ:2	0.192	0.394	0.195	0.396	−0.006
employ:3	0.056	0.229	0.055	0.228	0.002
employ:4	0.211	0.408	0.213	0.409	−0.005
employ:5	0.037	0.189	0.037	0.189	0.001

rmpw: Summary

Effects estimated: TE, NDE_0, NIE_1, NDE_1, NIE_0

Effect scale: marginal risk difference

Models: exposure and mediator

Variable types allowed:

- Exposure: binary
- Mediator: continuous, binary, count, nominal
- Outcome: continuous, binary, count

Multiple mediators? No

Single mediator code

```
library(rmpw)
results <- rmpw(data        = NSDUH_female,
                treatment   = "lgb",
                mediator    = "cig15",
                outcome     = "smoke",
```

TABLE 19.15
Balance table for NIE_1 estimand.

	tx.mn	tx.sd	ct.mn	ct.sd	std.eff.sz
age:1	0.349	0.477	0.349	0.477	0
age:2	0.23	0.421	0.231	0.421	−0.002
age:3	0.296	0.456	0.295	0.456	0.003
age:4	0.125	0.331	0.125	0.331	−0.002
race:1	0.589	0.492	0.588	0.492	0.002
race:2	0.134	0.341	0.134	0.341	0.001
race:3	0.177	0.381	0.178	0.382	−0.002
race:4	0.1	0.3	0.1	0.3	−0.001
educ:1	0.104	0.305	0.102	0.303	0.005
educ:2	0.23	0.421	0.229	0.42	0.004
educ:3	0.367	0.482	0.368	0.482	−0.003
educ:4	0.299	0.458	0.301	0.459	−0.005
income:1	0.213	0.409	0.213	0.409	0
income:2	0.311	0.463	0.31	0.463	0.002
income:3	0.149	0.356	0.15	0.357	−0.002
income:4	0.327	0.469	0.327	0.469	0
employ:1	0.504	0.5	0.506	0.5	−0.003
employ:2	0.192	0.394	0.191	0.393	0.004
employ:3	0.056	0.229	0.055	0.228	0.002
employ:4	0.211	0.408	0.211	0.408	−0.002
employ:5	0.037	0.189	0.037	0.189	0

```
                    propensity_x = c("age", "race", "educ",
                                     "income", "employ"),
                    outcome_x    = c("age", "race", "educ",
                                     "income", "employ"),
                    decomposition=0)
summary(results)
```

Note that the user can specify which decomposition is estimated. When `decomposition=0`, a single estimate for the NIE and NDE are generated (thereby assuming NDE_1=NDE_0 and NIE_1=NIE_0). But the package also includes decompositions based on VanderWeele's three-way decomposition [27]. When `decomposition = 1`, the total effect will be decomposed into the pure direct effect (NDE_0), total and pure indirect effects (NIE_1 and NIE_0), and the corresponding natural treatment-by-mediator interaction effect (NIE_1 - NIE_0). When `decomposition = 2`, the total effect will be decomposed into the pure indirect effect (NIE_0), the total and pure direct effects (NDE_1 and NDE_0), and the corresponding natural treatment-by-mediator interaction effect (NDE_1 - NDE_0).

Missing values are not allowed on the outcome or covariates. Standard errors are obtained using the two-step estimation method described in Bein et al. [28].

19.7.2 The `medflex` package

The `medflex` package [10] implements methods proposed by Lange et al. [29] and Vansteelandt et al. [8], using both a weighting-based approach (our focus) and an imputation-based approach. Weighting is implemented using Hong's RMPW method [17].

medflex: Summary

Effects estimated: TE, NDE_0, NIE_1, NDE_1, NIE_0

Effect scale: conditional odds ratio, marginal risk difference

Models: Mediator and outcome

Variable types allowed:

- Exposure: continuous, binary, count, nominal
- Mediator: continuous, binary, count, nominal
- Outcome: continuous, binary, count

Multiple mediators? No

Estimation of mediation effects via `medflex` requires specification of two models: (1) a regression model for the mediator and (2) a natural effects model for the counterfactual outcome. Using the `neWeight()` function, the user specifies the mediator model, which can be fitted using the `glm()` function (default) or the `vglm()` function from the `VGAM` package. When specifying the mediator model, the exposure variable should be listed as the first variable on the right-hand side followed by potential treatment-mediator confounders. Binary exposure variables need to be specified as factor variables. The `neWeight()` function returns an expanded version of the original dataset which includes observations corresponding to counterfactual values of the exposure. This expanded dataset includes two new counterfactual variables, named in the form of exposure0 and exposure1. Next, using the `neModel()` function, the user specifies the natural effects model. This model regresses the outcome on both counterfactual exposure variables created via `neWeight()`; potential confounders can also be included in the outcome model. The `neModel()` function estimates models using `glm()`. The specified dataset should be the expanded dataset created as the returned object from the `neWeight()` function. Use of `neEffdecomp()` returns estimates for the TE, NDE, and NIE. Standard errors can be obtained using the `summary()` function. By default, standard error estimation is performed based on the i.i.d. bootstrap. Note that `medflex` additionally allows the option of robust standard errors based on the sandwich estimator when models are estimated using `glm()` (specify `se = ''robust"`). The sample syntax provided below assumes a binary exposure. See the `medflex` documentation for the specific syntax to implement these models using a multinomial or continuous exposure.

Marginal risk difference code

```
library(medflex)
### Fit mediator model using neWeight() on original data
### neWeight() returned object is expanded dataset
expData <-  neWeight(cig15 ~ factor(lgb) + age + race +
                             educ + income + employ,
                     family  = binomial,
                     data    = NSDUH_female)
### Then fit natural effect model on expanded dataset
results <- neModel(smoke ~ lgb0 + lgb1,
                   expData = expData)
### summary() command will display estimates from natural effect model
    with SEs
```

```
summary(results)
### neEffdecomp() command will list TE, NDE, and NIE
neEffdecomp(results)
```

When estimating marginal risk differences, the estimates obtained from `medflex` are comparable to those from `rmpw`.

Conditional odds ratio code

```
library(medflex)
### Fit mediator model using neWeight() on original data
### neWeight() returned object is expanded dataset
expData <- neWeight(cig15 ~ factor(lgb) + age +
                               race + educ + income + employ,
                               family = binomial,
                               data = NSDUH_female)
### Then fit natural effect model on expanded dataset
results <- neModel(smoke ~ lgb0 + lgb1 + age +
                        race + educ + income + employ,
                        family = binomial,
                        expData = expData)
### summary() command will display estimates from natural effect model
    with SEs
summary(results)
### neEffdecomp() command will list TE, NDE, and NIE
neEffdecomp(results)
```

19.7.3 The `causalweight` package

The `causalweight` package [14] implements Huber's weighting approach [11] with the weights estimated using either logistic or probit regression.

`causalweight`: Summary

Effects estimated: TE, NDE_0, NIE_1, NDE_1, NIE_0

Effect scale: marginal risk difference

Models: Exposure and outcome

Variable types allowed:

- Exposure: binary
- Mediator: continuous, binary, count, nominal, ordinal
- Outcome: any type

Multiple mediators? Yes

Causal mediation analysis is implemented in the `causalweight` package using the function `medweight()`, as shown below:

Single mediator code

```
library(causalweight)
```

```
### Weight estimation using logistic regression
dat <- NSDUH_female
results <- medweight(y = dat$smoke,
                     d = dat$lgb,
                     m = dat$cig15,
                     x = cbind(dat$age,dat$race,
                         dat$educ,dat$income,dat$employ),
                     logit = TRUE)
summary(results)
```

The causalweight package returns the follow estimates: TE (labeled ATE), NDE_1 (labeled Dir.treat), NDE_0 (labeled Dir.control), NIE_1 (labeled Indir.treat), and NIE_0 (labeled Indir.control). By default, this package uses logistic regression to estimate weights (logit=TRUE); alternatively, probit regression can be specified (logit=FALSE). Standard error estimation is performed based on the i.i.d. bootstrap. By default, the causalweight package performs weight trimming. Specifically, the default is to discard observations with treatment probabilities given the covariates and mediator(s) smaller than 0.05 (5%) or larger than 0.95 (95%). To prevent weight trimming, specify the option trim=0. Missing values are not allowed for the outcome, exposure, mediator(s), or confounders.
Multiple mediator code

Implementation is very similar to the use of medweight() with a single mediator – however, the mediator m is now specified as a vector, cbind(dat$cig15, dat$alc15).

```
library(causalweight)
dat <- NSDUH_female
results <- medweight(y = dat$alc_cig_depend,
                    d = dat$lgb,
                    m = cbind(dat$cig15, dat$alc15),
                    x = cbind(dat$age, dat$race, dat$educ,
                          dat$income, dat$employ),
                 logit = TRUE)
```

19.7.4 The twangMediation package

The twangMediation package [12] implements Huber's weighting approach [11], using GBM as the default method for weight estimation (in contrast to logistic or probit regression). The twangMediation package builds on the twang package [30], which estimates propensity score weights (for binary, categorical, continuous, and time-varying treatments) using GBM.

twangMediation: Summary

Effects estimated: TE, NDE_0, NIE_1, NDE_1, NIE_0

Effect scale: marginal risk difference

Models: Exposure and outcome

Variable types allowed:

- Exposure: binary
- Mediator: continuous, binary, count, nominal, ordinal
- Outcome: any type

Multiple mediators? Yes

Estimation of mediation effects via `twangMediation` requires two steps: (1) estimation of propensity score weights for estimating the TE and (2) estimation of cross-world weights using Huber's approach. The first analytic step is to estimate propensity score weights used for weighting the counterfactual means, $E(Y_{(1,M_1)})$ and $E(Y_{(0,M_0)})$, (i.e., TE weights). While these weights can be estimated in any manner, we recommend estimating these weights with GBM using the `ps()` function in the `twang` package (as shown below) such that all weights are consistently estimated using GBM. Next, we use the `wgtmed()` function to estimate the cross-world weights and obtain the mediation estimates of interest. By default, `wgtmed()` estimates the cross-world weights using GBM (although logistic regression may also be specified).

Single mediator code

```
library(twang)
### Estimate propensity score weights for exposure
TEps <- ps(formula = lgb ~ age + race + educ + income
                     + employ,
           data     = NSDUH_female,
           n.trees = 7500,
           estimand= "ATE")
### Mediation estimands obtained using wgtmed() function
library(twangMediation)
results <- wgtmed(formula.med = cig15 ~ age + race + educ
                              + income + employ,
                  a_treatment              = "lgb",
                  y_outcome                = "smoke",
                  data                     = NSDUH_female,
                  method                   = "ps",
                  total_effect_ps          = TEps,
                  total_effect_stop_rule   = "ks.mean",
                  ps_version               = "gbm",
                  ps_n.trees               = 7500,
                  ps_stop.method           = "ks.mean")
```

The `twangMediation` package provides estimates of both direct effects, NDE_0 and NDE_1, as well as both indirect effects, NIE_0 and NIE_1. This package also provides robust balance diagnostics (as detailed in Section 19.6).

Multiple mediator code

The `twangMediation` package also estimates the joint effects of multiple mediators. The first step of estimating the propensity score weights for the exposure (lgb) is the same as in the prior example with a single mediator. However, in the second step using `wgtmed()`, both mediators are included on the left-hand side of the formula notation, separated by "+".

```
## In wgtmed function, specify multiple mediators on left-hand side
    separated by +
results <- wgtmed(cig15 + alc15 ~ age + race + educ +
                              income + employ,
                  a_treatment = "lgb",
                  y_outcome   = "alc_cig_depend",
                  data        = NSDUH_female,
                  method                 = "ps",
                  total_effect_ps        = TEps,
                  total_effect_stop_rule = "ks.mean",
                  ps_version             = "gbm",
                  ps_n.trees             = 6000,
                  ps_stop.method         = "ks.mean")
```

19.7.5 The `CMAverse` package

The `CMAverse` package is available on GitHub [18]. The `CMAverse` package implements six causal mediation analysis approaches using the `cmest()` function. Regarding weighting methods, the `CMAverse` package implements the IORW approach [19] and the weighting-based approach by VanderWeele and Vansteelandt [20]. Note that using this weighting-based approach with a single mediator is Albert's [23] approach. Using the `cmest()` function, the estimation method is specified in the `model=` argument.

IORW approach

`CMAverse`: IOWR summary

Effects estimated: TE, NDE_0, NIE_1

Effect scale: conditional odds ratio

Models: Exposure and outcome

Variable types allowed:

- Exposure: binary, nominal, ordinal
- Mediator: continuous, binary, count, nominal, ordinal
- Outcome: continuous, binary, count, nominal, ordinal, survival

Multiple mediators? Yes

Single mediator code

```
### Specify inverse odds ratio weighting approach with method = "iorw"
install.packages("devtools")
devtools::install_github("BS1125/CMAverse")
library(CMAverse)
results <- cmest(data       = NSDUH_female,
                 model       = "iorw",
                 outcome     = "smoke",
                 exposure    = "lgb",
                 mediator    = "cig15",
                 basec       = c("age", "race", "educ",
                                  "income", "employ"),
                 yreg        = "logistic",
                 ereg        = "logistic",
                 inference   ="bootstrap")
summary(results)
```

The outcome, exposure, and mediator are defined, and the pre-exposure confounders are specified in the `basec=` argument. The `yreg=` argument specifies the link function for the outcome regression model, and the `ereg=` argument specifies the link function for the exposure regression model. When the outcome is binary, the `CMAverse` package returns the following estimands: total effect odds ratio (labeled RTE), pure natural direct effect, NDE_0, odds ratio (labeled RPNDE), total natural indirect effect, NIE_1, odds ratio (labeled RTNIE), and proportion mediated (labeled PM). We note that this package does not provide the alternative decomposition of NDE_1 and NIE_0. By default, standard errors are calculated by bootstrapping.

Multiple mediator code

```
### Specify inverse odds ratio weighting approach with method = "iorw"
results <- cmest(data       = NSDUH_female,
                 model      = "iorw",
                 outcome    = "alc_cig_depend",
                 exposure   = "lgb",
                 mediator   = c("cig15", "alc15"),
                 basec      = c("age", "race", "educ",
                                "income", "employ"),
                 yreg       = "logistic",
                 ereg       = "logistic",
                 inference = "bootstrap")
summary(mult4)
```

Weighting-based approach (Albert's method) code

CMAverse: weighting-based approach summary

Effects estimated: $TE, NDE_0, NIE_1, NDE_1, NIE_0$

Effect scale: conditional odds ratio

Models: Exposure and outcome

Variable types allowed:

- Exposure: binary, nominal, ordinal
- Mediator: continuous, binary, count, nominal, ordinal
- Outcome: continuous, binary, count, nominal, ordinal

Multiple mediators? Yes

Single mediator code

```
### Specify weighting-based approach with model = "wb"
library(CMAverse)
results <- cmest(data          = NSDUH_female,
                 model         = "wb",
                 outcome       = "smoke",
                 exposure      = "lgb",
                 mediator      = "cig15",
                 mval          = list(0),
                 basec         = c("age", "race", "educ",
                                   "income", "employ"),
                 EMint         = FALSE,
                 yreg          = "logistic",
                 ereg          = "logistic",
                 inference     = "bootstrap")
summary(results)
```

The syntax is similar to that for the IORW method shown above. However, the `EMint=` argument is required to specify whether the outcome model should include an exposure-mediator interaction.

Additionally, the `mval=` argument is required, for each mediator variable, to specify a value at which the variable is controlled. When the outcome is a binary variable, the `CMAverse` package returns the following estimands: controlled direct effect odds ratio (labeled RCDE), pure natural direct effect, NDE_0, odds ratio (RPNDE), total natural direct effect, NDE_1, odds ratio (RTNDE), pure natural indirect effect, NIE_0, odds ratio (RPNIE), total natural indirect effect, NIE_1, odds ratio (RTNIE), total effect odds ratio (RTE), and overall proportion mediated (PM).
Multiple mediator code

```
### Specify weighting-based approach with model = "wb"
results <- cmest(data      = NSDUH_female,
                 model     = "wb",
                 outcome   = "alc_cig_depend",
                 exposure  = "lgb",
                 mediator  = c("cig15", "alc15"),
                 mval      = list(0,0),
                 basec     = c("age", "race", "educ",
                               "income", "employ"),
                 EMint     = FALSE,
                 yreg      = "logistic",
                 ereg      = "logistic",
                 inference = "bootstrap")
summary(results)
```

19.7.6 The `mediationClarity` package

The `mediationClarity` package, available on GitHub, implements a range of estimators of *marginal* natural (in)direct effects discussed in Nguyen et al. (2022) [16]. Some of these estimators use weighting as the primary way to estimate the effects; others use weighting as a way to induce robustness in strategies that otherwise rely primarily on regression. As examples, we introduce the package's implementation of (1) a pure weighting estimator, (2) an estimator that supplements weighting with a working regression model to reduce variance, and (3) an estimator that combines specific regression models with the same weighting scheme for triple robustness.

`mediationClarity`: summary

Effects estimated: TE, NDE_1, NIE_0, NDE_0, NIE_1

Effect scale: marginal risk difference, marginal risk or odds ratio

Models:

- weights are estimated via exposure models
- other models are specific to the estimator (the package covers a dozen plus estimators)

Variable types allowed:

- Exposure: binary
- Mediator: any type
- Outcome: any type for which one of the effect scales is relevant

Multiple mediators? Yes

To install and load the package, run this code.

```
install.packages("devtools")
devtools::install_github("trangnguyen74/mediationClarity")
library(mediationClarity)
```

We now present the code for estimation of the weights and for the three estimators of *marginal* natural (in)direct effects. We then briefly comment on using the weights to estimate *conditional* effects. All the code in this section is essentially the same regardless of the number of mediators.

Weights estimation

All three of the estimators highlighted rely on the same weighting scheme, which combines same-world and cross-world weighting. Same-world weighting weights the treated and control subsamples to mimic the full sample covariate distribution. Cross-world weighting weights the treated or control subsample (which one depends on the specific choice of TE composition) to mimic the full sample covariate distribution and to mimic the mediator given covariate distribution in the other subsample.

```
w.med <- weight_med(
    data        = NSDUH_female,
    cross.world = "10",
    a.c.form    = "lgb~age+race+educ+income+employ",
    a.cm.form   = "lgb~age+race+educ+income+employ+cig15+alc15",
    plot        = TRUE,
    c.std       = "age"
)
```

Function `weight_med()` estimates the weights and outputs the weighted data plus a weight distribution plot and a balance plot, using the method based on the third expression for the cross-world weights (see section 19.3.3). Argument `cross.world` accepts three values: `"10"` for the $(\text{NDE}_0, \text{NIE}_1)$ effect pair, `"01"` for the $(\text{NIE}_0, \text{NDE}_1)$ effect pair, and `"both"` for both pairs.

Currently the weights are estimated based on logit models for $\text{P}(A \mid C)$ and $\text{P}(A \mid C, M)$, whose formulas are specified in arguments `a.c.form` and `a.cm.form`, and the cross-world weight is computed using the formula in Equation 19.10. The models can be made more flexible to reduce misspecification and improve balance, for example, using spline terms of continuous variables or adding interaction terms among the variables. An alternative to estimate weights via direct optimization for balance (not using probability models) will be added later.

The following code extracts the output (weighted data and diagnostic plots).

```
wdat <- w.med$w.dat
w.med$plots$w.wt.distribution
w.med$plots$balance
```

The balance plot shows mean-balance for both (1) the covariates across the pseudo (i.e., weighted) treated, control, and cross-world samples and the full sample; and (2) the mediators between the pseudo cross-world and pseudo control samples if `cross.world = "10"` and between the pseudo cross-world and pseudo treated samples if `cross.world = "01"`. Arguments `c.std` and `m.std` indicate for which covariates and mediators to use standardized mean differences. This function (as well as the estimator functions below) can handle sampling weights. See documentation for details.

Pure weighting estimator

Function `estimate_wtd()` estimates the weights (calling `weight_med()`), uses the weights to compute estimates of potential outcome means, and contrasts the potential outcome means to estimate the effects. Confidence intervals are computed via bootstrapping.

```
est.wtd <- estimate_wtd(
    data          = NSDUH_female,
    cross.world   = "10",
    effect.scale  = "MD",
    a.c.form      = "lgb~age+race+educ+income+employ",
    a.cm.form     = "lgb~age+race+educ+income+employ+cig15+alc15",
    y.var         = "smoke",
    plot          = TRUE,
    c.std         = "age",
    boot.num      = 999,
    boot.method   = "cont-wt",
    boot.seed     = 77777
)
```

This function inherits weighting-related parameters from the `weight_med()` function. In addition, argument `effect.scale` accepts `"MD"` for mean/risk difference, `"MR"` for mean/risk/rate ratio, and `"OR"` for odds ratio. Argument `y.var` specifies the name of the outcome variable. Two bootstrap methods are available, the classic resampling bootstrap [31] and a recent method using continuous bootstrap weights [32]. These are specified by `boot.method = "resample"` and `boot.method = "cont-wt"`, respectively.

The following code extracts the output.

```
est.wtd$estimates
est.wtd$plots$w.wt.distribution
est.wtd$plots$key.balance
```

Weighting combined with a working regression model to improve precision

This estimator relies on weighting as the primary effect estimation strategy and uses a simple *working* regression model (not assumed to be correct) to improve estimation precision. It is motivated by the common practice of processing data to balance covariates and then fitting a simple regression model to the covariate-balanced data.

```
est.wtCadj <- estimate_wtCadj(
    data          = NSDUH_female,
    cross.world   = "10",
    effect.scale  = "MD",
    a.c.form      = "lgb~age+race+educ+income+employ",
    a.cm.form     = "lgb~age+race+educ+income+employ+cig15+alc15",
    y.var         = "smoke",
    y.link      = "logit",
    plot          = TRUE,
    c.std         = "age",
    boot.num      = 999,
    boot.method   = "cont-wt",
    boot.seed     = 77777
)
```

Function `estimate_wtCadj()` estimates the weights to form the pseudo samples (calling `weight_med()`); fits a simple working model (not assumed to be correct) to the weighted data, regressing the outcome on covariates and pseudo sample membership; and assembles estimates of the marginal effects using either model coefficients or average model predictions (depending on the type of outcome variable and the effect scale – see details in [16]).

This function's arguments are almost identical to those of `estimate_wtd()`. The only additional argument is `y.link`, which specifies the type of model for the working regression. No formula needs to be specified for this model. The output structure is exactly the same as that of `estimate_wtd()`.

Triply robust estimators combining weighting with conditional mean models

These estimators are termed the Ypred.R and NDEpred.R estimators (the latter only for additive effects) in [16]. We refer the reader to [16] for a proper presentation of these estimators, noting here that these estimators are robust in the sense that they do not depend on a specific model being correctly specified. These estimators are comprised of a pairing of a weight component and a conditional outcome or effect mean model – the estimator is consistent if either the weight component or the conditional mean model is correct. Below are the code for these two estimators.

```
est.Y2predR <- estimate_Y2predR(
    data          = NSDUH_female,
    cross.world   = "10",
    effect.scale  = "MD",
    a.c.form      = "lgb~age+race+educ+income+employ",
    a.cm.form     = "lgb~age+race+educ+income+employ+cig15+alc15",
    plot          = TRUE,
    c.std         = "age",

    y.c0.form     = "smoke~age+race+educ+income+employ",
    y.c1.form     = "smoke~age+race+educ+income+employ",
    y.cm1.form    = "smoke~age+race+educ+income+employ+cig15+alc15",
    y10.c.form    = "smoke~age+race+educ+income+employ",
    y.link        = "logit",

    boot.num     = 999,
    boot.method  = "cont-wt",
    boot.seed    = 77777
)
```

In function `estimate_Y2predR()`, in addition to the arguments inherited from `estimate_wtd()`, there is the `y.link` argument plus four model formula arguments: `y.c0.form`, `y.c1.form`, `y.cm1.form` and `y10.c.form`, for $\mathrm{E}[Y \mid C, A = 0]$, $\mathrm{E}[Y \mid C, A = 1]$, $\mathrm{E}[Y \mid C, M, A = 1]$ and $\mathrm{E}[Y_{1M_0} \mid C]$, respectively. The three arguments `y.c0.form`, `y.c1.form` and `y10.c.form`, if identical, can be replaced with a single shorthand argument, `y.c.form`.

```
est.NDEpredR <- estimate_NDEpredR(
    data          = NSDUH_female,
    cross.world   = "10",
    a.c.form      = "lgb~age+race+educ+income+employ",
    a.cm.form     = "lgb~age+race+educ+income+employ+cig15+alc15",
    plot          = TRUE,
    c.std         = "age",

    y.c0.form     = "smoke~age+race+educ+income+employ",
    y.c1.form     = "smoke~age+race+educ+income+employ",
    y.cm1.form    = "smoke~age+race+educ+income+employ+cig15+alc15",
    nde0.c.form   = "effect~age+race+educ+income+employ",
    y.link        = "logit",

    boot.num      = 999,
    boot.method   = "cont-wt",
    boot.seed     = 77777
```

```
)
```

The function `estimate_NDEpredR()` has very similar arguments to those of `estimate_Y2predR` The difference is that instead of using the model regressing the predicted cross-world potential outcome on covariates (`y.10.c.form`), it uses a model regressing a proxy of the natural direct effect on covariates (`nde0.c.form`). Also, this function does not have an `effect.scale` argument, because it only applies to additive effects. The output structure is the same between the two functions.

```
est.Y2predR$estimates
est.Y2predR$plots$w.wt.distribution
est.Y2predR$plots$balance
```

Estimating conditional effects using weighted data

While `mediationClarity` targets marginal effects, the `weight_med()` function can be used for a weighting first step (creating pseudo-treated, pseudo-control, and cross-world samples) before stacking these weighted samples and fitting a model regressing outcome on pseudo-sample membership and covariates to estimate conditional effects. For example, the code below estimates conditional effects defined on the odds ratio scale.

The starting point here is the data frame `wdat` (one of the outputs of `weight_med()`), where the three pseudo samples are stacked. Variable `samp` indicates which one: `"p11"` denotes the pseudo-treated sample, `"p00"` denotes the pseudo-control sample, and both `"p10"` and `"p01"` denote pseudo cross-world samples.

```
### Make 2 dummies A0, A1
wdat$A0 <- ifelse(wdat$samp=="p00", -1, 0)
wdat$A1 <- ifelse(wdat$samp=="p11",  1, 0)
### Fit a logit model
cond.logit <- glm(smoke~A0+A2+age+race+educ+employ,
                  data    = wdat,
                  weights = f.wt,
                  family  = binomial)
### Extract conditional ORs and CIs
exp(cbind(OR = coef(cond.logit),
          confint(cond.logit)))
```

The odds ratios associated with `A0` and `A1` estimate NDE_0 and NIE_1, respectively, if `weight_med()` was run with `cross.world = "10"`. They estimate NIE_0 and NDE_1, respectively, if `weight_med()` was run with `cross.world = "01"`.

19.7.7 Code for interventional (in)direct effects

In this example, `distress` is the binary mediator M and `cig15` is the intermediate confounder L. The outcome and pre-exposure covariates remain the same.

The first step is to fit the three relevant models: the propensity score model: $p(A \mid C)$, the mediator density given exposure, baseline covariates, and post-exposure confounder model: $p(M \mid C, A, L)$, and the mediator density given exposure and baseline covariates model: $p(M \mid C, A)$. For the two mediator models, we let exposure interact with all the predictors; this is important to make sure we estimate the mediator probability within each exposure condition well. The alternative is to fit mediator models stratified by exposure.

```
a.on.c <- glm(lgb ~ age+race+educ+income+employ,
              data = NSDUH_female, family = binomial)
m.on.ac <- glm(distress ~ age*lgb+race*lgb+educ*lgb+
                          income*lgb+employ*lgb,
```

```
                  data = NSDUH_female, family = binomial)
m.on.acl <- glm(distress ~ age*lgb+race*lgb+educ*lgb+
                           income*lgb+employ*lgb+cig15*lgb,
                data = NSDUH_female, family = binomial)
```

Based on these fitted models, we can compute the weights and form the corresponding pseudo samples. Here we compute w_{00} (for control units), and w_{11} and w_{10} (for treated units). If condition $(0, G_{1|C})$ is of interest, the code is easily modified to compute w_{01} (for control units).

```
### Compute the C-weight term
tmp <- NSDUH_female
ps <- predict(a.on.c, type = "response")
tmp$c.wt <- tmp$lgb / ps + (1-tmp$lgb) / (1-ps)
rm(ps, a.on.c)
### Prepare the 3 pseudo samples before computing M-weight
dat00 <- tmp[tmp$lgb==0,]
dat10 <- tmp[tmp$lgb==1,]
dat11 <- tmp[tmp$lgb==1,]
rm(tmp)
### Compute M-weight on pseudo samples 00 and 11
dat00$m.wt <-
    predict(m.on.ac,  newdata = dat00, type = "response") /
    predict(m.on.acl, newdata = dat00, type = "response")
dat11$m.wt <-
    predict(m.on.ac,  newdata = dat11, type = "response") /
    predict(m.on.acl, newdata = dat11, type = "response")
### Compute M-weight on pseudo sample 10
tmp <- dat10
tmp$lgb <- 0
dat10$m.wt <-
    predict(m.on.ac,  newdata = tmp,   type = "response") /
    predict(m.on.acl, newdata = dat10, type = "response")
rm(tmp, m.on.ac, m.on.acl)
### Multiply C-weight and M-weight
dat00$wt <- dat00$c.wt * dat00$m.wt
dat10$wt <- dat10$c.wt * dat10$m.wt
dat11$wt <- dat11$c.wt * dat11$m.wt
### Collect all weighted data
wdat <- rbind(cbind(dat11, samp = "p11"),
              cbind(dat00, samp = "p00"),
              cbind(dat10, samp = "p10"))
rm(dat11, dat00, dat10)
```

The result of this code is a dataset, `wdat`, that includes three pseudo samples corresponding to the three target conditions $(0, G_{0|C})$ (indicated by `samp = "p00"`), $(1, G_{1|C})$ (`samp = "p11"`), and $(1, G_{0|C})$ (`samp = "p10"`).

To use these data to estimate the IDE_0 and IIE_1, we use the same trick of dummy coding pseudo samples before fitting outcome regression used at the end of section 19.7.6 above. Note that here the coefficients of the dummy variables `A0` and `A1` estimate IDE_0 and IIE_1, respectively. When targeting the other pair of interventional (in)direct effects, the coefficients of `A0` and `A1` estimate IIE_0 and IDE_1. For completeness, the code follows.

```
### Make 2 dummies A0, A1
wdat$A0 <- ifelse(wdat$samp=="p00", -1, 0)
wdat$A1 <- ifelse(wdat$samp=="p11",  1, 0)
### Marginal additive effects: use either model below
```

```
#   both ok due to collapsibility of mean/risk difference;
#   second model may give smaller SE, CI
rd.mod1 <- lm(smoke ~ A0+A1,
              data = wdat, weights = wt)
rd.mod1 <- lm(smoke ~ A0+A1 + age+race+educ+employ,
              data = wdat, weights = wt)
### Conditional odds ratio effects:
or.mod <- glm(smoke ~ A0+A1 + age+race+educ+employ,
              data = wdat, weights = wt, family = binomial)
exp(cbind(OR = coef(or.mod),
          confint(or.mod)))
```

19.8 Conclusions

The focus of this chapter has been on weighting based estimators but it should be noted that there are other estimators (see Nguyen et al. (2022) [16] for an overview). In particular, a method proposed by Imai et al. (2010) [33] which is implemented in the R package `mediation`, an imputation-based approach introduced by Vansteelandt et al. (2012) [8] and implemented in the `medflex` package, and regression-based approaches which are covered in detail in VanderWeele (2015) [5] and most of which are implemented in the `CMAverse` package.

As weighting is a key component in many estimators, it is crucial to understand the target of the weighting components (the three conditions being contrasted, namely treated, control, and cross-world), the different ways to implement them, and how to check for the desired balance that should be achieved by the weighting. This book chapter and the code we cover here facilitate these important tasks.

References

[1] M. S. Schuler and R. L. Collins. Sexual minority substance use disparities: Bisexual women at elevated risk relative to other sexual minority groups. *Drug Alcohol Depend*, 206:107755, 2020.

[2] T. J. VanderWeele and S. Vansteelandt. Conceptual issues concerning mediation, interventions, and composition. *Statistics and Its Interface*, 2:457–468, 2009.

[3] J. M. Robins and S. Greenland. Identifiability and exchangeability for direct and indirect effects. *Epidemiology*, 3(2):143–55, 1992.

[4] Trang Quynh Nguyen, Ian Schmid, Elizabeth L. Ogburn, and Elizabeth A. Stuart. Clarifying causal mediation analysis for the applied researcher: Effect identification via three assumptions and five potential outcomes. 10, 246–279, 2022.

[5] T. J. VanderWeele. *Explanation in Causal Inference: Methods for Mediation and Interaction*. Oxford University Press, New York, 2015.

[6] J. Pearl. Direct and indirect effects. In *Proceedings of the Seventeenth Conference on Uncertainty in Artificial Intelligence*. Morgan Kaufman, 2001.

[7] G. Hong. Ratio of mediator probability weighting for estimating natural direct and indirect effects. In *Joint Statistical Meetings*. American Statistical Association, 2010.

[8] S. Vansteelandt, Maarten Bekaert, and Theis Lange. Imputation strategies for the estimation of natural direct and indirect effects. *Epidemiologic Methods*, 1(1):131–158, 2012.

[9] Xu Qin, Guanglei Hong, and Fan Yang. *rmpw: Causal Mediation Analysis Using Weighting Approach*, 2018. R package version 0.0.4.

[10] Johan Steen, Tom Loeys, Beatrijs Moerkerke, and Stijn Vansteelandt. medflex: An R package for flexible mediation analysis using natural effect models. *Journal of Statistical Software*, 76(11):1–46, 2017.

[11] Martin Huber. Identifying causal mechanisms (primarily) based on inverse probability weighting. *Journal of Applied Econometrics*, 29(6):920–943, 2014.

[12] Dan McCaffrey, Katherine Castellano, Donna Coffman, Brian Vegetabile, and Megan Schuler. *twangMediation: twang Causal Medition Modeling via Weighting*, 2021. R package version 1.0.

[13] B. K. Lee, J. Lessler, and E. A. Stuart. Improving propensity score weighting using machine learning. *Statistics in Medicine*, 29(3):337–346, 2010.

[14] Hugo Bodory and Martin Huber. *causalweight: Estimation Methods for Causal Inference Based on Inverse Probability Weighting*, 2021. R package version 1.0.2.

[15] Trang Quynh Nguyen. *mediationClarity: Estimation of Marginal Natural (in)direct Effects*, 2022. R package version 1.0.

[16] Trang Quynh Nguyen, Elizabeth L. Ogburn, Ian Schmid, Elizabeth B. Sarker, Noah Greifer, Ina M. Koning, and Elizabeth. A. Stuart. Causal mediation analysis: From simple to more robust strategies for estimation of marginal natural (in)direct effects. arXiv:2102.06048, 2022.

[17] G. Hong, Jonah Deutsch, and Heather D. Hill. Ratio-of-mediator-probability weighting for causal mediation analysis in the presence of treatment-by-mediator interaction. *Journal of Educational and Behavioral Statistics*, 40(3):307–340, 2015.

[18] B. Shi, C. Choirat, B. A. Coull, T. J. VanderWeele, and L. Valeri. Cmaverse: A suite of functions for reproducible causal mediation analyses. *Epidemiology*, 32(5):e20–e22, 2021.

[19] E. J. Tchetgen Tchetgen. Inverse odds ratio-weighted estimation for causal mediation analysis. *Statistics in Medicine*, 32(26):4567–80, 2013.

[20] T. J. VanderWeele and S. Vansteelandt. Mediation analysis with multiple mediators. *Epidemiologic Methods*, 2(1):95–115, 2014.

[21] R. M. Daniel, B. L. De Stavola, S. N. Cousens, and S. Vansteelandt. Causal mediation analysis with multiple mediators. *Biometrics*, 71:1–15, 2015.

[22] T. Lange, M. Rasmussen, and L. C. Thygesen. Assessing natural direct and indirect effects through multiple pathways. *American Journal of Epidemiology*, 179(4):513–8, 2014.

[23] J. M. Albert. Distribution-free mediation analysis for nonlinear models with confounding. *Epidemiology*, 23(6):879–88, 2012.

[24] Q. C. Nguyen, T. L. Osypuk, N. M. Schmidt, M. M. Glymour, and E. J. Tchetgen Tchetgen. Practical guidance for conducting mediation analysis with multiple mediators using inverse odds ratio weighting. *American Journal of Epidemiology*, 181(5):349–56, 2015.

[25] T. J. VanderWeele, S. Vansteelandt, and J. M. Robins. Effect decomposition in the presence of an exposure-induced mediator-outcome confounder. *Epidemiology*, 25(2):300–306, 2014.

[26] Trang Quynh Nguyen, Ian Schmid, and Elizabeth A. Stuart. Clarifying causal mediation analysis for the applied researcher: Defining effects based on what we want to learn. *Psychological Methods*, 26(2):255–271, 2021.

[27] T. J. VanderWeele. A three-way decomposition of a total effect into direct, indirect, and interactive effects. *Epidemiology*, 24(2):224–32, 2013.

[28] E. Bein, J. Deutsch, G. Hong, K. E. Porter, X. Qin, and C. Yang. Two-step estimation in ratio-of-mediator-probability weighted causal mediation analysis. *Statistics in Medicine*, 37(8):1304–1324, 2018.

[29] T. Lange, S. Vansteelandt, and M. Bekaert. A simple unified approach for estimating natural direct and indirect effects. *American Journal of Epidemiology*, 176(3):190–195, 2012.

[30] Matthew Cefalu, Greg Ridgeway, Dan McCaffrey, Andrew Morral, Beth Ann Griffin, and Lane Burgette. *twang: Toolkit for Weighting and Analysis of Nonequivalent Groups*, 2021. R package version 2.3.

[31] B. Efron. Bootstrap Methods: Another Look at the Jackknife. *The Annals of Statistics*, 7(1), January 1979. Publisher: Institute of Mathematical Statistics.

[32] Li Xu, Chris Gotwalt, Yili Hong, Caleb B. King, and William Q. Meeker. Applications of the fractional-random-weight bootstrap. *The American Statistician*, 74(4):345–358, October 2020.

[33] K. Imai, L. Keele, and D. Tingley. A general approach to causal mediation analysis. *Psychological Methods*, 15(4):309–334, 2010.

Part IV

Outcome Models, Machine Learning and Related Approaches

20

Machine Learning for Causal Inference

Jennifer Hill, George Perrett, Vincent Dorie

Hill, Perrett, and Dorie

CONTENTS

DOI: 10.1201/9781003102670-20

20.1 Introduction

Estimation of causal effects requires making comparisons across groups of observations exposed and not exposed to a a treatment or cause (intervention, program, drug, etc). To interpret differences between groups causally we need to ensure that they have been constructed in such a way that the comparisons are "fair." This can be accomplished though design, for instance, by allocating treatments to individuals randomly. However, more often researchers have access to observational data and are thus in the position of trying to create fair comparisons through post-hoc data restructuring or modeling. Many chapters in this book focus on the former approach (data restructuring). This chapter will focus on the latter (modeling) to illuminate what can be gained from such an approach. We illustrate the case for modeling the relationship between outcomes, covariates, and a treatment to estimate causal effects using a Bayesian machine learning algorithm known as Bayesian Additive Regression Trees (BART) [1–3].

20.2 Causal Foundations

This section introduces the building blocks necessary to understand what causal quantities represent conceptually and why they are so difficult to estimate empirically.

20.2.1 Fair comparisons

At a basic level, causal inference methods require fair comparisons. For instance, suppose we want to understand the effect of a nutritional intervention on cholesterol among women with high cholesterol levels. To do so, we have access to data on 200 women and 100 of them participated in the program. All of them had high cholesterol before the date that program started. If we simply compare average cholesterol levels between those who participated in the intervention and those who didn't one year after program onset, would it make sense to think of differences in average cholesterol levels as the causal effect of the program? After all, the individuals who participated in the intervention might have been different from those who didn't in any of a variety of ways even before the program began. For example, it might be that those who participated had higher levels of cholesterol on average. Or maybe they were more motivated to make lifestyle changes to lower their cholesterol. If differences like this exist, then it would be unfair to attribute any difference in mean outcomes to the treatment because we wouldn't be able to separate out what part of this difference was due to the program and what part of the difference was due to the baseline differences between groups.

20.2.2 Potential outcomes and causal quantities

We can conceptualize the assumptions required for causal inference as a formalization of this idea of fair comparisons. Critical building blocks in this formalization are *potential outcomes*. In the cholesterol example, the potential outcome Y_{0i} represents the cholesterol level we would see for individual i if they *did not* participate in the program. Y_{1i} represents the cholesterol level we would see for individual i if they *did* participate in the program. Potential outcomes allow us to formally define a causal effect for person i as the a comparison between them, $\tau_i = Y_{1i} - Y_{0i}$. This effect represents how different the cholesterol level would be for individual i if they *had* participated in the program compared to if they *had not*.

In practice, however, we can never measure both potential outcomes at the same point in time for the same person because we cannot observe that person both in the world where they received a treatment and the world where they did not. Therefore, if we denote the binary treatment received by individual i as Z_i, we can express the observed outcome, Y_i, as a function of the potential outcomes, $Y_i = Z_i Y_{1i} + (1 - Z_i)Y_{0i}$, such that Y_{1i} is revealed for those who receive the treatment and Y_{0i} is revealed for those who do not. We can perhaps imagine a situation where we could clone individual i to create individual j just at the moment of exposure to the treatment and have version i take the treatment and version j refrain. This would create a fair comparison when we compared their outcomes down the line because we would be assuming that both the individual and their clone would have had the same Y_0 and Y_1. Specifically we could assume $Y_{0i} = Y_{0j}$ and $Y_{1i} = Y_{1j}$. Therefore to estimate the causal effect for individual i, $\tau_i = Y_{1i} - Y_{0i}$, and even though their Y_{0i} would be missing, we could just substitute Y_{0j} which *is* observed (since individual j did not receive the treatment).

Most causal inference procedures try to mimic this situation but generally aim to estimate *average* treatment effects such as the mean of individual causal effects over a sample or population. We can express such an average effect generically as $\mathrm{E}[Y_{1i} - Y_{0i}]$. Some researchers may be interested instead in the average treatment effect for the type of individual who we observed to receive the treatment (or participate in the program). This quantity is referred to as the average effect of the treatment on the treated (ATT) and is formalized as $\mathrm{E}[Y_{1i} - Y_{0i} \mid Z_i = 1]$. This quantity may be of particular interest if we don't expect that our full control group would be eligible for or interested in the treatment. A reciprocal version of the ATT is the effect of the treatment on the controls, ATC, $\mathrm{E}[Y_{1i} - Y_{0i} \mid Z_i = 0]$, which captures the effect of the treatment for those we don't observe to take the treatment (or participate in the program). This may be of particular interest in situations where policy makers or practitioners would like to expand eligibility or incentivize different types of people to participate in a program or receive a treatment.

20.2.3 Assumptions

If our goal is to estimate the average causal effect for a group of individuals, we need to create fair comparisons for the group. What would be the most pristine way to accomplish this? Suspending disbelief for a moment, let's suppose we could clone everyone who was willing to participate in the study. Each of the original participants could be exposed to the treatment but none of the clones would be exposed. What are the implications for the assumptions needed for causal inference?

20.2.3.1 All confounders measured

In this hypothetical study with clones, one nice feature is that the everything about the original sample and the clones would be the same except for their treatment (and everything subsequent to the treatment). Since potential outcomes exist conceptually even prior to the treatment being administered, this implies that the distribution of potential outcomes would be the same across the treatment and the control groups,

$$p(Y_0, Y_1 \mid Z = 0) = p(Y_0, Y_1 \mid Z = 1).$$

This property is more commonly expressed as an independence statement

$$Y_0, Y_1 \perp Z.$$

While it is not possible to actually implement our hypothetical clone design, a completely randomized experiment does a good job of at least replicating this independence property because the randomization eliminates any *systematic* differences between the treatment and control groups. Unlike with our clones, the randomized experiment is still most useful for estimating average treatment effects. However, this does not guarantee that it will yield accurate *individual-level* treatment effect estimates.

Of course with small sample sizes a randomized experiment may still result in groups that differ simply by chance. While results will still be technically unbiased, any given treatment effect estimate may still be far from the truth[1]. Randomized block experiments, in which the randomization occurs within strata or blocks defined by covariates, can help to address this by ensuring perfect balance with respect to the blocking variables. Generally speaking, in randomized blocks experiments, independence is only achieved within blocks therefore these experiments satisfy a more specific assumption,

$$Y_0, Y_1 \perp Z \mid W,$$

where W denotes the blocks. For example, in a diet and exercise intervention we might seek to randomize individuals after grouping them based on their starting cholesterol levels.

Of course many questions can't be be addressed by randomized experiments due to any combination of logistical, financial, and ethical reasons. In that case researchers hope that the covariates they've measured essentially act like blocks in a randomized block experiment. That is, they hope that observations that have similar values on all their covariates have potential outcomes that are similar as well, regardless of their treatment assignment.

This assumption can be expressed formally as

$$Y_0, Y_1 \perp Z \mid X,$$

where X now denotes pre-treatment covariates. The intuition here is that for two groups (treatment and control) with the same values on all the pre-treatment variables, X, we assume they have, in effect, been randomized to the groups. This is basically the same assumption as is invoked by the randomized block experiment. The difference is that in a randomized block experiment the W are known and the assumption should hold in a pristine implementation of the design. In on observational study, on the other hand, researchers must make a leap of faith that their pre-treatment covariates, X, are sufficient to achieve this conditional independence. If X represents all confounders (informally, variables that predict both treatment and outcome), this assumption should be satisfied.

This assumption is referred to by many different names depending on the discipline and subfield. These include "ignorability," "selection on observables," "all confounders measured," "exchangeability," the "conditional independence assumption," and "no hidden bias" [5–8, 23]. In this chapter we will refer to this as the "all-confounders-measured" assumption.

Due to the critical role of this assumption in estimating unbiased treatment effects, the first step in many causal inference approaches is to try to ensure sure that is satisfied by including all *potential confounders* in X.[2] The second step, which is the focus of many of chapters in this book, is to figure out how to condition on these variables without making excessive additional assumptions. We'll illustrate some of the complications involved in this step in the next section with a hypothetical example, but first we discuss a related assumption.

20.2.3.2 Overlap

Another property of our hypothetical example with clones is that for each individual in our dataset there would be someone else with the exact same values of all pretreatment covariates (including confounders) but who received a different treatment. We might relax this to say that this neighbor

---

[1]For more discussion see [4].

[2]The idea of including all *possible* confounders is in tension with the desire to avoid "overfitting", discussed in Section 20.3.1: it cannot be fully known whether a variable is truly a confounder or a randomly correlated covariate, so that including additional but unrelated pre-treatment covariates may reduce generalizability. Another concern in including many potential confounders in a model is "bias amplification," which can occur when some true confounders are missing [10–12] and covariates are included that are strongly predictive of the treatment but not the outcome. However, this may be rare in real-world data, such that it is generally preferable to condition on most pre-treatment covariates [13]. Our advice would be to always include every pre-treatment covariate that is believed to be predictive of the outcome, particularly if it is also related to the treatment. If overfitting remains a concern use regularized/Bayesian models or perform a variable selection step.

would have to be in a sufficiently close neighborhood of the covariate space. Thus, by design, each individual would have an "empirical counterfactual" in the dataset. A more general formalization of this property is that, for every X and for $z \in \{0, 1\}$, $0 < \Pr(Z = z \mid X) < 1$. Conceptually we can think of this expression as requiring that in every neighborhood of the covariate (X) space spanned by our sample there has to be a positive (non-zero) probability of having both treated and control units.

In theory a completely randomized experiment should create overlap by design since the multivariate distributions of all pre-treatment variables (measured of not) should be the same across groups. Of course, in practice, if one were checking, it would be difficult to achieve perfect overlap even for a moderate number of variables simply due to sparsity [14]. However it turns out the overlap assumption is technically stronger than what we need to perform inference. A more precise requirement is that we have overlap with respect to our *true confounders*, X^C, as in $0 < \Pr(Z = z \mid X^C) < 1$. We'll return to this idea later in the chapter because BART affords some advantages with regard to this goal [15]. Since a completely randomized experiment has no confounders the requirement is satisfied trivially. A randomized block experiment would need to satisfy the overlap assumption with respect to its blocks.[3] Again this should be satisfied by design since randomized block experiments require a non-zero probability of assignment for each treatment variable in each block.

In an observational study, however, there is nothing to guarantee that overlap exists across treatment groups with respect to confounders. There may be certain type of people who will never participate in a program or will always be exposed to a treatment. If the overlap assumption does not hold, there may be some observations on our data set for which we simply don't have enough information about their counterfactual state to try to make inferences about them.

To push this to the extreme imagine what would happen in a study if all confounders are measured but overlap is violated. For example, suppose that we have a study where individuals are assigned to receive a treatment based on their age. Specifically, suppose that all individuals over age 50 receive the treatment and all individuals 50 and younger do not. Further suppose that within each group, treatment assignment has no impact on the potential outcomes; if the age restriction on treatment did not exist, the experimental design would be valid. Thus, even though all confounders are measured there is no overlap in the age distribution across treatment and control groups. In this situation arguably none of the observations would have empirical counterfactuals. If you wanted to understand the effect of the treatment for the individuals less than 50 you would be hampered by the absence of treated units in this age range to provide data on the missing y_1's. If you wanted to understand the effect of the treatment for individuals older than 50 you would be hampered by the absence of control units in this age range to provide data on the missing y_0's. Without further assumptions your only hope would be to focus on those observations closest to the threshold.[4]

20.2.3.3 SUTVA

While we won't devote much time to it in this chapter, one of the most important assumptions in causal inference is the Stable Unit Treatment Value Assumption (SUTVA) [18]. The basic idea is that we need to assume that each person's potential outcome is a function solely of their own treatment assignment, not the treatment assignment of anyone else. This portion of SUTVA is sometimes referred to as the "no-interference" assumption. This assumption also encapsulates the idea of "consistency." This can be formalized as the idea that $Y_{aj} = Y_j$ when $A_j = a$ [19]. This reflects the idea that if the observed value of Y for an individual j who received treatment equal what it would have been if the treatment was "set to" a. In other words, we assume that the manner in which treatment was set to a is irrelevant. This is sometimes referred to as the "no-multiple-versions-of-treatment" assumption [20].

[3] Not all blocks act as confounders, however, so this is actually also too strong a statement.

[4] For clever ideas of how you *can* estimate a causal effect in this situation, you can read about regression discontinuity designs [4, 16, 17].

20.3 Regression for Causal Inference

We illustrate the use of regression for causal inference by introducing a hypothetical example that we will augment to illustrate key features of the approach. This example imagines that a few years ago a large sporting equipment company released a new marathon running shoe called the hyperShoe with the claim that runners wearing the hyperShoe would be able to run faster without changing their fitness or running technique. In effect, the sporting company made the claim that using the hyperShoe caused reductions in marathon race times. Upon release, the hyperShoe was marketed to specific subset of runners. Thus, runners using the high performance shoes likely differ with respect to characteristics that also would be expected to predict their running performance. For instance, those who purchase the shoe are likely to be a different age and may be more serious about running. While the company claims that the high performance running shoes result in faster race times, skeptics argue that self-selection may explain the differences in performance. In other words if higher-performance runners are more likely to buy the shoes, then any differences in running times might simply be a result of the differences in the types of runners who tend to wear the shoes. In technical terms potential confounders need to be ruled out before reaching this causal conclusion.

To better understand the role of confounders, let's focus first on age, since it is easy to measure and is one of the strongest predictors of running time. Suspending disbelief for a moment, if we assumed that age were the *only* confounder when estimating the causal effect of the hyperShoe on running times, what would it mean to make fair comparisons? In this hypothetical, we would need to compare individuals of the same age but who differ with regard to whether they wore the hyperShoe. How does regression allow us to make this comparison?

Regression provides one way to condition on confounders in an attempt to create fair comparisons. Most students who take introductory statistics courses are taught that linear regression should not be used to draw causal conclusions. After all, they are told, and rightly so, "correlation is not causation." But that doesn't really get to the heart of the matter since alternatives like matching and weighting also just estimate (conditional) correlations or associations and also require additional assumptions to identify causal effects. It is critical to understand under what conditions a regression could recover a causal effect.

Let's start with a version of our hypothetical example in which age is linearly related to running time. Figure 20.1 displays what this might look like. Here lighter dots represent "treated" units (those with the hyperShoe) and darker squares represent control observations. The lighter line represents the relationship between Y_1 and age. The darker line represents the relationship between Y_0 and age. These lines are sometimes referred to as the "response surface". The vertical difference between the two lines represents, τ, the treatment effect at each level of age. Since the treatment effect is constant across age (reflected in the parallel lines), the expected treatment effect for any given person is the same as the average across the sample. The following regression equation, displayed by the dashed lines in the plot, provides a good fit to the data.

$$Y = 190.28 + .49X - 10.21Z + \epsilon,$$

where Y denotes running time, Z denotes use of the hyperShoe, and X denotes age. But is the coefficient on Z an estimate of the causal effect of the hyperShoe on running time?

Recall that in this example we are assuming that age is our only confounder. Thus the assumption of "all confounders measured" is satisfied. If, additionally, our parametric model is correct, we can say that

$$E[Y_0 \mid X] = E[Y_0 \mid X, Z = 0] = E[Y \mid X, Z = 0] = \beta_0 + \beta_1 X$$

and

$$E[Y_1 \mid X] = E[Y_1 \mid X, Z = 1] = E[Y \mid X, Z = 1] = \beta_0 + \beta_1 X + \tau.$$

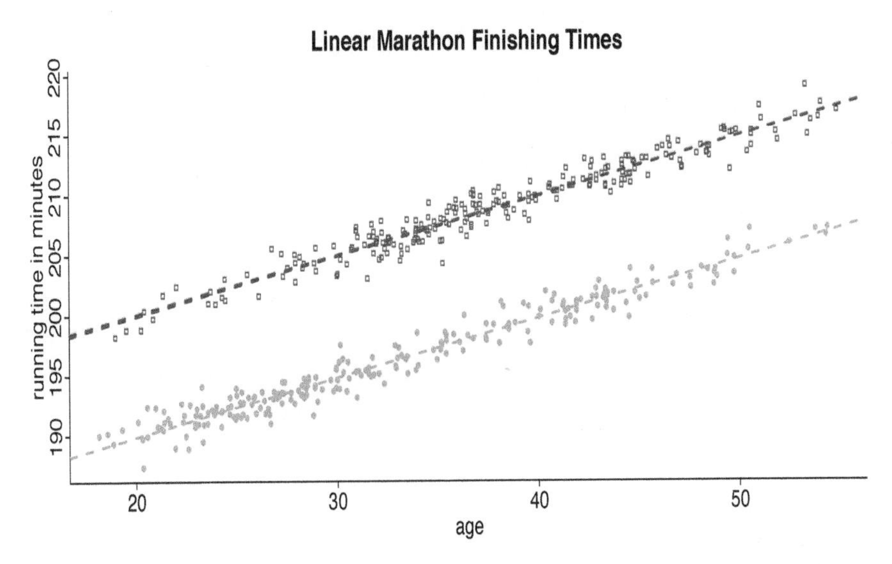

FIGURE 20.1
Hypothetical data on age and running times for a marathon. The solid line displays the true relationship between age and running time which, implausibly, is linear in this example. Lighter dots represent "treated" units – those with the hyperShoe – and darker squares represent control observations. The lighter line represents the relationship between Y_1 and age. The darker line represents the relationship between Y_0 and age.

That is, in this situation our regression model is also a causal model. Thus $\hat{\tau} = -10.21$ should be an unbiased estimate of the true treatment effect. In this case since we simulated the data we know that the true treatment effect for this example is -10. Interpreted causally, the hyperShoe decreases race time by 10 minutes.

In sum, if age is the only confounder *and* if the linear model is appropriate, then a linear regression model *can* recover the causal effect of the hyperShoe on running times. Unfortunately, neither of these assumptions is generally appropriate. We focus for the time being on the modeling component, and return to the foundational assumptions for causal inference in a later section.

20.3.1 Regression trees vs. linear regression

Unfortunately the real-world relationship between age and running time is not accurately represented in the hypothetical data presented in Figure 20.1 above. Figure 20.2 plots hypothetical age and marathon time data that better reflects the relationship documented in the literature [21, 22] as a blue line. The darker line, in contrast, represents what this relationship might look like if the hyperShoe was successful in reducing running times. Unlike in our previous example, a linear regression fit to the data (displayed with the dashed lines) fails to correctly represent the true relationship between age and running time for each potential outcome. The linear regression estimate of the causal effect is about -3.66, with a 95% confidence interval of $(-2.98, -4.34)$. However, the true ATT, ATE, and ATC for this sample are -3.85, -4.42, and -5.00, respectively. The linear regression confidence interval, which is best suited to estimate the ATE, captures the ATT but not the ATE or ATC. What's going on?

One obvious problem is that the true response surface is not linear! So we wouldn't expect a linear regression to represent it well. But there is another problem that exacerbates the issue. Not all parts of the of the "covariate space" (here, the range of the age) are well represented by both treated and control observations. Consequently, the model underperforms most severely in regions of the

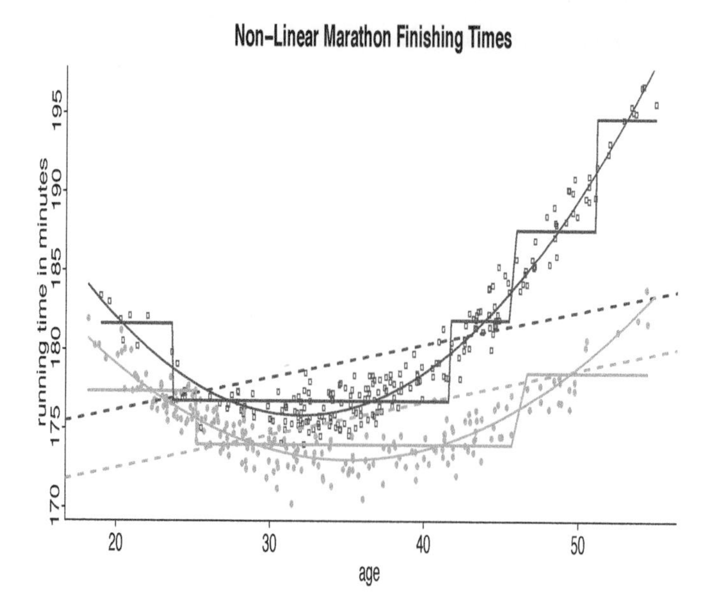

FIGURE 20.2
Hypothetical data on age and marathon running times in minutes. The solid curves (lighter for Y_1 and darker for Y_0) display the true relationship between the confounder and the potential outcomes. The dashed lines (with the same color mapping) represent a linear regression fit. The long-dash lines (with the same color mapping) represent the fit from a regression tree. Lighter dots are treated observations. Darker squares are control observations.

covariate space where counterfactual units are rare. While the treatment group is mostly strongly represented in the age range from 18 to 50, the control group more closely spans the full age range from about 20 to 55. Thus there are more empirical counterfactuals for the full treatment group than vice-versa.

Is there a simple way of providing a better fit to this response surface? Recall that regression is just a way to summarize information about how average outcomes of the response variable vary across subgroups defined by the covariates in our sample. In our current example we want to understand how marathon running times vary with subgroups defined by age. Linear regression does this in a way that places strong constraints on how these means are related to each other. Can we fit this relationship using a regression model that makes fewer assumptions?

A regression tree fit to these data would deconstruct the problem a bit differently than linear regression. Regression trees form subgroups within the dataset such that the within-subgroup variance in the outcome variable across subgroups is minimized (see chapter 9 of The Elements of Statistical Learning (ESL) by Hastie, Tibshirani, and Friedman (2009) [23]). The first step is to split the dataset into two subgroups. In a regression tree fit to our example data, the first split divides those individuals who are younger than 46 from those individuals older than 46. If we allow for further splits, the tree will continue to subdivide individuals by age and by who wore and did not wear the hyperShoe. This splitting process is repeated until a stopping condition is met, in our case until eight subgroups, or terminal nodes, are found. The tree with this stopping rule is displayed in the right panel of Figure 20.3 which shows the decision rules for each split and the mean in each terminal node. For instance the right most terminal node shows a mean of 195 which is the average outcome for individuals in our sample who did not receive the treatment (wear the shoe) and who are age 51 or above. The fit from this regression tree is summarized by the mean outcomes for the terminal node

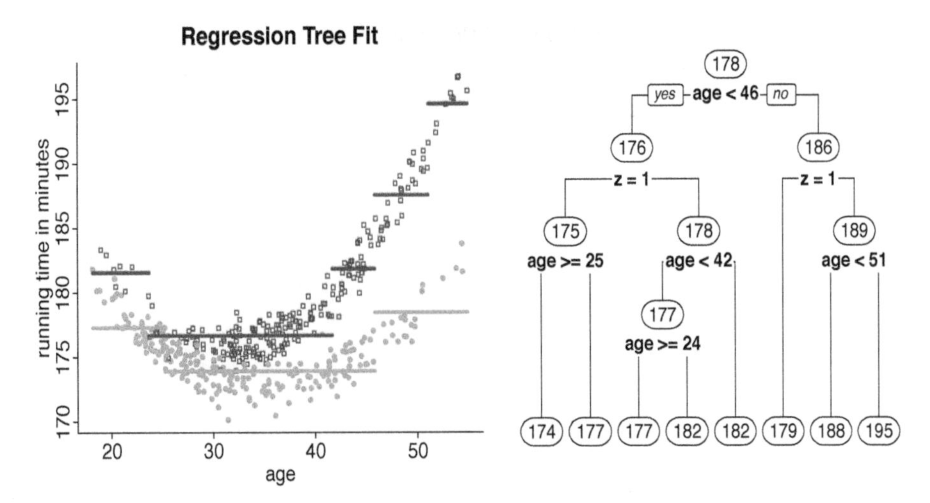

FIGURE 20.3
The left side of this figure displays the regression tree fit to the response surface which is represented by eight subgroup means. This means are displayed as horizontal line segments spanning the ages of the individuals in their subgroups and with darker colors for control observations and lighter for treated observations. The right side displays the branching and terminal nodes from the corresponding regression tree. In this tree "z" represents the binary treatment variable.

subgroups. These means are visualized by the horizontal lines (step function) in Figure 20.3. Notice how this model is better able to follow the curve of the response surface displayed on the left.

The flexibility of the regression tree fit is appealing. What are the downsides? We didn't mention above specifically how we decided to stop "growing the tree" (i.e. creating more subgroups). This is a tricky decision. If we stop too early we risk creating too crude a fit to the data. This is apparent in the fit in Figure 20.3, where it is often further from the observed data than we would like. On the other extreme we could allow the tree to grow so large that the fit yields a different "mean" for every observation. This model fit would be terrific at predicting the outcome within the current sample but is not likely to predict well at all in a new sample. This phenomenon is referred to as "overfitting." For a discussion of overfitting, see Chapter 7 of ESL.

In practice this tension is resolved by adjusting tuning parameters (also called hyperparameters in some fields) for the algorithm that govern features such as the number of observations required for a terminal node to be allowed to split, the minimum deviation within a terminal node to prevent further split, and the maximum depth for a tree. Typically these parameters are chosen via cross-validation (see chapters 7 and 9 of ESL). However cross-validation can only be directly used to understand to guard against overfit with regard to our observed data. It cannot be used to understand potential overfit with regard to unobserved counterfactuals.

There are several other downsides to regression trees. First, when we have multiple covariates, they aren't able to effectively capture additive effects well and tend to overemphasize multi-way interactions. Second, they don't directly estimate our uncertainty about our predictions or fit. This latter issue can be addressed using bootstrapping but that comes at a high computational cost. Finally, they tend to have a high variance – slightly different datasets can yield dramatically different trees.

20.3.2 Boosted regression trees

Boosted Regression Trees (chapter 10 of ESL) emerged as a way to address these issues of overfitting, difficulty in capturing additive structure, and overemphasis on high-level interactions. The idea

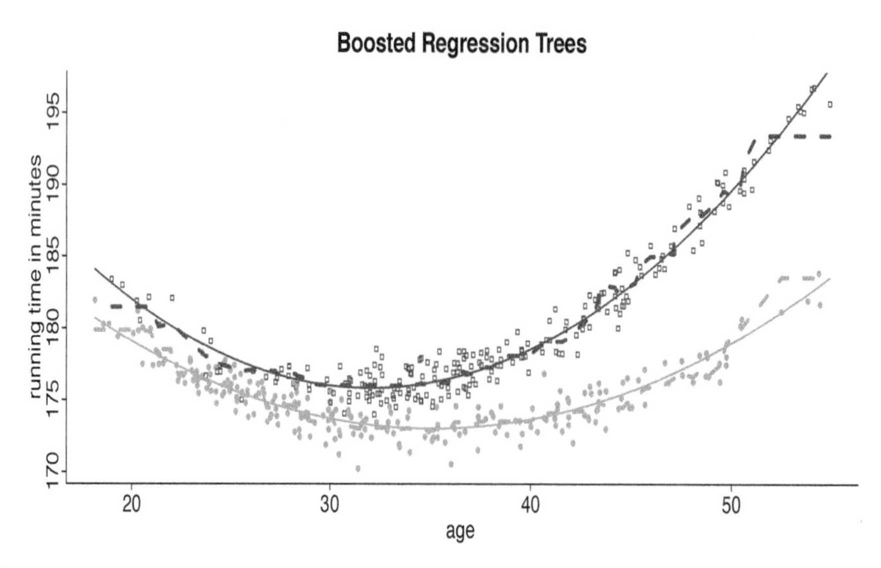

FIGURE 20.4
This figure displays the same response surface as in the previous figure (light for treated and Y(1) and dark for controls and Y(0)) but now with a boosted regression tree fit displayed as dashed lines.

is very clever! Instead of representing the fit to the data through a single regression tree, boosted regression trees represent the fit to the data through the sum of fits from multiple small regression trees. How does this work?

Consider our example above. A single tree fit would form subgroups based on treatment assignment and age as displayed in Figure 20.2. The residuals from that fit represent the variation in the outcome that has yet to be explained (which is a lot at this point!). What if we fit another tree to those residuals to try to explain more? Now new residuals can be formed by subtracting the fit from the second tree from the residuals from the previous step. This process can be repeated many times and at each step more of the unexplained variation in the outcome can be explained. Using this strategy, model complexity is controlled by using only small trees at each step, also known as "weak learners," tree predictions are combined using a weighted average or sum, reducing variance and forming an "ensemble," and by limiting the overall number of trees overfitting can be avoided.

Figure 20.4 displays a fit from a boosted regression tree with 100 trees. There is marked improvement to the fit of the response surface relative to the smaller tree in 20.3. However, note that the tree fit starts to depart from the true response surface when the treatment or control observations become relatively scarce. Traditional, machine-learning-style fits typically provide no way of being alerted to the increased uncertainty in this part of the covariate space.

While boosted regression trees were an important step forward as compared to standard regression trees, they still require choice of tuning parameters (additionally now including the number of trees included in the sum of trees) and fail to resolve the issues regarding uncertainty quantification.

20.4 Bayesian Additive Regression Trees

Bayesian Additive Regression Trees (BART) is a model that addresses some of the outstanding issues with boosted regression trees. It handles overfitting in a more principled, flexible, and data-driven way and also provides coherent uncertainty intervals.

The mean structure of BART is the same as the boosted regression tree. However, this structure is then embedded in a likelihood framework so that the fit of the model is considered to be a random variable and hence has a distribution. Given the importance of the treatment in this context we will write the BART for causal inference model explicitly incorporating the treatment variable in the notation. In addition, an error term is added to reflect deviations between the fit from the model and our observed values of the outcome (which reflects our uncertainty). We can formalize the model as

$$Y_i = f(\mathbf{x}_i, z) + \epsilon_i, \quad \epsilon_i \overset{\text{iid}}{\sim} N(0, \sigma^2), \tag{20.1}$$

$$f(\mathbf{x}, z) = g(\mathbf{x}, z, T_1, M_1) + g(\mathbf{x}, z, T_2, M_2) + \cdots + g(\mathbf{x}, z, T_m, M_m). \tag{20.2}$$

In this equation each of the g functions represents the fit from an individual tree, T_h represents the structure of the hth trees, and the corresponding $M_h = (\mu_{h1}, \mu_{h2}, \ldots \mu_{hb_h})'$ represents the set of subgroup means corresponding to the b_h terminal nodes of tree h.

20.4.1 BART prior

The key innovation in the model rests in the prior. Three elements of the BART model require prior specifications: the trees, the subgroup means, and the error term. While technical details can be found elsewhere (see BART: Bayesian Additive Regression Trees by Chipman, George, and McCulloch (CGM, 2010; [2]), we describe here the intuition behind and implications of the prior distributions.

20.4.1.1 Prior on the trees

The goal of the prior specification on the trees is to keep each of the individual trees relatively small. In fact, the default prior specification results in a prior probability for trees with only a single, terminal node of 0.05. Similarly, the prior probabilities for trees of sizes 2, 3, 4, and 5 or more terminal nodes are .55, .28, .09, .03, respectively. By expressing a preference for small trees, this prior avoids over-reliance on any single tree and also discourages high-level interactions. However, if the data suggest that larger trees are warranted the model will accommodate this. For instance in our example data a BART model constrained to have only a single tree in the response model has a median depth across iterations of 6. In a setting with more variables and a more complex structure, however, we might expect the average tree size to be larger.

20.4.1.2 Prior on the means

The prior on the means is also designed to help avoid overfitting. This is achieved by specifying a reasonable prior for the overall fit and then "reverse engineering" the implications for the means from individual trees. What is a reasonable prior for the overall fit? CGM achieve this by standardizing the outcome data so that it all lies between $-.5$ and $.5$. Given this, a reasonable prior for the expected response for the fit would assign high probability that the sum of all of the trees lies between these two numbers. The default prior sets this probability to 95%.

Remember, however, that the total expected value is the sum of m means from individual trees. The prior over these means can be derived from the overall prior. Suppose that we express the sum of the individual trees as $\text{N}(0, m\sigma_t^2)$, where σ_t^2 is the variance of the prior mean for any given tree. For this sum, by default, to have 95% probability of being within $-.5$ and $.5$ while treating the prior contribution of each tree as independent and identically distributed, σ_t has to be equal to $.5/2\sqrt{m}$, where $2 \approx 1.96 \approx \Phi^{-1}(0.975)$. Consequently, σ_t is a hyperparameter which determines the sensitivity of the nodes to the data. The default values work very well for continuous response variables; however, σ_t may require adjustment, crossvalidation, or the imposition of a hyperprior to avoid overfitting.

20.4.1.3 Prior on the error term

The prior on the error term should reflect the level of uncertainty we expect will remain after fitting the model. Rather than specifying an informative prior[5], CGM calibrate the prior using the residual error from a linear regression fit as a benchmark. Specifically the χ^2 prior assumes that there is 90% chance that the BART residual standard deviation will be less than that estimated by a standard linear regression.

Why does this make sense? Well suppose that the true model is linear. Then we would expect BART to have a similar residual standard deviation, though possibly slightly larger since it has to approximate a straight line with step functions. However, if the true response surface is nonlinear, then we would expect BART to have a smaller standard deviation due to the closer fit. In essence this prior assumes that in 90% of the examples where we will fit BART standard linear regression model will not provide a better fit, as defined by the residual standard deviation.

20.4.2 Gibbs sampler for BART

As a Bayesian algorithm, BART is fit by combining a statistical likelihood with priors on parameters to define a posterior distribution. Posteriors that do not have simple, named distributions are often summarized by sampling procedures. These samples can be drawn by defining a random walk whose stationary distribution is that of the target, that is by using a Markov chain Monte Carlo (MCMC) algorithm [24]. The algorithm starts with parameters initialized in some fashion – often drawn from their prior distribution or given reasonable, dispersed starting points – and proceeds by iteratively updating them from their current state to new ones in a stochastic fashion. The initial portion of such walks are often called "warm-up" and are not used in inference. During warm-up, the sampler progresses from the starting point to, ideally, the stationary distribution. Multiple, independent chains are often run as a diagnostic: when the random walks from many chains are all found to be in the same area of the posterior, the sampler is said to have "converged" to the stationary distribution.

In particular, BART is implemented as a Gibbs sampler, where the random walk proceeds by sequentially sampling from the conditional posteriors of individual parameters, given all of the others [25]. Since the likelihood for BART is Gaussian and the contribution of each tree is conditionally independent of all of the others, a Gibbs sampler for BART gives rise to an algorithm called "Bayesian backfitting": each tree can be randomly manipulated while fit against the residual of the response and every other tree. For details, see CGM. Finally, as a Gibbs sampler with a Gaussian likelihood, BART can be embedded in hierarchical models with components conditionally sampled, given different parameters. For example, see `stan4bart` [27], which fits multilevel models with both BART and so-called random effect mean structure. We discuss this extension in more detail below.

20.5 BART for Causal Inference

Just as a linear regression or regression trees could be used to estimate a causal effect, BART can be used as well. This section describes the steps involved in a basic implementation and available software.

[5] We avoid using the term "uninformative," as all priors incorporate some choice on where to distribute probability mass or density. For example, a so-called "flat" prior on a standard deviation expresses a strong preference for small valued variances.

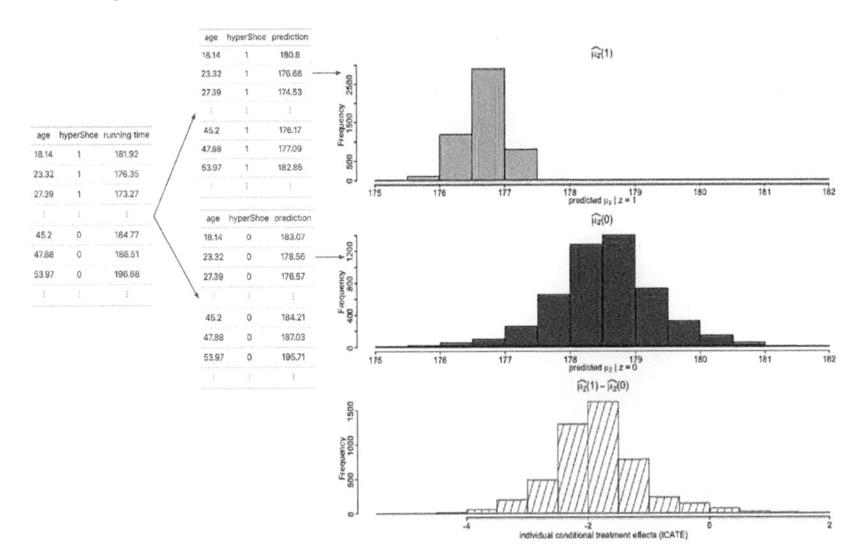

FIGURE 20.5
This figure displays how BART is used for causal inference. BART is fit to the original dataset (on the left). Predictions are made for two altered versions of that dataset (in the middle): one where all observations are assignment the treatment (top) and one where all observations are assigned to the control (bottom). These represent predictions of Y_0 and Y_1 for each person (last column of middle datasets). Posterior predictive intervals for each potential outcome for the 2nd individual in the dataset are displayed on the far right. The top plot is a histogram showing the empirical posterior distribution of $Y(1)$ for that individual. The darker histogram below it shows the empirical posterior distribution of $Y(0)$ for that individual. The difference between these distributions is the posterior distribution of the treatment effect for the 2nd individual (bottom-most histogram). These individual-level treatment effect distributions can be combined to estimate an average treatment effect for any of a variety of average treatment effects.

20.5.1 Basic implementation

A simple BART approach to causal inference implicitly involves several steps that are automated in current software (`bartCause`, described below). First, the algorithm is fit to the data using the model and fitting procedure described above. Second, predictions for the outcome are made for each observation in the original dataset, as well as a new "counterfactual" dataset. Intuitively we can think of that algorithm as imputing the missing potential outcomes in the dataset. In this way we can obtain predictions for each individual both for their observed treatment assignment and their counterfactual assignment.[6]

Since BART is a Bayesian algorithm, it provides not only a prediction of Y_0 and Y_1 for each person, but a full posterior predictive distribution for each of these potential outcomes. These distributions can be combined for a given individual, i, by differencing the two distributions to obtain a posterior distribution for $Y_{1i} - Y_{0i}$, individual i's causal effect. Figure 20.5 illustrates these steps and displays posterior distributions for the second observation in the dataset of our hypothetical example. Critically, posterior distributions for individual level causal effects can then be averaged to obtain the posterior distribution of the causal effect for the full sample or any subset thereof.[7]

[6] We describe it this way to make the intuition clear. In reality the computations only require that we fit on one dataset and make make predictions for that dataset and a version where the treatment assignment for all observations is flipped.

[7] Since these distributions condition on the covariates in our analysis sample, the average causal effect technically represents a conditional average causal effect rather than the sample average causal effect. Sample average effects can be obtained by

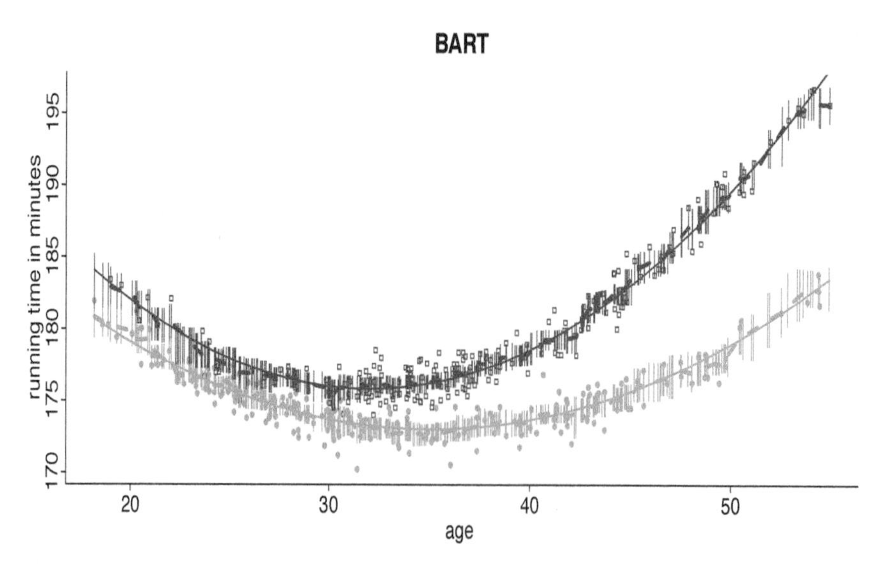

FIGURE 20.6
This figure displays the same response surface as in the previous figure (darker curve for treated and lighter curve for control) but now with a BART fit overlaid as dashed lines. 95% posterior uncertainty intervals (vertical lines) for all individuals (again with lighter lines for the treated and darker lines for the controls) are calculated by using normal approximations to their empirical marginal posteriors.

We present results of the BART fit to the data in our hypothetical example in Figure 20.6. This plot reproduces the true response surface and observations from Figure 20.4 above but instead overlays a BART fit. There are several ways to display this fit but we choose here to display two uncertainty intervals for each observation, i on the plot: (a) a 95% uncertainty interval for $E[Y_{0i} \mid X_i]$ and (b) a 95% uncertainty interval for $E[Y_{1i} \mid X_i]$.

20.5.2 Software: `bartCause`

A variety of packages currently exist in the R programming language to implement BART ([27, 29–35]), encompassing a multiplicity of applications and algorithmic adaptations. The R package bartCause allows for estimation of treatment effects using BART as described above and additionally accommodates the extensions discussed in the following section. It accepts as input the treatment and response variables as well as a list of confounders. It then creates a suitable counterfactual dataset to provide predictions of potential outcomes and their uncertainty.

The main function of the bartCause package is `bartc`. `bartc` is designed to be relatively similar to other model fitting functions in R, like `lm`, with the exception that it can fit *two* models instead of one. Consequently, it breaks the typical `formula` argument into three parts: `response` for the name of outcome variable, `treatment` for the name of the treatment, and `confounders` for the names of the covariates included to try to satisfy ignorability. The treatment and confounders are used to predict the outcome as described above. The confounders are also used in a model to predict the treatment as a way to flexibly estimate propensity scores. The propensity scores can be useful as an additional predictor in our outcome model or in calculating weights to help balance treatment and control groups.

---

using the observed factual outcome, and draws from the posterior predictive counterfactual distribution [28]. The uncertainty of population average effects can be obtained by using the posterior predictive distributions for both potential outcomes.

Additionally, bartCause accepts the standard model function fitting argument of a `data` object where it will look to resolve symbols, which can be a data frame or list with named elements. In other words, it takes inputs that are similar to the `lm()` and `glm()` functions in R. Since BART will look for a flexible, non-parametric relationship between the confounders and the treatment or response, the `confounders` themselves need merely be named. Deriving from the R model fitting syntax, they can be separated with a "+" sign, however, this is interpreted figuratively as "include this variable" and does not indicate a linear relationship. The `estimand` argument is also important to specify up front because the methodology differs for some approaches when targeting the `"ate"`, `"att"`, or `"atc."`

In the context of our illustrative example we could use the following command:

```
bartc_fit <- bartc(running_time, hyperShoe, age,
                   data = dat, estimand = "att",
                   seed = 0)
```

Once a model has been fit, callisng `summary` on it yields inferential results, including by default an estimate of the relevant population average treatment effect (in this case the population average effect of the treatment on the treated, PATT):

```
> summary(bartc_fit)
Call: bartCause::bartc(response = running_time, treatment = hyperShoe,
            confounders = age,  data = dat, estimand = "att",
            seed = 0)

Causal inference model fit by:
  model.rsp: bart
  model.trt: bart

Treatment effect (population average):
    estimate     sd ci.lower ci.upper
att   -3.866 0.1616   -4.182   -3.549
Estimates fit from 464 total observations
95% credible interval calculated by: normal approximation
  population TE approximated by: posterior predictive distribution
Result based on 500 posterior samples times 10 chains
```

In addition, bartCause includes a number of convenience plotting functions, all of which take a fitted model as their first argument.

- `plot_sigma`: for continuous responses only, produces by-chain trace plots of the residual standard deviation, where the x axis is the sample number and the y axis is the value; for use in assessing model convergence
- `plot_est`: by-chain trace plots of the estimand
- `plot_indiv`: histograms of individual-level quantities, including treatment effects and posterior means
- `plot_support`: scatter plots of individual-level quantities, with observations highlighted by the evidence of their common support, as discussed below

Finally, advanced users can access the posterior samples directly, using the `extract`, `fitted`, and `predict` generic functions. These can be useful for obtaining subgroup estimates, weighted averages, or for conducting additional diagnostics. bartCause can also be accessed now in a more user-friendly software package, thinkCausal, that additionally incorporates educational components to help the user understand the foundational concepts involved (`https://apsta.shinyapps.io/thinkCausal/`).

20.6 BART Extensions and Other Considerations for Causal Inference

The discussion above focused on average treatment effects for independent, identically distribution data where we assume that the all-confounders-measured and the overlap assumptions hold. What happens if these conditions aren't met or we want to estimate more complex treatment effects? This section briefly explores these issues.

20.6.1 Overlap, revisited

We revisit the overlap assumption to better the understand its implications vis-a-vis the BART approach to causal inference and other common causal inference strategies. To accurately estimate treatment effects, this identification strategy relies on its ability to accurately predict counterfactual values for the inferential observations (those represented in the estimand of interest). For instance, suppose the focus is on the effect of the treatment on the treated, $\mathrm{E}[Y_1 - Y_0 \mid Z = 1] = \mathrm{E}[Y_1 \mid Z = 1] - \mathrm{E}[Y_0 \mid Z = 1]$. We know a lot about $\mathrm{E}[Y_1 \mid Z = 1]$ because we observe Y_1 for all the members of the treatment group. However, we have no direct measures of $\mathrm{E}[Y_0 \mid Z = 1]$ because Y_0 is not observed for anyone in the treatment group. However, if all-confounders-measured holds, we can capitalize on the following equivalency: $\mathrm{E}[Y_0 \mid Z = 1, X = x] = \mathrm{E}[Y_0 \mid Z = 0, X = x]$. At a conceptual level, for each treatment group member, we use the information from control group observations that have the same values of X as the treatment group member to predict their Y_0. In practice, since it is unlikely that there are observations with exactly the same values on all variables, most matching methods matched on a lower-dimensional representation of the covariates, such as the propensity score. BART instead makes a prediction using its extremely flexible model.

In practice this means that we require an adequate number of observations that are sufficiently similar to a given treated observation to make a good guess about the missing Y_0. Failing this we may decide that it is inappropriate to try to make inferences about that particular observation.

How can we assess whether we have sufficient information to proceed? One option would be to check for each variable whether its distribution covers the same range across the two treatment groups, often summarized by a standardized difference in mean. However, it is possible that there is overlap only on the margins of each variable but not in their joint distribution.

A commonly advocated alternative is to focus only on the overlap in the propensity score [18,36]. This is certainly a simpler option and is justified by the theory if you have access to the true propensity score or a reasonably good prediction. However, this strategy places heavy emphasis on the propensity score estimation strategy. If the propensity score specification is incorrect, this can lead to misjudgements about whether or not overlap is satisfied. This problem can be compounded if the propensity score model includes covariates that are not actually confounders, as the additional variables might strongly predict the treatment – and hence degrade propensity score overlap – and yet have no influence on the outcome.

Hill and Su [15] discuss in more detail this idea that if the covariates in our model are a superset of the true confounders, then propensity score based approaches to identifying lack of overlap can lead to overly conservative judgments about which observations need to be excluded from an analysis. In contrast BART-based approaches to identifying observations that lack overlap [15] are geared toward discovering and mitigating threats due to lack of overlap from variables that are most likely to be true confounders. Thus it is more likely to yield estimates that satisfy common *causal* support [15].

We illustrate the approach in Figure 20.7 using a slightly modified version of our original example. In this scenario 14% of the treatment observations (those who used the hyperShoe) are younger than the youngest member of the control group (those who did not use the hyperShoe) and 11% of the control observations (those who did not use the hyperShoe) were older than the oldest member of the treatment group.

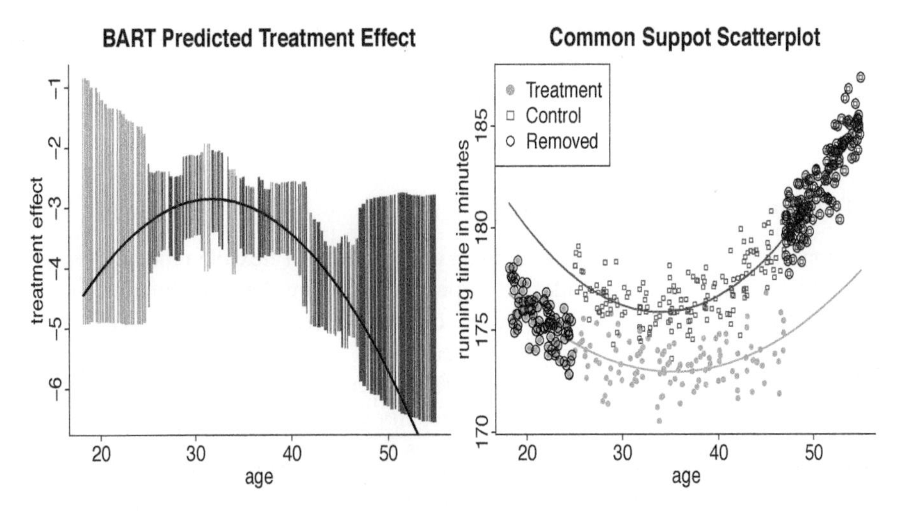

FIGURE 20.7
The left panel displays the average treatment effect as it varies with age as a solid black curve. The vertical lines represent 95% uncertainty intervals for the individual level treatment effect for each individual in the sample. The lighter lines correspond to treatment observations. The darker lines correspond to control observations. The right panel displays the response surface and observations again but circles the observations that BART would flag as being at high-risk of lacking common causal support. These assessments are related to the level of uncertainty in treatment effect estimation displayed in the left panel.

The left hand panel of Figure 20.7 displays a curve demonstrating how the true treatment effect varies with age. It also displays 95% uncertainty intervals for the individual-level treatment effects for each of the observations in the sample.[8] The right-hand panel displays the true response surface and observations.

A helpful feature of the BART predictions is that the uncertainty around them expands precipitously once we leave the range of common support. Standard BART measures of overlap [15], implemented in bartCause using the `commonSup.rule` and `commonSup.cut` arguments, would suggest that the circled units in the right-hand plot are sufficiently outside the range of sufficient overlap that we would not trust inferences about them. Therefore, we would discard those observations. The average treatment effect for the remaining observations is -3.92 and the BART estimate for those observations is -3.78.

The plot on the left also reminds us to be cautious in interpreting individual level treatment effects. While all of the intervals for the observations that would remain in our analysis still cover the true treatment effect (and some of the intervals for the discarded units cover as well!), some of them only just barely cover. Why is that? Because it's an exceptionally difficult inferential task to estimate a treatment effect specific to one covariate value, even when overlap exists in that neighborhood of the covariate space. When we estimate average effects we get to capitalize on the natural bias cancellation that occurs when adding up a bunch of slightly imprecise estimates, which leads to more accuracy for the average effects.

On the other hand this type of individual level prediction is much harder to do with most matching and weighting strategies that don't get to borrow strength across observations in the way

[8]Technically these are intervals of individual level *conditional* average treatment effects. For instance the interval displayed for a person who was 38 years old is just the interval for the average effect for anyone aged 38 in the sample. Formally, we can express each treatment effect as $\mathrm{E}[Y_{1i} - Y_{0i} \mid X_i]$, as distinct from, for example, $Y_{1i} - \mathrm{E}[Y_{0i} \mid X_i]$ (for a treatment observation) or $\mathrm{E}[Y_{1i} \mid X_i] - Y_{0i}$ (for a control observation).

that regression modeling approaches can. Even if we don't want to trust any given individual level prediction or interval we can still capitalize on the fact that they are estimated reasonably well to facilitate better understanding of trends in treatment effect heterogeneity, as is discussed in the next sections.

20.6.2 Treatment effect heterogeneity

The BART approach to causal inference, with its combination of flexible modeling embedded in a Bayesian likelihood framework, provides the opportunity for simultaneous inference on individual-level treatment effects as well as any of a variety of average treatment effects (ATE, ATC, ATT, or any subgroup effect defined by measured characteristics).

Hill [28] demonstrated the potential for BART to accurately detect individual-level treatment effects in scenarios where treatment effect heterogeneity is governed by observed covariates. Green and Kern [37] expanded on this to directly address the advantages of using BART to estimate treatment effect heterogeneity. Hahn et al. have since extended the BART framework to explicitly target treatment effects in a way that facilitates more accurate estimation of heterogeneous treatment effects [35]. A similar approach with a more formal decision-theoretic framework was developed by Sivaganesan and co-authors [38].

The BART output also provides a wealth of opportunities for summarizing the information in the posteriors. A simple but useful summary is a "waterfall" plot of point estimates and uncertainty intervals from posterior distributions for the CATE of every sample unit. Typically these plots order the treatment effects estimates and intervals by the magnitude of the treatment effect estimate (e.g., posterior mean or median).

An alternative to waterfall plots is demonstrated by Green and Kern [37] who present partial dependence plots [39] generated by estimating and averaging counterfactual treatment effect functions, which in turn are generated by manipulating potential effect moderators (instead of counterfactual predictions as in typical partial dependence plots). This approach can help us understand how the conditional average treatment effects vary with covariates.

Figure 20.8 provides another possibility. This figure plots data from an extension of our earlier example by considering the treatment effect of hyperShoe moderated by runners age and "mileage," (whether they run a high, moderate or low number of miles per week to prepare for the marathon). In the left panel of the figure we plot posterior distributions of the ATE associated with each of the three mileage groups. While there is some separation in the treatment effect distributions across these groups, there aren't strong differences between them. The right panel further distinguishes treatment effects by age and mileage group. Here we see that once we additionally condition on age we see much stronger differences in treatment effects across groups. Overall, older runners and runners who run more miles per week to prepare receive a greater benefit from the hyperShoe than younger runners and runners who run fewer miles per week.

20.6.3 Treatment effect moderation

The ability of BART to identify individual level treatment effects has additional benefits. As we saw in the previous example it allows us to explore what characteristics of observations are associated with variation in treatment effects. This is sometimes referred to as treatment effect moderation.

As another example, in recent work Carnegie et al. [40] explored treatment effect modification in an education example through a variety of graphical summaries. The left panel from Figure 1 of that paper displays a scatter plot of school-specific treatment effects versus a measure of average school-level fixed mindset beliefs. This plot was created to explore whether fixed mindset acted as a moderator of the treatment effect. However, unexpectedly, the plot revealed a cluster of schools with substantially different treatment effects. To better understand these differences a simple regression tree model was fit with estimated treatment effects (output from the BART fit) as the response and

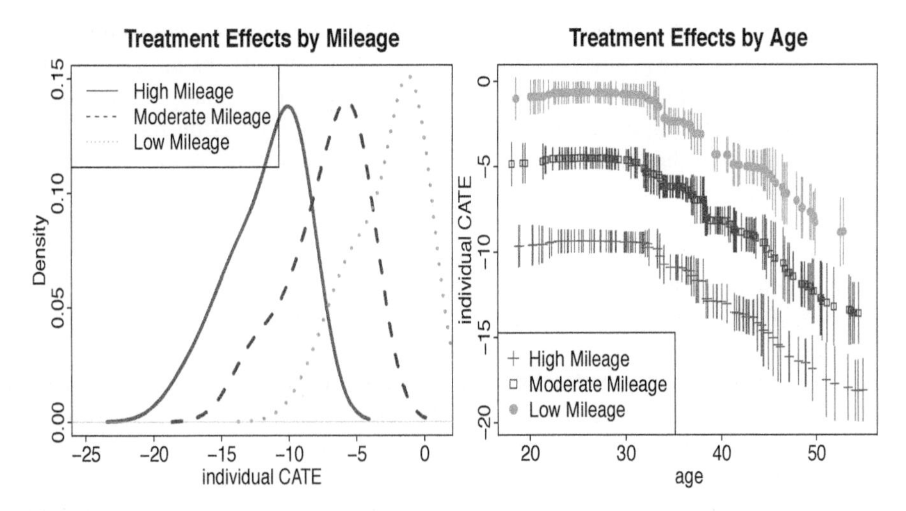

FIGURE 20.8
The left panel shows posterior distributions of the ATEs specific to "mileage" groups. While there is separation in the means of these distributions there is still a fair amount of overlap across these distributions. The right panel shows individual-level treatment effects as they vary with respect to levels of both mileage and age. If we hold age constant, we see more distinct separation in the treatment effect posterior distributions across mileage groups.

with the covariates as predictors. This fit revealed that the modifier creating this clustering was a measure of "urbanicity" of the schools. This is a simple and effective technique for discovering moderators and understanding more about treatment effect heterogeneity. The thinkCausal software mentioned previously automates implemention of this strategy (`https://apsta.shinyapps.io/thinkCausal/`).

Causal BART models also yield full posterior distributions over the outcomes under treatment and control. This can facilitate a formal decision theoretic treatment of optimal treatment selection for individuals by selecting the treatment rule that maximizes individual posterior expected utility. Logan et al. [41] use BART in this framework to estimate individualized treatment rules.

We note one caveat about these analyses. We recommend that exploration of treatment effect moderation and estimation of individual-level treatment effects is most plausible in the settings of randomized experiments. All-confounders-measured is a strong assumption in observational settings. Moreover, the overlap assumption is much stronger in observational settings, even if all-confounders-measured is satisfied. Satisfying these at an individual level is more difficult than at a group level. These complications compound in an observation setting whereas in a randomized experiment setting we mostly need to worry only about one of these issues.

20.6.4 Generalizability

When treatment effects vary across observations (e.g., students or schools) it is likely that the average treatment effect for a given sample will not be the same as that for another sample. How can we generalize the results from our original sample to a new sample or population that may represent a different distribution of treatment effects? For instance, suppose we find through a randomized experiment conducted in a dozen schools that an intervention was effective for lower performing students but had no impact on other students. If we want to have a sense of how that same intervention might impact a school with a different composition of lower and higher performing students, we would need to reweight (explicitly or implicitly) our estimate to mimic the population of interest.

Of course in more real-world applications treatment effects may vary based on a much larger number of individual-level (e.g. student) and group-level (e.g. school) characteristics. Thus simple reweighting strategies would be more complicated to implement, particularly if the treatment effect modifiers are not known at the outset. A BART approach capitalizes on its ability to estimate individual level effects. If the target population is simply made up a different compositions of the same types of observations that the algorithm was fit to originally,[9] then this task reduces to a prediction problem for a new sample of individuals (or groups) defined by these covariate values. If covariates are available for this new population then we can use BART to generate posterior distributions for both potential outcomes, which in turn can be used to generate posterior distributions for the treatment effect for each person in that new sample and any average effects based on groupings of these people.

If outcomes are additionally available (for instance we know test scores in the absence of an intervention but want to predict test scores (and thus effects) that would occur given exposure to the intervention then the prediction problem is less difficult because only half of the information is missing (the counterfactual outcome). BART-based strategies for generalizing treatment effects were discussed in [43], where BART was demonstrated to have superior performance over propensity score based approaches to generalization in the setting where all confounders are measured and we observe the covariates that modify the treatment effect. Generalization of average treatment effects from one sample or population to another requires an accurate portrayal of the way that treatment effects vary across subgroups. Thus it is no surprise that BART also demonstrates superior performance in this task.

20.6.5 Grouped data structures

When observations have a grouping structure (for example cluster randomized trials, repeated measurements, or individuals nested in schools or hospitals), it is important to explicitly include that structure in the treatment and response models. Failure to do so can omit a source of confounding, ignore correlation between observations, and lead to inaccurate measures of uncertainty. To address this, the bartCause package supports a wide variety of methods for grouped data.

The first decision to consider with grouped data is whether to include the grouping variables as so-called "fixed-effects" or "random-effects." These are also known as "unmodeled" or "modeled" parameters.[10] It is beyond the scope of this chapter to discuss the differences between fixed and random effects at length, although it's often the case that fixed effects are more appropriate when the researcher believes that there is unmeasured confounding at the group level. However, fixed effects can also lead to over-parameterization and noisy estimates in small subgroups. Random effects are typically more appropriate when there is a sizeable number of groups, they can be assumed to be independent of the treatment, and there are small groups that would benefit from being "partially pooled" toward the global average. For a longer discussion on fixed versus random effects in the context of causal inference, see this review chapter on multilevel modeling and causal inference [44] as well as research pointing to the potential dangers of using fixed effects for causal effects when all confounders are not measured [11].

Depending on the level of complexity of the grouping structure, bartCause has two supported interfaces. The first is for "varying intercepts," or models with a single grouping factor where the only predictor of interest is group membership itself. In that case, calling `bartc` and passing it the name of a grouping factor in the `group.by` argument will incorporate that variable. Further options include `use.ranef` to indicate the grouping factor should be random or fixed, and `group.effects`,

[9] Roughly speaking this translates into two assumptions: (1) that selection into each of the groups is ignorable with respect to the potential outcomes or the difference between them, and (2) that overlap exists between these groups. For more details see Stuart et al. [42].

[10] It should be noted that unless the foundational assumptions of Section 20.2.3 apply, the term "effect" is a misleading overstatement.

which determines if subgroup averages estimates should calculated as the result. Within the context of our hyperShoe example, suppose that observations were grouped by country, which might serve as a confounder through level of interest in running or funding available through a sports program. To fit a varying intercept model in bartCause and report subgroup average treatment effects:

```
mlm_fit <- bartc(running_time, hyperShoe, age,
                 data = dat, estimand = "att",
                 group.by = country, group.effects = TRUE,
                 seed = 0)
```

The result:

```
> summary(mlm_fit)
fitting treatment model via method 'bart'
fitting response model via method 'bart'
Call: bartCause::bartc(response = running_time, treatment = hyperShoe,
            confounders = age, data = dat, estimand = "att",
            group.by = country, group.effects = TRUE,
            seed = 0)

Causal inference model fit by:
  model.rsp: bart
  model.trt: bart

Treatment effect (population average):
           estimate     sd ci.lower ci.upper   n
country_1    -3.889 0.2615   -4.401   -3.377  42
country_2    -3.707 0.2466   -4.190   -3.223  49
country_3    -3.610 0.2438   -4.088   -3.132  49
country_4    -4.603 0.2709   -5.134   -4.072  42
country_5    -3.605 0.2353   -4.066   -3.144  52
total        -3.857 0.1592   -4.169   -3.545 234
Estimates fit from 464 total observations
95% credible interval calculated by: normal approximation
population TE approximated by: posterior predictive distribution
Result based on 800 posterior samples times 10 chains
```

More complicated grouping structures can be fit using the `parametric` argument of `bartc`. This argument accepts a full parametric equation that is added to the treatment and response models. Multilevel models, including nested or cross effects and varying slopes, are defined as in the lme4 package ([45]), using a vertical bar notation (`(var | group)`). For example, a varying intercept and slope model for our hyperShoe example that allowed a different coefficient for age by country would use the `parametric` argument of `(1 + age | country)`. More recently, extensions of the BART algorithm that accommodate parametric and multilevel structure have been developed in a package called stan4bart, available in R [27, 54].

20.6.6 Sensitivity to unmeasured confounding

The primary motivation behind using BART (or propensity scores, etc.) for causal inference is to avoid the bias incurred through misspecification of the response surface, ($E[Y(0)|X]$ and $E[Y(1)|X]$). However, the more difficult assumption to relax (in the absence of a controlled randomized or natural experiment) is the assumption that all confounders have been measured, in large part because this assumption is untestable.

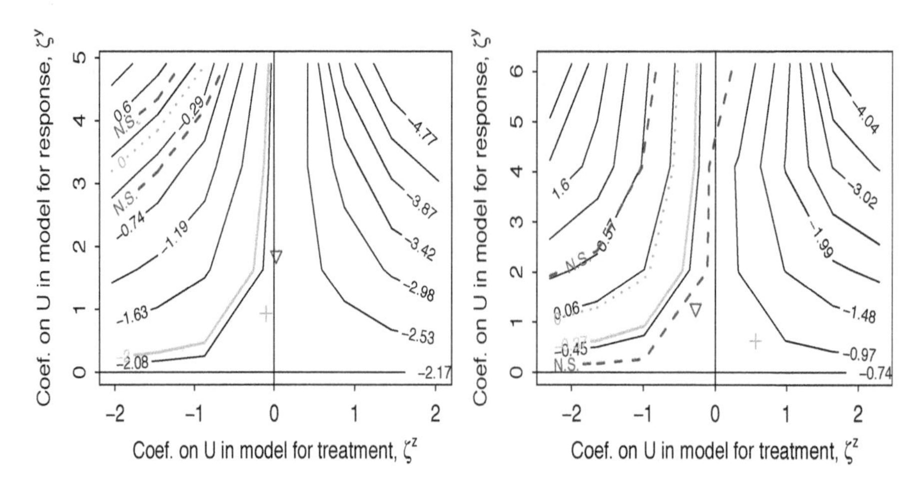

FIGURE 20.9
Contour lines corresponding to estimates of the treatment effect corrected for bias indicated by confounding levels at that location on the plot. The left plot displays an example that is not particularly sensitive to unobserved confounding. The right plot corresponds to an example that is more sensitive to unobserved confounding.

One approach to this is to evaluate the sensitivity of a study to potential unmeasured confounding across a variety of assumptions about the potential strength of that confounding. This allows the researcher to understand what level of confounding would be needed to substantively change the estimate of the treatment effect. For instance, what would it take to change the sign of this estimate or drive it to zero?

There is much work in this area but we will briefly focus on work by Hill and co-authors ([46]), extending earlier work by Carnegie, Harada, and Hill, [47] that allows BART to be incorporated into an existing sensitivity analysis framework. The original approach imposed a strict parametric model with the two parameters for to capture the role of an unobserved confounder; the extension relaxes the assumptions by allowing BART to model the response surface. This results in an easily interpretable framework for testing the potential impact of an unmeasured confounder that also limits the number of modeling assumptions. The performance of this approach was evaluated in a large-scale simulation setting and its usefulness was also demonstrated with high blood pressure data taken from the Third National Health and Nutrition Examination Survey [46].

To illustrate this approach we extended our original example to create two additional scenarios, both of which include an unobserved confounder. We conceive of this variable as a indicator of whether each runner had sponsorship for the race. Sponsorship is positively associated with the probability of having the hyperShoe and negatively associated with the running time (that is if you have sponsorship you are likely to have faster running times on average). In the first scenario the confounding was relatively weak and in the second it was relatively strong. Figure 20.9 displays results.

First consider the plot on the left. Each coordinate in the plot region represents a combination of sensitivity parameters. These parameters reflect the strength of association between our binary unobserved confounder, U (sponsorship), and the treatment (x-axis) as well as the strength of the unobserved confounder and the outcome (y-axis). In particular we can think of the parameter on the y axis as the difference in means in the outcome between groups with $U = 0$ and $U = 1$ after (non-parametrically) adjusting other the other covariates available; this parameter has been standardized to be represented in standard deviation units (with respect to the outcome variable). The parameter on the x-axis represents the coefficient on U in a probit regression model of Z on U and the other covariates. Each contour line in the plot reflects the set of such points (combinations of sensitivity parameters) that would result in a particular (standardized) estimate of the treatment effect, the magnitude of which is displayed on the line. The lighter dashed contour corresponds to a treatment effect estimate of 0 and the darker long-dashed contours show when the estimate would lose statistical significance. Finally, the plus sign and diamond represent that actual coefficients on the other two covariates in the model for these two equations (the triangle presents the estimate with a reversed sign so it is in a quadrant of the space represented by the plot). These help us to provide some context for the size of the coefficients and what might be considered a large magnitude for each of the sensitivity parameters.

Therefore, the left-hand plot, corresponding to the situation where the confounding is weak, accurately reflects that situation. It tells the researcher that a missing confounder would have to have a strong, negative relationship with the treatment and and exceptionally strong relationship with the outcome to drive the treatment effect estimate to zero; the sizes of strengths of those relationships would far exceed those of the observed confounders. On the other hand, the plot on the right displays a situation where the confounding is much stronger. In this case the results are much more sensitive to the unobserved confounder. The treatment effect estimates could be driven to zero with much more moderate associations between U and the treatment and outcome.

20.7 Evidence of Performance

Hill [28] first proposed use of BART in causal inference and provided simulation evidence of superior performance relative to simple propensity score approaches. Since then, many other papers have provided additional evidence of the advantages of BART-based methods to causal inference [35,37,43,48,49]. Although BART has proven to be particularly effective in settings with a single point in time binary treatment variable where all confounders measured holds, in principle any flexible regression model could be used in a similar manner and might have similar properties. The 2016 Causal Inference Data Analysis Challenge [50], associated with the annual Atlantic Causal Inference Conference (ACIC[11]), allowed researchers to submit any of a variety of causal inference algorithms to estimate causal effects across a variety of settings in this type of setting and indeed there were several promising machine-learning based submissions, including BART.[12]

20.8 Strengths and Limitations

This section outlines both strengths and limitations of using BART for causal inference. It also highlights directions for ongoing and future work towards ameliorating some of the existing shortcomings.

[11] ACIC now stands for American Causal Inference Conference.

[12] The 2017, 2018, and 2019 incarnations of these challenges also demonstrated superior performance for BART. The 2017 results are available as a technical report [51]. 2018 and 2019 results are currently unpublished, but were announced at each of the respective ACIC conferences.

20.8.1 Strengths

BART has three key features that contribute to its strengths: (1) flexible modeling strategy, (2) use of the outcome variable, and (3) the Bayesian framework. We discuss the advantages of a BART approach to causal inference by framing them in terms of their relationship to one or more of these features.

One of the biggest strengths of BART is that it allows for robust estimation of a wide variety of estimands, ranging from average treatment effects, to subgroup effects, to individual-level treatment effects. This versatility results as a combination for the flexible sum-of-trees model and the Bayesian framework which allows us to produce a full posterior distribution for each combination of observation and potential outcome. The ability to produce more robust estimates of individual treatment effects not only allows us to better understand treatment effect heterogeneity but also to explore what covariates moderate treatment effects. The fact that the BART modeling approach incorporates the outcome variable allows for more efficient estimation of these estimands as well.

Capitalizing on the information in the outcome variable also provides BART with a way of implicitly identifying which covariates are true confounders, based on which are most strongly predictive of the outcome. The algorithm then weights the contributions of those variables more strongly. In conjunction with the Bayesian framework which provides a principled strategy for estimating uncertainty, BART can identify observations without sufficient common causal support more easily than many other approaches to causal inference [15].

Additional advantages of the Bayesian framework include the ability to expand the model fairly easily to accommodate extensions. Currently extensions include the stan4bart multilevel BART model [32] and an amalgamation of the BART algorithm into the treatSens sensitivity analysis package, both described above. These are but two of many potential opportunities.

A further strength relates to the high performance of the default settings of most BART implementations. This yields an automated approach which allows the researcher to more easily pre-specify their model which, in turn, limits researcher degrees of freedom and makes the research more easily reproducible. This feature of BART makes it compatible with other "honest" approaches to causal inference because the researcher will not have the opportunity to adjust their model specification as a response to initial looks at treatment effect estimates. While many propensity score approaches also allow for this type of honesty because it is possible to choose a strategy without making use of the outcome variable, these strategies often still allow for many researcher degrees of freedom as the researcher searches for an optimal specification. The full modeling path can be difficult to reproduce for similar reasons.

20.8.2 Limitations and potential future directions

There are several limitations to the "vanilla" BART approach to causal inference. One of the most glaring is that it assumes normally distributed, independent, and identically distribution error terms. Extensions of this framework to incorporate group data structures (discussed above) help to weaken the independence assumption. Versions of BART that accommodate heteroskedastic errors address this limitation as well [52].

In addition, several different flavors of BART now exist to address the need to allow for a wider variety of distributional assumptions about the outcome and error term [53]. Refinements of these basic strategies to, for instance, use cross-validation to tune the additional hyperparameters required have been developed as well [46]. More work can be done to relax these assumptions and, ideally, allow the algorithm to automatically detect the right modeling choices.

This approach, like all causal inference approaches, assumes sufficient overlap. While BART has been used to develop some advances in the detection of observations that lack overlap in the confounder space, these approaches are still somewhat ad hoc and could be improved to target specific estimands (such as individual-level treatment effects).

Hahn et al. [35] have pointed out the potential for the most basic implementation of BART for causal inference to induce confounding through the regularization built into the prior. They have created an extension called Bayesian causal forests that provides a promising way to address this problem [35]. They also suggest that an approximate solution is to simply include a reasonable estimate of the propensity score as a covariate in the BART model.

Currently the standard BART approach is mostly useful for studies with a single binary treatment that occurs at one point in time. Many extensions to more complicated settings should be relatively straightforward but don't currently exist. Moreover, the primary implementations are in the R programming language. However stand-alone, user friendly software now exists that provides access to the software without requiring the user to program in R (`https://apsta.shinyapps.io/thinkCausal/`).

Finally, BART approaches to causal inference typically assume that all confounders have been measured. While BART has been incorporated into existing sensitivity analysis approaches [46], this doesn't entirely remove that problem. Use of these approaches requires humility in understanding what conclusions can reliably be drawn and transparency about the assumptions.

20.9 Conclusion

This chapter has introduced the basics of a causal inference approach that capitalizes on a Bayesian machine-learning-based algorithm, BART. It explains when and how flexible regression-based approaches to causal inference can be useful and has highlighted some potential advantages of these approaches relative to approaches that focus on data restructuring such as matching and weighting.

To our knowledge BART was the first machine-learning-based approach to causal inference introduced (with scholarly talks starting in 2005 and journal publication in 2011 [28]). However since that time several other machine-learning based approaches to causal inference have also been developed [55–59]. Many of these algorithms have similar desirable features. A distinguishing characteristic of BART is that the flexible model for the response surface is embedded within a Bayesian likelihood framework. This offers advantages with regard to uncertainty quantification, detection of observations that lack common causal support, simultaneous identification of a wide variety of causal estimands, and the ability for reasonably straightforward model extensions to accommodate features such as grouped data structures, sensitivity analysis, and varying distributional assumptions. Both the simple BART-causal implementation and many of the additional features described in this chapter are available in the `bartCause` package in R and the standalone thinkCausal software (`https://apsta.shinyapps.io/thinkCausal/`).

References

[1] Hugh Chipman, Edward George, and Robert McCulloch. Bayesian ensemble learning. In B. Schölkopf, J. Platt, and T. Hoffman, editors, *Advances in Neural Information Processing Systems 19*. MIT Press, Cambridge, MA, 2007.

[2] H. A. Chipman, E. I. George, and R. E. McCulloch. BART: Bayesian additive regression trees. *Annals of Applied Statistics*, 4(1):266–298, 2010.

[3] Jennifer Hill, Antonio Linero, and Jared Murray. Bayesian additive regression trees: A review and look forward. *Annual Review of Statistics and Its Application*, 7(1):251–278, 2020.

[4] Andrew Gelman, Jennifer Hill, and Aki Vehtari. *Regression and Other Stories*. Cambridge University Press, New York, 2020.

[5] Donald B. Rubin. Bayesian inference for causal effects: The role of randomization. *Annals of Statistics*, 6:34–58, 1978.

[6] B. S. Barnow, G. G. Cain, and A. S. Goldberger. Issues in the analysis of selectivity bias. In E. Stromsdorfer and G. Farkas, editors, *Evaluation Studies*, volume 5, pages 42–59. Sage, San Francisco, 1980.

[7] Sander Greenland and James M Robins. Identifiability, exchangeability, and epidemiological confounding. *International Journal of Epidemiology*, 15(3):413–419, 1986.

[8] Michael Lechner. Identification and estimation of causal effects of multiple treatments under the conditional independence assumption. In Michael Lechner and Friedhelm Pfeiffer, editors, *Econometric Evaluation of Labour Market Policies*, volume 13 of *ZEW Economic Studies*, pages 43–58. Physica-Verlag HD, 2001.

[9] Paul R. Rosenbaum. *Observational Studies*. Springer, New York, 2002.

[10] J. Pearl. On a class of bias-amplifying variables that endanger effect estimates. In *Proceedings of the Twenty-Sixth Conference on Uncertainty in Artificial Intelligence*, pages 425–432, Catalina Island, CA, 2010. Accessed 02/02/2016.

[11] J. Middleton, M. Scott, R. Diakow, and J. Hill. Bias amplification and bias unmasking. *Political Analysis*, 24:307–323, 2016.

[12] P. Steiner and Y. Kim. The mechanics of omitted variable bias: Bias amplification and cancellation of offsetting biases. *Journal of Causal Inference*, 4, 2016.

[13] P. Ding and L. Miratrix. To adjust or not to adjust? sensitivity analysis of m-bias and butterfly-bias. *Journal of Causal Inference*, 3:41–57, 2014.

[14] Alexander D'Amour, Peng Ding, Avi Feller, Lihua Lei, and Jasjeet Sekhon. Overlap in observational studies with high-dimensional covariates. *Journal of Econometrics*, 221(2):644–654, 2021.

[15] Jennifer Hill and Yu-Sung Su. Assessing lack of common support in causal inference using Bayesian nonparametrics: Implications for evaluating the effect of breastfeeding on children's cognitive outcomes. *Annals of Applied Statistics*, 7:1386–1420, 2013.

[16] Guido W. Imbens and Thomas Lemieux. Regression discontinuity designs: A guide to practice. *Journal of Econometrics*, 142(2):615–635, 2008. The regression discontinuity design: Theory and applications.

[17] Sebastian Calonico, Matias D. Cattaneo, Max H. Farrell, and Rocío Titiunik. Rdrobust: Software for regression-discontinuity designs. *The Stata Journal*, 17(2):372–404, 2017.

[18] Guido Imbens and Donald Rubin. *Causal Inference in Statistics, Social, and Biomedical Sciences*. Cambridge University Press, New York, 2015.

[19] James Robins. A new approach to causal inference in mortality studies with a sustained exposure period—application to control of the healthy worker survivor effect. *Mathematical Modelling*, 7(9):1393–1512, 1986.

[20] Luke Keele. The statistics of causal inference: A view from political methodology. *Political Analysis*, 23:313–35, 2015.

[21] B. Lara, J. Salinero, and J. Del Coso. The relationship between age and running time in elite marathoners is u-shaped. *Age (Dordr)*, 36(2):1003–1008, 2014.

[22] Niklas Lehto. Effects of age on marathon finishing time among male amateur runners in stockholm marathon 1979–2014. *Journal of Sport and Health Science*, 5(3):349–354, 2016.

[23] Trevor Hastie, Robert Tibshirani, and Jerome Friedman. *The Elements of Statistical Learning: Data Mining, Inference, and Prediction.* Springer-Verlag, New York, 2 edition, 2003.

[24] Luke Tierney. Markov Chains for Exploring Posterior Distributions. *The Annals of Statistics*, 22(4):1701 – 1728, 1994.

[25] George Casella and Edward I. George. Explaining the gibbs sampler. *The American Statistician*, 46(3):167–174, 1992.

[26] Vincent Dorie. *stan4bart: Bayesian Additive Regression Trees with Stan-Sampled Parametric Extensions*, 2021. R package version 0.0-1.

[27] Vincent Dorie. *stan4bart: Bayesian Additive Regression Trees with Stan-Sampled Parametric Extensions*, 2021. R package version 0.0-1.

[28] Jennifer Hill. Bayesian nonparametric modeling for causal inference. *Journal of Computational and Graphical Statistics*, 20(1):217–240, 2011.

[29] Hugh Chipman and Robert McCulloch. *BayesTree: Bayesian Additive Regression Trees*, 2016. R package version 0.3-1.4.

[30] Adam Kapelner and Justin Bleich. bartMachine: Machine learning with Bayesian additive regression trees. *Journal of Statistical Software*, 70(4):1–40, 2016.

[31] Vincent Dorie, Hugh Chipman, and Robert McCulloch. *dbarts: Discerete Bayesian Additive Regression Trees Sampler*, 2014. R package version 0.8-5.

[32] Bereket Kindo. *mpbart: Multinomial Probit Bayesian Additive Regression Trees*, 2016. R package version 0.2.

[33] Belinda Hernandez. *bartBMA: Bayesian Additive Regression Trees Using Bayesian Model Averaging*, 2020. R package version 1.0.

[34] Robert McCulloch, Matthew Pratola, and Hugh Chipman. *rbart: Bayesian Trees for Conditional Mean and Variance*, 2019. R package version 1.0.

[35] P. Richard Hahn, Jared S. Murray, and Carlos M. Carvalho. Bayesian Regression Tree Models for Causal Inference: Regularization, Confounding, and Heterogeneous Effects (with Discussion). *Bayesian Analysis*, 15(3):965–1056, 2020.

[36] Daniel Ho, Kosuke Imai, Gary King, and Elizabeth A. Stuart. Matchit: Nonparametric preprocessing for parametric causal inference. *Journal of Statistical Software*, 42(8):1–28, 2011.

[37] Holger L Kern, Elizabeth A Stuart, Jennifer Hill, and Donald P Green. Assessing methods for generalizing experimental impact estimates to target populations. *Journal of Research on Educational Effectiveness*, pages 1–25, 2016.

[38] Siva Sivaganesan, Peter Müller, and Bin Huang. Subgroup finding via bayesian additive regression trees. *Statistics in Medicine*, 36(15):2391–2403, 2017.

[39] Jerome H Friedman. Greedy function approximation: A gradient boosting machine. *Annals of Statistics*, 29(5):1189–1232, 2001.

[40] N. Carnegie, V. Dorie, and J. Hill. Examining treatment effect heterogeneity using BART. *Observational Studies*, 5:52–70, 2019.

[41] Brent R Logan, Rodney Sparapani, Robert E McCulloch, and Purushottam W Laud. Decision making and uncertainty quantification for individualized treatments using bayesian additive regression trees. *Statistical Methods in Medical Research*, page 0962280217746191, 2017.

[42] Elizabeth A Stuart, Stephen R Cole, Catherine P Bradshaw, and Philip J Leaf. The use of propensity scores to assess the generalizability of results from randomized trials. *Journal of the Royal Statistical Society A*, 174(2):369–386, 2011.

[43] Holger L Kern, Elizabeth A Stuart, Jennifer Hill, and Donald P Green. Assessing methods for generalizing experimental impact estimates to target populations. *Journal of Research on Educational Effectiveness*, pages 1–25, 2016.

[44] Jennifer Hill. Multilevel models and causal inference. In Marc A. Scott, Jeffrey S. Simonoff, and Brian D. Marx, editors, *The SAGE Handbook of Multilevel Modeling*. Sage Publications Ltd, 2013.

[45] Douglas Bates, Martin Mächler, Ben Bolker, and Steve Walker. Fitting linear mixed-effects models using lme4. *Journal of Statistical Software*, 67(1):1–48, 2015.

[46] Vincent Dorie, Masataka Harada, Nicole Carnegie, and Jennifer Hill. A flexible, interpretable framework for assessing sensitivity to unmeasured confounding. *Statistics in Medicine*, 35(20):3453–3470, 2016.

[47] Nicole Bohme Carnegie, Masataka Harada, and Jennifer Hill. Assessing sensitivity to unmeasured confounding using a simulated potential confounder. *Journal of Research on Educational Effectiveness*, 9:395–420, 2016.

[48] Jennifer L. Hill, Christopher Weiss, and Fuhua Zhai. Challenges with propensity score strategies in a high-dimensional setting and a potential alternative. *Multivariate Behavioral Research*, 46:477–513, 2011.

[49] T Wendling, K Jung, A Callahan, A Schuler, NH Shah, and B Gallego. Comparing methods for estimation of heterogeneous treatment effects using observational data from health care databases. *Statistics in Medicine*, 2018.

[50] V. Dorie, J. Hill, U. Shalit, M. Scott, and D. Cervone. Automated versus do-it-yourself methods for causal inference: Lessons learned from a data analysis competition. *Statistical Science*, 34(1):43–68, 2019.

[51] P. Richard Hahn, Vincent Dorie, and Jared S. Murray. Atlantic Causal Inference Conference (ACIC) Data Analysis Challenge 2017. *arXiv e-prints*, page arXiv:1905.09515, May 2019.

[52] M. T. Pratola, H. A. Chipman, E. I. George, and R. E. McCulloch. Heteroscedastic bart via multiplicative regression trees. *Journal of Computational and Graphical Statistics*, 29(2):405–417, 2020.

[53] Jared S. Murray. Log-linear bayesian additive regression trees for multinomial logistic and count regression models. *Journal of the American Statistical Association*, 116(534):756–769, 2021.

[54] Vincent Dorie, George Perrett, Jennifer L. Hill, and Benjamin Goodrich. Stan and bart for causal inference: Estimating heterogeneous treatment effects using the power of stan and the flexibility of machine learning. *Entropy*, 24(12):1782, Dec 2022.

[55] The H2O.ai team. *h2o: R Interface for H2O*, 2016. R package version 3.10.0.10.

[56] Erin LeDell. *h2oEnsemble: H2O Ensemble Learning*, 2016. R package version 0.1.8.

[57] Stefan Wager and Susan Athey. Estimation and inference of heterogeneous treatment effects using random forests. *Journal of the American Statistical Association*, 113(523):1228–1242, 2018.

[58] Sören R. Künzel, Jasjeet S. Sekhon, Peter J. Bickel, and Bin Yu. Metalearners for estimating heterogeneous treatment effects using machine learning. *Proceedings of the National Academy of Sciences*, 116(10):4156–4165, Feb 2019.

[59] Cheng Ju, Susan Gruber, Samuel D Lendle, Antoine Chambaz, Jessica M Franklin, Richard Wyss, Sebastian Schneeweiss, and Mark J van der Laan. Scalable collaborative targeted learning for high-dimensional data. *Statistical Methods in Medical Research*, 28(2):532–554, 2019. PMID: 28936917.

21

Treatment Heterogeneity with Survival Outcomes

Yizhe Xu, Nikolaos Ignatiadis, Erik Sverdrup, Scott Fleming, Stefan Wager, Nigam Shah

CONTENTS

DOI: 10.1201/9781003102670-21

Estimation of conditional average treatment effects (CATEs) plays an essential role in modern medicine by informing treatment decision-making at a patient level. Several metalearners have been proposed recently to estimate CATEs in an effective and flexible way by re-purposing predictive machine learning models for causal estimation. In this chapter we summarize the literature on metalearners and provide concrete guidance for their application for treatment heterogeneity estimation from randomized controlled trials' data with survival outcomes. The guidance we provide is supported by a comprehensive simulation study in which we vary the complexity of the underlying baseline risk and CATE functions, the magnitude of the heterogeneity in the treatment effect, the censoring mechanism, and the balance in treatment assignment. To demonstrate the applicability of our findings, we reanalyze the data from the Systolic Blood Pressure Intervention Trial (SPRINT) and the Action to Control Cardiovascular Risk in Diabetes (ACCORD) study. While recent literature reports the existence of heterogeneous effects of intensive blood pressure treatment with multiple treatment effect modifiers, our results suggest that many of these modifiers may be spurious discoveries. This chapter is accompanied by `survlearners`, an R package that provides well-documented implementations of the CATE estimation strategies described in this work, to allow easy use of our recommendations as well as reproduction of our numerical study.

21.1 Introduction

Healthcare decisions are commonly informed by a combination of *average treatment effects* (ATEs) from randomized controlled trials (RCTs), which do not account for fine-grained patient heterogeneity and *risk stratification* that identify those at the highest need for an intervention. For example, physicians assess patients' ten-year atherosclerotic cardiovascular disease (ASCVD) risk based on baseline covariates (such as age, blood pressure, and cholesterol levels) using the pooled cohort equations (PCEs) [1] and initiate statin treatment based on clinical practice guidelines [2]. Prioritizing treatment to patients who are at higher risk is a sensible starting point; however, may be suboptimal when baseline risk is not an appropriate surrogate for treatment effects. For example, a young patient with only one risk factor of elevated cholesterol level probably has a lower ASCVD risk than an older subject who has a normal cholesterol level but multiple other risk factors, such as high blood pressure and smoking, but the younger patient may benefit more from statin which reduces cholesterol [3,4]. Therefore, a key question in personalized care remains: *How* should we treat each *individual* patient?

Consequently, a key challenge in enabling precision medicine is to go beyond estimation of ATEs and risk stratification, and account for the varied patient response to treatment depending on factors such as patient characteristics, baseline risk, and sensitiveness to treatment [5]. Such heterogeneity in treatment effects (HTE) may be summarized by conditional average treatment effects (CATEs) as a function of subject-level baseline covariates. CATE estimation, however, is a difficult statistical task. Estimating CATEs, essentially requires estimating interactions of treatment and baseline covariates, and such interaction effects are often small compared to main effects that drive baseline risk. Conventional one-variable-at-a-time and subgroup analyses may produce false-positive results due to multiple testing and false-negative results due to insufficient power, i.e., small sample size in subgroups [6].

As we discuss more in the related work section below, a promising approach towards CATE estimation that addresses some of the above challenges is the development of estimation strategies that use flexible machine learning methods. A particularly convenient class of such methods are called metalearners [7,8]; The premise is that one can decompose the CATE estimation task into well-understood machine learning tasks (as would be conducted, e.g., for risk stratification). Statistical and domain expertise in risk modeling can thus be repurposed for a causal task, namely estimation of heterogeneous treatment effects.

In this chapter we build on the metalearners literature and provide concrete guidance for their usage in estimating treatment effect heterogeneity from RCT data with right-censored survival outcomes. Our contributions are as follows: 1) We provide an accessible summary of the mathematical underpinning of five popular metalearners (S-, T-, X-, M-, and R-learners) when combined with two popular machine learning strategies (Lasso and generalized random forests). While the described metalearners have been developed for uncensored continuous or binary data, we explain how they may be adapted to the survival setting through inverse probability of censoring weighting. 2) We provide R code in a package called `survlearners` [9] that demonstrates exactly how these methods may be implemented in practice and describe how machine learning models (e.g., risk models) are leveraged by each metalearner. 3) We conduct a comprehensive simulation study of the above CATE estimation strategies using several data generation processes (DGPs) in which we systematically vary the complexity of the baseline risk function, the complexity of the CATE function, the magnitude of the heterogeneity in treatment effects, the censoring mechanism, and the imbalance of treated and control units. 4) Based on the results of the simulation study, we summarize considerations that matter in choosing and applying metalearners and machine learning models for HTE estimation. 5) We apply our findings as a case study of HTE estimation on the systolic blood pressure intervention trial (SPRINT) [10] and the action to control cardiovascular risk in diabetes (ACCORD) trial [11].

21.1.1 Related work

Prior works have introduced metalearners for HTE estimation [7, 8, 12] and provided comprehensive tutorials and simulation benchmarks that explicate their usage [13]. Further works have conducted comprehensive simulation studies that compare both metalearners, as well as other machine learning approaches to HTE estimation [14–25]. More broadly, there have been substantial advances in non-parametric methodology for estimating HTEs [26–31] and in theoretical understanding thereof, e.g., with respect to minimax rates of estimation [32–34]. However, these studies are all conducted under a non-survival setting, while researchers in health care are often interested in right-censored, time-to-event outcomes. Our chapter fills this gap and provides a tutorial and benchmark of metalearners adapted to the survival analysis setting.

Several researchers [35–43] have developed machine learning based methods for estimating HTEs with data that have time-to-event outcomes (often called *survival data*). [44] conduct a comprehensive simulation study of the T-learner (one of the metalearners described in this work) combined with different predictive models for survival data. [45] reanalyze the SPRINT trial and in doing so extend the X-learner (a metalearner proposed by [7]) to the survival setting through inverse probability of censoring weights. However, we are not aware of a previous comprehensive tutorial and benchmark that demonstrates how to use a range of popular metalearners with survival data. Furthermore, while much of the related literature focuses on the more general setting with observational data under unconfoundedness, we focus exclusively on RCTs. In doing so, we hope to clarify the essential difficulties of HTE estimation [46] in the absence of confounding beyond censoring bias. We hope that our work will contribute to the principled estimation of HTEs from RCTs in medicine.

21.1.2 The PATH statement

The Predictive Approaches to Treatment effect Heterogeneity (PATH) Statement provides guidance for estimating and reporting treatment effect heterogeneity in clinical RCTs and was developed by a panel of multidisciplinary experts [47]. The PATH statement identifies two main approaches to predictive HTE estimation:

"PATH risk modeling": A predictive risk model is identified (or developed) and HTEs are estimated as a function of predicted risk in the RCT data.

"PATH effect modeling": A predictive model is developed for the outcome of interest with predictors that include the risk predictors, the treatment assignment, as well as interaction terms.

Our work is complementary to the PATH statement and provides further methodological guidance for HTE estimation in RCTs with survival outcomes. Concretely, the "PATH effect modeling" approach coincides with the "S-learner," a metalearner proposed in the machine learning literature and described in Section 21.3.1. Our numerical results and literature review confirm the caveats of effect modeling described in the PATH Statement. Other metalearners may be preferable in settings where those caveats apply. The PATH guidelines for "PATH risk modeling" approach are important for all the metalearners considered in this work. These guidelines advocate for the use of a parsimonious set of predictors for HTEs. A risk score (developed previously, or blinded to treatment assignment) is an important predictor for HTEs that is justified both mathematically and by clinical experience. In this work by using modern machine learning and regularization techniques, we allow for the possibility that X_i may be higher-dimensional. We do not provide guidance on how to choose predictors X_i based on domain expertise but describe methods for efficiently using any such HTE predictors chosen by the analyst (X_i may include, e.g., risk scores previously developed, but also additional predictors determined based on domain expertise).

21.1.3 Outline

The outline of this chapter is as follows. We define the CATE estimation problem in Section 21.2. In Section 21.3 we provide a brief tutorial on the use of metalearners and machine learning for estimating treatment heterogeneity in RCT with right-censored, time-to-event data. We describe our simulation study in Section 21.4 and summarize main takeaways in Section 21.5. We then present a case study on SPRINT and ACCORD in Section 21.6. Finally, we conclude with a discussion and future extensions in Section 21.7.

21.2 Problem Setup, Notation, and Assumptions

We discuss the problem of CATE estimation under the potential outcome framework in causal inference [48, 49]. Consider an RCT that provides N independent and identically distributed subject-level data, indexed by $i \in \mathcal{I} = \{1, 2, \dots, N\}$. Patient baseline covariates are denoted by $X_i = (X_{i1}, \dots, X_{id}) \in \mathbb{R}^d$ and the treatment variable is $W_i \in \{0, 1\}$. Let $T_i(w)$ be the potential survival time if subject i had been assigned the treatment w and C_i be the censoring time. Then, $T_i = W_i T_i(1) + (1 - W_i) T_i(0)$, and the observed follow-up time is $U_i = \min(T_i, C_i)$ with the corresponding non-censoring indicator $\Delta_i = I\{T_i \le C_i\}$. Our observed dataset is $\mathcal{O} = \{(X_i, W_i, U_i, \Delta_i) : i \in \mathcal{I}\}$. We also reserve the notation $\mathcal{I}_0 = \{i \in \mathcal{I} : W_i = 0\}$ for the index set of control units and $\mathcal{O}_0 = \{(X_i, 0, U_i, \Delta_i) : i \in \mathcal{I}_0\}$ for the control dataset. We define $\mathcal{I}_1$ and $\mathcal{O}_1$ analogously for the treated units. Occasionally we will explain concepts using R pseudocode in which case we will refer to the observed data as `X` ($N \times d$ matrix) and `W`, `U`, and `D` (length N vectors with i-th entry equal to W_i, U_i, resp. Δ_i).

The Conditional Average Treatment Effect (CATE) for the probability of survival beyond a pre-specified time t_0 is

$$\tau(x) = \mathrm{E}\, Y_i(1) - Y_i(0) \mid X_i = x, \tag{21.1}$$

where $Y_i(w) = \mathbb{1}\{T_i(w) > t_0\}$ is the indicator of survival beyond time t_0. We also write $Y_i = Y_i(W_i)$.

To ensure the CATE is identifiable, we make the following assumptions [50]; some of which already appeared implicitly in the notations above:

Consistency: The observed survival time in real data is the same as the potential outcome under the actual treatment assignment, i.e., $T_i = T_i(W_i)$.

RCT: The treatment assignment is randomized such that W_i is independent of $(X_i, T_i(1), T_i(0))$, i.e., $W_i \perp\!\!\!\perp (X_i, T_i(1), T_i(0))$ and $\mathbb{P}[W_i = 1] = e$ with known $0 < e < 1$.

Noninformative censoring: Censoring is independent of survival time conditional on treatment assignment and covariates, i.e., $C_i \perp\!\!\!\perp T_i \mid X_i, W_i$.

Positivity: There exists subjects who are at risk beyond the time horizon t_0, i.e., $\mathbb{P}[C_i > t_0 \mid X_i, W_i] \geq \eta_C$ for some $\eta_C > 0$.

Remark 21.1 (Observational studies). *As mentioned above, throughout this manuscript we focus our attention on RCT so as to provide a comprehensive discussion of issues involved in the estimation of HTEs in the absence of confounding beyond censoring bias. Nevertheless, conceptually, the metalearners we discuss are also applicable in the setting of observational studies under unconfoundedness. Concretely, we may replace the* RCT *assumption by the following two assumptions*

Unconfoundedness: *The potential survival times are independent of the treatment assignment W_i conditionally on baseline covariates, that is, $W_i \perp\!\!\!\perp (T_i(1), T_i(0))|X_i$.*

Overlap: *There exists $\eta \in (0,1)$ such that the propensity score $e(x) = \mathbb{P}[W_i = 1|X_i = x]$ satisfies $\eta \leq e(x) \leq 1 - \eta$ for all x in the support of X_i.*

We refer the interested readers to the original manuscripts introducing the different metalearners for an explanation of their application to observational studies. In short, for the methods we describe, it suffices to replace the treatment probability e (whenever it is used by a method) by $\widehat{e}(X_i)$ with $\widehat{e}(\cdot)$ is an estimate of the propensity score $e(\cdot) = \mathbb{P}[W_i = 1|X_i = \cdot]$. The statistical consequences of estimation error in $\widehat{e}(\cdot)$, as well as ways to make estimators robust to this error, are discussed further in [8] and [29].

21.3 Metalearners

Metalearners are specific meta-algorithms that leverage predictive models to solve the causal task of estimating treatment heterogeneity. Metalearners are motivated by the observation that predictive models are applied ubiquitously, that we have a good understanding about fitting models with strong out-of-bag predictive performance, and we know how to evaluate predictive models [51–53]. Metalearners repurpose this expertise to power the effective estimation of HTEs (a task which is less well-understood). Proposed metalearners build upon different predictive tasks and also combine these predictive models in distinct ways to estimate HTEs. In this Section we seek to provide a short, but instructive, introduction to commonly used metalearners in the context of CATE estimation (21.1) with survival data.

To emphasize the flexibility of metalearners in leveraging predictive models, we abstract away the concrete choice of predictive model for each task by introducing the notation

$$\mathcal{M}\left(\widetilde{Y} \sim \widetilde{X};\ \widetilde{\mathcal{O}}, [\widetilde{K}]\right), \tag{21.2}$$

to denote a generic prediction model that predicts $\widetilde{Y}$ as a function of covariates $\widetilde{X}$ based on the dataset $\widetilde{\mathcal{O}}$ with (optional) sample weights $\widetilde{K}$. Note that by default we assume that $\widetilde{K}_i = 1$ for all i, in which case we omit $\widetilde{K}$ from the notation and write $\mathcal{M}(\widetilde{Y} \sim \widetilde{X};\ \widetilde{\mathcal{O}})$. It will also be convenient to introduce notation for predictive models that give out-of-bag (oob)[1] predictions of $\widetilde{Y}$:

$$\mathcal{M}^{\text{oob}}\left(\widetilde{Y} \sim \widetilde{X};\ \widetilde{\mathcal{O}}, [\widetilde{K}]\right). \tag{21.3}$$

Below we will describe the high-level idea of each metalearner, followed by concrete examples for possible choices of $\mathcal{M}$. We first describe two metalearners – the S- and T-learners – for which we only need to predict the probability that $\{T_i > t_0\}$ as a function of certain covariates $\widetilde{X}_i$. We refer to this task as risk modeling, since

$$\mathbb{P}[T_i > t_0 \mid \widetilde{X}_i] = 1 - \mathbb{P}[T_i \leq t_0 \mid \widetilde{X}_i] = 1 - \text{Risk}(\widetilde{X}_i).$$

Standards for transparent reporting of predictive risk models [54] should be adhered to in the course of developing such models within the context of metalearners.

We emphasize that risk modeling, as defined in our work, has subtle differences from the "PATH risk modeling" approach (Subsection 21.1.2 and [47]) – to avoid any confusion we quote the latter throughout. One key difference is that, for us, risk models are always assumed to be developed on the training dataset $\mathcal{O}$ and not in a previous cohort ($\widetilde{\mathcal{O}}$ and $\mathcal{O}$ are the general and case-specific notations of a data set, respectively. This distinction also applies to the $\widetilde{Y}$, $\widetilde{X}$, and $\widetilde{K}$ in (21.2) throughout the chapter). Nevertheless, as we already alluded to in Subsection 21.1.2, the two frameworks are compatible with each other. We will further demonstrate their connection in our case-study reanalyzing the SPRINT RCT in Section 21.6: There we use the well-established Pooled Cohort Equations (PCE) [1] that predict the 10-year risk of a major ASCVD event as a function of predictors such as total cholesterol, blood pressure, age, etc. This gives us a new feature – the PCE score – which can then be used as a predictor for risk modeling within the SPRINT [10] cohort.

21.3.1 S-learner: modeling risk as a function of baseline covariates, treatment assignments, and their interactions

The "simplest" metalearner used in practice, called the *S-learner*, fits a *single* risk model using baseline covariates and the treatment variable. The S-learner builds upon the observation that under our assumptions, the CATE in (21.1) may be written as:

$$\tau(x) = \mu([x, 1]) - \mu([x, 0]) \text{ with } \mu([x, w]) = \mathbb{E}[Y_i \mid X_i = x, W_i = w]. \tag{21.4}$$

In the notation of (21.2), the S-learner proceeds as follows:

$$\widehat{\mu}(\cdot) = \mathcal{M}(Y \sim [X, W];\ \mathcal{O}), \quad \widehat{\tau}(x) = \widehat{\mu}([x, 1]) - \widehat{\mu}([x, 0]). \tag{21.5}$$

In words, the S-learner first learns a risk model $\widehat{\mu}(\cdot)$ as a function of X_i and W_i, i.e., the treatment assignment W_i is merely treated as "just another covariate" [27,28]. Then, given baseline covariates x, the S-learner applies the fitted model to $[x, 1]$ (that is, to the feature vector that appends $W_i = 1$ to x) to impute the response of the treated outcomes. The fitted model is then applied to $[x, 0]$ to impute the response of the control outcomes, and finally the CATE is estimated as the difference thereof.

In (21.4), any predictive model for $\mathbb{P}[T_i > t_0 \mid X_i, W_i]$ may be used, for example, a random survival forest.

Example 21.1 (S-learner with Random Survival Forest)**.** *A concrete example of a fully nonparametric model $\mathcal{M}(Y \sim [X, W];\ \mathcal{O})$ is the random survival forest of [55]. The basic idea of the random survival forest, and more generally, of generalized random forests (GRF) [26]) is the following: As*

[1] We use the term out-of-bag loosely also using it to refer to out-of-sample or out-of-fold predictions.

in the traditional random forest of [56], a collection of trees is grown. Each tree is grown based on a randomly drawn subsample of the training data and by recursively partitioning the covariate space. These trees are then used to adaptively weight [57] new test points. To be precise, let $\tilde{x}$ *be a test covariate (e.g.,* $\tilde{x} = [x, w]$ *for the S-learner), then the* i*-th data point in the training dataset is assigned weight* $\alpha_i(\tilde{x}) = \sum_{b=1}^{B} \mathbf{1}(\{\widetilde{X}_i \in L_b(\tilde{x})\})/(B||L_b(\tilde{x})|)$*, where* B *is the total number of trees,* $L_b(\tilde{x})$ *is the set of all training examples* i *such that* $\widetilde{X}_i$ *falls in the same leaf of the* b*-th tree as* $\tilde{x}$*, and* $\#|\cdot|$ *symbolizes the number of instances in a set. Then, given these weights* $\alpha_i(\tilde{x})$*, the Nelson-Aalen estimator with these weights is used to predict* $\mathbb{P}[T_i > t_0 \mid \widetilde{X}_i]$*. Using the random survival forest in the R package* `grf` *[58], the S-learner may be implemented as follows:*

```
library(grf)
m <- survival_forest(data.frame(X=X, W=W), U, D, prediction.type="Nelson-Aalen")
m1_x <- predict(m, data.frame(X=x, W=1), t0)$predictions
m0_x <- predict(m, data.frame(X=x, W=0), t0)$predictions
tau_x <- m1_x - m0_x
```

In the first line we load the `grf` *package. In the second line we fit the random survival forest based on the full dataset with augmented covariates for which we concatenate the baseline covariates* `X` *and treatment assignment vector* `W` *and with follow-up times* `U` *and event indicators* `D`*. Then, given test baseline covariates* `x`*, we impute (predict) the survival probability at* t_0 *(*`t0` *is a scalar that equals to* t_0*) using the fitted survival forest for the treated outcome (third line) and control outcome (fourth line), and finally we take their difference (fifth line).*

The random survival forest automatically captures interactions between baseline covariates X_i and treatment assignment W_i through the tree structure.[2] When a conventional regression is used as the predictive model, one needs to explicitly specify treatment-covariate interaction terms in order to model HTE:

Example 21.2 (S-learner with Cox-Lasso). *A commonly used risk model for survival data is given by the Cox proportional hazards (PH) model with Lasso penalization [59, 60]. Given covariates* $\widetilde{X}_i$*, the PH model assumes that the log-hazard is equal to* $\beta^\intercal \widetilde{X}_i$*, for an unknown coefficient vector* β *that is estimated by minimizing the negative log partial likelihood [61] plus the sum of absolute values of the coefficient* β_j *multiplied by the regularization parameter* $\lambda \geq 0$ *(i.e.,* $\lambda \cdot \sum_j |\beta_j|$*). One of the upshots of the Cox-Lasso is that it automatically performs shrinkage and variable selection;* λ *determines the sparsity of the solution (i.e., how many of the* β_j *are equal to zero). In the context of CATE estimation, the Cox-Lasso is typically fitted using the covariate vector* $\widetilde{X}_i = [X_i, W_i, W_i \cdot X_i]$*, that is, by explicitly including interaction terms* $W_i \cdot X_i$ *for the linear predictors. The S-learner with Cox-Lasso then takes the following form:*[3]

$$\widehat{\mu}(\cdot) = \mathcal{M}_{\lambda}^{\text{Cox-Lasso}}\left(Y \sim [X, W, W \cdot X];\ \mathcal{O}\right),\ \ \widehat{\tau}(x) = \widehat{\mu}([x, 1, 1 \cdot x]) - \widehat{\mu}([x, 0, 0 \cdot x]),$$

[2]There is a caveat to this claim: Since the treatment indicator is included in the same way as the other covariates, it is likely to be ignored in several trees that never split on it. This can cause the S-learner with Random Survival Forest to perform poorly in some situations.

[3]In this note we provide some more details about fitting the Cox-Lasso: Given a survival dataset $\widetilde{\mathcal{O}}$ indexed by $\widetilde{\mathcal{I}}$ and with covariates $\widetilde{X} \in \mathbb{R}^{\tilde{d}}$, and a tuning parameter $\lambda \geq 0$, the Cox-PH Lasso model for predicting the survival probability at t_0 is fitted as follows (assuming for simplicity that there are no ties in the observed event times U_i)

$$\widehat{\beta}_\lambda \in \arg\min_\beta \left\{ -\sum_{i \in \widetilde{\mathcal{I}}} \Delta_i \left[\widetilde{X}_i^\intercal \beta - \log \left(\sum_{j \in \widetilde{\mathcal{I}}\,:\, U_j \geq U_i} \exp(\widetilde{X}_j^\intercal \beta) \right) \right] + \lambda \sum_{j=1}^{\tilde{d}} |\beta_j| \right\},$$

$$\widehat{H}_\lambda(t_0) = \sum_{i \in \widetilde{\mathcal{I}}\,:\, U_i \leq t_0} \left(\Delta_i \Big/ \sum_{j \in \widetilde{\mathcal{I}}\,:\, U_j \geq U_i} \exp(\widetilde{X}_j^\intercal \widehat{\beta}_\lambda) \right).$$

where we make explicit in the notation that $\widehat{\lambda}$ *is typically chosen in a data-driven way by minimizing the cross-validated log partial likelihood [62, 63].*

A further challenge (beyond explicitly modeling interactions $W_i \cdot X_i$*) in applying the S-learner with Cox-Lasso is that there are many possible choices with respect to normalization of covariates, interaction terms, and to applying different penalty factors to different coefficients. In our implementation we do not apply any shrinkage on the coefficient of* W_i*. We discuss the Cox-Lasso model in more detail, as well as our normalization/shrinkage choices in Supplement A.1.*

21.3.2 T-learner: Risk modeling stratified by treatment

Another intuitive metalearner, called the *T-learner*, fits *two* risk models as a function of baseline covariates, separately for treatment and control arms. The T-learner relies on the following characterization of the CATE function under our assumptions:

$$\tau(x) = \mu_{(1)}(x) - \mu_{(0)}(x), \text{ where } \mu_{(w)}(x) = \mathbb{P}[T_i(w) > t_0 \mid X_i = x]. \tag{21.6}$$

In the notation of (21.2), the T-learner proceeds as follows:

$$\widehat{\mu}_{(1)}(\cdot) = \mathcal{M}(Y \sim X;\, \mathcal{O}_1), \quad \widehat{\mu}_{(0)}(\cdot) = \mathcal{M}(Y \sim X;\, \mathcal{O}_0), \quad \widehat{\tau}(x) = \widehat{\mu}_{(1)}(x) - \widehat{\mu}_{(0)}(x). \tag{21.7}$$

In words, the model $\widehat{\mu}_{(1)}(\cdot)$ is fitted using data only from treated subjects, while $\widehat{\mu}_{(0)}(\cdot)$ is fitted using data only from control subjects. $\widehat{\tau}(\cdot)$ is then computed as the difference of these two models. The risk models could again be, e.g., random survival forests, or Cox-Lasso models. The following example demonstrates how the T-learner with random survival forests may be implemented using the `grf` package.

Example 21.3 (T-learner with Random Survival Forest). *Following the code notation used in Example 21.1, the T-learner takes the following form:*

```
library(grf)
m1 <- survival_forest(X[W==1,], U[W==1], D[W==1], prediction.type="Nelson-Aalen")
m0 <- survival_forest(X[W==0,], U[W==0], D[W==0], prediction.type="Nelson-Aalen")
m1_x <- predict(m1, x, t0)$predictions
m0_x <- predict(m0, x, t0)$predictions
tau_x <- m1_x - m0_x
```

In the second line we fit the random survival forest with covariates `X` *only on the treated subjects through the subsetting* `W==1`*. In the third line we fit the same model on control subjects (subsetting* `W==0`*). Then we estimate the survival probability with covariates* `x` *under treatment (line 4) and control (line 5), and take the difference to estimate the CATE (line 6).*

The next example describes the T-learner with Cox-Lasso. In contrast to the S-learner from Example 21.2, here no special tuning or normalization is required when fitting the Cox-Lasso since the risks are modeled as functions of baseline covariates only.

Then given a test covariate $\widetilde{x}$, the prediction of the survival probability at t_0 is given by:

$$\exp\left\{-\widehat{H}_\lambda(t_0)\exp\left(\widetilde{x}^\intercal\widehat{\beta}_\lambda\right)\right\}.$$

In the main text (e.g., in Example 21.4), we use the following notation for the above predictive risk modeling procedure:

$$\mathcal{M}_\lambda^{\text{Cox-Lasso}}\left(Y \sim \widetilde{X};\widetilde{\mathcal{O}}\right).$$

Example 21.4 (T-learner with Cox-Lasso). *The CATE estimate is computed as follows:*

$$\widehat{\mu}_{(1)}(\cdot) = \mathcal{M}^{\text{Cox-Lasso}}_{\widehat{\lambda}_1}(Y \sim X;\ \mathcal{O}_1),\ \ \widehat{\mu}_{(0)}(\cdot) = \mathcal{M}^{\text{Cox-Lasso}}_{\widehat{\lambda}_0}(Y \sim X;\ \mathcal{O}_0),\ \ \widehat{\tau}(x) = \widehat{\mu}_{(1)}(x) - \widehat{\mu}_{(0)}(x).$$

Notice that the choice of penalty parameter $\widehat{\lambda}_1, \widehat{\lambda}_0$ is different for treated, resp. control groups. It may be chosen by cross-validation separately on treated subjects $\mathcal{O}_1$ and control subjects $\mathcal{O}_0$.

21.3.3 Metalearning by directly modeling treatment heterogeneity

A caveat of the S- and T-learners is that they target the statistical estimands $\mu([x,w])$ in (21.4), resp. $\mu_{(1)}(x), \mu_{(0)}(x)$ in (21.6), but not directly the CATE $\tau(x)$ in (21.1). As such they provide no direct way to control regularization of $\tau(x)$ and furthermore directly regularizing the risk models may lead to unintended suboptimal performance for the estimated CATE. For example, a major concern regarding the T-learner is that the two risk models $\widehat{\mu}_{(1)}(\cdot)$ and $\widehat{\mu}_{(0)}(\cdot)$ may use a different basis (predictors and their transformations) [14], e.g., in the T-learner with the Cox-Lasso, the two models may choose a different subset of covariates. We also refer to [7, Figure 1] for an iconic illustration of regularization-induced bias for the T-learner. In such cases, the difference between $\widehat{\mu}_{(1)}(\cdot)$ and $\widehat{\mu}_{(0)}(x)$ could be due to the discrepancy in model specification rather than true HTE.[4]

The metalearners we describe subsequently seek to address this shortcoming of the S- and T-learners by directly modeling the CATE. Suppose – as a thought experiment – that we could observe the individual treatment effects $Y_i(1) - Y_i(0)$ for all units i. Then we could directly learn the CATE by predictive modeling of this *oracle score*, denoted as $Y_i^{*,o} = Y_i(1) - Y_i(0)$, as a function of X_i and leverage assumed properties of the CATE function (e.g., smoothness, sparsity, linearity) through the choice of predictive model and regularization strategy.

There are two challenges in the above thought experiment: First, due to censoring we do not observe Y_i for the censored i; instead we observe (U_i, Δ_i). Second, by the fundamental problem of causal inference, we observe at most one potential outcome (each i is either treated or not), namely, $Y_i = Y_i(W_i)$. It turns out however, there exist metalearners that can handle the "missing data" issue and enable direct modeling of the CATE.

21.3.3.1 Censoring adjustments

As a first ingredient for the metalearners we describe below, we explain how one can account for censoring by Inverse Probability of Censoring Weights (IPCW) [64, 65].

We create a dataset with only *complete* cases, which are defined as subjects who had an event before t_0 or finished the follow-up until t_0 without developing an event, i.e., $\mathcal{I}_{\text{comp}} = \{i \in \mathcal{I} : \Delta_i = 1 \text{ or } U_i \geq t_0\}$ and $\mathcal{O}_{\text{comp}} = \{(X_i, W_i, U_i, \Delta_i) : i \in \mathcal{II}_{\text{comp}}\}$.[5] Next, we estimate the survival function of the censoring time as follows:[6]

$$\widehat{S^C}(\cdot) = \mathcal{M}^{\text{oob}}(\mathbb{1}\{C \geq u\} \sim [X, W];\ \mathcal{O}), \quad \text{where } S^C(u, x, w) = \mathbb{P}[C_i \geq u \mid X_i = x, W_i = w]. \tag{21.8}$$

Based on $\widehat{S^C}(\cdot)$, we then re-weight every subject $i \in \mathcal{I}_{\text{comp}}$ by $\widehat{K}_i$, an estimate of the inverse of

[4]The issue of regularization-induced bias becomes even more nuanced in the observational study setting of Remark 21.1, see e.g., [27].

[5]For these cases, the outcome Y_i is known. To see this, note first that if $\Delta_i = 1$, then we observe T_i and hence we observe $Y_i = 1(T_i > t_0)$. On the other hand, if the observation is censored ($\Delta_i = 0$), but $U_i \geq t_0$, then we also know that $T_i > U_i \geq t_0$, i.e., that $T_i > t_0$ and that $Y_i = 1$.

[6]We slightly abuse notation here since we define $S^C(u, x, w)$ as the probability of $\mathbb{1}\{C_i \geq u\}$ given X_i, W_i for all $u \geq 0$. In our implementations, however, we predict $S^C(u, x, w)$ with a strict inequality as $\mathbb{P}(C_i > u \mid X_i = x, W_i = w])$ as an approximation since the most commonly used survival models (including the ones we describe herein: Random Survival Forests, Cox-Lasso, Kaplan-Meier) follow the typical definition of a survival function, i.e., $S(t) = \mathbb{P}(T_i > t)$. We refer to [66] for a comprehensive discussion of survival models that can predict at all u, and survival models that cannot.

the probability of not being censored, namely:

$$\widehat{K}_i = 1 \Big/ \widehat{S^C}\left(\min\{U_i, t_0\}, X_i, W_i\right). \tag{21.9}$$

In view of (21.8), the $\widehat{K}_i$ are estimated out-of-bag and this helps to avoid overfitting. To estimate the survival function for censoring $S^C(u, x, w)$, we need to switch the roles of T_i and C_i by using the censoring indicator $1 - \Delta_i$ instead of Δ_i.

We first explain how to compute the $\widehat{K}_i$ with the Kaplan-Meier estimator, when censoring may be assumed to be completely independent, namely $C_i \perp\!\!\!\perp (T_i, X_i, W_i)$.

Example 21.5 (Censoring weights with Kaplan-Meier). *Let $\mathcal{F}_1, \ldots, \mathcal{F}_L$ be a partition of $\mathcal{I}$ into L folds (by default we set $L = 10$). Then, for $i \in \mathcal{F}_\ell$, the out-of-fold Kaplan-Meier estimator of K_i is equal to:*[7]

$$\widehat{K}_i = \left\{\widehat{S^C}_{\mathcal{F}_{-\ell}}(\min\{U_i, t_0\})\right\}^{-1} = \left\{\prod_{\substack{k: U_k \le \min\{U_i, t_0\} \\ k \in \mathcal{F}_{-\ell}}} \left(1 - \frac{1\{\Delta_k = 0\}}{\#\{j \in \mathcal{F}_{-\ell} : U_j > U_k\}}\right)\right\}^{-1},$$

where $\mathcal{F}_{-\ell} = \mathcal{U} \setminus \mathcal{F}_\ell$ is the set of all subjects outside fold $\mathcal{F}_\ell$ and $\Delta_k = \mathbb{1}\{\min(T_i, U_k) \le C_i\}$.

If censoring may depend on (X_i, W_i), then a more complicated model is required. For example, if one is willing to assume proportional hazards, then similarly to Examples 21.2 and 21.4, one could estimate $\widehat{K}_i$ by running the Cox-Lasso. A more nonparametric estimate is provided by random survival forests:

Example 21.6 (Censoring weights with Random Survival Forests). *Following the notation in Examples 21.3,21.1, the `grf` package may be used as follows for IPCW.*

```
library(grf)
cen <- survival_forest(data.frame(X,W), U, 1-D, prediction.type="Nelson-Aalen")
K <- 1/predict(cen, failure.times=pmin(U,t0), prediction.times="time")$predictions
```

In Line 2, we fit the forest with flipped event indicator `1-D` and covariates `X`, `W`, and in the last line we compute the vector of censoring weights `K`. We note that `survival_forest` in the `grf` package computes out-of-bag predictions by default (`compute.oob.predictions = TRUE`).

If we hypothetically had access to the oracle scores $Y_i^{*,o} = Y_i(1) - Y_i(0)$ for the uncensored samples $i \in \mathcal{I}_{\text{comp}}$ and weights $\widehat{K}_i$ as in (21.8), then we could estimate HTEs via weighted predictive modeling as $\widehat{\tau}(\cdot) = \mathcal{M}(Y^{*,o} \sim X;\ \mathcal{O}, \widehat{K})$ (see examples below). In the next subsections we describe three metalearners, the M-, X-, and R-learners that address the challenge that – even in the absence of censoring – the oracle scores $Y_i^{*,o} = Y_i(1) - Y_i(0)$ are not available, due to the fundamental challenge of causal inference.

The observation driving these methods is that for any given (observable) score Y_i^* with the property

$$\mathrm{E}\, Y_i^* \mid X_i = x = \tau(x) \ \text{ or } \ \mathbb{E}[Y_i^* \mid X_i = x] \approx \tau(x), \tag{21.10}$$

one can estimate $\widehat{\tau}(x)$ by predictive modeling of Y_i^* as a function of X_i. The oracle score $Y_i^{*,o} = Y_i(1) - Y_i(0)$ satisfies (21.10) by definition, however, it is not the only score with this property.

Remark 21.2 (Doubly robust censoring adjustments). *The IPCW (Inverse Probability of Censoring Weights) adjustment removes censoring bias, but it can be inefficient and unstable when the censoring*

[7] Assuming no ties for simplicity in the formula.

rate is high in a study and a majority of the censoring events happened before t_0. It also may be more sensitive to misspecification of the censoring model. In such cases, one can consider a doubly robust correction [67] similar to the augmented inverse-propensity weighting estimator of [68]. We do not pursue such a doubly robust correction here, because it is substantially more challenging to implement with general off-the-shelf predictive models. Case-by-case constructions are possible, e.g., the Causal Survival Forest (CSF) of [40] uses a doubly robust censoring adjustment, and we will compare to CSF in the simulation study.

21.3.3.2 M-learner

The *modified outcome method (M-learner)* [12, 14, 69, 70] leverages the aforementioned insight with the following score based on the [71] transformation / inverse propensity weighting (IPW):

$$Y_i^{*,M} = Y_i \left(\frac{W_i}{e} - \frac{1 - W_i}{1 - e} \right), \quad \mathbb{E}[Y_i^{*,M} \mid X_i = x] = \tau(x). \tag{21.11}$$

Then, the M-learner proceeds as follows:

$$\widehat{\tau}(x) = \mathcal{M}(Y^{*,M} \sim X;\ \mathcal{O}_{\text{comp}}, \widehat{K}), \quad \text{where } \widehat{K} \text{ is as in (21.9).} \tag{21.12}$$

The predictive model for (21.12) could be any predictive model. For example, a random (regression) forest could be used [56].

Example 21.7 (M-learner with Random Forest CATE modeling). *Let `K_hat` be a vector of censoring weights, derived as in Examples 21.5 or 21.6. Also let `e` be the treatment probability. Then the M-learner with random forest CATE modeling may be implemented as follows with the `regression_forest` function in the `grf` package.*

```
library(grf)
idx <- (D == 1) | (U >= t0)
Y_M <- (U > t0) * (W/e - (1-W)/(1-e))
tau_hat_forest <- regression_forest(X[idx,], Y_M[idx], sample.weights = K_hat[idx])
tau_x <- predict(tau_hat_forest, x)$predictions
```

In Line 2 we subset to the complete cases. In Line 3 we generate the IPW response in (21.11), *in Line 4 we fit the random forest and finally in Line 5 we extract the estimated CATE at* x.

In the context of survival data, [36] proposed a related M-learner approach that uses a single regression tree (rather than forest). Another alternative could be to fit the CATE with Lasso:

Example 21.8 (M-learner with Lasso CATE modeling). *If we seek to approximate $\tau(X)$ as a sparse linear function of X_i, we can use the Lasso [72] with squared error loss:*

$$(\widehat{\beta}_0, \widehat{\beta}_1) \in \mathit{arg\,min}_{\beta_0,\beta_1} \left\{ \sum_{i \in \mathcal{I}_{comp}} \widehat{K}_i \cdot \left(Y_i^{*,M} - \beta_0 - \beta_1^\mathsf{T} X_i \right)^2 + \widehat{\lambda}_\tau \sum_{j=1}^{d} |\beta_{1,j}| \right\}, \widehat{\tau}(x) = \widehat{\beta}_0 + \widehat{\beta}_1^\mathsf{T} x,$$

where $\mathcal{I}_{comp}$ is the set of complete observations, $\widehat{K}_i$ is the IPC-weight in (21.9), *and $Y_i^{*,M}$ is the M-learner (IPW) score in* (21.11). *$\widehat{\lambda}_\tau \geq 0$ may be chosen by, e.g., cross-validation on squared error loss.*

21.3.4 Modeling both the risk and treatment heterogeneity

A downside of the M-learner is the high variance (e.g., when $e \approx 0$, then the scores $Y_i^{*,M}$ can blow up for treated units). In this Section we describe the X- and R-learners that also directly model the

CATE. By including estimated risk models in the definition of the scores, these learners can estimate the CATE with lower variance.

21.3.4.1 X-learner

The X-learner [7] constructs scores that satisfy the approximate identity in (21.10) that build on the following observation:

$$Y_i^{*,X,1} = Y_i(1) - \mu_{(0)}(X_i), \quad \mathbb{E}[Y_i^{*,X,1} \mid X_i = x] = \tau(x), \tag{21.13}$$

where $\mu_{(0)}(x) = \mathbb{E}[Y_i(0) \mid X_i = x]$ is defined in (21.6). Since $\mu_{(0)}(x)$ is unknown, it needs to be estimated in a first stage (as in the T-learner (21.7)). The X-learner thus takes the following form

$$\widehat{\mu}_{(0)}(\cdot) = \mathcal{M}(Y \sim X;\ \mathcal{O}_0), \quad \widehat{\tau}_{(1)}(\cdot) = \mathcal{M}(Y - \widehat{\mu}_{(0)}(X) \sim X;\ \mathcal{O}_1 \cap \mathcal{O}_{\text{comp}}, \widehat{K}). \tag{21.14}$$

In words, $\widehat{\tau}_{(1)}$ is the CATE estimated using data from treated units, for which $Y_i(1)$ is observed, and then the unobserved $Y_i(0)$ is imputed as $\widehat{\mu}_{(0)}(X_i)$. The role of treatment and control groups in (21.13) may be switched[8] and hence analogously to (21.14) we could consider:

$$\widehat{\mu}_{(1)}(\cdot) = \mathcal{M}(Y \sim X;\ \mathcal{O}_1), \quad \widehat{\tau}_{(0)}(\cdot) = \mathcal{M}(\widehat{\mu}_{(1)}(X) - Y \sim X;\ \mathcal{O}_0 \cap \mathcal{O}_{\text{comp}}, \widehat{K}). \tag{21.15}$$

In a last stage, the X-learner combines the two CATE estimates as follows:

$$\widehat{\tau}(x) = (1 - e) \cdot \widehat{\tau}_{(1)}(x) + e \cdot \widehat{\tau}_{(0)}(x). \tag{21.16}$$

The intuition here is that we should upweight (21.14) if there are fewer treated units, and (21.15) if there are more treated units.

The two CATE models in (21.14) and (21.15) may be fitted using the same methods, described e.g., for the M-learner. We provide an example using random forests (analogous to Example 21.7):

Example 21.9 (X-learner with Random Forest CATE Model). *Suppose* `m1_hat`, *resp.* `m0_hat` *are vectors of length n with i-th entry equal to $\widehat{\mu}_{(1)}(X_i)$, resp. $\widehat{\mu}_{(0)}(X_i)$, with the models $\widehat{\mu}_{(1)}(\cdot), \widehat{\mu}_{(0)}(\cdot)$ fitted as in the T-learner (Subsection 21.3.2).[9] Furthermore, let* `K_hat` *be a vector of length n corresponding to IPC-Weights estimated as in Subsection 21.3.3.1. Then the X-learner that models the CATE using the random forest function in the* `grf` *package may be implemented as follows. First, we fit* (21.14) *by only retaining complete, treated cases (Line 2 below), constructing the estimated scores $Y_i^{*,X,1}$* (21.13) *(Line 3), fitting a random forest with IPCW (Line 4) and finally extracting the model prediction at* x*:*

```
library(grf)
idx_1 <- ((D == 1) | (U >= t0)) & (W == 1)
Y_X_1 <- (U > t0) - mu0_hat
tau_hat_1 <- regression_forest(X[idx_1,], Y_X_1[idx_1], sample.weights
= K_hat[idx_1])
tau_x_1 <- predict(tau_hat_1, x)$predictions
```

---

[8] That is, for $Y_i^{*,X,0} = \mu_{(1)}(X_i) - Y_i(0)$, it holds that $\mathbb{E}[Y_i^{*,X,0} \mid X_i = x] = \tau(x)$.

[9] For example, $\widehat{\mu}_{(1)}(\cdot), \widehat{\mu}_{(0)}(\cdot)$ could be Cox-Lasso models as in Example 21.4, or survival forests as in Example 21.3. In the latter case, the following code could be used for computing `m1_hat` and `m0_hat` using `grf`:

```
m1 <- survival_forest(X[W==1,], U[W==1], D[W==1], prediction.type="Nelson-Aalen")
m0 <- survival_forest(X[W==0,], U[W==0], D[W==0], prediction.type="Nelson-Aalen")
m1_hat <- predict(m1, X, t0)$predictions; m0_hat <- predict(m0, X, t0)$predictions
```

We compute `tau_x_0` (21.15) *analogously.*[10] *Finally, we compute* `tau_x <- (1-e)*tau_x_1 + e*tau_x_0` *to combine the two estimates as in* (21.16).

Other predictive models, e.g., the Lasso (as in Example 21.8) could also be used instead of random forests.

21.3.4.2 R-learner

The R-learner is a metalearner proposed by [8] that builds upon a characterization of the CATE in terms of residualization of W_i and Y_i [73]. To be concrete, we start by centering W_i and Y_i around their conditional expectation given X, that is, we consider $W_i - e$[11] and $Y_i - m(X_i)$ with $m(x) = \mathbb{E}[Y_i \mid X_i = x]$ equal to:

$$\begin{aligned} m(x) &= \mathbb{E}[Y_i \mid X_i = x, W_i = 1]\mathbb{P}[W_i = 1] + \mathbb{E}[Y_i \mid X_i = x, W_i = 0]\mathbb{P}[W_i = 0] \\ &= e\mu_{(1)}(x) + (1-e)\mu_{(0)}(x). \end{aligned} \tag{21.17}$$

With the above definition, the following calculations hold for the expectation of $Y_i - m(X_i)$ conditionally on $X_i = x, W_i = 1$:

$$\mathrm{E}\, Y_i - m(X_i) \mid X_i = x, W_i = 1 = \mu_{(1)}(x) - m(x) = (1-e)(\mu_{(1)}(x) - \mu_{(0)}(x)) = (1-e)\tau(x). \tag{21.18}$$

Similarly $\mathrm{E}\, Y_i - m(X_i) \mid X_i = x, W_i = 0 = -e\tau(x)$, and so, $\mathrm{E}\, Y_i - m(X_i) \mid X_i = x, W_i = w = (w-e)\tau(x)$. The preceding display enables characterization of the CATE $\tau(\cdot)$ through the loss-based representation [74],

$$\tau(\cdot) \in \arg\min_{\tilde{\tau}(\cdot)} \left\{ \mathbb{E}\left[(\{Y_i - m(X_i)\} - \{W_i - e\}\tilde{\tau}(X_i))^2 \right] \right\}. \tag{21.19}$$

(21.19) suggests the following CATE estimation strategy. First, we estimate the unknown $m(\cdot)$ out-of-bag (21.3):[12]

$$\widehat{\mu}_{(0)}(\cdot) = \mathcal{M}^{\text{oob}}(Y \sim X;\ \mathcal{O}_0),\ \widehat{\mu}_{(1)}(\cdot) = \mathcal{M}^{\text{oob}}(Y \sim X;\ \mathcal{O}_1),\quad \widehat{m}(\cdot) = e\widehat{\mu}_{(1)}(\cdot) + (1-e)\widehat{\mu}_{(0)}(\cdot). \tag{21.20}$$

The final procedure will be robust to step (21.20) in the following sense: Even if $\widehat{m}(\cdot)$ is not a good approximation to $m(\cdot)$, the R-learner may perform well. In fact, if $e = 0.5$ and we estimate $\widehat{m}(\cdot) \approx 0$ (even though $m(\cdot) \neq 0$), the R-learner predictions are very similar to the predictions of the M-learner.[13] On the other hand, when $\widehat{m}(\cdot) \approx m(\cdot)$, then centering by $\widehat{m}(X_i)$ helps stabilize the estimation compared to the M-learner.

---

[10]The code would look as follows:

```
idx_0 <- ((D == 1) | (U >= t0)) & (W == 0); Y_X_0 <- mu1_hat - (U > t0)
tau_hat_0 <- regression_forest(X[idx_0,], Y_X_0[idx_0], sample.weights =
 K_hat[idx_0]) tau_x_0 <- predict(tau_hat_0, x)$predictions
```

[11]Note that $\mathbb{E}[W \mid X = x] = \mathbb{E}[W] = \mathbb{P}[W = 1] = e$, due to our assumption that we are in the setting of an RCT.

[12]In the case without censoring, it suffices to fit a single predictive model by regressing Y_i on X_i, that is, $\widehat{m}(\cdot) = \mathcal{M}^{\text{oob}}(Y \sim X;\ \mathcal{O})$, and this is the approach suggested in [8]. However, under censoring that may depend on treatment assignment, fitting $\mathcal{M}^{\text{oob}}(Y \sim X;\ \mathcal{O})$ becomes challenging – for example, if we naively use a survival forest with covariates X, then the fitted model will typically be inconsistent for $m(\cdot)$. By fitting two separate predictive models to the two treatment arms, as outlined in (21.20), and taking their convex combination (weighted by the treatment probability), we overcome this challenge and may use general predictive models.

[13]The reason is that the following characterization of $\tau(\cdot)$ in lieu of Equation (21.19) also holds:

$$\tau(\cdot) \in \arg\min_{\tilde{\tau}(\cdot)} \left\{ \mathbb{E}\left[(Y_i - \{W_i - e\}\tilde{\tau}(X_i))^2 \right] \right\}.$$

We emphasize that we use out-of-bag estimates of $\widehat{m}(\cdot)$ in (21.20). If we fit $\widehat{\mu}_{(1)}(\cdot), \widehat{\mu}_{(0)}(\cdot)$ with `grf` survival forests, as in Example 21.3, then we obtain out-of-bag estimates by default. If the Cox-Lasso is used, as in Example 21.4, then we can get out-of-bag estimates by splitting $\mathcal{I}$ into 10 folds and using out-of-fold predictions.

Second, we let $\widehat{K}_i$ be IPC-weights as in (21.9) and $\mathcal{I}_{\text{comp}}$ be the index set of complete cases. Finally, we estimate $\tau(\cdot)$ by fitting a model that leads to small values of the R-learner loss $\sum_{i \in \mathcal{I}_{\text{comp}}} \widehat{K}_i \left(\{Y_i - \widehat{m}(X_i)\} - \{W_i - e\}\widehat{\tau}(X_i)\right)^2$. Such a fitting procedure is not directly accommodated by models of the form (21.2). Below, we will describe how we may achieve this task with general predictive models (21.2). However, before doing so, we provide some examples of procedures that directly operate on the R-learner loss. We start with the simplest case:

Example 21.10 (R-learner for estimating a constant treatment effect)**.** *Let $\widehat{m}(\cdot)$ be as in (21.20) and $\widehat{K}_i$ as in (21.9). Suppose there is no treatment heterogeneity, that is, $\tau(x) =$ constant for all x.*[14] *Then, estimating the constant $\widehat{\tau}$ with the R-learner loss boils down to fitting a weighted linear regression on the complete cases with response $Y_i - \widehat{m}(X_i)$, predictor $W_i - e$,* without *an intercept, and with weights $\widehat{K}_i$, and letting $\widehat{\tau}$ be equal to the slope of $W_i - e$ in the above regression.*

Generalizing the above example, if we seek to approximate $\tau(\cdot)$ as a sparse linear function of X_i, then we may use the Lasso (compare to Example 21.8):

Example 21.11 (R-learner with Lasso CATE modeling)**.** *Let $\widehat{m}(\cdot)$, $\widehat{K}_i$ be as in Example 21.10. The R-learner estimate of the CATE using the Lasso is the following:*

$$(\widehat{\beta}_0, \widehat{\beta}_1) \in \arg\min_{\beta_0, \beta_1} \left\{ \sum_{i \in \mathcal{I}_{comp}} \widehat{K}_i \cdot ((Y_i - \widehat{m}(X_i)) - (W_i - e)(\beta_0 + \beta_1^\intercal X_i))^2 + \widehat{\lambda}_\tau \sum_{j=1}^{d} |\beta_{1,j}| \right\},$$

$$\widehat{\tau}(x) = \widehat{\beta}_0 + \widehat{\beta}_1^\intercal x.$$

The above objective can be fitted with standard software for the Lasso (e.g., `glmnet` [77]) with the following specification. The response is equal to $Y_i - \widehat{m}(X_i)$, the covariates are $[W_i - e, (W_i - e)X_{i1}, \ldots, (W_i - e)X_{id}]$, the intercept is not included, and the coefficient of the first covariate $(W_i - e)$ is unpenalized.

More generally, we may fit the R-learner objective using methods for varying coefficient models [78]. The `causal_forest` function of the `grf` package enables direct modeling of $\tau(x)$ through the GRF framework of [26, Section 6]. For each x, the causal forest estimates $\widehat{\tau}(x)$ as in Example 21.10 with additional data-adaptive weighting localized around x and determined by the collection of tree splits [79, Section 1.3].

Example 21.12 (Causal Forest: R-learner with Random Forest CATE model)**.** *Suppose `m1_hat`, resp. `m0_hat` are vectors of length n with i-th entry equal to $\widehat{\mu}_{(1)}(X_i)$, estimated out-of-bag and let `K_hat` be a vector of length n corresponding to IPC weights estimated as in Subsection 21.3.3.1. Then the R-learner that models the CATE using the random forest function in the `grf` package may be implemented as follows.*

```
library(grf)
idx <- (D == 1) | (U >= t0)
m_hat <- mu1_hat * e + mu0_hat * (1 - e)
Y <- U > t0
tau_hat <- causal_forest(X[idx,], Y[idx], W[idx], Y.hat = m_hat[idx],
                         W.hat = e[idx], sample.weights = K_hat[idx])
tau_x <- predict(tau_hat, x)$predictions
```

[14]The procedure described in this example is valid also in the presence of HTEs and asymptotically recovers the overlap-weighted average treatment effect (see e.g., [75]). In the setting without censoring, the procedure is described also in [76].

In Line 2 we get the indices of complete cases. In Line 3 we combine the estimate of $\mu_{(1)}(\cdot), \mu_{(0)}(\cdot)$ *to get an estimate of* $m(\cdot)$ *as in* (21.20). *In Line 4 we compute the survival indicator* `Y`. *In Lines 5 and 6 we fit a causal forest that targets the R-learner objective. Note that here we subset only to complete cases given by* `idx`, *and we also specify the censoring weights* `K_hat`, *as well as the expected responses* `m_hat` *for* `Y`, *and* `e` *for* `W`. *Finally, in Line 7, we extract the causal forest estimate of the CATE at* `x`.

The preceding examples presented three approaches that directly operate on the R-learner loss function. It is possible, however, to cast R-learner based estimation of $\widehat{\tau}(\cdot)$ in the form (21.2). To do so, we rewrite (21.19) equivalently as:

$$\tau(\cdot) \in \arg\min_{\widetilde{\tau}(\cdot)} \left\{ \mathbb{E}\left[(W_i - e)^2 \left(\frac{Y_i - m(X_i)}{W_i - e} - \widetilde{\tau}(X_i) \right)^2 \right] \right\}. \tag{21.21}$$

This is a weighted least squares objective with weights $(W_i - e)^2$. Under unbalanced treatment assignment, e.g., when there are fewer treated units ($e < 0.5$), then the R-learner upweights the treated units compared to control units by the factor $(1-e)^2/e^2$. The upweighting of treated units by the R-learner is similar to the behavior of the X-learner (compare to (21.16)). In fact, the predictions of R-learner and X-learner (when used with the same predictive models) are almost identical in the case of strong imbalance ($e \approx 0$) [8].

(21.21) justifies the following equivalent implementation of the R-learner. Let $\widehat{m}(\cdot)$ be as in (21.20) and let $\widehat{K}_i$ be IPC-weights, then we may estimate $\widehat{\tau}(\cdot)$ as follows:

$$\widehat{Y}^{*,R} = \frac{Y - \widehat{m}(X)}{W - e}, \quad \widehat{\tau}(\cdot) = \mathcal{M}(\widehat{Y}^{*,R} \sim X;\ \mathcal{O}_{\text{comp}}, \widehat{K} \cdot (W - e)^2). \tag{21.22}$$

We make two observations: First, $\widehat{Y}^{*,R}$ approximates a score in the sense of (21.10).[15] Second, the weights used by predictive models are the product of $(W_i - e)^2$ and the IPC weights $\widehat{K}_i$.

To further illustrate how to apply (21.22), we revisit Example 21.11 and provide an equivalent way of implementing the R-learner with the Lasso.

Example 21.13 (Weighted representation of R-learner with Lasso)**.** *The estimator* $\widehat{\tau}(\cdot)$ *in Example 21.11 may be equivalently expressed as,*[16]

$$(\widehat{\beta}_0, \widehat{\beta}_1) \in \arg\min_{\beta_0, \beta_1} \left\{ \sum_{i \in \mathcal{I}_{comp}} \widehat{K}_i \cdot (W_i - e)^2 \cdot \left(\widehat{Y}_i^{*,R} - \beta_0 - \beta_1^{\intercal} X_i \right)^2 + \widehat{\lambda}_\tau \sum_{j=1}^{d} |\beta_{1,j}| \right\},$$

[15]Formally the score is as follows:

$$Y_i^{*,R} = \frac{Y_i - m(X_i)}{W_i - e}, \quad \mathbb{E}[Y_i^{*,R} \mid X_i = x] = \tau(x). \tag{21.23}$$

The U-learner by [7] is a metalearner, related to the R-learner, that directly operates on the score (21.23). The first step of the U-learner is identical to the first step of the R-learner, namely $\widehat{m}(\cdot)$ is estimated as in (21.20). The step (21.22), however, is replaced by the following step:

$$\widehat{Y}^{*,R} = \frac{Y - \widehat{m}(X)}{W - e}, \quad \widehat{\tau}(\cdot) = \mathcal{M}(\widehat{Y}^{*,R} \sim X;\ \mathcal{O}_{\text{comp}}, \widehat{K}).$$

The difference is in the weights; the U-learner directly uses the IPC-weights $\widehat{K}_i$, while the R-learner adjusts the weights as $\widehat{K}_i \cdot (W_i - e)^2$, i.e., by multiplying the IPC-weights by $(W_i - e)^2$. The U-learner performed subpar compared to the R-learner in the simulation studies of [8] and so we do not consider it in our benchmark.

[16]Examples 21.11 and 21.13 are formally equivalent as described in the main text. However, typical implementations of the Lasso first standardize all covariates to unit variance (e.g., option `standardize=TRUE` in the `glmnet` package [77]). In that case, one would get conflicting answers when implementing the R-learner as described in Example 21.11, respectively Example 21.13. In our implementation of the R-learner with the Lasso, we enable standardization and follow Example 21.13.

$$\widehat{\tau}(x) = \widehat{\beta}_0 + \widehat{\beta}_1^\intercal x,$$

where $\widehat{Y}_i^{*,R} = (Y_i - \widehat{m}(X_i))/(W_i - e)$. Note that this may be computed using standard Lasso software with response $\widehat{Y}_i^{*,R}$, covariates X_i, weights $\widehat{K}_i \cdot (W_i - e)^2$ and including an unpenalized intercept.

As a final example of the R-learner, we show how we may model the CATE with another popular machine learning method, namely XGBoost [80] through (21.22).

Example 21.14 (R-learner with XGBoost CATE model). *Suppose `m1_hat`, resp. `m0_hat` are vectors of length n with i-th entry equal to $\widehat{\mu}_{(1)}(X_i)$, estimated out-of-bag and let `K_hat` be a vector of length n corresponding to IPC-Weights estimated as in Subsection 21.3.3.1. Then the R-learner that models the CATE using the extreme gradient boosting function in the `xgboost` package may be implemented as follows.*

```
library(xgboost)
idx <- (D == 1) | (U >= t0)
m_hat <- mu1_hat * e + mu0_hat * (1 - e)
Y_R <- ((U > t0) - m_hat) / (W - e)
data <- data.frame(X = X[idx,], Y = Y_R[idx])
tau_hat <- xgboost(data = data, label = data$Y,
                   weight = K_hat[idx] * (W[idx] - e[idx])^2,
                   objective = "reg:linear", nrounds = 200)
tau_x <- predict(tau_hat, x)
```

In Line 1 we load the `xgboost` package. In Line 2 we get the indices of complete cases. In Lines 3 and 4 we construct the R-learner score (21.23). *In Lines 5–8 we fit the XGBoost model with sample weights `K_hat[idx]*(W[idx]-e[idx])^2`. Finally, in Line 9, we extract the XGBoost estimate of the CATE at x.*

21.3.5 Summary of metalearners

In this Section we described five different metalearners, namely the S, T, X, R, and M-learners. Table 21.1 provides a high-level summary/overview of the estimation strategy underlying each of these learners, and what models they need to fit.

In Section 21.4 we conduct a simulation study to benchmark concrete instantiations of all these metalearners along with different choices for the underlying predictive model. We implement these approaches as how they are typically used in applied research to reflect their performance in practice. Furthermore, these concrete methods are accompanied by code [9] that clarifies exact considerations required for their practical implementation.

21.4 Simulation Study

As we saw above, different metalearners utilize different estimation strategies, namely, risk modeling and/or direct modeling of treatment heterogeneity. Thus, certain combinations of metalearners and predictive models may be advantageous under particular data generating processes. In this section we conduct a comprehensive simulation study to explore conditions wherein different CATE estimation methods may perform well or poorly. We evaluate a plethora of CATE estimators (Subsection 21.4.1) using the rescaled root mean squared error and Kendall's τ (Subsection 21.4.2) under multiple data generating mechanisms (Subsection 21.4.3). All results are fully reproducible with the code accompanying this article [9].

TABLE 21.1
High-level overview of metalearners. For each learner (S, T, X, R, and M), we describe their requirements in fitting a risk model $\widehat{\mu}$, censoring model $\widehat{S^C}$, and/or CATE model $\widehat{\tau}$, as well as how these predictive models are combined by each meta-learner to produce CATE estimates.

<table>
<tr><th></th><th>Risk Model</th><th>Censoring Model</th><th>CATE Model</th></tr>
<tr><td>S</td><td>$\widehat{\mu}(\cdot) = \mathcal{M}(Y \sim [X, W];\ \mathcal{O})$</td><td>Not applicable</td><td>$\widehat{\tau}(x) = \widehat{\mu}([x,1]) - \widehat{\mu}([x,0])$</td></tr>
<tr><td>T</td><td rowspan="2">$\widehat{\mu}_{(1)}(\cdot) = \mathcal{M}(Y \sim X;\ \mathcal{O}_1)$
$\widehat{\mu}_{(0)}(\cdot) = \mathcal{M}(Y \sim X;\ \mathcal{O}_0)$</td><td>Not applicable</td><td>$\widehat{\tau}(x) = \widehat{\mu}_{(1)}(x) - \widehat{\mu}_{(0)}(x)$</td></tr>
<tr><td>X</td><td rowspan="3">$\widehat{S^C}(\cdot) = \mathcal{M}^{\text{oob}}(\mathbb{1}\{C \geq u\} \sim [X, W]; \mathcal{O})$
$\widehat{K} = 1/\widehat{S^C}(\min\{U, t_0\}, X, W)$</td><td>$Y^{*,X} = (1-W)(\widehat{\mu}_{(1)}(X) - Y) + W(Y - \widehat{\mu}_{(0)}(X))$
$\widehat{\tau}_{(1)}(x) = \mathcal{M}(Y^{*,X} \sim X;\ \mathcal{O}_1 \cap \mathcal{O}_{\text{comp}}, \widehat{K})$
$\widehat{\tau}_{(0)}(x) = \mathcal{M}(Y^{*,X} \sim X;\ \mathcal{O}_0 \cap \mathcal{O}_{\text{comp}}, \widehat{K})$
$\widehat{\tau}(x) = (1-e)\widehat{\tau}_{(1)}(x) + e\widehat{\tau}_{(0)}(x)$</td></tr>
<tr><td>R</td><td>$\widehat{\mu}_{(1)}(\cdot) = \mathcal{M}^{\text{oob}}(Y \sim X;\ \mathcal{O}_1)$
$\widehat{\mu}_{(0)}(\cdot) = \mathcal{M}^{\text{oob}}(Y \sim X;\ \mathcal{O}_0)$</td><td>$\widehat{m}(x) = e\widehat{\mu}_{(1)}(x) + (1-e)\widehat{\mu}_{(0)}(x)$
$Y^{*,R} = (Y - \widehat{m}(x))/(W - e)$
$\widehat{\tau}(x) = \mathcal{M}(Y^{*,R} \sim X;\ \mathcal{O}_{\text{comp}}, \widehat{K} \cdot (W-e)^2)$</td></tr>
<tr><td>M</td><td>Not applicable</td><td>$Y^{*,M} = WY/e + (1-W)Y/(1-e)$
$\widehat{\tau}(x) = \mathcal{M}(Y^{*,M} \sim X;\ \mathcal{O}_{\text{comp}}, \widehat{K})$</td></tr>
</table>

TABLE 21.2
Metalearner combinations considered in the simulation study. For risk and CATE models, we apply either the Cox-Lasso regression or the generalized random forest approach. For censoring models, we either employ the Kaplan-Meier estimator without covariates adjustment or the random survival forest method with variable adjustment.

Risk \ CATE Model	Lasso (L)	GRF (F)	None (*)
Cox-Lasso (L)	XLL, RLL	—	SL*, TL*
GRF Survival Forest (F)	XFL, RFL	XFF, RFF	SF*, TF*
None (*)	M*L	M*F	—

Censoring Model	
Kaplan-Meier	X, R, M
GRF Survival Forest	X, R, M
None	S, T

21.4.1 Estimators under comparison

We compare the following estimation strategies:

Metalearners: We implement the five metalearners (S,T,X,R,M) described in Section 21.3 (summarized in Table 21.1) where we vary the risk model (survival forest, Cox-Lasso) and the CATE model (random forest, Lasso). We use three-letter acronyms for each method, wherein the first letter corresponds to the meta-learner, the second to the risk model, and the third to the CATE model. For example, *XFF* is the X̲-learner with risk models fitted by GRF̲ Survival forests, and CATE fitted by GRF̲ Random forests. Table 21.2 presents the 12 combinations of metalearners we consider, and their acronyms. For the X,R,M metalearners that also require a censoring model, we use Kaplan-Meier IPCW by default (Example 21.5). In one of the simulation settings (Subsection 21.4.3.4) we also consider variants that use GRF Survival forest ICPW (Example 21.6) instead of Kaplan-Meier. Further details about the implementation are provided in Supplement A, and code implementing these metalearner combinations is available in the `survlearners` package [9].

Cox-proportional hazards model (CPH): This represents our baseline approach as it is very widely used in practice [81–83]. This method is the same as the S-learner in Example 21.2 but without Lasso penalization, i.e., $\lambda = 0$.

Causal Survival Forest (CSF): This estimator, proposed by [40], is similar to the RFF estimator we consider, with censoring model estimated with survival forests. The main difference, as discussed in more detail in Remark 21.2 is that instead of adjusting for censoring via IPCW, it implements a doubly robust adjustment. We use the CSF implementation in the R package `grf`.

21.4.2 Performance evaluation metrics

To assess the performance of each method, we considered two evaluation metrics:

Rescaled root mean squared error (RRMSE) $\mathbb{E}[(\hat{\tau}(X) - \tau(X))^2 \mid \hat{\tau}]^{1/2}/\mathrm{SD}(\tau(X))$*:* The root mean squared error in estimating the CATE, normalized by the standard deviation of the true CATE. The expectation is taken with respect to an independent test sample X with the same distribution as the X_i. This metric is informative when the goal is to quantify patient-level treatment effects.

Kendall's τ*:* The rank correlation coefficient between predicted and true CATEs. This metric may be more suitable than the rescaled root mean squared error when it is of interest to decide how to prioritize or allocate treatment.

The above metrics were computed in each Monte Carlo replicate of our simulations based on 5000 test samples. For each simulation setting we generated 100 Monte Carlo replicates and summarized the results through boxplots that show the $10\%, 25\%, 50\%, 75\%$, and 90% percentiles of the corresponding metric.[17]

21.4.3 Data generating processes

To conduct a comprehensive comparison, our starting point was a baseline data generating process (DGP) motivated by the *type 2* setup in [40]. We then systematically varied five different characteristics of that DGP with the goal of assessing how each variation impacts different methods. Table 21.3 provides a high-level overview of all the DGPs we considered; in total we constructed 22 distinct DGPs for method comparison.

The baseline DGP was as follows: We observe $n = 5,000$ independent samples in the training dataset. The baseline covariates $X_i = (X_{i1}, \ldots, X_{ip})$ are independent and identically distributed with $X_{ij} \sim \mathrm{N}(0,1)$ and $p = 25$. The treatment assignment was $W_i \sim \mathrm{Bernoulli}(e)$, where $e = 0.5$. The survival time T_i is drawn from a PH model with hazard function:

$$\lambda_T(t \mid X_i, W_i) = \exp\left(f_R(X_i) + f_\tau(X_i, W_i)\right) \cdot \sqrt{t}/2,\ f_R(X_i) = \beta_1 X_{i1},\ f_\tau(X_i, W_i) = (-0.5 - \gamma_1 X_{i2}) W_i, \tag{21.24}$$

where we set the coefficients $\beta_1 = 1$ and $\gamma_1 = 0.5$. The censoring time C_i is independent of (T_i, W_i, X_i) and follows a Weibull distribution with hazard function

$$\lambda_C = \kappa^{\rho - 1}, \tag{21.25}$$

where we set $\kappa = 4$ and $\rho = 2$ for the scale and shape parameters, respectively.

21.4.3.1 Complexity of the baseline risk function

The model for T_i considered in (21.24) is relatively simple: It is a well-specified Cox proportional hazards (PH) model with log-hazard that is linear in $(X_{i1}, W_i, W_i X_{i2})$. We evaluate how robust different CATE methods are to model misspecification and whether they can adapt to more complicated

[17] In particular, the lower and upper hinges of the boxplots we show correspond to 10%, resp. 90% percentiles.

TABLE 21.3
Data generating processes in the simulation study. DGPs were constructed by varying five characteristics of the underlying data structure. For each factor, there exists one base case and one to three variants. Risk and CATE models can be linear (Lin) or nonlinear (Nonlin) functions of p_R and p_τ number of covariates, respectively. The censoring mechanism can be independent, dependent of covariates, or even unbalanced between treatment arms.

	Base Case	Variant I	Variant II	Variant III
Model Complexity				
Baseline risk function (f_R)	Lin, $p_R = 1$	Lin, $p_R = 25$	Nonlin, $p_R = 1$	Nonlin, $p_R = 25$
Treatment-covariate interaction (f_τ)	Lin, $p_\tau = 1$	Lin, $p_\tau = 25$	Nonlin, $p_\tau = 1$	Nonlin, $p_\tau = 25$
Treatment Heterogeneity - $\text{sd}(\tau)/\text{sd}(\mu_0)$				
$f_R =$ Lin., $f_\tau =$ Lin.	0.50	0.90	0.19	–
$f_R =$ Nonlin., $f_\tau =$ Lin,	0.80	1.40	0.13	–
$f_R =$ Nonlin., $f_\tau =$ Nonlin.	0.40	0.65	0.13	–
Censoring				
Censoring rate ($\rho = 2$)	30% ($\kappa = 4$)	60% ($\kappa = 7$)	–	–
Censoring distribution ($\kappa = 4$)	Regular ($\rho = 2$)	Early ($\rho = 1$)	–	–
Censoring dependency	$C \perp\!\!\!\perp (X, W)$	$C \sim X_1 + X_2W$	$C \sim X_1 + X_2W + W$	–
Unbalanced Treatment Assignment (e)	0.50	0.08	–	–

models for T_i. To do so, we start by increasing the complexity of the baseline risk by including a larger number of predictors in f_R (21.24), by utilizing nonlinear f_R such as indicator functions, or both:

$$(\text{Lin.}, p_R = 25) : f_R(X_i) = \sum_{j=1}^{25} \beta_1 X_{ij}/\sqrt{p}, \quad (\text{Nonlin.}, p_R = 1) : \quad f_R(X_i) = \beta_1 \mathbb{1}\{X_{i1} > 0.5\},$$

$$(\text{Nonlin.}, p_R = 25) : \quad f_R(X_i) = \tilde{\beta}_1 \mathbb{1}\{X_{i1} > 0.5\} + \sum_{j=1}^{12} \tilde{\beta}_2 \mathbb{1}\{X_{i(2j)} > 0.5\} \mathbb{1}\{X_{i(2j+1)} > 0.5\}.$$

We set $\beta_1 = 1$ (as in the baseline DGP), and $\tilde{\beta}_1 = 0.99$, $\tilde{\beta}_2 = 0.33$. We note that the last specification $(\text{Nonlin.}, p_R = 25)$ also includes second-order interactions of baseline covariates in the log-hazard. Furthermore, in all cases, $X_i \in \mathbb{R}^{25}$, i.e., we do not change the dimension of the baseline covariates (e.g., in the case $p_R = 1$, only the first feature influences the baseline risk, yet this information is not "revealed" to the different methods).

21.4.3.2 Complexity of the CATE function

Treatment-covariate interaction terms directly determine HTEs. Hence we also increase the complexity of f_τ in (21.24) as follows:

$$\begin{aligned}
(\text{Lin.}, p_\tau = 25) : \quad & f_\tau(X_i) = \sum_{j=1}^{25} (-0.5 - \gamma_1 X_{ij}/\sqrt{p}) W_i, \\
(\text{Nonlin.}, p_\tau = 1) : \quad & f_\tau(X_i) = (-0.5 - \gamma_1 \mathbb{1}\{X_{i1} > 0.5\}) W_i, \\
(\text{Nonlin.}, p_\tau = 25) : \quad & f_\tau(X_i) = (-0.5 - \tilde{\gamma}_1 \mathbb{1}\{X_{i1} > 0.5\} \\
& \qquad - \sum_{j=1}^{12} \tilde{\gamma}_2 \mathbb{1}\{X_{i(2j)} > 0.5\} \mathbb{1}\{X_{i(2j+1)} > 0.5\}) W_i.
\end{aligned}$$

We set $\gamma_1 = 0.5$, $\tilde{\gamma}_1 = 0.99$ and $\tilde{\gamma}_2 = 0.33$. In varying both the baseline risk through f_R and the CATE complexity through f_τ, we only consider combinations so that f_R is at least as complex as

f_τ (that when $p_\tau = 25$, then also $p_R = 25$, and when f_τ is nonlinear, then we also take f_R to be nonlinear). This reflects the fact that the baseline risk could be arbitrarily complicated, but HTEs may be less so [7, 8].

21.4.3.3 Magnitude of treatment heterogeneity

The strength of treatment heterogeneity may also influence how easy or difficult it is to estimate CATEs. Intuitively, it may be easier for a model to detect the variation in treatment effects when the heterogeneity is strong versus weak. We simulate data under three of the DGPs above (fixing $p_R = 25, p_\tau = 1$ and varying the complexity, starting with $f_R =$ Lin., $f_\tau =$ Lin.). We then vary the parameter γ_1 in e.g.,(21.24): While the above simulations used $\gamma_1 = 0.5$, we also consider $\gamma_1 = 0$ and $\gamma_1 = 1$. Table 21.3 shows the treatment heterogeneity in each case, in which we measure it as the ratio of the standard deviation of true CATEs and the standard deviation of baseline risks. Note that the DGPs in the Variant II column correspond to zero treatment heterogeneity on the log-hazard, i.e., $\gamma_1 = 0$, and the heterogeneity we see is solely due to effect amplification [84]. Effect amplification can be described as individuals with higher baseline risks may experience larger risk reduction on the absolute risk difference scale. Effect amplification is a major reason for the "PATH risk modeling" approach to CATE estimation (Subection 21.1.2 and [47]) in which HTEs are modeled as a function of baseline risk.

21.4.3.4 Censoring mechanism

An ubiquitous challenge in working with survival outcomes is the presence of right-censoring. We examine the impact of censoring mechanism in three subcases: varied censoring rates and distributions, heterogeneous censoring, and unbalanced censoring.

We define the censoring rate as the proportion of complete cases, i.e., $\#\mathcal{I}_{\text{comp}}/N$. We construct different censoring rates by altering the scale parameter in (21.25). Compared to the 30% censoring rate in the baseline DGP, roughly 60% of the subjects are censored when $\kappa = 7$. We vary the censoring distribution by setting the shape parameter $\rho = 1$, which induces a slightly higher censoring rate (40%), but more importantly, a higher proportion of subjects being censored at early times of the follow-up than in the baseline DGP.

In the baseline case and its variants above, we generate censoring times independently of baseline covariates. We also consider *heterogeneous* (yet noninformative) censoring where we allow the scale parameter in the censoring time model to be a function of (X_i, W_i), namely,

$$\lambda_{C_i} = \kappa_i^\rho, \quad \kappa_i = \exp(0.5 + \alpha X_{i1} + \delta X_{i2} W_i),$$

where $\alpha = 2$ and $\delta = 2$. Under this setting, subjects' censorship depends on their characteristics and treatment type.

A more interesting scenario that builds on top of heterogeneous censoring is *unbalanced censoring*, that is, subjects in treated or untreated arms may be more likely to get censored. While κ_i already depends on W in the above setting, we make the censoring more unbalanced by also including a main effect of W_i.

$$\lambda_{C_i} = \kappa_i^\rho, \quad \kappa_i = \exp(1 + \alpha X_{i1} + \alpha W_i + \delta X_{2i} W_i).$$

Under this formulation, the censoring rate in the untreated arm is much higher than that in the treated arm (60% vs. 30%), which may happen e.g., if patients realize they are on the inactive treatment and drop early.

21.4.3.5 Unbalanced treatment assignment

Unbalanced treatment assignment may occur, e.g., when drugs are too expensive to be assigned to half of the cohort. A consequence of unbalanced treatment assignment is that there may be insufficient data under one treatment arm for accurate CATE estimation. The X-learner has been shown to be more reliable than T-learner under unbalanced treatment assignment [7]. To explore how robust all methods are to unbalanced treatment assignment, we create a scenario where only 8% of the patients are treated, i.e., $W_i \sim \text{Bernoulli}(e)$ with $e = 0.08$.

21.5 Simulation Results

21.5.1 Description of simulation results

21.5.1.1 Results under varying baseline risk and CATE complexity

We first discuss the performance of the 14 CATE estimation methods across different complexity of baseline risk functions and treatment-covariate interaction terms. The rows and columns in Figure 21.1 vary by the numbers of covariates used in main effects (p_R) and interaction terms ($p_\mathcal{T}$), and their functional forms, either linear (*Lin*) or nonlinear (*Nonlin*), respectively.

We first make the following observations regarding the linear-linear case (f_R = Lin., $f_\mathcal{T}$ = Lin., first column of Figure 21.1). Here, SL* and TL* outperform all other methods. The reason is that the Cox-Proportional hazards model is well-specified in this case. When either $p_R = 1$ or $p_L = 1$, SL* and TL* outperform CPH, i.e., the Lasso penalization helps. When all covariates are relevant ($p_R = 25, p_\mathcal{T} = 25$), then CPH, SL* and TL* perform similarly. When $p_R = 25, p_\mathcal{T} = 1$, SL* outperforms both TL* and CPH, i.e., it is beneficial to directly regularize the interaction effects. Finally, RLL and XLL perform quite well in absolute terms, although not as well as SL*, TL*. The reason is that RLL and XLL are misspecified: They model the CATE as linear when it is linear in terms of log-hazard ratios, but not in terms of survival probabilities.

When the baseline risk is nonlinear and the interaction term is linear (f_R = Nonlin., $f_\mathcal{T}$ = Lin), SL* performs the best, along with XLL, RFL, RLL. When both the baseline risk and the interaction function are nonlinear (f_R = Nonlin., $f_\mathcal{T}$ = Nonlin), the best performing methods are the ones that model risk/CATE using GRF (CSF, SF*, TF*, XFF, RFF, M*F). Method performance in terms of Kendall's τ correlation shows a similar pattern as the RRMSEs under the same settings (Supplemental Figure S1).

More broadly, this set of simulation suggests that it may be beneficial to use metalearners employing modern predictive models compared to CPH as soon as the true DGP is somewhat more complicated or has more structure compared to a dense, well-specified Cox-Proportional Hazard model. The R-learner always performs at least as well as the M-learner with the same choice of predictive model for the CATE (i.e., RFF outperforms M*F, and RLL, RFL outperform M*L). The R-learner appears to be relatively robust to the choice of model for the risk, i.e., RFL and RLL performed similarly across the settings considered (with a minor edge of RLL over RFL in the case where f_R = Lin., $f_\mathcal{T}$ = Lin.). The choice of CATE model matters more. The X-learner appears to be more sensitive to the specification of the risk model, and the performance of XFL and XLL sometimes deviates from each other.

21.5.1.2 Results under varying HTE magnitude

Figure 21.2 displays the performance under varying HTE magnitude across varying levels of model complexity (with $p_R = 25$ and $p_\mathcal{T} = 1$). All estimators show a performance drop when the treatment heterogeneity on the relative scale is zero ($\gamma = 0$). When the treatment heterogeneity gets larger

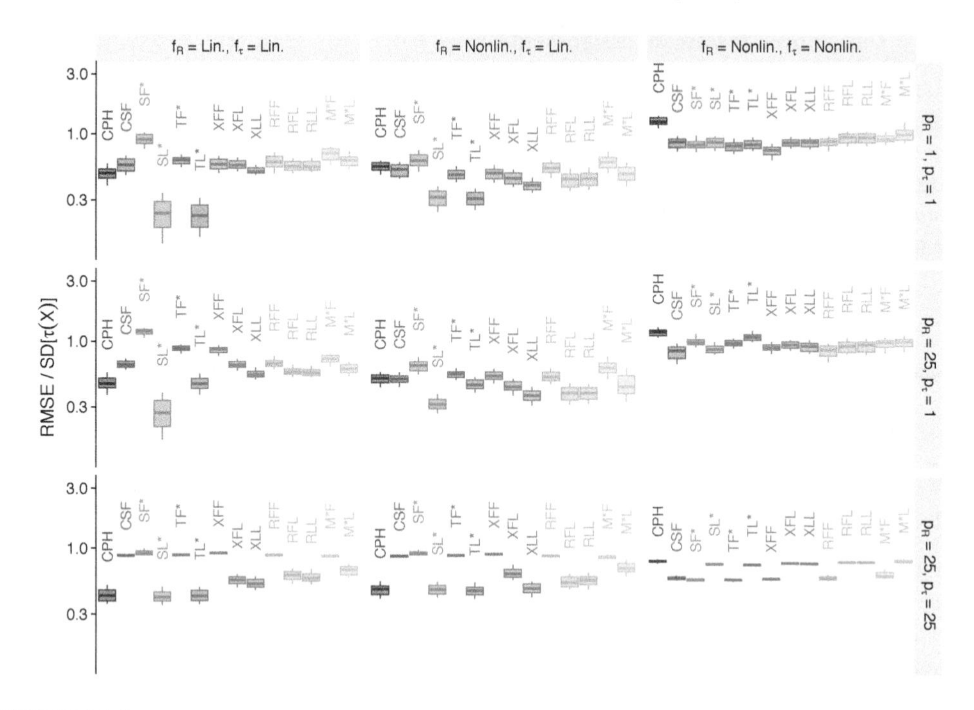

FIGURE 21.1
Rescaled root mean squared errors of metalearners across various levels of complexity of baseline risk functions (R) and treatment-covariate interactions (τ). The function forms (f_R and f_τ) vary between linear (Lin) and nonlinear (Nonlin) across columns, and the numbers of predictors (p_R and p_τ) vary across rows. We use 3-letter acronyms for each method, wherein the first letter corresponds to the meta-learner, the second to the risk model, and the third to the CATE model. For example, *XFF* is the X-learner with risk models fitted by GRF Survival forests, and CATE fitted by GRF Random forests.

($\gamma = 1$), most estimators yield smaller RRMSEs; more importantly, metalearners, such as RFF or XFF, that apply machine learning approaches that are misaligned with the underlying linear risk (Figure 21.2, f_R = Lin, f_τ = Lin) or CATE functional forms (Figure 21.2, f_R = Nonlin, f_τ = Lin) now perform similarly as the estimators whose predictive models match with the true functional forms. Moreover, methods show similar performance in terms of Kendall's τ correlation when $\gamma = 1$ and f_τ is linear (Figure S5).

21.5.1.3 Results under varying censoring mechanisms

We then summarize method performance under varied censoring mechanism. Figure 21.3 shows the RRMSE across different censoring models of all methods we have considered so far; while Figure S2 shows Kendall's τ. Figure 3 (resp. Figure S4) shows the RRMSE (resp. Kendall's τ) of all methods when the IPC-weights are estimated by a survival forest instead of Kaplan-Meier (this applies to the M-, X-, and R-learners). Finally, Figure 21.4 shows the ratio of the RRMSE of each of these methods (i.e., it contrasts the performance of the estimators with survival forest ICPW compared to Kaplan-Meier IPCW).

When the censoring is independent of baseline covariates and treatment (first three panels in above figures), then both Kaplan-Meier and survival forests are well-specified. Furthermore, all methods with both IPCW adjustment perform very similarly – even though a-priori one may have anticipated Kaplan-Meier IPCW to perform better because of reduced variance). On the other hand,

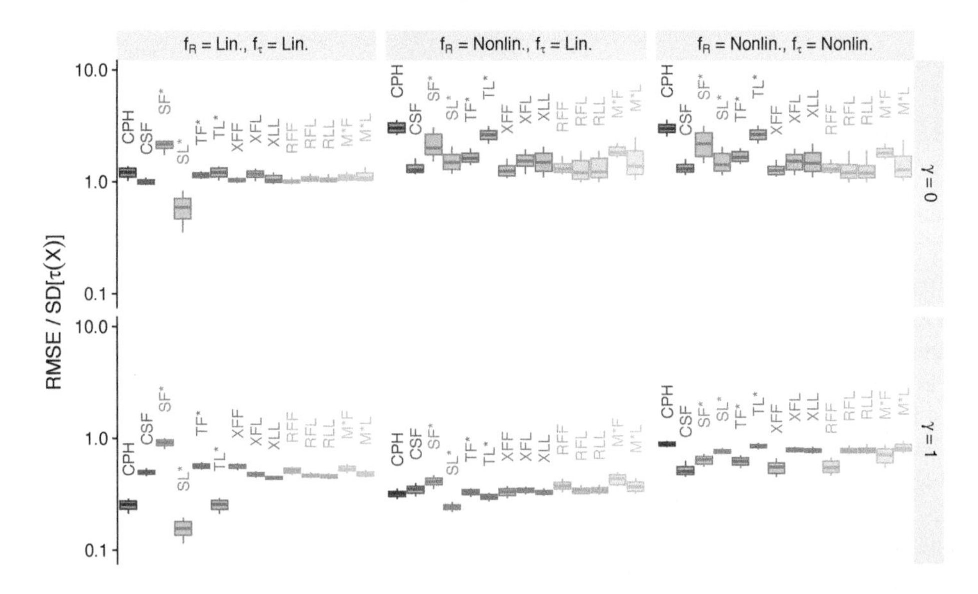

FIGURE 21.2
Rescaled root mean squared errors of metalearners under various levels of treatment heterogeneity. $\gamma = 0$ corresponds to zero treatment heterogeneity on the relative scale, and $\gamma = 1$ yields larger heterogeneity than in the base case (Table 21.3). The function forms vary across three combinations of linear and nonlinear. $p_R = 25$ and $p_\tau = 1$. Censoring is modeled using the Kaplan-Meier estimator. The metalearners are labeled in the same way as in Figure 21.1.

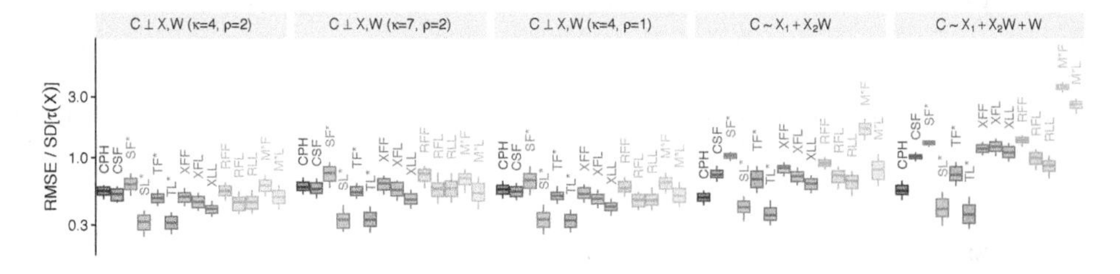

FIGURE 21.3
Rescaled root mean squared errors of metalearners under varied censoring mechanism. $C \perp\!\!\!\perp (X, W)$ and $C \sim (X, W)$ symbolizes censoring is *not* and is a function of covariates and treatment, respectively. κ and ρ are the scale and shape parameters in the hazard function of censoring. Censoring is modeled using the Kaplan-Meier estimator (except CSF). $p_R = 1$, $p_\tau = 1$, $f_R =$ Nonlin and $f_\tau =$ Lin. The metalearners are labeled in the same way as in Figure 21.1.

under heterogeneous censoring where $C \sim (X, W)$, the Kaplan-Meier estimator is misspecified. In that case, survival forest IPCW methods outperform Kaplan-Meier IPCW methods. This effect is particularly pronounced in the case of unbalanced heterogeneous censoring (panel 5). Furthermore, the X-learner appears to be more robust to misspecification of the censoring model compared to the R-learner. As we mentioned previously (also see Remark 21.2), CSF and RFF with survival forest ICPW are very similar methods: The main difference is that CSF accounts for censoring with a doubly robust augmentation term, and not just IPCW (as in RFF). In our simulations, CSF and RFF perform very similarly when censoring rates are not too high; however at high censoring levels, CSF outperforms RFF.

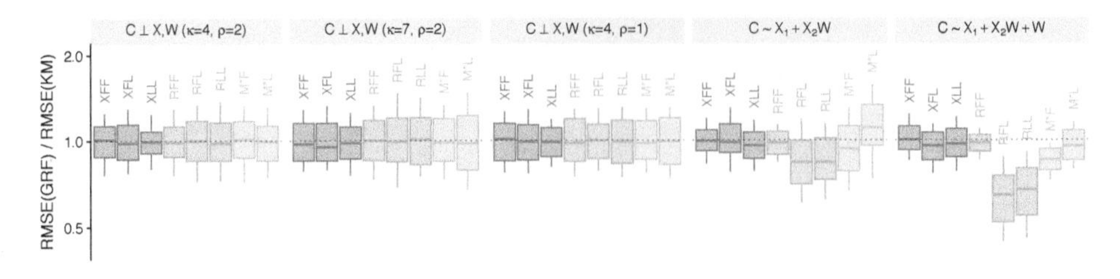

FIGURE 21.4
Ratios (log-scale) of rescaled root mean squared errors of metalearners under varied censoring mechanism. The contrast is formed between using a random survival forest model versus the Kaplan-Meier estimator for modeling censoring. $C \perp\!\!\!\perp (X, W)$ and $C \sim (X, W)$ symbolizes censoring is *not* and is a function of covariates and treatment, respectively. κ and ρ are the scale and shape parameters in the hazard function of censoring. $p_R = 1$, $p_\tau = 1$, $f_R =$ Nonlin and $f_\tau =$ Lin. The metalearners are labeled in the same way as in Figure 21.1.

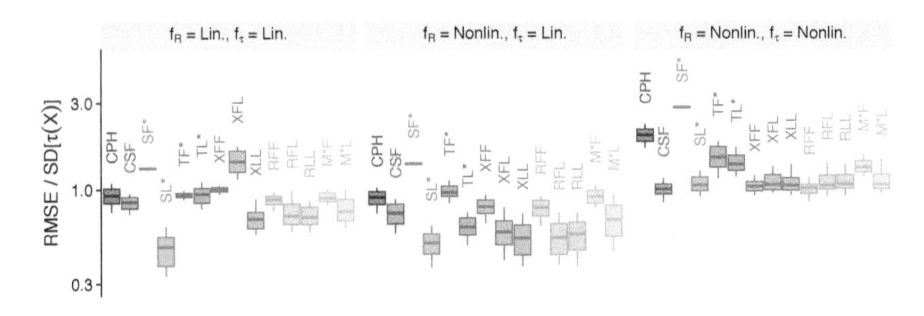

FIGURE 21.5
Rescaled root mean squared errors of metalearners unbalanced treatment assignment. Only 8% of subjects are treated. The function forms vary across three combinations of linear and nonlinear. $p_R = 25$ and $p_\tau = 1$. Censoring is modeled using the Kaplan-Meier estimator. The metalearners are labeled in the same way as in Figure 21.1.

21.5.1.4 Results under unbalanced treatment assignment

Lastly, Figures 21.5 and S6 show that CATE estimation becomes more challenging under unbalanced treatment assignment. When only 8% of the study samples are treated, all of the methods have smaller Kendalls' τ and increased median RRMSE with larger interquartile variation, and no estimator perform well when $f_R =$ Nonlin, $f_\tau =$ Nonlin. X- and R-learners, as well as CSF appear to be more robust to unbalanced treatment assignment than the other metalearners (especially when the predictive models for risk and CATE estimation are well-aligned with the true functional forms). In contrast, we observe larger performance drop in T-learners, which is consistent with earlier findings on the T-learner being sensitive to unbalanced treatment assignment [7].

21.5.2 Main takeaways from simulation study

In this section we summarize the major takeaways from our simulation study. We present a graphical summary of these takeaways in Figure 21.6.

First, we discuss the requirements for the risk model used with the different metalearners. The S- and T-learners require the risk models under both treatment arms to be well-estimated to provide

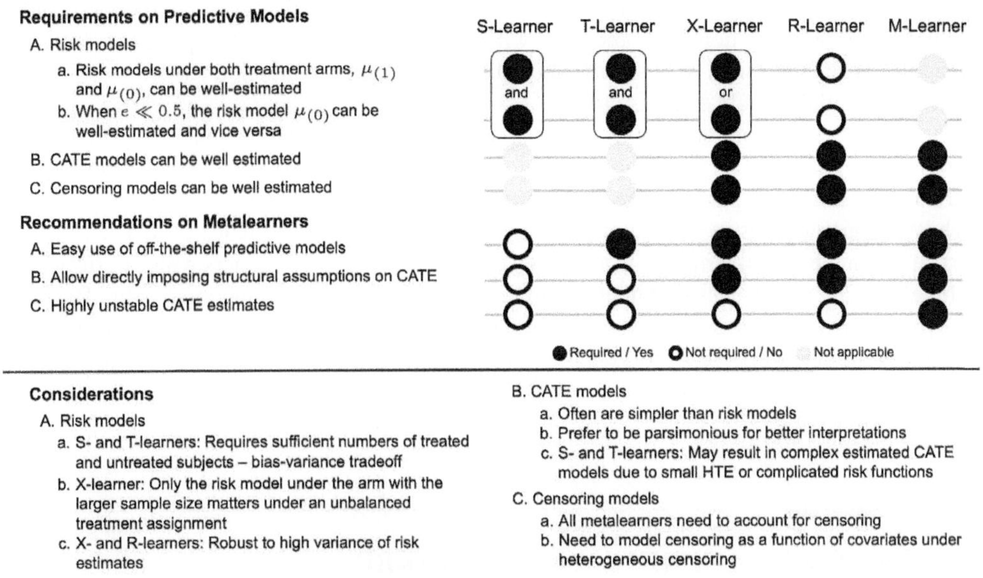

Requirements on Predictive Models	S-Learner	T-Learner	X-Learner	R-Learner	M-Learner
A. Risk models					
a. Risk models under both treatment arms, $\mu_{(1)}$ and $\mu_{(0)}$, can be well-estimated	Required / Yes (and)	Required / Yes (and)	Required / Yes (or)	Not required / No	Not applicable
b. When $e \ll 0.5$, the risk model $\mu_{(0)}$ can be well-estimated and vice versa	Required / Yes	Required / Yes	Required / Yes	Not required / No	Not applicable
B. CATE models can be well estimated	Not applicable	Not applicable	Required / Yes	Required / Yes	Required / Yes
C. Censoring models can be well estimated	Not applicable	Not applicable	Required / Yes	Required / Yes	Required / Yes
Recommendations on Metalearners					
A. Easy use of off-the-shelf predictive models	Not required / No	Required / Yes	Required / Yes	Required / Yes	Required / Yes
B. Allow directly imposing structural assumptions on CATE	Not required / No	Not required / No	Required / Yes	Required / Yes	Required / Yes
C. Highly unstable CATE estimates	Not required / No	Not required / No	Not required / No	Not required / No	Required / Yes

● Required / Yes ○ Not required / No Not applicable

Considerations

A. Risk models
- a. S- and T-learners: Requires sufficient numbers of treated and untreated subjects – bias-variance tradeoff
- b. X-learner: Only the risk model under the arm with the larger sample size matters under an unbalanced treatment assignment
- c. X- and R-learners: Robust to high variance of risk estimates

B. CATE models
- a. Often are simpler than risk models
- b. Prefer to be parsimonious for better interpretations
- c. S- and T-learners: May result in complex estimated CATE models due to small HTE or complicated risk functions

C. Censoring models
- a. All metalearners need to account for censoring
- b. Need to model censoring as a function of covariates under heterogeneous censoring

FIGURE 21.6
Main takeaways from the simulation study. For each listed item on the left, we summarize metalearners by assigning three types of labels: Required/Yes, Not required/No, and Not applicable, depending on the specific requirements or recommendations described. "*and*" indicates that both conditions need to be satisfied, and "*or*" means that only one of the two conditions is necessary.

accurate CATE estimates. Meeting this requirement is not easy in general and becomes particularly challenging under situations with unbalanced treatment assignment with a small number of subjects in one of the arms. The X- and R-learners are less sensitive to unbalanced treatment assignments. In the case of unbalanced treatment assignment, the X-learner performs well as long as the risk model for the arm with most subjects (typically, the control arm) is well-estimated. The R-learner only depends on the two risk models in order to decrease variance, and so is even more robust than the X-learner.

Second, we discuss the requirements for the CATE model. Estimating CATE functions is a crucial step for metalearners that directly model treatment effect heterogeneity, including M-, X-, and R-learners. The general intuition is that CATE functions are often simpler than risk functions, and we recommend applying parsimonious models to ensure the interpretability of CATE estimates. All three metalearners can flexibly estimate CATE functions by fitting a separate model, but M-learners tend to be unstable compared to X- and R-learners.

Third, we discuss the censoring model requirement. All approaches need to account for censoring one way or the other: S- and T-learners need to account for censoring in the process of fitting the risk models, while the other metalearners require explicit models for the censoring weights. When censoring functions are independent of baseline covariates and treatment, the Kaplan-Meier estimator is appropriate to use. However, the Kaplan-Meier estimator can induce performance degradation, when the censoring depends on baseline covariates or treatment. If it is unclear whether censoring is completely independent of treatment and covariates, we suggest to use random survival forests to model censoring as a function of relevant predictors – this choice appears to perform well even when the simpler Kaplan-Meier censoring model is correctly specified. When censoring rates are high we recommend applying the CSF [40] method instead of RFF.

We note that S-learners may perform well, however typically not when applied with off-the-shelf predictive models, but only with tailored models [14,27,85]. For instance, when used with flexible machine learning approaches such as random forests, S-learners do not give the treatment variable a special role (and so by default, some trees will not split on treatment assignment, even if HTEs are strong). When used with regression, analysts have to specify interaction terms, which typically requires substantial domain expertise [47].

To conclude, we recommend applying R- and X-learners for CATE estimation as strong default choices as any off-the-shelf machine learning models can be used. Besides, they allow imposing separate structural assumptions in the CATE and risk functions (stratified by treatment assignment), which is an important feature when these two functions are of different complexity. When background information on the control arm risk is available, analysts can apply X-learners with carefully chosen machine learning approaches that match with the possible underlying functional form. But if little is known about suitable risk models, we recommend implementing the R-learner with risk models estimated with survival forests (e.g., RFL or CSF) as default approaches as they are robust to misspecified risk models and provide stable CATE estimation.

21.6 Case Study on SPRINT and ACCORD-BP

The Systolic Blood Pressure Intervention Trial (SPRINT) is a multicenter RCT that evaluated the effectiveness of the intensive blood pressure (BP) treatment goal (systolic BP $<$ 120 mm Hg) as compared to the standard BP goal ($<$ 140 mm Hg) on reducing risks of cardiovascular disease (CVD). SPRINT recruited 9,361 participants and found that the intensive treatment reduced the risk of fatal and nonfatal CVD and all-cause mortality for patients at high risk of CVD events [10].

Several prior works have sought to estimate HTEs in SPRINT to improve BP control recommendations at a personalized level. For example, other authors have applied traditional subgroup analyses with either pre-specified subgroups [86, 87] or subgroups learned by machine learning methods [88, 89]. More closely related to our chapter, some authors have applied variants of metalearners for HTE estimation. In particular, [90] applied a S-learner with survival probabilities estimated by logistic regression (ignoring the censoring), [91] and [83] applied a S-learner along with the Cox elastic net (similar to Example 21.2, but with the elastic net [92] penalty instead of the Lasso penalty. [14] applied S-learners along with boosting and multivariate adaptive regression splines (MARS), modified to account for the treatment W_i in a specialized way. [45] developed a X-learner with random forests for both the Risk model and the CATE model and IPC-weights estimated by a Cox-PH model.

The prior works largely suggest the existence of heterogeneous effects of intensive BP treatment, along with diverging conclusions on potential treatment effect modifiers. However, in a unified analysis using the rank-weighted average treatment effect (RATE) method, [93] found no strong evidence of existing CATE estimators being able to learn generalizble HTEs in the SPRINT and ACCORD-BP trials.

The RATE can be used to evaluate the ability of any treatment prioritization rule to select patients who benefit more from the treatment than the average patient; when used to assess CATE estimators, the RATE evaluates the rule that prioritizes patients with the highest estimated CATE.

We build on this analysis by considering the behavior of a number of metalearners on these trials.

We first examine the performance of metalearners in estimating CATEs in SPRINT and the ACCORD-BP trial by conducting a global null analysis where, by construction, the CATE is zero for all subjects. Then, we analyze the CATEs in the SPRINT and ACCORD-BP data building upon insights leaned from our simulation study. Overall, our findings mirror those of [93], i.e., we do not find significant evidence of the ability of CATE estimators to detect treatment heterogeneity in these trials.

In all our analyses we seek to estimate the CATEs, defined as the difference in survival probabilities at the median follow-up time (i.e., 3.26 years). Our outcome is a composite of CVD events and deaths, which includes nonfatal myocardial infarction (MI), acute coronary syndrome not resulting in MI, nonfatal stroke, acute decompensated congestive heart failure, or CVD-related death. We identified 13 predictors from reviewing prior works [1,10,91], including age, sex, race black, systolic BP, diastolic BP, prior subclinical CVD, subclinical chronic kidney disease (CKD), number of antihypertensives, serum creatinine level, total cholesterol, high-density lipoprotein, triglycerides, and current smoking status. After retaining only subjects with no missing data on any covariate, the SPRINT sample includes 9,206 participants (98.3% of the original cohort)

In addition to the main predictor set (13 predictors) used above, we also consider a second, reduced predictor set with only two predictors. The first predictor is the 10-year probability of developing a major ASCVD event ("ASCVD risk") predicted from Pooled Cohort Equations (PCE) [1], which may be computed as a function of a subset of the aforementioned 13 predictors. Including the PCE score as a predictor is justified by the "PATH risk modeling" framework (Subsection 21.1.2 and [47]) according to which absolute treatment effects are expected to be larger for larger values of the PCE score. Such evidence was provided for the SPRINT trial by [94] who conducted subgroup analyses with subgroups stratified by quartiles of PCE scores. The second predictor is one of the 13 original predictors, namely the binary indicator of subclinical CKD. A subgroup analysis by CKD was pre-specified in the SPRINT RCT and considered, e.g., in [86,87].

Beyond the reanalysis of SPRINT, we also apply some of our analyses to the ACCORD-BP trial [11] that was conducted at 77 clinical cites across U.S. and Canada. ACCORD-BP also evaluated the effectiveness of intensive BP control as in SPRINT, but one major difference is that all subjects in ACCORD-BP are under type 2 diabetes mellitus. Moreover, ACCORD participants are slightly younger (mean age = 62.2 years) than SPRINT subjects (mean age = 67.9 years) and are followed for a longer time on average (mean follow-up time = 3.3 years for SPRINT vs. 4.7 years for ACCORD-BP). Our study sample contains 4,711 patient-level data (99.5% of the original data).

21.6.1 Global null analysis

A large challenge in evaluating CATE methods using real data is the lack of ground truth. One situation in which the true CATE is known, is under the global null, i.e., when all treatment effects are equal to 0. We implement the global null analysis by restricting our attention only to subjects in the control arm of SPRINT, and assign them an artificial and random treatment $Z \sim \text{Bernoulli}(0.5)$. Hence by construction Z is independent of $(T_i(1), T_i(0), X_i, C_i, W_i)$ and the CATEs must be null. We then estimate the CATE with respect to the new treatment Z using all the CATE approaches compared in Section 21.4. We repeat the same global null analysis by using all treated subjects in SPRINT, and for the treated and untreated participants in ACCORD-BP, separately.

In each global null analysis, we model censoring using two approaches: The Kaplan-Meier estimator or the random survival forest method via `grf`. Table 21.4 shows that some metalearners yielded much larger RMSEs than the others, which may indicate they are more likely to detect spurious HTE when it does not exist. The S-learner with GRF (SF*), M-learner with Lasso (M*L), and R-learners show consistently small RMSEs. The X-learner with GRF (XFF), which is methodologically similar to the method applied by [45] for HTE estimation in SPRINT, result in nontrivial RMSEs. When a survival forest model is used to estimate the censoring weights versus the Kaplan-Meier estimator, the RMSEs of most metalearners decrease, which may be an indication that the censoring mechanism in SPRINT depends on baseline covariates. Table S1 (Supplement C.1 shows similar results when the PCE scores and subclinical CKD are used as predictors.

Figure S7 compares the performance of different methods under the two considered choices of predictor sets. For the global null analysis conducted with untreated units on both SPRINT and ACCORD-BP, we make the following observation: The learners that applied a Lasso model for the CATE function have a decreased RMSE when the reduced predictor set is used. This observation is

TABLE 21.4
Performance assessment of metalearners in CATE estimation via global null analysis. By construction, the true CATEs in the modified SPRINT and ACCORD datasets are zero for all subjects. Original baseline covariates are used. The metalearners showed various amounts of overestimation of treatment heterogeneity (in terms of RMSEs), which may lead to false discoveries. The analysis was repeated for subjects under the standard and intensive arms and replicated to model censoring using a random survival forest and the Kaplan-Meier estimator, separately.

Standard				Intensive			
Survival Forest		Kaplan-Meier		Survival Forest		Kaplan-Meier	
Estimator	RMSE	Estimator	RMSE	Estimator	RMSE	Estimator	RMSE
SPRINT							
SF*	0.0005	SF*	0.0005	SF*	0.0003	SF*	0.0003
M*L	0.0008	M*L	0.0007	M*L	0.0021	M*L	0.0017
RFL	0.0011	RFL	0.0015	RLL	0.0024	RFL	0.0022
RLL	0.0021	RLL	0.0112	RFL	0.0027	RLL	0.0022
XLL	0.0085	XFF	0.0120	SL*	0.0090	XLL	0.0088
XFF	0.0114	RFF	0.0123	XFL	0.0094	SL*	0.0090
RFF	0.0118	TF*	0.0139	XLL	0.0127	XFL	0.0116
TF*	0.0139	CSF	0.0145	RFF	0.0136	RFF	0.0137
M*F	0.0141	M*F	0.0145	TF*	0.0141	TF*	0.0141
CSF	0.0145	SL*	0.0161	XFF	0.0146	CSF	0.0151
SL*	0.0161	XFL	0.0179	M*F	0.0147	XFF	0.0152
XFL	0.0176	XLL	0.0205	CSF	0.0151	M*F	0.0152
TL*	0.0351	TL*	0.0351	TL*	0.0280	TL*	0.0280
ACCORD-BP							
SF*	0.0010	SF*	0.0010	SF*	0.0013	SF*	0.0013
M*L	0.0103	M*L	0.0104	RLL	0.0026	RLL	0.0027
RFL	0.0106	RFL	0.0108	RFL	0.0034	RFL	0.0038
RLL	0.0121	RLL	0.0167	SL*	0.0047	SL*	0.0047
XLL	0.0132	XFF	0.0177	M*L	0.0062	M*L	0.0065
XFF	0.0178	XLL	0.0189	XLL	0.0144	XLL	0.0157
TF*	0.0204	TF*	0.0204	XFL	0.0283	XFL	0.0305
RFF	0.0212	RFF	0.0213	XFF	0.0329	XFF	0.0320
CSF	0.0220	CSF	0.0220	CSF	0.0336	CSF	0.0336
M*F	0.0253	XFL	0.0223	TL*	0.0338	TL*	0.0338
XFL	0.0265	M*F	0.0245	RFF	0.0347	RFF	0.0346
SL*	0.0291	SL*	0.0291	TF*	0.0367	TF*	0.0367
TL*	0.0296	TL*	0.0296	M*F	0.0438	M*F	0.0421

*The asterisk is part of the method's label, please refer to TABLE 21.2 where we introduce the abbreviations/labels for the methods we considered.

concordant with the "PATH risk modeling" recommendations [47]. In contrast, the learners that used GRF for the CATE function have a lower RMSE when the main predictor covariate set is used. A possible explanation is that the presence of multiple uninformative predictors drives different trees of the forest to split on different variables; upon averaging across trees the estimated CATEs are approximately zero (as they should be under the global null).

Figure 21.7 displays the relationship between the estimated CATEs in untreated SPRINT subjects and the ten-year PCE scores for a subset of the methods that use survival forest IPCW. We see

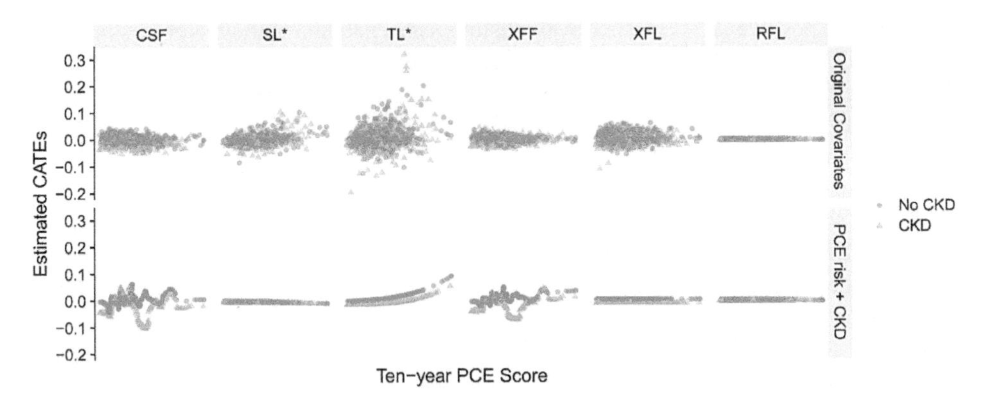

FIGURE 21.7
Scatter plot of ten-year PCE scores and estimated CATEs in SPRINT under a global null model. The CATE models were derived using 70% of the untreated patients with an artificial randomized treatment assignment then used to make predictions on the rest 30% test data. The censoring weights are estimated using a survival forest model. The analysis was replicated with the original covariates in SPRINT as the predictors (Row 1) and the estimated ten-year CVD risk (using pooled cohort equation) and subclinical CKD as the predictors (Row 2).

that RFL estimate constants CATEs (that is, all CATEs are equal to the estimated ATE) that are nearly zero. Under the global null, this behavior is desirable to avoid detection of spurious HTEs and showcases the benefit of R-learners in enabling direct modeling of CATEs and imposing, e.g., sparsity assumptions. Other approaches yielded non-zero CATEs with large variations. In particular, when the PCE scores and CKD are used as predictors, T-learner with Lasso (TL*) shows an increasing trend as a function of PCE scores for both subjects with or without prior subclinical CKD. This illustrates the regularization bias of the T-learner, as pointed out by [8], and explains the large RMSE of TL*. Figures S8–S14 are analogous to Figure 21.7 and present the other three settings (treated units in SPRINT, untreated and treated units in ACCORD-BP) for both Kaplan-Meier IPCW and survival forest IPCW.

21.6.2 CATE estimation in SPRINT and ACCORD

For our main SPRINT analysis, we applied five of the previously considered methods, namely, CSF, TL*, XFF, XFL, and RFL. We computed CATEs with respect to the actual (rather than null) treatment assignment in SPRINT and estimated the censoring weights using a random survival forest model. For internal validation, we randomly sampled 70% of the data as the training set and made predictions on the remaining testing set. For external validation, we employed the entire SPRINT data as the training set and made predictions on ACCORD-BP.

As ground truth CATEs are no longer available in this setting, we assess method performance in real data using two recently proposed metrics. RATEs [93] constitute a method for evaluating treatment prioritization rules, under which a subset of the population is prioritized for the allocated treatment. The RATE method quantifies the additional treatment benefits gained by using a prioritization rule to assign treatment relative to random treatment assignment measured as the area under the target operating curve (AUTOC). An AUTOC of zero for a prioritization rule indicates that using that rule would result in the same benefits as assigning treatment randomly within the population (i.e., average treatment effects), which further implies that either there is no treatment heterogeneity or the treatment prioritization rule does not effectively stratify patients in terms of treatment benefit. The RATE assessment can be considered as a discrimination metric for CATE estimates, as a prioritization rule with a large, positive AUTOC effectively distinguishes patients with greater treatment benefits from those with lesser treatment benefits by assigning them a high versus low treatment priority, respectively. The Expected Calibration Error for predictors of Treatment Heterogeneity (ECETH) [95] is another novel metric for quantifying the ℓ_2 calibration error of a CATE estimator. The calibration function of treatment effects is estimated using an AIPW (augmented inverse propensity weighted) score, which makes the metric robust to overfitting and high-dimensionality issues.

Table 21.5 shows that none of the methods detect significant treatment heterogeneity in SPRINT or ACCORD-BP. The insignificant results from the RATE evaluation suggest that there may be a lack of evidence for treatment heterogeneity. In Figure 21.8, the CATEs estimated using RFL also show an independent relationship with the PCE score when the original predictor set is used. But when the PCE score used as a predictor, the CATE estimates from all methods showed an overall increasing trend with the ten-year PCE score, consistent with e.g., the finding of [94]. Such a trend is less obvious in the external validation results (Figure S15), and all the resulting AUTOCs are nonpositive (under PCE + CKD), which suggest that the prioritization rules based on estimated CATEs in ACCORD-BP lead to similar treatment benefits as average treatment effects.

TABLE 21.5
Internal and external validation performance of CATE estimation in SPRINT and ACCORD-BP, respectively. None of the five metalearners showed significant AUTOCs, suggesting the lack of treatment heterogeneity of intensive BP therapy. RECETH is the square root of the default calibration error given by ECETH.

	Original Covariates		PCE + CKD	
Method	AUTOC	RECETH	AUTOC	RECETH
		Internal Validation		
SL*	−0.0043 (−0.0260, 0.0174)	0.0211 (0.0000, 0.0452)	−0.0002 (−0.0223, 0.0218)	0.0006 (0.0000, 0.0358)
TL*	−0.0002 (−0.0206, 0.0202)	0.0023 (0.0000, 0.0404)	0.0002 (−0.0209, 0.0212)	0.0142 (0.0000, 0.0431)
XFF	0.0048 (−0.0162, 0.0258)	0.0135 (0.0000, 0.0386)	−0.0144 (−0.0348, 0.0059)	0.0186 (0.0000, 0.0426)
XFL	−0.0053 (−0.0270, 0.0165)	0.0235 (0.0000, 0.0469)	−0.0019 (−0.0232, 0.0195)	0.0000 (0.0000, 0.0285)
RFL	0.0000 (0.0000, 0.0000)	0.0142 (0.0000, 0.0307)	0.0044 (−0.0161, 0.0248)	0.0000 (0.0000, 0.0301)
CSF	0.0012 (−0.0187, 0.0211)	0.0184 (0.0000, 0.0425)	−0.0172 (−0.0371, 0.0028)	0.0451 (0.0185, 0.0644)
		External Validation		
SL*	−0.0092 (−0.0281, 0.0097)	0.0077 (0.0000, 0.0332)	−0.0031 (−0.0201, 0.0138)	0.0000 (0.0000, 0.0277)
TL*	−0.0100 (−0.0274, 0.0073)	0.0242 (0.0000, 0.0421)	−0.0095 (−0.0261, 0.0072)	0.0215 (0.0000, 0.0406)
XFF	0.0169 (−0.0006, 0.0345)	0.0000 (0.0000, 0.0201)	−0.0088 (−0.0261, 0.0085)	0.0362 (0.0111, 0.0526)
XFL	−0.0081 (−0.0254, 0.0091)	0.0201 (0.0000, 0.0391)	−0.0095 (−0.0261, 0.0071)	0.0162 (0.0000, 0.0377)
RFL	0.0000 (0.0000, 0.0000)	0.0087 (0.0000, 0.0229)	0.0000 (0.0000, 0.0000)	0.0074 (0.0000, 0.0221)
CSF	0.0178 (0.0010, 0.0345)	0.0000 (0.0000, 0.0294)	−0.0111 (−0.0282, 0.0060)	0.0321 (0.0000, 0.0486)

21.7 Discussion

Given the increasing interest in personalized medicine, a number of advanced statistical methods have been developed for estimating CATEs, often referred to as personalized treatment effects. We focused on characterizing the empirical performance of metalearners in an RCT setting. An important direction for future work is to extend our tutorial and benchmark to the observational study setting where confounding plays a crucial role. In (21.1) we considered treatment effects in terms of difference in survival probabilities. CATEs can be also be defined on the relative scale such as hazard ratios and as the difference in restricted mean survival times. The latter is a popular estimand as it can be measured under any distribution of survival times and has a straightforward interpretation, that is, the expected life expectancy between an index date and a particular time horizon [96]. Further

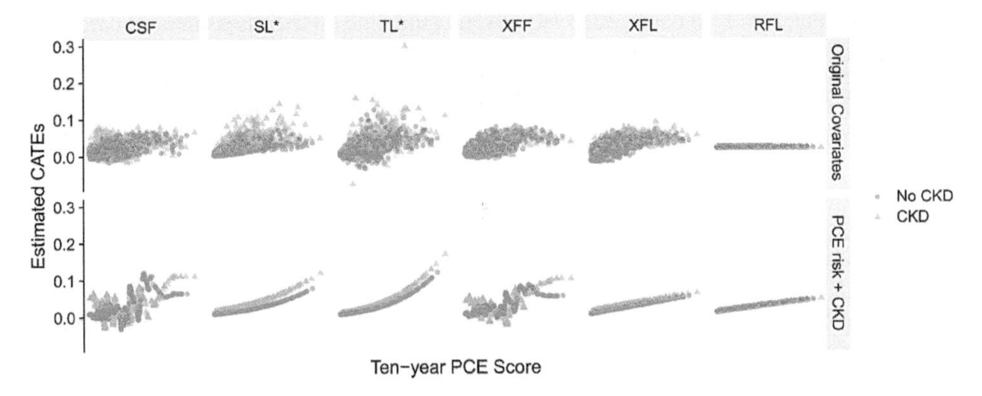

FIGURE 21.8
Scatter plot of ten-year PCE score and estimated CATEs in SPRINT. This figure is analogous to Figure 21.7, but with the full SPRINT cohort and using the true treatment assignment.

investigations on metalearners targeting such estimands may improve HTE estimation in clinical settings.

The current work aims to provide guidance on *when* and *how* to apply each approach for time-to-event outcomes in light of the specific characteristics of a dataset. We conducted comprehensive simulation studies to compare five state-of-the-art metalearners coupled with two predictive modeling approaches. We designed a spectrum of data generating processes to explore several important factors of a data structure and summarized their impacts on CATE estimation, as well as the weakness and strengths of each CATE estimator. Based on our findings from the simulation study, we highlighted main takeaways as a list of requirements, recommendations, and considerations for modeling CATEs, which provides practical guidance on how to identify appropriate CATE approaches for a given setting, as well as a strategy for designing CATE analyses. Finally, we reanalyzed the SPRINT and ACCORD-BP studies to demonstrate that some prior findings on heterogeneous effects of intensive blood pressure therapy are likely to be spurious, and we present a case study to demonstrate a proper way of estimating CATEs based on our current learning. To facilitate the implementation of our recommendations for all the CATE estimation approaches that we investigated and to enable the reproduction of our results, we created the R package survlearners [9] as an off-the-shelf tool for estimating heterogeneous treatment effects for survival outcomes.

Acknowledgments

This work was supported by R01 HL144555 from the National Heart, Lung, and Blood Institute (NHLBI).

References

[1] David C. Goff, Donald M. Lloyd-Jones, Glen Bennett, Sean Coady, Ralph B. D'Agostino, Raymond Gibbons, Philip Greenland, Daniel T. Lackland, Daniel Levy, Christopher J. O'Donnell, Jennifer G. Robinson, J. Sanford Schwartz, Susan T. Shero, Sidney C. Smith, Paul Sorlie, Neil J.

Stone, and Peter W. F. Wilson. 2013 ACC/AHA guideline on the assessment of cardiovascular risk. *Circulation*, 129(25_suppl_2):S49–S73, 2014.

[2] Donna K Arnett, Roger S Blumenthal, Michelle A Albert, Andrew B Buroker, Zachary D Goldberger, Ellen J Hahn, Cheryl Dennison Himmelfarb, Amit Khera, Donald Lloyd-Jones, J William McEvoy, et al. 2019 acc/aha guideline on the primary prevention of cardiovascular disease: a report of the american college of cardiology/american heart association task force on clinical practice guidelines. *Journal of the American College of Cardiology*, 74(10):e177–e232, 2019.

[3] Handrean Soran, Jonathan D Schofield, and Paul N Durrington. Cholesterol, not just cardiovascular risk, is important in deciding who should receive statin treatment. *European Heart Journal*, 36(43):2975–2983, 2015.

[4] George Thanassoulis, Michael J Pencina, and Allan D Sniderman. The benefit model for prevention of cardiovascular disease: an opportunity to harmonize guidelines. *JAMA Cardiology*, 2(11):1175–1176, 2017.

[5] RL Kravitz, N Duan, and J Braslow. Evidence-based medicine, heterogeneity of treatment effects, and the trouble with averages. *The Milbank Quarterly*, 82:661–687, 2004.

[6] Rui Wang, Stephen Lagakos, James Ware, David Hunter, and Jeffrey Drazen. Statistics in medicine — reporting of subgroup analyses in clinical trials. *The New England Journal of Medicine*, 357:2189–94, 12 2007.

[7] Sören R Künzel, Jasjeet S Sekhon, Peter J Bickel, and Bin Yu. Metalearners for estimating heterogeneous treatment effects using machine learning. *Proceedings of the national academy of sciences*, 116(10):4156–4165, 2019.

[8] Xinkun Nie and Stefan Wager. Quasi-oracle estimation of heterogeneous treatment effects. *Biometrika*, 108(2):299–319, 2021.

[9] Yizhe Xu, Nikolaos Ignatiadis, Erik Sverdrup, Scott Fleming, Stefan Wager, and Nigam Shah. *survlearners: Metalearners for Survival Data*, 2022. R package version 0.0.1.

[10] Jackson Wright, Jeff Williamson, Paul Whelton, Joni Snyder, Kaycee Sink, Michael Rocco, David Reboussin, Mahboob Rahman, Suzanne Oparil, Cora Lewis, Paul Kimmel, Karen Johnson, Goff Jr, Lawrence Fine, Jeffrey Cutler, William Cushman, Alfred Cheung, Walter Ambrosius, Mohammad Sabati, and Kasa Niesner. A randomized trial of intensive versus standard blood-pressure control. *New England Journal of Medicine*, 373:2103–2116, 2015.

[11] William Cushman, Gregory Evans, Robert Byington, David Jr, Richard Jr, Jeffrey Cutler, Denise Simons-Morton, Jan Basile, Marshall Corson, Jeffrey Probstfield, Lois Katz, Kevin Peterson, William Friedewald, John Buse, Thomas Bigger, Hertzel Gerstein, Faramarz Ismail-Beigi, and Elias Siraj. Effects of intensive blood-pressure control in type 2 diabetes mellitus. *New England Journal of Medicine*, 362:1575–1585, 04 2010.

[12] Susan Athey and Guido Imbens. Recursive partitioning for heterogeneous causal effects: Table 1. *Proceedings of the National Academy of Sciences*, 113:7353–7360, 07 2016.

[13] Alberto Caron, Gianluca Baio, and Ioanna Manolopoulou. Estimating individual treatment effects using non-parametric regression models: A review. *Journal of the Royal Statistical Society: Series A (Statistics in Society)*, 2022.

[14] Scott Powers, Junyang Qian, Kenneth Jung, Alejandro Schuler, Nigam Shah, Trevor Hastie, and Robert Tibshirani. Some methods for heterogeneous treatment effect estimation in high-dimensions. *Statistics in Medicine*, 37, 07 2017.

[15] T. Wendling, K. Jung, A. Callahan, A. Schuler, N. Shah, and Blanca Gallego. Comparing methods for estimation of heterogeneous treatment effects using observational data from health care databases. *Statistics in Medicine*, 37, 06 2018.

[16] Avi Feller, Jared Murray, Spencer Woody, and David Yeager. Assessing treatment effect variation in observational studies: Results from a data challenge. *Observational Studies*, 5:21–35, 01 2019.

[17] Yuta Saito, Hayato Sakata, and Kazuhide Nakata. Doubly robust prediction and evaluation methods improve uplift modeling for observational data. In *Proceedings of the 2019 SIAM International Conference on Data Mining*, pages 468–476. SIAM, 2019.

[18] Alicia Curth and Mihaela van der Schaar. Nonparametric estimation of heterogeneous treatment effects: From theory to learning algorithms. In *International Conference on Artificial Intelligence and Statistics*, pages 1810–1818. PMLR, 2021.

[19] Alicia Curth, David Svensson, Jim Weatherall, and Mihaela van der Schaar. Really doing great at estimating CATE? a critical look at ML benchmarking practices in treatment effect estimation. In *Thirty-fifth Conference on Neural Information Processing Systems Datasets and Benchmarks Track*, 2021.

[20] Daniel Jacob. CATE meets ML–The Conditional Average Treatment Effect and Machine learning. *arXiv preprint arXiv:2104.09935*, 2021.

[21] Michael C Knaus, Michael Lechner, and Anthony Strittmatter. Machine learning estimation of heterogeneous causal effects: Empirical monte carlo evidence. *The Econometrics Journal*, 24(1):134–161, 2021.

[22] Yaobin Ling, Pulakesh Upadhyaya, Luyao Chen, Xiaoqian Jiang, and Yejin Kim. Heterogeneous treatment effect estimation using machine learning for healthcare application: tutorial and benchmark. *arXiv preprint arXiv:2109.12769*, 2021.

[23] Andrea A. Naghi and Christian P. Wirths. Finite sample evaluation of causal machine learning methods: Guidelines for the applied researcher. Tinbergen Institute Discussion Paper TI 2021-090/III, 2021.

[24] Weijia Zhang, Jiuyong Li, and Lin Liu. A unified survey of treatment effect heterogeneity modelling and uplift modelling. *ACM Computing Surveys (CSUR)*, 54(8):1–36, 2021.

[25] Gabriel Okasa. Meta-Learners for estimation of causal effects: Finite sample cross-fit performance. *arXiv preprint arXiv:2201.12692*, 2022.

[26] Susan Athey, Julie Tibshirani, and Stefan Wager. Generalized random forests. *The Annals of Statistics*, 47(2):1148–1178, 2019.

[27] P Richard Hahn, Jared S Murray, and Carlos M Carvalho. Bayesian regression tree models for causal inference: Regularization, confounding, and heterogeneous effects (with discussion). *Bayesian Analysis*, 15(3):965–1056, 2020.

[28] JL Hill. Bayesian nonparametric modeling for causal inference. *Journal of Computational and Graphical Statistics*, 20(1):217–240, 2011.

[29] Dylan J Foster and Vasilis Syrgkanis. Orthogonal statistical learning. *arXiv preprint arXiv:1901.09036*, 2019.

[30] Stefan Wager and Susan Athey. Estimation and inference of heterogeneous treatment effects using random forests. *Journal of the American Statistical Association*, 113(523):1228–1242, 2018.

[31] Jiabei Yang and Jon Steingrimsson. Causal interaction trees: Finding subgroups with heterogeneous treatment effects in observational data. *Biometrics*, 2021.

[32] Zijun Gao and Yanjun Han. Minimax optimal nonparametric estimation of heterogeneous treatment effects. *Advances in Neural Information Processing Systems*, 33:21751–21762, 2020.

[33] Edward H Kennedy. Optimal doubly robust estimation of heterogeneous causal effects. *arXiv preprint arXiv:2004.14497*, 2020.

[34] Edward H Kennedy, Sivaraman Balakrishnan, and Larry Wasserman. Minimax rates for heterogeneous causal effect estimation. *arXiv preprint arXiv:2203.00837*, 2022.

[35] Szymon Jaroszewicz and Piotr Rzepakowski. Uplift modeling with survival data. In *ACM SIGKDD Workshop on Health Informatics (HI-KDD–14), New York City*, 2014.

[36] Weijia Zhang, Thuc Duy Le, Lin Liu, Zhi-Hua Zhou, and Jiuyong Li. Mining heterogeneous causal effects for personalized cancer treatment. *Bioinformatics*, 33(15):2372–2378, 2017.

[37] Nicholas C Henderson, Thomas A Louis, Gary L Rosner, and Ravi Varadhan. Individualized treatment effects with censored data via fully nonparametric bayesian accelerated failure time models. *Biostatistics*, 21(1):50–68, 2020.

[38] Sami Tabib and Denis Larocque. Non-parametric individual treatment effect estimation for survival data with random forests. *Bioinformatics*, 36(2):629–636, 2020.

[39] Jie Zhu and Blanca Gallego. Targeted estimation of heterogeneous treatment effect in observational survival analysis. *Journal of Biomedical Informatics*, 107:103474, 2020.

[40] Yifan Cui, Michael R Kosorok, Erik Sverdrup, Stefan Wager, and Ruoqing Zhu. Estimating heterogeneous treatment effects with right-censored data via causal survival forests. *arXiv preprint arXiv:2001.09887*, 2022.

[41] Paidamoyo Chapfuwa, Serge Assaad, Shuxi Zeng, Michael J Pencina, Lawrence Carin, and Ricardo Henao. Enabling counterfactual survival analysis with balanced representations. In *Proceedings of the Conference on Health, Inference, and Learning*, pages 133–145, 2021.

[42] Alicia Curth, Changhee Lee, and Mihaela van der Schaar. SurvITE: Learning heterogeneous treatment effects from time-to-event data. *Advances in Neural Information Processing Systems*, 34, 2021.

[43] Liangyuan Hu, Jung-Yi Lin, Keith Sigel, and Minal Kale. Estimating heterogeneous survival treatment effects of lung cancer screening approaches: A causal machine learning analysis. *Annals of Epidemiology*, 62:36–42, 2021.

[44] Liangyuan Hu, Jiayi Ji, and Fan Li. Estimating heterogeneous survival treatment effect in observational data using machine learning. *Statistics in Medicine*, 40(21):4691–4713, 2021.

[45] Tony Duan, Pranav Rajpurkar, Dillon Laird, Andrew Y Ng, and Sanjay Basu. Clinical value of predicting individual treatment effects for intensive blood pressure therapy: a machine learning experiment to estimate treatment effects from randomized trial data. *Circulation: Cardiovascular Quality and Outcomes*, 12(3):e005010, 2019.

[46] Jeroen Hoogland, Joanna IntHout, Michail Belias, Maroeska M Rovers, Richard D Riley, Frank E. Harrell Jr, Karel GM Moons, Thomas PA Debray, and Johannes B Reitsma. A tutorial on individualized treatment effect prediction from randomized trials with a binary endpoint. *Statistics in Medicine*, 40(26):5961–5981, 2021.

[47] David M. Kent, Jessica K. Paulus, David van Klaveren, Ralph B. D'Agostino, Steve Goodman, Rodney A. Hayward, John P. A. Ioannidis, Bray Patrick-Lake, Sally C. Morton, Michael J. Pencina, Gowri Raman, Joseph Ross, Harry P. Selker, Ravi Varadhan, Andrew Julian Vickers, John B. Wong, and Ewout Willem Steyerberg. The predictive approaches to treatment effect heterogeneity (PATH) statement. *Annals of Internal Medicine*, 172(1):35–45, 2020.

[48] J. Neyman. Sur les applications de la th'eorie des probabilit'es aux experiences agricoles: Essai des principes. *Roczniki Nauk Rolniczych*, 10, 01 1923.

[49] Donald B Rubin. Estimating causal effects of treatments in randomized and nonrandomized studies. *Journal of Educational Psychology*, 66(5):688, 1974.

[50] Paul Rosenbaum and Donald Rubin. The central role of the propensity score in observational studies for causal effects. *Biometrika*, 70:41–55, 04 1983.

[51] T. Hastie, R. Tibshirani, and J. Friedman. *The Elements of Statistical Learning: Data Mining, Inference, and Prediction, Second Edition*. Springer Series in Statistics. Springer New York, 2009.

[52] Hans van Houwelingen and Hein Putter. *Dynamic prediction in clinical survival analysis*. CRC Press, 2011.

[53] Frank E Harrell. *Regression modeling strategies: with applications to linear models, logistic and ordinal regression, and survival analysis*, volume 3. Springer, 2015.

[54] Gary S Collins, Johannes B Reitsma, Douglas G Altman, and Karel GM Moons. Transparent reporting of a multivariable prediction model for individual prognosis or diagnosis (TRIPOD): the TRIPOD statement. *Journal of British Surgery*, 102(3):148–158, 2015.

[55] Hemant Ishwaran, Udaya B Kogalur, Eugene H Blackstone, and Michael S Lauer. Random survival forests. *The Annals of Applied Statistics*, 2(3):841–860, 2008.

[56] L Breiman. Random forests. *Machine Learning*, pages 45: 5–32, 2001.

[57] Yi Lin and Yongho Jeon. Random forests and adaptive nearest neighbors. *Journal of the American Statistical Association*, 101(474):578–590, 2006.

[58] Julie Tibshirani, Susan Athey, Erik Sverdrup, and Stefan Wager. *grf: Generalized Random Forests*, 2022. R package version 2.1.0.

[59] Robert Tibshirani. The lasso method for variable selection in the Cox model. *Statistics in Medicine*, 16(4):385–395, 1997.

[60] Jelle J Goeman. l_1 penalized estimation in the Cox proportional hazards model. *Biometrical Journal*, 52(1):70–84, 2010.

[61] David R Cox. Regression models and life-tables. *Journal of the Royal Statistical Society: Series B (Methodological)*, 34(2):187–202, 1972.

[62] Hans C Van Houwelingen, Tako Bruinsma, Augustinus AM Hart, Laura J Van't Veer, and Lodewyk FA Wessels. Cross-validated Cox regression on microarray gene expression data. *Statistics in Medicine*, 25(18):3201–3216, 2006.

[63] Noah Simon, Jerome Friedman, Trevor Hastie, and Rob Tibshirani. Regularization paths for Cox's proportional hazards model via coordinate descent. *Journal of Statistical Software*, 39(5):1, 2011.

[64] Michael Kohler, Kinga Máthé, and Márta Pintér. Prediction from randomly right censored data. *Journal of Multivariate Analysis*, 80(1):73–100, 2002.

[65] Mark J Van der Laan and James M Robins. *Unified methods for censored longitudinal data and causality*, volume 5. Springer, 2003.

[66] Humza Haider, Bret Hoehn, Sarah Davis, and Russell Greiner. Effective ways to build and evaluate individual survival distributions. *Journal of Machine Learning Research*, 21(85):1–63, 2020.

[67] Anastasios A Tsiatis. Semiparametric theory and missing data. 2006.

[68] James Robins, A G Rotnitzky, and Lue Zhao. Estimation of regression coefficients when some regressors are not always observed. *Journal of The American Statistical Association*, 89:846–866, 09 1994.

[69] James Signorovitch. *Identifying informative biological markers in high-dimensional genomic data and clinical trials*. PhD thesis, Harvard University, 2007.

[70] Lu Tian, Ash Alizadeh, Andrew Gentles, and Robert Tibshirani. A simple method for estimating interactions between a treatment and a large number of covariates. *Journal of the American Statistical Association*, 109, 2014.

[71] Daniel G Horvitz and Donovan J Thompson. A generalization of sampling without replacement from a finite universe. *Journal of the American Statistical Association*, 47(260):663–685, 1952.

[72] Robert Tibshirani. Regression shrinkage and selection via the lasso. *Journal of the Royal Statistical Society. Series B (Methodological)*, 58(1):267–288, 1996.

[73] Peter Robinson. Root-n-consistent semiparametric regression. *Econometrica*, 56:931–54, 02 1988.

[74] James M Robins. Optimal structural nested models for optimal sequential decisions. In *Proceedings of the second Seattle Symposium in Biostatistics*, pages 189–326. Springer, 2004.

[75] Richard K Crump, V Joseph Hotz, Guido W Imbens, and Oscar A Mitnik. Dealing with limited overlap in estimation of average treatment effects. *Biometrika*, 96(1):187–199, 2009.

[76] Victor Chernozhukov, Denis Chetverikov, Mert Demirer, Esther Duflo, Christian Hansen, Whitney Newey, and James Robins. Double/debiased machine learning for treatment and structural parameters. *The Econometrics Journal*, 21(1):C1–C68, 2018.

[77] Jerome Friedman, Trevor Hastie, Rob Tibshirani, Balasubramanian Narasimhan, Kenneth Tay, Noah Simon, Junyang Qian, and James Yang. *glmnet: Lasso and Elastic-Net Regularized Generalized Linear Models*, 2022. R package version 4.1.4.

[78] Trevor Hastie and Robert Tibshirani. Varying-coefficient models. *Journal of the Royal Statistical Society: Series B (Methodological)*, 55(4):757–779, 1993.

[79] Susan Athey and Stefan Wager. Estimating treatment effects with causal forests: An application. *Observational Studies*, 5(2):37–51, 2019.

[80] Tianqi Chen and Carlos Guestrin. XGBoost: A scalable tree boosting system. In *Proceedings of the 22nd ACM SIGKDD International Conference on Knowledge Discovery and Data Mining*, pages 785–794, 2016.

[81] Aaron Baum, Joseph Scarpa, Emilie Bruzelius, Ronald Tamler, Sanjay Basu, and James Faghmous. Targeting weight loss interventions to reduce cardiovascular complications of type 2 diabetes: a machine learning-based post-hoc analysis of heterogeneous treatment effects in the look ahead trial. *The Lancet Diabetes and Endocrinology*, 5(10):808–815, 2017.

[82] Ann Lazar, Bernard Cole, Marco Bonetti, and Richard Gelber. Evaluation of treatment-effect heterogeneity using biomarkers measured on a continuous scale: Subpopulation treatment effect pattern plot. *Journal of Clinical Oncology: Official Journal of the American Society of Clinical Oncology*, 28:4539–44, 2010.

[83] Adam Bress, Tom Greene, Catherine Derington, Jincheng Shen, Yizhe Xu, Yiyi Zhang, Jian Ying, Brandon Bellows, William Cushman, Paul Whelton, Nicholas Pajewski, David Reboussin, Srinivasan Beddu, Rachel Hess, Jennifer Herrick, Zugui Zhang, Paul Kolm, Robert Yeh, Sanjay Basu, and Andrew Moran. Patient selection for intensive blood pressure management based on benefit and adverse events. *Journal of the American College of Cardiology*, 77:1977–1990, 04 2021.

[84] David Kent, Jason Nelson, Peter Rothwell, Douglas Altman, and Rodney Hayward. Risk and treatment effect heterogeneity: Re-analysis of individual participant data from 32 large clinical trials. *International Journal of Epidemiology*, 45:dyw118, 07 2016.

[85] Kosuke Imai and Marc Ratkovic. Estimating treatment effect heterogeneity in randomized program evaluation. *The Annals of Applied Statistics*, 7(1):443–470, 2013.

[86] Alfred K. Cheung, Mahboob Rahman, David M. Reboussin, Timothy E. Craven, Tom Greene, Paul L. Kimmel, William C. Cushman, Amret T. Hawfield, Karen C. Johnson, Cora E. Lewis, Suzanne Oparil, Michael V. Rocco, Kaycee M. Sink, Paul K. Whelton, Jackson T. Wright, Jan Basile, Srinivasan Beddhu, Udayan Bhatt, Tara I. Chang, Glenn M. Chertow, Michel Chonchol, Barry I. Freedman, William Haley, Joachim H. Ix, Lois A. Katz, Anthony A. Killeen, Vasilios Papademetriou, Ana C. Ricardo, Karen Servilla, Barry Wall, Dawn Wolfgram, and Jerry Yee. Effects of intensive BP control in CKD. *Journal of the American Society of Nephrology*, 28(9):2812–2823, 2017.

[87] Ara H Rostomian, Maxine C Tang, Jonathan Soverow, and Daniel R Sanchez. Heterogeneity of treatment effect in sprint by age and baseline comorbidities: The greatest impact of intensive blood pressure treatment is observed among younger patients without CKD or CVD and in older patients with CKD or CVD. *The Journal of Clinical Hypertension*, 22(9):1723–1726, 2020.

[88] Joseph Scarpa, Emilie Bruzelius, Patrick Doupe, Matthew Le, James Faghmous, and Aaron Baum. Assessment of risk of harm associated with intensive blood pressure management among patients with hypertension who smoke: a secondary analysis of the systolic blood pressure intervention trial. *JAMA network open*, 2(3):e190005–e190005, 2019.

[89] Yaqian Wu, Jianling Bai, Mingzhi Zhang, Fang Shao, Honggang Yi, Dongfang You, and Yang Zhao. Heterogeneity of treatment effects for intensive blood pressure therapy by individual components of FRS: An unsupervised data-driven subgroup analysis in SPRINT and ACCORD. *Frontiers in cardiovascular medicine*, 9, 2022.

[90] Krishna K Patel, Suzanne V Arnold, Paul S Chan, Yuanyuan Tang, Yashashwi Pokharel, Philip G Jones, and John A Spertus. Personalizing the intensity of blood pressure control: modeling the heterogeneity of risks and benefits from SPRINT (Systolic Blood Pressure Intervention Trial). *Circulation: Cardiovascular Quality and Outcomes*, 10(4):e003624, 2017.

[91] Sanjay Basu, Jeremy B Sussman, Joseph Rigdon, Lauren Steimle, Brian T Denton, and Rodney A Hayward. Benefit and harm of intensive blood pressure treatment: derivation and validation of risk models using data from the SPRINT and ACCORD trials. *PLoS Medicine*, 14(10):e1002410, 2017.

[92] Hui Zou and Trevor Hastie. Regularization and variable selection via the elastic net. *Journal of the Royal Statistical Society: Series B (Statistical Methodology)*, 67(2):301–320, 2005.

[93] Steve Yadlowsky, Scott Fleming, Nigam Shah, Emma Brunskill, and Stefan Wager. Evaluating treatment prioritization rules via rank-weighted average treatment effects. *arXiv preprint arXiv:2111.07966*, 2021.

[94] Robert A Phillips, Jiaqiong Xu, Leif E Peterson, Ryan M Arnold, Joseph A Diamond, and Adam E Schussheim. Impact of cardiovascular risk on the relative benefit and harm of intensive treatment of hypertension. *Journal of the American College of Cardiology*, 71(15):1601–1610, 2018.

[95] Yizhe Xu and Steve Yadlowsky. Calibration error for heterogeneous treatment effects. In *International Conference on Artificial Intelligence and Statistics*, pages 9280–9303. PMLR, 2022.

[96] Lu Tian, Hua Jin, Hajime Uno, Ying Lu, Bo Huang, Keaven Anderson, and Leejen Wei. On the empirical choice of the time window for restricted mean survival time. *Biometrics*, 76, 02 2020.

22

Why Machine Learning Cannot Ignore Maximum Likelihood Estimation

Mark J. van der Laan and Sherri Rose

CONTENTS

DOI: 10.1201/9781003102670-22

22.1 Introduction

The Avalanche of Machine Learning

The growth of machine learning as a field has been accelerating with increasing interest and publications across fields, including statistics, but predominantly in computer science. How can we parse this vast literature for developments that exemplify the necessary rigor? How many of these manuscripts incorporate foundational theory to allow for statistical inference? Which advances have the greatest potential for impact in practice? One could posit many answers to these queries. Here, we assert that one essential idea is for machine learning to integrate maximum likelihood for estimation of functional parameters, such as prediction functions and conditional densities.

The statistics literature proposed the familiar maximum likelihood estimators (MLEs) for parametric models and established that these estimators are asymptotically linear such that their $\sqrt{n}$-scaled and mean-zero-centered version is asymptotically normal [1]. This allowed the construction of confidence intervals and formal hypothesis testing. However, due to the restrictive form of these models, these parametric model-based MLEs target a projection of the true density on the parametric model that is hard to interpret. In response to this concern with misspecified parametric models, a rich statistical literature has studied so-called sieve-based or sieve MLEs involving specifying a sequence of parametric models that grow toward the desired true statistical model [2, 3]. Such a sieve will rely on a tuning parameter that will then need to be selected with some method such as cross-validation.

Simultaneously, and to a large degree independently from the statistics literature, the computer science literature developed a rich set of tools for constructing data-adaptive estimators of functional parameters, such as a density of the data, although this literature mostly focused on learning prediction functions. This has resulted in a wealth of machine learning algorithms using a variety of strategies to learn the target function from the data. In addition, a more recent literature in statistics established super learning based on cross-validation as a general approach to optimally combine a library of such candidate machine learning algorithms into an ensemble that performs asymptotically as well as an oracle ensemble under specified conditions [4, 5]. Although the latter framework is optimal for fast learning of the target function, it lacks formal statistical inference for smooth features of the target function and for the target function itself.

In this chapter, we will argue that, in order to preserve statistical inference, we should preference sieve MLEs as much as possible. If the target function is not a data density, one can use, more generally, minimum loss estimation with an appropriate loss function. For example, if the goal is to learn the conditional mean of an outcome, one could use the squared-error loss and use minimum least squares estimation. Therefore, in this chapter, the abbreviation MLE also represents the more general minimum loss estimator.

We will demonstrate the power of sieve MLEs with a particular sieve MLE that also relies on the least absolute shrinkage and selection operator (LASSO) [6] and is termed the highly adaptive LASSO minimum loss estimator (HAL-MLE) [7,8]. We will argue that the HAL-MLE is a particularly powerful sieve MLE, generally theoretically superior to other types of sieve MLEs. Moreover, for obtaining statistical inference for estimands that aim to equal or approximate a causal quantity defined in an underlying causal model, we discuss the combination of HAL-MLE with targeted maximum likelihood estimators (TMLEs) [9–11] in HAL-TMLEs as well as using undersmoothed HAL-MLE by itself as a powerful statistical approach [8, 10, 12, 13]. HALs can be used for the initial estimator in TMLEs as well as for nuisance functions representing the treatment and missingness/censoring mechanism. We will also clarify why many machine learning algorithms are not suited for statistical inference by having deviated from sieve MLEs. This chapter is a compact summary of recent and ongoing research; further background and references can be found in the literature cited. Sections 22.2–22.4 and 22.7 are more technical in nature, introducing core definitions and properties. Some readers may be interested in jumping to the material in Sections 22.5, 22.6, and 22.8 for technical but narrative discussions on contrasting HAL-MLE with other estimators as well as implementation of HAL-MLE and HAL-TMLE.

22.2 Nonparametric MLE and Sieve MLE

Before introducing the HAL-MLE, we describe the technical foundations of nonparametric and sieve MLEs.

22.2.1 Nonparametric MLE

We have observational unit on which we observe a random vector of measurements O with probability distribution P_0 (e.g., $O = (W, A, Y)$ with covariates W, intervention or treatment A, and outcome Y). Consider the case that we observe n i.i.d. copies $O_1, \dots, O_n$ from the common probability measure P_0 and suppose that we know that P_0 is an element of a set $\mathcal{M}$ of possible probability distributions, which is called the statistical model. Here, the underscore 0 indicates the true unknown probability measure. In addition, suppose that our target functional parameter is defined by a mapping $Q : \mathcal{M} \to D[0,1]^d$ from the probability measures into the space of d-variate real-valued cadlag functions [14].

Let $\mathcal{Q} = \{Q(P) : P \in \mathcal{M}\}$ be the parameter space for this functional parameter. Moreover, we assume that the target functional estimand $Q(P_0)$ can be characterized as the minimizer: $Q_0 = \arg\min_{Q \in \mathcal{Q}} P_0 L(Q)$, for some loss function $L(Q, O)$ also denoted with $L(Q)$ when viewed as a function of O. A first natural attempt for estimation of Q_0 would be to define the MLE over the entire parameter space:

$$\hat{Q}(P_n) = \arg\min_{Q \in \mathcal{Q}} P_n L(Q),$$

where we use empirical process notation $P_n f \equiv 1/n \sum_{i=1}^{n} f(O_i)$.

> However, for realistic statistical models $\mathcal{M}$, the parameter space is generally infinite dimensional and too flexible such that the minimizer of the empirical risk will overfit to the data and results in an ill-defined $\hat{Q}(P_n)$ or a statistically poor estimator.

22.2.2 Sieve MLE

A sieve MLE recognizes this need to regularize the nonparametric MLE and proposes a sequence $\mathcal{Q}_\lambda \subset \mathcal{Q}$ of subspaces indexed by some Euclidean-valued tuning parameter $\lambda \in \mathcal{I} \subset \mathbb{R}^k$ that measures the complexity of the space $\mathcal{Q}_\lambda$ [2, 3]. In addition, the sieve is able to approximate the target function such that for a sequence of values λ_j, $\mathcal{Q}_{\lambda_j}$ approximates the entire parameter space $\mathcal{Q}$ as j approximates infinity. These spaces are restricted in a manner so that an MLE: $Q_{n,\lambda} = \arg\min_{Q \in \mathcal{Q}(\lambda)} P_n L(Q)$, is well defined for all λ, or, at least, for a range $\mathcal{I}_n$ of λ values that approximates the complete index set $\mathcal{I}$ as sample size n grows. We can then define a cross-validation selector $\lambda_{n,cv} = \arg\min_\lambda 1/V \sum_v P^1_{n,v} L(\hat{Q}_\lambda(P_{n,v}))$ involving a V-fold sample splitting of the sample of n observations where the v-th sample split yields an empirical distribution $P^1_{n,v}$ and $P_{n,v}$ of the validation and training sample, respectively. Uniform bounds on this loss function implies that this cross-validation selector performs as well as the oracle selector: $\lambda_{0,n} = \arg\min_\lambda 1/V \sum_v P_0 L(\hat{Q}_\lambda(P_{n,v}))$ [15–17].

22.2.3 Score equations solved by sieve MLE

A key property of a sieve MLE is that it solves score equations. That is, one can construct a family of one dimensional paths $\{Q^h_{n,\lambda,\delta} : \delta\} \subset \mathcal{Q}$ through $Q_{n,\lambda}$ at $\delta = \delta_0 \equiv 0$ indexed by a direction $h \in \mathcal{H}$, and we have

$$0 = \frac{d}{d\delta_0} P_n L(Q^h_{n,\lambda,\delta_0}) \text{ for all paths } h \in \mathcal{H}.$$

We can define a path-specific score: $A_{Q_{n,\lambda}}(h) \equiv \frac{d}{d\delta_0} L(Q_{n,\lambda,\delta_0})$, for all h. This mapping $A_{Q_{n,\lambda}}(h)$ will generally be a linear operator in h and can therefore be extended to a linear operator on a Hilbert space generated or spanned by these directions $\mathcal{H}$. One will then also have that $P_n A_{Q_{n,\lambda}}(h) = 0$ for any h in the linear span of $\mathcal{H}$, thereby obtaining that $Q_{n,\lambda}$ solves a space of scores. For scores $S_{Q_{n,\lambda}}$ that can be approximated by scores $A_{Q_{n,\lambda}}(h)$ for certain h, this might provide the basis for showing that $P_n S_{Q_{n,\lambda}} = o_P(n^{-1/2})$ is solved up to the desired approximation. In particular, one can apply this class of score equations at $\lambda = \lambda_{n,cv}$ to obtain the score equations solved by the cross-validated sieve MLE $Q_{n,\lambda_{n,cv}}$.

22.3 Special Sieve MLE: HAL-MLE

We now introduce a special sieve MLE, the HAL-MLE. The HAL-MLE is generally theoretically superior to other types of sieve MLEs, with its statistical properties presented in the subsequent section.

22.3.1 Definition of HAL-MLE

A particular sieve MLE is given by the HAL-MLE. It starts with a representation $Q(x) = Q(0) + \sum_{s \subset \{1,\ldots,d\}} \int_{(0,x_s]} Q(du_s, 0_{-s})$, where $x_s = (x(j) : j \in s)$, $0_{-s} = (0(j) : j \notin s)$ and $Q(du_s, 0_{-s})$ is a generalized difference for an infinitesimal small cube $(u_s, u_s + du_s] \subset \mathbb{R}^{|s|}$ in the s-dimensional Euclidean space with respect to the measure generated by the s-specific section $Q_s(u_s) \equiv Q(u_s, 0_{-s})$ that sets the coordinates outside s equal to zero [7, 8, 18].

This representation shows that any cadlag function can be represented as a linear combination of indicator basis functions $x \rightarrow I_{u_s \leq x_s}$ indexed by knot point u_s, across all subsets s of $\{1, \ldots, d\}$ with coefficients given by $Q(du_s, 0_{-s})$. In particular, if Q is such that all its sections are like discrete cumulative distribution functions, then this representation of Q is literally a finite linear combination of these indicator basis functions. We note that each basis function $I_{u_s \leq x_s}$ is a tensor product $\prod_{j \in s} I(u_j \leq x_j)$ of univariate indictor basis functions $I(u_j \leq x_j)$, functions that jump from 0 to 1 at knot point u_j. Moreover, the L_1-norm of these coefficients in this representation is the so-called sectional variation norm:

$$\|Q\|_v^* =| Q(0) | + \sum_{s\subset\{1,\ldots,d\}} | Q(du_s, 0_{-s}) | .$$

This sectional variation norm of Q represents a measure of the complexity of Q. Functions that can be represented as $Q(x) = \sum_{j=1}^d Q_j(x_j)$ would correspond with $Q(du_s, 0_{-s}) = 0$ for all subsets s of size larger than or equal to 2. Such functions would have a dramatically smaller sectional variation norm than a function that requires high-level interactions. If the dimension $d = 1$, then this sectional variation norm is just the typical variation norm of a univariate function obtained by summing the absolute values of its changes across small intervals. For dimensional $d = 2$, this is defined the same way, but now one sums so-called generalized differences $Q(dx_1, dx_2) = Q(x_1+dx_1, x_2+dx_2)-Q(x_1+dx_1, x_2)-Q(x_1, x_2+dx_2)+Q(x_1, x_2)$, representing the measure Q assigns to the small two-dimensional cube $(x, x + dx]$. Similarly, one defines generalized differences for general d-dimensional function as the measure it would give to a d-dimensional cube $(x, x + dx]$, treating the function as a d-variate cumulative distribution function.

This measure of complexity now defines the sieve $\mathcal{Q}_\lambda = \{Q \in \mathcal{Q} : \|Q\|_v^* \leq \lambda\}$ as all cadlag functions with a sectional variation norm smaller than λ. The sieve-based MLE is then defined as: $Q_{n,\lambda} = \arg\min_{Q:\|Q\|_v^*\leq\lambda} P_n L(Q)$. Using a discrete support $\{u_{s,j} : j\}$ for each s-specific section $Q(u_s, 0_{0_{-s}})$, we can represent $Q = Q_\beta = \sum_{s,j} \beta(s,j)\phi_{s,j}$ with $\phi_{s,j}(x) = I(x_s \leq u_{s,j})$, where $u_{s,j}$ represent the knot points. Thus, we obtain a standard LASSO estimator:

$$\beta_{n,\lambda} = \arg \min_{\beta,\|\beta\|_1\leq\lambda} P_n L(Q_\beta),$$

and resulting $Q_{n,\lambda} = Q_{\beta_{n,\lambda}}$. As above, λ is then selected with the cross-validation selector $\lambda_{n,cv}$. This can generally be implemented with LASSO software implementations such as `glmnet` in `R` [19]. Additionally, `HAL9001` provides implementations for linear, logistic, Cox, and Poisson regression, which also provides HAL estimators of the conditional hazards and intensities [20].

22.3.2 Score equations solved by HAL-MLE

In order to define a set of score equations solved by this HAL-MLE, we can use as paths $\beta_\delta^h(s,j) = (1 + \delta h(s,j))\beta(s,j)$ for a vector h. In order to keep the L_1-norm constant along this path, we then enforce the constraint $r(h, \beta) = 0$, where

$$r(h, \beta) \equiv h(0) | \beta(0) | + \sum_{s,j} h(s,j) | \beta(s,j) | .$$

It can be verified that now $\|\beta_\delta^h\|_1 = \|\beta\|_1$ for δ in a neighborhood of 0. Therefore, we know that the HAL-MLE solves the score equation for each of these paths:

$$P_n A_{\beta_{n,\lambda}}(h) = 0 \text{ for all } h \text{ with } r(h, \beta_{n,\lambda}) = 0.$$

This shows that the HAL-MLE solves a class of score equations. Moreover, this result can be used to prove that

$$P_n \frac{d}{d\beta_{n,\lambda}(j)} L(Q_{\beta_{n,\lambda}}) = O_P(n^{-2/3}),$$

for all j for which $\beta_{n,\lambda}(j) \neq 0$. That is, the L_1-norm constrained HAL-MLE also solves the unconstrained scores solved by the MLE over the finite linear model $\{Q_\beta : \beta(j) = 0 \text{ if } \beta_{n,\lambda}(j) = 0\}$ implied by the fit $\beta_{n,\lambda}$. By selecting the L_1-norm λ to be larger, this set of score equations approximates any score, thereby establishing that the HAL-MLE, if slightly undersmoothed, will solve score equations up to $O_P(n^{-2/3})$ uniformly over all scores. For further details we refer prior work [13].

> This capability of the HAL-MLE to solve all score equations, even uniformly over a class that will contain any desired efficient influence curve of a target parameter, provides the fundamental basis for establishing its remarkably strong asymptotic statistical performance in estimation of smooth features of Q_0 as well as of Q_0 itself.

22.4 Statistical Properties of the HAL-MLE

Having introduced the special sieve MLE, the HAL-MLE Q_{n,λ_n}, we now further enumerate its statistical properties.

22.4.1 Rate of convergence

Firstly, $d_0(Q_{n,\lambda_n}, Q_0) = O_P(n^{-2/3}(\log n)^d)$, where $d_0(Q, Q_0) = P_0L(Q) - P_0L(Q_0)$ is the loss-based dissimilarity. This generally establishes that some L^2-norm of $Q_{n,\lambda_n} - Q_0$ will be $O_P(n^{-1/3}(\log n)^{d/2})$ [8,21].

22.4.2 Asymptotic efficiency for smooth target features of target function

Let $\Psi(Q_0)$ be a pathwise differentiable target feature of Q_0 and let $D^*_{Q(P),G(P)}$ be its canonical gradient at P, possibly also depending on a nuisance parameter $G(P)$. Let $R_{P_0}(Q, Q_0) = \Psi(Q) - \Psi(Q_0) + P_0 D^*_{Q,G_0}$ be its exact second-order remainder. By selecting λ_n to be large enough so that $P_n D^*_{Q_{n,\lambda_n},G_0} = o_P(n^{-1/2})$, we obtain that $\Psi(Q_{n,\lambda_n})$ is asymptotically linear with influence curve $D^*_{Q_0,G_0}$. That is, $\Psi(Q_n)$ is an asymptotically efficient estimator of $\Psi(Q_0)$. The only conditions are that $R_{P_0}(Q_{n,\lambda_n}, Q_0) = o_P(n^{-1/2})$, where one can use that $d_0(Q_{n,\lambda_n}, Q_0) = O_P(n^{-2/3}(\log n)^d)$. Generally speaking, the latter will imply that the exact remainder has the same rate of convergence. In fact, in double robust estimation problems, this exact remainder equals zero (because it is at G_0).

22.4.3 Global asymptotic efficiency

Moreover, now consider a very large class of pathwise differentiable target features $\Psi_t(Q_0)$ indexed by some $t \in [0,1]^m$, where $D^*_{t,Q,G}$ is the canonical gradient and $R_{P_0,t}(Q, Q_0)$ is the exact remainder. Then, by undersmoothing enough so that $\sup_t \mid P_n D^*_{t,Q_{n,\lambda_n},G_0} \mid = o_P(n^{-1/2})$, that is, we use global undersmoothing, we obtain that $\Psi_t(Q_n)$ is an asymptotically linear estimator of $\Psi_t(Q_0)$ having influence curve D^*_{t,Q_0,G_0}, with a remainder $R_n(t)$ that satisfies $\sup_t \mid R_n(t) \mid = o_P(n^{-1/2})$, so that $n^{1/2}(\Psi_t(Q_n) - \Psi_t(Q_0))$ converges weakly to a Gaussian process as a random function in function space. That is, $(\Psi_t(Q_n) : t \in [0,1]^m)$ is an asymptotically efficient estimator of $(\Psi_t(Q_0) : t \in [0,1]^m)$ in a sup-norm sense. In particular, it allows construction of simultaneous confidence bands. We note that this also implies that $Q_{n,h}(x) \equiv \int Q_n(u)K_h(u-x)du$, a kernel smoother of Q_n at x is an asymptotically efficient estimator of $Q_{0,h}(x) \equiv \int Q_0(u)K_h(u-x)du$, a kernel smoother of the true target function Q_0. In fact, this holds uniformly in all x, and for a fixed h

we have weak convergence of $n^{1/2}(Q_{n,h} - Q_{0,h})$ to a Gaussian process. That is, the undersmoothed HAL-MLE is an efficient estimator of the kernel smoothed functional of Q_0, for any h.

> *This may make one wonder if $Q_n(x)$ itself is not asymptotically normally distributed as well? While not a currently solved problem, we conjecture that, indeed, $(Q_n(x) - Q_0(x))/\sigma_n(x)$ converges weakly to a normal distribution, where the rate of convergence may be as good as $n^{-1/3}(\log n)^{d/2}$. If this results holds, then the HAL-MLE also allows formal statistical inference for the function itself!*

22.4.4 Nonparametric bootstrap for inference

It has been established that the nonparametric bootstrap consistently estimates the multivariate normal limit distribution of the plug-in HAL-MLE for a vector of target features [22]. In fact, it will consistently estimate the limit of a Gaussian process of the plug-in HAL-MLE for an infinite class of target features as discussed above. The approach is computationally feasible and still correct by only bootstrapping the LASSO-based fit of the model that was selected by the HAL-MLE; the bootstrap is only refitting a high-dimensional linear regression model with at most n basis functions (and generally much smaller than n). This allows one to obtain inference that also attains second-order behavior. In particular, obtaining the bootstrap distribution for each L_1-norm larger than or equal to the cross-validation selector of the L_1-norm and selecting the L_1-norm at which the confidence intervals plateau provides a robust *finite sample* inference procedure [22].

It will also be of interest to understand the behavior of the nonparametric bootstrap in estimating the distribution of the HAL estimator of the target function itself (say, at a point). Before formally addressing this, we need to first establish that the HAL-MLE is asymptotically normal as conjectured in a previous subsection.

22.4.5 Higher-order optimal for smooth target features

In recent work, higher-order TMLEs were proposed that target extra score equations beyond the first-order canonical gradient of the target feature, which are selected so that the exact second-order remainder for the plug-in TMLE will become a higher-order difference (say, third-order for the second order TMLE) [23]. Because an undersmoothed HAL-MLE also solves these extra score equations, it follows that the plug-in HAL-MLE, when using some undersmoothing, is not just asymptotically efficient but also has a reduced exact remainder analogue to the higher-order TMLE. (More on TMLEs in later sections.)

The key lesson is that the exact remainder in an expansion of a plug-in estimator can be represented as an expectation of a score: $R_0((Q,G),(Q_0,G_0)) = P_0 A_Q(h_{P,P_0})$, for some direction h_{P,P_0}. For example, in the nonparametric estimation of $E(Y_1) = E[E(Y \mid A = 1, W)]$, where $\bar{Q} = E(Y|A = 1, W)$ and $\bar{G}(W) = E(A \mid W)$, we have $R_0((Q,G),(Q_0,G_0)) = P_0(\bar{Q} - \bar{Q}_0)(\bar{G} - \bar{G}_0)/\bar{G}$, which can be written as $-E_0 I(A = 1)/\bar{G}_0(W)(\bar{G} - \bar{G}_0)/\bar{G}(Y - \bar{Q})$, which is an expectation of a score of the form $h_{G,G_0}(A,W)(Y - \bar{Q})$. The HAL-MLE $\bar{Q}_n$ approximately solves $P_n h_{G_n,G_0}(Y - \bar{Q}_n)$ for some approximation G_n of G_0, thereby reducing the remainder to $(P_n - P_0)h_{G_n,G_0}(Y - \bar{Q}_n)$. The latter is a term that is $o_P(n^{-1/2})$ – only relying on the consistency of G_n – not even a rate is required. Higher-order TMLE directly targets these scores to map the remainder in higher and higher order differences, but the undersmoothed HAL-MLE is implicitly doing the same (although not solving the score equations exactly).

> This demonstrates that solving score equations has enormous implications for first- and higher-order behavior of a plug-in estimator. Typical machine learning algorithms are generally not

tailored to solve score equations, and, thereby, will not be able to achieve such statistical performance for their plug-in estimator. In fact, most machine learning algorithms are not grounded in any asymptotic limit distribution theory.

22.5 Contrasting HAL-MLE with Other Estimators

One might wonder: what is particularly special about this sieve in the HAL-MLE relative to other sieve MLEs, such as those using Fourier basis or polynomial basis, wavelets, or other sequences of parametric models. Also, why might we prefer it over other general machine learning algorithms?

22.5.1 HAL-MLE vs. other sieve MLE

The simple answer is that these sieve MLEs generally do not have the (essentially) dimension-free/smoothness-free rate of convergence $n^{-1/3}(\log n)^{d/2}$, but instead their rates of convergence heavily rely on assumed smoothness. HAL-MLE uses a global complexity property, the sectional variation norm, rather than relying on local smoothness. The global bound on the sectional variation norm provides a class of cadlag functions $\mathcal{F}$ that has a remarkably nice covering number $\log N(\epsilon, \mathcal{F}, L^2) \lesssim 1/\epsilon(\log(1/\epsilon))^{2d-1}$, hardly depending on the dimension d. Due to this covering number, the HAL-MLE has this powerful rate of convergence – only assuming that the true target function is a cadlag function with finite sectional variation norm.

A related advantage of the special HAL sieve is that the union of all indicator basis functions is a small Donsker class, even though it is able to span any cadlag function. Most sets of basis functions include "high frequency" type basis functions that have a variation norm approximating infinity. As a consequence, these basis functions do not form a nice Donsker class. In particular, this implies that the HAL-MLE does not overfit, as long as the sectional variation norm is controlled. The fact that HAL-MLE itself is situated in this Donsker class also means that the efficient influence curves at HAL-MLE fits will fall in a similar size Donsker class. As a consequence, the Donsker class condition for asymptotic efficiency of plug-in MLE and TMLE is naturally satisfied when using the HAL-MLE, while other sieve-based estimators easily cause a violation of the Donsker class condition.

This same powerful property of the Donsker class spanned by these indicator basis functions also allows one to prove that nonparametric bootstrap works for the HAL-MLE, while, generally speaking, the nonparametric bootstrap generally fails to be consistent for machine learning algorithms.

Another appealing feature of the HAL sieve is that it is only indexed by the L_1-norm, while many sieve MLEs rely on a precise specification of the sequence of parametric models that grow in dimension. It should be expected that the choice of this sequence can have a real impact in practice. HAL-MLE does not rely on an ordering of basis functions, but rather it just relies on a complexity measure. For each value of the complexity measure, it includes all basis functions and represents an infinite dimensional class of functions rich enough to approximate any function with complexity satisfying this bound. A typical sieve MLE can only approximate the true target function, while the HAL-MLE includes the true target function when the sectional variation norm bound exceeds the sectional variation norm of the true target function.

As mentioned, the HAL-MLE solves the regular score equations from the data-adaptively selected HAL-model at rate $O_P(n^{-2/3})$. As a consequence, the HAL-MLE is able to uniformly solve the class of all score equations – only restricted by some sectional variation norm bound, where this bound can go to infinity as sample size increases. This strong capability for solving score equations appears to be unique for HAL-MLE relative to other sieve-based estimators. It may be mostly due to actually being an MLE over an infinite function class (for a particular variation norm bound). We also note that a parametric model-based sieve MLE would be forced to select a small dimension to avoid overfitting. However, the HAL-MLE adaptively selects such a model among all possible basis functions, and the dimension of this data-adaptively selected model will generally be larger. The latter is due to the HAL-MLE only being an L_1-regularized MLE for that adaptively selected parametric model, and thereby does not overfit the score equations, while the typical sieve MLE would represent a full MLE for the selected model.

By solving many more score equations *approximately* HAL-MLE can span a much larger space of scores than a sieve MLE that solves many fewer score equations perfectly.

22.5.2 HAL-MLE vs. general machine learning algorithms

Sieve MLEs solve score equations and, thereby, are able to approximately solve a class of score equations – possibly enough to approximate efficient influence curves of the target features of interest, and thus be asymptotically efficient for these target features. We discussed how these different sieve MLE can still differ in their performance in solving score equations, and that the HAL-MLE appears to have a unique strategy by choice of basis functions and measure of complexity. Thus, this gives it a distinct capability to solve essentially all score equations at a desired approximation error.

Many machine learning algorithms fail to be an MLE over any subspace of the parameter space. Such algorithms will have poor performance in solving score equations. As a consequence, they will not result in asymptotically linear plug-in estimators and will generally be overly biased and nonrobust.

22.6 Considerations for Implementing HAL-MLE

Having presented core definitions, properties, and comparisons, we now turn to some additional considerations for implementing HAL-MLEs.

22.6.1 HAL-MLE provides interpretable machine learning

The role of interpretable algorithms is a major consideration for many applications [24]. Is HAL-MLE interpretable? The HAL-MLE is a finite linear combination of indicator basis functions, so-called tensor products of zero-order splines, and the number of basis functions is generally significantly smaller than n. HAL-MLE has also been generalized to higher-order splines so that it is forced to have a certain smoothness. These higher-order spline HAL-MLE have the same statistical properties as reviewed above, as long as the true function satisfies the enforced level of smoothness. Such smooth HAL-MLE can be expected to result in even sparser fits, e.g., a smooth monotone function can be fitted with a few first-order splines (piecewise linear) while it needs relatively many knot points when fitting with a piecewise constant function. Either way, the HAL-MLE is a closed form object that can be evaluated and is thus interpretable. Therefore, the HAL-MLE has the potential to play a key role in interpretable machine learning.

22.6.2 Estimating the target function itself

HAL-MLE requires selecting the set of knot points and, thereby, the collection of spline basis functions. The largest set of knot points we have recommended (and suffices for obtaining the nonparametric rate of convergence $n^{-1/3}(\log n)^{d/2}$) is given by: $\{X_{s,i} : i = 1, \ldots, n, s \subset \{1, \ldots, d\}\}$, which corresponds (for continuous covariates) to $N = n(2^d - 1)$ basis functions. The LASSO-based fit will then select a relatively small subset ($<< n$) of this user-supplied collection.

Rather than selecting this full set of basis functions, one can incorporate model assumptions. This could include only selecting up to two-way tensor products, ranking the basis functions by their sparsity (i.e., proportion of 1's among the n observations) and selecting the top k, and specifying a specific additive model using standard `glm`-formula notation, such as $f(x_1, x_2, x_3) = f_1(x_1) + f_2(x_2, x_3) + f_3(x_3)$, and selecting the knot points accordingly. In particular, one could use some `glmnet` fit using main terms and standard interactions to decide on this type of additive model. In addition, one can specify that the coefficients of a certain set of basis functions should be nonnegative and others should be non-positive, thereby enforcing monotonicity of functions in the additive model. Finally, one can select among zero order and more generally k-th order spline basis functions, thereby specifying a desired smoothness of the HAL-MLE.

To reduce the computational burden of the implementation of HAL-MLE, one can randomly subset from a large set of basis functions or subset in a deterministic manner.

For example, for a continuous covariate X_j, rather than using as knot points $X_{j,i}$, $i = 1, \ldots, n$, one selects only five knot points – the five quantiles of the n observations $X_{j,i}$. In this manner, the number of basis functions will not grow linearly in n, it is now growing by a fixed factor 5. Similarly, the above restriction on the degree of the tensor products to only 2 reduces the 2^d-factor to d^2. Therefore, such choices reduces the total number of basis functions from $n(2^d - 1)$ to around $5 * d^2$ basis functions.

In certain cases, one might view some of these restrictions as a specification of the actual statistical model. For example, the statistical model might assume that Q_0 is an additive model including any one-way, two-way, and three-way function. However, in general, many of these model choices for the HAL-MLE will be hard to defend based on prior knowledge, although they might result in a statistically improved estimator relative to using the most nonparametric HAL-MLE. Therefore, we recommend building a super learner whose library contains a variety of such HAL-MLE specifications. In addition, by ranking these HAL-MLE fits by their complexity, one could implement a cross-validation scheme that implements and evaluates estimators from least complex to increasingly complex and stops when the cross-validated risk of the next HAL-MLE drops below the performance of the less complex previous HAL-MLE. In this manner, the resulting super learner avoids having to implement the highly computer intensive HAL-MLEs, except when they are really needed. Because the discrete super learner is asymptotically equivalent with the oracle selector, the resulting discrete super learner will perform as well as the best possible HAL-MLE, thereby also inheriting the rate of convergence of the most nonparametric HAL-MLE.

22.6.3 Using the HAL-MLE for smooth feature inference

By specifying the class of target features for which we want the HAL-MLE to be efficient, we could undersmooth the HAL-MLE until all the efficient score equations (one for each target feature) are solved at the desired level. In this case the above discrete super learner is not making the right trade-off. One might still be able to solve all the score equations under a restricted model choice, such as only including up to three-way tensor product basis functions. Thus, one might still rank various HAL-MLE specifications from least complex to maximal complex, but would now keep increasing complexity of the HAL-MLE according to this ranking until the desired class of score equations are solved at the desired level (i.e., $o_P(n^{-1/2})$ uniformly in all target features).

22.7 HAL and TMLE

Given the bevy of statistical properties for the HAL-MLE, one might wonder why one would need TMLE [9, 11]? However, the finite sample statistical properties of the HAL-MLE rely on solving the desired score equations up to a small enough error, even though the HAL-MLE has the statistical capacity of solving essentially all score equations up to an asymptotic rate $O_P(n^{-2/3})$. On the other hand, if the HAL-MLE is too undersmoothed, then its performance in estimating the target function deteriorates, and that can also have a detrimental effect on statistical performance. Targeted undersmoothing can deal with this to some degree by precisely finding the level of undersmoothing that solves the desired target equations at the preferred level (which could be $O_P(\sigma_n/(n^{1/2}\log n))$, where σ_n is a standard error of the efficient influence curve of the target feature). Nonetheless, in finite samples it might simply not be possible to achieve this goal.

22.7.1 Targeting a class of target features

Therefore, to achieve improved finite sample performance of the HAL-MLE for a desired target feature or class of target features, it follows to use TMLE with an initial HAL-MLE estimator [8]. In this manner the user has precise control over a set of target score equations while the HAL-MLE will provide important benefits by solving a large class of score equations and having a good rate of convergence.

Let Q_n^0 be a HAL-MLE, possibly a discrete super learner based on a library of different HAL-MLE specifications. Let $(\Psi_t(P_0) : t)$ be our class of target features and let $(D^*_{t,Q,G} : t)$ be the corresponding class of canonical gradients. We can then construct a universal least favorable path $\{Q^0_{n,\epsilon} : \epsilon\}$ through Q_n^0 at $\epsilon = 0$, possibly using an initial estimator G_n, and resulting MLE: $\epsilon_n = \arg\min_\epsilon P_n L(Q^0_{n,\epsilon})$, so that $\sup_t \mid P_n D^*_{t,Q^0_{n,\epsilon_n},G_n} \mid= o_P(n^{-1/2})$. Such a TMLE update step $Q_n^* = Q^0_{n,\epsilon_n}$ satisfies the identity:

$$\Psi_t(Q_n^*) - \Psi_t(Q_0) = (P_n - P_0)D^*_{t,Q_n^*,G_n} + R_{0,t}((Q_n^*, G_n), (Q_0, G_0)).$$

Because $d_0(Q_n^*, Q_0) = O_P(n^{-2/3}(\log n)^d)$, and assuming a HAL-MLE G_n also has $d_{01}(G_n, G_0) = O_P(n^{-2/3}(\log n)^d)$, under a strong positivity assumption, we will have:

$$\sup_t \mid R_{0,t}((Q_n^*, G_n), (Q_0, G_0)) \mid = O_P(n^{-2/3}(\log n)^d).$$

In addition, because HAL-MLEs fall in the well understood Donsker class of cadlag functions with a uniform bound on their sectional variation norm [21], it generally also follows that $\{D^*_{t,Q,G} : Q \in \mathcal{Q}, G \in \mathcal{Q}\}$ is a Donsker class with the same covering number rate as this cadlag function class. Therefore, we will also have:

$$(P_n - P_0)D^*_{t,Q_n^*,G_n} = (P_n - P_0)D^*_{t,Q_0,G_0} + O_P(n^{-2/3}(\log n)^d),$$

where the rate of the remainder would follow from using the empirical process finite sample asymptotic equicontinuity results [21]. As a consequence, we have

$$\Psi_t(Q_n^*) - \Psi_t(Q_0) = P_n D^*_{t,Q_0,G_0} + O_P(n^{-2/3}(\log n)^d).$$

In particular, it follows that $\Psi(Q_n^*)$ is an asymptotically efficient estimator of $\Psi(Q_0)$, possibly viewed as elements in function space endowed with a supremum norm [8].

Therefore, in great generality, we have that TMLEs that use HAL as an initial estimator are asymptotically efficient for the target features targeted by the TMLE assuming only that the nuisance functions are cadlag, have finite sectional variation norm, and a strong positivity assumption.

It is not required that the HAL-MLE is undersmoothed for TMLE. The solving of the score equations is carried out by the TMLE so that the HAL-MLE can be optimized for estimating Q_0 and G_0. In particular, one can now use the discrete super learner discussed above with a library of HAL-MLEs.

22.7.2 Score equation preserving TMLE update

Suppose that the HAL-MLE was undersmoothed and thereby was itself already a great estimator for these target features and also for other target features that were not targeted by the TMLE. In that case, could the TMLE destroy the score equations solved by the HAL-MLE and deteriorate some of the good performance of the initial HAL-MLE, such as its properties in estimation of other features or its higher-order properties in estimation of the actual target features targeted by the TMLE? That is, given an initial estimator has already succeeded in solving a large class of important score equations, making it potentially globally efficient and higher-order efficient for a large class of features, we should be worried about the TMLE not preserving these score equations during its TMLE update step solving the desired target score equations up to user supplied error.

This motivates us to generalize the TMLE update step to a TMLE update that is not only solving the target score equations but also preserves the score equations already solved by the initial estimator.

In particular, this is a motivation for using universal least favorable paths in the definition of the TMLE update, because such paths are only maximizing the likelihood in the direction of the target score equations, thereby not affecting any score equation orthogonal to these target score equations. However, in addition, one might do the following. We already specified the score equations exactly solved by the HAL-MLE above, one score for each coefficient that has a nonzero coefficient corresponding with a path that keeps the L_1-norm constant. Given this specified set of score equations and its linear span H_n of scores, we could compute the projection of the first-order canonical gradient $D^*_{t,Q,G}$ onto the space H_n spanned by these scores, and subtract it from $D^*_{t,Q,G}$, resulting in an orthogonalized $\tilde{D}^*_{t,Q,G} = D^*_{t,Q,G} - \Pi(D^*_{t,Q,G} \mid H_n)$.

One now defines the TMLE using the universal least favorable path based on this orthogonalized set of scores $(\tilde{D}^*_{t,Q,G} : t)$ rather than $(D^*_{t,Q,G} : t)$. In this case the TMLE update Q^*_n will still solve the score equations in H_n that were solved by the initial HAL-MLE Q^0_n, and, in addition, it will solve the score equations $P_n \tilde{D}^*_{t,Q^*_n,G_n} = 0$. As a consequence, it will also solve $P_n D^*_{t,Q^*_n,G_n} = 0$. So in this way, we have used the general TMLE to obtain a TMLE that not only targets a new set of score equations but also preserves the score equations already solved by the initial estimator Q^0_n.

> *In future work, we will study, implement, and evaluate this score equation preserving TMLE, and other variations of such score equation preserving TMLEs. The key message is that we will have further robustified the TMLE by not only solving the target score equations and preserving the rate of convergence of the initial estimator, but also preserving the score equations solved by the initial estimator with all its important additional statistical benefits.*

22.8 Considerations for Implementing HAL-TMLE

Causal inference relies on a general roadmap involving defining a causal model, causal quantity of interest, establishing an identification result, and thereby defining a target estimand of the data distribution [25]. These steps then dictate the statistical estimation problem in terms of data, statistical model, and target estimand [9, 26]. For estimation and statistical inference, the underlying causal model and its assumptions play no role, such that one now just proceeds with the roadmap of targeted learning given the statistical model and target estimand [9, 27]. That is, we can compute the canonical gradient of the target estimand, define an initial estimator, describe a TMLE update of this initial estimator, and then provide confidence intervals. In particular, the above plug-in HAL-MLE and score equation preserving TMLE using the HAL-MLE as initial estimator can now be applied. In addition, one can develop the higher-order TMLE using the HAL-MLE as initial estimator [23]. We now focus on additional considerations for implementing HAL-TMLEs.

22.8.1 Double robustness

Many causal inference estimation problems have a double robust structure in the sense that the exact remainder $R_0((Q,G),(Q_0,G_0))$ has a cross-product structure that can be naturally bounded by a product of $\|Q - Q_0\|$ and $\|G - G_0\|$ for certain L^2-norms (typically $L^2(P_0)$-norms), or, equivalently, in terms of $d_0^{1/2}(Q,Q_0)$ and $d_{01}^{1/2}(G,G_0)$. This structure implies that in randomized trials or observational studies where there is substantial knowledge of the censoring and treatment mechanism G_0, one can obtain excellent finite sample performance for TMLEs utilizing the resulting fast converging estimator G_n of G_0. In such a particular appeal due to its ability to remove all bias in the initial estimator Q^0_n when $G_n \approx G_0$.

22.8.2 HAL for treatment and censoring mechanisms

Using HAL for G_n (respecting the known model for G_0) has important benefits. In particular, even if Q_n is inconsistent, the TMLE will remain asymptotically linear if G_n is somewhat undersmoothed. In fact, if one uses a nonparametric HAL-MLE ignoring the model for G_0, and the full data model is nonparametric, then the TMLE will even be efficient, despite the inconsistency of Q_n. Similarly, the inverse probability of treatment and censoring weighted estimator using an undersmoothed HAL-MLE is asymptotically linear in these causal inference problems, and its efficiency is maximized by using an undersmoothed nonparametric HAL-MLE. On the other hand, in such a setting, if one would use other machine learning algorithms, including a super learner, these same estimators would lose their asymptotic linearity and not even converge at the desired $n^{-1/2}$-rate – failing to provide valid inference.

> Therefore, we learn that it is not only beneficial to use HAL-MLE as initial estimator of Q_0 due to above mentioned reasons, but it is also highly beneficial to use a HAL-MLE for G_n.

Let's consider the case where it is known that the treatment mechanism only depends on 2 confounders. By estimating G_0 with a model-based HAL-MLE, perhaps undersmoothed, the above arguments show important gains. However, the above arguments also state that ignoring the model for G_0 will even make it more efficient. Therefore, there is an important selection problem among candidate HAL-MLEs of G_0 that have varied complexity, ranging from the actual model to a fully nonparametric HAL-MLE. Cross-validation would then select the model-based HAL-MLE with high probability and thus be ignorant of the subtle bias-variance trade-offs at stake.

More generally, selecting among candidate estimators of G_0 based on the log-likelihood loss can be problematic when the positivity assumption is practically violated. For example, this could result in an estimator G_n that approaches zero, and, as a consequence, results in erratic TMLE updates. This is typically resolved by truncation, but one still needs to select the truncation level. Similarly, the adjustment set used by G_n might need to be tailored toward the MSE of the TMLE of the target estimand using G_n, rather than toward the estimation of G_0. For example, some baseline covariates might be highly predictive of treatment while not being real confounders (like instrumental variables), and it is well established that adjustment for such variables can easily increase both variance and bias [28–32]. Therefore, an important feature of causal inference problems is the targeted selection among candidate estimators of G_0.

22.8.3 Collaborative TMLE

TMLE provides a natural approach for such a selection by evaluating the performance of a candidate estimator of G_0 with the increase of the log-likelihood during the TMLE update step. The MLE of ϵ in the universal least favorable path through the initial estimator corresponds with fitting the target estimand itself. Specifically, it is the path along which the ratio of the squared change in target estimand by increase in log-likelihood is maximized, i.e., the slope in change of estimand per unit increase in log-likelihood is maximal. Therefore, this increase in log-likelihood during the TMLE update step indeed provides a targeted criterion for evaluating the performance of a candidate estimator G_n of G_0. After having selected an optimal choice with respect to this criterion, one can iterate the process by making the TMLE update the initial estimator and repeating this greedy selection, but now only selecting among candidate estimators that are more complex (say, higher log-likelihood) than the one selected at the previous step. In this manner we obtain a sequence of TMLE updates $Q^*_{n,k}$ of Q^0_n with corresponding increasingly complex $G_{n,k}$ estimators, $k = 1, \ldots, z$. And k can now be selected with cross-validation based on the loss function $L(Q)$ for Q_0, or a AIC/BIC type selector to avoid the double cross-validation. This approach can now be used, for example, to select the L_1-norm in a HAL-MLE, the truncation level, but also to select the maximal degree of the tensor products, or maximal sparsity of the basis functions, and so on [12,13,22,23,33]. Simulations have demonstrated that this collaborative (C-TMLE) procedure can dramatically outperform a TMLE that uses pure likelihood-based estimation of G_0 when there are positivity issues [12]. One can also let a HAL-MLE fit of Q_0 impact the basis functions included in the HAL-MLE of G_0, and use this

C-TMLE procedure to select the L_1-norm in the HAL-MLE of Q_0 and the resulting HAL-MLE of G_0 (as well as the truncation level). The latter type of estimator was termed the outcome adaptive HAL-TMLE [12], which built on prior work in outcome-adaptive LASSOs [34], providing a powerful tool for variable selection in G_n, and was shown to have strong practical performance.

22.9 Closing

There is a tendency for the machine learning literature to focus on piecemeal and small extensions of the "flashy" estimator of the moment. This was recently random forests, but is currently deep learning. However, the statistical theory and empirical process literature has offered strong guidance on the development of data-adaptive estimators for features of the data distribution that provide inference, while fully utilizing the knowledge of a statistical model.

In particular, for optimal robust (higher-order) estimation of target estimands, we need to solve specific first- and higher-order canonical gradients of the target estimand. Also, having the functional parameter of the data distribution needed for estimation of the target estimand as a member of a function class with a well behaving entropy integral (as implied by the covering number of the function class) provides good rates of convergence and allows for bootstrap based inference. HAL-MLE and HAL-TMLE appear to satisfy these key fundamental properties. It would be exciting for the general machine learning literature to build on these areas.

References

[1] Stephen Stigler. The epic story of maximum likelihood. *Statistical Science*, 22(4) 598–620, 2007.

[2] Whitney Newey. Convergence rates and asymptotic normality for series estimators. *Journal of Econometrics*, 79(1):147–168, 1997.

[3] Xiaotong Shen. On methods of sieves and penalization. *The Annals of Statistics*, 25(6) 2555–2591, 1997.

[4] Eric Polley, Sherri Rose, and Mark van der Laan. Super learning. In *Targeted Learning: Causal Inference for Observational and Experimental Data*, pages 43–66. Springer, 2011.

[5] Mark van der Laan, Eric Polley, and Alan Hubbard. Super learner. *Statistical Applications in Genetics and Molecular Biology*, 6(1), 2007.

[6] Robert Tibshirani. Regression shrinkage and selection via the lasso. *Journal of the Royal Statistical Society: Series B (Methodological)*, 58(1):267–288, 1996.

[7] David Benkeser and Mark van der Laan. The highly adaptive lasso estimator. In *2016 IEEE International Conference on Data Science and Advanced Analytics (DSAA)*, pages 689–696. IEEE, 2016.

[8] Mark van der Laan. A generally efficient targeted minimum loss based estimator based on the highly adaptive LASSO. *The International Journal of Biostatistics*, 13(2), 2017. DOI: 10.1515/ijb-2015-0097

[9] Mark van der Laan and Sherri Rose. *Targeted Learning: Causal Inference for Observational and Experimental Data*. Springer Science & Business Media, 2011.

[10] Mark van der Laan and Sherri Rose. *Targeted Learning in Data Science: Causal Inference for Complex Longitudinal Studies*. Springer, 2018.

[11] Mark van der Laan and Daniel Rubin. Targeted maximum likelihood learning. *The international Journal of Biostatistics*, 2(1), 2006.

[12] Cheng Ju, David Benkeser, and Mark van der Laan. Robust inference on the average treatment effect using the outcome highly adaptive LASSO. *Biometrics*, 76(1):109–118, 2020.

[13] Mark van der Laan, David Benkeser, and Weixin Cai. Efficient estimation of pathwise differentiable target parameters with the undersmoothed highly adaptive LASSO. *arXiv preprint arXiv:1908.05607*, 2019.

[14] Georg Neuhaus. On weak convergence of stochastic processes with multidimensional time parameter. *The Annals of Mathematical Statistics*, 42(4):1285–1295, 1971.

[15] Mark van der Laan and Sandrine Dudoit. Unified cross-validation methodology for selection among estimators and a general cross-validated adaptive epsilon-net estimator: Finite sample oracle inequalities and examples. *Technical Report 130, Division of Biostatistics, University of California, Berkeley*, 2003.

[16] Mark van der Laan, Sandrine Dudoit, and Aad van der Vaart. The cross-validated adaptive epsilon-net estimator. *Statistics & Decisions*, 24(3):373–395, 2006.

[17] Aad van der Vaart, Sandrine Dudoit, and Mark van der Laan. Oracle inequalities for multi-fold cross validation. *Statistics & Decisions*, 24(3):351–371, 2006.

[18] Richard Gill, Mark van der Laan, and Jon Wellner. Inefficient estimators of the bivariate survival function for three models. *Annales de l'IHP Probabilités et statistiques*, 31(3):545–597, 1995.

[19] Jerome Friedman, Trevor Hastie, Rob Tibshirani, Balasubramanian Narasimhan, Kenneth Tay, Noah Simon, and Junyang Qian. Package 'glmnet'. *CRAN R Repository*, 2021.

[20] Nima Hejazi, Jeremy Coyle, and Mark van der Laan. hal9001: Scalable highly adaptive lasso regression in R. *Journal of Open Source Software*, 5(53):2526, 2020.

[21] Aurélien Bibaut and Mark van der Laan. Fast rates for empirical risk minimization over cadlag functions with bounded sectional variation norm. *arXiv preprint arXiv:1907.09244*, 2019.

[22] Weixin Cai and Mark van der Laan. Nonparametric bootstrap inference for the targeted highly adaptive least absolute shrinkage and selection operator (LASSO) estimator. *The International Journal of Biostatistics*, 16(2), 2020. doi: 10.1515/ijb-2017-0070

[23] Mark van der Laan, Zeyi Wang, and Lars van der Laan. Higher order targeted maximum likelihood estimation. *arXiv preprint arXiv:2101.06290*, 2021.

[24] Cynthia Rudin. Stop explaining black box machine learning models for high stakes decisions and use interpretable models instead. *Nature Machine Intelligence*, 1(5):206–215, 2019.

[25] Judea Pearl. *Causality*. Cambridge University Press, 2009.

[26] Maya Petersen and Mark van der Laan. Causal models and learning from data: integrating causal modeling and statistical estimation. *Epidemiology*, 25(3):418, 2014.

[27] Megan Schuler and Sherri Rose. Targeted maximum likelihood estimation for causal inference in observational studies. *American Journal of Epidemiology*, 185(1):65–73, 2017.

[28] Sander Greenland. Invited commentary: variable selection versus shrinkage in the control of multiple confounders. *American Journal of Epidemiology*, 167(5):523–529, 2008.

[29] Jessica Myers, Jeremy Rassen, Joshua Gagne, Krista Huybrechts, Sebastian Schneeweiss, Kenneth Rothman, Marshall Joffe, and Robert Glynn. Effects of adjusting for instrumental variables on bias and precision of effect estimates. *American Journal of Epidemiology*, 174(11):1213–1222, 2011.

[30] Andrea Rotnitzky, Lingling Li, and Xiaochun Li. A note on overadjustment in inverse probability weighted estimation. *Biometrika*, 97(4):997–1001, 2010.

[31] Enrique Schisterman, Stephen Cole, and Robert Platt. Overadjustment bias and unnecessary adjustment in epidemiologic studies. *Epidemiology*, 20(4):488, 2009.

[32] Sebastian Schneeweiss, Jeremy Rassen, Robert Glynn, Jerry Avorn, Helen Mogun, and M. Alan Brookhart. High-dimensional propensity score adjustment in studies of treatment effects using health care claims data. *Epidemiology*, 20(4):512, 2009.

[33] Mark van der Laan, Antoine Chambaz, and Cheng Ju. C-tmle for continuous tuning. In *Targeted Learning in Data Science*, pages 143–161. Springer, 2018.

[34] Susan M Shortreed and Ashkan Ertefaie. Outcome-adaptive lasso: variable selection for causal inference. *Biometrics*, 73(4):1111–1122, 2017.

23

Bayesian Propensity Score Methods and Related Approaches for Confounding Adjustment

Joseph Antonelli

The University of Florida (USA)

CONTENTS

The goal of this chapter is to introduce readers to the different ways in which Bayesian inference has been successfully applied to causal inference problems. Our emphasis will be on approaches that utilize the propensity score, though we briefly discuss how other estimation approaches can be improved by using the Bayesian paradigm. As the chapter focuses on propensity scores, we will focus on the scenario where the treatment W is binary and the estimand of interest is the average treatment effect. Similar ideas apply to multilevel or continuous treatments as well as other estimands, and we will discuss some of these extensions toward the end of the chapter. We begin by introducing readers to the Bayesian paradigm and why it can be useful to adopt in the context of causal inference. Then we detail various ways in which the Bayesian paradigm has been used with propensity scores and doubly robust estimators, and we finish with extensions to Bayesian nonparametrics and other estimands.

DOI: 10.1201/9781003102670-23

23.1 Introduction to Bayesian Analysis

Here we provide a very brief review of the typical steps associated with a Bayesian analysis, and provide motivation for why it is a useful paradigm from which to perform inference. For more details and empirical examples, we recommend reading Bayesian Data Analysis [1]. The first step in any Bayesian analysis is to specify a joint probability model for the observed data as a function of unknown model parameters, i.e., the likelihood. We denote the likelihood of observing data $\boldsymbol{y}$ given a set of unknown parameters $\boldsymbol{\theta}$ by $p(\boldsymbol{y}|\boldsymbol{\theta})$. This is a key difference between Bayesian inference and much of the causal inference literature on weighting or matching, which are not necessarily rooted in likelihood-based inference. Once a likelihood has been defined, a prior distribution must be specified for the unknown parameters, which we denote by $p(\boldsymbol{\theta})$. This distribution is intended to reflect prior beliefs about how likely a particular value of $\boldsymbol{\theta}$ is. This prior knowledge can come from scientific expertise or from previous, related studies. In many cases, however, prior knowledge can be difficult to translate into a prior distribution, no prior knowledge exists on a particular parameter, or an analyst may wish to exclude any prior information and let the data alone provide information about the probability that $\boldsymbol{\theta}$ is equal to any particular value. These settings have led to a broad literature on non-informative or objective prior distributions [2,3]. There is no formal consensus on the definition of an objective Bayesian analysis, but loosely it is an analysis in which no subjective inputs from the analyst are placed into the prior distribution. Alternatively it can be thought of as a prior distribution that acknowledges the fact that the analyst has no prior knowledge on the value of $\boldsymbol{\theta}$. In practice, this commonly (though not always) amounts to prior distributions with very large variances or distributions that assign relatively equal weight to all possible values of a parameter.

Once the likelihood and prior distribution have been defined, the next step is to calculate the posterior distribution, which is defined as

$$p(\boldsymbol{\theta}|\boldsymbol{y}) = \frac{p(\boldsymbol{y}|\boldsymbol{\theta})p(\boldsymbol{\theta})}{p(\boldsymbol{y})} = \frac{p(\boldsymbol{y}|\boldsymbol{\theta})p(\boldsymbol{\theta})}{\int_{\boldsymbol{\theta}} p(\boldsymbol{y}|\boldsymbol{\theta})p(\boldsymbol{\theta})d\boldsymbol{\theta}} \propto p(\boldsymbol{y}|\boldsymbol{\theta})p(\boldsymbol{\theta}),$$

where $\propto$ implies that two quantities are proportional to each other and differ by a constant factor that does not depend on $\boldsymbol{\theta}$. The posterior distribution reflects our uncertainty in the unknown parameter $\boldsymbol{\theta}$ after observing data $\boldsymbol{y}$. We can see that the posterior distribution is a function of both the prior distribution and the likelihood and therefore combines information from both of these sources. Figure 23.1 illustrates the relationship between the prior distribution, likelihood, and posterior distribution. In the left panel, the prior distribution is relatively informative with a small variance. This leads to the prior distribution having a large influence on the resulting posterior distribution, which is pulled toward the prior distribution. In the middle and right panels, as we increase the variance of the prior distribution, the posterior looks closer to the likelihood, resulting in the likelihood and posterior being almost indistinguishable when the prior variance is large enough.

Once the posterior distribution is obtained, it is straightforward to perform inference for any function of $\boldsymbol{\theta}$. Point estimates of unknown quantities can be obtained using the corresponding posterior mean or median. $(1-\alpha)100\%$ credible intervals can be obtained in a number of ways. The most common of which is to construct a credible interval using the $\alpha/2$ and $1-\alpha/2$ quantiles of the posterior distribution for the quantity of interest. Unfortunately, the posterior distribution does not typically follow a known probability distribution and does not have a closed form solution that can be used to perform inference. Markov chain Monte Carlo (MCMC) methods are used to sample draws from the posterior distribution, and we can use these draws to approximate functionals of the posterior distribution. We won't discuss sampling considerations in this chapter, but it will be assumed throughout that we have obtained B draws from the posterior distribution of the unknown parameters, denoted by $\boldsymbol{\theta}^{(b)}$ for $b = 1, \ldots, B$. For additional information on computational issues and MCMC sampling, we point interested readers to existing textbooks on this topic [4,5]. Given

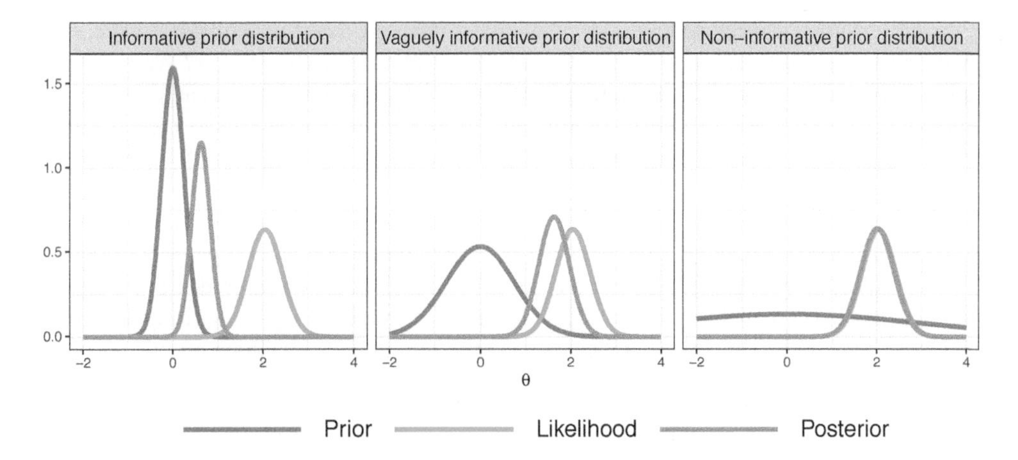

FIGURE 23.1
Interplay between the prior distribution, likelihood, and posterior distribution.

these posterior draws, the posterior mean can be approximated as $\frac{1}{B}\sum_{b=1}^{B}\boldsymbol{\theta}^{(b)}$, and analogous operations can be used to approximate other functionals such as the quantiles of the distribution for inference.

Another Bayesian quantity that is relevant to causal inference problems is the posterior predictive distribution, i.e., the distribution of a new, unobserved data point given the observed data. This can be defined as

$$p(\widetilde{y}|\boldsymbol{y}) = \int_{\boldsymbol{\theta}} p(\widetilde{y}|\boldsymbol{y},\boldsymbol{\theta})p(\boldsymbol{\theta}|\boldsymbol{y})d\boldsymbol{\theta} = \int_{\boldsymbol{\theta}} p(\widetilde{y}|\boldsymbol{\theta})p(\boldsymbol{\theta}|\boldsymbol{y})d\boldsymbol{\theta}.$$

This is useful for prediction as it allows the analyst to provide estimates and uncertainty assessments around predictions of new observations. This is useful for causal inference where we may be interested in predicting the outcome for an individual under a treatment level that is not observed.

23.1.1 Why use Bayesian methods for causal inference?

Before beginning with the discussion of how Bayesian inference can be used in conjuction with causal inference, we must first introduce a few elementary concepts in causal inference. Some concepts will be introduced throughout the chapter where relevant, but we must first define a few important quantities. Throughout, we let W be the treatment variable, which for now we assume to be binary. We let Y_1 be the potential outcome that we would have observed if, possibly contrary to fact, a subject was treated. Y_0 is the corresponding potential outcome under control, and the key problem is that we only ever get to observe at most one of these two quantities for a subject. Throughout most of the chapter we focus on estimating the average treatment effect, which is defined as $\Delta = E(Y_1 - Y_0)$. Estimating average potential outcomes, such as $E(Y_1)$, is a difficult problem because we only observe Y_1 for a non-random subset of the population given by those with $W = 1$. The set of individuals for which we observe Y_1 is driven by the propensity score, which is the probability of receiving treatment as a function of pre-treatment covariates $\boldsymbol{X}$ and is given by $P(W = 1|\boldsymbol{X})$. The propensity score is a key quantity in inferring causality from observational data, and one that we will utilize regularly throughout the chapter.

Now that we've introduced some basic causal inference concepts, we can discuss Bayesian inference and its application to causal inference. Bayesian inference is simply one approach to estimating unknown parameters, so a natural question to ask is why one might be interested in using it for causal inference problems. Bayesian inference has the added complication of sampling from the posterior distribution and can be more computationally intensive than frequentist inference, so it must provide something different than frequentist inference; otherwise why bother? We will see in this chapter that there are a number of desirable features that Bayesian inference can provide relatively easily that are useful in causal inference problems. Bayesian inference provides natural solutions to model averaging and variable selection, which is useful when the number of confounders is relatively large compared to the sample size. Sensitivity analysis, which is an important issue in causal inference, has natural solutions within the Bayesian paradigm where prior distributions on sensitivity parameters can be used to assess the robustness to key, untestable assumptions. Additionally, there are nice features of Bayesian inference that are not unique to causal inference problems, but are useful in general. These include the use of highly flexible Bayesian nonparametric approaches, easily constructing hierarchical models, and the ability to perform inference without relying on asymptotic approximations.

Part of this chapter focuses on propensity score estimation, and there has been some debate in the causal inference literature about whether an analysis that incorporates the propensity score can ever be fully Bayesian [6–9]. From our perspective, what determines whether an analysis is fully Bayesian or not is not well defined, nor is it clear why this would be advantageous. We do not attempt to clarify these arguments or claim that any of the methods discussed in this chapter are fully Bayesian. We believe that there are certain beneficial features of Bayesian methods that can be incorporated into causal analyses to improve their performance. Whether or not an analysis is fully Bayesian is irrelevant as long as the resulting inferential procedure is valid and the operating characteristics are well understood.

23.1.2 The progression of Bayesian causal inference

Before diving into the specific ways in which Bayesian methods have been used in causal inference, it is important to first step back and provide a timeline of what researchers have meant by the term Bayesian causal inference. In our view there are two distinct paradigms within Bayesian causal inference: 1) The explicit Bayesian modeling of potential outcomes, and 2) Bayesian modeling of commonly used causal inference tools such as propensity scores, outcome regression models, and related approaches. Bayesian causal inference began with the first of these two paradigms in the early years of causal inference [10]. The main idea is to posit a full probability model for the joint distribution of the potential outcomes and update the unknown parameters given the observed data, while acknowledging uncertainty or assessing sensitivity to non-identified parameters. The second paradigm takes a quite different approach, which is to extend a number of commonly used approaches to estimation in causal inference to the Bayesian paradigm. These approaches do not directly model the potential outcomes, but rather use Bayesian methods for various modeling tasks such as propensity score estimation or modeling the observed outcome conditional on the treatment and covariates. Bayesian methods have proven to be useful in certain aspects within this second paradigm, but also can complicate analyses substantially relative to frequentist counterparts. In this chapter we focus attention mostly on the second of these two paradigms to review the obstacles to Bayesian estimation in such settings, discuss approaches to avoiding such obstacles, and highlight the benefits that being Bayesian can provide. In Section 23.5 we will circle back to the first of these paradigms, partially due to renewed interest that has come about due to the increasing popularity of nonparametric Bayesian methodology for modeling potential outcome surfaces.

23.2 Bayesian Analysis of Propensity Scores

In this section we detail the various ways in which Bayesian inference can be combined with traditional propensity score analyses, and what can potentially be gained by utilizing the Bayesian paradigm. For a nice review of these ideas, we point readers to [11]. First we must specify the assumptions under which we can identify the causal effect in this setting. We will be assuming the stable unit treatment value assumption that states that the treatment status of one unit does not affect outcomes for other units and that there are not multiple versions of the treatment. Additionally, we assume unconfoundedness and positivity, which are defined as:

Positivity: $0 < P(W = 1|\boldsymbol{X} = \boldsymbol{x}) < 1$ with probability 1.

Unconfoundedness: $Y_w \perp\!\!\!\perp W|\boldsymbol{X}$ for $w = 0, 1$.

Before discussing Bayesian propensity score analysis, it is important to first clarify the different ways in which propensity scores can be used in an analysis, as this can greatly affect how easily the Bayesian paradigm can be incorporated. Throughout we will denote propensity scores by $e(\boldsymbol{X}, \boldsymbol{\alpha}) = P(W = 1|\boldsymbol{X}, \boldsymbol{\alpha})$, where we now include $\boldsymbol{\alpha}$ to represent unknown parameters of the propensity score model. While nonparametric models for the propensity score are possible and many of the same ideas will apply, we will focus on the following generalized linear model for the propensity score:

$$g_w^{-1}\Big(P(W = 1|\boldsymbol{X} = \boldsymbol{x}, \boldsymbol{\alpha})\Big) = \alpha_0 + \sum_{j=1}^{p} \alpha_j x_j \tag{23.1}$$

where $g_w(\cdot)$ is a standard link function such as the logistic or probit link functions. Once the propensity score is obtained, there are a myriad of ways in which they can be used to estimate the effect of the treatment. Matching on the propensity score is one common approach [12, 13], where two individuals are considered good matches if they have similar values of the propensity score. Inverse probability weighting is another estimator that estimates the average treatment effect as

$$\widehat{\Delta} = \frac{1}{n}\sum_{i=1}^{n}\left\{\frac{W_i Y_i}{e(\boldsymbol{X}_i, \boldsymbol{\alpha})} - \frac{(1 - W_i)Y_i}{1 - e(\boldsymbol{X}_i, \boldsymbol{\alpha})}\right\},$$

which is effectively a weighted average of the outcomes in the treated and control groups, where weights are used to construct a population that has no association between W and $\boldsymbol{X}$ and is therefore unconfounded [14]. Different numerators in this expression can be used to target different estimands [15], though similar ideas apply. A third approach, which we will focus on for much of this section, is to include the propensity score into an outcome regression model as follows:

$$g_y^{-1}\Big(E(Y|W = w, \boldsymbol{X} = \boldsymbol{x}, \boldsymbol{\beta})\Big) = \beta_0 + \beta_w w + h(e(\boldsymbol{x}, \boldsymbol{\alpha})),$$

where g_y is a standard link function. Here the $h(\cdot)$ function controls how the propensity score is included in the outcome model. It could be included linearly, using nonlinear functions of the propensity score, or through indicator functions that indicate membership into particular strata of the propensity score distribution [16]. Matching, IPW, and related estimators are not likelihood based and therefore don't have an immediate Bayesian formulation. For this reason, we will begin our discussion of Bayesian propensity score analysis with a focus on outcome regression as a method of estimating causal effects with propensity scores. We will discuss extensions of these ideas to non-likelihood-based estimators at the end of this section.

Now that our focus is on outcome regression models, we can discuss the Bayesian implementation of these ideas, which are discussed in detail in [17]. The two key components of any Bayesian analysis are the likelihood and prior distribution. Independent and non-informative prior distributions can be assigned for $\boldsymbol{\alpha}$ and $\boldsymbol{\beta}$ that are commonly used in Bayesian generalized linear models. Here, we need to model both the outcome and treatment data generating processes, and therefore the likelihood is given by

$$P(Y, W|\boldsymbol{X}, \boldsymbol{\alpha}, \boldsymbol{\beta}) = P(Y|W, \boldsymbol{X}, \boldsymbol{\alpha}, \boldsymbol{\beta})P(W|\boldsymbol{X}, \boldsymbol{\alpha}). \tag{23.2}$$

The likelihood factorizes into one component from the outcome model and one component from the treatment or propensity score model. One key distinction here is that $\boldsymbol{\alpha}$ is in both the likelihood for the treatment and for the outcome, despite the fact that these are the parameters of the propensity score model. The reason for this is that the propensity score, which is a function of $\boldsymbol{\alpha}$, is included in the outcome model. This is a crucial point that will make a Bayesian analysis of propensity scores more nuanced than traditional propensity score analysis. We will discuss the difficulties that arise from this complication in the following section, but first it is important to understand that this means the propensity score fit will be affected by the outcome since inference for $\boldsymbol{\alpha}$ will be based on $\boldsymbol{X}, W$, and Y. It has been argued that the design phase of a causal analysis, which the propensity score is one component of, should be separate from the analysis stage and should not use information from the outcome [18]. A traditional frequentist propensity score analysis would first estimate the MLE of $\boldsymbol{\alpha}$, using only information from the treatment, and then estimate the causal effect in a second stage conditional on these propensity score estimates. A fully Bayesian analysis on the other hand incorporates all information simultaneously to update parameters in a single analysis. While it may lead to additional complications in a propensity score analysis, there are certain reasons why one might wish to proceed with Bayesian inference here. One important reason is that it allows us to account for difficult sources of uncertainty, which we explore in Section 23.2.2. Another reason is that it permits the use of uniquely Bayesian tools such as Bayesian model averaging, which is beneficial when the number of confounders is large, or Bayesian nonparametric models that have been shown to work well in a wide range of settings.

23.2.1 Issues with model feedback

The likelihood in (23.2) is a product of two components or modules, and the parameter $\boldsymbol{\alpha}$ shows up in both modules. A common issue in Bayesian inference made up of distinct modules such as this is that misspecification of one model can contaminate inferences on the other model [19–21]. This problem is not unique to the propensity score analysis problem discussed here, but is potentially problematic for Bayesian analyses of treatment effects. A common approach to addressing this problem is to "cut" the feedback from one model into the other. To see this, let us consider how the $\boldsymbol{\alpha}$ parameter is updated in an MCMC sampler. Typically a Gibbs sampler is used and the parameters are updated from their full conditional distribution, given by

$$P(\boldsymbol{\alpha}|\boldsymbol{Y}, \boldsymbol{W}, \boldsymbol{X}, \boldsymbol{\beta}) \propto P(Y|W, \boldsymbol{X}, \boldsymbol{\alpha}, \boldsymbol{\beta}) P(W|\boldsymbol{X}, \boldsymbol{\alpha}) P(\boldsymbol{\alpha}).$$

Clearly if we sample from this distribution, the resulting $\boldsymbol{\alpha}$ values will be a function of both the treatment and outcome as both components of the likelihood are in this expression. Cutting the feedback would amount to an approximately Bayesian approach that simply updates from the modified full conditional distribution given by

$$P(W|\boldsymbol{X}, \boldsymbol{\alpha}) P(\boldsymbol{\alpha}).$$

This distribution will only use information from the treatment to update $\boldsymbol{\alpha}$ and will therefore be unaffected by any model misspecification from the outcome model component of the likelihood. The problem of model misspecification in Bayesian propensity score analysis was first considered in [22]. Their motivation was to better understand the potential impact of model misspecification in these situations, and they compared the fully Bayesian approach to the approximately Bayesian one that cuts feedback from the outcome model. They found similar results between the two approaches and suggested that authors use the fully Bayesian approach if they are confident in their outcome model or are able to use a flexible model that is less susceptible to misspecification.

More recent work [23] has identified a much larger issue inherent to a fully Bayesian analysis of propensity scores and an outcome regression that incorporates the propensity score. A central property of propensity scores that enables them to identify causal effects is the so-called balancing property [24], which states that

$$W \perp\!\!\!\perp \boldsymbol{X} | e(\boldsymbol{X}, \boldsymbol{\alpha})$$

In [23] the authors noted that it is not clear whether the propensity scores in a fully Bayesian analysis that incorporate the outcome will possess this property, and therefore, it is not clear how useful they are as a measure to remove confounding bias. Further, they showed that a fully Bayesian analysis can only lead to unbiased estimates of the treatment effect in extremely specific situations that are unlikely to hold. Intuitively, an outcome model that conditions on the propensity score is by definition misspecified as we don't typically think the true outcome process is a function of $e(\boldsymbol{X}, \boldsymbol{\alpha})$. Given that model feedback is problematic under model misspecification, this implies that model feedback will always be a problem for propensity score estimation. It can be shown that the only way in which this approach will provide unbiased estimates is if the true outcome model coefficients are a simple re-scaling of the propensity score coefficients, $\boldsymbol{\alpha}$. Clearly this is overly restrictive, and effectively renders the fully Bayesian approach as defined above useless.

In light of this, there are two ways to incorporate Bayesian analysis into propensity score analysis if an outcome model is used. The first is to use an approximately Bayesian approach that cuts the feedback and does not use information from the outcome when updating the propensity score parameters. Related two step Bayesian procedures to propensity score analysis are described in detail in [25]. While these ideas work, and can still propagate uncertainty from propensity score estimation into causal estimates, they are not always feasible. One key instance is when doing variable selection, which we cover in the following section, as we want the chosen set of variables to depend on both the treatment and outcome. [23] found that another way to solve the issue of feedback between the propensity score and outcome model is to include additional covariate adjustment in the outcome model as

$$g_y^{-1}(E(Y|W = w, \boldsymbol{X} = \boldsymbol{x}, \boldsymbol{\beta}) = \beta_0 + \beta_w w + h(e(\boldsymbol{x}, \boldsymbol{\alpha})) + \sum_{j=1}^{p} \beta_j x_j. \tag{23.3}$$

This additional covariate adjustment has been shown to work well in propensity score analyses [13, 26]. Not only does this address the issue of model feedback and allow users to utilize a fully Bayesian analysis of propensity scores, but it has also been shown to be doubly robust in the sense that only the additional covariate adjustment or propensity score need to be correctly specified in order to estimate causal effects. Lastly, as we will see in the following section, this approach will allow users to perform model averaging or variable selection in a Bayesian framework that utilizes both the treatment and outcome information, which is desirable for variable selection in causal inference problems [27–30].

23.2.2 Accounting for uncertainty in propensity score estimation

One of the main goals that motivated work on Bayesian analyses of propensity scores was to account for uncertainty in propensity score estimation. Intuitively, the propensity score parameters are unknown and must be estimated, and therefore, we need to properly account for uncertainty in estimation of these parameters [31]. An interesting result, however, is that for estimating the average treatment effect, the asymptotic variance of matching estimators [32] and IPW estimators [33] is smaller when using an estimated propensity score than the true propensity score. This rather surprising result would suggest that we can ignore uncertainty from propensity score estimation and still obtain valid inference. However, these results only hold for certain estimators of certain estimands and do not necessarily hold in general for all propensity score problems. Additionally, there are other sources of uncertainty that one might wish to account for in propensity score analysis that are not easily dealt with in traditional analyses such as model uncertainty [29], or uncertainty in which observations are dropped in matching analyses [34, 35].

Despite the work described above, there is not a clear consensus on the overall benefits of Bayesian analysis of propensity scores. Certainly there are benefits in being able to use some of the unique features of Bayesian inference, though in terms of uncertainty quantification the story is less clear. Many authors found that the Bayesian approaches provided similar interval coverage rates in simulations to their frequentist counterparts. Additionally, the approaches considered above were restricted to a specific type of estimation procedure, frequently an outcome model based procedure as this is the easiest to imbed in the Bayesian paradigm. Other estimators that are not likelihood based such as IPW estimators do not easily extend to Bayesian inference. Though work has been done to provide Bayesian versions of these estimators [7], there is some debate about their utility [8].

Recent progress has been made at bridging the gap between Bayesian inference and more general causal estimators that rely on propensity score approaches in [36]. This paper uses Bayesian inference for parameter estimation along with a broad class of causal estimators, and aims to provide an inferential strategy that works across all estimators. Using similar notation, we can define ν to be the output of the design stage of the study, or the output from a particular propensity score implementation. This can represent weights for an IPW estimator, a set of matched observations after matching on the propensity score, or a partition of the data into propensity score strata, among others. Letting Δ represent the treatment effect of interest, the goal is to find

$$P(\Delta|\boldsymbol{Y}, \boldsymbol{W}, \boldsymbol{X}) = \int_{\nu} P(\Delta|\boldsymbol{Y}, \boldsymbol{W}, \boldsymbol{X}, \nu) P(\nu|\boldsymbol{W}, \boldsymbol{X}) d\nu.$$

Note that the distribution of ν here does not depend on the outcome, following the principle that the outcome should not influence the design stage of a study. Interestingly, this posterior distribution can be decomposed even further as:

$$\int_{\nu} P(\Delta|\boldsymbol{Y}, \boldsymbol{W}, \boldsymbol{X}, \nu) \int_{\alpha} P(\nu|\boldsymbol{\alpha}, \boldsymbol{W}, \boldsymbol{X}) P(\boldsymbol{\alpha}|\boldsymbol{W}, \boldsymbol{X}) d\boldsymbol{\alpha} d\nu.$$

This suggests a sequential strategy to sampling from the posterior distribution of Δ. First, the propensity score parameters can be sampled from the posterior distribution $P(\boldsymbol{\alpha}|\boldsymbol{W}, \boldsymbol{X})$. Conditional on $\boldsymbol{\alpha}$, the design stage can be sampled from $P(\nu|\boldsymbol{\alpha}, \boldsymbol{W}, \boldsymbol{X})$. Lastly, the causal effect can be sampled from the posterior distribution of $P(\Delta|\boldsymbol{Y}, \boldsymbol{W}, \boldsymbol{X}, \nu)$. These three sources of uncertainty were referred to in [36] as analysis estimation uncertainty, design decision uncertainty, and design estimation uncertainty, respectively. In many cases, such as for IPW weights, there is no design decision uncertainty as the weights used in IPW are a deterministic function of $\boldsymbol{\alpha}$ and $\boldsymbol{X}$. Design decision uncertainty is more prevalent in matching estimators where there is uncertainty in the matched sets, and different replications of the same matching algorithm can lead to different analyses. If the final analysis strategy is likelihood based, such as through an outcome model, the analysis estimation uncertainty can be accounted for by the posterior distribution of the parameters in this model. If instead the analysis is done using non-likelihood based estimators such as IPW, doubly robust estimators, or matching, then this distribution can be replaced by the asymptotic distribution of that estimator leading to an approximately Bayesian analysis.

23.3 Covariate Selection in Propensity Score Models

In many cases the number of predictors available to us in an observational study is relatively large. From an identification perspective this can be a good thing as it makes the assumption of no unmeasured confounding more plausible than if fewer covariates were measured. From an estimation perspective, it complicates analyses for a number of reasons. The most commonly used approach to model selection in propensity score models is to simply include all available covariates in the propensity score model to ensure that all important confounders are included. Unfortunately, when the number of possible confounders is moderate or high-dimensional then this approach is inefficient at best, and infeasible at worst. In some cases the number of covariates is so large that traditional approaches to propensity score estimation do not apply, or they lead to severely overfit propensity scores that are effectively equal to 1 for treated subjects and 0 for control subjects. Many ad-hoc approaches to variable selection could be used in this setting such as stepwise regression approaches on the propensity score model, or simply looking at univariate correlations between predictors and the treatment or outcome [37]. High-dimensional models such as the lasso [38] and related models could be used to model the propensity score, but it has been noted in many cases that this can lead to substantial finite sample bias of the treatment effect [28–30, 39]. The crucial point is that confounders will be associated with both the treatment and outcome, and therefore the outcome should be used to help identify the important confounders [12, 40, 41]. The problem stems from the fact that a variable can be strongly associated with the outcome, and only weakly associated with the treatment, yet still induce non-negligible bias for the treatment effect. These are likely to be excluded by algorithms that select confounders based only on the treatment and ignore information from the outcome. An additional reason to incorporate the outcome into the confounder selection process is that variables associated with only the outcome can help improve the efficiency of resulting treatment effect estimates if they are included in the propensity score model. Note that this problem has spurred interest in model selection for causal inference [27, 30, 39, 42, 43], as well as interest in extending these ideas to the truly high-dimensional setting where $p > n$ [28, 44–50]. In this section we restrict attention to Bayesian approaches to model averaging and confounder selection in this context, but refer readers to the aforementioned papers for frequentist estimators in this setting.

23.3.1 The goal of Bayesian model averaging

One nice feature of estimating causal effects within the Bayesian framework is the ability to use uniquely Bayesian features, such as Bayesian model averaging. Instead of selecting a single model, Bayesian model averaging treats the choice of model as an unknown parameter that we wish to find the posterior distribution of [51, 52]. This leads to results which are averaged over all possible models, and model weights are assigned according to how strongly they are favored by the posterior distribution. Bayesian model averaging has been shown to improve upon analyses that condition on a single model, and it allows the analyst to account for uncertainty in model selection.

Due to these benefits, it stands to reason that Bayesian model averaging will be a useful tool for dealing with a large number of confounders when some form of confounder selection is required. As mentioned earlier, model selection is unique in causal inference as variables should be included or excluded based on their relative association with both the treatment and outcome, rather than just one of these variables. In this section we describe ideas that have been used to tailor Bayesian model averaging to causal inference problems by linking prior distributions for the treatment and outcome models. Before we discuss specific aspects of implementation for different estimators, we will first cover the overarching goal of all of these approaches.

Let M represent a set of variables that are chosen to be used in an analysis. For instance, if we choose to use covariates 1, 4, and 9 in an analysis, then $M = \{1, 4, 9\}$. Further, suppose that $\mathcal{M}^*$ is the set of all models M that satisfy the no unmeasured confounding assumption. This means that any model $M \in \mathcal{M}^*$ contains covariates such that the potential outcomes are independent of the treatment given these covariates. Using Bayesian model averaging, the posterior distribution of the causal effect can be written as

$$\begin{aligned} P(\Delta|\boldsymbol{Y},\boldsymbol{W},\boldsymbol{X}) &= \sum_{M} P(\Delta|M,\boldsymbol{Y},\boldsymbol{W},\boldsymbol{X})P(M|\boldsymbol{Y},\boldsymbol{W},\boldsymbol{X}) \\ &= \sum_{M\in\mathcal{M}^*} P(\Delta|M,\boldsymbol{Y},\boldsymbol{W},\boldsymbol{X})P(M|\boldsymbol{Y},\boldsymbol{W},\boldsymbol{X}) \\ &+ \sum_{M\notin\mathcal{M}^*} P(\Delta|M,\boldsymbol{Y},\boldsymbol{W},\boldsymbol{X})P(M|\boldsymbol{Y},\boldsymbol{W},\boldsymbol{X}) \end{aligned}$$

The main goal of Bayesian model averaging for causal inference is to assign as much posterior probability as possible to models in $\mathcal{M}^*$. Specifically, the goal is to obtain a posterior distribution such that $P(M \notin \mathcal{M}^*|\boldsymbol{Y},\boldsymbol{W},\boldsymbol{X}) \approx 0$ so that nearly all posterior probability is placed on models that satisfy the no unmeasured confounding assumption. For the rest of this section, we provide details on specific implementations of Bayesian model averaging in causal inference.

23.3.2 Bayesian model averaging in propensity score models

In the context of propensity scores, Bayesian model averaging was introduced in [29]. These authors used a fully Bayesian analysis of the propensity score and outcome regression models, which were defined previously in (23.1) and (23.3). The two key parameters of interest here are $\boldsymbol{\alpha}$ and $\boldsymbol{\beta}$, which correspond to the parameters of the propensity score model and the additional covariate adjustment in the outcome model, respectively. Each of these are $p-$dimensional, and we are working under the framework where the dimension of these needs to be reduced in some way. One way is to introduce spike-and-slab prior distributions of the following form

$$\begin{aligned} P(\alpha_j|\gamma_j) &\sim (1-\gamma_j)\delta_0 + \gamma_j\mathcal{N}(0,\sigma^2_\alpha) \\ P(\beta_j|\gamma_j) &\sim (1-\gamma_j)\delta_0 + \gamma_j\mathcal{N}(0,\sigma^2_\beta) \\ P(\gamma_j) &= p^{\gamma_j}(1-p)^{1-\gamma_j}, \end{aligned}$$

where δ_0 is a point mass at zero, and p is the prior probability of including a confounder into the models. This distribution is a mixture between a continuous density (the slab) and a discrete point mass at zero (the spike). The crucial parameter is $\gamma_j \in \{0,1\}$, which indicates whether a parameter is nonzero or not. A key difference here between standard spike-and-slab prior distribution implementations is that γ_j is shared between both models, which means that a confounder is either included in both models or excluded from both models. This induces two critical features for this model to perform well: 1) It eliminates model feedback as described in Section 23.2.1 and 2) it ensures that information from both the treatment and outcome is used when updating the posterior distribution of γ_j. Variables that have a strong association with either the treatment or outcome, but only a weak association with the other, are much more likely to be included using this strategy, which should increase the probability of including all the necessary confounders. This strategy has been shown to have strong empirical performance when the number of confounders is large, and interestingly can outperform (in terms of mean squared error) the approach of simply including all covariates even when the number of covariates is not prohibitively large.

23.3.3 Bayesian model averaging for related causal estimators

While the previous section detailed Bayesian model averaging for propensity score based estimation, similar ideas have been employed in similar contexts. One of the first works in this area was the Bayesian adjustment for confounding (BAC) prior distribution [27]. This approach estimates the treatment effect using an outcome model of the form:

$$E(Y|W = w, \boldsymbol{X} = \boldsymbol{x}) = \beta_0 + \beta_w w + \sum_{j=1}^{p} \beta_j x_j$$

Spike-and-slab prior distributions could simply be applied to the β_j coefficients, but this would ignore each variable's association with the treatment and could lead to substantial bias in finite samples. For this reason, they estimate both the outcome model above and the propensity score model defined in (23.1). Spike-and-slab prior distributions are then placed on the parameters of both models as

$$P(\alpha_j|\gamma_j^w) \sim (1 - \gamma_j^w)\delta_0 + \gamma_j^w \mathcal{N}(0, \sigma_\alpha^2)$$
$$P(\beta_j|\gamma_j^y) \sim (1 - \gamma_j^y)\delta_0 + \gamma_j^y \mathcal{N}(0, \sigma_\beta^2).$$

Note here that there are distinct γ_j^w and γ_j^y parameters indicating that the covariates in the treatment and outcome models can differ. To increase the likelihood that all confounders have $\gamma_j^y = 1$, the following prior distribution is used for the binary inclusion parameters:

$$P(\gamma_j^w = 0, \gamma_j^y = 0) = P(\gamma_j^w = 0, \gamma_j^y = 1) = P(\gamma_j^w = 1, \gamma_j^y = 1) = \frac{\omega}{3\omega + 1}$$
$$P(\gamma_j^w = 1, \gamma_j^y = 0) = \frac{1}{3\omega + 1},$$

where $\omega \geq 1$ is a tuning parameter that controls the degree of linkage between the propensity score and outcome models. If $\omega = 1$, then the prior distributions for γ_j^w and γ_j^y are independent; however, as ω increases the dependence between the two grows. It is easier to see this dependence by looking at the conditional odds implied by this prior distribution:

$$\frac{P(\gamma_j^y = 1|\gamma_j^w = 1)}{P(\gamma_j^y = 0|\gamma_j^w = 1)} = \omega, \quad \frac{P(\gamma_j^y = 1|\gamma_j^w = 0)}{P(\gamma_j^y = 0|\gamma_j^w = 0)} = 1, \quad \text{for } j = 1, \ldots, p$$

This shows that if a covariate is included in the treatment model, then it is far more likely to be included in the outcome model. This reduces the finite sample bias of the treatment effect by increasing the probability that all confounders are included into the outcome model. These ideas have been explored in a range of contexts in causal inference such as treatment effect heterogeneity [53], missing data [54], exposure-response curve estimation [55], and multiple exposures [56].

Now that Bayesian model averaging has been utilized in both propensity score and outcome regression models for estimating causal effects, it is natural to assume that they can be used for doubly robust estimation as well. We will cover doubly robust estimation in detail in the following section, but loosely it allows for consistent estimates of causal effects if either (but not necessarily both) of the treatment and outcome model are correctly specified. Clearly this is a desirable feature and was therefore adopted in [57], in which the authors utilized Bayesian model averaging within the context of doubly robust estimators. Let $\mathcal{M}^{om}$ and $\mathcal{M}^{ps}$ represent the model space for the outcome model and propensity score model, respectively. Further, let $\mathcal{M}_1^{ps}$ be the null model that includes no predictors in the propensity score. They assign a uniform prior on the space of outcome models, but use a prior distribution for the propensity score model space that is conditional on the outcome model, thereby linking the two models. The prior distribution for the propensity score model space is given by

$$\frac{P(\mathcal{M}_i^{ps}|\mathcal{M}_j^{om})}{P(\mathcal{M}_1^{ps}|\mathcal{M}_j^{om})} = \begin{cases} 1, & \mathcal{M}_i^{ps} \subset \mathcal{M}_j^{om} \\ \tau, & \text{otherwise.} \end{cases}$$

As with the BAC prior distribution, there is a tuning parameter $\tau \in [0,1]$ that controls the amount of linkage between the propensity score and outcome models. When $\tau = 0$, the propensity score can only include variables that are included in the outcome model. When τ is between 0 and 1, smaller weight is given to propensity score models that include terms that are not included in the outcome model. For each combination of the possible treatment and outcome models, the doubly robust estimator is calculated, and the overall estimator is a weighted average of these individual estimates given by the respective weights given to each model. Formally, their estimator is given by

$$\widehat{\Delta} = \sum_{i,j} w_{ij} \widehat{\Delta}_{ij},$$

where $w_{ij} = P(\mathcal{M}_i^{ps}, \mathcal{M}_j^{om} | \text{Data})$ is the posterior probability of propensity score model i and outcome model j, and $\widehat{\Delta}_{ij}$ is the doubly robust estimator using these two models. This model leverages Bayesian inference to provide model weights in a model averaged estimator, but does not use Bayesian inference in the estimation of the causal effect or to account for parameter uncertainty. In the following section we detail how doubly robust estimation can incorporate Bayesian tools to improve inference in a more general way than what was seen in [57].

23.4 Doubly Robust Estimation

Doubly robust estimation is a topic that has received substantial attention in the causal inference literature, and for good reason. Doubly robust estimators have a number of very desirable properties. The most obvious of which is the namesake property that says that consistent estimates of treatment effects can be obtained if only one of two models is correctly specified. Another benefit of certain doubly robust estimators that has received a great deal of attention is the property that fast $\sqrt{n}$ convergence rates can be obtained for estimators even when the nuisance parameters, such as those in a propensity score model or outcome regression model, converge at slower rates when both models are correctly specified. This final property allows for either 1) more flexible, nonparametric estimates of the nuisance functions or 2) high-dimensional covariate dimensions. There are a large number of doubly robust estimators in the literature that span a range of estimands such as the average treatment effect [58, 59], conditional average treatment effect [60], local average treatment effect [61], and continuous exposure-response curve [62], among others. Even within a particular estimand such as the ATE, there are a number of different estimators that can be shown to be doubly robust such as those based on weighting [59], matching [46, 63], covariate balancing propensity scores [64], or calibrated estimators of nuisance functions [65].

This chapter is about how Bayesian inference can be used to aid estimation of treatment effects, and nearly all of the estimators listed above do not have a natural Bayesian counterpart. For this reason, we will focus on the augmented IPW estimator, which we will define as

$$\Delta(\boldsymbol{D}, \boldsymbol{\Psi}) = \frac{1}{n} \sum_{i=1}^{n} \left\{ \frac{W_i(Y_i - m_{1i})}{p_{1i}} + m_{1i} - \frac{(1 - W_i)(Y_i - m_{0i})}{p_{0i}} - m_{0i} \right\}, \tag{23.4}$$

where $p_{wi} = P(W_i = w | \boldsymbol{X} = \boldsymbol{X}_i)$ and $m_{wi} = E(Y | W = w, \boldsymbol{X} = \boldsymbol{X}_i)$ are the treatment and outcome models, respectively. Note that we have written this estimator as a function of both $\boldsymbol{D} = (Y, W, \boldsymbol{X})$ and unknown parameters $\boldsymbol{\Psi}$, which encapture all unknown parameters of both the propensity score and outcome regression models. This is not a likelihood based estimator and therefore does not have a natural Bayesian formulation. Recent work in [9] aims to construct doubly robust estimators using Bayesian posterior predictive distributions and a change of measure using importance sampling, which they argue emits a natural Bayesian interpretation. In this chapter we will not address whether any of the following approaches are fully Bayesian, nor do we attempt to construct a fully Bayesian estimator. Rather we will highlight how certain Bayesian ideas can be very useful for certain aspects of doubly robust estimation, and how finite sample inference can be improved using Bayesian models for the nuisance parameters in a doubly robust estimator. In the previous section we saw how Bayesian model averaging could improve inference in doubly robust estimation, but there are a multitude of other useful properties that we would ideally be able to imbed within doubly robust estimation such as Bayesian nonparametrics, easy handling of missing data, or not needing to rely on asymptotic theory for inference.

While the doubly robust estimator described above does not have a Bayesian interpretation, both the propensity score and outcome regression models, which the doubly robust estimator is a function of, can be estimated from likelihood-based procedures and therefore within the Bayesian paradigm. Using the notation above, we can estimate $\boldsymbol{\Psi}$ within the Bayesian paradigm and obtain the posterior distribution of these parameters, which we denote by $P(\boldsymbol{\Psi}|\boldsymbol{D})$. There are two important questions that are left to be answered: 1) Once we have the posterior distribution of the propensity score and outcome model parameters, how do we construct point estimates and confidence intervals, and 2) why is this a useful pursuit? Do these estimators actually provide something that existing frequentist estimators do not, or are we simply performing frequentist-pursuit as some critics of Bayesian causal inference like to state? We answer both of these questions in what follows.

There are two possible estimators of Δ once the posterior distribution is obtained. The more common approach is the more common approach is to use an estimate of both p_{wi} and m_{wi}. We could use the posterior mean by setting $\widehat{p}_{wi} = E_{\boldsymbol{\Psi}|\boldsymbol{D}}[p_{wi}]$ and $\widehat{m}_{wi} = E_{\boldsymbol{\Psi}|\boldsymbol{D}}[m_{wi}]$. Then, we plug these values into (23.4) to estimate the average treatment effect. Using the notation above, this can be defined as $\widehat{\Delta} = \Delta(\widehat{\boldsymbol{\Psi}}, \boldsymbol{D})$ with $\widehat{\boldsymbol{\Psi}} = E_{\boldsymbol{\Psi}|\boldsymbol{D}}[\boldsymbol{\Psi}]$. This is the common approach taken in frequentist analyses, and while this strategy is a reasonable one, inference is more challenging. Either the bootstrap can be used to account for uncertainty in both stages of the estimator, or the estimator's asymptotic distribution is obtained from which inference can proceed. These approaches, however, may not be valid or may not perform well in finite samples with complex or high-dimensional models for the propensity score and outcome regression. It is natural to assume that the posterior distribution can be used to provide measures of uncertainty, but it is not clear how it can be used for this estimator. In light of this, a second approach can be use for estimating the ATE, which is to construct an estimator as

$$\widehat{\Delta} = E_{\boldsymbol{\Psi}|\boldsymbol{D}}[\Delta(\boldsymbol{\Psi}, \boldsymbol{D})], \tag{23.5}$$

which is the posterior mean of the $\Delta(\boldsymbol{\Psi}, \boldsymbol{D})$ function. Intuitively, for every posterior draw of the propensity score and outcome regression models we evaluate (23.4), and the mean of these values is our estimator. The posterior mean can be approximated using B posterior draws as

$$E_{\boldsymbol{\Psi}|\boldsymbol{D}}[\Delta(\boldsymbol{\Psi}, \boldsymbol{D})] \approx \frac{1}{B}\sum_{b=1}^{B} \Delta(\boldsymbol{\Psi}^{(b)}, \boldsymbol{D}),$$

This is also a somewhat obvious choice for an estimator, but we'll see that this estimator leads to a strategy for inference that works in difficult settings such as when the covariate space is high-dimensional, or when the models used to estimate the propensity score and outcome regression are highly flexible. The goal will be to construct an estimate of the variance of the effect estimate that accounts for all sources of uncertainty. The target variance is the variance of the sampling distribution of the estimator, defined by $\text{Var}_{\boldsymbol{D}} E_{\boldsymbol{\Psi}|\boldsymbol{D}}[\Delta(\boldsymbol{\Psi}, \boldsymbol{D})]$. There are two main sources of variability in this estimator: 1) the uncertainty in parameter estimation for the propensity score and outcome regression models and 2) sampling variability in $\boldsymbol{D}_i$ that is present even if we knew the true outcome and propensity score models. Following ideas seen in [50], we will show that the posterior distribution of model parameters can be combined with a simple resampling procedure to provide a variance estimator that is consistent when both the propensity score and outcome regression models are correctly specified and contract at sufficiently fast rates and is conservative in finite samples or under model misspecification.

Before defining our variance estimator, we must introduce additional notation. Let $\boldsymbol{D}^{(m)}$ be a resampled version of our original data $\boldsymbol{D}$, where resampling is done with replacement as in the nonparametric bootstrap. The variance estimator is defined as

$$\text{Var}_{\boldsymbol{D}^{(m)}}\{E_{\boldsymbol{\Psi}|\boldsymbol{D}}[\Delta(\boldsymbol{\Psi}, \boldsymbol{D}^{(m)})]\} + \text{Var}_{\boldsymbol{\Psi}|\boldsymbol{D}}[\Delta(\boldsymbol{\Psi}, \boldsymbol{D})]. \tag{23.6}$$

The first of these two terms resembles the true variance, except the outer variance is no longer with respect to $\boldsymbol{D}$, but is now with respect to $\boldsymbol{D}^{(m)}$. This is a crucial difference, however, as the first term does not account for variability due to parameter estimation. The inner expectation of the first term is with respect to the posterior distribution of $\boldsymbol{\Psi}$ given the observed data $\boldsymbol{D}$, not the resampled data $\boldsymbol{D}^{(m)}$. This means that this variance term does not account for variability that is caused by the fact that different data sets would lead to different posterior distributions. Ignoring this source of variability will likely lead to anti-conservative inference as our estimated variance will be smaller than the true variance of our estimator. To fix this issue, the second term is introduced, which is the variability of the estimator due to parameter uncertainty.

Computing this variance estimator is relatively straightforward once we have the posterior distribution of $\boldsymbol{\Psi}$. The second term, given by $\text{Var}_{\boldsymbol{\Psi}|\boldsymbol{D}}[\Delta(\boldsymbol{\Psi}, \boldsymbol{D})]$, is simply the variability across posterior samples of the doubly robust estimator in (23.4) evaluated at the observed data. To calculate the first term in (23.6), we can create M new data sets, $\boldsymbol{D}^{(1)}, \ldots, \boldsymbol{D}^{(M)}$, by sampling with replacement from the empirical distribution of the data. For each combination of resampled data set and bootstrap sample, we can calculate the doubly robust estimator defined in (23.4). This creates an $M \times B$ matrix of treatment effect estimates as given below.

$$\begin{pmatrix} \Delta(\boldsymbol{D}^{(1)}, \boldsymbol{\Psi}^{(1)}) & \Delta(\boldsymbol{D}^{(1)}, \boldsymbol{\Psi}^{(2)}) & \ldots & \Delta(\boldsymbol{D}^{(1)}, \boldsymbol{\Psi}^{(B)}) \\ \Delta(\boldsymbol{D}^{(2)}, \boldsymbol{\Psi}^{(1)}) & \ddots & & \Delta(\boldsymbol{D}^{(2)}, \boldsymbol{\Psi}^{(B)}) \\ \vdots & & \ddots & \vdots \\ \Delta(\boldsymbol{D}^{(M)}, \boldsymbol{\Psi}^{(1)}) & \Delta(\boldsymbol{D}^{(M)}, \boldsymbol{\Psi}^{(2)}) & \ldots & \Delta(\boldsymbol{D}^{(M)}, \boldsymbol{\Psi}^{(B)}) \end{pmatrix}$$

Once this matrix of estimates is obtained, we can take the mean within each row, which corresponds to the posterior mean of the estimator at each of the M resampled data sets. Taking the variance of these M estimators leads to an estimate of $\text{Var}_{\boldsymbol{D}^{(m)}}\{E_{\boldsymbol{\Psi}|\boldsymbol{D}}[\Delta(\boldsymbol{\Psi}, \boldsymbol{D}^{(m)})]\}$.

The variance estimator in (23.6) makes sense intuitively as it involves adding posterior variability to the first term, which was ignoring uncertainty from parameter estimation. However, it is not clear that the sum of these two terms lead to a valid variance estimator. Under general settings this estimate of the variance was shown to be conservative in that it gives estimates of the variance that are too large on average, leading to more conservative inference. This would seem potentially problematic at first as it is not known just how large this variance estimate could be, and it may lead to overly wide confidence intervals that lead to low statistical power. It was shown, however, that under certain conditions on the posterior distributions for the treatment and outcome models that this is a consistent variance estimator. In Bayesian statistics, asymptotics are frequently expressed in terms of posterior contraction rates instead of convergence rates used for point estimators. Posterior contraction rates detail the behavior of the entire posterior distribution instead of a simple measure of centrality such as the posterior mean or median. We say that the treatment and outcome models contract at rates ϵ_{nw} and ϵ_{ny} if the following holds:

(i) $\sup_{P_0} E_{P_0}\mathbb{P}_n\left(\frac{1}{\sqrt{n}}||\boldsymbol{p}_w - \boldsymbol{p}_w^*||_2 > M_w\epsilon_{nw} \mid \boldsymbol{D}\right) \rightarrow 0,$

(ii) $\sup_{P_0} E_{P_0}\mathbb{P}_n\left(\frac{1}{\sqrt{n}}||\boldsymbol{m}_w - \boldsymbol{m}_w^*||_2 > M_y\epsilon_{ny} \mid \boldsymbol{D}\right) \to 0,$

where M_w and M_y are constants, $\boldsymbol{p}_w = (p_{w1}, \ldots, p_{wn})$, $\boldsymbol{m}_w = (m_{w1}, \ldots, m_{wn})$, and $\boldsymbol{p}_w^*$ and $\boldsymbol{m}_w^*$ denote their unknown, true values.. Here, ϵ_{nw} and ϵ_{ny} determine how quickly the posterior distribution centers around the true values for the propensity score and outcome regression models. If we assume correctly specified parametric models, then these rates of contraction would be $n^{-1/2}$. For high-dimensional or nonparametric models, we expect slower rates of convergence where $\epsilon_{nw} \geq n^{-1/2}$ and $\epsilon_{ny} \geq n^{-1/2}$. It was shown that if both the treatment and outcome model posterior distributions contract at rates such that $\epsilon_{nw} \leq n^{-1/4}$ and $\epsilon_{ny} \leq n^{-1/4}$, i.e. faster rates than $n^{-1/4}$, then the variance estimator is consistent for the true variance. In addition to the variance being consistent, under these same assumptions, the estimator of the treatment effect defined in (23.5) is consistent at the $\sqrt{n}$ rate. The intuition behind this result is relatively straightforward. The first term in the variance estimator in (23.6) is effectively the variance of the doubly robust estimator that ignores parameter uncertainty. It is well known that one desirable feature of the doubly robust estimator is that its asymptotic variance is the same whether fixed parameters are used or not and therefore does not need to account for parameter uncertainty. This implies that the first term in (23.6) is asymptotically equivalent to the true variance of interest. The second term in (23.6), which amounts to uncertainty in the doubly robust estimator from parameter estimation, can be shown to be asymptotically negligible under the conditions above, and therefore the variance estimator is consistent.

These results are similar to those seen in the frequentist causal inference literature for semiparametric or high-dimensional models [44, 66], so a natural question might be to ask what Bayesian inference provided in this setting. The key to answering that question lies in the situations when the conditions of the theoretical results above do not hold. What happens in small sample sizes, or when one of the propensity score and outcome models is misspecified? In [50] it was shown that existing estimators that utilize asymptotic theory to provide inference, which ignore uncertainty from parameter estimation, perform poorly in small sample sizes or model misspecification in terms of interval coverage, obtaining levels well below the nominal level. The approach above, which uses the posterior distribution of $\boldsymbol{\Psi}$ to account for parameter uncertainty instead of ignoring it, performs well in these scenarios achieving coverage rates at or slightly above the nominal level. In simpler situations such as finite-dimensional parametric models for the two regression models and large sample sizes, the two approaches perform very similarly leading to valid inference on the treatment effect. In the more difficult settings of high dimensions, finite samples, or complex models for the two regression functions, the Bayesian approach provides a solution that works in general, and at worst is slightly conservative. Note again that this approach is not fully Bayesian, but rather Bayesian ideas are used to account for a difficult source of uncertainty that is commonly ignored.

23.5 Other Issues at the Intersection of Confounding Adjustment and Bayesian Analysis

So far we have discussed a variety of ways in which Bayesian inference can be combined with propensity scores to estimate treatment effects in observational studies. However, there are a number of other issues in causal estimation where Bayesian inference has proved to be quite useful. We do not have adequate space here to cover all such approaches, but we point readers to a number of interesting manuscripts for further reading. Bayesian inference can be particularly useful for problems such as principle stratification [67–69] or missing data [70], where the posterior distribution of the causal effect of interest can easily account for uncertainty in missing or latent quantities. The Bayesian approach has been used in a wide variety of other contexts such as mediation analysis [71, 72], sensitivity analysis for unmeasured confounders [73], estimation in panel data settings [74], estimation in regression discontinuity designs [75, 76], dealing with high-dimensional confounders [49, 77], and instrumental variable analysis [78], among others. In this chapter, we are focused on propensity score methods, and more generally the issue of confounding adjustment. Along these lines we discuss three main issues in this section: 1) sample estimands and uncertainty quantification, 2) the use of nonparametric Bayesian approaches for flexible confounding adjustment and effect estimation, and 3) treatment effect heterogeneity.

23.5.1 Sample estimands and fully Bayesian analysis of potential outcomes

One beneficial feature of the Bayesian paradigm when applied to causal inference problems is the straightforward estimation and uncertainty quantification of different estimands. So far we have only considered estimation of population average causal effects such as the average treatment effect defined by $E(Y_1 - Y_0)$. Another commonly used estimand is the sample average treatment effect, which is defined as

$$\text{SATE} = \frac{1}{n}\sum_{i=1}^{n}\{Y_{1i} - Y_{0i}\}.$$

Sample treatment effects differ from population based ones as they treat the potential outcomes as fixed quantities, while population estimands assume that the potential outcomes are random draws from a superpopulation. As is always the case, one of (Y_{0i}, Y_{1i}) is unknown for each individual, and the uncertainty in estimation of the SATE stems from the uncertainty in this missing value. Bayesian inference, however, readily allows for the imputation of this missing potential outcome if we model the joint distribution of the potential outcomes [10]. A simple approach is to posit a linear model for the potential outcomes using

$$\begin{pmatrix} Y_1 \\ Y_0 \end{pmatrix} = \mathcal{N}\left(\begin{pmatrix} \boldsymbol{X}\beta_1 \\ \boldsymbol{X}\beta_0 \end{pmatrix}, \begin{pmatrix} \sigma_1^2 & \rho\sigma_1\sigma_0 \\ \rho\sigma_1\sigma_0 & \sigma_0^2 \end{pmatrix} \right).$$

When updating all unknown parameters in an MCMC algorithm, the unknown potential outcome for each individual can be treated as an unknown parameter and updated from its full conditional distribution. For control individuals with $W_i = 0$, we update their missing potential outcome from the following distribution:

$$Y_{1i}|\cdot \sim \mathcal{N}\left(\boldsymbol{X}\beta_1 + \frac{\rho\sigma_1}{\sigma_0}(Y_i - \boldsymbol{X}\beta_0), \sigma_1^2(1-\rho^2) \right),$$

where an analogous distribution holds for treated individuals that require imputation of Y_{0i}. The main issue with this model is that Y_{0i} and Y_{1i} are never jointly observed and therefore there is no information in the data to inform the correlation between these two potential outcomes, denoted by ρ. One strategy is to vary ρ from 0 to 1 and assess how results vary as a function of the correlation between the potential outcomes. Another approach is to place a prior distribution on ρ that assigns probability to plausible values and allows the resulting posterior distribution to average over possible values of this correlation.

It is useful to take this approach from the Bayesian paradigm as it is easy to account for uncertainty in the imputations of the missing potential outcomes and to average over uncertainty about ρ, however, there are other potential benefits in this setting. If the dimension of the covariate space is relatively large then shrinkage priors or Bayesian model averaging can be used for β_0 and β_1. Additionally, the model above assumes both a linear association between the covariates and the mean of the potential outcomes, as well as normality of the potential outcomes. One or both of these assumptions can easily be alleviated using nonparametric Bayesian prior distributions, which do not make strong parametric assumptions about either the mean or distribution of the potential outcomes. These nonparametric Bayesian approaches have shown to be very effective in a variety of settings, and therefore we detail their usage in a variety of causal estimators in the following section.

23.5.2 Incorporating nonparametric Bayesian prior distributions

A common theme among Bayesian approaches in causal inference is the use of nonparametric Bayesian formulations. These models have been shown to work well in a vast array of settings, not just causal inference, and their impressive empirical performance has been carried over into causal inference in a number of distinct ways. Broadly speaking, nonparametric Bayes refers to situations in which a prior distribution is placed on an infinite dimensional parameter space, such as a function or a distribution. To illustrate the difference between traditional Bayesian inference and nonparametric Bayes, we can examine the simple problem of density estimation. Suppose we observe random variables Z_i for $i = 1, \ldots, n$ and we wish to estimate the their probability density function. The traditional Bayesian approach would be to assume a family of distributions governed by a finite set of parameters, and then assign prior distributions for these parameters. The simplest model would be the following:

$$\begin{aligned} Z_i|\mu, \sigma^2 &\sim \mathcal{N}(\mu, \sigma^2) \\ \mu &\sim \mathcal{N}(\mu_0, \sigma_0^2) \\ \sigma^2 &\sim \mathcal{IG}(a_0, b_0). \end{aligned}$$

Here we have assumed that Z_i follows a normal distribution with mean μ and variance σ^2, and then we assigned a normal prior distribution for the mean, and an inverse-gamma prior distribution for the variance. A nonparametric Bayesian alternative would be to assume

$$\begin{aligned} Z_i|G &\sim G \\ G &\sim DP(\alpha, M), \end{aligned}$$

where $DP(\alpha, M)$ represents a dirichlet process with centering parameter α and base distribution M. Here we could take M to be a normal distribution as above, but this formulation allows for deviations away from normality and the degree of flexibility is governed by α. We won't discuss this in further detail here, but the main idea is that Bayesian nonparametrics place prior distributions on infinite dimensional parameters so that parametric assumptions do not need to be made. The dirichlet process example above is simply one such Bayesian nonparametric solution to this problem, though many others exist. For an accessible introduction into Bayesian nonparametrics, we point readers to [79].

Now we can discuss how these ideas have been successfully applied in causal inference problems. First note that the nonparametric Bayesian ideas that follow can be applied to either propensity score estimation or outcome regression models. We choose to focus on outcome regression models here as they are far more commonly combined with Bayesian nonparametrics. Within outcome model estimation, there are two distinct ways in which Bayesian nonparametrics can be applied. The first is to use flexible nonparametric Bayesian prior distributions to improve estimation of the mean of the outcome regression surface, while the rest of the outcome distribution is allowed to be fully parametric. The second approach is to adopt a fully nonparametric solution to the entire data generating process rather than simply the mean of the outcome.

23.5.2.1 Modeling the mean function

Arguably the most common way in which Bayesian nonparametrics have influenced causal estimation is in flexible modeling of the outcome response surface. We can specify a parametric distribution for the outcome via

$$Y_i | W_i = w, \boldsymbol{X}_i = \boldsymbol{x} \sim \mathcal{N}(\mu(w, \boldsymbol{x}), \sigma^2),$$

and then utilize Bayesian nonparametrics for estimation of $\mu(w, \boldsymbol{x})$. The idea to utilize nonparametric Bayesian methods to estimate this function was first described in [80]. The key idea is that a number of estimands of interest, such as sample, population, or conditional treatment effects, can be obtained once we know $\mu(w, \boldsymbol{x})$. For this reason, they focused on building easy to use and flexible models for this unknown function. In this first paper the author utilized Bayesian additive regression trees (BART) [81] to model the mean function, which assumes that

$$\mu(w, \boldsymbol{x}) = \sum_{k=1}^{K} g(w, x; T_k, M_k).$$

Here, T_k corresponds to a particular tree structure or partition of the data, while M_k corresponds to the value of the responses in each terminal node of the tree given by T_k. Each of the individual trees is a weak-learner in that they do not predict the outcome well on their own, but the sum of these trees leads to a very strong predictive model. This approach was shown to work remarkably well at estimating causal effects. When the true regression models are nonlinear, the BART model greatly outperforms simple parametric models. When the true regression models are linear, the BART model performs nearly as well as an estimate of the true linear regression model. This is an overarching theme of Bayesian nonparametrics and why they can be so effective: they adapt well to complex situations, but also perform nearly as well as parametric models when the truth is indeed parametric.

Given a function $\mu(w, \boldsymbol{x})$, inference can proceed for a number of distinct estimands. Assuming unconfoundedness and consistency, the conditional average treatment effect, which we highlight in the following section, is given by

$$\begin{aligned} E(Y_1 - Y_0 | \boldsymbol{X} = \boldsymbol{x}) &= E(Y_1 | W = 1, \boldsymbol{X} = \boldsymbol{x}) - E(Y_0 | W = 0, \boldsymbol{X} = \boldsymbol{x}) \\ &= E(Y | W = 1, \boldsymbol{X} = \boldsymbol{x}) - E(Y | W = 0, \boldsymbol{X} = \boldsymbol{x}) \\ &= \mu(1, \boldsymbol{x}) - \mu(0, \boldsymbol{x}). \end{aligned}$$

Inference can proceed automatically using the relevant quantiles of the posterior distribution of $\mu(1, \boldsymbol{x}) - \mu(0, \boldsymbol{x})$. This shows one advantage of BART over more algorithmic tree-based approaches such as random forests [82]. Prior distributions are assigned to T_k and M_k and uncertainty is automatically captured by their posterior distribution. Under the same assumptions, the average treatment effect can be identified from the observed data as

$$\int_{\boldsymbol{x}} \mu(1, \boldsymbol{x}) - \mu(0, \boldsymbol{x}) f(\boldsymbol{x}) d\boldsymbol{x} \approx \frac{1}{n} \sum_{i=1}^{n} \mu(1, \boldsymbol{x}) - \mu(0, \boldsymbol{x})$$

Inference in this situation is slightly more nuanced, because we must also account for uncertainty in the distribution of the covariates, which we are approximating with the empirical distribution from our observed sample. The posterior distribution of $\mu(w, \boldsymbol{x})$ does not account for this uncertainty, but we can account for it using the Bayesian bootstrap [83]. Suppose we have B posterior draws from our model, given by $\mu^{(b)}(w, \boldsymbol{x})$. We can define weights as $\xi_i = u_i - u_{i-1}$ where $u_0 = 0, u_n = 1$ and $u_1, \ldots, u_{n-1}$ are the order statistics from $n - 1$ draws from a standard uniform distribution. We can do this for every posterior sample to obtain $\xi_i^{(b)}$ and create posterior draws of the average causal effect as

$$\frac{1}{n} \sum_{i=1}^{n} \xi_i^{(b)} \left[\mu^{(b)}(1, \boldsymbol{x}) - \mu^{(b)}(0, \boldsymbol{x}) \right] \quad \text{for } b = 1, \ldots, B.$$

Inference can then proceed in a traditional Bayesian framework once a posterior distribution is obtained. Note that if we were interested in sample average treatment effects, then we would not have to do this additional uncertainty assessment as this is only done to account for uncertainty from using a sample average to estimate a population expectation.

While BART has been used extensively for causal inference, it is not the only approach to estimating $\mu(w, \boldsymbol{x})$. A popular approach in Bayesian nonparametrics is to use the Gaussian process prior distribution, which amounts to specifying a prior for $\mu(w, \boldsymbol{x})$ as

$$\mu(w, \boldsymbol{x}) \sim \mathcal{GP}(\mu_0(\cdot), C(\cdot, \cdot)).$$

Here $\mu_0(\cdot)$ is the mean function of the Gaussian process, which is usually either assumed to be $\mu_0(w, \boldsymbol{x}) = 0$ or a linear function given by $\mu_0(w, \boldsymbol{x}) = \beta_0 + \beta_w w + \sum_{j=1}^{p} \beta_j x_j$. The Gaussian process allows for deviations away from this mean function, and this is dictated by the kernel function $C(\boldsymbol{z}, \boldsymbol{z}')$, where $\boldsymbol{z} = [w, \boldsymbol{x}']'$. This function describes how similar two covariate vectors are, and there are a number of choices for this function, such as

$$C(\boldsymbol{z}, \boldsymbol{z}') = \sigma^2 \exp\{-\delta \sum_{j=1}^{p+1} (z_j - z_j')^2\}.$$

The Gaussian process prior works under an assumption that similar covariate vectors should have similar values of the regression function $\mu(\cdot)$. The only real assumption on the regression function is that of smoothness, and the degree of smoothness is dictated by δ. An easier way to understand the Gaussian process is to see that the function evaluated at a finite set of locations follows a multivariate normal distribution. In our setting, we have that the prior distribution at our observed n data points is given by

$$(\mu(w_1, \boldsymbol{x}_1), \ldots, \mu(w_n, \boldsymbol{x}_n))' \sim \mathcal{N}((\mu_0(w_1, \boldsymbol{x}_1), \ldots, \mu_0(w_n, \boldsymbol{x}_n))', \Sigma)$$

where the (i, j) element of Σ is given by $C(\boldsymbol{z}_i, \boldsymbol{z}_j)$. While Gaussian processes are widely used and have very strong predictive performance, there are certain drawbacks relative to the BART model described earlier. The main drawback is the computational burden of Gaussian processes, which require the inversion of an $n \times n$ matrix in every MCMC sample. While this can be substantially alleviated through certain approximations [84–86], the BART model is extremely fast computationally. Additionally, the default specifications for BART perform remarkably well across a wide range of scenarios and therefore the model requires little to no tuning or prior expertise. This is likely one of the driving reasons for BART regularly performing well in causal inference data analysis competitions [87].

23.5.2.2 Modeling the joint distribution

The previous approaches focused on flexible modeling of the mean function for the outcome regression, but still assumed a normal distribution with fixed variance for the distribution of Y given the treatment and covariates. An alternative option, which we will see has some desirable features, instead models the entire distribution of the data in a nonparametric way. One of the simplest approaches to doing this, which is explained in the context of causal inference in [88], is the dirichlet process mixture model. For simplicity of notation, assume now that $\boldsymbol{Z}_i = [1, W_i, \boldsymbol{X}_i]$. This model assumes that

$$\begin{aligned} Y_i | \boldsymbol{Z}_i, \boldsymbol{\beta}_i &\sim p(Y_i | \boldsymbol{Z}_i, \boldsymbol{\beta}_i) \\ \boldsymbol{Z}_i &\sim p(\boldsymbol{Z}_i | \boldsymbol{\theta}_i) \\ \boldsymbol{\phi}_i &\sim G \\ G &\sim DP(\alpha, M), \end{aligned}$$

where $\phi_i = [\beta_i, \theta_i]$ and M is an appropriate base measure of the dirichlet process. Effectively, this is simply a mixture model where each individual has their own parameters that come from a distribution G. The dirichlet process prior on G necessarily leads to a discrete distribution on these parameters, where some individuals will have the same parameters as others. While this model is quite complex, extensions have been proposed and used in the causal inference literature such as the enriched dirichlet process model [89]. Of more interest is what advantages these models have over the nonparametric mean functions of the previous section. The first is that this specifies a model for the full joint distribution of the data, and therefore any missingness in the covariates can easily be addressed by imputing the missing values within the MCMC algorithm [70]. A second benefit is that we are no longer restricted to estimands that examine the mean of the potential outcome surface. Approaches such as these can easily estimate average treatment effects, but immediately extend to more complex estimands such as quantile treatment effects [90]. If we let $F_1(y)$ and $F_0(y)$ be the cumulative distribution functions for the potential outcomes under treatment and control, respectively, we can define quantile treatment effects as

$$F_1^{-1}(p) - F_0^{-1}(p).$$

Once the posterior distribution is obtained for the joint density of $(Y_i, \boldsymbol{Z}_i)$, these quantities can easily be obtained through appropriate standardization, and inference can be accounted for through quantiles of the posterior distribution.

23.5.3 Treatment effect heterogeneity

In this section our goal will be to estimate the conditional average treatment effect defined above to be

$$E(Y_1 - Y_0 | \boldsymbol{X} = \boldsymbol{x}) = \mu(1, \boldsymbol{x}) - \mu(0, \boldsymbol{x}).$$

Within the Bayesian paradigm, there are three natural ways to estimating this quantity. The first and most flexible approach is to separately model these two regression functions where only the treated individuals are used to estimate $\mu(1, \boldsymbol{x})$ and only the control individuals are used to estimate $\mu(0, \boldsymbol{x})$. This amounts to placing independent prior distributions on these two functions, and estimating independent posterior distributions. While this approach is sufficiently flexible to capture nearly any type of treatment effect heterogeneity, it is also highly inefficient. We are only interested in the difference between these two functions, rather than the individual functions themselves. However, the prior variability in the difference between these two functions is twice the prior variability of the individual functions themselves. This is not desirable, particularly because treatment effects are likely to be only moderately heterogeneous, and therefore, we believe a priori the variability of this difference should be small.

Another approach, which was described above and in [80], is to place a single BART prior distribution on $\mu(w, \boldsymbol{x})$, from which we can obtain either mean function and their difference. Recent work, however, has shown that a more carefully thought out prior distribution for this function that explicitly targets the estimand of interest can perform better [91]. The main idea is to parameterize this function as

$$\mu(w, \boldsymbol{x}) = f(\boldsymbol{x}) + \tau(\boldsymbol{x})w.$$

By parameterizing the function in this way, no flexibility has been lost in the resulting function, but now it has been separated into two separate functions: one corresponding to the prognostic portion of the covariates, and one corresponding to the treatment effect. This allows for separate prior distributions, and importantly, separate amounts of shrinkage or regularization applied to each of these two functions. Now it is straightforward to encode our prior beliefs about the degree of treatment effect heterogeneity into prior distributions, and it is clear how the model shrinks treatment effects towards an overall homogeneous treatment effect. Any approach can be used to estimate these functions, including the nonparametric ones discussed in the previous section, though the Bayesian causal forest (BCF) approach that introduced this idea was restricted to BART priors on both functions. To reflect the fact that they believe heterogeneity to typically be small to moderate in magnitude, they use a different BART prior for $\tau(\boldsymbol{x})$ that prioritizes smaller trees and simpler functions. In addition to the reparameterization of this model, a key insight of this paper is that overall estimation is improved by including an estimate of the propensity score into the prognostic function, $f(\boldsymbol{x})$. Specifically, they fit the following model:

$$\mu(w, \boldsymbol{x}) = f(\widehat{\pi}, \boldsymbol{x}) + \tau(\boldsymbol{x})w,$$

where $\widehat{\pi}$ is an initial, likely frequentist, estimate of the propensity score. It was shown that this can help to reduce bias that the authors refer to as regularization induced confounding bias. Essentially this is bias that occurs when regularization is applied in an outcome model without regard to the fact that the goal is treatment effect estimation and not prediction of the outcome. To see the utility of these different approaches, we simulated a data set with a constant treatment effect of 0.3 that does not vary by observed characteristics. We applied all three of the approaches considered here for estimating $E(Y_1 - Y_0|\boldsymbol{X} = \boldsymbol{X}_i)$ for $i = 1, \ldots, n$, and the results can be seen in Figure 23.2. We see that the approach that fits two separate models leads to substantial amounts of variability in the individual treatment effect estimates. The 95% credible intervals still work well in that they typically cover the true parameter, which is given by the grey line at 0.3. The BCF and single BART approaches perform similarly well, but the BCF approach has estimates that are closer to the truth and intervals that are generally smaller in width than the single BART approach. This is likely due to the explicit parameterization that allows them to enforce more shrinkage and simpler BART models for the function governing treatment effect heterogeneity.

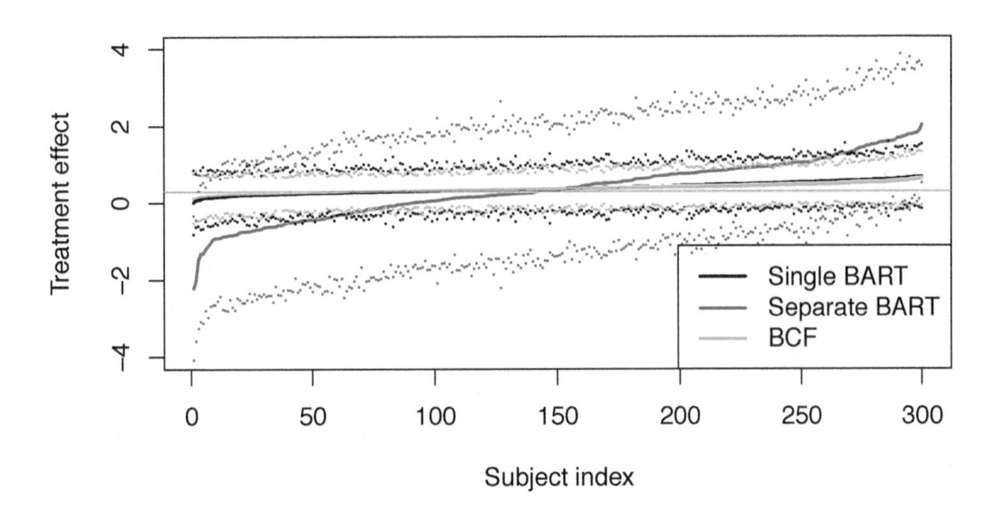

FIGURE 23.2
Conditional average treatment effect estimates evaluated at each $\boldsymbol{X}_i$ from the three models for treatment effect heterogeneity. The solid lines refer to posterior means, while the corresponding dots represent upper and lower 95% credible intervals.

Overall the Bayesian paradigm provided an elegant solution to this problem that provided a sufficient amount of flexibility for estimation of conditional average treatment effects while still being able to perform inference automatically on any estimands of interest. Future work could look into the best choice of prior distribution in this framework for estimating heterogeneous treatment effects. Recent work has shown that other priors such as Gaussian processes may perform better with respect to uncertainty quantification than BART, particularly when there is less propensity score overlap [92]. Other extensions of these ideas could involve non-binary treatments or higher-dimensional covariate spaces. The latter of these two is a difficult problem, particularly with respect to uncertainty quantification, but is one for which Bayesian inference is well-suited.

References

[1] A. Gelman, J. B. Carlin, H. S. Stern, D. B. Dunson, A. Vehtari, and D. B. Rubin, *Bayesian data analysis*. CRC Press, 2013.

[2] J. Berger, "The case for objective bayesian analysis," *Bayesian analysis*, vol. 1, no. 3, pp. 385–402, 2006.

[3] G. Consonni, D. Fouskakis, B. Liseo, and I. Ntzoufras, "Prior distributions for objective bayesian analysis," *Bayesian Analysis*, vol. 13, no. 2, pp. 627–679, 2018.

[4] W. R. Gilks, S. Richardson, and D. Spiegelhalter, *Markov chain Monte Carlo in practice*. CRC press, 1995.

[5] S. Brooks, A. Gelman, G. Jones, and X.-L. Meng, *Handbook of markov chain monte carlo*. CRC press, 2011.

[6] J. M. Robins and Y. Ritov, "Toward a curse of dimensionality appropriate (coda) asymptotic theory for semi-parametric models," *Statistics in Medicine*, vol. 16, no. 3, pp. 285–319, 1997.

[7] O. Saarela, D. A. Stephens, E. E. Moodie, and M. B. Klein, "On bayesian estimation of marginal structural models," *Biometrics*, vol. 71, no. 2, pp. 279–288, 2015.

[8] J. M. Robins, M. A. Hernán, and L. Wasserman, "On bayesian estimation of marginal structural models," *Biometrics*, vol. 71, no. 2, p. 296, 2015.

[9] O. Saarela, L. R. Belzile, and D. A. Stephens, "A bayesian view of doubly robust causal inference," *Biometrika*, vol. 103, no. 3, pp. 667–681, 2016.

[10] D. B. Rubin, "Bayesian inference for causal effects: The role of randomization," *Ann. Statist.*, vol. 6, no. 1, pp. 34–58, 1978.

[11] C. M. Zigler, "The central role of bayes' theorem for joint estimation of causal effects and propensity scores," *The American Statistician*, vol. 70, no. 1, pp. 47–54, 2016.

[12] D. B. Rubin and N. Thomas, "Matching using estimated propensity scores: relating theory to practice," *Biometrics*, pp. 249–264, 1996.

[13] E. A. Stuart, "Matching methods for causal inference: A review and a look forward," *Statistical Science: A Review Journal of the Institute of Mathematical Statistics*, vol. 25, no. 1, p. 1, 2010.

[14] J. M. Robins, M. A. Hernan, and B. Brumback, "Marginal structural models and causal inference in epidemiology," 2000.

[15] F. Li, K. L. Morgan, and A. M. Zaslavsky, "Balancing covariates via propensity score weighting," *Journal of the American Statistical Association*, vol. 113, no. 521, pp. 390–400, 2018.

[16] J. K. Lunceford and M. Davidian, "Stratification and weighting via the propensity score in estimation of causal treatment effects: A comparative study," *Statistics in Medicine*, vol. 23, no. 19, pp. 2937–2960, 2004.

[17] L. C. McCandless, P. Gustafson, and P. C. Austin, "Bayesian propensity score analysis for observational data," *Statistics in Medicine*, vol. 28, no. 1, pp. 94–112, 2009.

[18] D. B. Rubin *et al.*, "For objective causal inference, design trumps analysis," *Annals of Applied Statistics*, vol. 2, no. 3, pp. 808–840, 2008.

[19] D. Lunn, N. Best, D. Spiegelhalter, G. Graham, and B. Neuenschwander, "Combining mcmc with 'sequential'pkpd modelling," *Journal of Pharmacokinetics and Pharmacodynamics*, vol. 36, no. 1, p. 19, 2009.

[20] P. E. Jacob, L. M. Murray, C. C. Holmes, and C. P. Robert, "Better together? statistical learning in models made of modules," *arXiv preprint arXiv:1708.08719*, 2017.

[21] F. Liu, M. Bayarri, J. Berger, *et al.*, "Modularization in bayesian analysis, with emphasis on analysis of computer models," *Bayesian Analysis*, vol. 4, no. 1, pp. 119–150, 2009.

[22] L. C. McCandless, I. J. Douglas, S. J. Evans, and L. Smeeth, "Cutting feedback in bayesian regression adjustment for the propensity score," *The International Journal of Biostatistics*, vol. 6, no. 2, 2010.

[23] C. M. Zigler, K. Watts, R. W. Yeh, Y. Wang, B. A. Coull, and F. Dominici, "Model feedback in bayesian propensity score estimation," *Biometrics*, vol. 69, no. 1, pp. 263–273, 2013.

[24] P. R. Rosenbaum and D. B. Rubin, "The central role of the propensity score in observational studies for causal effects," *Biometrika*, vol. 70, no. 1, pp. 41–55, 1983.

[25] D. Kaplan and J. Chen, "A two-step bayesian approach for propensity score analysis: Simulations and case study," *Psychometrika*, vol. 77, no. 3, pp. 581–609, 2012.

[26] D. B. Rubin, "The use of propensity scores in applied bayesian inference," *Bayesian Statistics*, vol. 2, pp. 463–472, 1985.

[27] C. Wang, G. Parmigiani, and F. Dominici, "Bayesian effect estimation accounting for adjustment uncertainty," *Biometrics*, vol. 68, no. 3, pp. 661–671, 2012.

[28] A. Belloni, V. Chernozhukov, and C. Hansen, "Inference on treatment effects after selection among high-dimensional controls," *The Review of Economic Studies*, vol. 81, no. 2, pp. 608–650, 2014.

[29] C. M. Zigler and F. Dominici, "Uncertainty in propensity score estimation: Bayesian methods for variable selection and model-averaged causal effects," *Journal of the American Statistical Association*, vol. 109, no. 505, pp. 95–107, 2014.

[30] S. M. Shortreed and A. Ertefaie, "Outcome-adaptive lasso: Variable selection for causal inference," *Biometrics*, vol. 73, no. 4, pp. 1111–1122, 2017.

[31] A. Gelman and J. Hill, *Data analysis using regression and multilevel/hierarchical models*. Cambridge university press, 2006.

[32] A. Abadie and G. W. Imbens, "Matching on the estimated propensity score," *Econometrica*, vol. 84, no. 2, pp. 781–807, 2016.

[33] B. A. Brumback, "A note on using the estimated versus the known propensity score to estimate the average treatment effect," *Statistics & Probability Letters*, vol. 79, no. 4, pp. 537–542, 2009.

[34] C. M. Zigler and M. Cefalu, "Posterior predictive treatment assignment for estimating causal effects with limited overlap," *arXiv preprint arXiv:1710.08749*, 2017.

[35] R. M. Alvarez and I. Levin, "Uncertain neighbors: Bayesian propensity score matching for causal inference," *arXiv preprint arXiv:2105.02362*, 2021.

[36] S. X. Liao and C. M. Zigler, "Uncertainty in the design stage of two-stage bayesian propensity score analysis," *Statistics in Medicine*, vol. 39, no. 17, pp. 2265–2290, 2020.

[37] P. C. Austin, "A critical appraisal of propensity-score matching in the medical literature between 1996 and 2003," *Statistics in Medicine*, vol. 27, no. 12, pp. 2037–2049, 2008.

[38] R. Tibshirani, "Regression shrinkage and selection via the lasso," *Journal of the Royal Statistical Society: Series B (Statistical Methodology)*, vol. 58, no. 1, pp. 267–288, 1996.

[39] S. Vansteelandt, M. Bekaert, and G. Claeskens, "On model selection and model misspecification in causal inference," *Statistical Methods in Medical Research* , vol. 21, no. 1, pp. 7–30, 2012.

[40] D. B. Rubin, "Estimating causal effects from large data sets using propensity scores," *Annals of Internal Medicine*, vol. 127, no. 8_Part_2, pp. 757–763, 1997.

[41] M. A. Brookhart, S. Schneeweiss, K. J. Rothman, R. J. Glynn, J. Avorn, and T. Stürmer, "Variable selection for propensity score models," *American Journal of Epidemiology*, vol. 163, no. 12, pp. 1149–1156, 2006.

[42] S. Schneeweiss, J. A. Rassen, R. J. Glynn, J. Avorn, H. Mogun, and M. A. Brookhart, "High-dimensional propensity score adjustment in studies of treatment effects using health care claims data," *Epidemiology (Cambridge, Mass.)*, vol. 20, no. 4, p. 512, 2009.

[43] X. De Luna, I. Waernbaum, and T. S. Richardson, "Covariate selection for the nonparametric estimation of an average treatment effect," *Biometrika*, p. asr041, 2011.

[44] V. Chernozhukov, D. Chetverikov, M. Demirer, E. Duflo, C. Hansen, W. Newey, and J. Robins, "Double/debiased machine learning for treatment and structural parameters," *The Econometrics Journal*, vol. 21, pp. C1–C68, 01 2018.

[45] S. Athey, G. Imbens, and S. Wager, "Approximate residual balancing: debiased inference of average treatment effects in high dimensions," *Journal of the Royal Statistical Society Series B*, vol. 80, no. 4, pp. 597–623, 2018.

[46] J. Antonelli, M. Cefalu, N. Palmer, and D. Agniel, "Doubly robust matching estimators for high dimensional confounding adjustment," *Biometrics*, vol. 74, no. 4, pp. 1171–1179, 2018.

[47] J. Antonelli and M. Cefalu, "Averaging causal estimators in high dimensions," *Journal of Causal Inference*, vol. 8, no. 1, pp. 92–107, 2020.

[48] A. Ertefaie, M. Asgharian, and D. A. Stephens, "Variable selection in causal inference using a simultaneous penalization method," *Journal of Causal Inference*, vol. 6, no. 1, 2018.

[49] J. Antonelli, G. Parmigiani, and F. Dominici, "High-dimensional confounding adjustment using continuous spike and slab priors," *Bayesian Analysis*, vol. 14, no. 3, p. 805, 2019.

[50] J. Antonelli, G. Papadogeorgou, and F. Dominici, "Causal inference in high dimensions: A marriage between bayesian modeling and good frequentist properties," *Biometrics*, 2020.

[51] A. E. Raftery, D. Madigan, and J. A. Hoeting, "Bayesian model averaging for linear regression models," *Journal of the American Statistical Association*, vol. 92, no. 437, pp. 179–191, 1997.

[52] J. A. Hoeting, D. Madigan, A. E. Raftery, and C. T. Volinsky, "Bayesian model averaging: a tutorial," *Statistical Science*, pp. 382–401, 1999.

[53] C. Wang, F. Dominici, G. Parmigiani, and C. M. Zigler, "Accounting for uncertainty in confounder and effect modifier selection when estimating average causal effects in generalized linear models," *Biometrics*, vol. 71, no. 3, pp. 654–665, 2015. PMCID:PMC4575246.

[54] J. Antonelli, C. Zigler, and F. Dominici, "Guided bayesian imputation to adjust for confounding when combining heterogeneous data sources in comparative effectiveness research," *Biostatistics*, vol. 18, no. 3, pp. 553–568, 2017.

[55] G. Papadogeorgou, F. Dominici, *et al.*, "A causal exposure response function with local adjustment for confounding: Estimating health effects of exposure to low levels of ambient fine particulate matter," *Annals of Applied Statistics*, vol. 14, no. 2, pp. 850–871, 2020.

[56] A. Wilson, C. M. Zigler, C. J. Patel, and F. Dominici, "Model-averaged confounder adjustment for estimating multivariate exposure effects with linear regression," *Biometrics*, vol. 74, no. 3, pp. 1034–1044, 2018.

[57] M. Cefalu, F. Dominici, N. Arvold, and G. Parmigiani, "Model averaged double robust estimation," *Biometrics*, 2016.

[58] D. O. Scharfstein, A. Rotnitzky, and J. M. Robins, "Adjusting for nonignorable drop-out using semiparametric nonresponse models," *Journal of the American Statistical Association*, vol. 94, no. 448, pp. 1096–1120, 1999.

[59] H. Bang and J. M. Robins, "Doubly robust estimation in missing data and causal inference models," *Biometrics*, vol. 61, no. 4, pp. 962–973, 2005.

[60] V. Semenova and V. Chernozhukov, "Debiased machine learning of conditional average treatment effects and other causal functions," *arXiv preprint arXiv:1702.06240*, 2017.

[61] E. L. Ogburn, A. Rotnitzky, and J. M. Robins, "Doubly robust estimation of the local average treatment effect curve," *Journal of the Royal Statistical Society. Series B, Statistical methodology*, vol. 77, no. 2, p. 373, 2015.

[62] E. H. Kennedy, Z. Ma, M. D. McHugh, and D. S. Small, "Nonparametric methods for doubly robust estimation of continuous treatment effects," *Journal of the Royal Statistical Society. Series B, Statistical Methodology*, vol. 79, no. 4, p. 1229, 2017.

[63] S. Yang and Y. Zhang, "Multiply robust matching estimators of average and quantile treatment effects," *arXiv preprint arXiv:2001.06049*, 2020.

[64] Y. Ning, P. Sida, and K. Imai, "Robust estimation of causal effects via a high-dimensional covariate balancing propensity score," *Biometrika*, vol. 107, no. 3, pp. 533–554, 2020.

[65] Z. Tan *et al.*, "Model-assisted inference for treatment effects using regularized calibrated estimation with high-dimensional data," *Annals of Statistics*, vol. 48, no. 2, pp. 811–837, 2020.

[66] M. H. Farrell, "Robust inference on average treatment effects with possibly more covariates than observations," *Journal of Econometrics*, vol. 189, no. 1, pp. 1–23, 2015.

[67] F. Mealli and A. Mattei, “A refreshing account of principal stratification,” *The International Journal of Biostatistics*, vol. 8, no. 1, 2012.

[68] A. Mattei, F. Li, and F. Mealli, “Exploiting multiple outcomes in bayesian principal stratification analysis with application to the evaluation of a job training program,” *The Annals of Applied Statistics*, vol. 7, no. 4, pp. 2336–2360, 2013.

[69] L. Forastiere, F. Mealli, and T. J. VanderWeele, “Identification and estimation of causal mechanisms in clustered encouragement designs: Disentangling bed nets using bayesian principal stratification,” *Journal of the American Statistical Association*, vol. 111, no. 514, pp. 510–525, 2016.

[70] J. Roy, K. J. Lum, B. Zeldow, J. D. Dworkin, V. L. Re III, and M. J. Daniels, “Bayesian nonparametric generative models for causal inference with missing at random covariates,” *Biometrics*, vol. 74, no. 4, pp. 1193–1202, 2018.

[71] M. J. Daniels, J. A. Roy, C. Kim, J. W. Hogan, and M. G. Perri, “Bayesian inference for the causal effect of mediation,” *Biometrics*, vol. 68, no. 4, pp. 1028–1036, 2012.

[72] C. Kim, M. J. Daniels, B. H. Marcus, and J. A. Roy, “A framework for bayesian nonparametric inference for causal effects of mediation,” *Biometrics*, vol. 73, no. 2, pp. 401–409, 2017.

[73] L. C. McCandless, P. Gustafson, and A. Levy, “Bayesian sensitivity analysis for unmeasured confounding in observational studies,” *Statistics in Medicine*, vol. 26, no. 11, pp. 2331–2347, 2007.

[74] J. Antonelli and B. Beck, “Estimating heterogeneous causal effects in time series settings with staggered adoption: An application to neighborhood policing,” *arXiv e-prints*, pp. arXiv–2006, 2020.

[75] S. Chib and L. Jacobi, “Bayesian fuzzy regression discontinuity analysis and returns to compulsory schooling,” *Journal of Applied Econometrics*, vol. 31, no. 6, pp. 1026–1047, 2016.

[76] Z. Branson, M. Rischard, L. Bornn, and L. W. Miratrix, “A nonparametric bayesian methodology for regression discontinuity designs,” *Journal of Statistical Planning and Inference*, vol. 202, pp. 14–30, 2019.

[77] P. R. Hahn, C. M. Carvalho, D. Puelz, J. He, *et al.*, “Regularization and confounding in linear regression for treatment effect estimation,” *Bayesian Analysis*, vol. 13, no. 1, pp. 163–182, 2018.

[78] S. Adhikari, S. Rose, and S.-L. Normand, “Nonparametric bayesian instrumental variable analysis: Evaluating heterogeneous effects of coronary arterial access site strategies,” *Journal of the American Statistical Association*, vol. 115, no. 532, pp. 1635–1644, 2020.

[79] P. Müller, F. A. Quintana, A. Jara, and T. Hanson, *Bayesian nonparametric data analysis*. Springer, 2015.

[80] J. L. Hill, “Bayesian nonparametric modeling for causal inference,” *Journal of Computational and Graphical Statistics*, vol. 20, no. 1, pp. 217–240, 2011.

[81] H. A. Chipman, E. I. George, and R. E. McCulloch, “BART: Bayesian additive regression trees,” *The Annals of Applied Statistics*, vol. 4, no. 1, pp. 266–298, 2010.

[82] L. Breiman, “Random forests,” *Machine learning*, vol. 45, no. 1, pp. 5–32, 2001.

[83] D. B. Rubin, “The bayesian bootstrap,” *The annals of statistics*, pp. 130–134, 1981.

[84] R. B. Gramacy and H. K. H. Lee, "Bayesian treed gaussian process models with an application to computer modeling," *Journal of the American Statistical Association*, vol. 103, no. 483, pp. 1119–1130, 2008.

[85] S. Banerjee, A. E. Gelfand, A. O. Finley, and H. Sang, "Gaussian predictive process models for large spatial data sets," *Journal of the Royal Statistical Society: Series B (Statistical Methodology)*, vol. 70, no. 4, pp. 825–848, 2008.

[86] A. Banerjee, D. B. Dunson, and S. T. Tokdar, "Efficient gaussian process regression for large datasets," *Biometrika*, vol. 100, no. 1, pp. 75–89, 2013.

[87] V. Dorie, J. Hill, U. Shalit, M. Scott, and D. Cervone, "Automated versus do-it-yourself methods for causal inference: Lessons learned from a data analysis competition (with discussion)," *Statistical Science*, vol. 34, pp. 43–99, 02 2019.

[88] A. Oganisian and J. A. Roy, "A practical introduction to bayesian estimation of causal effects: Parametric and nonparametric approaches," *Statistics in Medicine*, vol. 40, no. 2, pp. 518–551, 2021.

[89] S. Wade, S. Mongelluzzo, and S. Petrone, "An enriched conjugate prior for bayesian nonparametric inference," *Bayesian Analysis*, vol. 6, no. 3, pp. 359–385, 2011.

[90] D. Xu, M. J. Daniels, and A. G. Winterstein, "A bayesian nonparametric approach to causal inference on quantiles," *Biometrics*, vol. 74, no. 3, pp. 986–996, 2018.

[91] P. R. Hahn, J. S. Murray, and C. M. Carvalho, "Bayesian regression tree models for causal inference: regularization, confounding, and heterogeneous effects," *Bayesian Analysis*, 2020.

[92] R. Papadogeorgou and F. Li, "Discussion of "Bayesian regression tree models for causal inference: Regularization, confounding, and heterogeneous effects"," *Bayesian Analysis*, vol. 15, no. 3, pp. 1007–1013, 2020.

Part V

Beyond Adjustments

24

How to Be a Good Critic of an Observational Study

Dylan S. Small

CONTENTS

In an observational study, an association between treatment and outcome is inescapably ambiguous – it could be explained by (i) a treatment effect or (ii) a bias in comparing the treatment and control group, e.g., due to differences between the treatment and control group prior to treatment in the absence of randomized assignment. However, this ambiguity can be reduced by investigating plausible biases and either (i) finding them wanting in explaining the association or (ii) finding that they do explain away the association. Good criticism of an observational study enhances what we learn from an observational study because it brings up plausible sources of biases and enables reducing ambiguity by investigating them. Bad criticism of an observational study shows doubt without enhancing understanding. This chapter is about how to be a good critic of an observational study.

As a case study, we will consider criticism of observational studies that claimed evidence for smoking causing lung cancer. We first provide background on smoking and lung cancer.

24.1 Smoking and Lung Cancer

Freedman [1] described the history of studies of causes of lung cancer up to the 1940s: "In the 1920s, physicians noticed a rapid increase of death rates from lung cancer. For many years, it was debated whether the increase was real or an artifact of improvement in diagnostics. (The lungs are inaccessible, and diagnosis is not easy.) By the 1940s, there was some agreement on the reality of

DOI: 10.1201/9781003102670-24

the increase, and the focus of the discussion shifted. What was the cause of the epidemic? Smoking was one theory. However, other experts thought that emissions from gas works were the cause. Still others believed that fumes from the tarring of roads were responsible."

There were early papers that reported an association between smoking and lung cancer and suggested a casual effect [2–4]. But two papers published in 1950, Wynder and Graham [5] in the US and Doll and Hill [6] in the UK, attracted more attention [1]. Many papers in the 1950s followed; see [7] for a review. In 1957 a Study Group appointed by the National Cancer Institute, the National Heart Institute, the American Cancer Society, and the American Heart Association, examined the scientific evidence on the effects of smoking on health and arrived at the following conclusion: "The sum total of scientific evidence establishes beyond reasonable doubt that cigarette smoking is a causative factor in the rapidly increasing incidence of human epidermoid carcinoma of the lung." [8] But the public was not convinced – cigarette sales were at an all time high in 1957 [9]. And there remained fierce scientific critics including the famous statisticians R.A. Fisher and Joseph Berkson. Motivated by what he saw as poor criticism in the debate over smoking and lung cancer, Bross [10] formulated ground rules for good statistical criticism.

24.2 Bross's Criterion for Good Criticism: Show that Counterhypothesis is Tenable

Consider an observational study that presents evidence in favor of the hypothesis that there is a treatment effect. A critic puts forth a counterhypothesis[1] there is no treatment effect and instead the results of the study can be explained by another factor, e.g., a preexisting difference between the treatment and control group. Bross [10] argued that a good critic should show that the counterhypothesis is in fact tenable. In the next two paragraphs we will contrast a criticism shown not to be tenable (first paragraph) and a criticism shown to be tenable (second paragraph).

One criticism of the hypothesis that smoking is a substantial cause of lung cancer was that the hypothesis was inconsistent with there being a higher rate of lung cancer among urban residents than rural residents [11–13]. These critics asserted that air pollution and exposure to industrial hazards were more likely to explain most of the increase in lung cancer than smoking, and that the higher air pollution and exposure to industrial hazards in urban areas explained the higher rate of lung cancer in urban areas. Evidence was cited that urban residents' lung cancer rate remained higher than rural residents' even after controlling for smoking history. This provides evidence that smoking is not the sole cause of lung cancer, but not that smoking has at most a negligible effect on lung cancer. Evidence for the latter would be that after controlling for place of residence, there is little difference between smokers and non-smokers in lung cancers. But Cornfield et al. [14] asserted, "Evidence now in hand weighs strongly against this last assertion," and continued, "Stocks and Campbell [15], in their report on lung-cancer mortality among persons in Liverpool, the suburban environs, and rural towns in North Wales, found that heavy smokers have higher lung-cancer rates when urban and rural males were studied separately. Mills and Porter [16] reported similar findings in Ohio. These results agree with the experience of the Hammond and Horn [17] study, which revealed markedly higher death rates for bronchogenic carcinoma among smokers regardless of whether they lived in cities or in rural areas. No contradictory observations are known to us." By Bross's criterion, the criticism of observational studies that suggested smoking caused lung cancer because they did not control for urban-rural residence was not good criticism because it failed to consider whether controlling for

[1] Bross [10] argued that it helpful for the counterhypothesis to be stated explicitly – just like it is important in the scientific method that the hypothesis be clearly stated so that it can be tested, good criticism of a scientific study should explicitly state a counterhypothesis so that the criticism can be evaluated using the scientific method.

urban-rural residence would make the counterhypothesis that smoking does not cause lung cancer tenable.

Postmenopausal hormone replacement therapy (HRT) is use of a drug containing female hormones to replace ones a woman's body no longer makes after menopause. Observational studies during the 1970s–1990s reported evidence that HRT protects against heart disease. For example, Stampfer et al. [18] published an observational study of nurses in the *New England Journal of Medicine* that compared postmenopausal women who had currently used HRT vs. those who had never used it and reported that after controlling for a number of measured covariates – including age, body mass index (BMI), and smoking – current use of HRT was associated with an estimated 70% reduction in the risk of coronary heart disease and a 95% confidence interval of 36% to 86% reduction in risk with a p-value for testing the null hypothesis of no effect of less than 0.002. Use of HRT was being advocated by many in the medical community and users of HRT could be thought of as complying with medical advice [19–23]. Pettiti [24] raised the concern that the observational studies of HRT might be biased because compliers with medical advice often have healthier lifestyles than noncompliers [25]. Pettiti considered the counterhypothesis that there was no causal effect of HRT on heart disease and the lower observed risk of heart disease was fully accounted for by this "compliance bias." To evaluate whether this counterhypothesis was tenable, Pettiti studied how much of a difference in heart disease risk there was among compliers with placebo vs. noncompliers with placebo in randomized trials of two drug treatments for coronary heart disease, beta blockers and clofibrate. Pettiti found that compliers had lower mortality (which was mostly due to heart disease) and that the reduction in mortality was similar to that of the reduction in mortality among users vs. non-users of HRT. Furthermore, Pettiti found that after adjusting for similar covariates as those adjusted for in the observational studies of HRT, the reduction in mortality among compliers remained the same. While it can be argued whether the bias among compliers vs. noncompliers in drug trials is comparable to that of compliers vs. noncompliers with medical advice about taking HRT, it is plausibly so, and Pettiti has at least made a serious case that her counterhypothesis is tenable. Pettiti's criticism was a good criticism.

Bross [10] describes several classes of what he considers bad criticism which do not satisfy his criterion of showing that the counterhypothesis is tenable – hit-and-run criticism, dogmatic criticism, speculative criticism, and tubular criticism. We review these next.

24.3 Bross's Types of Bad Criticism

24.3.1 Hit-and-run criticism

In hit-and-run criticism, the critic points out some real or fancied flaw and suggests that this automatically invalidates the author's conclusions. Since the critic makes no attempt to develop a tenable counterhypothesis, Bross [10] says the critic's performance is on a par with that of a proponent who glances at the data and then jumps to a conclusion. Bross gives two examples of hit-and-run criticism of Hammond and Horn's [26] prospective observational study of the effect of smoking on lung cancer. First, critics called attention to the possibility of misclassification of the cause of death on death certificates. "Most of these critics dropped the matter at that point (apparently under the impression that they had scored a hit)," Bross said. In fact, nondifferential misclassification of an outcome tends to underestimate the strength of relationship between an exposure and outcome [27]. Also Hammond and Horn [17] found that the association between smoking and lung cancer was greater for patients with a well-established diagnosis of lung cancer than for patients with less convincing evidence for a diagnosis of lung cancer. Thus, misclassification arguably enhanced rather than dented the evidence Hammond and Horn's study provided for smoking causing lung cancer.[2]

The second example of hit-and-run criticism that Bross [10] gives is Berkson's [30] paper that criticizes Hammond and Horn's [26] study on the grounds of selection bias. Hammond and Horn found a strong association between smoking and lung cancer and suggested that an active educational campaign be conducted to warn the public that smoking increases the risk of lung cancer. Berkson [30] asserted that "such a proposal seems poorly founded" because smokers in the study had equal or lower cancer death rates than the general U.S. population[3] and warned that "the operation of selective forces in statistical investigations can be a very subtle process." Berkson provided a model and hypothetical numerical example of how selection bias in Hammond and Horn's [26] study could produce a spurious (i.e., non-causal) association between smoking and lung cancer through the selection processes of (i) some seriously ill people at the start of the prospective study not participating in the study and (ii) smokers being more reluctant to participate in the study. While these selection processes may be present, Korteweg [31] showed that in Berkson's numerical example, if the arbitrarily chosen mortality rates in Berkson's hypothetical example are substituted by rates from Hammond and Horn [26] and U.S. official mortality statistics, "only a small part of the excess in death rates for lung cancer and for coronary disease with smokers can be explained as being spurious." Furthermore, under Berkson's model, if there was no causal effect of smoking on lung cancer and the association in Hammond and Horn's study arose entirely from selection bias, the association between smoking and lung cancer should diminish over the course of the study but in fact it increased [10, 14]. Berkson's criticism is like the lawyer's tactic of cross-examining an expert witness by asking a hypothetical question removed from the facts of the case. The lawyer asks the physician witness, "Would you do an MRI scan for a patient who makes repeated complaints of severe neck pain?" If there was no evidence of such complaints, a good reply by the witness would be "Yes, but in this case, the charts and the admission questionnaire shows no such complaints."

Another example of hit-and-run criticism was pointed out by Kodlin [32] concerning studies that showed an association between blood group and diseases such as gastric cancer, malaria and peptic ulcers. Alexander Wiener, a winner of the prestigious Lasker Prize, criticized these studies as "fallacious" and founded on a "bias in the collection of data" that in borderline cases of classifying a case as a certain disease or not, the investigator is subconsciously influenced by knowledge of the patient's blood-group [33, 34]. Wiener asserted that "such bias actually occurs in practice is proved by Billington [the author of a study that found an association between gastric cancer and blood group [35]]'s admission when he was questioned about possible bias when classifying a series of cases of carcinoma of the stomach according to site of the lesion" [34] In fact, Wiener asked "Can Dr. Billington exclude this possibility [bias from differential misclassification of borderline cases]?" [33] to which Dr. Billington replied "I admit I am unable to exclude the possibility." Hardly proof that such bias occurred! In fact, Wiener's counterhypothesis that bias from differential misclassification of borderline cases explains all the associations between blood group and disease does not seem tenable. It would require for certain diseases (e.g., peptic ulcers), a person with blood type O was more likely to be called a disease case when borderline but for other diseases (e.g., gastric cancer), a person with blood type A was more likely to be called a disease case. Furthermore, since the associations between blood group and disease have been repeatedly found, it would require that independent

[2]Sterling et al. [28] discussed the possibility that there is nondifferential misclassification of the cause of death where lung cancer is more likely to be listed as the cause of death for a smoker. Flanders [29] argued that the hypothesized bias from nondifferential misclassification would probably be of small magnitude and affect only a select subgroup of the many investigations of smoking and lung cancer so that consequently, "Even if the bias should prove real, current ideas about smoking and its adverse effects would change little."

[3]The study population consisted of white U.S. males from nine states and the follow-up period contained more low death summer months than high death winter months; Hammond and Horn [26] did not claim the study population was representative of the U.S. population. See Korteweg [31] for discussion.

teams, who are not necessarily particularly eager to confirm previous findings, are all biased in the same way [32].

24.3.2 Dogmatic criticism

Dogmatic criticism is criticism based on appeal to dogma rather than reason. Sir R.A. Fisher criticized the observational studies of smoking and lung cancer saying "The evidence linking cigarette smoking with lung cancer, standing by itself, is inconclusive, as it is apparently impossible to carry out properly controlled experiments with human material" [36]. This criticism is based on an appeal to dogma – the dogma that no observational study can provide strong evidence for causation, only a controlled experiment can.[4] Observational studies do have a potential bias compared to controlled, randomized experiments that individuals choose their exposure, and consequently treated and control individuals may differ systematically in ways other than exposure. But in some settings, this potential bias is not a plausible explanation for an association in an observational study and the observational study (or a series of observational studies) provides strong evidence for causation. John Snow provided strong evidence that contaminated drinking water causes cholera by comparing customers in London in 1854 of the Southwark and Vauxhall Water Company, whose water was heavily contaminated by sewage, to those of the Lambeth Company, whose water was purer, and finding that the cholera death rate was much higher among customers of the Southwark and Vauxhall Company [38]. Snow said:

> The mixing of the supply [of the Companies] is of the most intimate kind. The pipes of each Company go down all the streets, and into nearly all the courts and alleys...Each company supplies both rich and poor, both large houses and small; there is no difference either in the condition or occupation of the persons receiving the water of the different Companies...As there is no difference whatever in the houses or the people receiving the water supply of the two Water Companies, or in any of the physical conditions with which they are surrounded, it is obvious that no experiment could have been devised which would more thoroughly test the effect of water supply on the progress of cholera than this, which circumstances placed ready made before the observer.

Besides natural experiments like Snow's, other ways in which observational studies can provide strong evidence for causation include finding effects that are insensitive to plausible biases and confirming an extensive set of predictions about what will be observed which differ from the predictions made by competing theories [39, 40].

24.3.3 Speculative criticism

Speculative criticism is criticism which focuses on the effect something "might, could, or ought to" have on a study's interpretation in the absence of (enough) evidence that would decide an issue. Bross [10] argues that there is a place for speculation in criticism, but that speculation should be clearly labeled as such and should not enter the conclusions. Berkson [41] made the following speculative criticism of claims based on associations from observational studies that smoking causes lung cancer:

> One explanation is that the associations have a constitutional basis....The hypothesis is...persons who are nonsmokers, or relatively light smokers, are of a constitutional type that is biologically disposed to self-protective habits, and that this is correlated generally with constitutional resistance to mortal forces from disease. If 85 to 95 per cent of a population are smokers, then the small minority who are not smokers would appear, on the face of it, to be of some special type of constitution. It is not implausible that they should be on the

[4] Cook [37] discusses how some present day observers persist with this dogma.

> average relatively longevous, and this implies that death rates generally in this segment of the population will be relatively low. After all, the small group of persons who successfully resist the incessantly applied blandishments and reflex conditioning of the cigarette advertisers are a hardy lot, and, if they can withstand these assaults, they should have relatively little difficulty in fending off tuberculosis or even cancer! If it seems difficult to visualize how such a constitutional influence can carry over to manifest itself as a graded increase of death rate with a graded increase of intensity of smoking, then we must remember that we are wandering in a wilderness of unknowns. I do not profess to be able to track out the implications of the constitutional theory or to defend it, but it cannot be disposed of merely by fiat denial.

Bross credits Berkson for labeling this constitutional hypothesis for the association between smoking and lung cancer as speculative, but expresses concern that this speculative criticism appears to play an important role in Berkson's subsequent rejection of the claim that the observational studies provide strong evidence for smoking causing lung cancer. In the summary of his paper, Berkson states this rejection and restates the speculative criticism without the caution that it is speculative.

Bross anticipates arguments "that it is too stringent to require a critic to show that his substantive counterhypothesis is tenable because he is not actually asserting it but merely suggesting it as a possible line for future research. However, "I fail to see how a critic contributes to the scientific process if the suggested avenue for research is, in fact, a dead end road. Nor can I see how a critic can expect to point out a sensible direction for research unless he explores the tenability of his counterhypothesis – for example [in studies of smoking and lung cancer] whether his notion jibes with the incidence pattern for lung cancer."

Writing in 1959, Cornfield et al. [14] pointed out four ways in which the constitutional hypothesis does not entirely jibe with the incidence pattern for lung cancer: (i) the rapid rise in lung cancer in the first half of the 20th century; (ii) the carcinogenicity of tobacco tars for experimental mice; (iii) the existence of a large association of pipe and cigar tobacco smoking with cancers of the buccal cavity and larynx but not with cancers of the lung; and (iv) the reduced lung-cancer mortality among discontinued cigarette smokers. To explain (i) with the constitutional hypothesis, one could assert that the constitution of people got worse during the first half of the 20th Century[5]; to explain (ii), one could assert that there is a difference between mice and humans; to explain (iii), one could assert that the constitutions of pipe and cigar tobacco smokers tend to protect them against lung cancer but leaves them vulnerable to cancers of the buccal cavity and larynx; and to explain (iv), one could assert that the constitutions of cigarette smoking quitters is better than smokers. While it is possible that each of these assertions is true, all of them being true so that the constitutional hypothesis jibes with the data does not seem particularly tenable. Additional features of the incidence pattern for lung cancer that have arisen since 1959 are also hard for the constitutional hypothesis to explain. For example, prior to the 1960s, women were much less likely to smoke than men but in the 1960s, the tobacco industry sought to change that with products and advertising aimed at women. For instance, beginning in 1967, a time when the women's rights movement was gaining steam, the Virginia Slims brand was marketed under the slogan "You've come a long way baby" with ads that showed independent, stylish, confident and liberated women smoking. Women increased their smoking but men did not [42]. Thirty years later, Bailar and Gornik [43] wrote: "For lung cancer, death rates for women 55 or older have increased to almost four times the 1970 rate," but rates for males over 55 and rates for other cancer sites showed no such dramatic increase.[6]

[5]One could also augment the constitutional hypothesis with the hypothesis that some other factor that rose during the first half of the 20th century is a major cause of lung cancer

[6]This example was presented by Paul Rosenbaum in his IMS Medallion lecture at the 2020 Joint Statistical Meetings.

24.3.4 Tubular criticism

In tubular criticism, the critic has "tubular" (tunnel) vision and only sees the evidence against the causal hypothesis that a study purports to show, ignoring the evidence for it. Tubular criticism suffers from the same weakness as tubular advocacy in which the proponent of a causal hypothesis only sees the evidence favorable to the hypothesis. The philosopher Arthur Schopenhauer observed that "An adopted hypothesis gives us lynx-eyes for everything that confirms it and makes us blind to everything that contradicts it" [44].

As an example of tubular criticism, Bross [10] cites Berkson's [41]'s critique of the evidence on smoking causing lung cancer that "virtually all the evidence is obtained from statistical studies...We are not dealing with the results of laboratory experiments...Nor is the conclusion based on a synthesis, by a 'chain of reasoning,' of relevant scientific knowledge from many different sources." Berkson was unable to "see" the evidence from the carcinogenicity of tobacco tars for experimental mice [45], the lung tissue pathology in smokers [46, 47] and the consistency in support of smoking causing lung cancer from studies with different biases [39].

24.4 Less Stringent Criteria for Criticism

Several authors have suggested less stringent criteria for criticism than Bross (1960) [10].

24.4.1 Welcoming all criticism

Gelman [48] argues that "nearly all criticism has value." Gelman considers the following sorts of statistical criticism, aligned in roughly decreasing order of quality, and discusses the value of each to him as an author:

- A *thorough, comprehensive reassessment* where an expert goes to the trouble of (a) finding problems in an author's work, (b) demonstrating the errors were consequential, and (c) providing an alternative. This is a "good" criticism by Bross's criterion and is a clear step forward.
- A *narrow but precise correction.* If there is a mistake in data processing or analysis, or an alternative explanation for the findings has been missed, Gelman says he would like to know because "Even if it turns out that my error did not affect my main conclusions, it will be helpful to myself and to future researchers to fix the immediate problem."
- *Identification of a potential problem.* Gelman says, "What if someone criticizes one of my published papers by suggesting a problem without demonstrating its relevance? This can be annoying, but I don't see the problem with the publication of such a criticism: Readers should be made aware of this potential problem, and future researchers can explore it."
- *Confusion* – criticism that reveals a misunderstanding on the part of the critic. Gelman says this can have value too because it reveals the author has failed to communicate some point in his or her original article. "We can't hope to anticipate all possible misreadings of our work, but it is good to take advantage of opportunities to clarify." [48]
- *Hack jobs* – criticism that is motivated by a desire to muddy the waters rather than to get at the truth. Gelman says "Even a hack can make a good point, and hacks will use legitimate arguments where available." The problem Gelman says with hack criticism is not in the criticism itself but in the critical process, "a critic who aims at truth should welcome a strong response, while a hack will be motivated to avoid any productive resolution."

Gelman summarizes his argument by paraphrasing Al Smith's quote "The cure for the evils of democracy is more democracy," by "the ills of criticism can be cured by more criticism." That said, Gelman recognizes "that any system based on open exchange can be hijacked by hacks, trolls, and other insincere actors." How to have open exchange of ideas without it being hijacked by trolls and people seeking to just muddy the water is a major challenge. See [49] for some perspectives on possible approaches.

Gelman makes a good point that we can learn from any statistical criticism, whatever its source. The difficulty is that we all have limits on our time and cognitive capacity. It would be helpful if critics label the type of criticism they are providing so that authors and scientific community can prioritize their time in thinking about the criticism.

24.4.2 Welcoming more criticism

Bross argues that a critic should be required to check his or her counterhypothesis against available data and ensure that it is tenable. Law has a less stringent criterion for handling criticism. In legal cases expert witnesses often criticize hypotheses put forth by another witness. There are *Federal Rules of Evidence* that require that the expert be using "scientific methodology" [50], but experts are given a fair amount of latitude to express opinions and the decision maker, such as a judge or jury, is given the task of assigning different weights of evidence to different experts depending on their persuasiveness [51].

Reichardt [52] agrees with Bross that a critic should be required to check his or her counterhypothesis against available data. But Reichardt cautions that because data is always incomplete and can all too easily be misinterpreted, judging what is and what is not in agreement with data can be tricky. Reichardt argues that researchers and editors should be circumspect in applying Bross's rule (that a critic must show a counterhypothesis is in agreement with available data and tenable) and give somewhat more latitude to critics. As an example where Bross's rule for criticism might be too stringent, Reichardt considers Darwin's theory of evolution by natural selection – all organisms compete for resources and those that have an innate advantage prosper and pass on the advantage to their offspring, leading to continuing adaptation of a species to its environment. Darwin's theory was a counterhypothesis to intelligent design – "The hinges in the wings of an earwig, and the joints of its antennae, are as highly wrought, as if the Creator had nothing else to finish...The marks of design are too strong to be got over. Design must have had a designer. That designer must have been a person. That person is GOD" [53]. Darwin presented considerable evidence in support of his theory that species could adapt to their environment by natural selection without an intelligent designer, for example, common structures (a human arm, a cat's leg, a whale's flipper, and a bat's wing all are adapted to different purposes, but share the same bone structure, suggesting a common ancestor), nested geographic distributions (species that occur on islands, such as the finches of the Galapagos Islands, are often closely related but different than species on the nearby mainland despite considerable differences in the environment of the island, suggesting colonization and adaptation rather than independent creation of the species) and the fossil record (many fossilized species had been found, some of which did not exist any longer, suggesting a selection process). But there were holes in the evidence for Darwins' theory. Darwin's theory was that new species were continually evolving so that there should be intermediate forms scattered across the fossil record, but there were not [54]. Was Darwin's explanation for evolution in sufficient agreement with available data, and therefore sufficiently tenable, to permit publication according to Bross? Darwin argued that the lack of intermediate forms could be explained by the incompleteness of the fossil record: "Now let us turn to our richest museums, and what a paltry display we behold! That our collections are imperfect is admitted by every one...Only a small portion of the surface of the earth has been geologically explored, and no part with sufficient care" [55]. Today, many of the gaps in the fossil record have been filled, and hundreds of thousands of fossil organisms, found in well-dated rock sequences, represent successions of forms through time and manifest many evolutionary transitions [56].

Another interesting example Reichardt brings up to question whether Bross's criterion for a hypothesis or counterhypothesis to be tenable is too stringent is the work of Ingaz Semmelweis. Giving birth was perilous for mothers in the 1800s and childbed fever was one common cause of death. Childbed fever was thought to be due to multiple causes and the causes of illnesses were thought to be as unique as individuals themselves and determinable only on a case by case basis [52]. When Ignaz Semmelweis was hired as a physician at the Vienna General Hospital's First Obstetrical Clinic in 1846, as many as twenty percent of the women giving birth to a child in the hospital's First Clinic died from childbed fever; high death rates from childbed fever were common in many hospitals [52]. Semmelweis observed that the Vienna General Hospital's Second Obstetrical Clinic had around a 2% death rate from childbed fever. Semmelweis was severely troubled that his First Clinic had a much higher mortality rate than the Second Clinic. It "made me so miserable that life seemed worthless" [57]. Semmelweis was determined to find the cause of the difference. The First Clinic and Second Clinic admitted women on alternate days so differences in the humours of the body did not seem a likely explanation for the difference in death rates. Semmelweis started eliminating all possible differences between the two clinics, including even religious practices. The only major difference was the individuals who worked there. The First Clinic was the teaching service for medical students, while the Second Clinic had been selected in 1841 for the instruction of midwives only. Semmelweis uncovered a telling clue when his friend Jakob Kolletschka cut his finger while performing an autopsy and died from symptoms similar to childbed fever. Semmelweis hypothesized that childbed fever was caused by contamination from "cadaverous material" from doctors who often performed autopsies before serving on the obstetrics ward. This would explain the lower rate in the Second Clinic because midwives do not perform autopsies. Semmelweis instituted a policy that physicians wash their hands in a solution of chlorinated lime before examining patients. The First Clinic's death rate from childbed fever dropped 90% and became comparable to the Second Clinic.

Instead of being lauded as the "savior of mothers," as he is now called, Semmelweis was ridiculed and his ideas were rejected by the medical community [58]. An example is a paper by Carl Edvard Marius Levy, head of the Danish Maternity Institute at Copenhagen, who wrote:

> If Dr. Semmelweis had limited his opinion regarding infections from corpses to puerperal corpses, I would have been less disposed to denial than I am... the specific contagium seems to be of little importance to Dr. Semmelweis. Indeed it is so little considered that he does not even discuss the direct transmission of the disease from those who are ill to healthy persons lying nearby. He is concerned only with general infection from corpses without respect to the disease that led to death. In this respect his opinion seems improbable...a rapidly fatal putrid infection, even if the putrid matter is introduced directly into the blood, requires more than homeopathic doses of the poison. And, with due respect for the cleanliness of the Viennese students, it seems improbable that enough infective matter or vapor could be secluded around the fingernails to kill a patient...To prove his opinion, Dr. Semmelweis ordered chlorine washings to destroy every trace of cadaverous residue on the fingers. Would not the experiment have been simpler and more reliable if it had been arranged, at least during the experiment, that all anatomical work would be avoided?...In spite of these reservations, one must admit that the results of the experiment appear to support Dr. Semmelweis's opinion, but certainly one must admit no more. Everyone who has had the opportunity to observe the periodic variations in the mortality rate of maternity clinics will agree that his findings lack certain important confirmation...In the absence of more precise statistical information, it is conceivable that the results of the last seven months depend partially on periodic accidental factors...that insofar as they are laid out, his [Semmelweis'] views appear too unclear, his observations too volatile, his experiences too uncertain, to deduce scientific results therefrom.

Would application of Bross's criterion that a hypothesis or counterhypothesis should be tenable to be publishable have suggested that in light of the concerns raised by Levy, Semmelweis's findings should not have been published or his suggested policy of doctors washing their hands should not have been implemented? Levy raises some concerns that were legitimate at the time. Semmelweis's hypothesis that a small dose of invisible cadaverous particles cause childbed fever was less plausible before the acceptance of the germ theory of disease. Furthermore, Semmelweis had claimed at the time that Levy was writing that only cadaveric matter from corpses could cause childbed fever, a claim that was incorrect [59]. Also mortality rates of childbed fever fluctuated dramatically within hospitals and within towns more generally [59] and consequently the data Semmelweis had compiled at the time Levy was writing, shown in Figure 24.1, suggested handwashing was beneficial but perhaps did not make an overwhelming case[7]. However, Semmelweis did make a tenable case that handwashing was beneficial that should have been taken seriously. Levy's wholesale dismissal of Semmelweis's hypothesis – "his views appear too unclear, his observations too volatile, his experiences too uncertain, to deduce scientific results therefrom" – is based on hit-and-run criticism and tubular criticism. Other criticisms of Semmelweis were dogmatic. For example, some critics blamed Semmelweis's hypothesis that invisible particles from cadaverous material causing childbed fever on his Catholic faith. They said his idea that invisible particles could cause disease and death was simply a product of his Catholic superstition, and argued that the presence of Catholic priests bearing the Eucharist to dying patients was deeply frightening, and this fright induced child-bed fever. Semmelweis tested this theory by keeping priests out of one ward while admitting them to a second: no difference in illness or mortality was observed. Despite this, the critics continued to hold that Semmelweis's religion was the actual cause of the deadly disease [62].

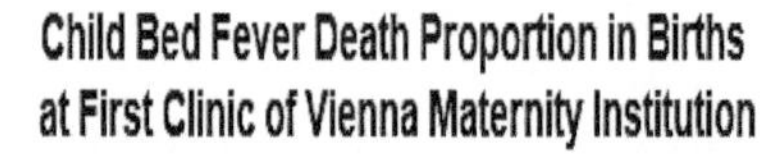

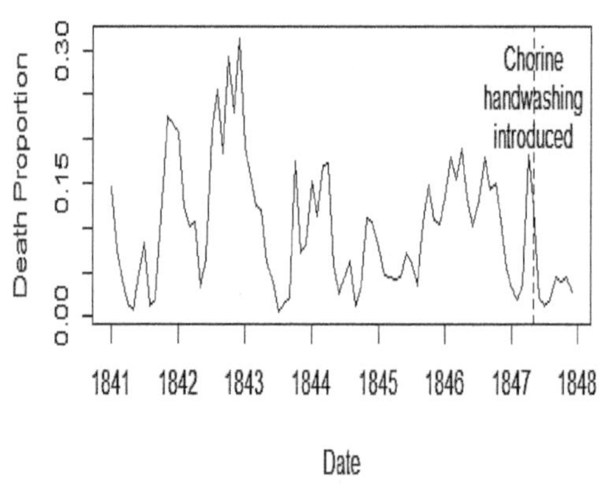

FIGURE 24.1
Data Semmelweis compiled on the monthly proportion of maternal deaths from child bed fever in births at the First Clinic of the Vienna Maternity Institution at the time Levy was writing his criticism of Semmelweis's hypothesis that chlorine handwashing prevented child bed fever deaths.

In welcoming more criticism, Reichardt also makes the good point that for a counterhypothesis to be tenable, it need not, by itself, account for all of the observed results: "I've seen instances where an estimate of a treatment effect, for example, is said to be immune to a rival hypothesis because the alternative explanation was insufficient to account for the entirety of the estimate. But more than one bias can be present. And perhaps together they could account for the whole treatment effect estimate. So the tenability of alternative explanations needs to be considered en masse rather than one at a time."

[7] Levy wrote his article based on a letter written in December 1847 by Heinrich Hermann Schwartz to Professor Gustav Adolph Michaelis, which Michaelis forwarded to Levy [60]. We have shown the data through December 1847. Data from [61]

Like Reichardt [52], the authors Ho [63], Rindskopf [64], Rosenbaum and Small [65] and Hill and Hoggatt [66] are all generally supportive of Bross's position that there should be standards for criticism – "requiring critics to do more than sling random criticisms without some backing for their statements was, and still is, a reasonable standard to meet" [64] – but they all express concerns that the standards not be too stringent as to stifle useful criticism. Ho [63] and Rindskopf [64] point out that the underlying data needed to do a reanalysis may not be publicly available. Making data publicly available should be encouraged. In contrast, dismissing criticism because a reanalysis was not done when the data is not made publicly available might encourage making data unavailable.

Rosenbaum and Small [65] argue that a criticism can advance understanding when it points out a logical inconsistency among data, a proponent's assumptions, and scientific knowledge from other sources, even if it does not make a definitive case that the proponent's hypothesis is wrong. For example, "Yang et al. [67] considered a plausible instrumental variable (IV) and used it to estimate a plausible beneficial treatment effect in one population in which the true treatment effect is unknown. They then applied the same IV to a second population in which current medical opinion holds that this same treatment confers no benefit, finding that this IV suggests a benefit in this second population also. Specifically, there is debate about whether delivery by caesarean section improves the survival of extremely premature infants, but current medical opinion holds that it is of no benefit for otherwise healthy but slightly premature infants. In contrast, the IV analysis suggested a substantial benefit for both types of infants. In light of this, there is logical incompatibility between four items: (i) the data, (ii) the claim that extremely premature infants benefit from delivery by caesarean section, (iii) the claim that otherwise healthy, slight premature infants do not benefit, (iv) the claim that the IV is valid in both groups of babies. Removal of any one of (i), (ii), (iii) or (iv) would remove the inconsistency, but there is no basis for removing one and accepting the others." This situation where several propositions are logically inconsistent so that they cannot all be true, yet at the present moment we are not in a position to identify which proposition(s) are false is called an aporia [68]. An aporia, though uncomfortable, is an advance in understanding: it can spur further investigation and further advances in understanding. Socrates, in Plato's *Meno* [69], thought that demonstrating an aporia in a curious person's thinking would spur discovery. Socrates said of a befuddled young interlocutor who he put in an aporia:

> At first he did not know what [he thought he knew], and he does not know even now: but at any rate he thought he knew then, and confidently answered as though he knew, and was aware of no difficulty; whereas now he feels the difficulty he is in, and besides not knowing does not think he knows...[W]e have certainly given him some assistance, it would seem, towards finding out the truth of the matter: for now he will push on in the search gladly, as lacking knowledge; whereas then he would have been only too ready to suppose he was right...[Having] been reduced to the perplexity of realizing that he did not know...he will go on and discover something

Hill and Hoggatt [66] point out that we should not have harsher standards for the critic than the proponent. If the proponent makes a claim based on an observational study which may be biased because of unmeasured confounders, we should not reflexively dismiss a criticism based on an alternative observational study just because the alternative observational study may be biased because of having its own unmeasured confounders. "If we are requiring that a counter-hypothesis be tenable, it seems the criteria should include a reasonable assessment of the plausibility of such assumptions. However, if we are comparing competing sets of untestable assumptions (corresponding to the proponent's original analysis and the critic's analysis in support of a counter-hypothesis) how should we assess which of the sets of assumptions are most plausible? Would it be better, for instance, to use an instrumental variables approach where the instrument is weak and the exclusion restriction is questionable or to use an observational study where we are uncertain that we have measured all confounders?" Hill and Hoggatt [66] propose sensitivity analyses as a way to tackle this problem.

24.4.3 Sensitivity analysis

A sensitivity analysis is an exploration of the sensitivity of estimates to violations of key assumptions of the analysis. The first sensitivity analysis for causal inference was done by Cornfield et al. [14] to assess the evidence on the effect of smoking causing lung cancer. Bross was an early developer of sensitivity analysis methods [70, 71]. There are many good references on sensitivity analysis, e.g., [72–79]. Hill and Hoggatt (2018) [66] propose that rather than making a binary decision about which counterhypotheses are tenable, a goal for critics (and proponents if they are acting as their own critics) could be to do sensitivity analyses that provide a range of estimates that are derived from different sets of assumptions.

Similarly, Rosenbaum and Small [65] argue that one can take a step "beyond criticism" and use sensitivity analysis to clarify what the research design combined with the data say about what would need to hold for the proponent or critic's claim to be true. For example, Fisher [36] criticized studies that showed an association between smoking and lung cancer and argued they do not provide strong evidence for a causal effect because he said the association could be explained by an unmeasured confounder such as a genotype that was associated with both smoking and lung cancer. Hammond [80] paired 36,975 heavy smokers to nonsmokers on the basis of age, race, nativity, rural versus urban residence, occupational exposures to dusts and fumes, religion, education, marital status, alcohol consumption, sleep duration, exercises, severe nervous tension, use of tranquilizers, current health, history of cancer other than skin cancer, and history of heart disease, stroke or high blood pressure. Of the 36,975 pairs, there were 122 pairs in which exactly one person died of lung cancer. Of these, there were 12 pairs in which the nonsmoker died of lung cancer and 110 pairs in which the heavy smoker died of lung cancer. Under the assumption of there being no unmeasured confounders, McNemar's test gives a p-value less than 0.0001. There would be strong evidence of smoking causing lung cancer if there were no unmeasured confounding. A sensitivity analysis shows that even if there were unmeasured confounding such that in a matched pair, one person had 5 times as high odds of being a smoker rather than the other person, there would still be significant evidence ($p = 0.03$) of smoking causing lung cancer but if the odds were 6 times as high, there would no longer be significant evidence ($p = 0.1$) [81]. The sensitivity analysis does not adjudicate an argument between a proponent and a critic about whether bias produced an association. Instead, it objectively clarified what is being said by a proponent who denies it is bias or a critic who asserts it is bias. The sensitivity analysis shows that to explain the association between heavy smoking and lung cancer as a bias, that bias would have to be enormous. This fact is less than we might like in that it does not definitively decide the argument between the proponent and the critic, but it is an important fact that advances scientific understanding nonetheless.

24.5 Evaluating Criticism

How should we evaluate statistical criticism? More generally, how should we evaluate the quality of evidence presented by a proponent or a critic? Rosenbaum [82] says:

> To discuss the quality of evidence provided by an empirical study one must first recognize that evidence is not proof, can never be proof, and is in no way inferior to proof. It is never reasonable to object that an empirical study has failed to prove its conclusions, though it may be reasonable, perhaps necessary, to raise doubts about the quality of its evidence. As expressed by Sir Karl Popper [83]: "If you insist on strict proof (or strict disproof) in the empirical sciences, you will never benefit from experience, and never learn from it how wrong you are."...

> Evidence may refute a theorem, not the theorem's logic, but its relevance...Evidence, unlike proof, is both a matter of degree and multifaceted. Useful evidence may resolve or shed light on certain issues while leaving other equally important issues entirely unresolved. This is but one of many ways that evidence differs from proof...
>
> Evidence, even extensive evidence, does not compel belief. Rather than being forced to a conclusion by evidence, a scientist is responsible and answerable for conclusions reached in light of evidence, responsible to his conscience and answerable to the community of scientists

Some of the criteria used for evaluating the credibility of news sources are also useful for evaluating the quality of evidence presented by a proponent or a critic. For evaluating news sources, [84] proposes the CRAAP test, a list of questions to ask about the news source related to Currency, Relevance, Authority, Accuracy, Purpose. For evaluating the quality of evidence, trying to directly evaluate accuracy is clearly an important consideration, but because there is typically uncertainty in reading a criticism about its accuracy, authority and purpose are also worth considering. Questions one should ask about authority include has the author demonstrated expertise on the topic? It is important though in evaluating authority to keep in mind that well known does not always mean authoritative and understanding of authority can itself be biased and leave out important voices [85]. One should try to be open to new voices. Of the self taught mathematical genius Srinivasa Ramanujan who was toiling away as a shipping clerk, Eysenck [86] wrote, "He tried to interest the leading professional mathematicians in his work, but failed for the most part. What he had to show them was too novel, too unfamiliar, and additionally presented in unusual ways; they could not be bothered." One mathematician G.H. Hardy took Ramanujan seriously; Hardy called their fruitful collaboration the "one romantic incident in my life" [87]. In evaluating the purpose of a criticism, one should ask, does the author have an agenda or bias, e.g., is the author part of a group whose interests would be affected by the research question? One should keep in mind though that as Gelman [48] mentions, even biased sources can make a good point and even "objective" sources may have a point of view that slips in [88]. In general, it is best to always try to think critically no matter the source. Thomas Jefferson emphasized the importance of critical thinking for the public when reading the news and it might equally apply to the scientific community when reading research [89]:

> The basis of our governments being the opinion of the people, the very first object should be to keep that right; and were it left to me to decide whether we should have a government without newspapers, or newspapers without a government, I should not hesitate a moment to prefer the latter. But I should mean that every man should receive those papers & be capable of reading them.

One useful criterion for evaluating statistical criticism is Bross's criterion that it should present a tenable counterhypothesis, along with modifications discussed above that relax it. Gastwirth [90] presents examples of how the use (lack of use) of Bross's criterion was helpful in producing fair (unfair) decisions in employment discrimination cases. Gastwirth also mentions an interesting public policy example about Reye syndrome where the use of Bross's criterion might have saved lives. Reye syndrome is a rapidly developing serious brain disease. Pediatric specialists had suspected that children receiving aspirin to alleviate symptoms of a cold or similar childhood disease might increase the risk of contracting Reye syndrome. In 1982, Halpin et al. [91] published a case control study in which aspirin use in cases of Reye syndrome was compared with that in controls who were in the case's class or homeroom and who were of the same sex, race, and age (± 1 year) and were recently absent with an acute illness or appeared ill to the teacher or school nurse. Controlling for the presence of fever, headache and sore throat, Halpin et al. estimated that aspirin use increased the odds of Reye syndrome 11.5 times ($p < 0.001$). Based on the findings by Halpin et al. and two earlier studies, the U.S. government initiated the process of warning the public, submitting the proposed warning and background studies to the Office of Information and Regulatory Analysis for review.

During the review period, The Aspirin Institute, which represented the interests of the industry, criticized the Halpin et al. study. The Institute argued that the association between aspirin use and Reye syndrome could be due the parents of cases who had Reye Syndrome having a stress-induced heightened recall of the medications they administered including aspirin. The Institute suggested that instead of the controls being children who had been absent from school with an illness, the controls be formed from children hospitalized for other diseases or who visited the emergency room since the parents of these children would under a more similar stress level as the cases. The government decided that another study should be conducted before warning the public about aspirin and Reye's syndrome. A Public Health Task Force was formed and planned the study during 1983 and a pilot study was undertaken during mid-February through May 1984. The data analysis was reviewed and made available in December 1984. The logistic regression model that controlled for fever and other symptoms using all control groups yielded an estimated odds ratio of 19.0 (lower 95% confidence limit: 4.8) and the two control groups suggested by the Aspirin Institute had the highest estimated odds ratios (28.5 for emergency room controls and 70.2 for inpatient controls) [92]. Citing these findings, on January 9, 1985, Health and Human Services Secretary Margaret Heckler, asked all aspirin manufacturers to begin using warning labels. The CDC reported that the number of cases of Reye syndrome reported to it dropped from 204 in 1984 before the warning to 98 in 1985 after the warning. See Gastwirth [90, 93] for further details and discussion of the aspirin-Reye syndrome story. Gastwirth [90] argues that had Bross's criterion for criticism been applied, the Aspirin Institute's criticism of the 1982 study – that an estimated odds ratio of 11.5 ($p < 0.001$) was due to recall bias from the caregiver being under stress rather than an effect of aspirin – should not have been considered tenable since no supporting evidence was provided for there being sufficient recall bias to create this high an estimated odds ratio and low a p-value. Consequently, Gastwirth [90] argues, the U.S. government should have issued a warning about aspirin causing Reye's syndrome in 1982 rather than 1985, and lives would have been saved. Gastwirth [93] makes a similar argument from a Bayesian perspective.

In interpreting statistical criticism, it is important to try to spot and avoid falling prey to "argument from ignorance" (*argumentum ad ignorantiam*), the trap of fallacious reasoning in which a premise is claimed to be true only because it has not been proven false or that it is false because it has not been proven true. Hit-and-run criticism (see Section 24.3.1) is an appeal to argument from ignorance. Other examples of argument from ignorance include the "margin of error folly" – if it could be (e.g., is within the margin of error), it is – or in a hypothesis testing context assuming that if a difference isn't significant, it is zero [50, 94]. One should not confuse a statement of "no evidence of an effect" with one of "evidence of no effect." In 2016, *US News and World Report* ran a story "Health Buzz: Flossing Doesn't Actually Work, Report Says" which reported on an Associated Press investigation that found that randomized trials provide only weak evidence of flossing's benefits. While this is true, the story's suggestion that flossing doesn't work wasn't justified. No well powered randomized trials of flossing's long terms effects have been conducted. A 2011 Cochrane review concluded, "Twelve trials were included in this review which reported data on two outcomes (dental plaque and gum disease). Trials were of poor quality and conclusions must be viewed as unreliable. The review showed that people who brush and floss regularly have less gum bleeding compared to toothbrushing alone. There was weak, very unreliable evidence of a possible small reduction in plaque. There was no information on other measurements such as tooth decay because the trials were not long enough and detecting early stage decay between teeth is difficult" [95]. Why haven't high quality studies of flossing's long term effects been conducted? For one thing, it's unlikely that an Institutional Review Board would approve as ethical a trial in which, for example, people don't floss for three years since flossing is widely believed by dentists to be effective [96]. Dr. Tim Iafolla, a dental health expert at the National Institute of Health said, "Every dentist in the country can look in someone's mouth and tell whether or not they floss" [97] Red or swollen gums that bleed easily are considered a clear sign that flossing and better dental habits are needed [97]. Another challenge in conducting a well powered, long run trial is that it would be difficult to monitor people's flossing habits over a long period and instead such a trial might need to rely on self report. And people tend to report what they think is the "right" answer when it comes to their health behaviors – e.g., say they are flossing regardless of whether they are – creating measurement error and reducing power [97]. "The fact that there hasn't been a huge population-based study of flossing doesn't mean that flossing's not effective," Dr. Iafolla said. "It simply suggests that large studies are difficult and expensive to conduct when you're monitoring health behaviors of any kind" [97]. Arguments from ignorance are sometimes made by supporting what is purported to be true not by direct evidence but by attacking an alternative possibility, e.g., a clinician might say "because the research results indicate a great deal of uncertainty about what to do, my expert judgment can do better in prescribing treatment than these results" [50]. Another way arguments from ignorance are sometimes made is from personal incredulity where because a person personally finds a premise unlikely or unbelievable, a claim is made that a premise can be assumed false or that another preferred but unproved premise is true instead [50].

A reasoning fallacy related to argument from ignorance is *falsum in uno, falsum in omnibus* (false in one thing, false in everything) implying that someone found to be wrong on one issue, must be wrong on all others. *Falsum in uno, falsum in omnibus* is a common law principle for judging the reliability of witnessess dating back from at least the Stuart Treason Trials in the late 1600s. Today, many jurisdictions have abandoned the principle as a formal rule of evidence and instead apply the rule as a "permissible inference that the jury may or may not draw" [98]. Judge Richard Posner drew a distinction between "the mistakes that witnesses make in all innocence...(witnesses are prone to fudge, to fumble, to misspeak, to misstate, to exaggerate)" and "slips that, whether or not they go to the core of the witness's testimony, show that the witness is a liar" [99]. In scientific arguments, hit-and-run criticism is often an attempt to discredit a proponent's hypothesis based on the equivalent of a fudge, fumble, misstatement, or exaggeration rather than something that goes to the core of the data supporting the proponent's hypothesis or the reliability of the proponent's work. For example, the Intergovernmental Panel on Climate Change (IPCC), which shared the Nobel Peace Prize in 2007, wrote an over a thousand page report with 676 authors assessing the evidence for climate change [100]. The IPCC had a procedure of only relying on data that has passed quality assurance mechanisms such as peer review. However, it was found that an estimate of how fast the glaciers in the Himalayas will melt – "if the present rate continues, the likelihood of them disappearing by the year 2035 and perhaps sooner is very high" – was based on a magazine writer's phone interview with an Indian scientist [101]. Newspaper headlines blared "World misled over Himalayan glacier meltdown" (*Sunday Times*, Jan. 7, 2010) and "UN report on glaciers melting is based on 'speculation" (*Daily Telegraph*, Jan. 7, 2010). The IPCC's chairman R.K. Pachauri argued that *falsum in uno, falsum in omnibus* should not be used to derail the IPCC's whole science based argument: "we slipped up on one number, I don't think it takes anything away from the overwhelming scientific evidence of what's happening with the climate of this earth" [101]. See Hubert and Wainer [50] for further discussion.

24.6 Self-Criticism

In his seminal paper on observational studies, Cochran [102] advocated that the proponent of a hypothesis should also try to be its critic:

> When summarizing the results of a study that shows an association consistent with the causal hypothesis, the investigator should always list and discuss all alternative explanations of his [sic] results (including different hypotheses and biases in the results) that occur to him. This advice may sound trite, but in practice is often neglected. A model is the section "Validity of

the results" by Doll and Hill (1952) [103], in which they present and discuss six alternative explanations of their results in a study.

While it is commonplace in scientific papers to report a study's limitations and weaknesses, Reichardt (2018) [52] argues that often such sections do not go to the heart of the limitations of a study:

> I've read reports where the limitations of a study include such obvious reflections as that the results should not be generalized beyond the population of participants and the outcome measures that were used. But the same reports ignore warnings of much more insidious concerns such as omitted variables and hidden biases.

One reason authors may be reluctant to acknowledge the weaknesses of a study is that the authors are afraid that acknowledging such weaknesses would disqualify their research from publication. Reviewers and editors should "explicitly disavow such disqualification when the research is otherwise of high quality – because such weaknesses are simply an inherent feature in some realms of research, such as observational studies. Even sensitivity analyses are not guaranteed to bracket the true sizes of treatment effects" [52]. To highlight the limitations and weaknesses of a study, Reichardt advocates that they should be put in their own section as Doll and Hill [103] did rather than in the discussion section.

In conducting observational studies, it may be good to keep in mind the great UCLA basketball coach John Wooden's advice [104]:

> You can't let praise or criticism get to you. It's a weakness to get caught up in either one.

References

[1] David Freedman. From association to causation: some remarks on the history of statistics. *Journal de la société française de statistique*, 140(3):5–32, 1999.

[2] Herbert L Lombard and Carl R Doering. Cancer studies in massachusetts: habits, characteristics and environment of individuals with and without cancer. *New England Journal of Medicine*, 198(10):481–487, 1928.

[3] Raymond Pearl. Tobacco smoking and longevity. *Science*, 87(2253):216–217, 1938.

[4] F.H. Muller. Tabakmissbrauch und lungcarcinom. *Zeitschrift fur Krebs forsuch*, 49:57–84, 1939.

[5] Ernest L Wynder and Evarts A Graham. Tobacco smoking as a possible etiologic factor in bronchiogenic carcinoma: a study of six hundred and eighty-four proved cases. *Journal of the American Medical Association*, 143(4):329–336, 1950.

[6] Richard Doll and A Bradford Hill. Smoking and carcinoma of the lung. *British Medical Journal*, 2(4682):739, 1950.

[7] Dean F Davies. A review of the evidence on the relationship between smoking and lung cancer. *Journal of Chronic Diseases*, 11(6):579–614, 1960.

[8] American Association for the Advancement of Science et al. Smoking and health: Joint report of the study group on smoking and health. *Science*, 125(3258):1129–1133, 1957.

[9] US Congress. False and misleading advertising (filter-tip cigarettes). In *Hearings before a Subcommittee of the Committee on Government Operations. Washington, DC: US House of Representatives, 85th Congress, First Session, July*, volume 18, pages 23–26, 1957.

[10] Irwin DJ Bross. Statistical criticism. *Cancer*, 13(2):394–400, 1960.

[11] Wilhelm C Hueper. *A quest into the environmental causes of cancer of the lung*. Number 452. US Department of Health, Education, and Welfare, Public Health Service, 1955.

[12] Paul Kotin. The role of atmospheric pollution in the pathogenesis of pulmonary cancer: A review. *Cancer Research*, 16(5):375–393, 1956.

[13] Ian Macdonald. Contributed comment: Chinks in the statistical armor. *CA: A Cancer Journal for Clinicians*, 8(2):70–70, 1958.

[14] Jerome Cornfield, William Haenszel, E Cuyler Hammond, Abraham M Lilienfeld, Michael B Shimkin, and Ernst L Wynder. Smoking and lung cancer: recent evidence and a discussion of some questions. *Journal of the National Cancer institute*, 22(1):173–203, 1959.

[15] Percy Stocks and John M Campbell. Lung cancer death rates among non-smokers and pipe and cigarette smokers. *British Medical Journal*, 2(4945):923, 1955.

[16] Clarence A Mills and Marjorie Mills Porter. Tobacco smoking, motor exhaust fumes, and general air pollution in relation to lung cancer incidence. *Cancer Research*, 17(10):981–990, 1957.

[17] EC Hammond and D Horn. Smoking and death rates. part i. total mortality. part ii. death rates by cause. *Journal of the American Medical Association*, 166:1159–1172, 1958.

[18] Meir J Stampfer, Walter C Willett, Graham A Colditz, Bernard Rosner, Frank E Speizer, and Charles H Hennekens. A prospective study of postmenopausal estrogen therapy and coronary heart disease. *New England Journal of Medicine*, 313(17):1044–1049, 1985.

[19] Veronica A Ravnikar. Compliance with hormone therapy. *American Journal of Obstetrics and Gynecology*, 156(5):1332–1334, 1987.

[20] C Lauritzen. Clinical use of oestrogens and progestogens. *Maturitas*, 12(3):199–214, 1990.

[21] PJ Ryan, R Harrison, GM Blake, and I Fogelman. Compliance with hormone replacement therapy (hrt) after screening for post menopausal osteoporosis. *BJOG: An International Journal of Obstetrics & Gynaecology*, 99(4):325–328, 1992.

[22] Kathryn A Martin and Mason W Freeman. Postmenopausal hormone-replacement therapy. *New England Journal of Medicine*, 328:1115–1117, 1993.

[23] Daniel M Witt and Tammy R Lousberg. Controversies surrounding estrogen use in postmenopausal women. *Annals of Pharmacotherapy*, 31(6):745–755, 1997.

[24] Diana B Petitti. Coronary heart disease and estrogen replacement therapy can compliance bias explain the results of observational studies? *Annals of Epidemiology*, 4(2):115–118, 1994.

[25] William H Shrank, Amanda R Patrick, and M Alan Brookhart. Healthy user and related biases in observational studies of preventive interventions: a primer for physicians. *Journal of General Internal Medicine*, 26(5):546–550, 2011.

[26] E Cuyler Hammond and Daniel Horn. The relationship between human smoking habits and death rates: a follow-up study of 187,766 men. *Journal of the American Medical Association*, 155(15):1316–1328, 1954.

[27] Irwin Bross. Misclassification in 2 x 2 tables. *Biometrics*, 10(4):478–486, 1954.

[28] Theodor D Sterling, Wilfred L Rosenbaum, and James J Weinkam. Bias in the attribution of lung cancer as cause of death and its possible consequences for calculating smoking-related risks. *Epidemiology*, pages 11–16, 1992.

[29] W Dana Flanders. Inaccuracies of death certificate information. *Epidemiology*, pages 3–5, 1992.

[30] Joseph Berkson et al. The statistical study of association between smoking and lung cancer. In *Proceedings of Staff Meetings of the Mayo Clinic*, volume 30, pages 319–48, 1955.

[31] R Korteweg. The significance of selection in prospective investigations into an association between smoking and lung cancer. *British Journal of Cancer*, 10(2):282, 1956.

[32] D Kodlin. Blood-groups and disease. *The Lancet*, 279(7243):1350–1351, 1962.

[33] Alexander S Wiener. Blood-groups and disease. *The Lancet*, 268(6956):1308, 1956.

[34] AlexanderS Wiener. Blood-groups and disease: A critical review. *The Lancet*, 279(7234): 813–816, 1962.

[35] BP Billington. Gastric cancer: relationships between abo blood-groups, site, and epidemiology. *The Lancet*, 268(6948):859–862, 1956.

[36] Ronald A Fisher. Cancer and smoking. *Nature*, 182(4635):596–596, 1958.

[37] Thomas D Cook. The inheritance bequeathed to william g. cochran that he willed forward and left for others to will forward again: The limits of observational studies that seek to mimic randomized experiments. *Observational Studies*, 1:141–164, 2015.

[38] John Snow. *On the mode of communication of cholera.* John Churchill, 1855.

[39] Paul R Rosenbaum. *Observation and experiment.* Harvard University Press, 2017.

[40] Charles S Reichardt. *Quasi-Experimentation: A guide to design and analysis.* Guilford Publications, 2019.

[41] Joseph Berkson. Smoking and lung cancer: some observations on two recent reports. *Journal of the American Statistical Association*, 53(281):28–38, 1958.

[42] David M Burns, Lora Lee, Larry Z Shen, Elizabeth Gilpin, H Dennis Tolley, Jerry Vaughn, Thomas G Shanks, et al. Cigarette smoking behavior in the united states. *Changes in Cigarette-Related Disease Risks and Their Implication for Prevention and Control. Smoking and Tobacco Control Monograph No*, 8:13–112, 1997.

[43] John C Bailar and Heather L Gornik. Cancer undefeated. *New England Journal of Medicine*, 336(22):1569–1574, 1997.

[44] Arthur Schopenhauer. The world as will and representation. volume 2, translated by david carus and richard aquila, 1844. translation published in 2011.

[45] Ernest L Wynder, Evarts A Graham, and Adele B Croninger. Experimental production of carcinoma with cigarette tar. *Cancer Research*, 13(12):855–864, 1953.

[46] Oscar Auerbach, J Brewster Gere, Jerome B Forman, Thomas G Petrick, Harold J Smolin, Gerald E Muehsam, Dicran Y Kassouny, and Arthur Purdy Stout. Changes in the bronchial epithelium in relation to smoking and cancer of the lung: a report of progress. *New England Journal of Medicine*, 256(3):97–104, 1957.

[47] Oscar Auerbach, AP Stout, E Cuyler Hammond, and Lawrence Garfinkel. Changes in bronchial epithelium in relation to cigarette smoking and in relation to lung cancer. *New England Journal of Medicine*, 265(6):253–267, 1961.

[48] Andrew Gelman. Learning from and responding to statistical criticism. *Observational Studies*, 4:32–33, 2018.

[49] Harrison Rainie, Janna Quitney Anderson, and Jonathan Albright. The future of free speech, trolls, anonymity and fake news online, 2017.

[50] Lawrence Hubert and Howard Wainer. *A statistical guide for the ethically perplexed.* CRC Press, 2012.

[51] W.B. Fairley and W.A. Huber. Statistical criticism and causality in *prima facie* proof of disparate impact discrimination. *Observational Studies*, 4:11–16, 2018.

[52] Charles S Reichardt. Another ground rule. *Observational Studies*, 4:57–60, 2018.

[53] William Paley. *Natural Theology: or, Evidences of the Existence and Attributes of the Deity, Collected from the Appearances of Nature.* Lincoln and Edmands, 1829.

[54] Bill Bryson. *A short history of nearly everything.* Broadway, 2004.

[55] Charles Darwin. *The origin of species.* John Murray, London, 1859.

[56] National Academy of Sciences. *Science and Creationism: A View from the National Academy of Sciences, Second Edition.* National Academies Press, 1999.

[57] Ignaz Semmelweis. *The etiology, concept, and prophylaxis of childbed fever.* Univ of Wisconsin Press, 1983.

[58] Caroline M De Costa. "the contagiousness of childbed fever": A short history of puerperal sepsis and its treatment. *Medical Journal of Australia*, 177(11):668–671, 2002.

[59] Dana Tulodziecki. Shattering the myth of semmelweis. *Philosophy of Science*, 80(5): 1065–1075, 2013.

[60] K Codell Carter and George S Tate. The earliest-known account of semmelweis's initiation of disinfection at vienna's allgemeines krankenhaus. *Bulletin of the History of Medicine*, 65(2):252–257, 1991.

[61] Flynn Tran. Kaggle contributor.

[62] Peter Gay. *The Cultivation of Hatred: The Bourgeois Experience: Victoria to Freud (The Bourgeois Experience: Victoria to Freud).* WW Norton & Company, 1993.

[63] Daniel E Ho. Judging statistical criticism. *Observational Studies*, 4:42–56, 2018.

[64] D Rindskopf. Statistical criticism, self-criticism and the scientific method. *Observational Studies*, 4:61–64, 2018.

[65] Paul R Rosenbaum and Dylan S Small. Beyond statistical criticism. *Observational Studies*, 4:34–41, 2018.

[66] J Hill and KJ Hoggatt. The tenability of counterhypotheses: A comment on bross' discussion of statistical criticism. *Observational Studies*, 4:34–41, 2018.

[67] Fan Yang, José R Zubizarreta, Dylan S Small, Scott Lorch, and Paul R Rosenbaum. Dissonant conclusions when testing the validity of an instrumental variable. *The American Statistician*, 68(4):253–263, 2014.

[68] Nicholas Rescher. *Aporetics: Rational deliberation in the face of inconsistency*. University of Pittsburgh Press, 2009.

[69] Plato, WRM Lamb, Robert Gregg Bury, Paul Shorey, and Harold North Fowler. *Plato in twelve volumes*. Heinemann, 1923.

[70] Irwin DJ Bross. Spurious effects from an extraneous variable. *Journal of Chronic Diseases*, 19(6):637–647, 1966.

[71] Irwin DJ Bross. Pertinency of an extraneous variable. *Journal of Chronic Diseases*, 20(7): 487–495, 1967.

[72] Paul R Rosenbaum. Sensitivity analysis for certain permutation inferences in matched observational studies. *Biometrika*, 74(1):13–26, 1987.

[73] Paul R Rosenbaum. *Design of observational studies*. Springer, 2010.

[74] James M Robins. Association, causation, and marginal structural models. *Synthese*, pages 151–179, 1999.

[75] Hyejin Ko, Joseph W Hogan, and Kenneth H Mayer. Estimating causal treatment effects from longitudinal hiv natural history studies using marginal structural models. *Biometrics*, 59(1):152–162, 2003.

[76] Bryan E Shepherd, Mary W Redman, and Donna P Ankerst. Does finasteride affect the severity of prostate cancer? a causal sensitivity analysis. *Journal of the American Statistical Association*, 103(484):1392–1404, 2008.

[77] Vincent Dorie, Masataka Harada, Nicole Bohme Carnegie, and Jennifer Hill. A flexible, interpretable framework for assessing sensitivity to unmeasured confounding. *Statistics in Medicine*, 35(20):3453–3470, 2016.

[78] Tyler J VanderWeele and Peng Ding. Sensitivity analysis in observational research: introducing the e-value. *Annals of Internal Medicine*, 167(4):268–274, 2017.

[79] Qingyuan Zhao, Dylan S Small, and Bhaswar B Bhattacharya. Sensitivity analysis for inverse probability weighting estimators via the percentile bootstrap. *Journal of the Royal Statistical Society: Series B (Statistical Methodology)*, 81(4):735–761, 2019.

[80] E Cuyler Hammond. Smoking in relation to mortality and morbidity. findings in first thirty-four months of follow-up in a prospective study started in 1959. *Journal of the National Cancer Institute*, 32(5):1161–1188, 1964.

[81] Paul R Rosenbaum. *Observational studies, 2nd Edition*. Springer, 2002.

[82] Paul R Rosenbaum. Choice as an alternative to control in observational studies. *Statistical Science*, pages 259–278, 1999.

[83] Karl Popper. *The logic of scientific discovery*. Harper and Row, 1968. English translation of Popper's 1934 *Logik der Forschung*.

[84] Chico Meriam Library, California State University. Evaluating information – applying the craap test, 2010.

[85] RutgersLibGuides. Evaluating news sources: Ask questions.

[86] Hans Jürgen Eysenck. *Genius: The natural history of creativity*, volume 12. Cambridge University Press, 1995.

[87] GH Hardy. Ramanujan: Twelve lectures on subjects suggested by his life and work, 1940.

[88] Kareem Abdul Jabbar. We may be a divided nation but we're united in not trusting the news media. November 8, 2020.

[89] Thomas Jefferson. *The Papers of Thomas Jefferson*, volume 11. Princeton University Press Princeton, 1950. Editor: Julian P. Boyd.

[90] JL Gastwirth. The potential usefulness of bross's principles of statistical criticism for the evaluation of statistical evidence in law and public policy. *Observational Studies*, 4:17–31, 2018.

[91] Thomas J Halpin, Francis J Holtzhauer, Robert J Campbell, Lois J Hall, Adolfo Correa-Villaseñor, Richard Lanese, Janet Rice, and Eugene S Hurwitz. Reye's syndrome and medication use. *Jama*, 248(6):687–691, 1982.

[92] Eugene S Hurwitz, Michael J Barrett, Dennis Bregman, Walter J Gunn, Lawrence B Schonberger, William R Fairweather, Joseph S Drage, John R LaMontagne, Richard A Kaslow, D Bruce Burlington, et al. Public health service study on reye's syndrome and medications: report of the pilot phase. *New England Journal of Medicine*, 313(14):849–857, 1985.

[93] Joseph L Gastwirth. Should law and public policy adopt 'practical causality'as the appropriate criteria for deciding product liability cases and public policy? *Law, Probability and Risk*, 12(3-4):169–188, 2013.

[94] David Rogosa. A school accountability case study: California api awards and the orange county register margin of error folly. *Defending standardized testing*, pages 205–226, 2005.

[95] Dario Sambunjak, Jason W Nickerson, Tina Poklepovic, Trevor M Johnson, Pauline Imai, Peter Tugwell, and Helen V Worthington. Flossing for the management of periodontal diseases and dental caries in adults. *Cochrane Database of Systematic Reviews*, (12), 2011.

[96] Jamie Holmes. Flossing and the art of scientific investigation.

[97] NIH News in Health. Don't toss the floss! the benefits of daily cleaning between teeth, November 2016.

[98] Corpus Juris Secundum. Witnesses, section 636.

[99] Richard Posner. Kadia v. gonzales, 501 f.3d 817, 821 (7th cir. 2007), 2007.

[100] Core Writing Team, RK Pachauri, and A Reisinger. Intergovernmental panel on climate change fourth assessment report. *IPCC, Geneva, Switzerland*, 2007.

[101] J. Raloff. Ipcc relied on unvetted himalaya melt figure.

[102] William G Cochran. The planning of observational studies of human populations. *Journal of the Royal Statistical Society. Series A (General)*, 128(2):234–266, 1965.

[103] Richard Doll and A Bradford Hill. Study of the aetiology of carcinoma of the lung. *British Medical Journal*, 2(4797):1271, 1952.

[104] ESPN.com staff. The wizard's wisdom: 'woodenisms'.

25

Sensitivity Analysis

C.B. Fogarty

CONTENTS

25.1 Why Conduct a Sensitivity Analysis?

The debate surrounding an observational study rarely centers around the K covariates that have been accounted for through matching, weighting, or some other form of adjustment; rather, a critic's counterclaim typically rests upon the existence of a nettlesome $K + 1$st covariate for which the researcher did not account, an unobserved covariate whose relationship with treatment assignment and the outcome of interest may explain away the study's purported finding. Should the totality of evidence assume the absence of hidden bias, as is common with studies proceeding under the usual

DOI: 10.1201/9781003102670-25

expedient of strong ignorability, the critic need merely suggest the existence of bias to cast doubt upon the posited causal mechanism. It is thus incumbent upon researchers not only to anticipate such criticism, but also to arm themselves with a suitable rejoinder.

A sensitivity analysis assesses the robustness of an observational study's conclusions to controlled departures from strong ignorability, determining the strength of hidden bias needed to materially alter its findings. Insensitivity instills confidence in the findings of an observational study, while skepticism is warranted should only a trifling degree of hidden bias be required. Through conducting a sensitivity analysis, a critic may no longer undermine confidence in an observational study by simply stating that hidden bias may exist; rather, the critic must specifically argue that the strength of a proposed lurking variable could realistically exceed the maximum degree of bias which a study could absorb while leaving the study's conclusions intact.

The first sensitivity analysis in an observational study was performed by [1] in an article assessing evidence for cigarette smoking causing lung cancer, quantifying the degree of imbalance on an unobserved covariate required to produce, in the absence of a causal relationship, an association between smoking and lung cancer to the extent observed. The approach presented therein possesses a few drawbacks: it does not account for sampling uncertainty; it does not allow for adjustment for measured confounders; and it is limited to binary outcome variables. This chapter describes an approach to sensitivity analysis first developed by Paul Rosenbaum and his collaborators appropriate for matched designs which overcomes these limitations. It then briefly presents a related approach by [2] appropriate when using inverse probability weighting.

Alternative frameworks for sensitivity analysis have been proposed through the years. See, among several, [3–12].

25.2 Sensitivity Analysis for Matched Designs

25.2.1 A model for biased assignments

There are N total individuals in the observational study who are placed into I matched sets using an optimal without-replacement matching algorithm, with n_i total individuals in the ith of I matched sets. Let W_{ij} be the binary treatment indicator for the ijth individual, let y_{1ij} and y_{0ij} be the potential outcomes for this individual under treatment and control (assumed univariate initially), and let $\mathbf{x}_{ij}$ and u_{ij} be the observed covariate and unobserved covariate for this individual. We assume that ith matched set contains 1 treated individual and $n_i - 1$ control individuals, such that $\sum_{j=1}^{n_i} W_{ij} = 1$; see [23, §4, Problem 12] for the straightforward modifications required to accommodate full matching. Let $\mathcal{F} = \{(y_{0ij}, y_{1ij}, \mathbf{x}_{ij}, u_{ij}) : i = 1, ..., I; j = 1, .., n_i\}$ be a set containing the potential outcomes, observed covariates, and unobserved covariates for the individuals in the study. In what follows, we will consider inference conditional upon $\mathcal{F}$, without assuming a particular generative model for the potential outcomes or the covariates.

In the population before matching, we assume that individual ij received treatment independently of individual $i'j'$ for distinct units ij and $i'j'$ with probability $\pi_{ij} = \text{pr}(W_{ij} = 1 \mid y_{0ij}, y_{1ij}, \mathbf{x}_{ij}, u_{ij})$, such that

$$\pi_{ij} = \text{pr}(W_{ij} = 1 \mid y_{0ij}, y_{1ij}, \mathbf{x}_{ij}, u_{ij});$$
$$\text{pr}(\mathbf{W} = \mathbf{w} \mid \mathcal{F}) = \prod_{i=1}^{I}\prod_{j=1}^{n_i} \pi_{ij}^{w_{ij}}(1 - \pi_{ij})^{(1-w_{ij})},$$

where $\mathbf{W}$ is the lexicographically ordered vector of treatment assignments $\mathbf{W} = (W_{11}, W_{12}, ..., W_{In_I})^T$, and the analogous notation holds for other boldfaced quantities in what follows.

In a sensitivity analysis we imagine that treatment assignment is not strongly ignorable given the observed covariates $\mathbf{x}$ alone, but that treatment would be strongly ignorable given $(\mathbf{x}, u)$, i.e.

$$0 < \pi_{ij} = \text{pr}(W_{ij} = 1 \mid \mathbf{x}_{ij}, u_{ij}) < 1. \tag{25.1}$$

Observe that (25.1) along with assuming that $\text{pr}(W_{ij} = 1 \mid \mathbf{x}_{ij}, u_{ij}) = \text{pr}(W_{ij} = 1 \mid \mathbf{x}_{ij})$ imply that $0 < \text{pr}(W_{ij} = 1 \mid y_{0ij}, y_{1ij}, \mathbf{x}_{ij}) = \text{pr}(W_{ij} = 1 \mid \mathbf{x}_{ij}) < 1$, such that (25.1) and irrelevance of u for treatment assignment given $\mathbf{x}$ provides strong ignorability given $\mathbf{x}$ alone.

The model for a sensitivity analysis introduced in [22] bounds the odds ratio of π_{ij} and π_{ik} for two individuals j, k in the same matched set i,

$$\frac{1}{\Gamma} \leq \frac{\pi_{ij}(1-\pi_{ik})}{\pi_{ik}(1-\pi_{ij})} \leq \Gamma \;\; (i = 1, ..., I; j, k = 1, .., n_i). \tag{25.2}$$

This can be equivalently expressed in terms of a logit model for π_{ij},

$$\log\left(\frac{\pi_{ij}}{1-\pi_{ij}}\right) = \kappa(\mathbf{x}_{ij}) + \gamma u_{ij}, \tag{25.3}$$

where $\gamma = \log(\Gamma)$, $\kappa(\mathbf{x}_{ij}) = \kappa(\mathbf{x}_{ik})$ for all $i = 1, ..., I; j, k = 1, ..., n_i$, and $0 \leq u_{ij} \leq 1$ for $i = 1, ..., I; j = 1, ..., n_i$; see [15] for a proof of this equivalance. That is, imagining a logit form with a scalar unmeasured covariate bounded between 0 and 1 as in (25.3) imposes the same restrictions as does (25.2), which makes no reference to the dimension of the unmeasured covariate.

Attention may be returned to the matched structure at hand by conditioning upon the event that the observed treatment assignment satisfies the matched design. Let Ω be the set of treatment assignments $\mathbf{w}$ adhering to the matched design, i.e. satisfying $\sum_{j=1}^{n_i} w_{ij} = 1$ for all i, and let $\mathcal{W}$ be the event that $\mathbf{W} \in \Omega$. When (25.3) holds at Γ, the conditional distribution for $\mathbf{W} \mid \mathcal{F}, \mathcal{W}$ may be expressed as

$$\begin{aligned} \text{pr}(\mathbf{W} = \mathbf{w} \mid \mathcal{F}, \mathcal{W}) &= \frac{\exp(\gamma \mathbf{w}^T \mathbf{u})}{\sum_{\mathbf{b}\in\Omega} \exp(\gamma \mathbf{b}^T \mathbf{u})}. \\ &= \prod_{i=1}^{I} \frac{\exp(\gamma \sum_{j=1}^{n_i} w_{ij} u_{ij})}{\sum_{j=1}^{n_i} \exp(\gamma u_{ij})}. \end{aligned} \tag{25.4}$$

We see that through conditioning upon $\mathcal{W}$, we have removed dependence upon the nuisance parameters $\kappa(\mathbf{x})$, but that dependence upon u remains. At $\Gamma = 1 \Leftrightarrow \gamma = 0$, $\text{pr}(W_{ij} = 1 \mid y_{0ij}, y_{1ij}, \mathbf{x}_{ij}, u_{ij}) = \text{pr}(W_{ij} = 1 \mid \mathbf{x}_{ij})$, such that the study is free of hidden bias. In this case, the conditional distribution (25.4) is precisely the distribution for treatment assignments in a finely stratified experiment with 1 treated individual and $n_i - 1$ controls in each stratum [16, 17]. This reflects mathematically the adherence of matching to the advice attributed to H.F. Dorn that "the planner of an observational study should always ask himself the question, 'how would the study be conducted if it were possible to do it by controlled experimentation?'" [18, p. 236]. At $\Gamma = 1$, matching provides a reference distribution for inference on treatment effects that aligns with what we would have attained through a finely stratified experiment.

Taking $\Gamma > 1$ in (25.2) allows for departures from equal assignment probabilities within a matched set, be they due to the impact of hidden bias or to residual discrepancies on the basis of observed covariates. We attain a family of conditional treatment assignment distributions (25.4) for each $\Gamma > 1$, indexed by the length-N vector of unmeasured confounders $\mathbf{u} \in \mathcal{U}$, where $\mathcal{U}$ is the N-dimensional unit cube.

25.2.2 Randomization distributions for test statistics

Let $G(\mathbf{W},\mathbf{Y})$ be an arbitrary function of both the vector of treatment assignments $\mathbf{W}$ and the vector of observed responses $\mathbf{Y}$, where $Y_{ij} = W_{ij}y_{1ij} + (1 - W_{ij})y_{0ij}$. For any $\mathbf{u} \in \mathcal{U}$, when (25.2) is assumed to hold at Γ, the randomization distribution for $G(\mathbf{W},\mathbf{Y})$ takes the form

$$\text{pr}\{G(\mathbf{W},\mathbf{Y}) \leq \mathbf{k} \mid \mathcal{F}, \mathcal{W}\} = \frac{\sum_{\mathbf{b}\in\Omega} \exp(\gamma \mathbf{b}^T\mathbf{u}) 1\{G(\mathbf{b},\mathbf{Y}) \leq \mathbf{k}\}}{\sum_{\mathbf{b}\in\Omega} \exp(\gamma \mathbf{b}^T\mathbf{u})}, \quad (25.5)$$

with the inequality $G(\mathbf{W},\mathbf{Y}) \leq \mathbf{k}$ interpreted coordinate-wise in the multivariate case and $1\{A\}$ being an indicator that the event A occurred. The randomization distribution in (25.5) is generally unknown to the practitioner for all $\mathbf{k}$ without further assumptions. For any value of $\Gamma \geq 1$, (25.8) depends upon values for the potential outcomes that are unknown: if $W_{ij} = 1$, then y_{0ij} is unknown, and if $W_{ij} = 0$, then y_{1ij} is unknown. As a result, the value $G(\mathbf{w},\mathbf{Y})$ that would be observed under treatment assignment $\mathbf{w} \in \Omega$ is generally unknown for all $\mathbf{w} \in \Omega$ spare the observed assignment $\mathbf{W}$ without additional assumptions. For $\Gamma > 1$, (25.8) further depends upon the unknown vector of unmeasured confounders $\mathbf{u} \in \mathcal{U}$.

25.2.3 Reference distributions for sharp null hypotheses

A sharp null hypothesis is a hypothesis about the potential outcomes under treatment and control which equates a function of the potential outcomes under treatment to a function of the potential outcomes under control for each individual,

$$H_{sharp} : f_{1ij}(y_{1ij}) = f_{0ij}(y_{0ij}) \;\; (i = 1,..,I; j = 1,...,n_i). \quad (25.6)$$

In so doing the observed outcome Y_{ij} imputes the missing value for the function of the potential outcomes under the treatment assignment that was not observed. Perhaps the most famous sharp null hypothesis is Fisher's sharp null hypothesis of no treatment effect for any individual in the study, which can be tested by choosing $f_{0ij}(y_{0ij}) = y_{0ij}$ and $f_{1ij}(y_{1ij}) = y_{1ij}$. The supposition that the treatment effect is constant at some value τ_0 for all individuals can be tested by setting $f_{0ij}(y_{0ij}) = y_{0ij}$ and $f_{1ij}(y_{1ij}) = y_{1ij} - \tau_0$. Other choices of $f_{1ij}(\cdot)$ and $f_{0ij}(\cdot)$ can yield tests allowing for subject-specific causal effects such as dilated treatment effects, displacement effects and tobit effects; see [23, §5] and [19, §§2.4-2.5] for an overview.

From our data alone we observe

$$F_{ij} = W_{ij}f_{1ij}(y_{1ij}) + (1 - W_{ij})f_{0ij}(y_{0ij}) \quad (25.7)$$

Under the assumption of the sharp null hypothesis (25.6), we further have that $\mathbf{F} = \mathbf{f_0} = \mathbf{f_1}$, where $\mathbf{F}$, $\mathbf{f_0}$, and $\mathbf{f_1}$ are the lexicographically ordered vectors of length N containing F_{ij}, $f_{0ij}(y_{ij})$, and $f_{1ij}(y_{1ij})$ in their entries. Let $T(\mathbf{W},\mathbf{F})$ be a scalar-valued test statistic, and suppose that larger values for the test statistic reflect evidence against the null hypothesis. The right-tail probability for $T(\mathbf{W},\mathbf{F})$ is

$$\text{pr}\{T(\mathbf{W},\mathbf{F}) \geq t \mid \mathcal{F}, \mathcal{W}\} = \frac{\sum_{\mathbf{b}\in\Omega} \exp(\gamma \mathbf{b}^T\mathbf{u}) 1\{T(\mathbf{b},\mathbf{F}) \geq t\}}{\sum_{\mathbf{b}\in\Omega} \exp(\gamma \mathbf{b}^T\mathbf{u})},$$

which under the sharp null (25.6) may be expressed as

$$\text{pr}\{T(\mathbf{W},\mathbf{F}) \geq t \mid \mathcal{F}, \mathcal{W}, H_{sharp}\} = \frac{\sum_{\mathbf{b}\in\Omega} \exp(\gamma \mathbf{b}^T\mathbf{u}) 1\{T(\mathbf{b},\mathbf{f_0}) \geq t\}}{\sum_{\mathbf{b}\in\Omega} \exp(\gamma \mathbf{b}^T\mathbf{u})}. \quad (25.8)$$

25.3 Sensitivity Analysis for Sharp Null Hypotheses

When $\Gamma = 1 \Leftrightarrow \gamma = 0$, the reference distribution (25.8) does not depend upon the unknown $\mathbf{u}$. Hence, p-values constructed with (25.8) may be used to test the sharp null hypothesis (25.6), providing inference whose Type I error rate may be controlled at α regardless of the sample size. At $\Gamma > 1$, (25.8) is no longer known due to its dependence on $\mathbf{u}$. In a sensitivity analysis we instead construct bounds on the possible p-values produced by (25.8) when assuming that the sensitivity model (25.2) holds at some value Γ. We then increase the value of Γ until the worst-case bound on the p-value exceeds α. This changepoint value for Γ, called the *sensitivity value* in [20], provides a measure of robustness for the rejection of the sharp null hypothesis, describing the minimum degree of bias required to overturn a rejection of the sharp null (25.6). The sensitivity value is closely related to the E-value described in [12], differing primarily in that the sensitivity value also takes sampling uncertainty into account.

25.3.1 Bounds on p-values for sum statistics

In what follows we will assume that the test statistics under consideration take the form

$$T(\mathbf{W}, \mathbf{F}) = \mathbf{W}^T\mathbf{q}, \tag{25.9}$$

for some vector $\mathbf{q} = \mathbf{q}(\mathbf{F})$, referred to as a *sum statistic* . Most common statistics take this form, and a few examples for testing Fisher's sharp null of no effect follow to help build intuition. In paired observational studies, McNemar's test statistic with binary outcomes takes $q_{ij} = Y_{ij}$, while the difference in means with any outcome variable type can be attained by choosing $q_{ij} = (Y_{ij} - Y_{ij'})/I$ in a paired design. To recover Wilcoxon's signed rank statistic, let d_i be the ranks of $|Y_{i1} - Y_{i2}|$ from 1 to I, and let $q_{ij} = d_i 1\{F_{ij} > F_{ij'}\}$. With multiple controls, $q_{ij} = \sum_{j' \neq j}(Y_{ij} - Y_{ij'})/\{I(n_i - 1)\}$ returns the treated minus control difference in means. The aligned rank test of [21] first forms aligned responses in each stratum i as $Y_{ij} - \sum_{j'=1}^{n_i} Y_{ij'}/n_i$, ranks the aligned responses from 1 to N (temporarily ignoring the stratification), and then sets q_{ij} equal to the rank for the ijth aligned response accomplishes this.

When $T(\mathbf{W}, \mathbf{F}) = \mathbf{W}^T\mathbf{q}(\mathbf{F})$ is a sum statistic, intuition suggests that (25.8) will be larger when large values for q_{ij} correspond to higher treatment assignment probabilities in a matched set, and will be smaller when large values for q_{ij} correspond to lower treatment assignment probabilities. In matched pair designs we can construct random variables T_Γ^- and T_Γ^+ whose upper tail probabilities bound (25.8) from below and from above for any t at any value of Γ in (25.2). These random variables accord with the above intuition: let T_Γ^+ have the distribution (25.8) with $u_{ij} = 1\{q_{ij} \geq q_{ij'}\}$, and let T_Γ^- have the distribution (25.8) with $u_{ij} = 1\{q_{ij} \leq q_{ij'}\}$ for $i = 1, ..., I$ and $j \neq j' = 1, 2$. In each matched set, T_Γ^+ gives the largest possible probability, $\Gamma/(1+\Gamma)$, to the larger value of $\{q_{i1}, q_{i2}\}$ for $i = 1, ..., I$, and T_Γ^- gives the lowest possible probability, $1/(1+\Gamma)$, to the larger value of $\{q_{i1}, q_{i2}\}$ for $i = 1, ..., I$. Then, for any t and any sample size, if (25.2) holds at Γ,

$$\text{pr}(T_\Gamma^- \geq t) \leq \text{pr}(\mathbf{W}^T\mathbf{q} \geq t \mid \mathcal{F}, \mathcal{W}, H_{sharp}) \leq \text{pr}(T_\Gamma^+ \geq t). \tag{25.10}$$

For a subclass of sum statistics called *sign-score* statistics, we can similarly construct bounding random variables T_Γ^- and T_Γ^+ such that (25.10) holds after matching with multiple controls; see [23, §4.4] for details. When using other sum statistics after matching with multiple controls, [28] show that for any t, $\text{pr}(\mathbf{W}^T\mathbf{q} \geq t \mid \mathcal{F}, \mathcal{W}, H_{sharp})$ is minimized or maximized for distinct vectors $\mathbf{u}$ in $\mathcal{U}^-$ and $\mathcal{U}^+$ respectively, where both $\mathcal{U}^-$ and $\mathcal{U}^+$ are sets containing $\prod_{i=1}^{I}(n_i - 1)$ binary vectors. Unlike in the paired case, however, the values for $\mathbf{u}$ attaining the lower and upper bounds may vary as a function of t. More troublesome, finding these exact bounds becomes computationally intractable due to the requirement of enumerating the $\prod_{i=1}^{I}(n_i - 1)$ elements of $\mathcal{U}^-$ and $\mathcal{U}^+$.

Owing to these computational limitations, [25] turned to an asymptotic approximation to provide upper and lower bounds on pr$(\mathbf{W}^T\mathbf{q} \geq t \mid \mathcal{F}, \mathcal{W}, H_{sharp})$ based upon a normal approximation for the distribution of $\mathbf{W}^T\mathbf{q}$. These bounds are not valid for all t, but are valid for the values of t relevant to conducting inference: the lower bound is valid any t less than the smallest possible expectation for $\mathbf{W}^T\mathbf{q}$, and the upper bound is valid for any t greater than the largest possible expectation for $\mathbf{W}^T\mathbf{q}$. As this random variable may be expressed as the sum of I independent random variables, the approximation is justified under mild regularity conditions on the constants $\mathbf{q}$ ensuring that the Lindeberg-Feller central limit theorem holds. We will consider the problem of upper bounding pr$(\mathbf{W}^T\mathbf{q} \geq t \mid \mathcal{F}, \mathcal{W}, H_{sharp})$ when (25.2) is assumed to hold at Γ by constructing a normal random variable $\tilde{T}_\Gamma^+$ with a suitable mean and variance; the procedure for finding the lower bound $\tilde{T}_\Gamma^-$ is analogous. Roughly stated, we proceed by finding the vector $\mathbf{u}$ which maximizes the expected value of $\mathbf{W}^T\mathbf{q} \mid \mathcal{F}, \mathcal{W}$ when (25.2) is assumed to hold at Γ; if multiple vectors $\mathbf{u}$ yield the same maximal expectation, we choose the one which maximizes the variance. Then, we simply compute the probability that a normal random variable with this mean and variance exceeds a given value t larger than the maximal expectation. Importantly, the mean and variance for this normal variable may be pieced together *separately* on a stratum-by-stratum basis, requiring an optimization over $n_i - 1$ candidate solutions for each matched set i rather than requiring global optimization over $\prod_{i=1}^{I}(n_i - 1)$ binary vectors. [25] refer to this optimization as possessing *asymptotic separability*.

To proceed, rearrange the values q_{ij} in each matched set i such that $q_{i1} \leq q_{i2} \leq ... \leq q_{in_i}$. Let $\mathcal{U}_i^+$ denote the collection of $n_i - 1$ binary vectors for the ith stratum of the form $u_{i1} = ... = u_{ia_i} = 0$ and $u_{ia_i+1} = ... = u_{in_i} = 1$ for some $a_i = 1, ..., n_i - 1$. Let μ_i be the largest possible expectation for $\mathbf{u}_i \in \mathcal{U}_i^+$,

$$\begin{aligned}\mu_i^+ &= \max_{\mathbf{u}_i \in \mathcal{U}_i^+} \frac{\sum_{j=1}^{n_i} \exp(\gamma u_{ij}) q_{ij}}{\sum_{j=1}^{n_i} \exp(\gamma u_{ij})} \\ &= \max_{a_i = 1,\ldots,n_i-1} \frac{\sum_{i=1}^{a_i} q_{ij} + \Gamma \sum_{j=a_i+1}^{n_i} q_{ij}}{a_i + \Gamma(n_i - a_i)}\end{aligned}$$

Let $\mathcal{A}_i$ be set of values for a_i attaining the maximal expectation μ_i^+, and define the corresponding variance ν_i^+ as

$$\nu_i^+ = \max_{a_i \in \mathcal{A}_i} \frac{\sum_{j=1}^{a_i} q_{ij}^2 + \Gamma \sum_{j=a_i+1}^{n_i} q_{ij}^2}{a_i + \Gamma(n_i - a_i)} - (\mu_i^+)^2$$

By independence of conditional treatment assignments across matched sets, the expectation and variance of the asymptotically bounding normal random variable $\tilde{T}_\Gamma$ are

$$E(\tilde{T}_\Gamma^+) = \sum_{i=1}^{I} \mu_i^+; \ \text{var}(\tilde{T}_\Gamma^+) = \sum_{i=1}^{I} \nu_i^+. \tag{25.11}$$

The asymptotic upper bound for pr$(\mathbf{W}^T\mathbf{q} \geq t \mid \mathcal{F}, \mathcal{W})$ returned by this procedure whenever $t \geq E(\tilde{T}_\Gamma^+)$ is

$$1 - \Phi\left\{\frac{t - E(\tilde{T}_\Gamma^+)}{\sqrt{\text{var}(\tilde{T}_\Gamma^+)}}\right\}, \tag{25.12}$$

where $\Phi(\cdot)$ is the cumulative distribution function of the standard normal. As seen through its construction, this asymptotic approximation reduces an optimization problem with $\prod_{i=1}^{I}(n_i - 1)$ candidate solutions to I tractable optimization problems which may be solved in isolation, each requiring enumeration of only $n_i - 1$ candidate solutions.

To find the moments $E(\tilde{T}_\Gamma^-)$ and var$(\tilde{T}_\Gamma^-)$ needed for the asymptotic lower bound, simply replace q_{ij} with $\tilde{q}_{ij} = -q_{ij}$ for all ij and follow the above procedure, finding expectations $\tilde{\mu}_i^+$ and $\tilde{\nu}_i^+$ for all $i = 1, .., I$ for the test statistic $\mathbf{W}^T\tilde{\mathbf{q}}$. Then, take $E(\tilde{T}_\Gamma^-) = -\sum_{i=1}^{I}\tilde{\mu}_i^+$ and var$(\tilde{T}_\Gamma^-) = \sum_{i=1}^{I}\tilde{\nu}_i^+$. For any $t \leq E(\tilde{T}_\Gamma^-)$, pr$(\mathbf{W}^T\mathbf{q} \geq t \mid \mathcal{F}, \mathcal{W}, H_{sharp})$ is asymptotically lower bounded by (25.12) with $\tilde{T}_\Gamma^+$ replaced by $\tilde{T}_\Gamma^-$.

25.3.2 Sensitivity analysis for point estimates

Beyond providing upper and lower bounds on p-values under sharp null hypotheses, we can further construct intervals for point estimates under various models of effects and confidence intervals accounting for both sampling uncertainty and the impact of hidden bias at different values of Γ in (25.2). We will focus on point estimates and confidence intervals for the unknown treatment effect when a constant, or additive, treatment effect model is assumed; that said, the core ideas presented herein extend naturally to other models for effects such as multiplicative and tobit effects. An additive treatment effect model states that the treatment effect for each individual in the study equals some value τ, i.e.

$$y_{1ij} = y_{0ij} + \tau \quad (i = 1, ..., I; j = 1, .., n_i). \tag{25.13}$$

Consider first testing the null hypothesis that $\tau = \tau_0$ in (25.13). By setting $f_{0ij}(y_{0ij}) = y_{0ij}$ and $f_{1ij}(y_{1ij}) = y_{1ij} - \tau_0$ in (25.6) for all ij, we have that F_{ij} in (25.7) are the adjusted responses $Y_{ij} - W_{ij}\tau_0$, and that $\mathbf{F} = \mathbf{f_0} = \mathbf{f_1}$. Therefore, we are entitled to the reference distribution (25.8) for conducting inference, and the methods for sensitivity analysis described in §25.3 may be deployed whenever $T(\mathbf{W}, \mathbf{Y} - \mathbf{W}\tau_0)$ is a sum statistic. As we now describe, this facilitates sensitivity analyses for point estimates and confidence intervals. We assume moving forwards that $T(\mathbf{w}, \mathbf{Y} - \mathbf{w}\tau)$ is both a sum statistic and is a monotone decreasing function of τ for any $\mathbf{Y}$ and any $\mathbf{w} \in \Omega$. The latter condition simply requires that the statistic T measures the size of the treatment effect. We will also assume that the potential outcomes are continuous, as a constant treatment effect model makes little sense with discrete outcomes.

In [24], Hodges and Lehmann consider a general approach to producing point estimates based upon hypothesis tests. Their idea, loosely stated, is to find the value of τ_0 such that the adjusted responses $\mathbf{Y} - \mathbf{W}\tau_0$ appear to be exactly free of a treatment effect as measured by the statistic T. When $T(\mathbf{W}, \mathbf{Y} - \mathbf{W}\tau_0)$ is continuous as a function of τ_0, this is accomplished by finding the value of τ_0 such that the observed value of $T(\mathbf{W}, \mathbf{Y} - \mathbf{W}\tau_0)$ exactly equals its expectation when assuming (25.13) holds at τ_0. The approach can be modified to accomodate discontinuous statistics T such as rank statistics as follows. Letting $M_{\Gamma,\mathbf{u},\tau}$ be the expectation of $T(\mathbf{W}, \mathbf{Y} - \mathbf{W}\tau_0)$ when the treatment effect is constant at τ_0 and (25.2) holds at Γ with vector of hidden bias $\mathbf{u}$, the Hodges-Lehmann estimator $\hat{\tau}_{HL}$ is

$$\begin{aligned} \hat{\tau}_{HL} = [&\inf\{\tau : T(\mathbf{W}, \mathbf{Y} - \mathbf{W}\tau) < M_{\Gamma,\mathbf{u},\tau}\} \\ &+ \sup\{\tau : T(\mathbf{W}, \mathbf{Y} - \mathbf{W}\tau) > M_{\Gamma,\mathbf{u},\tau}\}]/2, \end{aligned} \tag{25.14}$$

i.e. the average of the smallest value of τ such that the test statistic falls below its expectation and the largest τ such that the test statistic falls above its expectation.

At $\Gamma = 1$ in (25.2) there is a single Hodges-Lehmann estimator for each choice of test statistic. For $\Gamma > 1$, the single estimator instead becomes an interval of estimates: each value for the vector of hidden bias $\mathbf{u}$ can potentially provide a unique value for $M_{\Gamma,\mathbf{u},\tau}$. This interval of point estimates is straightforward to construct given our assumption of a sum statistic that is monotone decreasing in τ, as we can leverage steps involved in the procedure of §25.3.1 for constructing asymptotic tail bounds. Recalling that $T(\mathbf{W}, \mathbf{Y} - \mathbf{W}\tau)$ declines with τ, to find the lower bound for the interval of Hodges-Lehmann estimates we compute (25.14) replacing $M_{\Gamma,\mathbf{u},\tau}$ with $E(\tilde{T}^{+}_{\Gamma,\tau})$, the largest possible expectation for $T(\mathbf{W}, \mathbf{Y} - \mathbf{W}\tau)$ as given by (25.11). For the upper bound, we instead compute (25.14) when $M_{\Gamma,\mathbf{u},\tau}$ is replaced by $E(\tilde{T}^{-}_{\Gamma,\tau})$, the smallest possible expectation for $T(\mathbf{W}, \mathbf{Y} - \mathbf{W}\tau)$.

25.3.3 Sensitivity analysis for confidence intervals

To form a confidence interval for τ when (25.2) holds at Γ, we use the duality between hypothesis tests and confidence intervals: for any $\mathbf{u}$ and Γ, the test statistic $T(\mathbf{W}, \mathbf{Y} - \mathbf{W}\tau_0)$ may be used to form a two-sided $100(1-\alpha)\%$ by taking the set of values τ_0 such that the null hypothesis that $\tau = \tau_0$ in (25.13) cannot be rejected at level $\alpha/2$ when testing with both a greater-than and a less-than alternative. The union of these sets over $\mathbf{u} \in \mathcal{U}$ is then called the *sensitivity interval* when (25.2) holds at Γ. A value of τ is in the sensitivity interval whenever there exists a vector of hidden bias $\mathbf{u}$ such that τ cannot be rejected when (25.2) is assumed to hold at Γ. To find the upper-bound of this sensitivity interval at a given Γ, we find the largest value τ such that the worst-case (largest) left-tail p-value falls above $\alpha/2$. For the lower bound, we instead find the smallest value τ such that the largest possible right-tail p-value falls above $\alpha/2$. Using the asymptotic approach described in 25.3.1, the endpoints of the $100(1-\alpha)\%$ sensitivity interval when (25.2) holds at Γ are L_Γ and U_Γ, where

$$L_\Gamma = \inf\left\{\tau : \frac{T(\mathbf{W}, \mathbf{Y} - \mathbf{W}\tau) - E(\tilde{T}^{+}_{\Gamma,\tau})}{\text{var}(\tilde{T}^{+}_{\Gamma,\tau})} \leq \Phi^{-1}(1-\alpha/2)\right\};$$

$$U_\Gamma = \sup\left\{\tau : \frac{T(\mathbf{W}, \mathbf{Y} - \mathbf{W}\tau) - E(\tilde{T}^{-}_{\Gamma,\tau})}{\text{var}(\tilde{T}^{-}_{\Gamma,\tau})} \geq \Phi^{-1}(\alpha/2)\right\},$$

and $\tilde{T}^{+}_{\Gamma,\tau}$ and $\tilde{T}^{-}_{\Gamma,\tau}$ are the asymptotically upper and lower bounding normal random variables for $T(\mathbf{W}, \mathbf{Y} - \mathbf{W}\tau)$ returned through the procedure in §25.3.1.

25.4 Design Sensitivity

25.4.1 Bias trumps variance

To this point we have not considered the particular form of the test statistic $T(\mathbf{W}, \mathbf{F})$ used when conducting the sensitivity analysis. In practice the particular choice of test statistic can substantially impact the sensitivity analysis, with different test statistics potentially furnishing markedly different perceptions of how robust the rejection of the sharp null is to unmeasured confounding. In this section we develop general intuition for the features of a test statistic which are most important for improved performance in a sensitivity analysis.

Suppose that we use a sum statistic $T = \mathbf{W}^T\mathbf{q}$ to perform inference for a sharp null of the form (25.6). Hidden bias may or may not exist, and the sharp null may or may not actually hold. Let $T_I = T/I$, and let μ and σ^2/I be the true, but unknowable, mean and variance for T_I. Under mild conditions a central limit theorem will apply to the statistic $I^{1/2}(T - \mu)/\sigma$, such that for any k

$$\text{pr}\left\{\frac{I^{1/2}(T-\mu)}{\sigma} \geq k\right\} \to \Phi(k).$$

Because the data are observational, the researcher conducts a sensitivity analysis using T as a test statistic when assuming that the sharp null holds. For any value of Γ in (25.2) the practitioner uses the method in §25.3.1 to calculate the moments of a normal random variable which provides, asymptotically, an upper bound on the upper tail probability of T under the assumption that the sharp null holds. Let $\mu_\Gamma = E(\tilde{T}_\Gamma^+)/I$ and $\sigma_\Gamma^2/I = \text{var}(\tilde{T}_\Gamma^+)/I^2$ be the returned expectation and variance for $T_I = T/I$ by the procedure in (25.3.1). If α is the desired level of the procedure, using a normal approximation the sensitivity analysis rejects at Γ if

$$\frac{I^{1/2}(T_I - \mu_\Gamma)}{\sigma_\Gamma} \geq k_\alpha,$$

with $k_\alpha = \Phi^{-1}(1-\alpha)$. This rejection occurs with probability

$$\text{pr}\left\{\frac{I^{1/2}(T_I - \mu_\Gamma)}{\sigma_\Gamma} \geq k_\alpha\right\} = \text{pr}\left\{\frac{I^{1/2}(T_I - \mu)}{\sigma} \geq \frac{k_\alpha\sigma_\Gamma + I^{1/2}(\mu_\Gamma - \mu)}{\sigma}\right\} \quad (25.15)$$
$$\to 1 - \Phi\left\{\frac{k_\alpha\sigma_\Gamma + I^{1/2}(\mu_\Gamma - \mu)}{\sigma}\right\},$$

which tends to 1 if $\mu > \mu_\Gamma$ and 0 otherwise. That is to say, because bias due to unmeasured confounding is $O(1)$, whether or not the sensitivity analysis rejects depends solely upon whether or not the worst-case expectation at Γ under the null, μ_Γ, exceeds the true value of the test statistic's expectation, μ, with neither σ nor σ_Γ playing a role. The value of Γ for which $\mu = \mu_\Gamma$ is called the *design sensitivity* and is denoted by $\tilde{\Gamma}$. It is thus desirable to use test statistics with larger values for the design sensitivity when the alternative is true, as these test statistics provide evidence which is more robust to adversarially aligned hidden bias.

25.4.2 A favorable reality unknown to the practitioner

While the design sensitivity may be calculated under any assumed situation, it is common in practice to proceed with calculations under the following *favorable setting*. Imagine that assignment was actually strongly ignorable given $\mathbf{x}$, such that (25.2) actually holds at $\Gamma = 1$, and that the treatment effect is truly positive. Given that our data are observational and treatment assignment is outside of our control, a critic could always counter that a large effect estimate is merely due to bias from lurking variables. Hence, even in this favorable situation of a ignorable treatment assignment given $\mathbf{x}$ and a positive treatment effect, we would hope that our inferences would prove robust to moderate degrees of hidden bias to protect ourselves against such criticism. In this favorable setting there is thus no hidden bias, there is truly an effect, but the practitioner, blind to this reality, hopes that her inferences perform well under the stress of a sensitivity analysis. Rejections of the null hypothesis of no effect are warranted and desired, and we would prefer to use test statistics with larger values for the design sensitivity. Under this setting, we assess our test's ability to discriminate between (1) no treatment effect and hidden bias; and (2) a treatment effect without hidden bias.

Design sensitivities computed under this favorable situation have many uses when considering the design and analysis of observational studies [25]. They can help quantify otherwise informal advice about design decisions which can improve the quality of an observational study [26, 27]; can inform how many matched controls are matched to a given treated individual [28]; and can help illuminate whether estimators designed with an eye towards efficiency perform well when considering robustness to hidden bias [18].

25.4.3 An illustration: design sensitivities for m-statistics

We will now illustrate the usefulness of design sensitivity by comparing members of a class of statistics known as m-statistics. M-statistics were introduced by [30] as the testing analogue of Huber's robust m-estimates, wherein the influence of any particular observation is controlled and limited through the choice of an estimating equation. For simplicity we restrict attention to matched pairs in this illustration; the concepts extend to matching with multiple controls without issue, albeit with messier formulae. We closely follow the presentation in [28, 31]. Suppose we are interested in performing a sensitivity analysis for the null hypothesis that the additive treatment effect model (25.13) holds with effect τ_0. Let $\hat{\tau}_i = (W_{i1} - W_{i2})(Y_{i1} - Y_{i2})$ be the treated-minus-control difference in means in the ith pair. We will consider using an m-statistic to conduct inference, which takes the form

$$T(\mathbf{W}, \mathbf{Y} - \mathbf{W}\tau_0) = \sum_{i=1}^{I} \psi\left(\frac{\hat{\tau}_i - \tau_0}{h_{\tau_0}}\right),$$

where ψ is an odd function, $\psi(x) = -\psi(-x)$, which is nonnegative for $x > 0$; and h_{τ_0} is a scaling factor, typically taken to be a quantile of the absolute differences $|\hat{\tau}_i - \tau_0|$. For any member of this class, if (25.2) holds at Γ then under the null the right-tail probabilities for T are bounded above and below by variables T_Γ^+ and T_Γ^- with an intuitive construction as in (25.10): let T_Γ^+ be the sum of I independent random variables that take the value $\psi_i = \psi(|\hat{\tau}_i - \tau_0|/h_{\tau_0})$ with probability $\Gamma/(1+\Gamma)$ and $-\psi_i$ otherwise. Likewise, let T_Γ^- be the sum of I independent random variables that take the value ψ_i with probability $1/(1+\Gamma)$ and $-\psi_i$ otherwise. In each pair, T_Γ^+ thus puts the largest possible mass on ψ_i, the positive value, while T_Γ^- puts the largest mass on $-\psi_i$ in each pair.

From this construction we see that the expectation and variance for T_Γ^+ take the form

$$E(T_\Gamma^+) = \frac{\Gamma - 1}{1 + \Gamma} \sum_{i=1}^{I} \psi_i$$

$$\text{var}(T_\Gamma^+) = \frac{4\Gamma}{(1+\Gamma)^2} \sum_{i=1}^{I} \psi_i^2,$$

such that under mild regularity conditions on ψ_i, asymptotically we can reject the null hypothesis with a greater-than alternative at level α whenever

$$\frac{T - \frac{\Gamma-1}{1+\Gamma}\sum_{i=1}^{I}\psi_i}{\sqrt{\frac{4\Gamma}{(1+\Gamma)^2}\sum_{i=1}^{I}\psi_i^2}} \geq \Phi^{-1}(1-\alpha).$$

Until this point inference has been performed conditional upon $\mathcal{F}$, and a generative model for $\mathcal{F}$ has been neither required nor assumed. For calculating design sensitivity it is convenient to assume a superpopulation model, imagining that the responses themselves are themselves drawn from a distribution. We imagine we are the favorable situation of an effect equal to $\tau > \tau_0$ and no bias described in §25.4.2, and imagine that $\hat{\tau}_i$ are generated through

$$\hat{\tau}_i = \epsilon_i + \tau, \tag{25.16}$$

where ϵ_i are drawn *iid* from a distribution P with mean zero and finite variance. From the discussion in §25.4.1, we see that the design sensitivity will be the value Γ such that $E(T) = \frac{\Gamma-1}{1+\Gamma}\sum_{i=1}^{I} E\{\psi(|\hat{\tau}_i - \tau_0|/\eta\}$, where η is the probability limit of h_{τ_0}. Solving for Γ, we find that

$$\tilde{\Gamma} = \frac{E\left\{\psi\left(\frac{|\hat{\tau}_i - \tau_0|}{\eta}\right)\right\} + E\left\{\psi\left(\frac{\hat{\tau}_i - \tau_0}{\eta}\right)\right\}}{E\left\{ß\psi\left(\frac{|\hat{\tau}_i - \tau_0|}{\eta}\right)\right\} - E\left\{\psi\left(\frac{\hat{\tau}_i - \tau_0}{\eta}\right)\right\}}$$

TABLE 25.1
Design sensitivities for three m-statistics under different generative models and different magnitudes of treatment effect.

	Normal			$t_{df=3}$			Laplace		
τ	ψ_t	ψ_{hu}	ψ_{in}	ψ_t	ψ_{hu}	ψ_{in}	ψ_t	ψ_{hu}	ψ_{in}
1/4	1.9	1.8	1.9	2.2	2.2	2.4	2.0	2.0	2.0
1/2	3.5	3.3	4.0	4.5	4.8	5.5	3.9	3.8	4.1
3/4	6.7	6.2	8.9	8.7	9.6	11.5	7.1	7.0	8.1
1	13.0	12.2	22.3	15.6	17.5	21.5	12.6	12.4	15.9

$$= \frac{\int_0^\infty \psi\left(|y|/\eta\right)\, dF(y)}{\int_{-\infty}^0 \psi\left(|y|/\eta\right) dF(y)},$$

where $F(\cdot)$ is the distribution function of $\hat{\tau}_i - \tau_0$, and η is the median of $F(\cdot)$.

We now compare design sensitivities for three m-statistics under different distributional assumption for ϵ_i. The competing test statistics use h_{τ_0} equal to the median value of $|\hat{\tau}_i - \tau_0|$, and have ψ functions

$$\begin{aligned}\psi_t(x) &= x\\ \psi_{hu}(x) &= \text{sgn}(x)\min(|x|, 2)\\ \psi_{in}(x) &= (4/3)\text{sgn}(x)\max\{0, \min(2, |x|) - 1/2\}.\end{aligned}$$

The function $\psi_t(\cdot)$ simply returns the permutational t-test based upon the treated-minus-control difference in means. We consider performing "outer trimming" with ψ_{hu}, which uses Huber's weighting function with weights leveling off at 2, hence limiting the impact of outlying points. We also consider "inner trimming" with ψ_{in}, which levels off at 2 (outer trimming) but is modified such that any points between 0 and 1/2 have no influence (inner trimming) [28]. For each of these ψ functions, we compare results when $\tau_0 = 0, \tau = 1/4, 1/2, 3/4, 1$ and (a) ϵ_i are *iid* standard normal; (b) $\sqrt{3}\epsilon_i$ are t-distributed with 3 degrees of freedom; and (c) $\sqrt{2}\epsilon_i$ are Laplace, or double exponential, distributed. The scalings of ϵ_i in (b) and (c) ensure that ϵ_i have variance 1 in all three settings.

Table 25.1 shows the results, with the columns with normal errors replicating the first set of columns in Table 3 of [28]. We see first that the design sensitivity increases with τ in each column, illustrating the intuitive fact that larger treatment effects are more difficult to explain away as the result of hidden bias. Observe next that for all three distributions, the function ψ_{in} produces the largest design sensitivities. This may be particularly striking when considering the results with normally distributed errors, where it is known that the test ψ_t has the largest Pitman efficiency in the context of inference at $\Gamma = 1$. This highlights that the considerations when designing a test statistic that performs well in a sensitivity analysis deviate from those when performing inference in a randomized experiment or under strong ignorability.

25.5 Multiple Comparisons

Developments to this point have focused on sensitivity analyses leveraging a single test statistic. Oftentimes in observational studies, methods for combining multiple test statistics are either desired or required. One example of this is described in [32], where a method is discussed for combining multiple distinct test statistics for testing the *same* sharp null hypothesis in paired designs, in so doing providing a sensitivity analysis with improved robustness properties. Another common scenario is when the researcher wants to investigate the potential causal effects of a single treatment on multiple outcome variables, with distinct test statistics for each outcome variable under consideration.

Suppose that there are K outcome variables, with $\mathbf{y}_0$, $\mathbf{y}_1$, and $\mathbf{Y}$ now representing $N \times K$ matrices of potential outcomes under control, under treatment, and observed outcomes respectively. Suppose that for each outcome, there is a sharp null hypothesis of the form (25.6) under consideration, denoted as $H_1, ..., H_K$. From these K outcomes, we form L sum statistics, with $T_\ell = \mathbf{W}^T \mathbf{q}_\ell$ for $\ell = 1, ..., L$ and $\mathbf{q}_\ell = \mathbf{q}_\ell(\mathbf{F})$ being a length-N vector formed as a function of the matrix $\mathbf{F}$ whose form is determined by the sharp null hypotheses (25.6) under consideration for each outcome variable. We first consider a level-α sensitivity analysis for the global null hypothesis that all K of these hypotheses are true,

$$H_0 : \bigwedge_{k=1}^{K} H_k, \tag{25.17}$$

while assuming that (25.2) holds at Γ. An extremely simple way to conduct a sensitivity analysis for (25.17) would be to first separately conduct L sensitivity analyses, one for each test statistic, and to record the worst-case p-value for the ℓth statistic, call it $P_{\Gamma,\ell}$. Then, we could simply employ a Bonferroni correction and reject the null if $\min_{\ell=1,...,L} P_{\Gamma,\ell} \leq \alpha/L$. While straightforward to implement, this approach ends up being unduly conservative because it allows the worst-case vector of hidden bias $\mathbf{u}$ used to furnish the worst-case p-value to vary from one test statistic to next. If the sensitivity model holds at Γ, there is a single vector $\mathbf{u} \in \mathcal{U}$ which determines the true, but unknowable, conditional assignment distribution for $\mathbf{W}$ given by (25.4). This in turn determines the reference distribution (25.8) for each test statistic under the sharp null. Substantially more powerful tests of (25.17) can be attained by enforcing the requirement that $\mathbf{u}$ cannot vary from one null to the next.

25.5.1 Multivariate sensitivity analysis as a two-person game

We now present a general approach to combining test statistics for use in sensitivity analysis, following closely the presentation in [33]. Define $\varrho_{ij} = \text{pr}(W_{ij} = 1 \mid \mathcal{F}, \mathcal{W})$ to be the conditional assignment probabilities for the ijth individual given the matched design. Under the global null (25.17) and recalling that our test statistics are of the form $T_\ell = \mathbf{W}^T \mathbf{q}_\ell$ with $\mathbf{q}_\ell$ fixed under the global null, the expectation $\mu(\boldsymbol{\varrho})$ and covariance $\Sigma(\boldsymbol{\varrho})$ for the vector of test statistics $\mathbf{T} = (T_1, \ldots, T_L)^T$ are

$$\mu(\boldsymbol{\varrho})_\ell = \sum_{i=1}^{I} \sum_{j=1}^{n_i} q_{ij\ell} \varrho_{ij}$$

$$\Sigma(\boldsymbol{\varrho})_{\ell,\ell'} = \sum_{i=1}^{I} \left\{ \sum_{j=1}^{n_i} q_{ij\ell} q_{ij\ell'} \varrho_{ij} - \left(\sum_{j=1}^{n_i} q_{ij\ell} \varrho_{ij} \right) \left(\sum_{j=1}^{n_i} q_{ij\ell'} \varrho_{ij} \right) \right\};$$

see [33–35] for details. For a given vector of probabilities $\boldsymbol{\varrho}$, under suitable regularity conditions on the constants $q_{ij\ell}$ the distribution of T is asymptotically multivariate normal through an application of the Cramér-Wold device. That is, for any fixed nonzero vector $\boldsymbol{\lambda} = (\lambda_1, ..., \lambda_L)^T$ the standardized deviate

$$\frac{\boldsymbol{\lambda}^T \{\mathbf{T} - \mu(\boldsymbol{\varrho})\}}{\{\boldsymbol{\lambda}^T \Sigma(\boldsymbol{\varrho}) \boldsymbol{\lambda}\}^{1/2}}$$

is asymptotically standard normal.

The constraints imposed by the sensitivity model (25.2) holding at Γ on ϱ can be represented by a polyhedral set. For a particular Γ this set, call it $\mathcal{P}_\Gamma$, contains vectors ϱ such that

(i) $\varrho_{ij} \geq 0$ $(i = 1, ..., I; j = 1, ..., n_i)$;

(ii) $\sum_{i=1}^{n_i} \varrho_{ij} = 1$ $(i = 1, ..., I)$;

(iii) $s_i \leq \varrho_{ij} \leq \Gamma s_i$ for some s_i $(i = 1, ..., I; j = 1, ..., n_i)$.

Conditions (i) and (ii) simply reflect that ϱ_{ij} are probabilities, while the s_i terms in (iii) represent $\sum_{j=1}^{n_i} \exp(\gamma u_{ij})$ and arise from applying a Charnes-Cooper transformation [36].

Let $\mathbf{t} = (t_1, ..., t_L)^T$ be the observed vector of test statistics. For fixed $\boldsymbol{\lambda}$, a large-sample sensitivity analysis for the global sharp null could be achieved by minimizing the standardized deviate $\boldsymbol{\lambda}^T\{\mathbf{t} - \mu(\varrho)\}/\{\boldsymbol{\lambda}^T\Sigma(\varrho)\boldsymbol{\lambda}\}^{1/2}$ subject to $\varrho \in \mathcal{P}_\Gamma$, and assessing whether the minimal objective value exceeds the appropriate critical value from a standard normal. Taking $\boldsymbol{\lambda}$ as a vector of all zeroes except for a 1 in the ℓth coordinate returns a univariate sensitivity analysis using the statistic T_ℓ. In general, any pre-specified value of $\boldsymbol{\lambda}$ amounts to a univariate sensitivity analysis with modified test statistic $T_{\boldsymbol{\lambda}} = \mathbf{W}^T \left(\sum_{\ell=1}^{L} \lambda_\ell \mathbf{q}_\ell\right)$.

Rather than fixing $\boldsymbol{\lambda}$, consider instead the two-person game

$$a^*_{\Gamma,\Lambda} = \min_{\varrho \in \mathcal{P}_\Gamma} \sup_{\boldsymbol{\lambda} \in \Lambda} \frac{\boldsymbol{\lambda}^T\{\mathbf{t} - \mu(\varrho)\}}{\{\boldsymbol{\lambda}^T\Sigma(\varrho)\boldsymbol{\lambda}\}^{1/2}}, \tag{25.18}$$

where Λ is some subset of $\mathbb{R}^L$ without the zero vector. By allowing for maximization over Λ, the practitioner is allowed additional flexibility to choose a linear combination of test statistics which is more robust to the impact of hidden bias, all the while imposing the condition that the vector of unmeasured covariates $\mathbf{u}$ cannot vary from one test statistic to the next. Setting $\Lambda = \{\mathbf{e}_\ell\}$ where e_ℓ is a vector with a 1 in the ℓth coordinate and zeroes elsewhere returns a univariate sensitivity analysis for the kth outcome with a greater-than alternative, while $-\mathbf{e}_\ell$ would return the less-than alternative. When the test statistics T_ℓ are rank tests, setting $\Lambda = \{\mathbf{1}_L\}$ where 1_L is a vector containing L ones returns the coherent rank test of [37]. When $\Lambda = \{\mathbf{e}_1, ..., \mathbf{e}_L\}$, the collection of standard basis vectors, (25.18) returns the method of [35] with greater-than alternatives, and $\Lambda = \{\pm\mathbf{e}_1, ..., \pm\mathbf{e}_L\}$ gives the same method with two-sided alternatives. The method of [34] amounts to a choice of $\Lambda = \mathbb{R}^L \setminus \{\mathbf{0}_L\}$, i.e. all possible linear combinations except the vector $\mathbf{0}_L$ containing L zeroes, while the method of [33] takes Λ equal to the nonnegative orthant, again excluding the zero vector.

The problem (25.18) is not convex, making the problem challenging to solve in practice; however, consider replacing it with the modified optimization problem

$$b^*_{\Gamma,\Lambda} = \min_{\varrho \in \mathcal{P}_\Gamma} \sup_{\boldsymbol{\lambda} \in \Lambda} \max\left[0, \frac{\boldsymbol{\lambda}^T\{\mathbf{t} - \mu(\varrho)\}}{\{\boldsymbol{\lambda}^T\Sigma(\varrho)\boldsymbol{\lambda}\}^{1/2}}\right]^2. \tag{25.19}$$

This replaces negative values for the standardized deviate with zero, and then takes the square of the result. Note that this does not preclude directional testing (for instance, testing a less-than alternative for a single hypothesis) due to flexibility in designing the elements contained within Λ. The benefit of this transformation is that the function $g(\varrho) = \sup_{\boldsymbol{\lambda} \in \Lambda} \max[0, \boldsymbol{\lambda}^T\{\mathbf{t} - \mu(\varrho)\}/\{\boldsymbol{\lambda}^T\Sigma(\varrho)\boldsymbol{\lambda}\}^{1/2}]^2$ is convex in ϱ for any set Λ not containing the zero vector. This allows for efficient minimization over the polyhedral set $\mathcal{P}_\Gamma$ using methods such as projected subgradient descent, made practicable by the fact that the constraints in $\mathcal{P}_\Gamma$ are blockwise in nature, with distinct blocks of constraints for each matched set and with no constraints spanning across multiple matched sets.

25.5.2 Testing the intersection null

The sensitivity analysis at Γ concludes by comparing $b^*_{\Gamma,\Lambda}$ to an appropriate critical value $c_{\alpha,\Lambda}$, rejecting if $b^*_{\Gamma,\Lambda}$ in (25.19) falls above the critical value. The required critical value depends on the structure of Λ, through which it is seen that additional flexibility in the set Λ comes with the cost of a larger critical value. When Λ is a singleton, the critical value is simply the square of the usual critical value from the standard normal. If Λ is instead a finite set with $|\Lambda| > 1$, we could, in general, proceed using a Bonferroni correction, which would inflate the critical value. Improvements may be attained in paired designs when all L test statistics are based upon the same outcome variable as described in [32], or in the general case by optimizing for the worst-case correlation between test statistics for a given value of Γ as described in [38]. When $\Lambda = \mathbb{R}^L$ excluding the zero vector, [34] applies a result on quadratic forms of multivariate normals to show that we must instead use a critical value from a χ^2_L distribution when conducting inference through (25.19). This result underpins Scheffé's method for multiplicity control while comparing all linear contrasts of a multivariate normal [39].

25.5.3 Testing individual nulls

A class of methods adhering to the sequential rejection principle [40] control for multiple comparisons by using tests for intersections of null hypotheses as building blocks. Examples include closed testing [36], hierarchical testing [42], and the inheritance procedure [43] among others. When using procedures of this form, each intersection null hypothesis may be tested using the methodology outlined in §§25.5.1–25.5.2. For a general discussion of how tests of intersection nulls may be used with any procedure using the sequential rejection principle in a sensitivity analysis, see [35, §5].

We now describe how to proceed with closed testing. Suppose that we desire inference for K null hypotheses while controlling the Type I error rate at α. For each $k = 1, ..., K$, let $\mathcal{H}_k$ be the set of subsets of $\{1, ..., K\}$ containing the index k. The closure principle states that we can reject H_k while controlling the familywise error rate at α if we can reject all possible intersection nulls containing the kth hypothesis using a level α test for each intersection null. That is, we can reject H_k if we can reject using a level α test

$$\bigwedge_{a \in \mathcal{K}} H_a \quad \text{for all } \mathcal{K} \in \mathcal{H}_k.$$

For concreteness, consider $K = 3$. Then, we can reject H_1 if we can reject the null hypotheses $H_1 \wedge H_2 \wedge H_3$, $H_1 \wedge H_2$, $H_1 \wedge H_3$, and H_1 using tests that are each level α. When conducting a sensitivity analysis at Γ, each required test of an intersection null may be performed using the optimization problem (25.19).

25.5.4 Design sensitivity with multiple comparisons

There are two opposing forces at play when determining how rich a set Λ to consider in (25.19). On the one hand, additional flexibility afforded by a richer set Λ allows the researcher to mitigate the impact of hidden bias on the magnitude of the test statistic. On the other, by making Λ more flexible the resulting test statistic must exceed a larger critical value in order to reject the global null (25.17). This larger critical value impacts not only the power of tests for the global null, but also the power for tests of individual null hypotheses when procedures leveraging tests of intersections of null hypotheses are deployed for familywise error control.

Initial insight into the roles of these competing forces can be gleaned by inspecting the calculations leading to (25.15) when using a single test statistic. Importantly, observe that the critical value k_α plays no role, asymptotically, in determining the design sensitivity of a given test. As a result, for a given test statistic we would have attained the same design sensitivity had we used an overly conservative (i.e. larger than necessary) critical value. Through this, we see that the critical value takes a back seat in design sensitivity calculations, with the determining factor instead being the robustness of a chosen test statistic, as measured by the true expectation of the test statistic staying above worst-case expectation under the null for larger values of Γ in (25.2).

We now turn to calculations under the favorable setting described in §25.4.2 to formalize this intuition in the context of multiple comparisons. For a given generative model for the potential outcomes, let $\tilde{\Gamma}_{\boldsymbol{\lambda}}$ be the design sensitivity for the linear combination of test statistics $T_{\boldsymbol{\lambda}} = \mathbf{W}^T(\sum_{\ell=1}^{L} \boldsymbol{\lambda}_\ell \mathbf{q}_\ell)$. Now, consider the design sensitivity when using $b^*_{\Gamma,\Lambda}$ in (25.19) as the test statistic and rejecting based upon an appropriate critical value $c_{\alpha,\Lambda}$. First, the design sensitivity for this test of the intersection null satisfies

$$\tilde{\Gamma}_\Lambda \geq \max_{\boldsymbol{\lambda}\in\Lambda} \tilde{\Gamma}_{\boldsymbol{\lambda}}.$$

In words, the design sensitivity for testing the intersection null using $b^*_{\Gamma,\Lambda}$ as a test statistic is at least as large as that of the most robust linear combination of test statistics; see [38, Theorem 2]. Stated another way, imagine the researcher had oracle access to $\tilde{\boldsymbol{\lambda}} = \arg\max_{\boldsymbol{\lambda}\in\Lambda} \tilde{\Gamma}_{\boldsymbol{\lambda}}$ and performed a sensitivity analysis with this most robust linear combination $T_{\tilde{\boldsymbol{\lambda}}} = \mathbf{W}^T(\sum_{\ell=1}^{L} \tilde{\boldsymbol{\lambda}}_\ell \mathbf{q}_\ell)$, rather than choosing the linear combination based upon the data. In this case the researcher would not have to pay a price for multiple comparisons, and could simply use a critical value from a standard normal for inference. The above result shows that there is no loss, and a potential for gain, in design sensitivity from adaptively choosing the linear combination based upon the observed data, as the sensitivity analysis using (25.19) with a suitable critical value has design sensitivity no smaller than $\max_{\boldsymbol{\lambda}\in\Lambda} \tilde{\Gamma}_{\boldsymbol{\lambda}}$.

Next, it is straightforward to show that whenever $\Lambda_2 \subseteq \Lambda_1$, the design sensitivity for a test using Λ_2 in (25.19) is no larger than that of a test using Λ_1; see [33, Theorem 2] for a proof. Hence, the richer the set Λ, the larger the design sensitivity, and hence asymptotically the benefits of a larger set Λ outweigh the costs of a larger critical value. This has implications for testing individual null hypotheses through familywise error control procedures built upon tests of intersections of hypotheses. Consider closed testing as an illustration. When proceeding with closed testing, we reject the individual null H_ℓ only when all intersections of null hypotheses containing H_ℓ can be rejected. As a concrete example, the test for $H_1 \wedge H_2 \wedge H_3$ might take Λ_{123} to be all of $\mathbb{R}^3$ without the zero vector, while the test of $H_1 \wedge H_2$ may be thought of as constructing Λ_{12} by setting the third coordinate of $\boldsymbol{\lambda}$ equal to zero, but otherwise allowing the first two coordinates to range over all of $\mathbb{R}^2$ without the zero vector, such that $\Lambda_{12} \subseteq \Lambda_{123}$. The design sensitivities for intersections of hypotheses will be no smaller than those for their individual component hypotheses. As a result, the design sensitivities for testing individual nulls after closed testing will equal the design sensitivities for testing individual nulls had we not accounted for multiple comparisons at all.

25.5.5 The power of a sensitivity analysis with multiple comparisons and moderate sample sizes

Design sensitivity calculations are asymptotic, imagining the limiting power of a sensitivity analysis as $I \to \infty$. In small and moderate sample sizes, issues such as the null variance of the chosen test statistic and the critical value used to perform inference play a larger role in determining whether or not the null may be rejected for a given Γ in (25.2). How well do the insights gleaned from design sensitivity with multiple comparisons translate to small and moderate samples?

When discussing the power of a sensitivity analysis, we again imagine we are in the favorable setting in §25.4.2 of a treatment effect and no hidden bias. For a given sample size I and a given value of Γ, we assess the probability of correctly rejecting the null hypothesis of no treatment effect. This probability will decrease as a function of Γ for fixed I, and as I increases the power as a function of Γ will converge to a step function with the point of discontinuity equal to the design sensitivity.

Through a simulation study, we now investigate the loss in power from controlling for multiple comparisons. We imagine we have a paired observational study with $K = 3$ outcome variables. The treated-minus-control differences in outcomes in each pair $(\hat{\tau}_{i1}, \hat{\tau}_{i2}, \hat{\tau}_{i3})^T$, $i = 1, .., I$, are *iid* distributed as exchangeable normals with common mean 0.2, marginal variances 1, and correlation 0.5. For each outcome variable k, we apply Huber's ψ with truncation at 2.5 to $\hat{\tau}_{ik}/h_0$, where h_0 is the median of $|\hat{\tau}_i|$ and the ψ function is

$$\psi(x) = \text{sgn}(x) \max(|x|, 2.5).$$

For a range of values of Γ, we consider power for rejecting both the global null of no effect on any of the three outcome variables; and rejecting the null of no effect for the first outcome variable while controlling the familywise error rate at $\alpha = 0.05$. The procedures for sensitivity analysis we will compare are

1. Use closed testing. For each subset $\mathcal{K} \subseteq \{1, 2, 3\}$, compute $b^*_{\Gamma,\Lambda_{\mathcal{K}}}$ in (25.19) with $\Lambda_{\mathcal{K}} = \{\pm \mathbf{e}_k, k \in \mathcal{K}\}$ in (25.19). Reject the intersection null for all outcomes in $\mathcal{K}$ if $1 - \Phi(b^*_{\Gamma,\Lambda}) \geq \alpha/|\Lambda_{\mathcal{K}}|$, recovering the procedure in [35].

2. Combine individual sensitivity analyses using the techniques for univariate outcomes applied separately to each outcome variable in a closed testing procedure. This amounts to using a Bonferroni correction on the individual p-values for testing the global null hypothesis, and using Holm-Bonferroni to assess individual null hypotheses. Letting $P_{\Gamma,k}$ be the worst-case two-sided p-value for the kth outcome, reject the global null if $\min_{k=1,2,3} P_{\Gamma,k} \leq \alpha/3$. For each subset $\mathcal{K} \subseteq \{1, 2, 3\}$, reject the intersection null if $\min_{k \in \mathcal{K}} P_{\Gamma,k} \leq \alpha/|\mathcal{K}|$

Both procedures will control the familywise error rate asymptotically. From the discussion in §25.5.3, we know that approach 2 will be unduly conservative, as it allows different patterns of hidden bias for each outcome variable when testing intersection nulls.

As a baseline method, we additionally consider the following modification to procedure 2

2'. Conduct individual sensitivity analyses as in procedure 2. Reject the global null if the smallest two-sided p-value is below α, and reject an individual null if its two-sided p-value is below α.

Note that procedure 2' does **not** account for multiple comparisons: it simply rejects each individual test if its two-sided p-value falls below α. Procedure 2' does not control the familywise error rate, giving it an unfair advantage in the simulations that follow. When testing individual null hypotheses, procedure 2' would provide the best possible power for a given choice of test statistic. We thus investigate how close procedures 1 and 2 come to attaining this optimal power for testing nulls on individual outcoes while, unlike procedure 2', providing familywise error control.

Figures 25.1 and 25.2 show the power of these procedures for the global null hypothesis and for rejecting the null for the first outcome variable, respectively, as a function of Γ for $I = 250, 500, 1000$, and 2000 when testing at $\alpha = 0.05$. We see in both figures that procedure 1 uniformly dominates procedure 2 in terms of power, due to the conservativeness of combining individual sensitivity analyses; see [35] for additional discussion. In Figure 25.1, we see that procedure 2' initially performs best for testing the overall null for $I = 250$ and $I = 500$. That is, despite procedure 1 having superior design sensitivity, the use of a larger critical value to control the familywise error rate impacts performance in small samples. At $I = 1000$ and beyond procedure 1 has superior power to both procedure 2 and 2'. This is in spite of the fact that procedure 2' **does not** provide familywise error control and is once again a reflection of the conservativeness of combining individual sensitivity analyses while allowing different patterns of hidden bias for each outcome. By $I = 2000$ we begin to see convergence of the power curves to step functions dicatated by the procedures' design sensitivities, wherein it is reflected that procedure 1 has the superior design sensitivity.

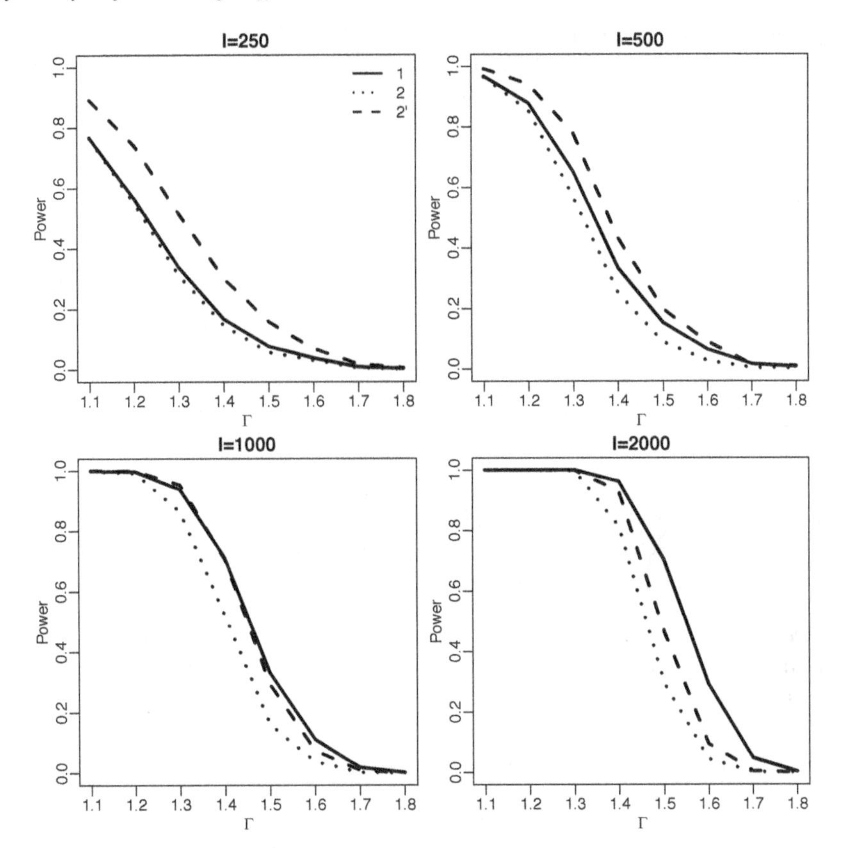

FIGURE 25.1
Power of a sensitivity analysis for testing the global null hypothesis of no effect for $K = 3$ outcomes as a function of Γ at $I = 250, 500, 1000, 2000$ using procedures 1,2, and 2'. Only procedures 1 and 2 provide familywise error control.

When testing the sharp null of no effect for the first outcome, at $I = 250$ and $I = 500$ the power of procedure 1 lags that of procedure 2' (which does not account for multiple comparisons). Note that the gap decreases as the sample size increases, and by $I = 1000$ the power profiles for procedures 1 and 2' are in near perfect alignment. The improvements provided by procedure 1 when testing intersection null hypotheses trickle down to improving power for testing individual nulls after closed testing, essentially providing the same power for testing individual null hypotheses as what would have been attained had we not controlled for multiple comparisons at all. Observational studies with $I = 1000$ and above are commonplace in practice: observational data is cheap and plentiful but prone to bias. Once we account for hidden bias in sensitivity analysis, we see that in large enough samples there is no further loss for controlling for multiple comparisons when closed testing is used in conjunction with a suitable method for combining test statistics. Rather, it is primarily a statistic's robustness to hidden bias that governs the sensitivity analysis.

25.6 Sensitivity Analysis for Average Effects

A common critique of the approach to sensitivity analysis described to this point is that the methods provide tests for *sharp* null hypotheses, i.e. hypotheses which impute the missing values for the

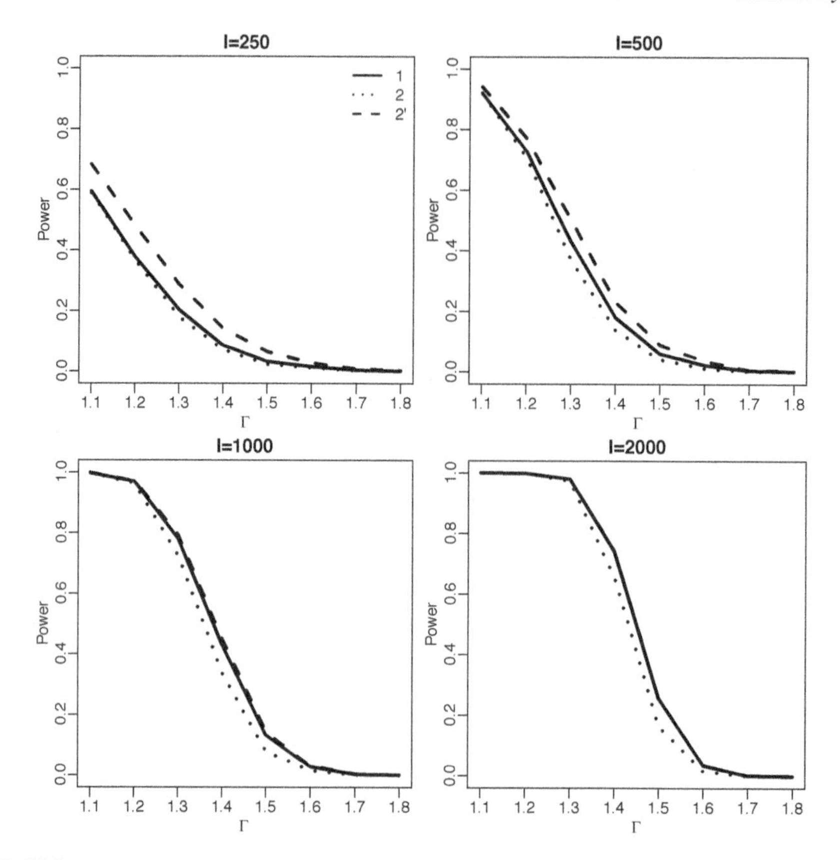

FIGURE 25.2
Power of a sensitivity analysis for testing the null of no effect for the first outcome as a function of Γ at $I = 250, 500, 1000, 2000$. Only procedures 1 and 2 provide familywise error control.

potential outcomes. Sharp nulls are viewed by some as overly restrctive. Moreover, a particular fear is that sensitivity analyses conducted assuming Fisher's sharp null of no effect at all may paint an overly optimistic picture of the study's sensitivity to hidden bias if effects are instead heterogeneous but average to zero, with unmeasured confounding conspiring with the unidentified aspects of the constant effects model to render the analysis assuming constant effects inadequate.

In this section we focus attention on matched pair designs, and consider an approach for sensitivity analysis for the average of the $N = 2I$ treatment effects in the matched sample, $\bar{\tau} = N^{-1}\sum_{i=1}^{I}\sum_{j=1}^{n_i}\tau_{ij}$, while allowing for heterogeneity in the individual treatment effects. The procedure, developed in [24], provide a sensitivity analysis that is simultaneously (i) asymptotically valid for inference on the average treatment effect; and (ii) valid for any sample size if treatment effects are constant at the hypothesized value.

Assume that the model (25.2) holds at Γ and consider testing the weak null hypothesis

$$H_{weak} : \bar{\tau} = \bar{\tau}_0 \tag{25.20}$$

with a greater-than alternative. Let $\hat{\tau}_i = (W_{i1} - W_{i2})(Y_{i1} - Y_{i2})$ be the treated-minus-control difference in means in the ith of I pairs. Define $D_{\Gamma i}$ as

$$D_{\Gamma i} = \hat{\tau}_i - \bar{\tau}_0 - \left(\frac{\Gamma - 1}{1 + \Gamma}\right)|\hat{\tau}_i - \bar{\tau}_0|, \tag{25.21}$$

and let $\bar{D}_\Gamma = I^{-1}\sum_{i=1}^{I} D_{\Gamma i}$. The term $\{(\Gamma-1)/(1+\Gamma)\}|\hat{\tau}_i - \bar{\tau}_0|$ in (25.21) is the largest possible expectation for $\hat{\tau}_i - \tau_0$ at Γ when treatment effects are assumed to be *constant* at $\bar{\tau}_0$, such that $E(\bar{D}_{\Gamma i} \mid \mathcal{F}, \mathcal{W}) \leq 0$ under constant effects. Perhaps surprisingly, this centered variable $D_{\Gamma i}$ continues to be useful for sensitivity analysis even when effects are heterogeneous.

Define $\text{se}(\bar{D}_\Gamma)$ as the usual standard error estimate for a paired design,

$$\text{se}(\bar{D}_\Gamma) = \sqrt{\frac{1}{I(I-1)}\sum_{i=1}^{I}(D_{\Gamma i} - \bar{D}_\Gamma)^2}.$$

Consider now the test statistic

$$S_\Gamma(\mathbf{W}, \mathbf{Y} - \mathbf{W}\bar{\tau}_0) = \max\left\{0, \frac{\bar{D}_\Gamma}{\text{se}(\bar{D}_\Gamma)}\right\},$$

which studentizes $\bar{D}_\Gamma$ by the standard error $\text{se}(\bar{D}_\Gamma)$, and replaces negative values for the standardized deviate by zero.

Despite effects being heterogeneous under (25.20), consider using the worst-case distribution for $S_\Gamma(\mathbf{W}, \mathbf{Y} - \mathbf{W}\bar{\tau}_0)$ under the assumption of *constant* effects as a reference distribution for conducting inference. More precisely, let $\mathbf{a} = \mathbf{Y} - \mathbf{W}\bar{\tau}_0$ be the observed vector of adjusted responses, let s_Γ^{obs} be the observed value for the test statistic, let $u_{ij} = 1\{a_{ij} \geq a_{ij'}\}$ for $j \neq j'$, $i = 1, ..., I$, and use as a p-value

$$p_\Gamma = \frac{\sum_{\mathbf{b}\in\Omega}\exp(\gamma \mathbf{b}^T\mathbf{u})1\{S_\Gamma(\mathbf{b}, \mathbf{a}) \geq s_\Gamma^{obs}\}}{\sum_{\mathbf{b}\in\Omega}\exp(\gamma \mathbf{b}^T\mathbf{u})}. \tag{25.22}$$

[24] shows that under under mild regularity conditions, rejecting the weak null hypothesis (25.20) when p_Γ falls at or below α yields a sensitivity analysis whose Type I error rate is, asymptotically, no larger than α if (25.2) holds at Γ and the weak null (25.20) holds. Furthermore, if treatment effects are constant at $\bar{\tau}_0$ as in (25.13), using p_Γ as a p-value produces a sensitivity analysis whose Type I error rate is less than or equal to α at any sample size when (25.2) holds at Γ.

If we are indifferent to maintaining finite-sample Type I error control under constant effects, the following large-sample procedure also produces an asymptotically valid sensitivity analysis for any Γ: rather than constructing a p-value using (25.20), simply reject the weak null (25.20) when

$$\frac{\bar{D}_\Gamma}{\text{se}(\bar{D}_\Gamma)} \geq \Phi^{-1}(1-\alpha). \tag{25.23}$$

In the usual way the procedure may be inverted to produce sensitivity intervals for the sample average treatment effect at any Γ.

Note that the procedure using (25.22) as a p-value differs from a sensitivity analysis based solely upon the average of the treated-minus-control difference in means due to the use of studentization. [24] shows that while the a sensitivity analysis based upon the difference in means and assuming constant effects uses a reference distribution whose expectation bounds that of $I^{-1}\sum_{i=1}^{I}(\hat{\tau}_i - \bar{\tau}_0)$ even when effects are heterogeneous, the resulting reference distribution may have too small a variance when (25.2) holds at $\Gamma > 1$. As a result, the sensitivity analysis using the unstudentized difference in means and assuming constant effects might be anti-conservative when conducted at $\Gamma > 1$. That said, by virtue of the reference distribution properly bounding the worst-case expectation it is shown that this sensitivity analysis will be asymptotically valid if conducted at $\Gamma + \epsilon$ for any $\epsilon > 0$, as in that case the candidate worst-case expectation will exceed the true expectation if (25.2) actually holds at Γ. This provides assurances that sensitivity analyses based upon the difference in means without studentization are unlikely to be unduly optimistic if interpreted as providing inference for the average treatment effect.

In the particular case of binary outcomes in a paired observational study, [45] studies the performance of sensitivity analyses for McNemar's test,

$$T(\mathbf{W}, \mathbf{Y}) = \mathbf{W}^T\mathbf{Y},$$

valid under the sharp null hypothesis of no effect, when effects are actually heterogeneous but average to zero. They show that sensitivity analyses based upon McNemar's statistic are asymptotically valid as sensitivity analyses of the null of no sample average treatment effect. Using terminology common in causal inference with binary outcomes, sensitivity analyses using McNemar's test under the assumption of Fisher's sharp null maintain asymptotic validity as tests that the sample causal risk difference is zero, or equivalently that the sample causal risk ratio is 1, when (25.2) is assumed to hold at Γ. This provides both a useful fortification and further interpretations for sensitivity analyses which have used this statistic in the past: any sensitivity analysis using McNemar's test may be interpreted as a asymptotically valid sensitivity analysis testing that the average effect is zero, while additionally providing an exact sensitivity analysis for testing Fisher's sharp null.

25.7 Sensitivity Analysis with Instrumental Variables

Instrumental variable designs are commonly used when treatment assignment is not believed to be strongly ignorable given $\mathbf{x}$. An instrumental variable may be viewed as a haphazard nudge or encouragement toward receiving a particular level of the treatment without fully determining the level of treatment received. Crucial to the use of instrumental variables is the assumption that there are no hidden covariates which simultaneously impact the candidate instrumental variable and the outcome. A sensitivity analysis in an instrumental variable design tests departures from this assumption. We now described how the previously described methods for sensitivity analysis for both sharp null hypotheses and assuming heterogeneous effects may be adopted for use in matched instrumental variable designs with binary instruments.

The value of the instrumental variable randomly assigned to the ijth individual is W_{ij}. Each individual has two potential values for the level of treatment actually received, d_{0ij} and d_{1ij}. Likewise, each individual has two potential outcomes y_{0ij} and y_{1ij}. The treatment level actually received is $D_{ij} = (1 - W_{ij})d_{0ij} + W_{ij}d_{1ij}$, and the observed outcome is $Y_{ij} = (1 - W_{ij})y_{0ij} + W_{ij}y_{1ij}$. We again use (25.2) to model departures from random assignment, with the notion of strong ignorability given $(\mathbf{x}, u)$ modified to $\pi_{ij} = \text{pr}(W_{ij} = 1 \mid y_{0ij}, y_{1ij}, d_{0ij}, d_{1ij}, \mathbf{x}_{ij}, u_{ij}) = \text{pr}(W_{ij} = 1 \mid \mathbf{x}_{ij}, u_{ij})$.

The proportional dose model states that the difference in potential outcomes is proportional to the differences in the received treatment levels for all individuals,

$$H_{prop} : y_{1ij} - y_{0ij} = \beta(d_{1ij} - d_{0ij}). \;\; (i = 1, ..., I; j = 1, ..., n_i). \tag{25.24}$$

This model is described in further detail in [46–48] and [23, §5.4].

Suppose we want to conduct a sensitivity analysis for the null that (25.24) holds at $\beta = \beta_0$. Observe that under the null hypothesis, $y_{1ij} - d_{1ij}\beta_0 = y_{0ij} - d_{0ij}\beta_0$. Therefore, $Y_{ij} - D_{ij}\beta_0$ is fixed across randomizations under this null hypothesis. Defining $\mathbf{f}_0 = \mathbf{Y} - \mathbf{D}\beta_0$, we are entitled to the reference distribution (25.8) for any $\mathbf{u} \in \mathcal{U}$. Therefore, any of the methods for sensitivity analysis described for sharp null hypotheses may be used. Similarly, ranges of Hodges-Lehmann estimates and sensitivity intervals for β in (25.24) follow from the discussion in §§25.3.2 - 25.3.3

Should the assumption of proportional doses prove unpalatable, in paired designs we may instead conduct a sensitivity analysis for a parameter known as the effect ratio [49],

$$\beta = \frac{N^{-1}\sum_{i=1}^{I}\sum_{j=1}^{2}(y_{1ij} - y_{0ij})}{N^{-1}\sum_{i=1}^{I}\sum_{j=1}^{2}(d_{1ij} - d_{0ij})}, \tag{25.25}$$

where we assume that $\sum_{i=1}^{n}\sum_{j=1}^{2}(d_{1ij} - d_{0ij}) \neq 0$. The effect ratio is the ratio of two treatment effects. In the numerator we have the effect of the encouragement W_{ij} on the outcome variable, while in the denominator we have the effect of the encouragement of the level of the treatment actually received.

The method to be presented will provide a valid sensitivity analysis for the effect ratio β when interpreted as the ratio of two treatment effects. That said, in instrumental variable studies it is common to make additional assumptions about the relationships between the potential levels of treatment received and the potential outcomes. In defining an instrumental variable, [50] require that the variable satisfies the *exclusion restriction*: the instrument can only affect the outcome by influencing the treatment received, stated formally as $d_{wij} = d_{w'ij} \Rightarrow y_{wij} = y_{w'ij}$ for $w, w' = 0, 1$. Observe that this assumption holds under the proportional dose model (25.24). Typically, IV studies further invoke a monotonicity assumption of the following form: for all individuals, $d_{w'ij} \geq d_{wij}$ for $w' \geq w$. When the potential treatment levels d_{wij} are binary variables reflecting whether or not the treatment would actually be received, the exclusion restriction and monotonicity imply that the causal estimand β may then be interpreted as the sample average treatment effect among compliers, those individuals for whom $d_{0ij} = 0$ and $d_{1ij} = 1$. These assumptions are not needed for inference on the effect ratio β; rather, by making these assumptions, we confer additional interpretations unto β.

Suppose we wants to conduct a sensitivity analysis for the null that the effect ratio in (25.25) equals some value β_0 without imposing the proportional dose model (25.24). The methods for sensitivity analysis for the sample average treatment effect described in §25.6 may also be used for the effect ratio through a simple redefinition of $D_{\Gamma i}$ in (25.21). First let $\hat{\zeta}_i$ be the treated-minus-control difference in the adjusted responses $Y_{ij} - \beta_0 D_{ij}$,

$$\begin{aligned}\hat{\zeta}_i &= (W_{i1} - W_{i2})\{Y_{i1} - Y_{i2} - \beta_0(D_{i1} - D_{i2})\}\\ &= W_{i1}\{Y_{i1} - Y_{i2} - \beta_0(D_{i1} - D_{i2})\} + W_{i2}\{Y_{i2} - Y_{i1} - \beta_0(D_{i2} - D_{i1})\}.\end{aligned}$$

Using these differences, redefine $D_{\Gamma i}$ as

$$D_{\Gamma i} = \hat{\zeta}_i - \left(\frac{\Gamma - 1}{1 + \Gamma}\right)|\hat{\zeta}_i|. \tag{25.26}$$

With this change in definition of $D_{\Gamma i}$, the developments in §25.6 may be used without further modification. The p-value in (25.22) provides a sensitivity analysis that is simulataneously asymptotically valid for testing the null that the effect ratio in (25.25) equals β_0, and is further valid for any sample size when the proportional dose model (25.24) holds at β_0. Should we be indifferent to performance under the proportional dose model, an alternative asymptotically valid approach for inference on the effect ratio can be attained by rejecting the null based upon (25.23). The latter approach extends the proposal of [49] for inference on the effect ratio at $\Gamma = 1$ to providing inference for any value of Γ in (25.2).

25.8 Sensitivity Analysis after Inverse Probability Weighting

Under a model for departures from strong ignorability closely related to (25.2), sensitivity analyses can also be performed for weighted estimators of average treatment effects. In what follows, we will assume that the observed data $(Y_{0i}, Y_{1i}, W_i, \mathbf{X}_i)$, $i = 1, .., N$, are drawn *iid* from a common distribution P. As in (25.1) we imagine that strong ignorability holds given $(\mathbf{X}, U)$ for some unmeasured covariate U. Let $\pi(\mathbf{X}, U) = \text{pr}(W = 1 \mid \mathbf{X}, U)$, and let $e(\mathbf{X}) = \text{pr}(W = 1 \mid \mathbf{X})$ [51] introduced a model facilitating sensitivity analysis called the *marginal sensitivity model*, which bounds the odds ratio of $\pi(\mathbf{X}, U)$ and $e(\mathbf{X})$,

$$\frac{1}{\Lambda} \leq \frac{\pi(\mathbf{X}, U)\{1 - e(\mathbf{X})\}}{e(\mathbf{X})\{1 - \pi(\mathbf{X}, U)\}} \leq \Lambda. \tag{25.27}$$

Note that this presentation differs slightly from that presented in [51] and [2], who choose U to be one of the potential outcomes; this distinction has no practical consequences. Further note that Λ in (25.27) has no connection with Λ as used in §25.5. Equivalently, we imagine that the following logit model holds:

$$\log\left\{\frac{\text{pr}(W = 1 \mid \mathbf{X} = \mathbf{x}, U = u)}{\text{pr}(W = 0 \mid \mathbf{X} = \mathbf{x}, U = u)}\right\} = \log\left\{\frac{e(\mathbf{x})}{1 - e(\mathbf{x})}\right\} + \lambda(2u - 1), \tag{25.28}$$

where $\lambda = \log(\Lambda)$ and $0 \leq u \leq 1$. The models (25.27) and (25.28) are equivalent in the sense that there is a model of the form (25.27) describing assignment probabilities if and only if (25.28) is satisfied. To see that (25.27) implies (25.28), for any U define for $\Lambda > 1$

$$U^* = \log\left\{\frac{\pi(\mathbf{X}, U)\{1 - e(\mathbf{X})\}}{e(\mathbf{X})\{1 - \pi(\mathbf{X}, U)\}}\right\}/(2\lambda) + 1/2, \tag{25.29}$$

and set $U^* = 0$ when $\Lambda = 1$. Observe that $0 \leq U^* \leq 1$ when (25.27) holds at Λ. Moreover, observe that $(\mathbf{X}, U^*)$ is a balancing score in the terminology of [52], as by inspection of (25.29) we can express $\pi(\mathbf{X}, U)$ as a function of X and U^* for any fixed Λ. By the assumption of strong ignorability given $(\mathbf{X}, U)$, this implies by Theorem 3 of [52] that $\text{pr}(W = 1 \mid \mathbf{X}, U^*, Y_1, Y_0) = \text{pr}(W = 1 \mid \mathbf{X}, U^*) = \text{pr}(W = 1 \mid \mathbf{X}, U)$. Therefore, both (25.28) and (25.27) hold when we replace U with U^*. Showing that (25.28) implies (25.27) is straightforward, and the proof is omitted.

Compare the logit form (25.28) with that used in sensitivity analysis for matched designs,

$$\log\left\{\frac{\text{pr}(W = 1 \mid \mathbf{X} = \mathbf{x}, U = u)}{\text{pr}(W = 0 \mid \mathbf{X} = \mathbf{x}, U = u)}\right\} = \kappa(\mathbf{x}) + \gamma u, \tag{25.30}$$

with $0 \leq u \leq 1$ and $\kappa(\mathbf{x}_{ij}) = \kappa(\mathbf{x}_{ij'})$ for each $i = 1, ..., I$, $j, j' = 1, ..., n_i$. If individuals in the same matched set have the same covariates $\mathbf{x}_{ij}$, then (25.28) holding at Λ implies that (25.30) holds at $\Gamma = \Lambda^2$ setting $\kappa(\mathbf{x}_{ij}) = \log\{e(\mathbf{x})/(1 - e(\mathbf{x}))\} - \lambda$, such that the previously described methods for sensitivity analysis when (25.2) holds at Γ are also applicable whenever (25.27) is assumed to hold at $\sqrt{\Gamma}$. The motivation for using this slightly different model comes from the different approaches to dealing with nuisance parameters in matching versus weighting. When matching, by conditioning on the matched structure we remove dependence upon the parameters $\kappa(\mathbf{x})$ under the model (25.30) whenever all individuals in the same matched set have the same value for $\kappa(\mathbf{x})$. When using a weighting estimator such as inverse probability weighting, we instead use a plug-in estimate for the nuisance parameter. The model (25.30) does not specify a particular form for $\kappa(\mathbf{x})$. As a result, we cannot estimate $\kappa(\mathbf{x})$ from the observed data. The modification (25.28) elides this difficulty by relating $\kappa(\mathbf{x})$ to the propensity score $e(\mathbf{x})$, such that we may proceed with a plug-in estimator for the propensity score.

Under this modification, [2] describe an approach for sensitivity analysis for stabilized inverse probability weighted estimators and stabilized augmented inverse probability weighted estimators. For now, imagine that the propensity score $e(\mathbf{x})$ is known to the researcher. For each $i = 1, .., N$, let $e_i = e(\mathbf{x}_i)$ and let $g_i = \log\{e_i/(1-e_i)\}$ be the log odds of the assignment probabilities given $\mathbf{x}_i$, i.e. the log odds transform of the propensity score. If (25.28) holds at Λ, we have that

$$\frac{1}{1+\Lambda\exp(-g_i)} \leq \text{pr}(W_i = 1 \mid \mathbf{X}_i = \mathbf{x}_i, U_i = u_i) \leq \frac{1}{1+\Lambda^{-1}\exp(-g_i)},$$

where $\text{pr}(W_i = 1 \mid \mathbf{X}_i = \mathbf{x}_i, U_i = u_i) = [1+\exp\{\lambda(2u_i-1)\}\exp(-g_i)]^{-1}$. Let $z_i = \exp\{\lambda(2u_i-1)\}$, and consider the following optimization problem for finding the upper and lower bounds on the possible SIPW estimators of the population average treatment effect, $E(Y_{1i} - Y_{0i})$, when (25.27) holds at Λ:

$$\begin{aligned} &\text{max or min} \quad \frac{\sum_{i=1}^N W_i Y_i\{1+z_i\exp(-g_i)\}}{\sum_{i=1}^N W_i\{1+z_i\exp(-g_i)\}} - \frac{\sum_{i=1}^N (1-W_i)Y_i\{1+z_i\exp(g_i)\}}{\sum_{i=1}^N (1-W_i)\{1+z_i\exp(g_i)\}} \\ &\text{subject to} \quad \Lambda^{-1} \leq z_i \leq \Lambda \quad (i = 1, ..., N) \end{aligned} \tag{25.31}$$

Problem (25.31) is a fractional linear program, and may be converted to a standard linear program through the use of a Charnes-Cooper transformation [36]. As a result, the problem may be solved efficiently using off-the-shelf solvers for linear programs.

Let $\hat{\tau}_{\Lambda,\min}$ and $\hat{\tau}_{\Lambda,\max}$ be the optimal objective values when performing either minimization or maximization in (25.31). [2] additionally describe how $100(1-\alpha)\%$ sensitivity intervals may be attained through the use of the percentile bootstrap. In short, in the bth of B bootstrap iterations we solve the optimization problems in (25.31) for the bootstrap sample, storing these values as $\hat{\tau}^{(b)}_{\Lambda,\min}$ and $\hat{\tau}^{(b)}_{\Lambda,\max}$. The lower bound of sensitivity interval is then the $\alpha/2$ quantile of $\{\hat{\tau}^{(b)}_{\Lambda,\min} : b = 1, .., B\}$, and the upper bound equal to the $(1-\alpha/2)$ quantile $\{\hat{\tau}^{(b)}_{\Lambda,\max} : b = 1, ..., B\}$. In practice the propensity scores e_i must be estimated for each individual, with the resulting estimators needing to be sufficiently smooth such that the bootstrap sensitivity intervals maintain their asymptotic validity. Estimation introduces the possibility of model misspecification, particularly when parametric models such as logistic regression are used. [2] consider a slight modification to the model (25.27) that instead bounds the odds ratio of $\pi(\mathbf{X}, U)$ and the best parametric approximation to $e(\mathbf{X})$ between Λ^{-1} and Λ. This modification allows for discrepancies due to both unmeasured confounding and misspecification of the propensity score model and is similar in spirit to the model (25.2), which allowed for differences in assignment probabilities for two individuals in the same matched set due to both hidden bias and discrepancies on observed covariates.

25.9 Additional Reading

For the interested reader, below is a non-exhaustive list of additional topics on sensitivity analysis not covered in this chapter.

- **Attributable effects** are a class of estimands which allow for heterogeneous treatment effects. With binary outcomes, the attributable effect is the number of events which occurred only because of the treatment. [53] introduces sensitivity analysis for attributable effects in matched designs, highlighting their use with binary outcomes and for assessing displacement effects with continuous outcomes; see also [23, §2.5].

- When conducting a sensitivity analysis using the model (25.2), we bound the impact that hidden bias may have on the assignment probabilities for two individuals in the same matched set. The model imposes no restrictions on the relationship between hidden bias and the potential outcomes; indeed, the potential outcomes are fixed quantities through conditioning. When performing a sensitivity analysis, the worst-case vector of unmeasured covariates bounding the upper tail probabilities for a given test statistic generally has near perfect correlation with the potential outcomes. This may seem implausible and overly pessimistic. An **amplification** of a sensitivity analysis maps the one-parameter sensitivity analysis considered here to a set of two-parameter sensitivity analysis separately bounding the impact of hidden bias on the treatment and the potential outcomes. No new calculations are required; rather, for each Γ we are entitled to an alternative set of interpretations of the strength of biases under consideration. See [30] and [19, §3.6] for further discussion.

- **Evidence factors** are tests of the same null hypothesis about treatment effects which may be treated as statistically independent, by virtue of either being truly independent or producing p-values whose joint distribution stochastically dominates that of the multivariate uniform distribution. Evidence factors are, through their nature, susceptible to distinct patterns of hidden bias. With evidence factors a given counterargument may overturn the finding of a given test, yet have little bearing on the findings of another. For a comprehensive overview of evidence factors and their role in strengthening evidence rather than replicating biases, see [21].

- There is much qualitative advice on what can be done to improve the design and analysis of observational studies. Through a sensitivity analysis, we can often quantify the benefits of utilizing quasi-experimental devices. Some examples are **coherence**, or **multiple operationalism** [26, 33, 35, 37]; and the incorporation of either a **negative control outcome** (i.e. an outcome known to be unaffected by the treatment) [56] or an outcome with a **known direction of effect** [23, §6] to rule out certain patterns of hiddden bias.

- The choice of test statistic plays a large role in the sensitivity value returned by a sensitivity analysis. While a unifying theory for choosing test statistics for optimal performance in a sensitivity analysis does not exist, many statistics have been designed which have superior performance to the usual choices such as the difference in means or Wilcoxon's signed rank test. Certain m-tests, described in §25.4.3, provide one example. Members of the class of **u-statistics** provide yet another. See [49] for details

- Other metrics beyond design sensitivity exist for guiding the choice of a test statistic. **Bahadur efficiency**, introduced in [58] and developed for sensitivity analysis in [26], provides comparisons between test statistics below the minimum of their design sensitivities, providing of an assessment of which test statistic best distinguishes a moderately large treatment effect from a degree of bias below the minimal design sensitivities.

- For worked examples of sensitivity analyses, see [60, §9] and [23, §4] among several resources. Most of the references cited in this chapter also have data examples accompanying them. For a more hands-on and visual tutorial of how to conduct a sensitivity analysis, the following `shiny` app, created by Paul Rosebaum, provides an interactive illustration: `https://rosenbap.shinyapps.io/learnsenShiny/`

25.10 Software

Many `R` packages and scripts implement the methods described in this chapter. Below are descriptions of some of these resources.

- Testing sharp null hypotheses, §25.3
  - The R packages sensitivitymv and sensitivityfull, available on CRAN, facilitate sensitivity analyses when using m-statistics.
  - The R package senU, available on CRAN, provides sensitivity analysis for u-statistics
  - The R package sensitivity2x2xk, available on CRAN, implements sensitivity analysis for McNemar's test (paired studies) and the Mantel-Haenszel test (general matched designs) when outcomes are binary.
  - The R package senstrat, available on CRAN, allows for more customization in the choice of test statistic when conducting a sensitivity analysis. The user can provide particular values for the elements of $\mathbf{q}$ in (25.9), rather than restricting attention to a given class of test statistics. This method does not use the asymptotically separable algorithm described in §25.3.1, but instead uses a clever Taylor approximation described in [27] to form an upper-bound on the worst-case p-value.
  - The R script sensitivitySimple.R, available at https://github.com/colinbfogarty/SensitivitySimple, provides an alternative approach which solves exactly the optimization problem which [27] approximates by solving a quadratic integer program. This script also allows for customization of the vector $\mathbf{q}$ in (25.9). The method is described in Section F of the supplementary material for [62].
- Sensitivity analysis with multiple outcomes, §25.5
  - The R script multiCompareFunctions, available at https://github.com/colinbfogarty/SensitivityMultipleComparisons, provides functions to conduct a sensitivity analysis when Λ in (25.18) consists of standard basis vectors and/or negative 1 times a standard basis vector, thus implementing the method described in [35]. The script reproducescript.R reproduces the data analysis and simulation studies in that paper, and is available at the same web address.
  - The R script chiBarSquaredTest.R, available at https://github.com/PeterLCohen/Directed-Multivariate-Sensitivity-Analysis, implements a multivariate sensitivity analysis when Λ in (25.18) equals the non-negative orthant. Scripts walking through examples using real and synthetic data sets are also available at the same web address.
- Sensitivity analysis with heterogeneous treatment effects, §§25.6-25.7
  - The R script StudentizedSensitivity.R, available at available at https://github.com/colinbfogarty/StudentizedSensitivitySATE, implements the method for sensitivity analysis for the sample average treatment effect in paired observational studies described in [24].
  - The R script ERSensitivity.R, available at https://github.com/colinbfogarty/EffectRatioHeterogeneous, implements the sensitivity analysis for the effect ratio described in §25.7. The script IV_Sens_Example.R, available at the same web address, illustrates the method with a simulated data set.
- Sensitivity analysis for inverse probability weighted estimators, §25.8
  - The R script bootsens.R, available at https://github.com/qingyuanzhao/bootsens, implements the sensitivity analysis for IPW estimators of [2] described in §25.8.

References

[1] Jerome Cornfield, William Haenszel, E Cuyler Hammond, Abraham M Lilienfeld, Michael B Shimkin, and Ernst L Wynder. Smoking and lung cancer: Recent evidence and a discussion of some questions. *Journal of the National Cancer Institute*, 22:173–203, 1959.

[2] Qingyuan Zhao, Dylan S Small, and Bhaswar B Bhattacharya. Sensitivity analysis for inverse probability weighting estimators via the percentile bootstrap. *Journal of the Royal Statistical Society: Series B (Statistical Methodology)*, 81(4):735–761, 2019.

[3] Sue M Marcus. Using omitted variable bias to assess uncertainty in the estimation of an aids education treatment effect. *Journal of Educational and Behavioral Statistics*, 22(2):193–201, 1997.

[4] James M Robins, Andrea Rotnitzky, and Daniel O Scharfstein. Sensitivity analysis for selection bias and unmeasured confounding in missing data and causal inference models. In *Statistical models in epidemiology, the environment, and clinical trials*, pages 1–94. Springer, 2000.

[5] Guido W Imbens. Sensitivity to exogeneity assumptions in program evaluation. *American Economic Review*, 93(2):126–132, 2003.

[6] Binbing Yu and Joseph L Gastwirth. Sensitivity analysis for trend tests: application to the risk of radiation exposure. *Biostatistics*, 6(2):201–209, 2005.

[7] Liansheng Wang and Abba M Krieger. Causal conclusions are most sensitive to unobserved binary covariates. *Statistics in Medicine*, 25(13):2257–2271, 2006.

[8] Brian L Egleston, Daniel O Scharfstein, and Ellen MacKenzie. On estimation of the survivor average causal effect in observational studies when important confounders are missing due to death. *Biometrics*, 65(2):497–504, 2009.

[9] Carrie A Hosman, Ben B Hansen, and Paul W Holland. The sensitivity of linear regression coefficients' confidence limits to the omission of a confounder. *The Annals of Applied Statistics*, 4(2):849–870, 2010.

[10] José R Zubizarreta, Magdalena Cerdá, and Paul R Rosenbaum. Effect of the 2010 Chilean earthquake on posttraumatic stress reducing sensitivity to unmeasured bias through study design. *Epidemiology*, 24(1):79–87, 2013.

[11] Weiwei Liu, S Janet Kuramoto, and Elizabeth A Stuart. An introduction to sensitivity analysis for unobserved confounding in nonexperimental prevention research. *Prevention Science*, 14(6):570–580, 2013.

[12] Tyler J VanderWeele and Peng Ding. Sensitivity analysis in observational research: introducing the e-value. *Annals of Internal Medicine*, 167(4):268–274, 2017.

[13] Paul R Rosenbaum. *Observational studies*. Springer, New York, 2002.

[14] Paul R. Rosenbaum. Sensitivity analysis for certain permutation inferences in matched observational studies. *Biometrika*, 74(1):13–26, 1987.

[15] Paul R Rosenbaum. Quantiles in nonrandom samples and observational studies. *Journal of the American Statistical Association*, 90(432):1424–1431, 1995.

[16] Colin B Fogarty. On mitigating the analytical limitations of finely stratified experiments. *Journal of the Royal Statistical Society: Series B (Statistical Methodology)*, 80:1035–1056, 2018.

[17] Nicole E Pashley and Luke W Miratrix. Insights on variance estimation for blocked and matched pairs designs. *Journal of Educational and Behavioral Statistics*, 46(3):271–296, 2021.

[18] William G Cochran. The planning of observational studies of human populations. *Journal of the Royal Statistical Society. Series A (General)*, 128(2):234–266, 1965.

[19] Paul R Rosenbaum. *Design of observational studies*. Springer, New York, 2010.

[20] Qingyuan Zhao. On sensitivity value of pair-matched observational studies. *Journal of the American Statistical Association*, 114(526):713–722. 2019.

[21] Joseph L Hodges and Erich L Lehmann. Rank methods for combination of independent experiments in analysis of variance. *The Annals of Mathematical Statistics*, 33(2):482–497, 1962.

[22] Paul R Rosenbaum and Abba M Krieger. Sensitivity of two-sample permutation inferences in observational studies. *Journal of the American Statistical Association*, 85(410):493–498, 1990.

[23] Joseph L Gastwirth, Abba M Krieger, and Paul R Rosenbaum. Asymptotic separability in sensitivity analysis. *Journal of the Royal Statistical Society: Series B (Statistical Methodology)*, 62(3):545–555, 2000.

[24] Joseph L Hodges and Erich L Lehmann. Estimates of location based on rank tests. *The Annals of Mathematical Statistics*, 34(2):598–611, 1963.

[25] Elizabeth A Stuart and David B Hanna. Commentary: Should epidemiologists be more sensitive to design sensitivity? *Epidemiology*, 24(1):88–89, 2013.

[26] Paul R Rosenbaum. Design sensitivity in observational studies. *Biometrika*, 91(1):153–164, 2004.

[27] Paul R Rosenbaum. Heterogeneity and causality. *The American Statistician*, 59(1):147–152, 2005.

[28] Paul R Rosenbaum. Impact of multiple matched controls on design sensitivity in observational studies. *Biometrics*, 69(1):118–127, 2013.

[29] Paul R Rosenbaum. Design sensitivity and efficiency in observational studies. *Journal of the American Statistical Association*, 105(490):692–702, 2010.

[30] JS Maritz. A note on exact robust confidence intervals for location. *Biometrika*, 66(1):163–170, 1979.

[31] Paul R Rosenbaum. Sensitivity analysis for M-estimates, tests, and confidence intervals in matched observational studies. *Biometrics*, 63(2):456–464, 2007.

[32] Paul R Rosenbaum. Testing one hypothesis twice in observational studies. *Biometrika*, 99(4):763–774, 2012.

[33] Peter L Cohen, Matt A Olson, and Colin B Fogarty. Multivariate one-sided testing in matched observational studies as an adversarial game. *Biometrika*, 107(4):809–825, 2020.

[34] Paul R Rosenbaum. Using Scheffé projections for multiple outcomes in an observational study of smoking and periodontal disease. *The Annals of Applied Statistics*, 10(3):1447–1471, 2016.

[35] Colin B Fogarty and Dylan S Small. Sensitivity analysis for multiple comparisons in matched observational studies through quadratically constrained linear programming. *Journal of the American Statistical Association*, 111(516):1820–1830, 2016.

[36] Abraham Charnes and William W Cooper. Programming with linear fractional functionals. *Naval Research Logistics Quarterly*, 9(3-4):181–186, 1962.

[37] Paul R Rosenbaum. Signed rank statistics for coherent predictions. *Biometrics*, pages 556–566, 1997.

[38] Siyu Heng, Hyunseung Kang, Dylan S Small, and Colin B Fogarty. Increasing power for observational studies of aberrant response: An adaptive approach. *Journal of the Royal Statistical Society: Series B (Statistical Methodology)*, 2021.

[39] Henry Scheffé. A method for judging all contrasts in the analysis of variance. *Biometrika*, 40(1-2):87–110, 1953.

[40] Jelle J Goeman and Aldo Solari. The sequential rejection principle of familywise error control. *Annals of Statistics*, 38(6):3782–3810, 2010.

[41] Ruth Marcus, Peritz Eric, and K Ruben Gabriel. On closed testing procedures with special reference to ordered analysis of variance. *Biometrika*, 63(3):655–660, 1976.

[42] Nicolai Meinshausen. Hierarchical testing of variable importance. *Biometrika*, 95(2):265–278, 2008.

[43] Jelle J Goeman and Livio Finos. The inheritance procedure: multiple testing of tree-structured hypotheses. *Statistical Applications in Genetics and Molecular Biology*, 11(1):1–18, 2012.

[44] Colin B Fogarty. Studentized sensitivity analysis for the sample average treatment effect in paired observational studies. *Journal of the American Statistical Association*, 115(531):1518–1530, 2020.

[45] Colin B Fogarty, Kwonsang Lee, Rachel R Kelz, and Luke J Keele. Biased encouragements and heterogeneous effects in an instrumental variable study of emergency general surgical outcomes. *Journal of the American Statistical Association*, pages 1–12, 2021.

[46] Paul R Rosenbaum. Identification of causal effects using instrumental variables: Comment. *Journal of the American Statistical Association*, 91(434):465–468, 1996.

[47] Paul R Rosenbaum. Choice as an alternative to control in observational studies. *Statistical Science*, 14(3):259–278, 1999.

[48] Guido W Imbens and Paul R Rosenbaum. Robust, accurate confidence intervals with a weak instrument: quarter of birth and education. *Journal of the Royal Statistical Society: Series A (Statistics in Society)*, 168(1):109–126, 2005.

[49] Mike Baiocchi, Dylan S Small, Scott Lorch, and Paul R Rosenbaum. Building a stronger instrument in an observational study of perinatal care for premature infants. *Journal of the American Statistical Association*, 105(492):1285–1296, 2010.

[50] Joshua D Angrist, Guido W Imbens, and Donald B Rubin. Identification of causal effects using instrumental variables. *Journal of the American Statistical Association*, 91(434):444–455, 1996.

[51] Zhiqiang Tan. A distributional approach for causal inference using propensity scores. *Journal of the American Statistical Association*, 101(476):1619–1637, 2006.

[52] Paul R Rosenbaum and Donald B Rubin. The central role of the propensity score in observational studies for causal effects. *Biometrika*, 70(1):41–55, 1983.

[53] Paul R Rosenbaum. Attributing effects to treatment in matched observational studies. *Journal of the American Statistical Association*, 97(457):183–192, 2002.

[54] Paul R Rosenbaum and Jeffrey H Silber. Amplification of sensitivity analysis in matched observational studies. *Journal of the American Statistical Association*, 104(488), 2009.

[55] Paul R Rosenbaum. *Replication and evidence factors in observational studies*. Chapman and Hall/CRC, 2021.

[56] Paul R Rosenbaum. Sensitivity analyses informed by tests for bias in observational studies. *Biometrics*, 2021.

[57] Paul R Rosenbaum. A new u-statistic with superior design sensitivity in matched observational studies. *Biometrics*, 67(3):1017–1027, 2011.

[58] Raghu Raj Bahadur. Stochastic comparison of tests. *The Annals of Mathematical Statistics*, 31(2):276–295, 1960.

[59] Paul R Rosenbaum. Bahadur efficiency of sensitivity analyses in observational studies. *Journal of the American Statistical Association*, 110(509):205–217, 2015.

[60] Paul R Rosenbaum. *Observation and experiment: An introduction to causal inference*. Harvard University Press, 2017.

[61] Paul R Rosenbaum. Sensitivity analysis for stratified comparisons in an observational study of the effect of smoking on homocysteine levels. *The Annals of Applied Statistics*, 12(4):2312–2334, 2018.

[62] Colin B Fogarty, Pixu Shi, Mark E Mikkelsen, and Dylan S Small. Randomization inference and sensitivity analysis for composite null hypotheses with binary outcomes in matched observational studies. *Journal of the American Statistical Association*, 112(517):321–331, 2017.

26

Evidence Factors

Bikram Karmakar

Department of Statistics, University of Florida

CONTENTS

26.1 Introduction

There is nothing thought to be known more precisely in the recorded human knowledge than the charge of a single electron. The National Institute of Standards and Technology reports the numerical absolute value of the elementary charge e as $1.602\ 176\ 634 \times 10^{-19}$ coulomb (C), where in the standard error[1] field it reports *(exact)* [1]. This exact determination was made official in 2018, and before that, there was still a standard error of the order of 10^{-27} C associated with its determination.

[1]In metrology, conventionally, this is called *standard uncertainty*.

DOI: 10.1201/9781003102670-26

It is quite remarkable that in the short period from a Friday in 1897 when J. J. Thompson announced his discovery of the electron to the audience at the Royal Institution of Great Britain, to 2018, we have been able to determine this minuscule number accurately. It is helpful to look at some history regarding the efforts in determining e. Robert A. Millikan designed the famous oil-drop experiment and published his determination of e in 1913 in a seminal paper [2]. He reported his estimate of e with a standard error that was several times smaller than anything know before. Millikan was awarded the 1923 Nobel prize in Physics "for his work on the elementary charge of electricity and on the photoelectric effect."

Using his newly developed apparatus Millikan had been able to calculate e as $1.5924 \pm 0.003 \times 10^{-19}$ C.[2] This determination was corroborated independently in several replications between 1913 and 1923 using the oil drop experiment. During this time, Millikan further refined the apparatus and his calculations and collected more data points to improve this estimate. In his Nobel lecture in 1924, he reported a more than 40% improved standard error of 0.0017×10^{-19} C [3]. During the lecture he notes:

> After ten years of work in other laboratories in checking the methods and the results obtained in connection with the oil-drop investigation published ... there is practically universal concurrence upon their correctness...

If we look closely, though, we see that the true value of e is more than 3.25 standard error larger than Millikan's first estimate and more than 5.75 standard error larger than Millikan's improved estimate. Thus, in statistical terms, the estimate was biased. It was biased the first time it was published, and it was biased in every replication of the experiment during that ten years. The replicated results using the oil-drop experiment only provided a higher certainty about this biased estimate.

How did we correct this bias? Did future oil-drop investigations suddenly remove the bias in the estimate? No. Did much larger sample sizes remove the bias in the estimate? No. Did refinements of the statistical methods remove the bias in the estimate? No.

In 1924 Karl Manne Siegbahn was awarded the Nobel prize in Physics for his work in the field of X-ray spectroscopy. Using this technology, in his thesis, Erik Bäcklin (1928) was able to find an alternate method to calculate e. This value was higher than Millikan's determination. But this was not a confirmation that Millikan's determination was biased. Far from that, this would be considered an extreme determination among several determinations of the value of e that provided surprising concurrence.

That same year, Arthur S. Eddington published a purely theoretical determination of e based on the newly developed Pauli's exclusion principle and General Relativity [4]. His calculation also gave a higher estimate than Millikan's determination. In the paper Eddington writes:

> the discrepancy is about three times the probable error attributed to [Millikan's] value, I cannot persuade myself that the fault lies with the theory.

He then writes:

> I have learnt of a new determination of e by Siegbahn .... I understand that a higher value is obtained which would closely confirm the present theory.

Today we know that Millikan's determination was biased because of an erroneous value for the coefficient of the air's inner friction in the oil-drop experiment. Replications of the experiment or refinement of the calculations could not correct this bias. Only when different experiments were conducted did the bias start become clear. Logically, it was also possible that the determination from the oil-drop experiments was correct, and the determinations by Siegbahn and Eddington were biased.

[2]Millikan reported in statC; 1 C = 2997924580 statC.

After all, each of these methods used fairly new technologies that would be open to criticism at the time. It is not the many replications or variations of one method but the corroboration of the results from a multitude of independent methods that are susceptible to different biases that strengthened our belief regarding the value of e.

26.1.1 Evidence factors

In causal inference from observational studies, the principle source of bias stems from the fact that the treatment and control groups may not be comparable even after adjustments for the observed pre-treatment covariates. Thus, when the aim is to strengthen evidence from an observational study, there is little to be gained by just repeating the study with a larger sample size or reanalyzing the data with a different statistical method that replaces one untestable and unverifiable assumption with another.

Instead, an *evidence factor analysis* performs multiple and nearly independent analyses, also known as *factors*, each depending on assumptions that do not completely overlap. When such two or more nearly independent analyses for a causal hypothesis provide supportive evidence, the evidence for a causal effect is strengthened. This is because neither bias nor statistical error that might invalidate one piece of the evidence can invalidate supportive evidence from the other factors.

Further, the factors possess the property that even when one of them is (perhaps infinitely) biased, the other factors are not affected. This property is formalized mathematically by showing that the factors are independent under sensitivity analyses that relax (not replace) the assumptions of no bias in the factors.

Thus, evidence for a causal effect is strengthened if multiple evidence factors support a causal effect and such evidence is further robust to a moderate degree of unmeasured biases in different directions.

All the relevant code for this chapter can be found at `https://github.com/bikram12345k/efchapter`.

26.2 Evidence Factors in Different Study Designs

Evidence factors are often available in observational studies but are overlooked. In this section we illustrate the variety of studies where an evidence factor analysis is possible. This section builds intuition, and the following sections provide some of the mathematical details.

26.2.1 Treatments with doses

Studies that are used to investigate possible treatment effects often have treatments with doses. These doses either are apparent in the study or can be constructed by additional knowledge about the causal problem.

Example 26.1. *Consider studying the effect of radiation exposure on solid cancer incidence. The Life Span Study is a long-running longitudinal study that investigates the long-term health effects of radiation exposure among survivors of the atomic bombings in Hiroshima and Nagasaki. The study includes survivors who were in the cities during the bombings and "not-in-city (NIC) residents" who were residents of the cities but were fortunate enough not to be in the city during the incident. We only work with the data from Hiroshima to reduce heterogeneity.*[3]

[3]Data were obtained from the Radiation Effects Research Foundation; a public interest foundation funded by the Japanese Ministry of Health, Labour and Welfare and the U.S. Department of Energy. The views of the author do not necessarily reflect those of the two governments.

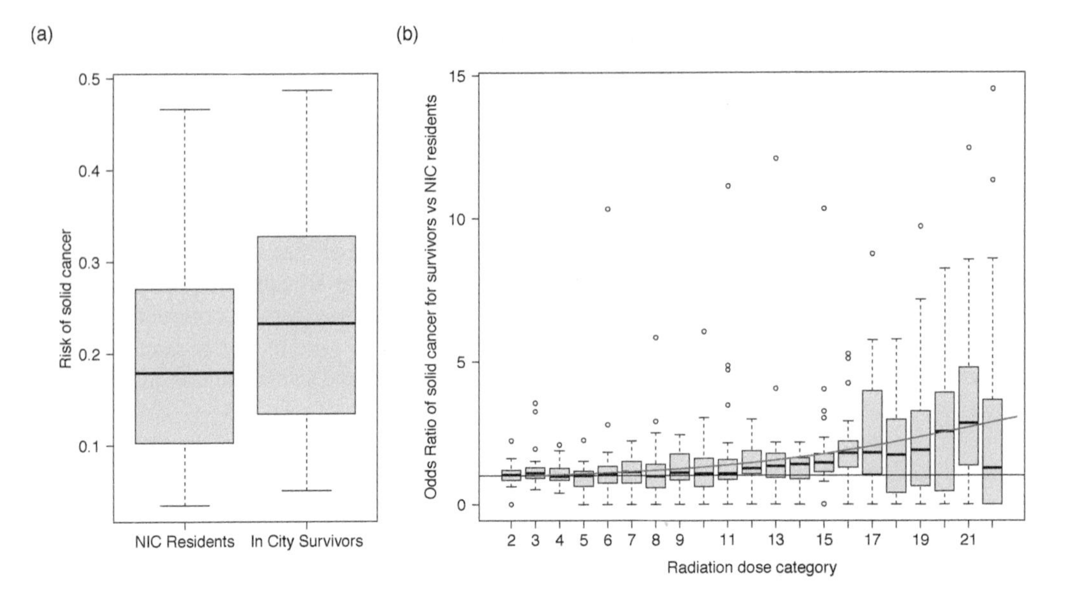

FIGURE 26.1
Radiation exposure and solid cancer incidence adjusted for age and sex. Panel (b) only shows in city survivors. Dose categories are 2 for 0–5 mGy, 3 for 5–20 mGy, 4 for 20–40 mGy, ... 20 for 1750–2000 mGy, 21 for 2000–2500 mGy, and 22 for 2500-3000 mGy colon dose, see www.rerf.or.jp.

Figure 26.1 shows two plots based on this study. Individuals are stratified into 30 strata based on age and sex; see details in [5]. Panel (a) of Figure 26.1 plots the risks of solid cancer for the NIC residents and survivors. The risk tends to be higher for the survivors relative to NIC residents. The Mantel–Haenszel test rejects the null hypothesis of no effect of radiation exposure on solid cancer in favor of a carcinogenic effect of radiation exposure; one-sided p-value 2.35×10^{-10}. Note that this comparison is valid under the assumption that the survivors and NIC residents are similar in all characteristics other than their radiation exposure. This assumption could be violated, for example, if not-in-city residents were better educated or employed.

We could redo this test with a different test statistic that provides increased power for one-sided alternatives in contingency tables; see examples in [6]. But if in fact there is no carcinogenic effect of radiation and the previous analysis is biased, this second test does not serve to weaken the effect of the bias in our evidence. In this section we are interested in structures of evidence factor analyses. Hence, we will not focus on the choice of test statistics. In Section 26.6 we provide some discussion on appropriate choices of test statistics.

Figure 26.1(b) looks at a different aspect of the effect of radiation. This time, the plot is of estimated odds ratios of cancer for the survivors versus NIC residents plotted against dose categories of the survivors. The pattern apparent in this plot is that the odds ratio increases with a higher radiation dose. There are a few important facts about this pattern. First, this pattern would appear if the null hypothesis of no effect was false and the carcinogenic effect gets worse with higher radiation dose. Kendall's test for correlation gives a one-sided p-value of 5.26×10^{-14}. Second, we are not entitled to this p-value without assumptions. This analysis could also be biased because the hypocentre was close to the urban center so that the survivors who were exposed to higher doses tended to be located in more urban areas; also, high-dose survivors might have been comparatively healthy to have survived a high dose. But, this bias acts differently from the one that worried us in the first analysis. Third, Figure 26.1(a) gives no indication that such a pattern would exist in Figure 26.1(b). In a sense, these two pieces of evidence complement each other. This is in contrast to two

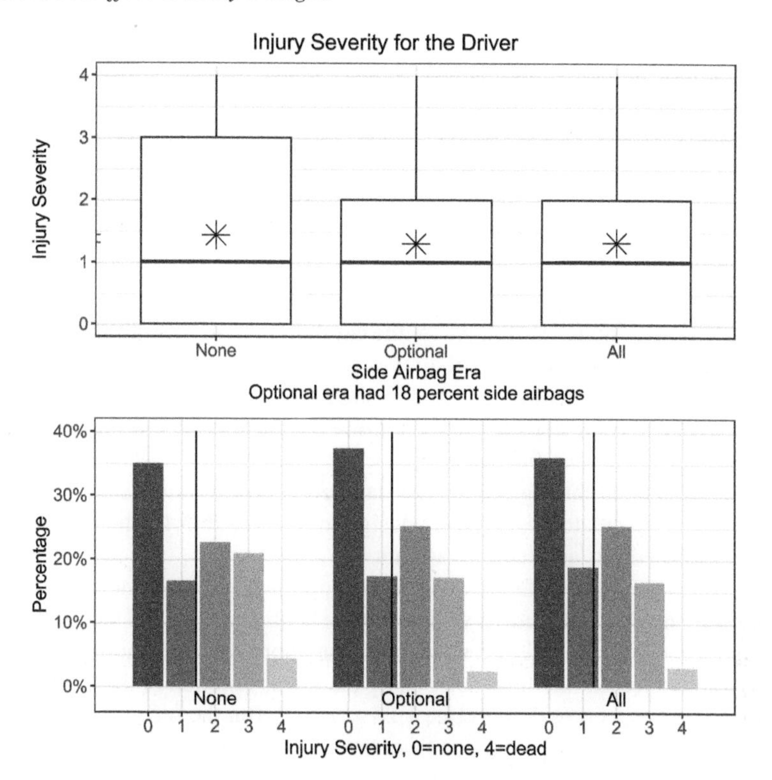

FIGURE 26.2
Injury severity in matched crashes across three ears of side-airbag availability, None, Optional, and All. The star in the boxplot and the vertical line in the barplot represent the corresponding mean. This is Figure 3 of [7].

tests on the same Figure 26.1(b), say with Kendall's test and Pearson's correlation test, which would be highly correlated and similarly biased. Stated a bit more formally, one could derive the asymptotic joint distribution of the two test statistics using a bi-variate Central Limit Theorem. The asymptotic correlation between the two tests will be 0, indicating that they are asymptotically independent. Figure 26.2(a) and (b) thus illustrate two factors in an evidence factor analysis of the carcinogenic effect of radiation from the Life Span Study data.

In the above example the doses of the treatment were easily conceived. The p-values were very small. We discuss in Section 26.4 how to leverage the properties of evidence factors analyses to make stronger conclusions. The example below is different from the previous example: it does not come with an obvious treatment with doses, and the comparisons do not unambiguously agree with each other. How do we create the factors? What do we learn from their evidence factor analysis?

Example 26.2. *Adoption of new safety technology in cars happens gradually. There is an earlier period where cars would not have had the technology and a later period where most cars will have the safety technology by default. In between, there is often a transition period where the technology will be available for purchase optionally. There are many car manufacturers; some may follow a faster adoption than others. This timeline may also vary across different vehicle models by the same manufacturer.*

Consider in this context studying the effectiveness of side-airbags during a crash. A simple com/parison of injuries to drivers of cars with side-airbags and without side-air bags is not an appealing comparison because side airbags are only one difference between these vehicles and their drivers. Perhaps Volvos (which tended to have side-air bags earlier than others) attract drivers concerned with safety, with the possible consequence that Volvos are driven differently from, say, Dodge Chargers.

A more attractive comparison can be made that avoids this bias. We can compare crashes of the same makes and models of cars across eras that differ in side-airbag availability. This comparison was made in [7] using the U.S. NHTSA's U.S. Fatality Analysis Reporting System (FARS) records. We avoid some discussion regarding possible selection bias and its resolution in this data set which could occur since FARS only has records of fatal crashes; see [7] Section 2.2. The analysis created matched sets of three vehicle types of vehicles with each set consisting of cars with the same make and model, one from the "none" period where the car did not have side-airbags, either one or three from the "optimal" period where side-airbag was available only as an optional purchase and one from the "all" period. The matching adjusted for measured characteristics of the driver, e.g., age, gender and if belted, and characteristics of the crash, e.g., the direction of impact. The outcome is a measure of severity of the injury, 0 for uninjured to 4 for death.

This type of matching of three groups is not standard in observational study designs. We discuss the algorithm we used to create this matched design in Section 26.5.1. The matched design in this study had 2,375 matched sets and total 9,084 cars.

This study has two evidence factors: (i) all-and-optional-versus-none-era and (ii) all-versus-optional. Note that only 18% of studied vehicles in the "optional" era had side airbags. Figure 26.2 reproduces Figure 3 of [7]. This figure shows that the "none" period had higher injuries. This is also seen in a stratified test for the first factor using the `senm` function of the `sensitivitymult` package in `R` which gives a one-sided p-value 1.26×10^{-06}. But the second factor gives a one-sided p-value 0.5801. This indicates the evidence from the first factor is not plausibly an effect caused by side-airbags, because there is no significant difference between the "all" and "optional" eras. Yet, in the "optional" era, only 18% of owners had purchased vehicles with side-airbags. If we had not looked at two factors, the comparison of the "none" era to the other eras might mistakenly have been taken to be evidence for effects caused by side-airbags.

Other comparisons could also be of interest in this design. Comparison of the "all" era to the "none" era also provides a signal of reduced injury when side-airbags were standard in cars. But, this does not complement the first analysis from our previous discussion. These two analyses are not independent; thus they do not form evidence factors. We continue this discussion in Section 26.4.2.

26.2.2 Case–control studies

In case–control studies, patients with (cases) or without (controls) an out- come of interest are compared in terms of their exposure to treatment. Case–control studies are particularly useful for an outcome that is relatively rare in the population. In a case–control study there is often a choice of how to define a case. A "broad" definition of a case may have higher sensitivity but less specificity, whereas a "narrow" definition of a case may have higher specificity but less sensitivity.

In a case–control study with narrow and broad cases, we expect that if the exposure has an effect and our theory that the narrow cases are more likely to be caused by the exposure than the more heterogeneous broad cases is correct and also there is no unmeasured confounding, then (a) the exposure should have a larger association with narrow cases than marginal cases, i.e., cases that are broad but not narrow and (b) the exposure should have an association with broad cases compared to controls.

We can test these patterns using evidence factor analysis where the two factors are formed by comparing (i) narrowly defined cases to other cases, and (ii) broad cases to controls in their exposure rates. The relevant methodological development for this analysis is presented in [8].

26.2.3 Nonreactive exposure and its reactive dose

In studies of the human population some exposures are external, they are "nonreactive," whereas a dose of an exposure might also incorporate an individual's reaction to a nonreactive exposure. A nonreactive exposure can be an individual's exposure to a pollutant, or an individual taking a vaccine shot. A reactive dose is an outcome within the individual, e.g., level of the pollutant's metabolites in the blood or level of the vaccine's related antibodies. These levels will vary according to the individual's metabolism or blood cells' reactions to the vaccine; on biology in general.

A reactive dose, although personal, is not necessarily preferable to its nonreactive counterpart for making causal inference about the exposure on a disease outcome. This is because many physiological processes that might be involved in an individual's internalization of the external exposure may also influence or be influenced by the disease process [9], which would introduce a confounding bias. Such reasons have led some to suggest caution in using exposure biomarkers to infer a causal effect and propose using the nonreactive exposure instead [10].

An evidence factor analysis in such a study can look at two things – the nonreactive and reactive exposure – at once while keeping the analyses separate from each other [11].

Example 26.3. *[12] and other research suggest that cigarette smoking has a causal effect on periodontal disease. We reconsider this question using data from the 2003-2004 National Health and Nutrition Examination Survey (NHANES). We defined a smoker as a person who has smoked every day for the last 30 days and a nonsmoker as a person who has smoked less than 100 cigarettes in their life.*

We adjusted for age (in 10 yrs groups), sex, education (high-school graduate or less), and race (Hispanic, Black, or other) by creating a stratification where the strata are identified jointly by the categories of these five covariates.

Following [12], we calculate the outcome, periodontal disease, as the proportion of periodontal sites that had a pocket depth or loss of clinical attachment of ≥ 4 mm. Figure 26.3 plots several aspects of the smoking status, serum cotinine level (a biomarker for smoking), and periodontal disease.

Figure 26.3(a) plots the disease for the smokers and non-smokers. This comparison is our first factor. For the other two plots, which only plot smokers, we calculated the value of the cotinine level and periodontal disease for the smokers relative to their reference nonsmokers in the same stratum. We defined the relative cotinine level as the smoker's cotinine level minus the average cotinine level for the nonsmokers from its stratum, and the relative periodontal disease as the smoker's disease outcome divided by the sum of the average disease for its reference nonsmokers and overall difference in the average disease for smokers from nonsmokers.

Figure 26.3(c) shows that the boxplot for relative cotinine level is mostly on the positive axis since smoking increases cotinine level. We note that, for the nonsmokers, the first and third quartile of cotinine are 0.016 and 0.216, respectively. Therefore, we see a much large variability in cotinine levels for the smokers. It is possible that physiological processes that internalize smoking exposure to serum cotinine also have variabilities. We also note that, some nonsmokers have cotinine levels of 50, 100, 200, or even 500, indicating perhaps some mislabeling of the smoking status in the survey.

Figure 26.3(b) shows the other two factors. On the horizontal axis we have three equal-sized groups based on the relative cotinine levels. Factor 2 asks if periodontal disease increases from a medium relative to a high relative cotinine level. Factor 3 asks if periodontal disease increases from a low to a medium or high relative cotinine level.

We use val Elteren's rank-based tests for stratified data [13] for the first factor and Wilcoxon's rank-sum tests for the other two factors. The p-values are calculated as $< 1.0 \times 10^{-16}$, 0.0024 and 0.066 for these three factors, respectively. We shall revisit this example to present additional analyses in Section 26.3.3.

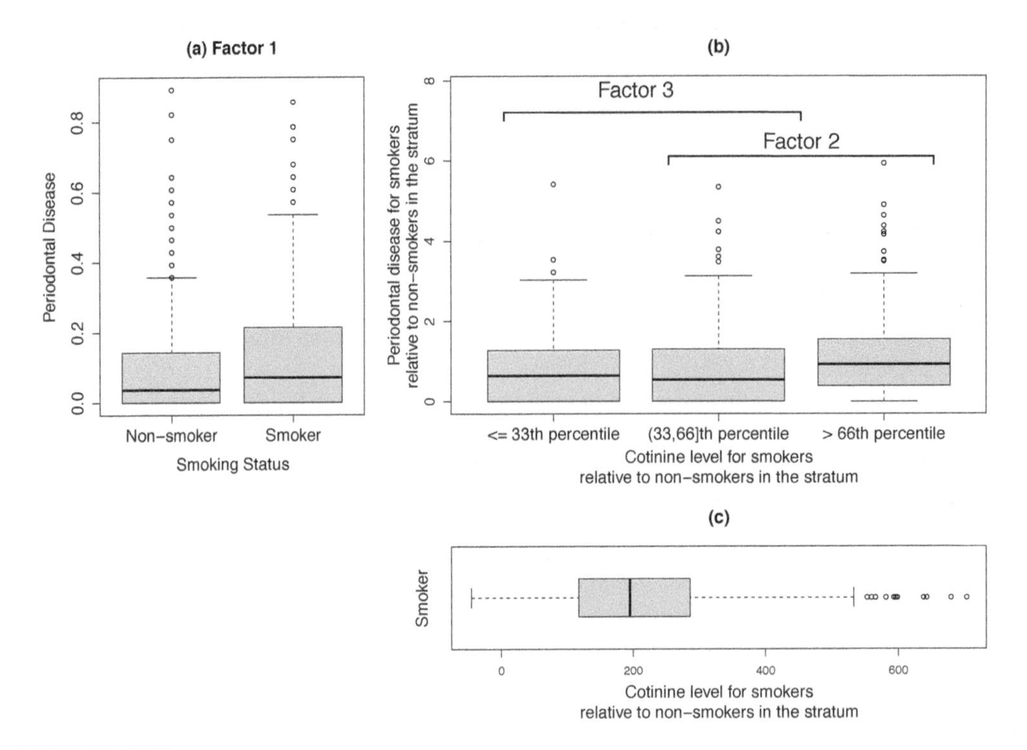

FIGURE 26.3
Covariate adjusted periodontal disease for smokers and nonsmokers (a). Smokers' relative periodontal disease for groups of relative cotinine levels (b). Smokers' relative cotinine level to nonsmokers (c).

26.2.4 Studies with possibly invalid instrument(s)

An instrumental variable, a "random nudge" to accept the treatment, provides a unique opportunity to separate a meaningful causal effect from bias. Valid usage of an instrumental variable is subject to a few conditions. There must be no direct effect of the instrument on the outcome, i.e., it must satisfy the exclusion restriction assumption, and no unmeasured confounders between the instrument and the outcome. However, in practice, these assumptions are often violated even after controlling for observed covariates and are untestable without solid scientific knowledge.

Depending on the context, other instruments can be used for the same question. If multiple instruments tend to induce bias in similar directions then there is less to be gained in terms of separating causal effect from bias by using multiple instruments because one bias may affect multiple results in the same direction. However, if the instrumental variable analyses could produce separate pieces of evidence, we can obtain randomness in bias from multiple instruments even without replicating studies.

In a study with multiple candidate instruments we can construct evidence factors where each factor is valid when the corresponding instrument is conditionally valid given the other instruments. This evidence factor analysis thus relaxes the exclusion restriction on the instruments. Two liberally defined exclusion restrictions have been considered in the literature. The first is the ordered partial exclusion restriction and second is the unordered partial exclusion restriction. Consider K candidate instruments. Let $\mathcal{K} \subseteq \{1, \ldots, K\}$ and $k_{\min}$ be the smallest index in $\mathcal{K}$.

Definition 26.1. *([14]) The ordered partial exclusion restriction holds for $\mathcal{K}$ if, with the values of the first $k-1$ instruments fixed by conditioning, the potential outcomes of the units are specified by the exposure status of the units.*

Definition 26.2. *([15]) The unordered partial exclusion restriction holds for $\mathcal{K}$ if, with the values of the instruments not in $\mathcal{K}$ fixed by conditioning, the potential outcomes of the units are specified by the exposure status of the units.*

The unordered partial exclusion restriction is less restrictive than its ordered counterpart. For example, with $K = 3$ and $\mathcal{K} = \{2\}$, the unordered partial exclusion restriction holds for $\mathcal{K}$ if the third instrument directly affects the outcome, but the ordered partial exclusion restriction does not hold. We refer the readers to [15] for a detailed discussion of these two restrictions.

If the ordered partial restriction holds for $\mathcal{K}$, and further there are no unmeasured confounders in the instruments in $\mathcal{K}$ then we can create $|\mathcal{K}|$ (the cardinality of $\mathcal{K}$) evidence factors using the reinforced design developed in [14]. In this method, K comparisons are made where the kth step performs an instrumental variables analysis using the kth instrument after stratifying on the previous $k - 1$ instruments. If $\mathcal{K}$ is known, one could only conduct $|\mathcal{K}|$ analyses that stratify on the first $k_{\text{min}} - 1$ instruments. But it is unlikely in practice that $\mathcal{K}$ would be known. In that case, a planned analysis can synthesize the results from the K comparisons to provide useful detail regarding the strength of the evidence in the presence of invalid instruments. Planned analyses of evidence factors are discussed in Section 26.4.

If the unordered partial exclusion restriction holds for $\mathcal{K}$, then a reinforced design could fail to give valid evidence factors. For example, suppose $K = 3$, $\mathcal{K} = \{2\}$ and the third instrument directly affects the outcome. In this case the reinforced design does not work. Instead, we can construct evidence factors for this situation using the balanced blocking method of [15]. This method creates ex-post-facto strata where the empirical distribution of the instruments in each stratum are made jointly independent. For example, for $K = 2$ binary instruments, a stratum may look like (a) or (b) in Table 26.1, but not Table 26.1(c) where the second instrument tends to be higher for higher values of the first instrument.

TABLE 26.1
Examples of balanced blocks (a) and (b), and an unbalanced block (c).

(a)

Unit	Z_1	Z_2
1	0	0
2	0	1
3	1	0
4	1	1

(b)

Unit	Z_1	Z_2
1	0	0
2	0	1
3	0	1
4	1	0
5	1	1
6	1	1

(c)

Unit	Z_1	Z_2
1	0	0
2	0	1
3	1	0
4	1	1
5	1	1
6	1	1

A few remarks on differences between a reinforced design and balanced blocking follow. (i) Balanced blocking requires the creation of a special design, which is not required in reinforced design, where we balance the distribution of the instruments against each other. If the instruments are highly imbalanced, balanced blocking may require throwing away many data points, or it may be impossible to create a balanced blocking. The latter happens, for example, when the instruments are nested. For nested instruments, [15] develop the mutual stratification method for evidence factor analysis. (ii) The analysis in balanced blocking does not require conditional analyses on other instruments that are central to a reinforced design. Instead, one can analyze each instrument marginally. Marginal analyses often improve the interpretability of the inference; see Section 3.5 of [15]. (iii) The factors from a reinforced design are nearly independent in the finite sample (technical details in Section 26.3.3), but the factors from a balanced blocking method are only asymptotically independent.

Both reinforced design and balanced blocking ask for novel design tools. It turns out, both these design problems are NP-hard. Intuitively, this means it is not always possible to compute the optimal designs fast. Section 26.5 gives algorithms that run fast to create designs that are close to the optimal designs.

Example 4. Does Catholic schooling have a positive impact in terms of future earnings relative to public schooling? [16] were the first to study this question by directly comparing public high schools to private Catholic high schools. Following the study, controversy arose partly due to methodological reasons. It is easy to see that parents' education, income and social status can confound this analysis. So, we should adjust for them. But, parents' and child's commitment to education and ambitions are not easy to measure accurately and thus are not easy to adjust for, which could bias the direct comparison.

In the literature at least two instrumental variables have been used to study this question. The first uses the fact that geographic proximity to Catholic schools nudges a child to attend a Catholic school. Thus, if the answer to the above question was yes, then we should see higher income for the individuals raised closer to Catholic schools relative to others who were otherwise comparable. The second uses the fact that being raised in a Catholic family also has a strong influence on the decision to attend a Catholic school. But the literature also has many arguments and counter-arguments regarding the validity of these candidate instruments.

[14] present an evidence factor analysis where rather than choosing one comparison, there are three factors from the two candidate binary instruments (raised in an urban/rural area and a Catholic/other household) above and the direct comparison. They use a reinforced design where the comparisons are made in the order, first, proximity to Catholic schools, then, raised in a Catholic family, and finally, the direct comparison. This analysis provides nearly independent inferences that provide valid inference for the question under an ordered partial exclusion restriction.

Using the same data from the Wisconsin Longitudinal Study of students graduating high school 1957 (see details in [14]), we present an evidence factor analysis using the balanced blocking method. Our outcome measure is yearly wages in 1974. We adjust for students' IQ, parents' education, income, occupational socioeconomic index (SEI) score and occupational prestige score. We create a balanced blocking that adjusts for these covariates, where every balanced block includes 5 individuals each for four possible pairs of values of the two instruments. We discuss the algorithm for how this is achieved in Section 26.5.3.

Table 26.2 is our balance table showing the summary of the covariates before and after balanced blocking. An absolute standardized difference on the table has the same interpretation as in a matched pairs design but requires a different definition for a stratified design which is given in Section 26.6.

Pearson's χ^2 statistic for an association between the urban/rural instrument and attending Catholic school is 27.65, and between the religion instrument and attending Catholic school is 204.38. Both instruments are therefore strongly associated with the treatment.

There are several choices of test statistics for our three comparisons. We choose the stratified Wilcoxon's rank-sum statistic because it is robust to extreme values of the outcome, and the stratum sizes are equal. We have three factors in this study. The first compares the outcomes for urban vs rural using the strata created by the balanced blocks. The second compares the outcomes for raised Catholic vs other using the strata created by the balance blocks. The third compares the outcomes for Catholic vs public schooling using strata created by the balance blocks and the two binary instruments.

Table 26.3 reports the results from these analyses. The point estimates and confidence intervals are calculated in the usual way by inverting the test ([17], section 2.4) under the assumption of a Tobit effect, see Section 2.4.5 of [18]. Our three factors agree in finding that there is no significant effect of Catholic schooling vs public schooling. Since these comparisons are nearly independent, more can be said from these results. For example, what is the evidence by combining these factors? What evidence remains if we believe one of these factors is biased? These are discussed in Section 26.4.1.

TABLE 26.2
Covariate balance before and after balanced blocking for example 4. We report the means for each covariate at the two levels of the instruments and then the absolute standardized difference of the covariate between the levels, e.g., 0.107 is the absolute standardized difference in IQ score between the urban and rural groups before balanced blocking.

	Before		After		Before		After	
	Rural	Urban	Rural	Urban	Catholic	Other	Catholic	Other
IQ	99.7	103.1	102.2	103.1	102.8	102.6	102.8	102.5
	0.107		0.030		0.032		0.009	
Father's education	10.1	11	10.7	11	11	10.8	11	10.7
	0.154		0.050		0.006		0.046	
Mother's education	10.6	10.9	10.6	10.9	10.9	10.6	10.9	10.6
	0.047		0.040		0.052		0.053	
Parental income ($100)	56	76.5	68.1	73.3	74.5	71.6	72.2	69.2
	0.176		0.047		0.029		0.027	
Occupational SEI score	27.4	35.8	32.4	35.8	35.1	33.4	34.9	33.2
	0.198		0.083		0.033		0.040	
Occupational prestige score	59	63.2	61.7	63.2	62.9	62.2	62.8	62.1
	0.191		0.068		0.037		0.035	

TABLE 26.3
Evidence factors in the study for the effect of Catholic schooling on yearly wage ($) in 1974.

	Urban/Rural	Raised Catholic/Other	Schooling Public/Catholic
p-value	.0640	.1398	.1608
Point estimate	4,946	1,145	419
95% Confidence interval	(−1,671, 25,833)	(−945, 3,586)	(−697, 1,759)

A brief remark: In our analysis, we made two adjustments for the covariates. Once in balanced blocking where we created blocks with the units close in their covariate values. Then, we used a covariance adjustment method that used the residuals from a robust linear regression of the outcome on the covariates and block levels as inputs for the stratified Wilcoxon's tests; this method is discussed in [19].

26.3 Structure of Evidence Factors

A single theory does not describe all the ways we created our factors in the examples from Section 26.2. It is nonetheless helpful to see one type of construction of evidence factors. This is done in Section 26.3.1. The present section also reviews the fundamentals of sensitivity analysis in Section 26.3.2 and describes joint sensitivity analyses with evidence factors in Section 26.3.3.

26.3.1 A simple construction of exactly independent comparisons

Consider n units and three treatment groups of sizes n_1, n_2 and n_3, respectively. Let Z_{i1} be the indicator that unit i is in the second or third treatment group and Z_{i2} be the indicator that unit i is in the third treatment group. Consider three potential outcomes for each unit i depending on the treatment group. Denote them by Y_{i1}, Y_{i2} and Y_{i3} for treatment group 1, 2, and 3, respectively. Let $\mathcal{F} = \{Y_{ig} : 1 \leq i \leq n, g = 1, 2, 3\}$.

We want to test the null hypothesis of no treatment effect which states that the potential outcomes of the units do not vary with the group, i.e., $Y_{i1} = Y_{i2} = Y_{i3}$ for all i. Consider two comparisons. The first compares treatment group one to the other two treatment groups. The second compares the second and third treatment groups. We can do these comparisons using Wilcoxon's rank-sum tests. These two comparisons use the same data set. Yet, we show below that for a completely random treatment/group assignment, they are exactly independent under the null hypothesis.

The $2n$ indicators Z_{i1}'s and Z_{i2}'s define a treatment assignment. At this point, it is convenient to change our perspective a bit. Let σ be a vector of size n, where the first n_1 elements of σ are the indices of the units assigned to treatment group 1, following n_2 elements are the indices of the units assigned to treatment group 2, and the last n_3 elements are the indices assigned to treatment group 3. Then, σ is a permutation of $\{1, \ldots, n\}$. But, if any of n_1, n_2 or n_3 is more than 1, many such σ are possible. To pick one, we define the following.

Definition 26.3. *Call a permutation σ of $\{1, \ldots, n\}$ (n_1, n_2, n_3)-increasing permutation if the first n_1 indices of σ are in increasing order, the next n_2 indices are in increasing order, and the last n_3 indices are in increasing order.*

Given the $2n$ indicators, we define the random vector σ as its corresponding (n_1, n_2, n_3)-increasing permutation. Let $\sigma(i)$ denote the ith element of σ.

To simplify notation, suppose the null hypothesis is true. Then we can write our two Wilcoxon's rank-sum statistics using the vector $(Y_{\sigma(1)1}, \ldots, Y_{\sigma(n)1})$. Assume there are no ties. Define $(R_1^\sigma, \ldots, R_n^\sigma)$ to be the ranks of the elements in this vector and $(\tilde{R}_1^\sigma, \ldots, \tilde{R}_{n_2+n_3}^\sigma)$ to be the ranks of the last $n_2 + n_3$ elements in this vector among themselves. Then the test statistics are $T_1 = \sum_{i=1}^{n_2+n_3} R_{n_1+i}^\sigma$ and $T_2 = \sum_{i=1}^{n_3} \tilde{R}_{n_2+i}^\sigma$.

Proposition 26.1. *Consider a completely randomized treatment assignment. When the null is true, given $\mathcal{F}$, T_1 and T_2 are independent.*

Proof A completely randomized design implies that all possible choices of (n_1, n_2, n_3)-increasing permutations are equally likely. Notice that $T_1 = n(n+1)/2 - \sum_{i=1}^{n_1} R_i^\sigma$. Then the theorem follows from the following two facts.

First, let $(R'_1, \ldots, R'_n)$ be the ranking of the elements of the vector $(Y_{11}, \ldots, Y_{1n})$ among themselves. Then $R_i^\sigma = R'_{\sigma(i)}$. Hence, given $\mathcal{F}$, when there are no ties, all possible choices of the vector $(R_1^\sigma, \ldots, R_n^\sigma)$ are equally likely.

Second, there is a one to one correspondence between the vectors $(R_1^\sigma, \ldots, R_n^\sigma)$ and $(R_1^\sigma, \ldots, R_{n_1}^\sigma, \tilde{R}_1^\sigma, \ldots, \tilde{R}_{n_2+n_3}^\sigma)$. We state this result without proof here. It can easily be verified in small instances.

Hence, all values of $(R_1^\sigma, \ldots, R_{n_1}^\sigma, \tilde{R}_1^\sigma, \ldots, \tilde{R}_{n_2+n_3}^\sigma)$ are equally likely in the product set $\{(r_1, \ldots, r_{n_1}) : 1 \leq r_i \leq n, r_i\text{s distinct integers}\} \times \{(\tilde{r}_1, \ldots, \tilde{r}_{n_2+n_3}) : 1 \leq \tilde{r}_i \leq n_2 + n_3, \tilde{r}_i\text{s distinct integers}\}$. Hence, the result follows. □

The above proposition shows that we can create exactly independent tests for a null hypothesis in the same study. It is easy to see that we do not need to use Wilcoxon's rank-sum statistics for the tests. Any function of $(R_1^\sigma, \ldots, R_{n_1}^\sigma)$ could be used for T_1 and any function of $(\tilde{R}_1^\sigma, \ldots, \tilde{R}_{n_2+n_3}^\sigma)$ could be used for T_2.

Alternatively, exactly independent comparisons can also be created by sample splitting. However, performing independent comparisons from two or more split samples is not a good strategy for creating evidence factors. First, it reduces the effective sample size of analysis. Second, more importantly, two analyses using similar statistical methods but from separate splits do not protect us from a variety of unmeasured biases.

The representation of a treatment assignment as a random permutation comes in quite handy in one kind of construction of evidence factors where, as we did here, the factors are created from a factorization of the treatment assignment. A general construction along this line is given in [20], which is further extended in Theorem 9 of [21].

Ranks are somewhat special, in the sense that they can provide exactly independent analyses. But many common pairs of test statistics will not have this property. It turns out that for an evidence factor analysis we do not need exact independence; a sense of near independence is sufficient. Near-independence of tests is formally defined in Definition 26.5. Also, nearly independent comparisons are easier to achieve. Further, nearly independent comparisons can be combined as if they were independent to report combined evidence against the null that can only have a conservative type-I error. Thus, pieces of evidence from two nearly independent comparisons do not repeat each other.

26.3.2 Brief introduction to sensitivity analysis for unmeasured confounders

To be evidence factors we require more than just (near) independence of the factors when there are no unmeasured confounders. Since we do not rule out the possibility of biases, we want the (near) independence of the factors to remain even when one comparison is biased. In this way one factor can even be infinitely biased, thus useless for our inference, and still, we can use the evidence from the other factor that does not repeat the bias in the former one. To make this concrete we briefly discuss sensitivity analysis for unmeasured confounders in observational studies.

Consider a treatment-control analysis. Suppose we have a set of measured covariates for the analysis, and these covariates are adjusted using a matched pairs design. If there are no unmeasured biases, then in these matched pairs the probability is 1/2 that the first unit in the pair receives the treatment. In a sensitivity analysis for unmeasured confounders, this typically unrealistic assumption of no unmeasured confounders is relaxed gradually, not removed or replaced with another assumption. We use Rosenbaum's model for sensitivity analysis [22, 23]. Sensitivity analysis under this model is well developed for various observational study structures and its properties have been well studied; see [24–29].

The aforementioned relaxation is parametrized with a sensitivity parameter. For our matched pairs design let this parameter be Γ. A value of $\Gamma \geq 1$ says that the probability that the first unit in the pair receives treatment is in the interval $[1/(\Gamma+1), \Gamma/(\Gamma+1)]$. Thus, when $\Gamma = 1$, we get back our no unmeasured confounding assumption. For a value $\Gamma > 1$, we have an interval of values of the treatment assignment probability. This interval becomes larger with larger values of Γ. The size of the interval, equivalently the value of Γ, depicts the amount of influence we are assuming for the unmeasured confounder.

This parameter Γ has a useful and intuitive interpretation. Let p denote the probability of the first unit in the pair receiving treatment. We have $1/(\Gamma+1) \leq p \leq \Gamma/(\Gamma+1)$. Thus, for Γ equals to 2, we get that the maximum possible odds of the first unit being treated rather than the second unit, $p/(1-p) = 2$. Meaning, a $\Gamma = 2$ indicates a bias level where for a pair of units that are similar in the observed covariates, one unit can have twice the odds receiving the treatment. Thus $\Gamma = 2$ is not a negligible level of bias.

Recall that a confounder is associated with both the treatment and the outcome. In the above, Γ only bounds the treatment assignment and allows for any association with the outcome. Inherently, Γ indicates the dependence of the treatment assignment on the potential outcome after adjustment for the observed covariates have been made. The parameter Γ has another interpretation that separates the two associations of the confounder and treatment and the confounder and outcome [30]. Interpreted this way, a bias level of $\Gamma = 2$ could be created by an unmeasured confounder that increases the odds of getting the treatment 5 times and increases the chance of a positive difference in the treatment minus control outcome for the pair by threefold. Again, these effects are not small.

What do we do after specifying a value of Γ? If $\Gamma = 1$, we know there are no unmeasured confounders. Thus, in a matched pairs design we know that the probability of a unit in a pair receiving the treatment is a toss of a fair coin. We can use this randomized treatment assignment distribution to get the null distribution of our test statistic under the null. This way, we can calculate an exact p-value for our test [23]. All the p-values we reported in Section 26.2 are calculated this way but under different randomization distributions.

If we specify $\Gamma > 1$, we do not know the exact distribution of the treatment assignment. Rather, a family of treatment assignment distributions is possible. Then, a sensitivity analysis p-value is calculated as an upper bound of all possible p-values that are possible under this family.

In example 3, when $\Gamma = 1$ the comparison of the smokers to nonsmokers on periodontal disease gave a p-value of $< 1.0 \times 10^{-16}$ for van Elteren's test. To assess the sensitivity of this inference, let $\Gamma = 2$. We calculate the sensitivity analysis p-value for the test (this can be done using the `R` package `senstrat`) as 0.0003. This number is larger than the p-value under $\Gamma = 1$, which is expected. But it is still smaller than the nominal level 0.05. Thus, a bias of $\Gamma = 2$ is not enough to refute a rejection decision for the null hypothesis. A large enough bias level can invalidate any inference made under the assumption of no unmeasured confounding. At $\Gamma = 2.75$, the sensitivity analysis p-value is .1707, and at $\Gamma = 3$ the sensitivity analysis p-value is .3937.

26.3.3 Evidence factors: definition and integrating evidence using joint sensitivity analyses

Now consider K factors and their corresponding sensitivity parameters $\Gamma_1, \ldots, \Gamma_K$. Let $\overline{P}_{k,\Gamma_k}$ denote the p-value upper bound from comparison k at bias $\Gamma_k \geq 1$. We have the following Definition 26.5 for evidence factors.

Definition 26.4. *A vector of p-values* $(P_1, ..., P_K) \in [0,1]^K$ *is stochastically larger than the uniform when for all coordinate-wise non-decreasing bounded function* $g : [0,1]^K \to \mathbb{R}$, *we have, under the null,*

$$E\{g(P_1, \ldots, P_K)\} \geq E\{g(U_1, \ldots, U_K)\}, \tag{26.1}$$

where $U_1, \ldots, U_K$ *are i.i.d. uniform[0,1] random variables.*

Definition 26.5. *Multiple analyses are evidence factors when (i) bias that invalidates one analysis does not necessarily bias other analyses; and (ii) for any* $(\Gamma_1, \ldots, \Gamma_K)$ *the p-value upper bounds* $(\overline{P}_{1,\Gamma_1}, \ldots, \overline{P}_{K,\Gamma_K})$ *are stochastically larger than the uniform.*

Consider a simple implication of two factors satisfying the definition. Define a non-decreasing bounded function g on $[0,1]^2$ as $g(p_1, p_2) = 1$ if and only if $p_1 \geq \alpha_1$ or $p_2 \geq \alpha_2$ for some specified values of α_1 and α_2 in $(0, 1]$. Then the definition applied to two factors for $\Gamma_1 = \Gamma_2 = 1$ implies

$$\Pr(\overline{P}_{1,1} < \alpha_1 \text{ and } \overline{P}_{2,1} < \alpha_2) \leq \Pr(U_1 < \alpha_1 \text{ and } U_2 < \alpha_2) = \alpha_1\alpha_2.$$

Assume $\overline{P}_{1,1}$ has a uniform distribution, which will happen in many common situations for a large sample. Then, the above implies that, under the null, $\Pr(\overline{P}_{2,1} < \alpha_2 \mid \overline{P}_{1,1} < \alpha_1) \leq \alpha_2$. Thus, if we had a testing procedure that rejected the null if $\overline{P}_{1,1} < \alpha_1$, then a decision to reject the null using the first factor will not inflate the type-I error for the second factor. Thus, the comparisons are nearly independent in the sense of (ii) in Definition 26.5.

Another important implication of the definition is the following result. Consider any method that combines K independent p-values into a valid p-value. If the method is coordinatewise non-decreasing, then it can be applied to $(\overline{P}_{1,\Gamma_1}, \ldots, \overline{P}_{K,\Gamma_K})$ to calculate a p-value that is valid for testing the null hypothesis when the biases in the factors are at most $\Gamma_1, \ldots, \Gamma_K$. Hence, we can combine the factors as if they were independent.

Many methods are available that combine independent p-values. Fisher's method is a popular choice [31]. This method calculates a new test statistic as -2 times the sum of the logarithms of the K p-values. It is easy to see that if the p-values were independent and uniformly distributed on [0,1] this statistic will have a χ^2-distribution with $2K$ degrees of freedom. Fisher's combination method calculates the combined p-values using this null distribution. Specifically, Fisher's combination of $\overline{P}_{k,\Gamma_k}$'s is

$$\Pr(\chi^2_{2K} > -2\sum_{k=1}^{K} \log \overline{P}_{k,\Gamma_k}).$$

By the near independence property of the factors, this is a valid p-value when the biases are at most Γ_ks. [32] gives a comprehensive, although slightly outdated, summary of various combination methods. Using such methods we can produce joint sensitivity analyses with several factors.

Two questions: Which combination method should we use? What do you get extra from looking at joint sensitivity analyses of the factors rather than at the individual factors?

Consider the first question. Some combination methods are better for joint sensitivity analysis. An attractive choice is the truncated product method of [33]. It uses a threshold κ where $0 < \kappa \leq 1$ to define the statistic $-2\sum_{k:\overline{P}_{k,\Gamma_k}} \log \overline{P}_{k,\Gamma_k}$. The combined p-value is calculated by comparing the value of this statistic to its null distribution, which has an analytic form [33]. The truncated product method de-emphasizes larger ($> \kappa$) p-values by treating them all as 1. When $\kappa = 1$ we get back Fisher's combination test. Since $\overline{P}_{k,\Gamma_k}$'s are p-value upper bounds, their exact numeric values are of little interest beyond a certain value. This fact makes the truncated product method appealing in joint sensitivity analyses.

Although Fisher's and the truncated product method have equivalent asymptotic power performance (see Proposition 7 of [34] with $k = 1$), in practical situations, the truncated product method often shows higher sensitivity to unmeasured confounders [35]. Typically, the suggested value of κ is .1 or .2.

For the second question, consider example 3 again.

Example 3 (continued). Recall the three factors from Figure 26.3. Notice that not everything about Figure 26.3(a) is independent of Figure 26.3(b). There are smokers on the second boxplot of 26.3(a) and their periodontal disease are on the vertical axis, and in 26.3(b), we again have the periodontal disease of the smokers on the vertical axis. Yet, the three comparisons in Figure 26.3(a) and (b) are nearly independent.

Assuming no unmeasured biases, i.e., $\Gamma_1 = \Gamma_2 = \Gamma_3 = 1$, the p-values from the three comparisons were $< 1.0 \times 10^{-16}$, .0024 and .066, respectively. Considered separately, they provide different impressions regarding the strength of evidence from the factors. Factor 3 is not significant at the 5% significance level. But taken together with factor 2, the combined p-value of the two is significant; combined p-value .0013 using the truncated product method with $\kappa = .2$.

The factors can also be susceptible to different levels of biases. In Table 26.4 we report joint sensitivity analyses of the factors with varied values of Γ_1, Γ_2, and Γ_3. [5] show that this table, despite having many p-values, does not require a multiplicity correction. Thus, we can read off the table at the same time that at the 5% significance level, we have evidence to reject the null when $\Gamma_1 = 2$ and $\Gamma_2 = \Gamma_3 = 1.5$, and that the inference is sensitive to bias at $\Gamma_1 = 2.25$ and $\Gamma_2 = \Gamma_3 = 1.75$. We can also see that if factor 1 is infinitely biased, the evidence remains if the other factors are free of biases. Also, if factors 2 and 3 are infinitely biased, we still have evidence to reject the null for $\Gamma_1 \leq 2$.

The analysis in Table 26.4 combines all three factors. After reading this table, can we see the table for factors 1 and 2, for the individual factors? Can we also consider the comparison of high vs low relative cotinine levels in Figure 26.3(b)? These questions are answered in the following section.

TABLE 26.4
Joint sensitivity analyses of the evidence factors for smoking and periodontal disease. The p-value upper bounds of the factors are combined using the truncated product method with $\kappa = .2$.

	$\Gamma_2 = \Gamma_3$							
Γ_1	1	1.25	1.5	1.75	2	2.25	2.5	∞
1	.0000	.0000	.0000	.0000	.0000	.0000	.0000	.0000
1.25	.0000	.0000	.0000	.0000	.0000	.0000	.0000	.0000
1.5	.0000	.0000	.0000	.0000	.0000	.0000	.0000	.0000
1.75	.0000	.0000	.0000	.0002	.0002	.0002	.0002	.0002
2	.0000	.0004	.0012	.0071	.0071	.0071	.0071	.0071
2.25	.0001	.0044	.0128	.0593	.0593	.0593	.0593	.0593
2.5	.0004	.0200	.0522	.1896	.1896	.1896	.1896	.1896
2.75	.0012	.0494	.1200	.4317	.4317	.4317	.4317	.4317
∞	.0047	.1448	.3059	1.0000	1.0000	1.0000	1.0000	1.0000

26.4 Planned Analyses of Evidence Factors

Meta-analysis is the process of synthesizing independent analyses of the same hypothesis. As noted in the previous section, the nearly independent comparisons from an evidence factor analysis can be combined using meta-analytic tools. But, unlike in a meta-analysis, because we know that the data will be available, we do not need to wait for the analyses to be completed and their results reported to plan ways we can integrate the evidence from evidence factors.

This allows us to plan ahead and plan smart. We can often get additional details regarding the strength of evidence while being aware of the required control for the overall type-I error. Below we present two such methods of planned analyses.

26.4.1 Partial conjunctions with several factors

Consider K evidence factors with p-value upper bounds from their sensitivity analyses given by $(\overline{P}_{1,\Gamma_1}, \ldots, \overline{P}_{K,\Gamma_K})$ for $\Gamma_k \geq 1$, $k = 1, \ldots, K$.

Fix k, and let $H_{0;k,\Gamma_k}$ denote the intersection of the null hypothesis and that the bias in factor k is at most Γ_k. Let $\bigwedge$ denote the "and" operation between hypotheses, and $\bigvee$ denote the "" operation between hypotheses. Then, the combined sensitivity analysis as we discussed in Section 26.3.3 tests for the global intersection hypothesis $\bigwedge_{1 \leq k \leq K} H_{0;k,\Gamma_k}$.

Consider now analyses that are concerned with evidence from a subset of the factors. For example, if we are interested in the evidence from the first two factors when biases are at most Γ_1 and Γ_2, respectively, then a combination of those two factors tests the conjunction hypothesis $H_{0;1,\Gamma_1} \bigwedge H_{0;2,\Gamma_2}$. It will make sense to combine these two factors if there are arguments saying the other factors could be highly biased. But there are $2^K - 1$ possible combinations, and it is possible to provide arguments for these many combinations.

A partial conjunction hypothesis does not postulate which factors are likely less biased but instead postulates how many are likely less biased. In that sense define

$$H^{k|K}_{0;\Gamma_1,\ldots,\Gamma_K} = \bigvee_{l=K-k+1}^{K} \bigwedge_{t \in \{t_1,\ldots,t_l\}, 1 \leq t_1 < \cdots < t_l \leq K} H_{0;t,\Gamma_t}$$

for a kth order partial conjunction hypothesis among the K factors. In other words, $H^{k|K}_{0;\Gamma_1,\ldots,\Gamma_K}$ says that at most $k-1$ of the K factors are valid at their specified bias levels. A partial conjunction can be defined for each of the order $k = 1, \ldots, K$. The Kth order partial conjunction is the global intersection hypothesis from before. Thus, testing several of these partial conjunctions could provide more comprehensive evidence in a causal inference method.

A primary reason to focus on the K partial conjunctions rather than the $2^K - 1$ combinations is that we can guarantee a family-wise error rate control without any multiplicity correction. To see this, first define $Pr^{k|K}_{\Gamma_1,\ldots,\Gamma_K}$ as the combination of largest $K - k + 1$ of the p-value upper bounds in $(\overline{P}_{1,\Gamma_1}, \ldots, \overline{P}_{K,\Gamma_K})$.

We have the following results for type-I error control. These results are proved in [34].

1. For fixed k and $\Gamma_1, \ldots, \Gamma_K$, $P^{k|K}_{\Gamma_1,\ldots,\Gamma_k}$ is a valid p-value for testing $H^{k|K}_{0;\Gamma_1,\ldots,\Gamma_K}$ for any coordinatewise nondecreasing combination method.

2. Fix k and any set $\mathfrak{J}$ of values for $(\Gamma_1, \ldots, \Gamma_K)$. The testing procedure that rejects $H^{k|K}_{\Gamma_1,\ldots,\Gamma_K}$ if $P^{k|K}_{\Gamma_1,\ldots,\Gamma_K} < \alpha$ has a familywise error rate of at most α for the set of hypotheses $\{H^{k|K}_{\Gamma_1,\ldots,\Gamma_K} : (\Gamma_1, \ldots, \Gamma_K) \in \mathfrak{J}\}$.

3. The same testing procedure as above also provides a familywise error of at most α for the set of hypotheses $\{H^{k|K}_{\Gamma_1,\ldots,\Gamma_K} : 1 \leq k \leq K; (\Gamma_1, \ldots, \Gamma_K) \in \mathfrak{J}\}$ under an additional mild ordering condition on the combination methods, see [34] Proposition 3.

These results allow us to look at the sensitivity analyses of partial conjunctions of all orders in an evidence factor analysis without having to pay for an inflated type-I error due to multiplicity.

After rejecting a partial conjunction hypothesis, it could be of interest to test the individual hypotheses, asking if at least k of the K hypotheses are false, that is, if $H^{k|K}_{0;\Gamma_1,\ldots,\Gamma_K}$ is rejected, which of the individual hypotheses are false? This question can be answered by comparing each factor t at an adjusted level $\alpha/(K - k)$ or by adjusting the corresponding p-value by $(K - k)\overline{P}_{k,\Gamma_k}$.

A special case of the above deserves attention. Suppose $K = 2$ and we reject the hypothesis that at least one of them is false by comparing the combination of the two factors to level α. Since $K = 2, k = 1$, hence $K - k = 1$, and $H^{1|K}$ has been rejected, the individual p-value upper bounds can be considered without further adjustments. This is equivalent to closed testing of two evidence factors [5,36]. Closed testing of more than two factors is also possible, but with $2^K - 1$ conjunctions it could get complicated. The partial conjunction method has a universal structure in this sense.

There is a family of combination methods that is optimal in the sense of having the largest Bahadur slopes [37]. Fisher's combination and the truncated product combination for p-values belong to this family.

Example 4 (continued). In Table 26.3 the p-values were calculated assuming no bias from unmeasured confounding. Since the evidence is separated into three nearly independent analyses, combining them will strengthen our evidence. Fisher's method for p-value combination applied to the three factors gives a value of 0.0237. Although the individual factors are not statistically significant at level 0.05, this small combined p-value tells us that there is only a 2.37% probability that we would have seen these three relatively small p-values, .0640, .1398, and .1608, for the factors if the null were true. We can also get an estimate and a confidence interval for the effect by inverting this combined p-value. They are: effect estimate $\$1,265$ and 95% confidence interval $(\$120, \$5,303)$. Thus, the combined p-value supports hypothesis of a positive effect the when all the factors are free of any bias.

What evidence remains if at most one of them is invalid? Using the partial conjunction analysis this may be calculated as Fisher's combination of the two largest numbers, i.e., the combination of .1398 and .1608, which is .0714.

If we assume small bias levels $\Gamma_1 = \Gamma_2 = \Gamma_3 = 1.1$, the p-values from the three factors become .2099, .3572 and .2895, respectively, which when combined is no longer significant. Thus, our evidence factor analysis gives very weak evidence regarding a positive effect of Catholic schooling vs public schooling that does not hold up if either the factor that most favors this hypothesis is invalid or there are small biases.

The technical results in this section tell us that we do not have to pick one of these results to report worrying that we might incur additional type-I error; we can report all of them at the same time.

Here is a note regarding possible violation of the exclusion restriction and bias from unmeasured confounding in an instrument. It is not always possible to separate these two. For instance, higher ambition might be interpreted as a confounder so that conditioning on it renders the urban/rural instrument valid, but without measurements of ambition, the instrument is invalid because of the failure of the no unmeasured confounders assumption. Or, one might argue differently that living in urban areas comes along with ambition, resulting in a direct effect on earnings when ambitions affect earnings. It is not possible within the study to separate the two.

26.4.2 Incorporating comparisons that are not factors

Some interesting comparisons may not be factors. In example 2 the factors are (i) none-versus-others eras and (ii) optional-versus-all eras. We also mentioned the pure treatment-control comparison of the all to none era. This comparison does not fit with the previous two factors in an evidence factor analysis.

Let $\mathcal{H} \subseteq \{H_{0;k,\Gamma_k} : 1 \leq k \leq K, \Gamma \geq 1\} \cup \{H^{k|K}_{\Gamma_1,\ldots,\Gamma_K} : 1 \leq k \leq K, \Gamma_k \geq 1\}$ be the set of hypotheses we plan to test in an evidence factor analysis. In Section 26.4.1 we discussed methods that provide familywise error rate control for $\mathcal{H}$. With $\mathcal{H}$ we can incorporate a comparison H_0' using a budgeting strategy on the total type-I error α. This method has three steps.

Let $0 < \alpha' \leq \alpha$. We test H_0' at level α'. Next, test $\mathcal{H}$ with a desired familywise error rate control of α if H_0' is rejected or $\alpha - \alpha'$ if H_0' is not rejected. In the last step, if all hypotheses in $\mathcal{H}$ are rejected, test H_0' again at level α. One can show that in this procedure the familywise error rate for the family $\mathcal{H} \cup H_0'$ is controlled at α.

The special case of $\alpha' = \alpha$ proposes the incompatible comparison as the primary analysis and the evidence factor analysis as the secondary analysis which is only attempted if the primary hypothesis is rejected. When $\alpha' = \alpha$, this procedure does not require any adjustments in the primary analysis. Thus, $\alpha' = \alpha$ could be preferred if we believe the test for H_0' will have a high statistical power among all the hypotheses. But, the risk of using H_0' as the primary analysis is that, if it is not rejected, we do not see anything from the evidence factor analysis.

We may also consider a family of hypotheses $\mathcal{H}'$ instead of a single hypothesis H_0'. For example, $\mathcal{H}'$ could be the all-versus-none comparison with its own sensitivity analysis. The above procedure still provides a familywise error rate control for $\mathcal{H} \cup \mathcal{H}'$ if we test $\mathcal{H}'$ with a familywise error rate control at α' or at α in the first and the third step, respectively.

Example 2 (continued). Consider the case with the incompatible all-versus-none era as a primary hypothesis. The one-sided p-value, when there is no bias, for this analysis is 5.27×10^{-5}. Thus, we would proceed to the evidence factor analysis that, in our previous discussion, raised doubt that side-airbags are effective in reducing injury in crashes.

We can also consider a sensitivity analysis of the all-versus-none era comparison at bias level $\Gamma = 1.15$. The p-value upper bound is .0988. If this were our primary analysis, i.e., $\alpha' = \alpha$, we would fail to reject the null and see that the analysis is very sensitive to bias. But, as the hypothesis is not rejected with $\alpha' = \alpha$, we would not see the evidence factors. If instead $\alpha' = \alpha/2 = .025$, we will see the sensitivity of the all-versus-none comparison at level 1.15 and see the evidence factor analysis with remaining $\alpha/2 = 0.025$. The evidence factor analysis with two factors here can be done using the closed testing as discussed in Section 26.4.1.

26.5 Algorithmic Tools for Designs of Evidence Factors

The importance of carefully designing an observational study is now well established [18, 38, 39]. An appropriate design is governed by the future laid out plan for statistical analysis.

A treatment-control analysis is conceptually simple and requires relatively simple design elements. Still, there are many variations: in designs, in their structures and in their goals to balance for pre-treatment covariates. Commonly used design structures for a treatment-control analysis are: matched pairs, one treated and a fixed number of controls in each stratum, and one treated or one control design (or, full matching). Various balancing goals include: exact matching, fine, near-fine, and refined balancing for categorical variables. This is not a comprehensive list of the variety of matching methods for treatment-control observational studies. But most of these design problems (including the ones listed above) can be solved to the optimal design using *efficient algorithms*. The efficiency of an algorithm here is measured in terms of its computational complexity. We provide some details regarding computational complexity to define what is an efficient algorithm.

The computational complexity of an algorithm is a valuable measure of how fast the algorithm is expected to run to solve the problem in a data set. But this expectation is mostly only a guess. It is often more useful to compare the computational complexities of two algorithms that solve the same problem. If one has a better computational complexity than the other, it is unlikely that we would not always use the former algorithm (there are exceptions; the simplex algorithm for solving linear programming has a very poor worst case computational complexity and yet it is used widely, although there are efficient alternatives).

Another use of computational complexity is in characterizing the difficulties of problems. While a big literature exists on different classifications and sub-classifications, we shall use the P versus NP-hard classification. Suppose the size of the problem is parametrized by $a_1, a_2, \ldots, a_l$. Then the problem belongs to the P class if we can find an algorithm to solve this problem that has a worst case time complexity of $O(f(a_1, \ldots, a_l))$ where f is any polynomial in its arguments. Otherwise, the problem is NP-hard. To avoid these technical details, we simply call an algorithm *efficient* if it has a polynomial computational complexity.

We saw in the previous sections that with evidence factors we have more elaborate plans for analyses. In contrast to treatment-contrast analyses, it turns out, most design problems for evidence factor analyses are NP-hard. As we cannot find a polynomial-time algorithm to solve these problems, there are two ways out. First, use a heuristic algorithm that seems to work well. Heuristic algorithms often have an evolutionary path. Someone proposes a heuristic algorithm; it is then tested widely in many instances of the problem; some edge instances are observed where the algorithm does badly; a modification is proposed to improve the performance of the algorithm for some or all these edge instances; the algorithm thus evolves and so on. The second way to approach an NP-hard problem is to find an *approximation algorithm*.

Definition 26.6. *[40] An approximation algorithm for an optimization problem is a polynomial-time algorithm that for all instances of the problem produces a solution whose value is within a factor of* α *if the optimal solution for some constant* α.

An approximation algorithm ensures efficiency at the cost of sacrificing a bit of optimality. Unlike heuristic algorithms, there are no edge instances in an approximation algorithm. We have a guarantee that the solution produced is not too far from the optimal solution, more specifically, by a factor α. Between two approximation algorithms, typically, the lower the α the better.

Below we discuss some design problems specific to evidence factors, although, they may appear in other instances. We shall discuss heuristic and approximation algorithms for these problems.

26.5.1 Matching with three or more groups

Consider n units in three groups $\mathcal{I}_1, \mathcal{I}_2$ and $\mathcal{I}_3$. Suppose set $\mathcal{I}_1$ has the smallest cardinality. For any two units u and u' from two different groups let $d_{u,u'}$ denote a non-negative, symmetric distance measure defined using pre-treatment covariates. Suppose d satisfies the inequality

$$d_{u,u''} \leq d_{u,u'} + d_{u',u''} \tag{26.2}$$

where u, u' and u'' are from different groups. We are interested in matched sets of these three groups. Specifically, given positive integers κ_1 and κ_2 we want to create $|\mathcal{I}_1|$ many disjoint sets (here indexed by i) of the form $(u_i, u'_{i1}, \ldots, u'_{i\kappa_1}, u''_{i1}, \ldots, u''_{i\kappa_2})$ where u is from $\mathcal{I}_1$, u''s are from $\mathcal{I}_2$ and u'''s are from $\mathcal{I}_3$ that attempt to minimize

$$\frac{1}{\kappa_2}\sum_{i'=1}^{\kappa_1} d_{u_i,u'_{ii'}} + \frac{1}{\kappa_1}\sum_{i''=1}^{\kappa_2} d_{u_i,u''_{ii''}} + \frac{1}{\kappa_1\kappa_2}\sum_{i'=1}^{\kappa_1}\sum_{i''=1}^{\kappa_2} d_{u'_{ii'},u''_{ii''}}. \tag{26.3}$$

In the side-airbag example (example 2), the three groups were the cars from three eras of side-airbag availability. Conceptually, there we solved this problem many times for each make and model. However, operationally, this is done in one large problem with all cars that imposed exact matching on make and model. We defined the distances based on the pre-treatment covariates that we mentioned in our earlier discussion. We skip some details regarding the calculations and only give the key ideas. Let $\mathbf{x}_u$ denote the vector of covariates for unit u. Then one can define $d_{u,u'} = \sqrt{(\mathbf{x}_u - \mathbf{x}_{u'})^\top \Sigma^{-1}(\mathbf{x}_u - \mathbf{x}_{u'})}$, where Σ is the covariance matrix combining all the groups. The is the Mahalanobis distance of the covariates. The squared root is needed in defining d's to satisfy (26.2). In our matching, we used a robust version of the Mahalanobis distance that uses ranks of the covariates instead of the actual values, hence less sensitive to extreme covariate values, but also satisfies (26.2); see [7] and Chapter 8 of [18].

It turns out even a simpler version of the problem (26.3) where all three groups have equal size and $\kappa_1 = \kappa_2 = 1$ is NP-hard [41]. But in our design problems we are unlikely to get equal-sized groups, and since some groups can be very large, it is natural to ask if we can match more than one unit from the group in our matched sets. Obviously, we cannot have κ_1 and κ_2 too large. Otherwise, there will not be any feasible designs.

[7] give an approximation algorithm for this problem with approximation factor $\mathfrak{a} = 2$. Given κ_1 and κ_2, this algorithm has a worst case computational complexity that is of the order $O(n^3)$. This algorithm is implemented in the `R` package `approxmatch`. The function `tripletmatching` in that package takes as input the distances and group labels of the units to create an approximately optimal design. The package also has a function `multigrp_dist_struc` that provides several ways to calculate the distances from the covariates that satisfy (26.2).

The `R` function for three group matching can do more than just minimize (26.3). If there are categorical variables, it can be asked to solve the design problem approximately while also maintaining a near-fine balance of the categorical variables; details are in [7]. Intuitively, a near-fine balance requires making the marginal distributions of the categorical variables as close to equal as possible across the groups in the matched design.

The algorithm has two steps that are relatively easy to understand, although we skip the details. It starts with matching the first group to the second group by minimizing $\frac{1}{\kappa_2}\sum_{i'=1}^{\kappa_1} d_{u_i,u'_{ii'}}$. This way the partial matched sets are created. Then, in the next step it completes the matched sets by choosing the rest of the units by minimizing $\frac{1}{\kappa_1}\sum_{i''=1}^{\kappa_2} d_{u_i,u''_{ii''}} + \frac{1}{\kappa_1\kappa_2}\sum_{i'=1}^{\kappa_1}\sum_{i''=1}^{\kappa_2} d_{u'_{ii'},u''_{ii''}}$. These two steps, individually, can be solved optimally as each is a matching problem between two groups. But together, the three-group matched sets design is not optimal because we may regret some of the partial matches in the first step between the first two groups if we had seen the third group. But it can be shown that because of (26.2), this regret is not large, and the algorithm has an approximation factor 2.

The same package `approxmatch` also has a function `kwaytmatching` that can create matched sets for more than 3 groups. All the features of `tripletmatching` extend to this function. This time, for K groups, we have a $(K-1)$-approximation algorithm. As the problem with more groups is more difficult, the worst case approximation performance is also worse. Given K, the computational complexity of the algorithm does not change; it is $O(n^3)$.

26.5.2 Stratification

In some instances, we want to adjust for covariates without necessarily disturbing the distribution of the treatments, because the distributions of the treatments might be informative. Such is the case in the reinforced design with multiple instruments [14]. The design problem, in this case, is different from the matching problem of Section 26.5.1. This time we want to stratify ungrouped/unlabeled data based on the covariate values. In example 3 we had a few categorical variables. Thus, we stratified based on joint categories of the variable, e.g., female, between 50 and 59, white, high school graduate defines a stratum. In the Catholic schooling example, we had many variables, both continuous and categorical. Thus, we cannot stratify this way.

Consider n units and distances d_{ij} between units. We want to divide these n units into non-overlapping sets of units of size k each so that the within-set differences are minimized. This is an attempt to stratify the data into strata of size k that are homogeneous in the covariates. The problem is important even outside of the context of observational studies. In experimental designs with multiple treatments, or one treatment with multiple factors, this stratification can be used to create a blocked design.

This problem is also NP-hard for any $k \geq 3$. For $k = 2$, it is equivalent to a non-bipartite matching problem that can be solved efficiently [42]; this algorithm is implemented in the `R` package `nbpMatching` [43].

For general k, there have been some heuristic algorithms for this problem. Moore [44] proposed a heuristic algorithm that works as follows. Let $n = mk$. This algorithm first creates m pairs greedily, finding one pair at a time which is the best from the available units. Then, if $k = 3$, it greedily matches one unit to each pair. For larger k, the method proceeds similarly until a stratification is created. [45] proposed a randomized heuristic algorithm for the problem (implemented in the `R` package `blockingChallenge`[4]). It first randomly chooses m units as template units for the m strata. The remaining $m(k-1)$ units are optimally assigned to these m templates at a ratio $(k-1):1$ to create the stratification. This method attempts many sets of random templates and chooses the best of the created stratification.

An approximation algorithm for this design problem is developed in [46][5]. The objective function is the maximum within strata distance (There are certain advantages to using this objective instead of a sum of within strata distances. Briefly, it guarantees that the average imbalance for any treatment assignment cannot exceed this maximum; minimizing the sums or averages does not provide such a guarantee.). For the special case where $k = 2^J$ for some positive integer J, this algorithm calls a non-bipartite matching J times. The first call creates $m2^{J-1}$ strata of size 2, i.e., pairs, such that the maximum paired distance is minimized. Each subsequent call halves the number of strata by pairing the previous set of blocks so that the maximum of the paired within block distances is minimized in this local problem. This procedure needs some modification if k is not a power of 2.

[46] gives a theoretical study of this algorithm. Theorem 3 from their paper shows that when d_{ij}'s satisfy the triangle inequality, this is an approximation algorithm with approximation factor

[4]available from `https://github.com/bikram12345k/blockingChallenge`

[5]An implementation of this algorithm is available from `https://github.com/bikram12345k/BlockingAlgo`

$\mathfrak{a} = (k-1)$. The algorithm has a computational complexity of $O(n^3)$. Further, Proposition 4 of the paper shows that no other approximation algorithm can have a better constant in general.

26.5.3 Balanced blocking

Consider n units and two binary instruments Z_{i1} and Z_{i2} for unit i. A balanced blocking creates m blocks where block b say has $n_{b(z_1,z_2)}$ units from $Z_1 = z_1$ and $Z_2 = z_2$ which satisfy

$$n_{b(0,0)}/n_{b(0,1)} = n_{b(1,0)}/n_{b(1,1)}. \tag{26.4}$$

The general balanced blocking problem is therefore to find non-overlapping blocks from the n units by optimizing over the $n_{b(z_1,z_2)}$ values under the constraint (26.4) while minimizing the total/maximum within-block average distances. There has not been any algorithm, approximation or heuristic, that solves this design problem. Thus a careful study of the design problem for balanced blocking remains incomplete.

In our example 4, we solved a slightly easier problem. We pre-specified the numbers $n_{b(z_1,z_2)}$'s to be 5 for all blocks b and all (z_1, z_2) values. Hence, (26.4) is satisfied by this choice. We used a combination of the algorithms from Section 26.5.1 and 26.5.2 to create our design heuristically.

We first defined four groups based on the values of $(Z_1, Z_2) \in \{0,1\}^2$. Next, using our approximation algorithm for matching with multiple groups, we created an intermediate design of quadruples, matched sets where the groups are represented 1:1:1:1. We used a squared root rank based on Mahalanobis distance based on the covariates in Table 26.2. Additionally, we used a near-fine balancing constraint on indicators for missing values of the covariates. Next, we used this quadruples structure as an input to a stratification algorithm from Section 26.5.2. This time we defined the distance between two quadruples as the total of 4×4 distances between the units from the quadruples. The stratification problem is solved with $k = 5$ to create our balanced blocked design. The number 5 was chosen by trying out a few choices of k with an eye on the covariate balance after balanced blocking.

26.6 Supplemental Notes

This section provides some additional remarks regarding the use of evidence factors.

26.6.1 Absolute standardized difference for stratified designs

The standardized mean difference is a commonly used measure for imbalance in the covariates across groups in the data. Typically, the standardized mean difference in the covariate is measured before and after matching on two groups to quantify the imbalances of the covariate between the groups and give an understanding of whether the matching was able to improve the imbalance. But in a stratified design, the usual definition of standardized mean difference is not very useful. Because this measure only considers the marginal distributions of the covariate across groups, before and after matching. Suppose the stratification kept all the units in the data from before stratification. In that case, the marginal distribution of the covariate does not change, and this standardized difference also does not change and does not capture whether reasonably homogeneous strata were created.

We propose the following definition of absolute standardized difference for stratified designs. Before stratification, perform a one-way ANOVA of the covariate on the group labels as the factors. Define the before stratification absolute standardized difference as the squared root of the ratio of sum of squares of the groups and sum of squares of the residuals. We caution that this definition does not give the same number for the absolute standardized mean difference as in the usual definition. But in our experiments (which are not reported here) this proposed definition captures similar imbalances in the data that the usual standardized mean difference captures. After stratification, we calculate our absolute standardized differences in two steps. First, perform a two-way ANOVA with groups as the first factor and the stratum levels as the second factor. Take the ratio of the sum of squares for the groups from this ANOVA to the sum of squares of the residuals from the previous ANOVA. Finally, calculate the absolute standardized difference as the squared root of this ratio after multiplying it by the ratio of sample size before stratification over the sample size after stratification.

The standardized differences in Table 26.2 were calculated using this process. The sample size before stratification was 4,449. After stratification, the sample size was 2,980 in the form of 149 strata of 20 units in each balanced block.

26.6.2 Guided choice of test statistics

As we have highlighted at a few places in this chapter, although there might be multiple choices for test statistics (say a t-test and a Wilcoxon's test), not a lot may be gained by using multiple of them in the same comparison. But, in an observational study some statistics could be preferred over others.

In planning a clinical trial the test statistic is chosen based on its expected good power performance under likely distributions of the outcome and the treatment effects. In an observational study, in addition to the power to detect a treatment effect, a second criterion of the power of sensitivity analysis for unmeasured confounders should be used. A better test should more likely report insensitivity to larger unmeasured confounders when there is a treatment effect. As in planning a clinical trial, in planning an observational study, we can evaluate the tests in simulated data according to these two criteria.

Existing literate has investigated the performances of many test statistics in observational studies [47–51]. Although there is not a complete guide to picking the one best test statistic, it is known that some statistics are better than others. For example, Wilcoxon's statistic is known to exaggerate sensitivity to bias. Alternative rank-based statistics from the family of U-statistic by [49] have comparable power to Wilcoxon's statistic but demonstrate larger insensitivity to bias.

26.6.3 Evidence factors and triangulation

Triangulation is an old idea that has received attention recently in the discussion of causal inference. Triangulation has been defined, somewhat loosely, as a method where a causal inference is strengthened by integrating results from several approaches that are different in key sources of potential bias [52, 53]. One can see similarities between this definition and the key points of evidence factor analysis. In a sense, an evidence factor analysis provides a formalization of the key aspects of triangulation.

One key difference between the two is the emphasis in an evidence factor analysis of the factors' statistical independence or near independence. In triangulation the independence aspect is not explicit, perhaps since there is often an understanding that different studies are used for the different pieces used to triangulate the causal effect. Using evidence factors we have shown that independent or nearly independent analyses can be created in a single study that are susceptible to different biases. Separate comparisons in a single study can be important in observational studies where even a single study tends to have several challenges, e.g., measurement, missing data, selection bias. Thus, it is a potent strategy for causal inference if more can be achieved from a single study by better design and analysis, with an appropriate care for the challenges. At the same time, the definition of evidence factors does not restrict itself to a single study.

Triangulation can be used to make inferences regarding a causal hypothesis rather than a specific treatment effect. The same is possible with evidence factors as well. [34] develop such a strategy where tools of planned evidence factor analysis are used to provide a comprehensive assessment of the evidence that corroborates a causal hypothesis.

26.7 Acknowledgment

Bikram Karmakar was partly supported by NSF Grant DMS-2015250.

References

[1] Eite Tiesinga, Peter J Mohr, David B Newell, and Barry N Taylor. Codata recommended values of the fundamental physical constants: 2018. *Journal of Physical and Chemical Reference Data*, 50(3):033105, 2021.

[2] Robert Andrews Millikan. On the elementary electrical charge and the avogadro constant. *Physical Review*, 2(2):109, 1913.

[3] Robert Andrews Millikan. *The electron and the light-quant from the experimental point of view.* Stockholm: Imprimerie Royale. P. A. Norstedt & Fils, 1925, Nobel Lecture delivered on May 23 1924. Available from https://www.nobelprize.org/uploads/2018/06/millikan-lecture.pdf.

[4] Arthur Stanley Eddington. The charge of an electron. *Proceedings of the Royal Society of London. Series A, Containing Papers of a Mathematical and Physical Character*, 122(789):358–369, 1929.

[5] Bikram Karmakar, Benjamin French, and Dylan S Small. Integrating the evidence from evidence factors in observational studies. *Biometrika*, 106(2):353–367, 2019.

[6] Vance Berger and Harold Sackrowitz. Improving tests for superior treatment in contingency tables. *Journal of the American Statistical Association*, 92(438):700–705, 1997.

[7] Bikram Karmakar, Dylan S Small, and Paul R Rosenbaum. Using approximation algorithms to build evidence factors and related designs for observational studies. *Journal of Computational and Graphical Statistics*, 28(3):698–709, 2019.

[8] Bikram Karmakar, Chyke A Doubeni, and Dylan S Small. Evidence factors in a case-control study with application to the effect of flexible sigmoidoscopy screening on colorectal cancer. *The Annals of Applied Statistics*, 14(2):829–849, 2020.

[9] David A Savitz and Gregory A Wellenius. Invited commentary: exposure biomarkers indicate more than just exposure. *American journal of epidemiology*, 187(4):803–805, 2018.

[10] Marc G Weisskopf and Thomas F Webster. Trade-offs of personal vs. more proxy exposure measures in environmental epidemiology. *Epidemiology (Cambridge, Mass.)*, 28(5):635, 2017.

[11] Bikram Karmakar, Dylan S Small, and Paul R Rosenbaum. Using evidence factors to clarify exposure biomarkers. *American Journal of Epidemiology*, 189(3):243–249, 2020.

[12] Scott L Tomar and Samira Asma. Smoking-attributable periodontitis in the united states: findings from nhanes iii. *Journal of Periodontology*, 71(5):743–751, 2000.

[13] PH Van Elteren. On the combination of independent two sample tests of wilcoxon. *Bull Inst Intern Staist*, 37:351–361, 1960.

[14] Bikram Karmakar, Dylan S Small, and Paul R Rosenbaum. Reinforced designs: Multiple instruments plus control groups as evidence factors in an observational study of the effectiveness of catholic schools. *Journal of the American Statistical Association*, 116(533):82–92, 2021.

[15] Anqi Zhao, Youjin Lee, Dylan S Small, and Bikram Karmakar. Evidence factors from multiple, possibly invalid, instrumental variables. *The Annals of Statistics*, 50(3):1266–1296, 2022.

[16] James Coleman, Thomas Hoffer, and Sally Kilgore. Cognitive outcomes in public and private schools. *Sociology of education*, pages 65–76, 1982.

[17] Erich Leo Lehmann and Howard J D'Abrera. *Nonparametrics: Statistical methods based on ranks.* Holden-day, 1975.

[18] Paul R Rosenbaum. *Design of observational studies*, volume 10. Springer, 2010.

[19] Paul R Rosenbaum. Covariance adjustment in randomized experiments and observational studies. *Statistical Science*, 17(3):286–327, 2002.

[20] Paul R Rosenbaum. The general structure of evidence factors in observational studies. *Statistical science*, 32(4):514–530, 2017.

[21] Paul R Rosenbaum. *Replication and Evidence Factors in Observational Studies.* Chapman and Hall/CRC, 2021.

[22] Paul R Rosenbaum. Sensitivity analysis for certain permutation inferences in matched observational studies. *Biometrika*, 74(1):13–26, 1987.

[23] Paul R Rosenbaum. *Observational studies.* Springer, 2002.

[24] Colin B Fogarty. Studentized sensitivity analysis for the sample average treatment effect in paired observational studies. *Journal of the American Statistical Association*, 115(531):1518–1530, 2020.

[25] Joseph L Gastwirth, Abba M Krieger, and Paul R Rosenbaum. Asymptotic separability in sensitivity analysis. *Journal of the Royal Statistical Society: Series B (Statistical Methodology)*, 62(3):545–555, 2000.

[26] Paul R Rosenbaum. Bahadur efficiency of sensitivity analyses in observational studies. *Journal of the American Statistical Association*, 110(509):205–217, 2015.

[27] Paul R Rosenbaum. Sensitivity analysis for stratified comparisons in an observational study of the effect of smoking on homocysteine levels. *The Annals of Applied Statistics*, 12(4):2312–2334, 2018.

[28] Paul R Rosenbaum and Abba M Krieger. Sensitivity of two-sample permutation inferences in observational studies. *Journal of the American Statistical Association*, 85(410):493–498, 1990.

[29] Paul R Rosenbaum and Dylan S Small. An adaptive mantel–haenszel test for sensitivity analysis in observational studies. *Biometrics*, 73(2):422–430, 2017.

[30] Paul R Rosenbaum and Jeffrey H Silber. Amplification of sensitivity analysis in matched observational studies. *Journal of the American Statistical Association*, 104(488):1398–1405, 2009.

[31] Ronald Aylmer Fisher. Statistical methods for research workers. In *Breakthroughs in statistics*, pages 66–70. Springer, 1992.

[32] Betsy Jane Becker. Combining significance levels. *The handbook of research synthesis*, pages 215–230, 1994.

[33] Dmitri V Zaykin, Lev A Zhivotovsky, Peter H Westfall, and Bruce S Weir. Truncated product method for combining p-values. *Genetic Epidemiology: The Official Publication of the International Genetic Epidemiology Society*, 22(2):170–185, 2002.

[34] Bikram Karmakar and Dylan S Small. Assessment of the extent of corroboration of an elaborate theory of a causal hypothesis using partial conjunctions of evidence factors. *The Annals of Statistics*, 48(6):3283–3311, 2020.

[35] Jesse Y Hsu, Dylan S Small, and Paul R Rosenbaum. Effect modification and design sensitivity in observational studies. *Journal of the American Statistical Association*, 108(501):135–148, 2013.

[36] Ruth Marcus, Peritz Eric, and K Ruben Gabriel. On closed testing procedures with special reference to ordered analysis of variance. *Biometrika*, 63(3):655–660, 1976.

[37] Bikram Karmakar. Improved power of multiple sensitivity analyses in observational studies using smoothed truncated product method. 2021. Unpublished manuscript.

[38] Donald B Rubin. The design versus the analysis of observational studies for causal effects: parallels with the design of randomized trials. *Statistics in Medicine*, 26(1):20–36, 2007.

[39] Donald B Rubin. For objective causal inference, design trumps analysis. *The Annals of Applied Statistics*, 2(3):808–840, 2008.

[40] David P Williamson and David B Shmoys. *The design of approximation algorithms*. Cambridge University Press, 2011.

[41] Yves Crama and Frits CR Spieksma. Approximation algorithms for three-dimensional assignment problems with triangle inequalities. *European Journal of Operational Research*, 60(3):273–279, 1992.

[42] Robert Greevy, Bo Lu, Jeffrey H Silber, and Paul Rosenbaum. Optimal multivariate matching before randomization. *Biostatistics*, 5(2):263–275, 2004.

[43] Cole Beck, Bo Lu, and Robert Greevy. nbpmatching: Functions for optimal non-bipartite matching, 2016. R package version 1.5.1.

[44] Ryan T Moore. Multivariate continuous blocking to improve political science experiments. *Political Analysis*, 20(4):460–479, 2012.

[45] Bikram Karmakar. blockingchallenge: Create blocks or strata which are similar within., 2018. R package version 1.0.

[46] Bikram Karmakar. An approximation algorithm for blocking of an experimental design. *Journal of the Royal Statistical Society - Series B*, 2022. doi: 10.1111/rssb.12545.

[47] Steven R Howard and Samuel D Pimentel. The uniform general signed rank test and its design sensitivity. *Biometrika*, 108(2):381–396, 2021.

[48] Paul R Rosenbaum. Design sensitivity and efficiency in observational studies. *Journal of the American Statistical Association*, 105(490):692–702, 2010.

[49] Paul R Rosenbaum. A new u-statistic with superior design sensitivity in matched observational studies. *Biometrics*, 67(3):1017–1027, 2011.

[50] Paul R Rosenbaum. An exact adaptive test with superior design sensitivity in an observational study of treatments for ovarian cancer. *The Annals of Applied Statistics*, 6(1):83–105, 2012.

[51] Paul R Rosenbaum. Weighted m-statistics with superior design sensitivity in matched observational studies with multiple controls. *Journal of the American Statistical Association*, 109(507):1145–1158, 2014.

[52] Debbie A Lawlor, Kate Tilling, and George Davey Smith. Triangulation in aetiological epidemiology. *International Journal of Epidemiology*, 45(6):1866–1886, 2016.

[53] Neil Pearce, Jan Vandenbroucke, and Deborah A Lawlor. Causal inference in environmental epidemiology: old and new. *Epidemiology (Cambridge, Mass.)*, 30(3):311, 2019.

Index

Note: Locators in *italics* represent figures and **bold** indicate tables in the text

N